Vector Formulas

$$\mathbf{A} \cdot (\mathbf{B} \times \mathbf{C}) = (\mathbf{A} \times \mathbf{B}) \cdot \mathbf{C} = \mathbf{C} \cdot (\mathbf{A} \times \mathbf{B}) = (\mathbf{C} \times \mathbf{A}) \cdot \mathbf{B} = \mathbf{B} \cdot (\mathbf{C} \times \mathbf{A})$$

$$\mathbf{A} \times (\mathbf{B} \times \mathbf{C}) = \mathbf{B}(\mathbf{A} \cdot \mathbf{C}) - \mathbf{C}(\mathbf{A} \cdot \mathbf{B})$$

$$(\mathbf{A} \times \mathbf{B}) \cdot (\mathbf{C} \times \mathbf{D}) = (\mathbf{A} \cdot \mathbf{C})(\mathbf{B} \cdot \mathbf{D}) - (\mathbf{A} \cdot \mathbf{D})(\mathbf{B} \cdot \mathbf{C})$$

Derivatives of Sums

$$\nabla (f + g) = \nabla f + \nabla g$$

$$\nabla \cdot (\mathbf{A} + \mathbf{B}) = \nabla \cdot \mathbf{A} + \nabla \cdot \mathbf{B}$$

$$\nabla \times (\mathbf{A} + \mathbf{B}) = \nabla \times \mathbf{A} + \nabla \times \mathbf{B}$$

Derivatives of Products

$$\nabla (fg) = f \nabla g + g \nabla f$$

$$\nabla (\mathbf{A} \cdot \mathbf{B}) = \mathbf{A} \times (\nabla \times \mathbf{B}) + \mathbf{B} \times (\nabla \times \mathbf{A}) + (\mathbf{A} \cdot \nabla) \mathbf{B} + (\mathbf{B} \cdot \nabla) \mathbf{A}$$

$$\nabla \cdot (f\mathbf{A}) = f (\nabla \cdot \mathbf{A}) + \mathbf{A} \cdot (\nabla f)$$

$$\nabla \cdot (\mathbf{A} \times \mathbf{B}) = \mathbf{B} \cdot (\nabla \times \mathbf{A}) - \mathbf{A} \cdot (\nabla \times \mathbf{B})$$

$$\nabla \times (f\mathbf{A}) = f (\nabla \times \mathbf{A}) - \mathbf{A} \times (\nabla f)$$

$$\nabla \times (\mathbf{A} \times \mathbf{B}) = \mathbf{A} (\nabla \cdot \mathbf{B}) - \mathbf{B} (\nabla \cdot \mathbf{A}) + (\mathbf{B} \cdot \nabla) \mathbf{A} - (\mathbf{A} \cdot \nabla) \mathbf{B}$$

Second Derivatives

$$\nabla \times (\nabla \times \mathbf{A}) = \nabla (\nabla \cdot \mathbf{A}) - \nabla^2 \mathbf{A}$$

$$\nabla \cdot (\nabla \times \mathbf{A}) = 0$$

$$\nabla \times (\nabla f) = 0$$

Integral Theorems

$$\int_V (\nabla \cdot \mathbf{A}) \, dV = \oint_S \mathbf{A} \cdot \hat{\mathbf{n}} \, dS \qquad \text{Gauss's (divergence) Theorem}$$

$$\int_S (\nabla \times \mathbf{A}) \cdot \hat{\mathbf{n}} \, dS = \oint_C \mathbf{A} \cdot d\boldsymbol{\ell} \qquad \text{Stokes's (curl) Theorem}$$

$$\int_{\mathbf{a}}^{\mathbf{b}} (\nabla f) \cdot d\boldsymbol{\ell} = f(\mathbf{b}) - f(\mathbf{a})$$

$$\int_V \left(f \nabla^2 g - g \nabla^2 f \right) dV = \oint_S (f \nabla g - g \nabla f) \cdot \hat{\mathbf{n}} \, dS \qquad \text{Green's Theorem}$$

ELECTROMAGNETISM

ELECTROMAGNETISM

Gerald L. Pollack
Michigan State University

Daniel R. Stump
Michigan State University

San Francisco   Boston   New York
Cape Town   Hong Kong   London   Madrid   Mexico City
Montreal   Munich   Paris   Singapore   Sydney   Tokyo   Toronto

Acquisitions Editor: Adam Black
Project Editor: Nancy Gee
Cover Designer: Blakeley Kim
Marketing Manager: Christy Lawrence
Managing Editor: Joan Marsh
Manufacturing Buyer: Vivian McDougal
Project Coordination: Integre Technical Publishing Co., Inc.

Library of Congress Cataloging-in-Publication Data
Pollack, Gerald L., 1933–
    Electromagnetism / Gerald L. Pollack and Daniel R. Stump.
      p. cm.
    Includes index.
    ISBN 0-8053-8567-3
    1. Electromagnetism. I. Stump, Daniel R. II. Title.
QC760.P65 2002
537-dc21   2001045977

9 10 — IBT — 08

www.aw.com/physics

Contents

6 ■ Electrostatics and Dielectrics   186

7 ■ Electric Currents   222

10 ■ Electromagnetic Induction   355

11 ■ The Maxwell Equations   397

Preface

This is an intermediate-level textbook on electricity and magnetism. It is intended to be used for a two- or one-semester course for students of physics, engineering, mathematics, and other sciences, who have already had a one-year introductory physics course with calculus.

The book is flexible enough to be used in several ways: (1) The traditional two-semester course would cover electrostatics and magnetostatics in the first semester using Chapters 1-8; and then magnetic materials and time-dependent fields in the second semester using Chapters 9-15. (2) An instructor teaching a one-semester course could cover all the basic principles of electromagnetism by using Chapters 1-3 and 6-11; there might also be time for a few examples from Chapters 4 and 5. (3) An interesting alternative approach in a two-semester course would be to go over the basic principles of Chapters 1-3 and 6-11 in the first semester, and then applications and advanced topics in the second semester based on Chapters 4,5, and 12-15.

The total material in the book is more than could be realistically covered by any instructor, even in two semesters. Instructors are encouraged to pick and choose based on their own judgment of what is important. Electricity and magnetism is a wonderfully interesting subject, but to students at the intermediate level its physical concepts are non-intuitive, and the associated mathematical techniques are new and challenging. Therefore it's important in teaching this subject to avoid the kind of heroic pace which will tire out all but the strongest students and instructors. The general principle that in teaching it's better to uncover a little than to cover a lot, applies to this subject of course.

The order of presentation of subjects is the traditional one: electrostatics first, then magnetism, electrodynamics and Maxwell's equations, relativity, and radiation. Chapter 2 is an introductory treatment of vector calculus, which should help students acquire the necessary mathematical armamentarium. Our experience in teaching this subject is that at the outset of the course most students do not know vector calculus well enough to study electromagnetic field theory, so it's important to help them gain the necessary mastery. Chapter 2 is sophisticated in places, and it is not necessary to comprehend all of it before starting on Chapter 3; the student can return to Chapter 2 when additional mathematical skill is needed. Students might also read a specialized book on vector calculus (e.g., one of the two references at the end of Chapter 2) while studying Chapter 2.

We have given an extensive treatment of electrostatics, in Chapters 3-6. The topics treated later in the book are more interesting than electrostatics to many students and instructors, so there's a tendency to hurry into them. But our experience is that time invested studying electrostatics pays dividends later on, because students acquire in electrostatics mathematical skills and confidence that are needed for other areas of electromagnetism.

Many good textbooks have been written on electromagnetism, and thousands of students have learned the subject from them. The two authors have taught this subject to hundreds of students over several decades, using some of those earlier books. It was from those many interactions with students, as well as with our colleagues and teachers, all remembered with pleasure, that we were led to write this book.

What is special about this book? For the most part it is a traditional, even conventional, exposition of electromagnetism, but we have also done three things we believe are important, and not stressed quite enough in other textbooks. First, we have tried to show how the mathematical principles that students are studying are used in modern technology—i.e., in real applications that students encounter in science and everyday life—applications such as cellular phones, optical fibers, magnetic resonance imaging, and charged particle accelerators and detectors. How is Faraday's Law related to the electricity in a wall socket? How can we calculate the interaction between radio waves and the ionosphere? Although it is necessary to study highly idealized, academic examples in field theory—e.g., the magnetic field of an infinitely long wire carrying a constant current, or the electric field in a spherical capacitor—students should also learn that the theory describes real physical phenomena and devices. The ideal cases are not the whole story.

Second, we have included in the text many worked-out mathematical examples in each area, including some examples that go beyond the elementary, exactly solvable, ones, and other examples that require multi-step analysis such as the use of the superposition principle. This book is based on a two-semester course that we have taught at Michigan State University. The course is taken by senior undergraduate students, and some first-year graduate students who are not yet enrolled in the graduate-level course. Our experience is that for students to master the intermediate level they must study more than just the simplest cases. Even to "uncover a little" requires that the instructor show a variety of examples. Each chapter starts at an elementary level, with topics the student is likely to know from an introductory physics course. The discussion then leads into the junior/senior-level material which is the heart of the book. At the end of several chapters we've introduced an interesting, more advanced, subject; we hope this will inspire students to future further study of the subject by indicating what lies beyond the horizon.

Third, we have included a number of computer calculations, both in the text and in the end of chapter exercises. Computer software that integrates analysis, numerical calculation, and graphics, e.g., Mathematica, Maple, Mathcad, Matlab, Excel, etc., can be used for these calculations. Students who are comfortable with Fortran or other computer languages can do the exercises by writing their own programs. Much of current physics, both experimental and theoretical, is done with computers, so today's students need this experience. On the other hand, the computer cannot replace the understanding of theoretical principles. We regard the computer exercises as an important, but not dominant, part of the book.

An essential part of this subject is learning how to do problems. There is a tendency (maybe it's even a tradition) for some textbooks on this subject—Smythe's book being the most dramatic example—to give many very difficult problems, on which the student spends uncounted and often frustrating hours. Many practicing physicists, the authors included, have been brought up in this draconian school. We believe however that the principle of "all things in moderation" should be applied to E&M problems. At the end of each chapter we have given a number of exercises of various degrees of difficulty, but mostly of only moderate difficulty, which are intended to help students understand the subject. Hints and answers are given in many of the exercises.

In writing *Electromagnetism* we had in mind the learning needs of present-day students, who are in some ways ready for a deeper understanding of the subject than we, their instructors, were at their age. They have had sophisticated courses in mathematics, even if they are still learning to apply this mathematics; we therefore use advanced mathematics freely, but also give generous explanations, so that students can exercise these valuable, newly acquired, skills. Mastering the subject at the intermediate level is not easy, and we encourage students to discuss what they are learning with other students and with their instructors at every opportunity.[1] Learning and doing physics has a social component. One learns much more from such discussions than by reading a textbook alone.

We believe it is especially important to meet the needs of those students who will go on to further study of electrodynamics, for example in graduate school in physics or electrical engineering, or who will use these principles in industry or engineering. Among these are the men and women who will write the next generation of books on this and related subjects, invent and develop new applications and (who knows?) discover new principles. To this end we try to extend the students' knowledge to a high enough level that they will be adequately prepared for working in J. D. Jackson's *Classical Electrodynamics*, or similar advanced books.

We would like to thank our students for listening to so many of our lectures on electromagnetism. We also owe a great debt to our own teachers of this subject, among them, R. P. Feynman, M. Firebaugh, R. C. Garth, D. L. Huber, F. E. Low, W. Mais, and W. R. Smythe.

---

[1] "Many will range far and wide and knowledge will increase." *Daniel* 12.4.

Encouragement for the Student

This Section together with Chapter 1 might appropriately be called "Loomings", in the sense that "Loomings" is the title of Chapter 1 of *Moby Dick*[2], because the purpose here, as there, is to give background and indicate an approach to a large subject. In any case, natural *looming*, which is observed in the atmosphere over sea or land, is an optical, *i.e.*, electromagnetic, phenomenon. It is caused by variation of the index of refraction of air with temperature and height. Looming makes objects like ships and lighthouses look larger and more threatening than they really are. There is something like that effect when one first approaches the study of electromagnetism, and is confronted by a very large subject. But, as in optical looming, the size is not really so large. After all, there are only 4 Maxwell equations.

Although learning electromagnetism is a great intellectual challenge – mastering any subject is difficult of course – there's nothing in it that should deter serious students. On the contrary, in our study of the subject we will encounter many interesting ideas, physical phenomena, and mathematical techniques; and many of these things have great beauty! It's a challenge less dangerous than the voyage of the *Pequod*. A good thing about the electromagnetism voyage is that it won't end.

In this Section the authors offer some suggestions to smooth the way of learning.

How to study this book. In learning a new subject, especially one with a lot of mathematical content, it's a good idea to take it in small increments. In this book we've tried to limit the ideas/page, *i.e.,* the idea density, to a value for which 3-5 pages per learning session is about right. It is good practice to read the material before going to class. This first reading can be casual, without paper and pencil. Then the instructor's presentation will be more understandable. But then it is necessary to reread the material, this time slowly and carefully with paper and pencil at hand. Try to fill in the intermediate steps in the calculations. When you get mired down, which is inevitable in theoretical physics, don't spend an inordinate amount of time struggling with a particular step. Pick up the argument wherever you can and go on.

Worked-out examples. The Examples in the text illustrate application of the principles to classic problems. They range from elementary to sophisticated. The

[2]Herman Melville, *Moby Dick*, 1851.

latter introduce new ideas. We suggest that you read each example line by line, so that you are able to repeat the calculation on your own and apply the ideas to other problems.

Formal language.   Informal language plays an essential role in learning and exchanging information. We all use it most of the time. But more formal language, which is used in this and similar books for proofs, derivations, and statements of problems, is also necessary. It is important not to be intimidated by the formalism, whose purpose is to avoid confusion or ambiguity. You'll get used to it, and after reading a few chapters you'll feel comfortable with it.

Exercises.   In the end, most of what you learn will have come from doing exercises yourself.[3] Because so much time and effort must go into this process, the experience should be as pleasant as possible. Therefore, exercises are provided at the end of each chapter. Ideally, you should try to understand and solve all the end-of-chapter exercises.

A common student difficulty in using textbooks is "to understand the material" but not be able to do the problems. There's a gap between what is learned from the text and what is needed to do the exercises. There is no easy fix for this difficulty, but we have tried some things to help bridge the gap: Some exercises are footnoted in the text where they are related to the text. Some exercises are grouped together by the appropriate section of the chapter. Of course the *best* exercises require material from many sections, and a wider perspective. Those exercises are more like the problems that arise in research in science and engineering.

The end-of-chapter exercises include a range of difficulty, but we have tried to omit the very difficult kind that only a few students can solve, and then only after a Herculean effort.[4] Although it is crucial to work on the exercises yourself, it is also useful to discuss them with classmates, instructors, or interested bystanders. There is a social aspect of science, and you will learn a lot by interactions with other scientists. Please eschew the "method of multiple books" – looking for the solution of an exercise in another book. That takes more time than just figuring it out yourself.

If you get stuck on some problems, console yourself with the thought that this happens to everyone. The knowledge you are seeking has taken hundreds of years to accrue, developed by thousands of people who made the same mistakes and false starts as you. So, if you feel hopelessly stuck, then close the book, contemplate the problem at leisure, talk it over with some other people, and try again tomorrow.[5]

---

[3] Most physicists will say: "Everything I know I taught myself!" What they mean is that they worked out a lot of problems, by the sweat of their own brows, and that's how they learned the subject.

[4] If you want to try some really difficult problems, they can be found in more advanced books like *Smythe* or *Jackson*.

[5] You can take encouragement from the Midrash Rabba (Fifth century) which says: "There isn't a blade of grass in the world below which doesn't have a star and an angel above that strike it and tell it, 'Grow! Grow!' "

Introductory textbooks. This book is intended for students who have already taken an introductory college-level course on physics. All introductory textbooks cover electricity and magnetism, in quite a lot of detail, including the mathematical theory. The textbook used in your introductory course should be used as a reference book when you encounter a difficult topic in this intermediate-level book. (In the same way, when you study electromagnetism in a graduate course, you will refer back to this book when you encounter a difficult topic in the advanced-level graduate text.) Reading over the material at a slightly lower level of sophistication can be very helpful in understanding the new ideas.

A list of good introductory textbooks is given at the end of Chapter 1.

CHAPTER

1

History and Perspective

"The theory I propose may therefore be called a theory of the Electromagnetic Field, because it has to do with the space in the neighborhood of the electric or magnetic bodies."
James Clerk Maxwell, Transactions of the Royal Society, 1864.

The electromagnetic interaction is one of the fundamental interactions of the physical world. It is the basic interaction in atoms and molecules. In nature it manifests itself in sunshine, lightning, rainbows, and many other phenomena, so that its study is as old as the attempts to understand such common observations. In technology the study of electromagnetism is as new as techniques for communicating with NASA's planetary probes, electromagnetic medical imaging devices, computer electronics, and other modern marvels. The subject will no doubt be renewed by the understanding and future applications developed by serious students of science and engineering, like you, for whom this book is intended.

Electromagnetism is important in *all* areas of physics. Section 1.2 of this chapter describes the place of electromagnetism in the "Standard Model" of the fundamental interactions. But we begin now with a brief history of our subject.

1.1 ■ BRIEF HISTORY OF THE SCIENCE OF ELECTROMAGNETISM

Electric and magnetic phenomena have been known for millenia. The earliest examples were forces produced by static electricity and by ferromagnetism. Of course there are other phenomena that we recognize today to be of electromagnetic origin, which have been observed since the beginning of time. For example, lightning is an electric discharge. Also, light consists of electromagnetic waves, or, in the quantum theory, photons. But before the scientific revolution it was not recognized that these varied phenomena have a common origin. During the 19th century a unified theory was constructed, describing electricity, magnetism, and optics, based on the *electromagnetic field*.

The word "electric" was coined by William Gilbert[1] as a name for the phenomena of static electricity. The origin of the word is from the Greek "electron" ($\eta\lambda\epsilon\kappa\tau\rho o\nu$) which means *amber*. Amber is a fossilized plant resin, an electrical

[1] William Gilbert was born in 1544 in Colchester, England, and died in 1603.

insulator which, when rubbed with animal fur, acquires the ability to attract small bits of matter such as straw or paper. This primitive example of static electricity was known to the philosophers of ancient Greece.

The word "magnetic" was also coined by Gilbert as a name for the phenomena of ferromagnetism. The origin of this word is the Greek place-name "Magnesia," which was a region of ancient Greece where magnetic iron ore (magnetite) occurs in the Earth. The forces between natural magnets were also known to the philosophers of ancient Greece.

These earliest observed electric and magnetic phenomena must have seemed interesting (in the way of a mystery) to philosophers, but they had no *practical* importance. In contrast, the scientific study of electricity and magnetism during the past 400 years has dispelled much of the mystery, and created technological power that the ancients could not have imagined.

The study of electricity and magnetism was an early part of the scientific revolution. In 1600 Gilbert published an important book, *De Magnete*, one of the first scientific works on electricity and magnetism. It was Gilbert who demonstrated that a compass needle points north because the Earth itself is a large magnet exerting a magnetic force on the magnetized needle. Regarding electricity, Gilbert discovered that many substances besides amber exhibit static electricity from friction, including glass, sulphur, sealing wax, and gemstones.

Static Electricity and Current Electricity

The science of electricity progressed significantly during the 18th century, because of two technical advances. Electrostatic devices were invented that create and store large amounts of charge, i.e., static electricity. Also, electrochemical cells were developed, the forerunners of today's batteries, to create controlled electric currents.

DuFay[2] discovered that there are in fact two kinds of static electricity. He named these forms *resinous*, produced by rubbing such materials as amber; and *vitreous*, produced by rubbing glass. Today we call these negative and positive charge, respectively. The theoretical models that were developed at that time to explain static electricity were based on the assumption that electricity is some kind of "fluid" distinct from matter. Two models were in competition, the one-fluid model and the two-fluid model. Franklin[3] invented the one-fluid model, which supposed that neutral matter possesses a certain amount of the electric fluid, and positive charge is an excess of the fluid, negative charge a deficiency. On the other side, the two-fluid model assumed that there are two different fluids—one positive, the other negative—and neutral matter contains equal amounts of the two. The modern understanding of electricity emerged much later, with the discovery of the electron by Thomson[4] in 1897: Charge is a property of matter, residing in the elementary particles—electrons and protons—that compose atoms.

[2]Charles Francois DuFay, b. 1698, d. 1739
[3]Benjamin Franklin, b. 1706, d. 1790
[4]J. J. Thomson, b. 1856, d. 1940

The definitive experiment on the force between charges was done by Coulomb[5] in 1785. He devised a delicate torsion balance for measuring small forces accurately. Using this balance, he determined that the force between small charged objects is inversely proportional to the distance squared, and proportional to the charges, $F \propto q_1 q_2 / r^2$.

Electric current—the flow of electric charge—occurs in discharges of static electricity. Lightning is an example in nature. But accurate experiments with current were not possible until the invention of the battery, which provides a source of continuous current. A battery is a chemical reactor that produces a constant electromotive force between two electrodes. The battery was developed by Volta,[6] following the discovery by Galvani[7] that if different metals are in contact with an aqueous salt solution then there is an electric current. Galvani's original discovery was in animal tissue (frogs' legs to be precise) and he believed that galvanic effects were at least partly biological. But Volta showed that current could be produced with purely physical samples. By constructing a "pile" of metal plates, alternating between different metals, immersed in acid or salt solution, large currents can be produced. This discovery made available a new source of electricity, in the form of continuous current. The first suggestion that the forces holding atoms and molecules together are electric forces came from Carlisle and Nicholson[8] in 1800 who passed a current through salt water and obtained hydrogen and oxygen gases. Current electricity was a necessary prerequisite for the advances in the understanding of electricity and magnetism that followed in the 19th century.

Unification of Electricity and Magnetism

In 1820, during a public lecture on electric and magnetic effects, Oersted[9] discovered that electric current in a metal wire produces deflection of a compass needle. This accidental discovery was the first indication that electricity and magnetism are related. In modern terms, Oersted had found that an electric current produces a magnetic field. Within a few days of the discovery, Biot and Savart[10] had reproduced the effect in their laboratory, and found that the strength of the magnetic field varies inversely with the distance from the wire. As to direction, by tracing the magnetic force in the region of the wire they showed that the magnetic field *curls around* the current.

Ampère[11] studied in great detail how electric currents produce magnetic effects. His ultimate result was an equation for the action of one current element

---

[5] Charles Augustin de Coulomb, b. 1736, d. 1806
[6] Alessandro Volta, b. 1745, d. 1827
[7] Luigi Galvani, b. 1737, d. 1798
[8] Anthony Carlisle (1768–1840) and William Nicholson (1753–1815) built the first battery in Britain, made from British coins and wet cardboard.
[9] Hans Christian Oersted, b. 1777, d. 1851
[10] Jean-Baptiste Biot (1744–1862) and Felix Savart (1791–1841)
[11] André Marie Ampère, b. 1775, d. 1836

$I_1 d\ell_1$ on another $I_2 d\ell_2$; the force on the first element has the form

$$dF_1 = C I_1 d\ell_1 \times \left(I_2 d\ell_2 \times \hat{r} \right) / r^2$$

where C is a constant that depends on the system of units. ($\hat{r}$ is the unit vector in the direction from $d\ell_2$ to $d\ell_1$ and r is the distance between them.) The student may recall that the cross product of two vectors points in a direction perpendicular to both. Ampère's result, which involves two cross products, anticipates that magnetic fields and forces may be a little tricky to figure out.

If electric current produces magnetism, then a natural question is whether magnetism can produce an electric current. Faraday[12] began experiments on electricity and magnetism in 1821, and he discovered electromagnetic induction in 1831. Magnetism does produce an electromotive force when the magnetic field is changing in time. The same effect was discovered at about the same time by Henry[13] in America.

After these observations relating electricity and magnetism, the challenge was to develop a theory that could account for all the effects in a unified way. The theory that finally succeeded was *field theory*.

The Electromagnetic Field

Faraday, from his years of experimentation with electric and magnetic phenomena, had developed a theoretical picture in which the phenomena were created by something *outside* the charges and currents. Faraday was self-educated, and more interested in new experimental results than advanced mathematical theories. His thinking was not limited by the existing theories of action at a distance that had been used to describe the forces between charges and between currents since the time of Coulomb and Ampère, in analogy with Newton's theory of gravity. Faraday called his alternative idea "the lines of force," by which he meant a physical entity that extends throughout space, outside the charge or current, and affects other charges or currents. This field concept is the opposite of action at a distance: The force on a charge or current element is due to the field *at the location* of the charge or current element.

Faraday's concept of the lines of force was put into precise mathematical form by Maxwell.[14] Maxwell published his equations in 1864, and we use the same equations to describe electromagnetism today. Maxwell coined the phrase "the electromagnetic field" as the name for his theory, based on the hypothesis that

---

[12]Michael Faraday (1791–1867) is often regarded as the greatest experimentalist in the history of physics. From humble origins Faraday rose to fame in science, and his biography is truly inspirational. Electromagnetic induction, discovered by Faraday, is the basis for the electric power we use every day.

[13]Joseph Henry (1797–1878) was a teacher in Albany, and his teaching duties were so heavy that he could only do research in August. He later became the first director of the Smithsonian Institution in Washington.

[14]James Clerk Maxwell was born in Edinburgh in 1831 and died in Cambridge in 1879. His field equations, published in 1864, are the subject of many physics textbooks, including this one.

the cause of electromagnetic phenomena is found in the space surrounding the charges. He was guided by fluid mechanics and elasticity—each of which is a mechanical system in a medium—and regarded the electromagnetic field as a measure of stresses and strains in a medium called the "aether." Today, however, we ascribe physical reality to the field itself, and deny the existence of an aether.

To construct a mathematically consistent set of field equations, Maxwell had to postulate the existence of a new electromagnetic effect, not known from experiments at that time: the displacement current. The resulting equations have wave solutions. When Maxwell calculated the wave speed, from measured electric and magnetic force parameters, he found that the wave speed has the same value as the measured speed of light. Thus this *mathematical theory* revealed that light is electromagnetic. Maxwell's equations form a unified theory of electricity, magnetism, and optics. Electric and magnetic effects are described by Maxwell's equations.

1.2 ■ ELECTROMAGNETISM IN THE STANDARD MODEL

The subject of this book is *classical electromagnetism*. Electromagnetism is one of the fundamental interactions of nature. But it is important to realize that it is a part of a larger picture. The purpose of this section is to describe briefly how electromagnetism fits into our current knowledge of physical forces. The current theory of the fundamental interactions is called the *Standard Model*. In this model, all forces and particle reactions derive from four basic interactions: electromagnetic and weak, chromodynamic, and gravity.

The standard model for the *electroweak interactions* is a field theory. The two types of weak interaction (charged and neutral) together with electromagnetism are all described by a unified field. The complete theory is consistent with the principles of relativity, and has interesting symmetries among the interactions. Electromagnetism is part of this unified theory.

Chromodynamics, or *quantum chromodynamics* (QCD), is the theory of the interactions between quarks and gluons—the constituents of protons and neutrons. The strong forces exerted by protons and neutrons in an atomic nucleus are a consequence of QCD interactions between the constituent quarks. QCD, too, is a field theory. It is a remarkable discovery, established during the final quarter of the 20th century, that the basic interactions of elementary particles (except gravity) are described by field theories analogous to electromagnetism.[15]

The modern theory of gravity is Einstein's general relativity, in which spacetime is non-Euclidean. An important unsolved problem in theoretical physics is to develop a fully unified theory, combining the electroweak and chromodynamic field theories and gravity.

---

[15] In the modern jargon the electroweak and chromodynamic theories are *gauge theories*, i.e., possessing a special symmetry. The electromagnetic field theory also possesses a gauge symmetry, which we'll learn about in Chapter 11.

Elementary Charges, Photons, and QED

The basic phenomena of electromagnetism, from which all electromagnetic effects derive, are interactions involving elementary charged particles and photons. These particles are so small (in energy) that they can only be described accurately by quantum theory. The quantum theory of electromagnetism—*quantum electrodynamics*, or QED—has been tested by experiments more precisely than any other theory in physics.

The elementary charges in matter are the electron, which has negative charge $-e$ where $e = 1.6 \times 10^{-19}$ C; and the quarks, which have fractional charge $\frac{2}{3}e$ (*u* type) and $-\frac{1}{3}e$ (*d* type). A proton has three valence quarks *uud*; its charge is $+e$, i.e., the sum of the valence quark charges, because charge obeys the superposition principle. A neutron has valence quarks *udd*, so its charge is 0.

What Is Electric Charge?   Charge is a property of the elementary particles of matter. The property belongs to electrons and quarks. Therefore charge is *quantized*, in units of $e/3$. In ordinary matter quarks are confined in bound states with integer charge—protons and neutrons—by QCD forces. So, in ordinary matter charge is quantized in units of e. The value of the electric charge of an elementary particle is a *coupling constant*: It determines the strength of interaction between the particle and the electromagnetic field.

The electromagnetic field is also quantized. The smallest (in energy) excitation of the field is a single photon. A photon is a packet of field with energy $E = h\nu$ and momentum $p = h\nu/c$, where ν is the frequency of oscillation of the field, and $h = 6.63 \times 10^{-34}$ Js is Planck's constant. The energy and momentum are related by $E = pc$, which is the same as for a massless particle in special relativity. Light has both particle-like and wave-like behavior. A beam of light is a stream of photons; but light undergoes interference and diffraction like waves. It seems hard to comprehend this dual nature of light, but the particle and wave aspects of light are reconciled in the quantum theory QED. Classical electromagnetism describes the wave aspect of light, and we'll study electromagnetic waves in detail in Chapters 11–15.

The Classical Limit

Classical electromagnetism is the science of electromagnetic phenomena in macroscopic systems.

In a macroscopic system, the energy, charge, and field strength are so large that the quantization, i.e., *discreteness,* of these quantities is irrelevant. For example, consider a wire carrying current $I = 1$ A. The rate at which electrons pass a point in the wire is $I/e = 6 \times 10^{18}$ per second, which is so large that the discreteness of the current in electrons is unobservable. Or, consider a radio transmitter with frequency $\nu = 1$ MHz and power $P = 1$ kW. The rate of photon production is $P/h\nu = 1.5 \times 10^{30}$ per second; again, the fact that the radio wave consists of discrete photons is unobservable because the number of photons is so large. Such macroscopic systems are described accurately by the classical theory.

Classical electromagnetism is the limit of QED for large systems, i.e., systems that involve many elementary charges or many photons. In QED the electromagnetic field is subject to quantum uncertainty, due to emission and absorption of single photons. The classical field is the *mean field*, in the sense of quantum theory, the average over the quantum probability distribution of field configurations. In a macroscopic system the uncertainty of the field is very small compared to the mean field, as illustrated by the previous two examples.

The mean field of QED satisfies the classical field equations of Maxwell. Therefore Maxwell's classical field theory is an accurate description of electromagnetic effects for macroscopic charged systems and waves.

Note that, of the fundamental interactions, only electromagnetism has both quantum and classical realms accessible to experiment. Gravity is so weak that on the quantum scale it is immeasurably small. The weak and chromodynamic interactions have very short range, so they do not have observable macroscopic effects that could be described by classical theories. Electromagnetism, *uniquely*, follows quantum theory on the atomic scale, and classical theory in the macroscopic limit.

This book is concerned with the classical electromagnetic field.

Conservation of Charge

That charge is quantized is not important in classical electromagnetism, because the quantum of charge e is so small. That charge is conserved is *essential*.

One statement of conservation of charge is that the net charge of an isolated system is constant. A more far-reaching statement is that if the charge of any sample of matter changes by Δq, then the charge of matter in contact with the sample must change by $-\Delta q$. This conservation law is ultimately based on experimental observations. It is a basic postulate of classical electrodynamics and a consequence of QED. The particles that carry charge, namely electrons and quarks, are themselves conserved, so the sum of their charges is constant. Creation or annihilation of particle-antiparticle pairs can occur in relativistic collisions, but still the charge is conserved because particle and antiparticle have equal but opposite charges. All interactions of the Standard Model respect this conservation law.[16]

FURTHER READING

On the history of the science of electromagnetism:

Edmund T. Whittaker, *A History of the Theories of Aether and Electricity* (Harper, New York, 1960).

Gerrit Verschuur, *Hidden Attraction; The History and Mystery of Magnetism* (Oxford University Press, New York, 1993).

Richard Feynman, *QED; The Strange Theory of Light and Matter* (Princeton University Press, New Jersey, 1985).

[16]In QED, the gauge symmetry implies conservation of charge through Noether's theorem.

E. Kashy and S. McGrayne, "Electricity and Magnetism", Encyclopaedia Britannica, 15th ed., Vol. 18, pp. 159–194 (1995). A very good overview of the subject.

Introductory textbooks:

R. P. Feynman, M. L. Sands, and R. B. Leighton, *The Feynman Lectures on Physics*, Volume 2 (Addison Wesley, Reading, 1994). This book and its companion volumes are characterized by Feynman's insights and flair for presentation, but they require diligent study.

P. M. Fishbane, S. Gasiorowicz, and S. T. Thornton, *Physics for Scientists and Engineers*, 2nd ed. (Prentice Hall, New Jersey, 1996)

D. Halliday, R. Resnick, and J. Walker, *Fundamentals of Physics*, 6th ed. (John Wiley, New York, 2001).

E. M. Purcell, *Electricity and Magnetism*, Berkeley Physics Course Volume 2, 2nd ed. (McGraw-Hill, New York, 1985).

R. Wolfson and J. M. Pasachoff, *Physics*, 3rd ed. (Addison-Wesley, Reading, 1999)

CHAPTER

2

Vector Calculus

*"Mathematics is distinguished from all other sciences,
except only ethics, in standing in no need of ethics."*

From the essay "The Essence of Mathematics"
by Charles Sanders Peirce (1839–1914).

Electromagnetic phenomena are caused by electric and magnetic fields, $\mathbf{E}(\mathbf{x}, t)$ and $\mathbf{B}(\mathbf{x}, t)$. A field is a physical entity that fills a volume of space. $\mathbf{E}(\mathbf{x}, t)$ and $\mathbf{B}(\mathbf{x}, t)$ are vector functions of position $\mathbf{x}$ and time t. The most basic equations of electrodynamics, the Maxwell equations, are partial differential equations that state how the fields change in space and time. The body of mathematical knowledge that we need to describe the physics of fields is *vector calculus*.

The purpose of this chapter is to review some important results in vector calculus. Anyone studying electromagnetism at the level of this book will probably already have some knowledge of this mathematics. Several good textbooks on vector calculus, such as Refs. [1] and [2], present theorems, proofs, interesting examples and exercises. This chapter summarizes the important results.

Vector calculus is an important tool in studying electricity and magnetism. Maxwell and the other early developers of this subject did not have the advantage of using today's vector calculus, and a study of their work shows how difficult and tedious the mathematics could be without it. Much of what is needed to use vector calculus is discussed in this chapter, and additional techniques will be introduced later when needed in specific applications. Our treatment includes the use of *suffix notation* (also called index notation), a modern approach, for some proofs and calculations. Suffix notation is an excellent method, but, frankly, it takes some getting used to at first. It is not necessary to master everything in the chapter the first time through it. Mastery will come with experience. It may be useful to revisit this material as needs arise.

A Comment on Notation

In calculations with vectors, it is essential to distinguish between different types of quantities: vectors, vector components, unit vectors, and scalars. All are different, and one must adopt notations that identify the type of each term in a mathematical expression. Table 2.1 shows the notations used in this book. A physicist must habitually use a notation that identifies which kind of quantity is represented by a

TABLE 2.1   Notations

Quantity	Notation	Examples
vector	boldface	$\mathbf{x}$, $\mathbf{E}$
vector	suffix notation	x_i, E_i
unit vector	hat $\hat{}$	$\hat{\mathbf{i}}$, $\hat{\mathbf{r}}$, $\hat{\mathbf{n}}$
vector component	plainface and subscript	E_x, B_θ
vector component	subscript with vector in ( )	$(\mathbf{E})_x$, $(\mathbf{B})_\theta$
vector magnitude	plainface or vector in $|\ |$	$|\mathbf{E}| = E$
scalar	plainface	t, U

symbol. In handwritten equations a vector must be identified as such, e.g., by an arrow over the symbol; for example, $E(\vec{x}, t)$ for $\mathbf{E}(\mathbf{x}, t)$. Every vector must have its arrow, and every unit vector its hat!

Getting the notation right is truly important! In complicated mathematical problems, an error of notation is a common cause of a failed calculation.

2.1 ■ VECTOR ALGEBRA

2.1.1 ▦ Definitions

In elementary mathematics a vector is defined as a quantity with a magnitude and a direction. For our purposes, we need to be more precise. The most basic example of a vector is the position vector $\mathbf{x}$ of a point P in space, with respect to a chosen origin.

In this book the symbol $\mathbf{x}$ is used to denote a position in space, or a position vector. Another position in the same system might be called $\mathbf{x}'$. In other courses the reader may have used the symbols $\mathbf{r}$ and $\mathbf{r}'$ for this purpose. So, for example, a function of position in our notation is denoted $F(\mathbf{x})$. We use $\mathbf{x}$, rather than $\mathbf{r}$, so that there is no chance of confusing the vector $\mathbf{x}$ with its magnitude $r \equiv |\mathbf{x}|$. Also, we prefer the symbol $\mathbf{x}$ for the position vector because $\mathbf{r}$ is often used to denote a *relative vector*, such as the vector from one charge to another. We use $\mathbf{x}$ for the position with respect to a fixed origin in space so that there is no chance of confusing $\mathbf{x}$ with a relative vector $\mathbf{r}$.[1]

In terms of Cartesian coordinates, the position vector $\mathbf{x}$ of a point P is written as

$$\mathbf{x} = x\,\hat{\mathbf{i}} + y\,\hat{\mathbf{j}} + z\hat{\mathbf{k}}. \tag{2.1}$$

Here $\hat{\mathbf{i}}$, $\hat{\mathbf{j}}$, and $\hat{\mathbf{k}}$ are the unit direction vectors of a set of Cartesian axes; and x, y, and z are distances from the origin along the axes, as illustrated in Fig. 2.1. The components x, y, z of $\mathbf{x}$ are *lengths*; $\hat{\mathbf{i}}$, $\hat{\mathbf{j}}$, $\hat{\mathbf{k}}$ are dimensionless.

---

[1] The notation $\mathbf{x}$ for position is also used in the famous graduate textbook on electrodynamics by J. D. Jackson.

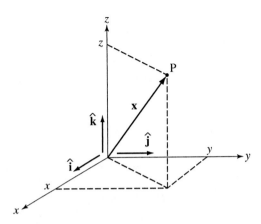

FIGURE 2.1   Cartesian coordinates. x is the position vector of the point P; the components of **x** are the coordinates of P. We write $\mathbf{x} = x\hat{\mathbf{i}} + y\hat{\mathbf{j}} + z\hat{\mathbf{k}}$.

If the coordinate axes are rotated, but the point P is left fixed in space, then the direction vectors change, $\hat{\mathbf{i}}, \hat{\mathbf{j}}, \hat{\mathbf{k}} \rightarrow \hat{\mathbf{i}}', \hat{\mathbf{j}}', \hat{\mathbf{k}}'$; and the coordinates of P on the new axes are different from those on the original axes, $x, y, z \rightarrow x', y', z'$. The vector **x** does not change, because P is fixed in space, so

$$\mathbf{x} = x'\hat{\mathbf{i}}' + y'\hat{\mathbf{j}}' + z'\hat{\mathbf{k}}'. \tag{2.2}$$

For example, for a rotation by angle θ about $\hat{\mathbf{k}}$, as shown in Fig. 2.2, the transformation of axis directions is

$$\hat{\mathbf{i}}' = \hat{\mathbf{i}} \cos\theta + \hat{\mathbf{j}} \sin\theta \tag{2.3}$$

$$\hat{\mathbf{j}}' = -\hat{\mathbf{i}} \sin\theta + \hat{\mathbf{j}} \cos\theta \tag{2.4}$$

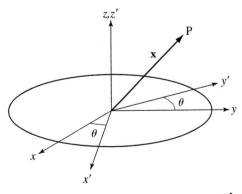

FIGURE 2.2   Rotation of coordinate axes by angle θ about axis $\hat{\mathbf{k}}$. The point P is fixed as the axes rotate.

$$\hat{\mathbf{k}}' = \hat{\mathbf{k}};\tag{2.5}$$

and the transformation of coordinates, which follows from (2.2), is

$$x' = \hat{\mathbf{i}}' \cdot \mathbf{x} = x\cos\theta + y\sin\theta\tag{2.6}$$

$$y' = \hat{\mathbf{j}}' \cdot \mathbf{x} = -x\sin\theta + y\cos\theta\tag{2.7}$$

$$z' = \hat{\mathbf{k}}' \cdot \mathbf{x} = z.\tag{2.8}$$

(For example, to derive (2.6), x' is the dot product of (2.2) and $\hat{\mathbf{i}}'$; the right-hand side of (2.6) is obtained from the dot product of (2.1) and (2.3).) The coordinates transform linearly, so in general we may write the transformation as a matrix equation

$$\begin{pmatrix} x' \\ y' \\ z' \end{pmatrix} = R \begin{pmatrix} x \\ y \\ z \end{pmatrix}\tag{2.9}$$

where R is a 3×3 matrix.[2] The matrix R depends on the angle and axis of rotation, but not on P. That is, the components of all position vectors transform by the same rotation matrix.

The definition of a *scalar* is a quantity that does not change under rotation of the coordinate axes. For example, the charge density $\rho(\mathbf{x})$ is a scalar function. The charge per unit volume at the point P is invariant under rotations of the coordinate axes. The definition of a *vector* is a quantity with three components that transform under rotation in the same way as the coordinates of a point. Let $\mathbf{A}$ be a vector with components A_x, A_y, and A_z in the original coordinate system. Then the components of $\mathbf{A}$ with respect to the rotated axes must be

$$\begin{pmatrix} A'_x \\ A'_y \\ A'_z \end{pmatrix} = R \begin{pmatrix} A_x \\ A_y \\ A_z \end{pmatrix}\tag{2.10}$$

where R is the same 3×3 matrix as in (2.9). Equation (2.10) is the defining property of a *vector*.

The electric and magnetic fields

$$\mathbf{E} = E_x\hat{\mathbf{i}} + E_y\hat{\mathbf{j}} + E_z\hat{\mathbf{k}}$$

$$\mathbf{B} = B_x\hat{\mathbf{i}} + B_y\hat{\mathbf{j}} + B_z\hat{\mathbf{k}}$$

are examples of vectors. Just having three components is not enough to make a quantity a vector. For example, $\hat{\mathbf{i}}\,E_x + \hat{\mathbf{j}}\,E_y + \hat{\mathbf{k}}B_z$ is not a vector. The three components must be geometrically related such that (2.10) holds for coordinate rotations.

[2]Exercise 5 at the end of this chapter.

2.1.2 ■ Addition and Multiplication of Vectors

The product of a scalar α times a vector $\mathbf{A}$ is a vector $\alpha\mathbf{A}$ whose components are αA_x, αA_y, and αA_z. The sum of any two vectors $\mathbf{A}$ and $\mathbf{B}$ is a vector $\mathbf{A} + \mathbf{B}$, whose components are $A_x + B_x$, $A_y + B_y$, and $A_z + B_z$. It is straightforward to verify that $\alpha\mathbf{A}$ and $\mathbf{A} + \mathbf{B}$ satisfy the definition of a vector (2.10), as does any linear combination of vectors, $\alpha\mathbf{A} + \beta\mathbf{B}$.

We also define multiplication of vectors. Indeed there are two multiplicative operations, dot product and cross product. The *dot product*, or scalar product, of two vectors $\mathbf{A}$ and $\mathbf{B}$ is a scalar $\mathbf{A} \cdot \mathbf{B}$, defined by

$$\mathbf{A} \cdot \mathbf{B} = A_x B_x + A_y B_y + A_z B_z = \sum_{i=1}^{3} A_i B_i. \qquad (2.11)$$

(In the last expression the values of $i = 1, 2, 3$ correspond to x, y, z; i.e., $A_1 = A_x$, $A_2 = A_y$, and $A_3 = A_z$. We shall use this notation on many occasions.) $\mathbf{A} \cdot \mathbf{B}$ is a scalar because it does not change under the transformation (2.9):

$$\mathbf{A}' \cdot \mathbf{B}' = (R\mathbf{A}) \cdot (R\mathbf{B}) = \begin{pmatrix} A_x & A_y & A_z \end{pmatrix} R^T R \begin{pmatrix} B_x \\ B_y \\ B_z \end{pmatrix} = \mathbf{A} \cdot \mathbf{B}. \qquad (2.12)$$

The third equality (which proves the claim that $\mathbf{A} \cdot \mathbf{B}$ is invariant) is true because any rotation matrix is an *orthogonal matrix*; that is, $R^T = R^{-1}$. (R^T is the transpose of R, and R^{-1} is the inverse of R.) We will discuss the geometric meaning of the dot product in Sec. 2.1.4, from which it will be obvious that $\mathbf{A} \cdot \mathbf{B}$ is a scalar.

The *cross product*, or vector product, of two vectors $\mathbf{A}$ and $\mathbf{B}$ is a vector $\mathbf{A} \times \mathbf{B}$. The ith component of $\mathbf{A} \times \mathbf{B}$ is

$$(\mathbf{A} \times \mathbf{B})_i = A_j B_k - A_k B_j \qquad (2.13)$$

where (ijk) is a cyclic permutation of (123).[3] (Again, each of the indices i, j, and k takes a value 1, 2, or 3, which correspond to x, y, or z respectively.) A handy way to calculate a cross product is to use the determinant formula

$$\mathbf{A} \times \mathbf{B} = \begin{vmatrix} \hat{\mathbf{i}} & \hat{\mathbf{j}} & \hat{\mathbf{k}} \\ A_x & A_y & A_z \\ B_x & B_y & B_z \end{vmatrix}. \qquad (2.14)$$

Another formula, which may look difficult at first sight but is very powerful, is

$$(\mathbf{A} \times \mathbf{B})_i = \sum_{j=1}^{3}\sum_{k=1}^{3} \epsilon_{ijk} A_j B_k \qquad (2.15)$$

where ϵ_{ijk} is the *Levi-Civita alternating tensor*. The value of ϵ_{123} is 1, and ϵ_{ijk} is antisymmetric under exchange of any two indices. (ϵ_{ijk} is sometimes called the

---

[3]The cyclic permutations of (123) are (123), (231), and (312).

completely antisymmetric tensor for three dimensions.) The antisymmetry implies that ϵ_{ijk} is 0 unless i, j, and k are all different. The nonzero components of ϵ_{ijk} are

$$\epsilon_{123} = \epsilon_{231} = \epsilon_{312} = 1,$$

$$\epsilon_{213} = \epsilon_{132} = \epsilon_{321} = -1. \tag{2.16}$$

Later we will use the Levi-Civita tensor to derive useful vector identities involving cross products.

It can be shown that $\mathbf{A} \times \mathbf{B}$ satisfies the definition (2.10) of a vector,

$$(\mathbf{A} \times \mathbf{B})' = \mathbf{A}' \times \mathbf{B}' = R\,(\mathbf{A} \times \mathbf{B}) \tag{2.17}$$

for any rotation R. The geometric meaning of $\mathbf{A} \times \mathbf{B}$ will be discussed in Sec. 2.1.4.

An important property of the cross product is that $\mathbf{B} \times \mathbf{A} = -\mathbf{A} \times \mathbf{B}$. This *noncommutativity* of the cross product is obvious from the definition (2.13), or from the antisymmetry of ϵ_{ijk} in (2.15). Also, by setting $\mathbf{B} = \mathbf{A}$ we see that the cross product of a vector with itself is 0, $\mathbf{A} \times \mathbf{A} = 0$.

2.1.3 ■ Vector Product Identities

Cross products and dot products often appear in multiplicity in calculations. There are important identities that relate such multiple products, of which we shall discuss two. These identities are used often in vector field theory.

One identity, true for all vectors, is

$$\mathbf{A} \cdot (\mathbf{B} \times \mathbf{C}) = (\mathbf{A} \times \mathbf{B}) \cdot \mathbf{C}. \tag{2.18}$$

The most straightforward (but rather laborious!) way to verify this equation is to write out each side in components, and verify that the six terms are the same on both sides. There is a much easier way to prove the result, and other product identities, but it requires an understanding of *suffix notation*. We need to make a brief digression on suffix notation.[4]

On Suffix Notation
In suffix notation, the vector $\mathbf{A}$ is denoted A_i. The notation is potentially dangerous, because A_i stands for two different things: the ith component of the vector, and the vector itself. The context must be used to decide which is meant. We use suffix notation rarely in this book, mainly just in proofs of vector identities. There is one branch of theoretical physics where suffix notation is *always* used for vectors, namely, the theory of relativity. This notation for vectors has become more widespread in physics in recent years, because of its prevalence in relativistic calculations.

[4]Suffix notation is explained in detail in Ref. [1].

The dot product of two vectors is

$$\mathbf{A} \cdot \mathbf{B} = A_x B_x + A_y B_y + A_z B_z. \tag{2.19}$$

In suffix notation, this is written very succinctly as

$$\mathbf{A} \cdot \mathbf{B} = A_i B_i, \tag{2.20}$$

making use of the *Einstein summation convention*: Any repeated suffix is understood to be summed from 1 to 3. So, $A_i B_i$ means $\sum_{i=1}^{3} A_i B_i$. The sum is just understood. A repeated suffix is a "dummy index," because it is summed over all of its values; it does not have a particular value. For this reason $A_j B_j$ is *exactly the same quantity* as $A_i B_i$; both are equal to the sum in (2.19). The summation convention, invented by Einstein, is a wonderfully simplifying notation if we are careful and understand its usage.

The cross product of two vectors is

$$\mathbf{A} \times \mathbf{B} = \left(A_y B_z - A_z B_y \right) \hat{\mathbf{i}} + \left(A_z B_x - A_x B_z \right) \hat{\mathbf{j}} + \left(A_x B_y - A_y B_x \right) \hat{\mathbf{k}}. \tag{2.21}$$

Suffix notation is more compact,

$$(\mathbf{A} \times \mathbf{B})_i = \epsilon_{ijk} A_j B_k \tag{2.22}$$

where j and k are summed.

Now, to prove (2.18) using suffix notation, note that the left-hand side of the equation is

$$A_i \left(\mathbf{B} \times \mathbf{C} \right)_i = \epsilon_{ijk} A_i B_j C_k \tag{2.23}$$

where again we have used the summation convention; on the left-hand side of (2.23) i is summed, and on the right-hand side i, j, and k are all summed. The right-hand side of (2.18) is

$$(\mathbf{A} \times \mathbf{B})_k \, C_k = \epsilon_{kij} A_i B_j C_k \tag{2.24}$$

where again any repeated indices are summed from 1 to 3. But $\epsilon_{kij} = \epsilon_{ijk}$ for any i, j, k by the complete antisymmetry of the Levi-Civita tensor;

$$\epsilon_{kij} = -\epsilon_{ikj} = +\epsilon_{ijk}.$$

Thus the two sides of (2.18) are equal, and the identity is proven.

Another important vector identity is the *double cross product identity*

$$\mathbf{A} \times (\mathbf{B} \times \mathbf{C}) = \mathbf{B} \left(\mathbf{A} \cdot \mathbf{C} \right) - \mathbf{C} \left(\mathbf{A} \cdot \mathbf{B} \right). \tag{2.25}$$

This identity is used over and over again in field theory. To prove it by writing out both sides of the equation in components would be quite tedious. Instead, we shall prove it using suffix notation.

The left-hand side is a vector; in suffix notation, it is $[\mathbf{A} \times (\mathbf{B} \times \mathbf{C})]_i$. Now, convert the first cross using (2.15), and then the second by the same equation,

$$[\mathbf{A} \times (\mathbf{B} \times \mathbf{C})]_i = \epsilon_{ijk} A_j (\mathbf{B} \times \mathbf{C})_k = \epsilon_{ijk}\epsilon_{k\ell m} A_j B_\ell C_m. \qquad (2.26)$$

Again the repeated indices ($jk\ell m$) are summed from 1 to 3; the second expression is a double sum, and the third a quadruple sum. (Note that in the second application of (2.15) it was necessary to introduce new dummy indices ℓ and m because j and k were already in use.) An identity of the Levi-Civita tensor is

$$\epsilon_{ijk}\epsilon_{k\ell m} = \delta_{i\ell}\delta_{jm} - \delta_{im}\delta_{j\ell} \qquad (2.27)$$

where δ_{ij} is the Kronecker delta tensor

$$\delta_{ij} = \begin{cases} 1 & \text{if } i = j \\ 0 & \text{if } i \neq j. \end{cases} \qquad (2.28)$$

The proof of (2.27) is an exercise.[5] A suffix sum involving the Kronecker delta may be eliminated by putting the summed suffix equal to the other suffix of the δ; for example, $\delta_{ij}u_j = u_i$. Using this trick for the sums over ℓ and m in (2.26) we find that the right-hand side of (2.26) is

$$A_j\delta_{i\ell}B_\ell\delta_{jm}C_m - A_j\delta_{j\ell}B_\ell\delta_{im}C_m = A_j B_i C_j - A_j B_j C_i \qquad (2.29)$$

(sum on j implied) which is, in suffix notation, the vector on the right-hand side of (2.25). Hence the double cross product identity is proven.

Having proven that (2.18) and (2.25) are true for any vectors, we may use these identities freely in calculations. We don't need to rederive the result each time we use it. Students often find it difficult to know when to use a vector identity, or to remember the identities. Practicing physicists may not remember all the identities either, but when an identity is needed it can be derived "on the fly" (after plenty of practice!) using the Levi-Civita tensor.

2.1.4 ■ Geometric Meanings

Our discussion of vectors has been purely algebraic so far. Vectors also have a geometric aspect. The geometric meanings of the dot and cross products are contained in the equations

$$\mathbf{A} \cdot \mathbf{B} = |\mathbf{A}||\mathbf{B}| \cos\theta, \qquad (2.30)$$

$$\mathbf{A} \times \mathbf{B} = |\mathbf{A}||\mathbf{B}| \sin\theta \, \hat{\mathbf{n}}, \qquad (2.31)$$

where θ is the angle between $\mathbf{A}$ and $\mathbf{B}$. Our notation is that $|\mathbf{A}|$ denotes the magnitude of $\mathbf{A}$,

$$|\mathbf{A}| = \sqrt{A_x^2 + A_y^2 + A_z^2} = (A_k B_k)^{1/2}. \qquad (2.32)$$

---

[5] See Exercise 3.

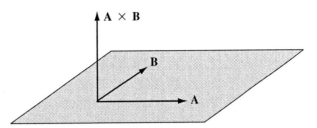

FIGURE 2.3   Direction of the cross product. $\mathbf{A} \times \mathbf{B}$ is perpendicular to both $\mathbf{A}$ and $\mathbf{B}$, *i.e.,* to the plane spanned by $\mathbf{A}$ and $\mathbf{B}$, in the sense given by the right-hand rule.

For example, for the position vector $\mathbf{x}$ the magnitude $|\mathbf{x}|$ is the distance from the origin to P.

$\mathbf{A} \cdot \mathbf{B}$ is the projection of $\mathbf{A}$ on $\mathbf{B}$ times the magnitude of $\mathbf{B}$, or the projection of $\mathbf{B}$ on $\mathbf{A}$ times the magnitude of $\mathbf{A}$. If $\mathbf{A}$ and $\mathbf{B}$ are parallel, then $\mathbf{A} \cdot \mathbf{B}$ is just the product of their magnitudes. If $\mathbf{A}$ and $\mathbf{B}$ are perpendicular, then $\mathbf{A} \cdot \mathbf{B}$ is 0. Because lengths of vectors and angles between vectors are invariant under coordinate rotations, it is obvious that $\mathbf{A} \cdot \mathbf{B}$ is a scalar.

In (2.31) $\hat{\mathbf{n}}$ is the unit vector perpendicular to the plane spanned by $\mathbf{A}$ and $\mathbf{B}$, in the sense defined by the right-hand rule, as shown in Fig. 2.3: If the fingers of your right hand curl from $\mathbf{A}$ to $\mathbf{B}$ then the thumb points in the direction of $\mathbf{A} \times \mathbf{B}$. The cross product is perpendicular to both $\mathbf{A}$ and $\mathbf{B}$. The factor $\hat{\mathbf{n}}$ in (2.31) is *crucial*; without that factor (2.31) would be nonsensical, because the left-hand side is a vector, but the right-hand side (without the $\hat{\mathbf{n}}$) would be a scalar. A vector cannot equal a scalar! When doing calculations with cross products, it is important not to lose the vector. Again, the geometric statement (2.31) makes it obvious that $\mathbf{A} \times \mathbf{B}$ transforms as a vector under coordinate rotations.

EXAMPLE 1   Let $\mathbf{A}$ be the vector $\hat{\mathbf{i}} + \hat{\mathbf{j}}$ in the xy plane, and $\mathbf{B}$ the vector $\hat{\mathbf{j}} + \hat{\mathbf{k}}$ in the yz plane, shown in Fig. 2.4. Find the magnitude and direction of $\mathbf{A} \times \mathbf{B}$.

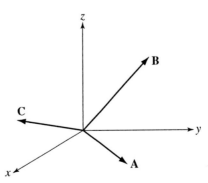

FIGURE 2.4   Example 1. $\mathbf{C}$ is $\mathbf{A} \times \mathbf{B}$. The vectors shown are $\mathbf{A} = \hat{\mathbf{i}} + \hat{\mathbf{j}}$, $\mathbf{B} = \hat{\mathbf{j}} + \hat{\mathbf{k}}$, and $\mathbf{C} = \hat{\mathbf{i}} - \hat{\mathbf{j}} + \hat{\mathbf{k}}$. The direction of $\mathbf{C}$ is perpendicular to both $\mathbf{A}$ and $\mathbf{B}$.

We could calculate $\mathbf{A} \times \mathbf{B}$ from the determinant formula, but for this example it is simple enough to expand the product, as

$$\mathbf{A} \times \mathbf{B} = (\hat{\mathbf{i}} + \hat{\mathbf{j}}) \times (\hat{\mathbf{j}} + \hat{\mathbf{k}})$$
$$= \hat{\mathbf{i}} \times \hat{\mathbf{j}} + \hat{\mathbf{i}} \times \hat{\mathbf{k}} + \hat{\mathbf{j}} \times \hat{\mathbf{j}} + \hat{\mathbf{j}} \times \hat{\mathbf{k}}$$
$$= \hat{\mathbf{k}} - \hat{\mathbf{j}} + \hat{\mathbf{i}}. \tag{2.33}$$

The magnitude of $\hat{\mathbf{i}} - \hat{\mathbf{j}} + \hat{\mathbf{k}}$ is $\sqrt{3}$. This magnitude agrees with (2.31); the magnitudes of $\mathbf{A}$ and $\mathbf{B}$ are both $\sqrt{2}$, and the angle θ between $\mathbf{A}$ and $\mathbf{B}$ is 60 degrees (because $\cos\theta = \mathbf{A} \cdot \mathbf{B}/(AB) = 1/2$) so $AB\sin\theta = \sqrt{3}$. The direction of $\hat{\mathbf{i}} - \hat{\mathbf{j}} + \hat{\mathbf{k}}$ is indicated in Fig. 2.4 by $\mathbf{C}$, perpendicular to both $\hat{\mathbf{i}} + \hat{\mathbf{j}}$ and $\hat{\mathbf{j}} + \hat{\mathbf{k}}$.

2.2 ■ VECTOR DIFFERENTIAL OPERATORS

2.2.1 ■ Gradient of a Scalar Function

The fields of electromagnetism are functions of position $\mathbf{x}$, so there are three spatial partial derivatives: $\partial/\partial x$, $\partial/\partial y$, and $\partial/\partial z$. The *gradient operator* ∇, called "del," combines the three partial derivatives into a vector operator. The action of ∇ on a scalar function $f(\mathbf{x})$ produces a vector function ∇f. Using a Cartesian coordinate system ∇f is

$$\nabla f = \hat{\mathbf{i}}\frac{\partial f}{\partial x} + \hat{\mathbf{j}}\frac{\partial f}{\partial y} + \hat{\mathbf{k}}\frac{\partial f}{\partial z}. \tag{2.34}$$

To prove that ∇f is a vector, i.e., that it satisfies the definition (2.10), we must determine how its components transform under a rotation of the Cartesian axes. The ith component is $(\nabla f)_i = \partial f/\partial x_i$. The transformed component is

$$\left(\nabla' f\right)_i = \frac{\partial f}{\partial x_i'} = \sum_{j=1}^{3} \frac{\partial x_j}{\partial x_i'}\frac{\partial f}{\partial x_j}, \tag{2.35}$$

where the second equality is by the chain rule of differentiation. Now, the transformation (2.9) of the coordinates implies $x_j = (R^{-1})_{jk}x_k'$, with a sum over k implied by the Einstein summation convention. Therefore,

$$\frac{\partial x_j}{\partial x_i'} = \left(R^{-1}\right)_{ji} = R_{ij}, \tag{2.36}$$

where R_{ij} is the ij component of the matrix R in (2.9), and we have used the fact that $R^{-1} = R^T$ for any rotation matrix. Thus ∇f transforms as a vector: $(\nabla' f)_i = R_{ij}(\nabla f)_j$ (sum over j implied).

Equation (2.34) is the formula for calculating ∇f in a Cartesian coordinate system. But in field theory it is often more useful to use curvilinear coordinates, like cylindrical or spherical polar coordinates. The generalization of (2.34) to curvilinear coordinates is nontrivial, as we shall see. So we need a *coordinate-independent definition* of the gradient. Consider the change of $f(\mathbf{x})$ resulting from an infinitesimal displacement $d\mathbf{x}$. That is, $d\mathbf{x}$ is the vector

$$d\mathbf{x} = \hat{\mathbf{i}} \, dx + \hat{\mathbf{j}} \, dy + \hat{\mathbf{k}} \, dz. \tag{2.37}$$

The change of f from $\mathbf{x}$ to $\mathbf{x} + d\mathbf{x}$ is

$$df = f(\mathbf{x} + d\mathbf{x}) - f(\mathbf{x}) = dx\frac{\partial f}{\partial x} + dy\frac{\partial f}{\partial y} + dz\frac{\partial f}{\partial z} = d\mathbf{x} \cdot \nabla f. \tag{2.38}$$

The third expression in (2.38) is Cartesian, but the fourth form is coordinate independent. Therefore the differential relation $df = d\mathbf{x} \cdot \nabla f$ is the definition of ∇f. We shall use this definition later to express ∇f in curvilinear coordinates.

An important theorem is that the direction of ∇f at a point $\mathbf{x}$ is perpendicular to the surface of constant f that includes the point $\mathbf{x}$. To prove the theorem, note that for any displacement $d\mathbf{x}$ along the surface of constant f, df is 0. But then by (2.38) ∇f is orthogonal to $d\mathbf{x}$. Since $d\mathbf{x}$ is an arbitrary displacement on the surface, ∇f must point in the direction perpendicular to the surface.

For an arbitrary displacement $d\mathbf{x}$ the change of f is $df = |\nabla f||d\mathbf{x}| \cos\theta$ where θ is the angle between ∇f and $d\mathbf{x}$. The change of f is maximum if $\theta = 0$. In other words, ∇f points in the direction of the maximum increase of $f(\mathbf{x})$. The magnitude of ∇f is the rate of change of f in that direction.

2.2.2 ■ Divergence of a Vector Function

The operator ∇ acts algebraically as a vector. The *divergence* of a vector function $\mathbf{F}(\mathbf{x})$, denoted $\nabla \cdot \mathbf{F}$ and read as "del dot F", is a scalar function. In Cartesian coordinates the divergence is given by[6]

$$\nabla \cdot \mathbf{F} = \frac{\partial F_x}{\partial x} + \frac{\partial F_y}{\partial y} + \frac{\partial F_z}{\partial z}. \tag{2.39}$$

It can be shown that $\nabla \cdot \mathbf{F}$ is invariant under rotation of the coordinate system.

Again, we need a coordinate-independent definition of the divergence, that can be used to derive $\nabla \cdot \mathbf{F}$ in curvilinear coordinates. The definition follows from the "divergence" property of this operator: $\nabla \cdot \mathbf{F}$ is a measure of how the vector function $\mathbf{F}(\mathbf{x})$ diverges, i.e., *spreads out* from $\mathbf{x}$. Let dV be an infinitesimal cubic volume centered at $\mathbf{x}$, of size $\epsilon \times \epsilon \times \epsilon$, aligned with the Cartesian directions $\hat{\mathbf{i}}$, $\hat{\mathbf{j}}$, and $\hat{\mathbf{k}}$ as shown in Fig. 2.5. The flux of $\mathbf{F}$ outward through the boundary surface dS of dV is the sum of six terms, from the six faces of the cube. The area of each face is ϵ^2; and for opposite faces the outward normals are $+\hat{\mathbf{e}}_i$ and $-\hat{\mathbf{e}}_i$, where $\hat{\mathbf{e}}_1$, $\hat{\mathbf{e}}_2$, $\hat{\mathbf{e}}_3$ denote the Cartesian directions $\hat{\mathbf{i}}$, $\hat{\mathbf{j}}$, $\hat{\mathbf{k}}$. Thus, treating ϵ as an infinitesimal length, the flux through dS is

---

[6] In suffix notation, $\nabla \cdot \mathbf{F} = \partial F_i / \partial x_i$.

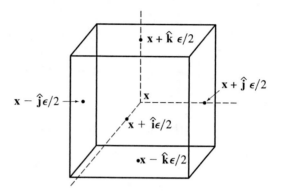

FIGURE 2.5   An infinitesimal cube. We consider the outward flux of a vector field to derive a coordinate-independent definition of divergence.

$$\oint_{dS} \mathbf{F} \cdot d\mathbf{A} = \sum_{i=1}^{3} \left[F_i(\mathbf{x} + \epsilon \hat{\mathbf{e}}_i /2) - F_i(\mathbf{x} - \epsilon \hat{\mathbf{e}}_i /2) \right] \epsilon^2$$

$$= \sum_{i=1}^{3} \frac{\partial F_i(\mathbf{x})}{\partial x_i} \epsilon^3 = (\mathbf{\nabla} \cdot \mathbf{F})\epsilon^3. \tag{2.40}$$

Comparing the last and first expressions, equation (2.40) says that the divergence is equal to the *flux per unit volume through an infinitesimal closed surface*. This statement serves as a coordinate-independent definition of the divergence; letting S be the boundary of V,

$$\mathbf{\nabla} \cdot \mathbf{F} = \lim_{V \to 0} \frac{1}{V} \oint_S \mathbf{F} \cdot d\mathbf{A}. \tag{2.41}$$

Another important differential operator is the Laplacian ∇^2, read as "del squared". Its action on a scalar function $f(\mathbf{x})$ is defined by

$$\nabla^2 f = \mathbf{\nabla} \cdot (\mathbf{\nabla} f). \tag{2.42}$$

In words, the Laplacian is the divergence of the gradient. It is a straightforward exercise to show that in Cartesian coordinates

$$\nabla^2 f = \frac{\partial^2 f}{\partial x^2} + \frac{\partial^2 f}{\partial y^2} + \frac{\partial^2 f}{\partial z^2}. \tag{2.43}$$

2.2.3 ■ Curl of a Vector Function

If $\mathbf{\nabla} \cdot \mathbf{F}$ is a scalar, then naturally $\mathbf{\nabla} \times \mathbf{F}$ ("del cross F") is a vector. It is called the *curl* of **F**. In Cartesian coordinates the curl may be written as a determinant operator,

$$\nabla \times \mathbf{F} = \begin{vmatrix} \hat{\mathbf{i}} & \hat{\mathbf{j}} & \hat{\mathbf{k}} \\ \partial/\partial x & \partial/\partial y & \partial/\partial z \\ F_x & F_y & F_z \end{vmatrix}. \tag{2.44}$$

Or, equivalently, in terms of the Levi-Civita tensor, the ith component of $\nabla \times \mathbf{F}$ is

$$(\nabla \times \mathbf{F})_i = \epsilon_{ijk}\frac{\partial F_k}{\partial x_j} \tag{2.45}$$

(where the sum over j and k from 1 to 3 is implied).[7]

The curl is a measure of vorticity, i.e., how the vector function curls around the point $\mathbf{x}$. Let dS be an infinitesimal square of size $\epsilon \times \epsilon$ centered at $\mathbf{x}$, aligned with the Cartesian directions $\hat{\mathbf{e}}_i$ and $\hat{\mathbf{e}}_j$. (Remember, $\hat{\mathbf{e}}_1, \hat{\mathbf{e}}_2, \hat{\mathbf{e}}_3$ stand for $\hat{\mathbf{i}}, \hat{\mathbf{j}}, \hat{\mathbf{k}}$.) The line integral of $\mathbf{F}(\mathbf{x})$, counterclockwise around the perimeter $dP(ij)$ of the square, illustrated in Fig. 2.6, is

$$\oint_{dP(ij)} \mathbf{F} \cdot d\boldsymbol{\ell} = \epsilon F_i(\mathbf{x} - \epsilon\hat{\mathbf{e}}_j/2) + \epsilon F_j(\mathbf{x} + \epsilon\hat{\mathbf{e}}_i/2)$$

$$- \epsilon F_i(\mathbf{x} + \epsilon\hat{\mathbf{e}}_j/2) - \epsilon F_j(\mathbf{x} - \epsilon\hat{\mathbf{e}}_i/2)$$

$$= \left(\frac{\partial F_j}{\partial x_i} - \frac{\partial F_i}{\partial x_j}\right)\epsilon^2 = (\nabla \times \mathbf{F})_k\,\epsilon^2 \tag{2.46}$$

where (ijk) is a cyclic permutation of (123). The four terms in the second expression come from the four sides of the square, starting at the bottom and proceeding counterclockwise, and again ϵ is treated as an infinitesimal. Comparing the last and first expressions, (2.46) says that the curl of $\mathbf{F}$ is equal to the *circulation of* $\mathbf{F}$

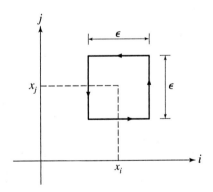

FIGURE 2.6 An infinitesimal square. We consider the counterclockwise circulation of a vector field to derive a coordinate-independent definition of curl.

[7]The student should beware that (2.44) and (2.45) are not correct for cylindrical or spherical coordinates. The correct expressions for curvilinear coordinates are derived in Sec. 2.4.

per unit area around an infinitesimal loop. (By the circulation we mean $\oint \mathbf{F} \cdot d\boldsymbol{\ell}$.) This property provides a coordinate-independent definition of the curl. To define the component of $\nabla \times \mathbf{F}$ in a specified direction $\hat{\mathbf{n}}$, let C denote a planar closed curve with normal vector $\hat{\mathbf{n}}$ and area A; then

$$\hat{\mathbf{n}} \cdot (\nabla \times \mathbf{F}) = \lim_{A \to 0} \frac{1}{A} \oint_C \mathbf{F} \cdot d\boldsymbol{\ell}. \tag{2.47}$$

As an example of the direction of $\nabla \times \mathbf{F}$, if the curve C in (2.47) lies in the xy plane and is traversed counterclockwise, then the left-hand side of (2.47) is the $+z$ component of the curl.

EXAMPLE 2   Consider the scalar function $f(\mathbf{x}) = Cz$. Calculate ∇f and interpret the result geometrically.

The gradient is $\nabla f = \hat{\mathbf{k}} df/dz = C\hat{\mathbf{k}}$, a uniform vector field independent of $\mathbf{x}$. In general the gradient points in the direction of the maximum increase of the function, which in this case is the z direction. The magnitude of the gradient is the slope of the function in that direction, which in this case is the constant C.

EXAMPLE 3   Consider the vector function $\mathbf{F}(\mathbf{x}) = Cx\,\hat{\mathbf{i}}$. The only nonzero component of $\mathbf{F}$ is $F_x = Cx$. Calculate the divergence $\nabla \cdot \mathbf{F}$ and interpret the result geometrically.

Figure 2.7 shows the vectors at a few points in the xy plane, assuming $C > 0$. The divergence at an arbitrary point $\mathbf{x}$ is $\nabla \cdot \mathbf{F} = dF_x/dx = C$. In general $\nabla \cdot \mathbf{F}$ describes the net flux of $\mathbf{F}$ away from $\mathbf{x}$. In Fig. 2.7 the net flux is the same at all points, and positive. For example, on a small surface surrounding $\mathbf{x}$ in the region $x > 0$, the flow away from $\mathbf{x}$ on the right side is greater than the flow toward $\mathbf{x}$ on

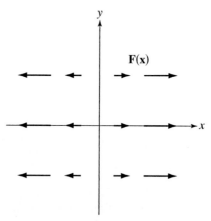

FIGURE 2.7   Example 3. A few vectors of the vector function $\mathbf{F}(\mathbf{x}) = Cx\,\hat{\mathbf{i}}$ are shown, illustrating a vector function with nonzero divergence.

the left. For **x** in the region $x < 0$ the divergence is also positive because there is a flux away from **x**.

EXAMPLE 4 Consider the vector function $\mathbf{G(x)} = Cx\,\hat{\mathbf{j}}$. The only nonzero component of **G** is $G_y = Cx$. Calculate the curl $\nabla \times \mathbf{G}$ and interpret the result geometrically.

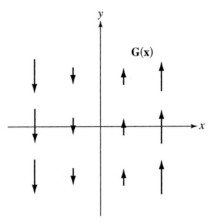

FIGURE 2.8 **Example 4.** A few vectors of the vector function $\mathbf{G(x)} = Cx\,\hat{\mathbf{j}}$ are shown, illustrating a vector function with nonzero curl.

Figure 2.8 shows the vectors at a few points in the xy plane, assuming $C > 0$. The curl at an arbitrary point **x** is $\nabla \times \mathbf{G} = \hat{\mathbf{k}}dG_y/dx = C\hat{\mathbf{k}}$. In general $\nabla \times \mathbf{G}$ describes the circulation of **G** around **x**. In Fig. 2.8 there is a counterclockwise circulation around the z axis, independent of **x**. For example, consider a small loop in the region $x > 0$. The flow in the $+y$ direction on the right side of the loop is greater than flow in the $+y$ direction on the left side, so there is a net circulation around the loop.

2.2.4 ■ Del Identities

We often encounter ∇ acting on products of functions. Also, an expression may contain multiple derivatives. It is important to derive some general identities for expressions of this kind, so that when such an expression arises in a calculation it can immediately be simplified by substitution from the general rule. Table 2.2 lists some useful identities involving ∇.

The identities in Table 2.2 can be derived by expanding them in Cartesian components, or more compactly using suffix notation. For example, consider the first identity in the table. In Cartesian components it is

TABLE 2.2   Del identities. Here f and g denote scalar functions of $\mathbf{x}$, and $\mathbf{F}$ and $\mathbf{G}$ denote vector functions of $\mathbf{x}$.

Derivatives of products:

$$\mathbf{\nabla}(fg) \qquad = f\mathbf{\nabla}g + g\mathbf{\nabla}f$$
$$\mathbf{\nabla}\cdot(g\mathbf{F}) \qquad = g\mathbf{\nabla}\cdot\mathbf{F} + \mathbf{\nabla}g\cdot\mathbf{F}$$
$$\mathbf{\nabla}\times(g\mathbf{F}) \qquad = g\mathbf{\nabla}\times\mathbf{F} + \mathbf{\nabla}g\times\mathbf{F}$$
$$\mathbf{\nabla}\cdot(\mathbf{F}\times\mathbf{G}) = (\mathbf{\nabla}\times\mathbf{F})\cdot\mathbf{G} - \mathbf{F}\cdot(\mathbf{\nabla}\times\mathbf{G})$$
$$\mathbf{\nabla}\times(\mathbf{F}\times\mathbf{G}) = (\mathbf{G}\cdot\mathbf{\nabla})\mathbf{F} - (\mathbf{F}\cdot\mathbf{\nabla})\mathbf{G} + \mathbf{F}(\mathbf{\nabla}\cdot\mathbf{G}) - \mathbf{G}(\mathbf{\nabla}\cdot\mathbf{F})$$

Products of derivatives:

$$\mathbf{\nabla}\times(\mathbf{\nabla}f) \qquad = 0$$
$$\mathbf{\nabla}\cdot(\mathbf{\nabla}\times\mathbf{F}) \qquad = 0$$
$$\mathbf{\nabla}\times(\mathbf{\nabla}\times\mathbf{F}) = \mathbf{\nabla}(\mathbf{\nabla}\cdot\mathbf{F}) - \nabla^2\mathbf{F}$$

$$\mathbf{\nabla}(fg) = \hat{\mathbf{i}}\frac{\partial}{\partial x}(fg) + \hat{\mathbf{j}}\frac{\partial}{\partial y}(fg) + \hat{\mathbf{k}}\frac{\partial}{\partial z}(fg)$$

$$= f\left(\hat{\mathbf{i}}\frac{\partial g}{\partial x} + \hat{\mathbf{j}}\frac{\partial g}{\partial y} + \hat{\mathbf{k}}\frac{\partial g}{\partial z}\right) + g\left(\hat{\mathbf{i}}\frac{\partial f}{\partial x} + \hat{\mathbf{j}}\frac{\partial f}{\partial y} + \hat{\mathbf{k}}\frac{\partial f}{\partial z}\right)$$

$$= f\mathbf{\nabla}g + g\mathbf{\nabla}f.$$

More compactly, in suffix notation $\mathbf{\nabla}f$ is $\partial f/\partial x_i$, so the identity is

$$\frac{\partial(fg)}{\partial x_i} = f\frac{\partial g}{\partial x_i} + \frac{\partial f}{\partial x_i}g; \qquad (2.48)$$

but this is just the Leibniz rule for differentiating a product.

Or, consider the second identity. Writing both sides of the identity in suffix notation, the identity is

$$\frac{\partial(gF_i)}{\partial x_i} = g\frac{\partial F_i}{\partial x_i} + \frac{\partial g}{\partial x_i}F_i. \qquad (2.49)$$

In each term i is summed from 1 to 3 in accord with the summation convention. The equation is again just the Leibniz rule, so the identity is proven.[8] These two quick derivations illustrate the power of the suffix notation.

As an example involving two derivatives, consider $\mathbf{\nabla}\times(\mathbf{\nabla}f)$. In suffix notation this vector is

$$(\mathbf{\nabla}\times\mathbf{\nabla}f)_i = \epsilon_{ijk}\frac{\partial^2 f}{\partial x_j \partial x_k}.$$

[8] As an exercise the reader should prove the identity again by writing out the Cartesian components explicitly.

The factor $\partial^2 f/\partial x_j \partial x_k$ is symmetric under exchange of j and k, whereas the factor ϵ_{ijk} is antisymmetric, and this implies that the right-hand side is 0: Interchanging jk in the first step below produces a minus sign from ϵ_{ijk}. But j and k are dummy indices, summed from 1 to 3; so *renaming* the dummy indices ($k \rightarrow j$ and $j \rightarrow k$) in the second step we have

$$\epsilon_{ijk} \frac{\partial^2 f}{\partial x_j \partial x_k} = -\epsilon_{ikj} \frac{\partial^2 f}{\partial x_k \partial x_j} = -\epsilon_{ijk} \frac{\partial^2 f}{\partial x_j \partial x_k}.$$

That is, the left-hand side is equal to its negative, so it must be 0. (See also Exercise 4.) Hence the identity $\nabla \times \nabla f = 0$. In words, *the curl of the gradient of a scalar function is identically* 0. The identity is true no matter what the function $f(\mathbf{x})$ is. Similarly, consider $\nabla \cdot (\nabla \times \mathbf{F})$. In suffix notation this scalar is

$$\nabla \cdot (\nabla \times \mathbf{F}) = \epsilon_{ijk} \frac{\partial^2 F_k}{\partial x_i \partial x_j};$$

again we have a suffix sum of a symmetric and an antisymmetric tensor, which is 0,[9] so $\nabla \cdot (\nabla \times \mathbf{F}) = 0$. *The divergence of the curl of a vector function is identically zero.* This and the previous identity are important in electromagnetic theory, because they lead to the introduction of scalar and vector potentials for the electric and magnetic fields.

We have used suffix notation, the Einstein summation convention, and the Levi-Civita tensor to derive useful identities, but these techniques will not be needed for many calculations in this book, because we can simply apply the relevant identity from Table 2.2 when needed.

2.3 ■ INTEGRAL THEOREMS

The purpose of this section is to describe two theorems involving *integrals* in vector calculus—Gauss's theorem and Stokes's theorem—that are very important in electromagnetic field theory.

In our study of electromagnetism we will frequently encounter *surface integrals* and *line integrals*,[10] i.e., integrals of the form $\int_S f \, dA$ and $\int_C f \, d\ell$ where S is a surface with area element dA and C a curve with length element $d\ell$. The *flux* of a vector field $\mathbf{F}$ through S is the surface integral of $\hat{\mathbf{n}} \cdot \mathbf{F}(\mathbf{x})$ where $\hat{\mathbf{n}}$ is the unit normal vector at the point $\mathbf{x}$ on S. Positive flux means $\mathbf{F}$ points through S in the sense of $\hat{\mathbf{n}}$. The *circulation* of $\mathbf{F}$ around a loop C is the line integral of $\hat{\mathbf{t}} \cdot \mathbf{F}(\mathbf{x})$ where $\hat{\mathbf{t}}$ is the unit tangent vector at the point $\mathbf{x}$ on C. Positive circulation means $\mathbf{F}$ loops around C in the sense of $\hat{\mathbf{t}}$.

---

[9]See Exercise 4.
[10]References [1] and [2] at the end of this chapter, and any of the introductory physics textbooks listed at the end of Chap. 1, have good discussions on surface integrals and line integrals.

2.3.1 ■ Gauss's Theorem

Gauss's theorem, also called the divergence theorem, states that the flux of a vector quantity outward through a *closed surface* S is equal to the integral of the divergence of the function in the enclosed volume V,

$$\oint_S \mathbf{F} \cdot d\mathbf{A} = \int_V \mathbf{\nabla} \cdot \mathbf{F} d^3 x. \qquad (2.50)$$

$d\mathbf{A}$ is a directed area element on S; for an infinitesimal patch on S the direction of $d\mathbf{A}$ is normal to S and the magnitude is the area dA. Also, $d^3 x$ is a volume element in V.[11] The "O-integral" notation $\oint_S$ indicates that the integration region S is a closed surface, i.e., a surface with no boundary curve, like a sphere or cube.

We have already seen an example of Gauss's theorem, for an infinitesimal surface, in (2.40). The mathematical proof for an arbitrary surface can be found in math books, such as Refs. [1] and [2]. The basic idea of the proof is this: Subdivide the volume V into infinitesimal volume elements. Apply (2.40) to each subvolume, i.e., convert the integral of $\mathbf{\nabla} \cdot \mathbf{F}$ in the subvolume to the surface flux of $\mathbf{F}$ through the boundary of the subvolume. Then note that for each subvolume-boundary-surface in the interior of V there are two flux integrals, which cancel in the total integral; because each interior surface belongs to the boundary of two adjacent subvolumes, with the flux in opposite directions from the two subvolumes. This point is illustrated in Fig. 2.9. All that is left after canceling the internal fluxes is the flux through the outer surface S; hence (2.50).

Gauss's theorem is reminiscent of the fundamental theorem of calculus,

$$g(b) - g(a) = \int_a^b g'(x)\, dx$$

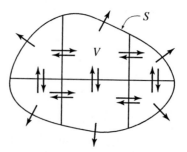

FIGURE 2.9   Illustration of the proof of Gauss's theorem. A volume V is subdivided into very small subvolumes. The integral of $\mathbf{\nabla} \cdot \mathbf{F}$ in V is the sum of integrals in the subvolumes. Each subvolume integral is equal to the flux of $\mathbf{F}$ outward through its boundary surface. The arrows indicate outward normals $d\mathbf{A}$. The flux integrals over interior surfaces cancel in pairs, leaving just the flux outward through the boundary surface S.

[11]We could use dV to denote the volume element, but that is *potentially* confusing because V also denotes the *potential*! The notation $d^3 x$ emphasizes that it is a 3D element.

for any function $g(x)$. In words, the integral of the derivative is determined by the function at the boundary points of the range of integration. Similarly in (2.50) the integral of the divergence is determined by the function at the boundary S of the region of integration V. But in (2.50) the region of integration is a volume and the boundary is a closed surface, over which $\mathbf{F} \cdot \hat{\mathbf{n}}$ is integrated.

Gauss's theorem may appear abstract at first sight and this discussion of it has been quite brief. But it will become more familiar in Chapter 3 where we will use it often. We'll also gain good physical insights into its meaning as we apply it to electric and magnetic systems throughout this book.

2.3.2 ■ Stokes's Theorem

Stokes's theorem states that the circulation of a vector function around a closed curve C is equal to the flux of vorticity[12] through any surface S bounded by C,

$$\oint_C \mathbf{F} \cdot d\boldsymbol{\ell} = \int_S (\nabla \times \mathbf{F}) \cdot d\mathbf{A}. \qquad (2.51)$$

$d\boldsymbol{\ell}$ is a length element of C tangent to C; that is, $d\boldsymbol{\ell} = \hat{\mathbf{t}}d\ell$ where $\hat{\mathbf{t}}$ is the unit tangent vector. Also, $d\mathbf{A}$ is a directed area element normal to S; that is, $d\mathbf{A} = \hat{\mathbf{n}}dA$ where $\hat{\mathbf{n}}$ is the unit normal vector. The direction of $d\mathbf{A}$ is coordinated with that of $d\boldsymbol{\ell}$ by the right-hand rule: With the fingers curling in the direction of $d\boldsymbol{\ell}$, the thumb points in the direction of $d\mathbf{A}$. The "O-integral" notation indicates that C is a closed curve, i.e., a loop. However, the surface integral is over an *open surface* S whose boundary curve is C.

We have already seen an example of Stokes's theorem, for an infinitesimal loop, in (2.46). The proof for an arbitrary curve can be found in Refs. [1] and [2]. The basic idea is this: Subdivide the surface S into infinitesimal area elements. Apply (2.46) to each subarea, i.e., convert the flux integral of $\nabla \times \mathbf{F}$ through the subarea to the line integral of $\mathbf{F}$ around the boundary of the subarea. Then note that the line integrals along the subarea boundary segments in the interior of S cancel in pairs, because each interior segment belongs to the boundary of two adjacent subareas, with the line integral in opposite directions for the two subareas. This point is illustrated in Fig. 2.10. All that is left after canceling the interior line integrals is the line integral around the boundary curve C; hence (2.51). It is obvious from the proof that S in (2.51) can be *any* surface bounded by C.[13]

EXAMPLE 5   Consider the vector function $\mathbf{F}(\mathbf{x}) = \mathbf{x}$. Calculate the divergence and curl of $\mathbf{F}(\mathbf{x})$, the flux through a sphere centered at the origin, and the circulation around any circle on the sphere. Verify that Gauss's theorem and Stokes's theorem are obeyed.

[12]"Vorticity" in this context is just another word for $\nabla \times \mathbf{F}$.
[13]See Exercise 11.

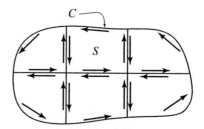

FIGURE 2.10   Illustration of the proof of Stokes's theorem. C is a closed curve. A surface S bounded by C is subdivided into very small subareas. The flux of $\mathbf{\nabla} \times \mathbf{F}$ over S is the sum of integrals over the subareas. Each subarea integral is equal to the circulation of $\mathbf{F}$ around its boundary curve. The arrows indicate the tangent vectors $d\boldsymbol{\ell}$. The line integrals of $\mathbf{F}$ along interior segments cancel in pairs, leaving just the circulation around C.

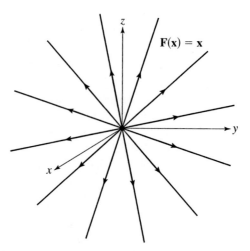

FIGURE 2.11   Example 5. Tangent curves of the vector function $\mathbf{F}(\mathbf{x}) = \mathbf{x}$.

    Using Cartesian coordinates, it is straightforward to show that $\mathbf{\nabla} \cdot \mathbf{F} = 3$ and $\mathbf{\nabla} \times \mathbf{F} = 0$. Figure 2.11 shows the tangent curves[14] of $\mathbf{F}(\mathbf{x})$, which are radial straight lines. The curves diverge, that is, *they spread out* from any point $\mathbf{x}$; but there is no curl, that is, *vorticity*, at any $\mathbf{x}$. The flux of $\mathbf{F}$ through a sphere of radius r centered at the origin is $\oint \mathbf{F} \cdot d\mathbf{A} = rA = 4\pi r^3$ where A is the surface area of the sphere. This result agrees with Gauss's theorem, because $\int \mathbf{\nabla} \cdot \mathbf{F} d^3x = 3V = 4\pi r^3$ where V is the volume of the sphere. The circulation of $\mathbf{F}$ around any circle on the sphere is 0, in agreement with Stokes's theorem, because $\mathbf{F}$ is everywhere orthogonal to the circle, so $\mathbf{F} \cdot d\boldsymbol{\ell} = 0$.

---

[14]The *tangent curves* of a vector function are a family of curves everywhere tangent to the vectors. For example, the tangent curves of the electric field are the electric field lines.

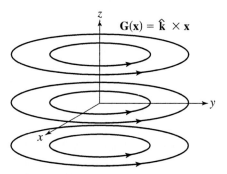

FIGURE 2.12   Example 6. Tangent curves of the vector function $\mathbf{G}(\mathbf{x}) = \hat{\mathbf{k}} \times \mathbf{x}$.

EXAMPLE 6   Consider the vector function $\mathbf{G}(\mathbf{x}) = \hat{\mathbf{k}} \times \mathbf{x}$. In cylindrical coordinates, $\mathbf{G}(\mathbf{x}) = \hat{\boldsymbol{\phi}} r$ where $r = \sqrt{x^2 + y^2}$. Calculate the curl and divergence, the circulation around a circle parallel to the xy plane centered at the z axis, and the flux through a sphere centered at the origin. Verify that Stokes's theorem and Gauss's theorem are obeyed.

Using Cartesian coordinates, in which $\mathbf{G}(\mathbf{x}) = -y\,\hat{\mathbf{i}} + x\,\hat{\mathbf{j}}$, it is straightforward to show that $\nabla \times \mathbf{G} = 2\hat{\mathbf{k}}$ and $\nabla \cdot \mathbf{G} = 0$. Figure 2.12 shows the tangent curves of $\mathbf{G}(\mathbf{x})$, which are circles around the z axis. There is vorticity but no divergence at any point. The circulation of $\mathbf{G}$ around a circle of radius r is $\oint \mathbf{G} \cdot d\boldsymbol{\ell} = rC = 2\pi r^2$ where C is the circumference of the circle. This result agrees with Stokes's theorem, because $\int \nabla \times \mathbf{G} \cdot d\mathbf{A} = 2A = 2\pi r^2$ where A is the area inside the circle. The flux of $\mathbf{G}$ through a sphere centered at the origin is 0, in agreement with Gauss's theorem, because $\mathbf{G}$ is everywhere tangent to the sphere, so $\mathbf{G} \cdot d\mathbf{A} = 0$.

2.3.3 ▨ Vector Calculus in Fluid Mechanics

Vector calculus is the language of electromagnetic field theory. This mathematics is also used in other areas of theoretical physics, notably fluid mechanics. The basic parameters for fluid motion are density, pressure, and velocity. The density $\rho(\mathbf{x}, t)$ and pressure $p(\mathbf{x}, t)$ are scalar functions. The fluid velocity $\mathbf{v}(\mathbf{x}, t)$ is a vector function, that depends on position within the fluid. The analysis of fluid motion relies on vector calculus.

The divergence and curl operators describe aspects of the fluid motion. For example, conservation of mass, which holds at every point in the fluid, implies the equation

$$\frac{\partial \rho}{\partial t} + \nabla \cdot (\rho \mathbf{v}) = 0, \tag{2.52}$$

which is called the continuity equation. The divergence of $\rho\mathbf{v}$ at $\mathbf{x}$ is the outward mass flux per unit volume there, by the definition (2.41) of divergence; therefore $\nabla \cdot (\rho\mathbf{v})$ must equal $-\partial\rho/\partial t$ because mass is conserved. Or, applying Gauss's theorem to (2.52), the mass flux outward through any closed surface S equals the rate of decrease of mass in the enclosed volume V:

$$\oint_S \rho\mathbf{v} \cdot d\mathbf{A} = -\frac{d}{dt} \int_V \rho d^3x.$$

As another example, the *vorticity* at a point is defined by $\nabla \times \mathbf{v}(\mathbf{x})$. If there is a whirlpool, i.e., the fluid swirls around some point, then $\oint_C \mathbf{v} \cdot d\boldsymbol{\ell}$ is nonzero if C surrounds the point. By Stokes's theorem, the vorticity is nonzero in the neighborhood of the whirlpool; $\mathbf{v}$ curls around a point of vorticity.

When studying electromagnetism, we will develop a geometric intuition for the direction and magnitude of fields. Understanding the vector differential operators, $\nabla\cdot$ and $\nabla\times$, and their relation to flux and circulation integrals, is a necessary step in that direction. It may sometimes help to picture $\mathbf{E}$ or $\mathbf{B}$ as a "flow" (even if the fields are static). The divergence ($\nabla\cdot$) and curl ($\nabla\times$) can be visualized in fluid flow which may help us to comprehend their meaning in field theory.

2.4 ■ CURVILINEAR COORDINATES

In problems with cylindrical or spherical symmetry it is more convenient to use appropriate curvilinear coordinates than Cartesian coordinates. Therefore, we will derive the formulas for calculating gradient, divergence, and curl in curvilinear coordinates. It is a common error to apply a Cartesian equation, (2.34), (2.39), (2.43), or (2.44), for cylindrical or spherical coordinates. As we shall see, these equations must be modified for non-Cartesian coordinates.

2.4.1 ■ General Derivations

In general terms, let u_1, u_2, u_3 denote three coordinates (distances or angles) that specify the points in three dimensions. The corresponding unit vectors $\hat{\mathbf{e}}_1, \hat{\mathbf{e}}_2, \hat{\mathbf{e}}_3$, point in the directions of independent positive displacements of u_1, u_2, u_3, respectively. These unit vectors are assumed to form an orthogonal triad at each point in space. The infinitesimal displacement $d\mathbf{s}$ in space that results from changing the coordinates by du_1, du_2, du_3 can always be written in the form

$$d\mathbf{s} = \hat{\mathbf{e}}_1 h_1 du_1 + \hat{\mathbf{e}}_2 h_2 du_2 + \hat{\mathbf{e}}_3 h_3 du_3. \tag{2.53}$$

Here h_1, h_2, h_3 are *scale factors* that relate distance (ds_i) to change of coordinate (du_i); in other words h_i is the ratio ds_i/du_i, for $i = 1, 2, 3$. The scale factors are, in general, functions of position (u_1, u_2, u_3). Also, the scale factor for a distance coordinate is dimensionless, while that for an angle coordinate has units of length. The simplest case is Cartesian coordinates, for which $u_1 = x$, $u_2 = y$, and $u_3 = z$. The scale factors are $h_1 = h_2 = h_3 = 1$ in this case. But next we will derive equations for grad, div and curl in general coordinates. And then we will write them specifically for cylindrical and spherical coordinates.

Gradient

The gradient of a scalar function f is defined by the differential relation $df = \nabla f \cdot d\mathbf{s}$. Viewing f as a function of (u_1, u_2, u_3), we have

$$df = \frac{\partial f}{\partial u_1} du_1 + \frac{\partial f}{\partial u_2} du_2 + \frac{\partial f}{\partial u_3} du_3, \tag{2.54}$$

and

$$\nabla f \cdot d\mathbf{s} = (\nabla f)_1 \, h_1 du_1 + (\nabla f)_2 \, h_2 du_2 + (\nabla f)_3 \, h_3 du_3. \tag{2.55}$$

Equating the coefficients of the independent differentials du_1, du_2, du_3, we find that the component of ∇f along $\hat{\mathbf{e}}_i$ is

$$(\nabla f)_i = \frac{1}{h_i} \frac{\partial f}{\partial u_i}. \tag{2.56}$$

(There is no sum over i here; this is the ith component of the vector.) Note that this result differs from the Cartesian formula (2.34) unless $h_1 = h_2 = h_3 = 1$, which is the condition for Cartesian coordinates.

Divergence

To derive the formula for the divergence of a vector function $\mathbf{F}(u_1, u_2, u_3)$, we apply the definition (2.41) to the infinitesimal cubic volume defined by displacements du_1, du_2, du_3 (see Fig. 2.13):

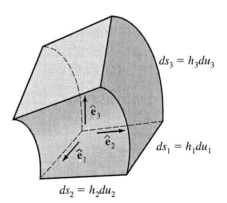

FIGURE 2.13   An infinitesimal curvilinear volume. The volume integral of $\nabla \cdot \mathbf{F}$ is the total outward flux of $\mathbf{F}$ through the six faces. ds_1, ds_2, and ds_3 are the elements of length in the three orthogonal directions.

$$\mathbf{\nabla} \cdot \mathbf{F} \, dV = \left[F_1 h_2 h_3 |_{u_1+du_1} - F_1 h_2 h_3 |_{u_1} \right] du_2 du_3$$
$$+ \left[F_2 h_3 h_1 |_{u_2+du_2} - F_2 h_3 h_1 |_{u_2} \right] du_3 du_1$$
$$+ \left[F_3 h_1 h_2 |_{u_3+du_3} - F_3 h_1 h_2 |_{u_3} \right] du_1 du_2 \qquad (2.57)$$

where dV is the infinitesimal volume

$$dV = (h_1 du_1)(h_2 du_2)(h_3 du_3) . \qquad (2.58)$$

The right-hand side of (2.57) is the outward flux through the boundary surface of dV, consisting of 6 terms from the 6 faces of the cube. F_1, F_2, F_3 are the projections of $\mathbf{F}$ onto the directions $\hat{\mathbf{e}}_1, \hat{\mathbf{e}}_2, \hat{\mathbf{e}}_3$. For example, the flux of $\mathbf{F}$ through a face with normal $\hat{\mathbf{e}}_1$ is $\pm F_1 (h_2 du_2)(h_3 du_3)$, with whichever sign corresponds to outward flux. Adding up the six fluxes gives (2.57); this equation is illustrated in Fig. 2.13. Dividing out the differentials du_1, du_2, du_3, we find

$$\mathbf{\nabla} \cdot \mathbf{F} = \frac{1}{h_1 h_2 h_3} \left[\frac{\partial}{\partial u_1} (F_1 h_2 h_3) + \frac{\partial}{\partial u_2} (F_2 h_3 h_1) + \frac{\partial}{\partial u_3} (F_3 h_1 h_2) \right] . \quad (2.59)$$

Again, this differs from the Cartesian formula (2.39) in general. Recall that h_1, h_2, h_3 can be functions of position. For example, $h_2 h_3$ may have a different value at $u_1 + du_1$ than at u_1. The products of scale factors must be included in the derivatives in (2.59).

Laplacian
The Laplacian operator $\mathbf{\nabla}^2$, which we will encounter in electrostatics, is defined by the relation $\mathbf{\nabla}^2 f = \mathbf{\nabla} \cdot (\mathbf{\nabla} f)$. Combining (2.56) and (2.59) we have for general curvilinear coordinates

$$\mathbf{\nabla}^2 f = \frac{1}{h_1 h_2 h_3} \sum_{(ijk)} \frac{\partial}{\partial u_i} \left(\frac{h_j h_k}{h_i} \frac{\partial f}{\partial u_i} \right) \qquad (2.60)$$

where the sum is over the three cyclic permutations of (123).

Curl
Curl, being a vector operator, is a bit tricky. To derive the formula for the curl of a vector function $\mathbf{F}(u_1, u_2, u_3)$, let's start by applying (2.47), which is the infinitesimal version of Stokes's theorem, to the infinitesimal rectangular area produced by displacements du_1 and du_2 (see Fig. 2.14):

$$(\mathbf{\nabla} \times \mathbf{F})_3 (h_1 du_1)(h_2 du_2) = \left(F_2 h_2 |_{u_1+du_1} - F_2 h_2 |_{u_1} \right) du_2$$
$$- \left(F_1 h_1 |_{u_2+du_2} - F_1 h_1 |_{u_2} \right) du_1 . \quad (2.61)$$

The right-hand side is the circulation integral, consisting of four terms from the four sides of the rectangular area, as illustrated in Fig. 2.14. Thus, canceling the

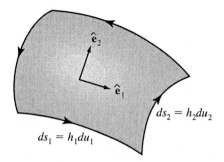

FIGURE 2.14   An infinitesimal curvilinear loop. The surface integral of $\hat{\mathbf{n}} \cdot (\nabla \times \mathbf{F})$ is the sum of the four line integrals.

displacements $du_1 du_2$,

$$(\nabla \times \mathbf{F})_3 = \frac{1}{h_1 h_2} \left[\frac{\partial}{\partial u_1} (F_2 h_2) - \frac{\partial}{\partial u_2} (F_1 h_1) \right]. \tag{2.62}$$

The other components are similar; replace (123) by (231) or (312). We may express the curl as a determinant,

$$\nabla \times \mathbf{F} = \begin{vmatrix} \hat{\mathbf{e}}_1 / h_2 h_3 & \hat{\mathbf{e}}_2 / h_3 h_1 & \hat{\mathbf{e}}_3 / h_1 h_2 \\ \partial / \partial u_1 & \partial / \partial u_2 & \partial / \partial u_3 \\ h_1 F_1 & h_2 F_2 & h_3 F_3 \end{vmatrix}. \tag{2.63}$$

To use (2.63), expand the determinant in terms of minors of the first row, taking care to use the correct signs. The scale factor and component of $\mathbf{F}$ must be differentiated together. For example, one term in the expansion is

$$\frac{\hat{\mathbf{e}}_1}{h_2 h_3} \frac{\partial}{\partial u_2} (h_3 F_3),$$

and there are five other terms.

2.4.2 ■ Cartesian, Cylindrical, and Spherical Coordinates

Cartesian Coordinates

Cartesian coordinates are $u_1 = x$, $u_2 = y$, and $u_3 = z$. The scale factors h_1, h_2, h_3 are all 1. The position vector of the point with coordinates (x, y, z) is

$$\mathbf{x} = x\,\hat{\mathbf{i}} + y\,\hat{\mathbf{j}} + z\hat{\mathbf{k}}.$$

Then the above equations for gradient, divergence, and curl reduce to the Cartesian equations.

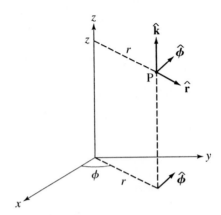

FIGURE 2.15   **Cylindrical coordinates** r, ϕ, z. The unit vectors $\hat{\mathbf{r}}$, $\hat{\boldsymbol{\phi}}$, and $\hat{\mathbf{k}}$ at P point in the directions of increasing r, ϕ, and z, respectively.

Cylindrical Coordinates

Cylindrical coordinates are $u_1 = r$, $u_2 = \phi$, and $u_3 = z$, as defined in Fig. 2.15. The formulas for differential operators in cylindrical coordinates are recorded in Table 2.3. It is also important to know how to write the position vector $\mathbf{x}$ in cylin-

TABLE 2.3   **Cylindrical coordinates.**

position	$\mathbf{x} = r\hat{\mathbf{r}} + z\hat{\mathbf{k}}$
displacement	$d\mathbf{s} = \hat{\mathbf{r}}dr + \hat{\boldsymbol{\phi}}rd\phi + \hat{\mathbf{k}}dz$
scale factors	$h_r = 1$, $h_\phi = r$, $h_z = 1$
volume element	$dV = r\,dr\,d\phi\,dz$
gradient	$\nabla f = \hat{\mathbf{r}}\dfrac{\partial f}{\partial r} + \hat{\boldsymbol{\phi}}\dfrac{1}{r}\dfrac{\partial f}{\partial \phi} + \hat{\mathbf{k}}\dfrac{\partial f}{\partial z}$
divergence	$\nabla \cdot \mathbf{F} = \dfrac{1}{r}\dfrac{\partial}{\partial r}(rF_r) + \dfrac{1}{r}\dfrac{\partial F_\phi}{\partial \phi} + \dfrac{\partial F_z}{\partial z}$
Laplacian	$\nabla^2 f = \dfrac{1}{r}\dfrac{\partial}{\partial r}\left(r\dfrac{\partial f}{\partial r}\right) + \dfrac{1}{r^2}\dfrac{\partial^2 f}{\partial \phi^2} + \dfrac{\partial^2 f}{\partial z^2}$
curl	$(\nabla \times \mathbf{F})_r = \dfrac{1}{r}\dfrac{\partial F_z}{\partial \phi} - \dfrac{\partial F_\phi}{\partial z}$
	$(\nabla \times \mathbf{F})_\phi = \dfrac{\partial F_r}{\partial z} - \dfrac{\partial F_z}{\partial r}$
	$(\nabla \times \mathbf{F})_z = \dfrac{1}{r}\dfrac{\partial}{\partial r}\left(rF_\phi\right) - \dfrac{1}{r}\dfrac{\partial F_r}{\partial \phi}$

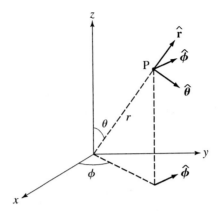

FIGURE 2.16 **Spherical polar coordinates** r, θ, ϕ. The unit vectors $\hat{\mathbf{r}}, \hat{\boldsymbol{\theta}}$, and $\hat{\boldsymbol{\phi}}$ at P point in the directions of increasing r, θ, and ϕ, respectively.

drical coordinates. Looking at Fig. 2.15 we see that

$$\mathbf{x} = r\hat{\mathbf{r}} + z\hat{\mathbf{k}}. \tag{2.64}$$

We may also express (x, y, z) in (r, ϕ, z):

$$x = r\cos\phi, \quad y = r\sin\phi, \quad z = z. \tag{2.65}$$

Spherical Coordinates
Spherical polar coordinates are $u_1 = r$, $u_2 = \theta$, and $u_3 = \phi$, as defined in Fig. 2.16. The formulas for differential operators in spherical coordinates are recorded in Table 2.4. Again, it is important to know the position vector in spherical coordinates,[15]

$$\mathbf{x} = r\hat{\mathbf{r}}. \tag{2.66}$$

We may also express (x, y, z) in (r, θ, ϕ),

$$x = r\sin\theta\cos\phi, \quad y = r\sin\theta\sin\phi, \quad z = r\cos\theta. \tag{2.67}$$

EXAMPLE 7 The unit basis vectors $\hat{\mathbf{r}}, \hat{\boldsymbol{\theta}}, \hat{\boldsymbol{\phi}}$ of spherical coordinates are vector functions of the position $\mathbf{x}$. For each of these vector functions calculate the divergence and curl, and interpret the results geometrically.

1. *Radial.* For the function $\mathbf{F}(\mathbf{x}) = \hat{\mathbf{r}}$, the components are $F_r = 1$, $F_\theta = F_\phi = 0$. Using the divergence and curl in spherical coordinates (Table 2.4) we find

---

[15]Failure to appreciate this little equation is the source of much confusion. A common mistake is to write $r\hat{\mathbf{r}} + \theta\hat{\boldsymbol{\theta}} + \phi\hat{\boldsymbol{\phi}}$, which doesn't even have consistent units!

TABLE 2.4   **Spherical coordinates.**

position	$\mathbf{x} = r\hat{\mathbf{r}}$
displacement	$d\mathbf{s} = \hat{\mathbf{r}}dr + \hat{\boldsymbol{\theta}}rd\theta + \hat{\boldsymbol{\phi}}r\sin\theta d\phi$
scale factors	$h_r = 1, h_\theta = r, h_\phi = r\sin\theta$
volume element	$dV = r^2 dr \sin\theta d\theta d\phi$
gradient	$\nabla f = \hat{\mathbf{r}}\dfrac{\partial f}{\partial r} + \hat{\boldsymbol{\theta}}\dfrac{1}{r}\dfrac{\partial f}{\partial \theta} + \hat{\boldsymbol{\phi}}\dfrac{1}{r\sin\theta}\dfrac{\partial f}{\partial \phi}$
divergence	$\nabla \cdot \mathbf{F} = \dfrac{1}{r^2}\dfrac{\partial}{\partial r}\left(r^2 F_r\right) + \dfrac{1}{r\sin\theta}\dfrac{\partial}{\partial \theta}(\sin\theta\, F_\theta) + \dfrac{1}{r\sin\theta}\dfrac{\partial F_\phi}{\partial \phi}$
Laplacian	$\nabla^2 f = \dfrac{1}{r^2}\dfrac{\partial}{\partial r}\left(r^2\dfrac{\partial f}{\partial r}\right) + \dfrac{1}{r^2\sin\theta}\dfrac{\partial}{\partial \theta}\left(\sin\theta\dfrac{\partial f}{\partial \theta}\right) + \dfrac{1}{r^2\sin^2\theta}\dfrac{\partial^2 f}{\partial \phi^2}$
curl	$(\nabla \times \mathbf{F})_r = \dfrac{1}{r\sin\theta}\left[\dfrac{\partial}{\partial \theta}(\sin\theta\, F_\phi) - \dfrac{\partial F_\theta}{\partial \phi}\right]$
	$(\nabla \times \mathbf{F})_\theta = \dfrac{1}{r}\left[\dfrac{1}{\sin\theta}\dfrac{\partial F_r}{\partial \phi} - \dfrac{\partial}{\partial r}(r F_\phi)\right]$
	$(\nabla \times \mathbf{F})_\phi = \dfrac{1}{r}\left[\dfrac{\partial}{\partial r}(r F_\theta) - \dfrac{\partial F_r}{\partial \theta}\right]$

$\nabla \cdot \hat{\mathbf{r}} = 2/r$ and $\nabla \times \hat{\mathbf{r}} = 0$. These results make sense geometrically, because the tangent curves of $\hat{\mathbf{r}}$ are radial lines, which diverge (spread out) but have no vorticity.

2. *Polar angle.* For the function $\mathbf{F}(\mathbf{x}) = \hat{\boldsymbol{\theta}}$, the components are $F_\theta = 1$, $F_r = F_\phi = 0$. The divergence and curl of this vector field are $\nabla \cdot \hat{\boldsymbol{\theta}} = (\cot\theta)/r$ and $\nabla \times \hat{\boldsymbol{\theta}} = \hat{\boldsymbol{\phi}}/r$. These results make sense geometrically. The tangent curves of $\hat{\boldsymbol{\theta}}$ are circles with constant longitude. These curves diverge at all points except the equator ($\theta = \pi/2$) where $\nabla \cdot \hat{\boldsymbol{\theta}} = 0$, and diverge most strongly near the poles ($\theta = 0$ or π) where $\nabla \cdot \hat{\boldsymbol{\theta}} = \pm\infty$. Also, they curl around the direction $\hat{\boldsymbol{\phi}}$ (draw a picture that shows this!), which explains why $\nabla \times \hat{\boldsymbol{\theta}} \propto \hat{\boldsymbol{\phi}}$.

3. *Azimuthal.* For the function $\mathbf{F}(\mathbf{x}) = \hat{\boldsymbol{\phi}}$, the components are $F_\phi = 1$, $F_r = F_\theta = 0$. The divergence and curl of this vector field are $\nabla \cdot \hat{\boldsymbol{\phi}} = 0$ and $\nabla \times \hat{\boldsymbol{\phi}} = (\hat{\mathbf{r}}\cot\theta - \hat{\boldsymbol{\theta}})/r = \hat{\mathbf{k}}/(r\sin\theta)$. (The last result follows from the relation $\hat{\mathbf{k}} = \hat{\mathbf{r}}\cos\theta - \hat{\boldsymbol{\theta}}\sin\theta$, between Cartesian and polar coordinates.) These equations make sense geometrically. The tangent curves of $\hat{\boldsymbol{\phi}}$ are circles with constant latitude. The tangent curves of $\hat{\boldsymbol{\phi}}$ curl around $\hat{\mathbf{k}}$, but do not diverge at any point.

2.5 ■ THE HELMHOLTZ THEOREM

We end this chapter with a theorem in vector calculus that requires an understanding of the basic concepts defined earlier.

Preliminaries

A vector function (or *field*) $\mathbf{F}(\mathbf{x})$ that has zero curl, $\nabla \times \mathbf{F} = 0$, is called *irrotational*. For instance, the electric field of any collection of static charges is irrotational. A vector function $\mathbf{G}(\mathbf{x})$ that has zero divergence, $\nabla \cdot \mathbf{G} = 0$, is called *solenoidal*. For instance, any magnetic field is solenoidal.

Any vector function $\mathbf{H}(\mathbf{x})$ can be written as the sum of an irrotational function $\mathbf{F}(\mathbf{x})$ and a solenoidal function $\mathbf{G}(\mathbf{x})$. The functions $\mathbf{F}$ and $\mathbf{G}$ are not necessarily unique. However, in some cases, if suitable boundary conditions are imposed, then the decomposition $\mathbf{H} = \mathbf{F} + \mathbf{G}$ is unique.

An irrotational field ($\nabla \times \mathbf{F} = 0$) can be written as a gradient, $\mathbf{F} = -\nabla \psi$. (We proved in Sec. 2.2.4 that the curl of $\nabla \psi$ is 0.) A solenoidal field ($\nabla \cdot \mathbf{G} = 0$) can be written as a curl, $\mathbf{G} = \nabla \times \mathbf{A}$. (We proved in Sec. 2.2.4 that the divergence of $\nabla \times \mathbf{A}$ is 0.) So, the decomposition of an arbitrary vector function $\mathbf{H}(\mathbf{x})$ into irrotational and solenoidal parts becomes $\mathbf{H}(\mathbf{x}) = -\nabla \psi + \nabla \times \mathbf{A}$.

The Helmholtz Theorem

The preliminaries bring us to *the Helmholtz theorem*, a general result in vector calculus with important applications in theoretical physics. Let $\mathbf{H}(\mathbf{x})$ be differentiable at all points in space, with divergence $\nabla \cdot \mathbf{H} \equiv d(\mathbf{x})$ and curl $\nabla \times \mathbf{H} \equiv \mathbf{c}(\mathbf{x})$. If $d(\mathbf{x})$ and $\mathbf{c}(\mathbf{x})$ approach 0 faster than r^{-2} as $r \to 0$, and $\mathbf{H}(\mathbf{x})$ approaches 0, then $\mathbf{H}(\mathbf{x}) = -\nabla \psi + \nabla \times \mathbf{A}$ where

$$\psi(\mathbf{x}) = \int \frac{d(\mathbf{x}')d^3x'}{4\pi |\mathbf{x} - \mathbf{x}'|}, \tag{2.68}$$

$$\mathbf{A}(\mathbf{x}) = \int \frac{\mathbf{c}(\mathbf{x}')d^3x'}{4\pi |\mathbf{x} - \mathbf{x}'|}. \tag{2.69}$$

The integration region is all of space. The proof of the theorem is given in Appendix B, but it requires some techniques that will be developed in Chapter 3.

The theorem has important applications in electromagnetic field theory, and also in other areas of theoretical physics. It implies that if the divergence and curl of a vector function (field) are known, then the function can be determined uniquely (under the assumptions of the theorem) by (2.68) and (2.69).

FURTHER READING

Either of the two books below provide a brief but sufficient introduction to vector calculus.

1. Paul C. Matthews, *Vector Calculus* (Springer-Verlag, Berlin, 1998). This book is especially recommended for learning suffix notation, which is important in advanced areas of theoretical physics and useful for deriving vector identities.

2. H. M. Schey, *Div, grad, curl, and all that: an informal text on vector calculus* (Norton, New York, 1992).

EXERCISES

Sec. 2.1. Vector Algebra

2.1. Determine the angle, in degrees, between a diagonal of a cube and an adjacent edge. (Hint: Use the unit cube $0 \leq x, y, z \leq 1$; one diagonal is along $\hat{\mathbf{i}} + \hat{\mathbf{j}} + \hat{\mathbf{k}}$, and an adjacent edge is along $\hat{\mathbf{k}}$.)

2.2. (a) Prove that $|\mathbf{A} \times \mathbf{B}|$ is equal to the area of the parallelogram defined by $\mathbf{A}$ and $\mathbf{B}$.

(b) Why is the equation $\mathbf{A} \times \mathbf{B} = AB \sin\theta$ false?

(c) Prove that $\mathbf{A} \cdot (\mathbf{B} \times \mathbf{C})$ is equal to the determinant

$$\begin{vmatrix} A_x & A_y & A_z \\ B_x & B_y & B_z \\ C_x & C_y & C_z \end{vmatrix}.$$

(d) Show that $\mathbf{A} \cdot (\mathbf{B} \times \mathbf{C})$ is equal to the volume of the parallelepiped defined by $\mathbf{A}, \mathbf{B}$, and $\mathbf{C}$. From this result prove that $\mathbf{A} \cdot (\mathbf{B} \times \mathbf{C}) = (\mathbf{A} \times \mathbf{B}) \cdot \mathbf{C}$.

2.3. (a) Prove equation (2.27). (Hint: Convince yourself that (2.27) has 81 components in general. However, it is sufficient to consider just those with $i = 1$ by symmetry. Then examine various combinations of $j\ell m$, and show that the left and right hand sides of (2.27) are equal in all cases.)

(b) Prove the identity

$$(\mathbf{A} \times \mathbf{B}) \cdot (\mathbf{C} \times \mathbf{D}) = (\mathbf{A} \cdot \mathbf{C})(\mathbf{B} \cdot \mathbf{D}) - (\mathbf{A} \cdot \mathbf{D})(\mathbf{B} \cdot \mathbf{C}).$$

2.4. Prove that if $S_{ij} = S_{ji}$ and $A_{ij} = -A_{ji}$, then $S_{ij}A_{ij} = 0$. The summation convention is understood. In words, the contraction of a symmetric tensor and an antisymmetric tensor is 0. ("Contraction" is a bit of jargon from tensors; it means the sum over dummy indices.)

2.5. (a) Write the 3×3 rotation matrix R for the rotation by angle θ about the z axis.

(b) Verify that the result of (a) is an orthogonal matrix; that is, $R^{-1} = R^T$.

Sec. 2.2. Vector Differential Operators

2.6. Prove these identities:

(a) $\nabla \cdot (g\mathbf{F}) = g\nabla \cdot \mathbf{F} + \nabla g \cdot \mathbf{F}$

(b) $\nabla \times (g\mathbf{F}) = g\nabla \times \mathbf{F} + \nabla g \times \mathbf{F}$

(c) $\nabla \cdot (\mathbf{F} \times \mathbf{G}) = \mathbf{G} \cdot (\nabla \times \mathbf{F}) - \mathbf{F} \cdot (\nabla \times \mathbf{G})$

(d) $\nabla \times (\nabla \times \mathbf{F}) = \nabla (\nabla \cdot \mathbf{F}) - \nabla^2\mathbf{F}$.

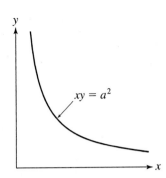

FIGURE 2.17   Exercise 7. A certain scalar function is given in the region bounded by the axes and the hyperbola.

2.7. Consider the function $f(x, y) = Cxy$ in the region bounded by the half-planes

$$\text{half plane } 1 = xz \text{ plane } (y = 0) \text{ with } x \geq 0,$$

$$\text{half plane } 2 = yz \text{ plane } (x = 0) \text{ with } y \geq 0,$$

and by the hyperbolic sheet $xy = a^2$, as shown in Fig. 2.17. (z is unbounded.)

(a) Show that $f(x, y)$ satisfies Laplace's equation, $\nabla^2 f = 0$.

(b) Calculate the boundary values of f on the three surfaces.

(c) Calculate ∇f, and sketch the tangent curves of the vector function ∇f.

2.8. (a) Calculate $\nabla \Phi$ where $\Phi(\mathbf{x}) = \mathbf{p} \cdot \mathbf{x}/r^3$. Here $\mathbf{p}$ is a constant vector, and $r = |\mathbf{x}|$. (Hint: Φ is fg where $f = \mathbf{p} \cdot \mathbf{x}$ and $g = r^{-3}$.) Sketch the tangent curves of the vector function $\nabla \Phi$, for $\mathbf{p} = p\hat{\mathbf{k}}$.

(b) Calculate $\nabla \times \mathbf{A}$ where $\mathbf{A}(\mathbf{x}) = \mathbf{m} \times \mathbf{x}/r^3$, and $\mathbf{m}$ is a constant vector. Sketch the tangent curves of $\nabla \times \mathbf{A}$, for $\mathbf{m} = m\hat{\mathbf{k}}$.

This is a hard problem, but important. The tangent curves of $\nabla \Phi$ and $\nabla \times \mathbf{A}$ look the same. Φ has the form of the scalar potential of a pointlike electric dipole, and $\mathbf{A}$ has the form of the vector potential of a pointlike magnetic dipole (apart from constant factors).

Sec. 2.3. Integral Theorems

2.9. (a) Sketch a picture of the vector field $\mathbf{F}(\mathbf{x}) = \hat{\mathbf{k}} \times \mathbf{x}$.

(b) Calculate directly the line integral of $\hat{\mathbf{k}} \times \mathbf{x}$ around a circle in the xy plane, centered at 0 with radius a.

(c) Calculate the line integral by using Stokes's theorem.

2.10. (a) Sketch a picture of the vector field $\mathbf{F}(\mathbf{x}) = \mathbf{x}$.

(b) Calculate directly the flux of $\mathbf{F}(\mathbf{x})$ outward through the surface of the unit cube defined by $0 \leq x, y, z \leq 1$.

(c) Calculate the flux by using Gauss's theorem.

2.11. (a) Use Gauss's theorem to prove that

$$\int_{S_1} (\nabla \times \mathbf{F}) \cdot d\mathbf{A} = - \int_{S_2} (\nabla \times \mathbf{F}) \cdot d\mathbf{A}$$

where S_1 and S_2 are open surfaces with the same boundary curve C. ($d\mathbf{A}$ points outward from the volume enclosed by $S_1 \cup S_2$.)

(b) Prove the same result from Stokes's theorem.

Sec. 2.4. Curvilinear Coordinates

2.12. Calculate the gradient ∇f and the Laplacian $\nabla^2 f$ of these scalar functions:

(a) $f = x^2 + y^2 + z^2$,

(b) $f = \left(x^2 + y^2 + z^2\right)^{1/2}$,

(c) $f = \left(x^2 + y^2 + z^2\right)^{-1/2}$.

Do each calculation two ways: using Cartesian coordinates, and using spherical polar coordinates. You must get the same answer either way.

2.13. Calculate the divergence and curl of these vector functions:
(a) $\mathbf{x}$, (b) $\mathbf{x}/r$, where $r = |\mathbf{x}|$, and (c) $\hat{\mathbf{k}} \times \mathbf{x}$.
Do each calculation two ways: using Cartesian coordinates, and using spherical polar coordinates.

2.14. A useful mathematical skill is to be able to expand the unit vectors of one system of coordinates in the unit vectors of another system. For Cartesian coordinates (x, y, z) and spherical polar coordinates (r, θ, ϕ) use Fig. 2.16 to prove:

(a)

$$\hat{\mathbf{i}} = \hat{\mathbf{r}} \sin\theta \cos\phi + \hat{\boldsymbol{\theta}} \cos\theta \cos\phi - \hat{\boldsymbol{\phi}} \sin\phi$$

$$\hat{\mathbf{j}} = \hat{\mathbf{r}} \sin\theta \sin\phi + \hat{\boldsymbol{\theta}} \cos\theta \sin\phi + \hat{\boldsymbol{\phi}} \cos\phi$$

$$\hat{\mathbf{k}} = \hat{\mathbf{r}} \cos\theta - \hat{\boldsymbol{\theta}} \sin\theta$$

(b)

$$\hat{\mathbf{r}} = \hat{\mathbf{i}} \sin\theta \cos\phi + \hat{\mathbf{j}} \sin\theta \sin\phi + \hat{\mathbf{k}} \cos\theta$$

$$\hat{\boldsymbol{\theta}} = \hat{\mathbf{i}} \cos\theta \cos\phi + \hat{\mathbf{j}} \cos\theta \sin\phi - \hat{\mathbf{k}} \sin\theta$$

$$\hat{\boldsymbol{\phi}} = -\hat{\mathbf{i}} \sin\phi + \hat{\mathbf{j}} \cos\phi$$

Sec. 2.5. The Helmholtz Theorem

2.15. Consider fluid flow in the xy plane directed radially away from the origin, so the fluid velocity is $\mathbf{v} = v_r \hat{\mathbf{r}}$. [The point $(0, 0, 0)$ is a fluid source.] Suppose that as a function of distance r from the origin, $v_r(r) = Cr^p$, where C and p are constants.

(a) For what values of C and p can $\mathbf{v}$ be expressed as a gradient, $\mathbf{v} = \nabla \psi$?

(b) Find $\psi(\mathbf{x})$ for arbitrary C and p. Treat the case $p = -1$ separately.

2.16. It is possible to construct vector functions that have both divergence and curl equal to 0. Consider

$$\mathbf{F}(x, y) = \frac{-y\hat{\mathbf{i}} + x\hat{\mathbf{j}}}{x^2 + y^2}$$

which is defined for the domain $\mathcal{D}$ of all points $(x, y) \neq (0, 0)$.

(a) Show that $\nabla \cdot \mathbf{F} = 0$ and $\nabla \times \mathbf{F} = 0$.

(b) Now calculate $\oint \mathbf{F} \cdot d\boldsymbol{\ell}$ for the unit circle in the xy plane. The integral is $\neq 0$. Why is this *not* a counterexample to Stokes's theorem?

2.17. Let $\mathbf{H}(\mathbf{x}) = x^2 y\,\hat{\mathbf{i}} + y^2 z\,\hat{\mathbf{j}} + z^2 x\hat{\mathbf{k}}$. Find an irrotational function $\mathbf{F}(\mathbf{x})$ and a solenoidal function $\mathbf{G}(\mathbf{x})$ such that $\mathbf{H} = \mathbf{F} + \mathbf{G}$.

2.18. $\mathbf{F} = -\nabla\psi$ is irrotational, $\mathbf{G} = \nabla \times \mathbf{A}$ is solenoidal, and $\mathbf{H} = \mathbf{F} + \mathbf{G}$. Show that ψ and $\mathbf{A}$ satisfy the equations

$$\nabla^2 \psi = -\nabla \cdot \mathbf{H},$$

$$\nabla (\nabla \cdot \mathbf{A}) - \nabla^2 \mathbf{A} = \nabla \times \mathbf{H}.$$

[The solutions to these equations are (2.68) and (2.69).]

General Exercises

2.19. Consider the vector function $\mathbf{F}(\mathbf{x}) = C\mathbf{x}$, where C is a constant.

(a) What is the divergence of $\mathbf{F}(\mathbf{x})$?

(b) What is the flux of $\mathbf{F}$ through a cubic surface of size $\ell \times \ell \times \ell$ centered at the origin?

2.20. Consider the vector function $\mathbf{G}(\mathbf{x}) = \hat{\mathbf{k}} \times \mathbf{x}$.

(a) Sketch the tangent curves of $\mathbf{G}(\mathbf{x})$.

(b) What is the circulation $\oint \mathbf{G} \cdot d\boldsymbol{\ell}$ around the unit circle C in the xy plane?

2.21. A vector function $\mathbf{F}(\mathbf{x})$ has the form $\mathbf{F}(\mathbf{x}) = f(r)\mathbf{x}$, where $r = |\mathbf{x}|$.

(a) Prove that $\nabla \times \mathbf{F} = 0$.

(b) Now suppose also $\nabla \cdot \mathbf{F} = 0$. What is the most general form allowed for $f(r)$?

2.22. Consider the scalar function

$$h(\mathbf{x}) = (\mathbf{x} \times \mathbf{A}) \cdot (\mathbf{x} \times \mathbf{B})$$

where $\mathbf{x} = x\hat{\mathbf{i}} + y\hat{\mathbf{j}} + z\hat{\mathbf{k}}$ and $\mathbf{A}$ and $\mathbf{B}$ are constant vectors. Show that the gradient of $h(\mathbf{x})$ is

$$\nabla h = \mathbf{A} \times (\mathbf{x} \times \mathbf{B}) + \mathbf{B} \times (\mathbf{x} \times \mathbf{A}).$$

Also, verify this general result for the special case $\mathbf{A} = a\,\hat{\mathbf{i}}$ and $\mathbf{B} = b\,\hat{\mathbf{j}}$.

2.23. As usual, $\mathbf{x} = x\hat{\mathbf{i}} + y\hat{\mathbf{j}} + z\hat{\mathbf{k}}$ and $r = |\mathbf{x}|$. Let $\mathbf{A}$ and $\mathbf{B}$ be arbitrary constant vectors. Prove these two results:

(a) $\mathbf{A} \cdot \nabla \left(\dfrac{1}{r} \right) = -\dfrac{\mathbf{A} \cdot \mathbf{x}}{r^3}$

(b) $\mathbf{B} \cdot \nabla \left[\mathbf{A} \cdot \nabla \left(\dfrac{1}{r} \right) \right] = \dfrac{3(\mathbf{A} \cdot \mathbf{x})(\mathbf{B} \cdot \mathbf{x})}{r^5} - \dfrac{\mathbf{A} \cdot \mathbf{B}}{r^3}$

2.24. Consider the vector function $\mathbf{F}(\mathbf{x}) = \hat{\boldsymbol{\phi}}$. Calculate $\oint_C \mathbf{F} \cdot d\boldsymbol{\ell}$ and $\int_H \nabla \times \mathbf{F} \cdot d\mathbf{A}$, where C is the circle of radius r in the xy plane centered at the origin, and H is the hemisphere above the xy plane with boundary curve C. (Take the integral over C to be counterclockwise.) Also, calculate $\int_D \nabla \times \mathbf{F} \cdot d\mathbf{A}$ where D is the disk in the xy plane bounded by C. Verify Stokes's theorem in both cases.

2.25. Consider the vector function

$$\mathbf{F}(\mathbf{x}) = \frac{\mathbf{x}}{(r^2 + \epsilon^2)^{3/2}} = \frac{r\hat{\mathbf{r}}}{(r^2 + \epsilon^2)^{3/2}}.$$

(r is the spherical radial coordinate.) Calculate the divergence of $\mathbf{F}(\mathbf{x})$, and sketch a plot of the divergence as a function of r, for $\epsilon \ll 1$, ≈ 1, and $\gg 1$. Calculate the flux of $\mathbf{F}$ outward through a sphere of radius R centered at the origin, and verify that it is equal to the integral of the divergence inside the sphere. Show that the flux is 4π, independent of R, in the limit $\epsilon \to 0$.

The limit $\epsilon \to 0$ is very singular! The function $\nabla \cdot \mathbf{F}$ approaches an infinitesimally narrow function of r with volume integral 4π. The mathematics of this example is important in electromagnetism. In the limit $\epsilon \to 0$, $\mathbf{F}(\mathbf{x})$ has the form of the electric field of a point charge at rest at the origin.

2.26. Consider the scalar function $\Phi(\mathbf{x}) = \mathbf{C} \cdot \mathbf{x}/r^3$, where $\mathbf{C}$ is a constant vector. In a coordinate system with $\mathbf{C} = C\hat{\mathbf{k}}$, Φ may be written as

$$\Phi = \frac{Cz}{(x^2 + y^2 + z^2)^{3/2}} = \frac{C \cos\theta}{r^2}$$

using Cartesian and spherical polar coordinates, respectively.

(a) Calculate $-\nabla\Phi$, which is the vector field $\mathbf{E}(\mathbf{x})$. You may use either Cartesian or polar coordinates.

(b) Determine the Cartesian direction of $\mathbf{E}$ at a point on the x axis.

(c) Determine the Cartesian direction of $\mathbf{E}$ at a point on the z axis.

(d) Sketch the tangent curves of $\mathbf{E}$.

2.27. Consider the vector function $\mathbf{F}(\mathbf{x}) = \mathbf{x}/r^3$. (As usual, r is the magnitude of $\mathbf{x}$.)

(a) Calculate the flux of $\mathbf{F}$ through a sphere of radius a, centered at the origin.

(b) Calculate the flux of $\mathbf{F}$ through a planar disk of radius a, parallel to the xy plane and centered at the z axis at $z = H$.

(c) Determine the numerical value of the result of (b) for $H = a$.

2.28. (a) Calculate the divergence of the function

$$\mathbf{G}(\mathbf{x}) = \mathbf{x}e^{-\alpha r}.$$

(b) Sketch a graph of $\nabla \cdot \mathbf{G}$ as a function of r.

(c) Determine the volume integral of $\mathbf{V} \cdot \mathbf{G}$ inside a sphere of radius a. (Hint: Use Gauss's theorem.)

(d) What is the volume integral of $\mathbf{V} \cdot \mathbf{G}$ over the infinite volume?

2.29. Prove *Green's theorem*: If $f(\mathbf{x})$ and $g(\mathbf{x})$ are any two scalar functions, then

$$\oint_S (f\nabla g - g\nabla f) \cdot d\mathbf{A} = \int_V \left(f\nabla^2 g - g\nabla^2 f \right) d^3x$$

where S is a closed surface and V the enclosed volume.

Computer Exercises

At the end of most chapters there are a few exercises that require the use of a computer. Convenient software for these exercises is Mathematica or Maple, because these programs integrate analytic calculations, numerical calculations, and graphics. However, other programs, e.g., Matlab, Mathcad, or old-fashioned programming languages like Fortran can be used if Maple and Mathematica are not available. Some of the simpler exercises can probably be done with a graphing calculator, although that is a relatively limited technology.

To get started doing computer calculations in field theory, here is an exercise in elementary functional analysis.

2.30. Consider a charged line segment with constant charge per unit length λ and length 2ℓ, located on the z axis from $z = -\ell$ to $z = +\ell$. The electrostatic potential throughout the space around the line segment is the function

$$V(x, y, z) = \frac{\lambda}{4\pi\epsilon_0} \ln\left\{ \frac{z + \ell + \sqrt{x^2 + y^2 + (z+\ell)^2}}{z - \ell + \sqrt{x^2 + y^2 + (z-\ell)^2}} \right\}.$$

This function looks pretty complicated, so we use a computer to study it.

(a) Let ℓ be the unit of length. Plot $V(x, 0, 0)$ in units of $\Lambda \equiv \lambda/(4\pi\epsilon_0)$ for x in the domain $(0, 5)$.

(b) The electric field is $\mathbf{E}(\mathbf{x}) = -\nabla V$. Plot $E_x(x, 0, 0)$ in units of Λ/ℓ for x in the domain $(0, 5)$. (If Mathematica or Maple is used, the derivative can be calculated analytically by the program. The result may be reduced to a simple form.)

(c) Make a log-log plot of $E_x(x, 0, 0)$ vs x, over a sufficiently large range of x, and show that the functional dependence is x^{-1} for $x \ll \ell$ and x^{-2} for $x \gg \ell$. We are often interested in such asymptotic expansions in field theory.

CHAPTER
3

Basic Principles of Electrostatics

The theory of electromagnetism encompasses all electric, magnetic, and optical phenomena. To learn such a far-reaching theory we must start someplace. We start with *electrostatics*, which is conceptually the simplest part of the full theory. Many of the theoretical ideas and mathematical techniques that we will acquire in studying electrostatics will be used again in the other parts of the theory.

Everybody who comes to this subject has already had a lot of experience with electricity and magnetism, both from applications, such as electric lights, motors, telephones, radio, and television, as well as from natural phenomena, such as light, lightning, static discharges, and Earth's magnetic field. Often the motivation for studying electromagnetism is a desire to understand the physics underlying these experiences. It is therefore natural to expect a book on electromagnetism to start with discussions of such familiar devices and phenomena. However, those discussions must be postponed until later. We must start with *electrostatics* to lay the foundation for later understanding of these more complicated things. Devices are made by people, so it's fair to say they are understood; but some natural phenomena are only partly understood and remain the subjects of current research.

Electrostatics is the science of the interactions between electric charges and the electric field in static systems, i.e., systems that do not change in time. Strictly, the charges must be at rest, and the fields must be constant. Practically, we also use the equations of electrostatics for systems that change slowly in time, a rather forgiving approximation called the quasistatic approximation. As we'll see, this approximation permits us to apply electrostatic principles to many electromagnetic devices and phenomena.

3.1 ■ COULOMB'S LAW

All of electrostatics can be derived from two principles, Coulomb's law and the principle of superposition. Coulomb's law, which is a formula for the force between two point-like charges, is based directly on experiment. The first accurate and detailed measurements of the force between charges were made by Coulomb in 1785. From these experiments, with a very precise torsion balance, Coulomb discovered that the force is inversely proportional to the square of the distance, and in the direction of the line joining the charges. Also, the force is proportional to the product of the charges. Writing an equation that expresses these facts, the force on a charge q_1 due to a charge q_2 is

$$\mathbf{F}_1 = K\frac{q_1 q_2}{r^2}\hat{\mathbf{r}} \tag{3.1}$$

where r is the distance between the charges and $\hat{\mathbf{r}}$ is the unit vector pointing from q_2 to q_1. If we let $\mathbf{r}$ be the vector from q_2 to q_1, then $r = |\mathbf{r}|$ and $\hat{\mathbf{r}} = \mathbf{r}/r$. For q_1 and q_2 with the same sign the force is repulsive, in the direction of $\hat{\mathbf{r}}$; for charges of opposite signs the force is attractive.

It is useful to introduce another notation for the force equation, that displays explicitly the two positions. Let $\mathbf{x}_1$ and $\mathbf{x}_2$ denote the positions of q_1 and q_2 with respect to some origin in space, as shown in Fig. 3.1. Then the relative vector $\mathbf{r}$ from q_2 to q_1 is $\mathbf{r} = \mathbf{x}_1 - \mathbf{x}_2$, and the force on q_1 is

$$\mathbf{F}_1 = Kq_1 q_2 \frac{\mathbf{x}_1 - \mathbf{x}_2}{|\mathbf{x}_1 - \mathbf{x}_2|^3}. \tag{3.2}$$

The constant of proportionality K depends on the system of units. In this book we use Standard International (SI) units, also known as rationalized MKSA units (for meter, kilogram, second, Ampere). The table below lists the units of some important quantities.[1]

Quantity	Unit	Abbreviation
length	meter	m
mass	kilogram	kg
time	second	s
current	ampere	A
force	newton	N
charge	coulomb	C
potential	volt	V
magnetic field	tesla	T

In SI units the value of K in (3.2) is defined to be

$$K = \frac{1}{4\pi\epsilon_0} = 10^{-7}c^2\,\frac{\text{N s}^2}{\text{C}^2} = 8.99 \times 10^9\,\frac{\text{N m}^2}{\text{C}^2} \tag{3.3}$$

where c is the speed of light and ϵ_0 is called the permittivity of the vacuum.[2] The unit of charge, the coulomb (C), is defined by $1\,\text{C} = 1\,\text{A s}$, where the unit of current, the ampere (A), is the fundamental electric unit. The ampere is defined in terms of the magnetic force between current-carrying wires, as we'll learn in

[1] Other systems of electric and magnetic units exist, and the field equations depend on the choice of units. When reading the literature it is necessary to identify which system of units is being used. Appendix A compares SI and Gaussian units.

[2] In the rest of the book we will never use K, but always write $1/4\pi\epsilon_0$ explicitly.

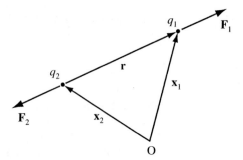

FIGURE 3.1   Interacting charges. The forces are equal but opposite. For the force directions shown, q_1 and q_2 have the same sign.

Chapter 8. The coulomb is defined such that $1/(4\pi\epsilon_0)$ has the value in (3.3). The numerical value is the force in newtons between two charges each of 1 C that are 1 m apart.

Coulomb's law (3.1), the first principle of electrostatics, is an experimental fact.

3.1.1 ▧ The Superposition Principle

The superposition principle merely states that the force on a charge q due to a set of charges $\{q_1, q_2, q_3, \ldots, q_N\}$ is the sum of the individual Coulomb forces

$$\mathbf{F}_q = \sum_{k=1}^{N} \frac{qq_k}{4\pi\epsilon_0} \frac{\mathbf{x} - \mathbf{x}_k}{|\mathbf{x} - \mathbf{x}_k|^3} \tag{3.4}$$

where $\mathbf{x}$ is the position of q, and $\mathbf{x}_k$ that of q_k.

3.2 ▧ THE ELECTRIC FIELD

3.2.1 ▧ Definition

The electric field $\mathbf{E}(\mathbf{x})$ is a vector-valued function of position. That is, there is a field vector at every point in space.

The definition of $\mathbf{E}(\mathbf{x})$ is the force per unit charge that would be exerted on a small "test charge" q if it were located at $\mathbf{x}$, in the limit $q \to 0$

$$\mathbf{E}(\mathbf{x}) = \lim_{q \to 0} \frac{\mathbf{F}}{q}. \tag{3.5}$$

The test charge is taken to be small so that it does not affect the other charges. Because the force $\mathbf{F}$ on q is proportional to q, the electric field is independent of the test charge. This definition provides a technique in principle for measuring the defined quantity $\mathbf{E}(\mathbf{x})$: Take a small charge to $\mathbf{x}$, and measure $\mathbf{F}/q$.

The subject of this chapter is the interaction between charges and the field. The definition (3.5) tells us one aspect of the interaction: The electric field exerts a force on a charged particle q, given by

$$\mathbf{F}_q = q\mathbf{E}(\mathbf{x}) \tag{3.6}$$

where $\mathbf{x}$ is the position of q. This equation states how the field affects a charge. Next we describe the other aspect of the interaction—how charges affect the field.

3.2.2 ■ Charge as the Source of E

We regard $\mathbf{E}(\mathbf{x})$ as a real physical entity, not just a mathematical construction. From the way it is defined above one might think that $\mathbf{E}$ is just a mathematical technique for describing the electric force, but that is the wrong way to think of it. We know that *light* consists of fluctuating electric and magnetic fields, and no one would dispute the statement that light is a real physical entity, not just a mathematical construction! Field theory was developed by Faraday and Maxwell as an alternative to "action at a distance." The field is something real that extends throughout a volume of space and exerts forces on charges in the space. Now the question is, where does this entity, the field, come from? What is its *source*?

Electric charge is one source of electric field. In electrostatics, charge is the only source.[3] Comparing (3.5) and (3.4) shows immediately that the field created by a set of static point charges $\{q_1, q_2, \ldots, q_N\}$ is

$$\mathbf{E}(\mathbf{x}) = \sum_{k=1}^{N} \frac{q_k}{4\pi\epsilon_0} \frac{\mathbf{x} - \mathbf{x}_k}{|\mathbf{x} - \mathbf{x}_k|^3}. \tag{3.7}$$

We refer to $\mathbf{x}$ as the field point and to the $\mathbf{x}_k$'s as source points. The vector $(\mathbf{x} - \mathbf{x}_k)/|\mathbf{x} - \mathbf{x}_k|$ is the *unit vector* in the direction from the source point toward the field point.

The field as a function of position $\mathbf{x}$, due to a single point charge q at position $\mathbf{x}_q$, is

$$\mathbf{E}(\mathbf{x}) = \frac{q}{4\pi\epsilon_0} \frac{\hat{\mathbf{r}}}{r^2}. \tag{3.8}$$

where $\mathbf{r} = \mathbf{x} - \mathbf{x}_q$. Charge acting as a source of $\mathbf{E}$ is the other aspect of the *interaction* between charges and the electric field. Equation (3.8) states how a charge affects the field, and (3.6) states how the field affects a charge.

EXAMPLE 1   Figure 3.2 shows six identical charges, one at each vertex of a regular hexagon in the xy plane centered at the origin.

---

[3] In Chapter 10 we will learn that another source of $\mathbf{E}$ is a magnetic field that changes with time. But that is a dynamic, rather than static, phenomenon.

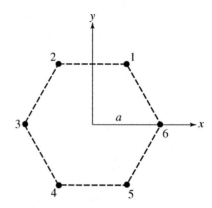

FIGURE 3.2   Example 1. Six charges at the corners of a regular hexagon.

1. What is the electric field at any field point **x** in the xy plane?
   Let the charges be numbered with the index k from 1 to 6. Then the position
   of the kth charge is

$$\mathbf{x}'_k = \hat{\mathbf{i}}\, a \cos \frac{k\pi}{3} + \hat{\mathbf{j}}\, a \sin \frac{k\pi}{3}. \tag{3.9}$$

The field for arbitrary **x** is then obtained from (3.7) as

$$\mathbf{E}(\mathbf{x}) = \frac{q}{4\pi\epsilon_0} \sum_{k=1}^{6} \frac{(x - a\cos k\pi/3)\,\hat{\mathbf{i}} + (y - a\sin k\pi/3)\,\hat{\mathbf{j}}}{\left[(x - a\cos k\pi/3)^2 + (y - a\sin k\pi/3)^2 \right]^{3/2}}, \tag{3.10}$$

where a is the distance from the origin to any of the charges. The field is
singular, *i.e.,* goes to infinity, at the six charges, but is finite everywhere else.

2. What is the field on the x axis?
   We set $y = 0$ in (3.10) and obtain

$$E_x(x, 0) = \frac{q}{4\pi\epsilon_0} \sum_{k=1}^{6} \frac{(x - a\cos k\pi/3)}{\left[x^2 - 2ax\cos k\pi/3 + a^2 \right]^{3/2}}, \tag{3.11}$$

$$E_y(x, 0) = \frac{-qa}{4\pi\epsilon_0} \sum_{k=1}^{6} \frac{\sin k\pi/3}{\left[x^2 - 2ax\cos k\pi/3 + a^2 \right]^{3/2}} = 0. \tag{3.12}$$

It is clear that $E_y(x, 0)$ must be zero, because of the symmetric positions of
the charges; in the sum in (3.12) the terms with $k = 3$ and 6 are 0, and the
other terms cancel in pairs.

   At the origin, **E** is 0, which is easily seen from the equations or just physically because the fields due to charges on opposite vertices cancel.

3. As an example of *asymptotic approximation*, what is the field on the x axis far from the origin, accurate through order x^{-4}?

The y component is zero on the x axis. For $x \gg a$, the x component must be calculated by expanding (3.11) in a power series in the small quantity a/x, using the method of Taylor series. The result is

$$E_x(x, 0) \approx \frac{1}{4\pi\epsilon_0} \left\{ \frac{6q}{x^2} + \frac{9qa^2}{2x^4} \right\}. \tag{3.13}$$

The leading term is, as expected, the same as for a point charge $6q$ at the origin.

4. If the 6th charge is removed from the group, what is the field at the origin for the remaining 5 charges ($k = 1, 2, \ldots, 5$)?

We could start from (3.11) but sum only from $k = 1$ to 5. However, the superposition principle enables us to answer the question more easily if we realize that the field of charges 1 to 5 is the same as the field of charges 1 to 6 *superposed* on the field of a charge $-q$ at $(a, 0)$. But the field at the origin due to the original 6 charges is 0, so the field of charges 1 to 5 is (note the sign!)

$$\mathbf{E}_{5\text{charges}}(0, 0) = \frac{+q\,\hat{\mathbf{i}}}{4\pi\epsilon_0 a^2}. \tag{3.14}$$

3.2.3 ■ Field of a Charge Continuum

Charge resides in elementary particles—electrons and protons—so any field may be attributed to discrete, point-like sources, as in (3.7). However, these *microscopic* charges are incredibly small, and much smaller than the relevant length scales of any macroscopic system. In a macroscopic system it is a good approximation to replace the discrete elementary charges by a continuous distribution of charge. We denote by $\rho(\mathbf{x})$ the *volume charge density* (charge per unit volume) as a function of position $\mathbf{x}$, defined with respect to volumes that are small compared to the whole system, but large compared to the distances between individual particles. That is, the density $\rho(\mathbf{x}')$ at a source point $\mathbf{x}'$ is the total charge in a small volume $\delta V'$ around $\mathbf{x}'$, divided by $\delta V'$. By a "small volume" we mean small on the macroscopic scale; but the number of elementary particles in $\delta V'$ is very large. Making this continuum approximation, the sum in (3.7) may be replaced by an integral over the volume of the charge distribution

$$\mathbf{E}(\mathbf{x}) = \frac{1}{4\pi\epsilon_0} \int \frac{\mathbf{x} - \mathbf{x}'}{|\mathbf{x} - \mathbf{x}'|^3} \rho(\mathbf{x}') d^3 x'. \tag{3.15}$$

Equation (3.15) is a fundamental equation of electrostatics. Given the charge density $\rho(\mathbf{x}')$ throughout the source, (3.15) determines the electric field $\mathbf{E}(\mathbf{x})$ at the "field point" **x**. In (3.15) we use the convention that primed coordinates represent source points, and unprimed coordinates represent field points.

We might write the equation (3.15) more succinctly as

$$\mathbf{E} = \frac{1}{4\pi\epsilon_0} \int \frac{\hat{\mathbf{r}}\,dq'}{r^2}, \tag{3.16}$$

which, however, has the disadvantage that it does not display explicitly the field point and the source point. The meaning of $\hat{\mathbf{r}}$ and r is not manifest, and one must *remember* that r is the distance between the source point and the field point, and $\hat{\mathbf{r}}$ the direction vector.

In (3.15) we consider the charge to be distributed in three dimensions, with volume density $\rho(\mathbf{x}')$ at $\mathbf{x}'$. That is, $\rho(\mathbf{x}')d^3x'$ is a small charge element dq'. If the charge is distributed over a 2-dimensional surface, then dq' is $\sigma(\mathbf{x}')dA'$ where dA' is an area element and $\sigma(\mathbf{x}')$ is the surface density, which has units C/m^2. If the charge is distributed over a 1-dimensional curve or line, then dq' is $\lambda(\mathbf{x}')d\ell'$, where $d\ell'$ is an infinitesimal line segment and $\lambda(\mathbf{x}')$ is the linear density, which has units C/m. In these latter cases the field $\mathbf{E}(\mathbf{x})$ would be expressed as a surface integral, or a line integral, over the charge distribution.

Whether one should use (3.15) or (3.7) depends on the system of interest. To describe elementary particles within a molecule (3.7) is appropriate. To describe the field of a macroscopic charge distribution, for example a van de Graaff generator, (3.15) is appropriate. In this book we are mainly concerned with classical systems, which are necessarily much larger than an atom, so the continuum approximation is more often relevant here. The next three examples illustrate the use of (3.15) to calculate the electric field.

EXAMPLE 2   What is the electric field on the midplane of a uniformly charged thin wire of length 2ℓ? The charge per unit length is λ.

Figure 3.3 shows the wire extending from $(0, 0, -\ell)$ to $(0, 0, +\ell)$ along the z axis. Because of axial symmetry it is sufficient to find the electric field on the x axis. As shown, the charge element $\lambda dz'$ produces field $d\mathbf{E}$ at the point $(x, 0, 0)$. Because the distribution is symmetric about $z = 0$, the resultant field due to the entire wire will be in the x direction. From (3.15) the x component of $\mathbf{E}$ is

$$E_x(x, 0, 0) = \frac{1}{4\pi\epsilon_0} \int_{-\ell}^{\ell} \frac{x\lambda dz'}{\left(x^2 + z'^2\right)^{3/2}} = \frac{\lambda\ell}{2\pi\epsilon_0 x\sqrt{x^2 + \ell^2}}. \tag{3.17}$$

We might derive this result in another, more geometrical way. The contribution dE_x due to $dq' = \lambda dz'$ is $\hat{\mathbf{i}} \cdot d\mathbf{E} = dE\cos\theta$, where $\cos\theta = x/\sqrt{x^2 + z'^2}$. Then E_x is $\int \cos\theta dE$, leading again to (3.17).

Note that the field in the midplane is radial. Generalizing to any point on the xy plane, at distance r from the z axis,

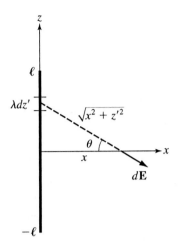

FIGURE 3.3   Example 2. A charged wire. $(0, 0, z')$ is a source point and $(x, 0, 0)$ is a field point.

$$\mathbf{E}(r) = \frac{\lambda \ell \hat{\mathbf{r}}}{2\pi\epsilon_0 r \sqrt{r^2 + \ell^2}}, \tag{3.18}$$

where $\hat{\mathbf{r}}$ is the radial unit vector in the xy plane.

It is interesting to consider the limiting behavior of this field for far points and near points. Far from the line charge, i.e., $r \gg \ell$, the field is $E_r = \lambda\ell/(2\pi\epsilon_0 r^2)$. As we'd expect, this is the same as the field of a point charge $q = 2\ell\lambda$ at the origin; from far away the line charge looks, to a first approximation, like a point. Near the wire, i.e., $r \ll \ell$, the field is $E_r = \lambda/(2\pi\epsilon_0 r)$.

If the line charge is infinitely long, then the field on the x axis is obtained from (3.17) by extending the integral from $-\infty$ to $+\infty$. Evaluating that integral gives, for any point in the midplane,

$$\mathbf{E}(r) = \frac{\lambda \hat{\mathbf{r}}}{2\pi\epsilon_0 r}. \tag{3.19}$$

This is the same result that we found for a finite line charge for points near the line. From close enough the finite line looks, to a first approximation, infinitely long.

EXAMPLE 3   What is the electric field on the axis of a circular loop of uniformly charged thin wire with total charge q? Let a be the radius of the wire circle.

In Figure 3.4 the wire is in the xy plane centered at O. The charge element dq produces field $d\mathbf{E}$ at the point $(0, 0, z)$, as shown. Because the charge distribution is axially symmetric, the resultant field of the entire wire is in the z direction, and $dE_z = dE \cos\xi = dE(z/\sqrt{a^2 + z^2})$. Each charge element makes the same

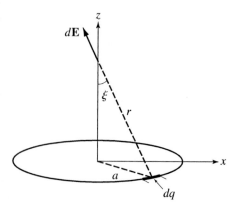

FIGURE 3.4   Example 3. A circular loop of wire. $d\mathbf{E}$ is the field of the elemental charge dq.

contribution to dE_z because they are all at the same distance $\sqrt{a^2 + z^2}$ from the field point. Therefore the integrand in (3.15) is a constant, and we have

$$E_z(0, 0, z) = \frac{qz}{4\pi\epsilon_0 \left(a^2 + z^2\right)^{3/2}}. \tag{3.20}$$

A charged disk, or a charged plane, can be built up from elemental annuli, so (3.20) can be integrated over the loop radius to find the field of a disk or plane.[4]

EXAMPLE 4   What is the electric field $\mathbf{E}(\mathbf{x})$ due to a uniformly charged spherical shell of radius R with total charge Q? The shell thickness is negligible, and the surface charge density is $\sigma = Q/(4\pi R^2)$.

Figure 3.5 shows the shell with its center at O. Because of the symmetry the field will be spherically symmetric, $\mathbf{E}(\mathbf{x}) = E_r(r)\hat{\mathbf{r}}$, where $\hat{\mathbf{r}}$ is the unit radial vector. It is sufficient to find the field at P on the z axis, where $r = z$ and $\hat{\mathbf{r}} = \hat{\mathbf{k}}$.

Let the shell be built up from elemental annuli, as shown in Fig. 3.5. On the z axis we know the field of an annular ring from Example 3. Consider now the annular ring at polar angle θ.[5] Its radius is $R\sin\theta$. Its width is $Rd\theta$, so it carries charge $dQ = \sigma \, 2\pi R^2 \sin\theta d\theta$. The perpendicular distance from this annular ring to P is $r - R\cos\theta$. Thus, using (3.20) for the annuli and integrating over the entire sphere from $\theta = 0$ to π,

$$E_r(r) = \frac{\sigma R^2}{2\epsilon_0} \int_0^{\pi} \frac{(r - R\cos\theta)\sin\theta d\theta}{\left[R^2\sin^2\theta + (r - R\cos\theta)^2\right]^{3/2}}. \tag{3.21}$$

---

[4] See Exercise 6.

[5] We often use the convention that source-point coordinates are primed. Here, however, because E_r does not depend on angular coordinates, there is no ambiguity in letting θ (without a prime) be the angular coordinate of an elemental ring in the source. Because of tricky notational details like this, it is important to make a sketch with the variables labeled for any field calculation.

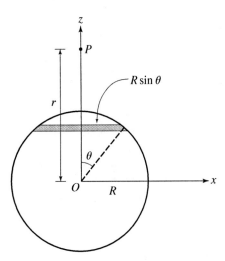

FIGURE 3.5 **Example 4.** A spherical shell. The elemental ring has radius $R \sin\theta$ and area $dA = 2\pi R^2 \sin\theta d\theta$.

Changing the variable of integration to $u = \cos\theta$ gives an integral that can be evaluated from a table of integrals, or from an analytic computer program; the result is

$$E_r(r) = \frac{\sigma R^2}{2\epsilon_0} \int_{-1}^{1} \frac{(r - Ru)du}{\left(R^2 + r^2 - 2rRu\right)^{3/2}} = \frac{\sigma R^2}{2\epsilon_0 r^2} \left[\frac{r - R}{\sqrt{(r - R)^2}} + 1 \right]. \quad (3.22)$$

Equation (3.22) is valid both inside and outside the sphere if the positive value is taken for the radical $\sqrt{(r - R)^2}$. That is, outside the sphere $(r > R)$ the radical is $r - R$, and inside the sphere $(r < R)$ the radical is $R - r$. Therefore the electric field is

$$\mathbf{E}(\mathbf{x}) = \begin{cases} \dfrac{Q\hat{\mathbf{r}}}{4\pi\epsilon_0 r^2} & \text{for } r > R, \\ 0 & \text{for } r < R. \end{cases} \quad (3.23)$$

The final result is remarkably simple. The field outside the spherical shell is the same as if all the charge were concentrated at its center. And the field inside the spherical shell is 0, because the vector sum of the Coulomb fields from all the surface elements is 0.

It is not too hard to extend the conclusions of the last paragraph to *any* finite spherically symmetric charge distribution, i.e., for *any* charge density $\rho(r')$ that depends only on radial distance. Any such charge distribution can be considered to be made up of elemental spherical shells. The field outside the distribution is the same as if all the charge were at the center, because that is true of every elemental shell. Also, the field at a point inside the charged region, say at distance r from the center, depends only on the charge $\rho(r')$ at points $r' < r$, because the elemental shells with $r' > r$ have zero field at r.

An interesting historical note is that the gravitational field problem analogous to the problem we have just discussed, namely to find the field outside and inside a spherical distribution of mass, stymied Isaac Newton as he was developing integral calculus. Fortunately for the development of science, he did eventually solve it.

The field inside a spherical shell of charge is 0, i.e., the vector sum of the Coulomb fields of the surface elements is 0, because the Coulomb field is inversely proportional to the square of the distance, $d\mathbf{E} \propto \hat{\mathbf{r}}/r^2$. If the field obeyed any other power law, then $\mathbf{E}$ would not be 0 inside the sphere. (It would be 0 at the exact center of the sphere by symmetry, but not throughout the interior.) This special property of the inverse square law provides the most accurate experimental test of Coulomb's Law, which will be discussed in Chapter 4.

We will return to Examples 2, 3 and 4 later, to illustrate *other methods* for calculating the electric field. The direct evaluation of the field, by integrating over the charge distribution, is often *not* the easiest way to determine the field. In fact, in many examples it is the *hardest way* to do the calculation!

3.3 ■ CURL AND DIVERGENCE OF E

From (3.15) we can derive two *differential relations*, that must be satisfied by any static electric field, for the curl and divergence of $\mathbf{E}(\mathbf{x})$. We will show that

$$\nabla \times \mathbf{E} = 0, \tag{3.24}$$

and

$$\nabla \cdot \mathbf{E} = \rho/\epsilon_0. \tag{3.25}$$

Equation (3.25) is called *Gauss's Law*, in the form of a differential equation, and it is true for any electric field including time-dependent fields. Equation (3.24) is true for the field from any distribution of static charge, such as all those treated in this chapter. Later, when we consider electromagnetic systems that vary in time, we will replace (3.24) by the equation for Faraday's Law.

To prove these relations, we first derive an alternate form of (3.15). Note that

$$\frac{\mathbf{x} - \mathbf{x}'}{|\mathbf{x} - \mathbf{x}'|^3} = -\nabla \frac{1}{|\mathbf{x} - \mathbf{x}'|}. \tag{3.26}$$

To verify this equation is a straightforward exercise in vector calculus: The gradient on the right-hand side is, by the chain rule of differentiation,

$$\nabla \frac{1}{|\mathbf{x} - \mathbf{x}'|} = \frac{-1}{|\mathbf{x} - \mathbf{x}'|^2} \overline{\nabla |\mathbf{x} - \mathbf{x}'|}; \tag{3.27}$$

and for any Cartesian component i,[6]

$$\nabla_i |\mathbf{x} - \mathbf{x}'| = \frac{\partial}{\partial x_i} \sqrt{(x_1 - x_1')^2 + (x_2 - x_2')^2 + (x_3 - x_3')^2} \tag{3.28}$$

$$= \frac{1}{2|\mathbf{x} - \mathbf{x}'|} 2(x_i - x_i'). \tag{3.29}$$

Combining these results yields (3.26). So, replace $\mathbf{r}/r^3$ in (3.15) by $-\nabla(1/r)$, where $\mathbf{r} = \mathbf{x} - \mathbf{x}'$. The gradient with respect to $\mathbf{x}$ can be pulled out of the integral over $\mathbf{x}'$, so

$$\mathbf{E}(\mathbf{x}) = -\nabla \frac{1}{4\pi \epsilon_0} \int \frac{\rho(\mathbf{x}') d^3 x'}{|\mathbf{x} - \mathbf{x}'|}. \tag{3.30}$$

In this alternate form of (3.15), $\mathbf{E}(\mathbf{x})$ is written as the gradient of a scalar function. That $\nabla \times \mathbf{E}$ is 0 follows immediately, because we proved in Chapter 2 that the curl of the gradient of any scalar function is identically 0. Hence (3.24) is proven. In electrostatics the field is irrotational.

We have written $\mathbf{E}(\mathbf{x})$ in terms of a scalar function $V(\mathbf{x})$ as

$$\mathbf{E}(\mathbf{x}) = -\nabla V(\mathbf{x}) \tag{3.31}$$

where, as we see from (3.30),

$$V(\mathbf{x}) = \frac{1}{4\pi \epsilon_0} \int \frac{\rho(\mathbf{x}') d^3 x'}{|\mathbf{x} - \mathbf{x}'|}. \tag{3.32}$$

The very important function $V(\mathbf{x})$ is called the *electric potential*. We will have much more to say about $V(\mathbf{x})$ in Secs. 3.5 and 3.6. According to (3.25) $V(\mathbf{x})$ satisfies the equation

$$-\nabla^2 V = \rho/\epsilon_0, \tag{3.33}$$

which is called *Poisson's equation*.

For (3.32) the charge is distributed in a 3-dimensional volume. If the charge is distributed over a 2-dimensional surface, or along a 1-dimensional curve, then $\rho d^3 x'$ would be replaced by $\sigma d A'$ or $\lambda d\ell'$, respectively. If the charge distribution consists of isolated point charges, then the potential is

$$V(\mathbf{x}) = \frac{1}{4\pi \epsilon_0} \sum_{k=1}^{N} \frac{q_k}{|\mathbf{x} - \mathbf{x}_k|}. \tag{3.34}$$

---

[6]Here $i = 1, 2, 3$ correspond to x, y, z; that is, $x_1 = x, x_2 = y, x_3 = z$.

The potential function may be used to solve problems where the charge distribution is known, in the following way: Calculate $V(\mathbf{x})$ from (3.32) or (3.34), or by solving Poisson's equation; then obtain the field as $\mathbf{E} = -\nabla V$. This approach is often easier to accomplish than evaluating the vector integral in (3.15).

The proof of (3.25) is addressed in Secs. 3.4 and 3.5.

3.3.1 ■ Field Theory Versus Action at a Distance

Action at a distance assumes that two charges, separated by any distance, exert forces on each other directly. Field theory is a different concept. Charges create the field. The field fills the volume, determined by the *local relations* (3.25) and (3.24) at every point in the volume. The force on a charge is exerted by the field at the position of the charge. All the action is local.

3.3.2 ■ Boundary Conditions of the Electrostatic Field

In later chapters we will study systems in which there are interfaces between different materials, e.g., between insulator and conductor, or between different dielectrics. It will be important to know how the field changes at such a surface of discontinuity. We can determine the boundary conditions from general considerations, based on the relations (3.24) and (3.25).

Let S be an arbitrary surface in space. For example, one might think of the interface between different materials, or a surface separating two parts of an electrostatic system. *Because* $\nabla \times \mathbf{E} = 0$, *the tangential components* $\mathbf{E}_t$ *of the electric field must be continuous across S*. To prove this statement, consider an infinitesimal rectangular loop cutting through the surface, as in Fig. 3.6(a). By Stokes's theorem the line integral of $\mathbf{E}$ around the loop is 0, because $\nabla \times \mathbf{E} = 0$. In the limit that the two segments on opposite sides of S approach S, the line integral approaches

$$\mathbf{E}_{t2} \cdot d\boldsymbol{\ell}_2 + \mathbf{E}_{t1} \cdot d\boldsymbol{\ell}_1 = (\mathbf{E}_{t2} - \mathbf{E}_{t1}) \cdot d\mathbf{s}$$

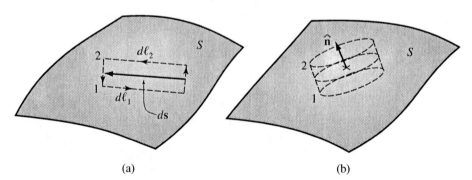

(a)   (b)

FIGURE 3.6 Proof of boundary conditions. (a) An infinitesimal loop cutting through a surface S. Region 1 is below the surface and region 2 is above. $d\mathbf{s}$ is a small displacement on the surface. (b) An infinitesimal pill box cutting through S. $\hat{\mathbf{n}}$ is the unit vector normal to the surface, pointing from region 1 to region 2.

where $\mathbf{E}_{t1}$ and $\mathbf{E}_{t2}$ are tangential components on opposite sides of S and $d\mathbf{s}$ is the segment on the surface. There is no contribution from the segments perpendicular to S because their lengths shrink to 0 in the limit (see Fig. 3.6(a)). Thus, because the line integral is 0, $\mathbf{E}_{t1}$ must be equal to $\mathbf{E}_{t2}$. That is, $\mathbf{E}_t$ is continuous across S.

Gauss's Law implies that the normal component E_n of the electric field is continuous across S unless there is a surface charge, in which case the discontinuity of E_n is σ/ϵ_0. To prove this statement, consider an infinitesimal pill box cutting through the surface, as in Fig. 3.6(b). By Gauss's divergence theorem the flux of $\mathbf{E}$ out of the pill box is $Q_{\text{enclosed}}/\epsilon_0$, because $\mathbf{V} \cdot \mathbf{E} = \rho/\epsilon_0$. In the limit that the two faces on opposite sides of S approach S, the charge enclosed approaches $\sigma\, dA$ where dA is the area of those faces. The flux approaches $E_{n2}\, dA - E_{n1}\, dA$, where E_{n1} and E_{n2} are the normal components on opposite sides of S. (The cylindrical area does not contribute to the flux in the limit that the height shrinks to 0.) Thus the discontinuity of E_n is $E_{n2} - E_{n1} = \sigma/\epsilon_0$.

We shall encounter many examples of these boundary conditions.

3.4 ■ THE INTEGRAL FORM OF GAUSS'S LAW

3.4.1 ■ Flux and Charge

We have stated the differential equation (3.25) that relates divergence and density at every point throughout the volume. That equation is called the *differential form of Gauss's Law*.

The *integral form of Gauss's Law* is a relation between the flux of $\mathbf{E}$ through any closed surface S, and the total charge Q enclosed by S.[7] We may derive the integral relation from the differential relation, by applying Gauss's divergence theorem. Consider the integral of $\mathbf{V} \cdot \mathbf{E}$ in an arbitrary volume V. By (3.25) this integral is Q/ϵ_0. But by Gauss's theorem the integral of $\mathbf{V} \cdot \mathbf{E}$ in V equals the flux of $\mathbf{E}$ through the surface S of V. That is,

$$\oint_S \mathbf{E} \cdot d\mathbf{A} = \frac{1}{\epsilon_0} \int_V \rho(\mathbf{x}) d^3 x = \frac{Q_{\text{enclosed}}}{\epsilon_0}, \tag{3.35}$$

which is the integral form of Gauss's Law. Equations (3.25) and (3.35) are mathematically equivalent statements of the same physical principle—that charge causes divergence of the electric field.

3.4.2 ■ Proof of Gauss's Law

Gauss's Law (3.35) is often used in physics to calculate electric fields, so it is important to prove that it comes from experiment—from Coulomb's law. The proof is an interesting exercise in analytic geometry. To get the basic idea, let's first prove it for the very special case of a point charge q and a sphere S of radius

---

[7] The integral form of Gauss's Law is used in elementary textbooks of physics, and the reader may find it helpful to review the discussion in such a book.

R around q. The electric field at a point $\mathbf{x}$ on the sphere is

$$\mathbf{E}(\mathbf{x}) = \frac{q\hat{\mathbf{r}}}{4\pi\epsilon_0 R^2}. \qquad (3.36)$$

The area element $d\mathbf{A}$ at $\mathbf{x}$ is $d\mathbf{A} = \hat{\mathbf{r}}R^2 d\Omega$, where $d\Omega$ is the solid angle subtended by $d\mathbf{A}$. Thus the flux of $\mathbf{E}$ out through S is

$$\oint_S \mathbf{E} \cdot d\mathbf{A} = \frac{q}{4\pi\epsilon_0} \oint d\Omega = \frac{q}{\epsilon_0}, \qquad (3.37)$$

which agrees with Gauss's Law (3.35) for this special case.[8] Note that the flux does not depend on the radius R of the sphere, because as R increases the area increases as R^2 while the field decreases as $1/R^2$.

Gauss's Law is true for *any* charge distribution and *any* surface, so we must generalize the above proof. So now, let S be an arbitrary closed surface, and again compute the flux out through S of the field due to a point charge q enclosed by S. The area element at a point $\mathbf{x}$ on S is

$$d\mathbf{A} = \frac{r^2 d\Omega}{|\cos\theta|}\hat{\mathbf{n}}, \qquad (3.38)$$

where $\hat{\mathbf{n}}$ is the unit outward normal at $\mathbf{x}$, $d\Omega$ is the solid angle subtended by $d\mathbf{A}$, and θ is the angle between $\hat{\mathbf{n}}$ and $\hat{\mathbf{r}}$. (The unit vector $\hat{\mathbf{r}}$ points radially away from q.) Figure 3.7 shows the geometry. A projection factor $1/|\cos\theta|$ is in (3.38) because $d\mathbf{A}$, which is normal to S, is not generally parallel to the radial direction. Imagine a beam of light shining on a surface at an oblique angle; the illuminated area on the surface would be larger than the cross section of the beam by the same projection factor $1/|\cos\theta|$.

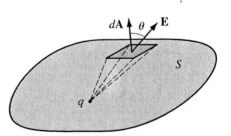

FIGURE 3.7   Proof of Gauss's Law. The solid angle subtended by the area element $d\mathbf{A}$ is $d\Omega = \hat{\mathbf{r}} \cdot d\mathbf{A}/r^2$. The electric field $\mathbf{E}$ due to the point charge q is in the direction of $\hat{\mathbf{r}}$.

---

[8]The differential solid angle in spherical coordinates is $d\Omega = \sin\theta\, d\theta\, d\phi$. The integrated solid angle, over all angles θ and ϕ, is 4π.

The flux through $d\mathbf{A}$ is

$$\mathbf{E} \cdot d\mathbf{A} = \frac{q}{4\pi\epsilon_0 r^2}\cos\theta\frac{r^2 d\Omega}{|\cos\theta|} = \pm\frac{q d\Omega}{4\pi\epsilon_0} \qquad (3.39)$$

where r is the distance from q to $d\mathbf{A}$. The sign depends on the relative directions of $\hat{\mathbf{r}}$ and $\hat{\mathbf{n}}$, and is positive if q is enclosed by S. Finally, the flux through the whole surface is proportional to the integral over all solid angles, $\oint d\Omega = 4\pi$. Hence $\oint \mathbf{E} \cdot d\mathbf{A} = q/\epsilon_0$. This proves that Gauss's Law follows from Coulomb's law.

We have assumed that q is enclosed by S. On the other hand, if q is outside S, then the flux through $d\mathbf{A}$ is negative over part of the surface S (assuming q is positive), because $d\mathbf{A}$ is the *outward* normal. On the part of the surface closer to q, the radial vector has a *negative* projection on $\hat{\mathbf{n}}$, so there the flux is negative. On the part of the surface farther from q, the radial vector has a *positive* projection on $\hat{\mathbf{n}}$, so there the flux is positive. In this case the solid angle integral is 0 because the positive and negative sections of S subtend the same solid angle from q, so the net flux is 0. This result is also consistent with Gauss's Law because q is not an enclosed charge.

If there are many charges present, or a charge continuum, then Gauss's Law is still true by the superposition principle. The electric field is the superposition of fields due to individual charges, or charge elements, and Gauss's Law (3.35) holds for each charge in the system so it holds for the sum of charges.

3.4.3 ■ Calculations Based on Gauss's Law

For charge distributions with a high degree of symmetry, it is often possible to determine the field $\mathbf{E}(\mathbf{x})$ by direct application of (3.35), taking into account the symmetry. We will consider three examples of this technique, with spherical, cylindrical, and planar symmetry.[9]

EXAMPLE 5   What is the electric field due to a uniformly charged sphere?

Let ρ be the charge density for $r \leq a$, where a is the radius of the sphere; uniformly charged means that ρ is constant, so the total charge is $Q = \frac{4}{3}\pi a^3 \rho$. By the spherical symmetry of the problem, $\mathbf{E}(\mathbf{x})$ must be in the radial direction ("diverging from the charge") and the magnitude $|\mathbf{E}(\mathbf{x})|$ can only depend on the distance $r = |\mathbf{x}|$ from the center. That is, in spherical coordinates $\mathbf{E}(\mathbf{x})$ must have the form

$$\mathbf{E}(\mathbf{x}) = E(r)\hat{\mathbf{r}}. \qquad (3.40)$$

We want to find $\mathbf{E}(\mathbf{x})$ both inside and outside the sphere of charge. To find the external field apply (3.35) to the spherical Gaussian surface of radius $r > a$ shown dashed in Fig. 3.8(a). The flux is $\Phi_E = 4\pi r^2 E(r)$. For this case the charge en-

---

[9]When studying these examples, please sketch a picture of the charge distribution, field, and Gaussian surface for each calculation.

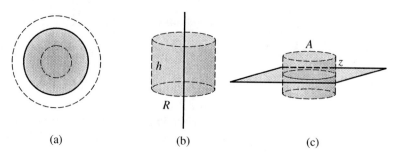

FIGURE 3.8   Gaussian surfaces for three symmetries. (a) Spherical symmetry; (b) cylindrical symmetry; (c) planar symmetry. The Gaussian surfaces are dashed.

closed is Q, so

$$E(r) = \frac{Q}{4\pi\epsilon_0 r^2} \quad \text{for} \quad r > a. \tag{3.41}$$

This is the same as the field for a point charge Q at the origin. The same result was obtained in Example 4 by integrating the Coulomb field. Generalizing this analysis, we see again that for *any* spherically symmetric charge distribution of finite radius, the electric field outside the charge is the same as if all the charge were concentrated at the center.

To find the internal field, apply (3.35) to the spherical Gaussian surface with $r < a$ shown dashed in Fig. 3.8(a). The flux is again $4\pi r^2 E(r)$, but in this case the charge enclosed is $\frac{4}{3}\pi r^3 \rho = Qr^3/a^3$; so

$$E(r) = \frac{Qr}{4\pi\epsilon_0 a^3} \quad \text{for} \quad r \le a. \tag{3.42}$$

Inside the sphere the field decreases as r decreases, and is 0 at the center. The reason is that the field at r depends only on the charge *inside r*. In other words, the contribution to the field at radius r from the charge density *outside* radius r is 0.

In Example 4 we found that the field inside a uniformly charged spherical shell is 0, by integrating the Coulomb field. Gauss's Law provides a simpler proof of this result. For a spherical shell the field must be radial, by symmetry. But there is no charge enclosed by a Gaussian surface drawn inside the shell. Therefore the field inside must be 0.

We have obtained from Gauss's Law the same results that we obtained earlier by integrating the Coulomb field. But it is important to appreciate that the calculations by Gauss's law are *much easier* than evaluating the integral.

EXAMPLE 6   What is the electric field of a uniformly charged line? The line is a mathematical idealization, both because the length is infinite and the diameter is infinitesimal. But the result will be a good approximation for the field of a long charged wire for points near the wire far from the ends, i.e., for $r \ll L$ where L is the length of the wire. Approximating the finite wire as an infinite line is an example of what is called "neglecting the end effects."

Let λ be the charge per unit length. By the cylindrical symmetry of the problem, and the fact that $\mathbf{E}$ diverges from charge sources, $\mathbf{E}(\mathbf{x})$ must have the form

$$\mathbf{E}(\mathbf{x}) = E(r)\hat{\mathbf{r}}, \tag{3.43}$$

where r is the perpendicular distance to the line, and $\hat{\mathbf{r}}$ the radial unit vector of cylindrical coordinates. Now apply (3.35) to a cylindrical Gaussian surface of radius R and height h, shown in Fig. 3.8(b). (This Gaussian surface respects the cylindrical symmetry.) The flux through the cylinder surface is $\Phi_E = 2\pi R h E(R)$, and the charge enclosed is λh, so

$$E(R) = \frac{\lambda}{2\pi \epsilon_0 R}. \tag{3.44}$$

This result agrees with (3.19), obtained in Example 2 by integrating the Coulomb field, but the calculation from Gauss's Law is much easier.

EXAMPLE 7 What is the electric field of a uniformly charged plane? The infinite plane is another mathematical idealization, but the result is a good approximation for the field of a charged plate far from the edges.

Let σ be the charge per unit area, and take the z axis to be perpendicular to the plane. By the planar symmetry, and the fact that $\mathbf{E}$ diverges from charge sources, $\mathbf{E}(\mathbf{x})$ must have the form

$$\mathbf{E}(\mathbf{x}) = \begin{cases} E(z)\hat{\mathbf{k}} & \text{for} \quad z > 0 \\ -E(|z|)\hat{\mathbf{k}} & \text{for} \quad z < 0, \end{cases} \tag{3.45}$$

pointing away from the plane in both regions. Now apply (3.35) to a cylindrical Gaussian surface whose end faces are at $\pm z$ and have area A, shown in Fig. 3.8(c). (This Gaussian surface respects the planar symmetry.) There is flux of $\mathbf{E}$ through the end faces, equal to $\Phi_E = 2AE(z)$, and the charge enclosed is σA, so

$$E(z) = \frac{\sigma}{2\epsilon_0}. \tag{3.46}$$

Note that the electric field vector is *discontinuous* at the charged surface, being $+\hat{\mathbf{k}}\sigma/(2\epsilon_0)$ above and $-\hat{\mathbf{k}}\sigma/(2\epsilon_0)$ below, with discontinuity $\hat{\mathbf{k}}\sigma/\epsilon_0$. The discontinuity agrees with the boundary condition derived in Sec. 3.3. The field is independent of the height z above the plane because we have considered the field of an infinite plane. For a finite plate, far from the edges, the field is approximately constant for z much less than the size L of the plate; but for $z > L$ the field decreases with z.

Gauss's Law is always true, even for time-dependent problems. However, it is useful for calculating **E** only if the symmetry is high, as in these examples. Although limited, the Gaussian-surface trick is powerful. For the spherical case, as in Example 5, Gauss's Law will give **E** for any spherically symmetric density $\rho(r)$. For the cylindrical case, Gauss's Law will give **E** if the charge density depends only on the cylindrical radial coordinate r. If ρ varies with z, e.g., for a charge density of finite extent in z, the method cannot be used. In the planar case, Gauss's Law will give **E** only if the surface charge density σ is constant. Of course any solutions can be superposed; for example, one can find the field for *two planes* with different charge densities using the result of Example 7. Also, the solution for a highly symmetric system may be a useful approximation to a nonsymmetric case; for example, finite parallel plates may be approximated by infinite parallel planes if edge effects are neglected.

3.5 ■ GREEN'S FUNCTION AND THE DIRAC DELTA FUNCTION

The previous section shows how the integral form of Gauss's Law can be used to analyze an electric field. However, in many ways the differential form (3.25) is even more useful. Or, writing $\mathbf{E} = -\nabla V$, where $V(\mathbf{x})$ is the electric potential, the function $V(\mathbf{x})$ satisfies Poisson's equation (3.33) which may be solved by various methods.

To develop an appreciation for the differential form of Gauss's Law, it is useful to study an important modern viewpoint, based on the Green's function method. We need the fact that $1/4\pi|\mathbf{x} - \mathbf{x}'|$ is the Green's function of the differential operator $-\nabla^2$. In this method, a point source is described by setting the density function equal to a Dirac delta function. A brief mathematical excursion is in order.

3.5.1 ■ The Dirac delta Function

The Dirac delta function $\delta(x)$ is a generalized function with the following defining property: *for every continuous function $f(x)$,*

$$\int_{-\infty}^{\infty} \delta(x) f(x) \, dx = f(0). \tag{3.47}$$

This looks innocent enough, but in fact $\delta(x)$ is an extremely singular function. To satisfy the definition, $\delta(x)$ must be 0 for all values of x not equal to 0. (Proof: Consider the class of functions $g(x)$ such that $g(x) \geq 0$ for all x, and $g(0) = 0$. Then $\int \delta(x)g(x)dx = 0$. The only way this can be true for every $g(x)$ in the class is if $\delta(x) = 0$ for all $x \neq 0$.) However, the integral of $\delta(x)$ is 1,

$$\int_{-\infty}^{\infty} \delta(x) \, dx = 1. \tag{3.48}$$

(Proof: Let $f(x)$ be the constant function 1 in (3.47).) Thus the graph of $\delta(x)$ is an infinitely sharp spike at $x = 0$.

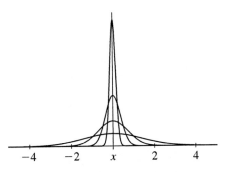

FIGURE 3.9   **The Dirac delta function** is the limit of a sequence of sharply peaked functions. In the limit, the width → 0 and the integral → 1.

To justify rigorously the existence of such a mathematical object requires a careful analysis in advanced mathematics. In fact, mathematicians call $\delta(x)$ a "generalized function" or "Schwartz distribution," to distinguish it from ordinary functions. But we will adopt a more intuitive attitude toward the delta function. It is perhaps most easily understood as the limit of a sequence of more and more sharply peaked functions, as illustrated in Fig. 3.9. The delta function was invented by a physicist (Paul Dirac) and is widely used in many branches of mathematical physics.

The 3-dimensional delta function $\delta^3(\mathbf{x})$ is defined analogously: For any continuous function $f(\mathbf{x})$,

$$\int \delta^3(\mathbf{x}) f(\mathbf{x}) d^3x = f(\mathbf{0}). \tag{3.49}$$

Since d^3x is just $dx\, dy\, dz$, the 3D delta function is the product of 1D delta functions for each Cartesian coordinate

$$\delta^3(\mathbf{x}) = \delta(x)\delta(y)\delta(z). \tag{3.50}$$

Thus $\delta^3(\mathbf{x})$ is 0 for $\mathbf{x} \neq 0$; its integral is 1, $\int \delta^3(\mathbf{x})d^3x = 1$. If x is a spatial coordinate, then dx has units of length, and $\delta(x)$ has units of $(\text{length})^{-1}$ because $\int \delta(x)dx = 1$. The units of $\delta^3(\mathbf{x})$ are $(\text{length})^{-3}$.

The delta function usually appears in an integral. We evaluate integrals involving the delta function by changing the variable of integration (if necessary) to convert the integral to the form (3.47), or in three dimensions (3.49). For example, consider the integral of $\delta(x - y)f(y)$ with respect to y; changing the variable of integration from y to $\xi \equiv x - y$ we find

$$\int_{-\infty}^{\infty} \delta(x - y)f(y)dy = \int_{-\infty}^{\infty} \delta(\xi)f(x - \xi)d\xi = f(x) \tag{3.51}$$

where the second equality follows from (3.47). Similarly, for three dimensions,

$$\int \delta^3(\mathbf{x} - \mathbf{y}) f(\mathbf{y}) d^3 y = f(\mathbf{x}), \tag{3.52}$$

which is an identity that we will use later. The meaning of (3.52) should be clear: $\delta^3(\mathbf{x})$ is a sharp spike at $\mathbf{x} = 0$, so $\delta^3(\mathbf{x} - \mathbf{y})$ is a spike at $\mathbf{y} = \mathbf{x}$; integrating over $\mathbf{y}$ picks out the value of $f(\mathbf{y})$ at $\mathbf{y} = \mathbf{x}$.

The Green's Function of $-\nabla^2$

Green's function techniques, which have wide usage in physics, provide a way to write the solution of a partial differential equation in closed form, as an integral over distributed sources. In electrostatics the partial differential equation is Poisson's equation (3.33), and the source is the charge density. The physical reasoning underlying the mathematical technique is that any distributed source can be considered as a sum or integral over elemental sources. Therefore, one first finds the potential function of a point source of unit strength located at an arbitrary point $\mathbf{x}'$; that potential is the Green's function. Then the potential function for the whole charge distribution is the integral of the Green's function over the source density.

The Green's function of $-\nabla^2$ is the function $G(\mathbf{x} - \mathbf{x}')$ defined by the equation

$$-\nabla^2 G(\mathbf{x} - \mathbf{x}') = \delta^3(\mathbf{x} - \mathbf{x}'), \tag{3.53}$$

along with the boundary condition that G approaches 0 at infinity.[10] We shall prove the following theorem:

$$G(\mathbf{x} - \mathbf{x}') = \frac{1}{4\pi |\mathbf{x} - \mathbf{x}'|}. \tag{3.54}$$

Proof: First, it is sufficient to let $\mathbf{x}'$ be at the origin, because all the relevant functions depend only on $\mathbf{x} - \mathbf{x}'$, the relative position of $\mathbf{x}$ and $\mathbf{x}'$. So, what we need to prove is that

$$-\nabla^2 \frac{1}{|\mathbf{x}|} = 4\pi \delta^3(\mathbf{x}). \tag{3.55}$$

For all $\mathbf{x}$ with $|\mathbf{x}| \neq 0$ the left-hand side of (3.55) is 0: Using spherical coordinates,

$$\nabla^2 \left(\frac{1}{r}\right) = \frac{d^2}{dr^2}\left(\frac{1}{r}\right) + \frac{2}{r}\frac{d}{dr}\left(\frac{1}{r}\right) = 0 \tag{3.56}$$

for $r \neq 0$. (This may also be proven in Cartesian coordinates.) For the point $|\mathbf{x}| = 0$, $\nabla^2(1/r)$ is not well-defined in the usual sense, because $1/r$ is singular at $r = 0$. To discover what happens at the origin, consider the integral of $\nabla^2(1/r)$ in a spherical volume V of any radius around the origin; using Gauss's theorem,

$$\int_V \nabla^2 \left(\frac{1}{r}\right) d^3 x = \oint_S \nabla \left(\frac{1}{r}\right) \cdot \hat{\mathbf{r}} dA \tag{3.57}$$

---

[10]We shall see in Sec. 3.6 that $G(\mathbf{x} - \mathbf{x}')/\epsilon_0$ is the electrostatic potential of a unit charge at $\mathbf{x}'$, i.e., charge density $\delta^3(\mathbf{x} - \mathbf{x}')$.

where S is the spherical boundary of V. But $\hat{\mathbf{r}} \cdot \nabla(1/r)$ is $-1/r^2$, and $dA = r^2 d\Omega$; therefore the right-hand side of (3.57) is $-\oint_S d\Omega = -4\pi$. So we have shown that $\nabla^2(1/r)$ is 0 for $r \neq 0$, but its integral over the volume of any sphere is -4π. These are precisely the characteristics of the delta function, summarized in (3.49); hence equation (3.55) is proven.

An alternative proof of (3.55), somewhat more comfortable for people who avoid ∞ in calculations, can be provided. This analysis does not use the singular function $1/r$ directly, but regards $1/r$ as the limit of $1/\sqrt{r^2 + \epsilon^2}$ as $\epsilon \rightarrow 0$. Denote by $g(r, \epsilon)$ the latter function, which is nowhere infinite as long as $\epsilon \neq 0$. Now, it is straightforward to calculate $-\nabla^2 g$,

$$-\nabla^2 g = \frac{3\epsilon^2}{(r^2 + \epsilon^2)^{5/2}}.$$

This function becomes more and more sharply peaked at $r = 0$ as ϵ approaches 0. Note that $-\nabla^2 g$ tends to 0 if $r \neq 0$, and to ∞ if $r = 0$, as $\epsilon \rightarrow 0$. The total volume integral of $-\nabla^2 g$ is 4π, independent of ϵ; the integral can be evaluated either using Gauss's theorem, or more directly by integrating over spherical polar coordinates. Thus $-\nabla^2 g$ approaches $4\pi \delta^3(\mathbf{x})$ as $\epsilon \rightarrow 0$, proving (3.55) since $g(r, 0) = 1/r$.

The reader may feel uneasy about this very singular mathematics. But Green's function methods are common in mathematical physics, and the delta function is a modern tool of theoretical physics. In Sec. 3.6 we will use the Green's function (3.54)—as the potential of a unit point source—to solve Poisson's equation.

3.5.2 ▓ Another Proof of Gauss's Law

After these mathematical developments, we are prepared to calculate the divergence of $\mathbf{E}$ directly from (3.30). Knowing the Dirac delta function, the calculation is extremely simple: really just a formality. From (3.30) and (3.54),

$$\nabla \cdot \mathbf{E} = \frac{1}{\epsilon_0} \int \left[-\nabla^2 G(\mathbf{x} - \mathbf{x}') \right] \rho(\mathbf{x}') d^3 x'; \tag{3.58}$$

but the quantity in square brackets is $\delta^3(\mathbf{x} - \mathbf{x}')$, so by the identity (3.52), $\nabla \cdot \mathbf{E} = \rho(\mathbf{x})/\epsilon_0$, which is the divergence equation (3.25), Gauss's Law of electricity.

3.6 ▪ THE ELECTRIC POTENTIAL

3.6.1 ▓ Definition and Construction

A general problem of electrostatics, which we are now prepared to consider, is to find $\mathbf{E}(\mathbf{x})$ everywhere for an *arbitrary* charge distribution $\rho(\mathbf{x}')$, if $\rho(\mathbf{x}')$ is known *a priori*. This is an important milepost on our path to learning how to solve the *most general problem* in electrostatics, which includes conductors and dielectrics in the system, discussed in Chapters 4 through 6.

For all of these cases the solution for $\mathbf{E}(\mathbf{x})$ should be unique. This raises a mathematical question: What charges, potentials, and boundary conditions must be specified in order that the function $\mathbf{E}(\mathbf{x})$ that solves the field equations is a unique solution? The answer must be consistent with the physical condition of uniqueness: Once the charges, batteries, conductors, and dielectrics are in place, then nature determines a unique field $\mathbf{E}$, which could be measured with suitable apparatus.

For electrostatics the electric field $\mathbf{E}(\mathbf{x})$ is *irrotational*; that is, $\boldsymbol{\nabla} \times \mathbf{E} = 0$. It follows that we may construct the *electric potential* $V(\mathbf{x})$, a scalar function such that

$$\mathbf{E}(\mathbf{x}) = -\boldsymbol{\nabla} V(\mathbf{x}). \tag{3.59}$$

We proved in Chapter 2 that the curl of the gradient of any scalar function is 0, so (3.59) guarantees that $\mathbf{E}$ is irrotational. The electric potential was introduced into the theory of electrostatics by Green.[11] It is an important mathematical technique in field theory. The SI unit of electric potential is the volt (V) defined by 1 V=1 J/C; that is, voltage has the units of work per unit charge.

To justify (3.59) we must address two issues—the existence and uniqueness of such a function $V(\mathbf{x})$. As to existence, we have already seen that $\mathbf{E}(\mathbf{x})$ can be written as a gradient in (3.30), and indeed from that equation we have written a formula for $V(\mathbf{x})$ of a static charge density,

$$V(\mathbf{x}) = \frac{1}{4\pi \epsilon_0} \int \frac{\rho(\mathbf{x}')d^3x'}{|\mathbf{x} - \mathbf{x}'|}. \tag{3.60}$$

This proves the existence of $V(\mathbf{x})$, by construction. For example, for a point charge q at the origin the electrostatic potential is

$$V(\mathbf{x}) = \frac{q}{4\pi \epsilon_0 r}. \tag{3.61}$$

One can easily verify (3.59) for this basic example.

But it is instructive to construct $V(\mathbf{x})$ in another way. Let $-V(\mathbf{x})$ be the line integral of $\mathbf{E}$ along a curve Γ from a reference point $\mathbf{x}_0$ to $\mathbf{x}$

$$V(\mathbf{x}) = -\int_\Gamma \mathbf{E} \cdot d\boldsymbol{\ell}. \tag{3.62}$$

This equation says that $V(\mathbf{x})$ is equal to the work that must be done by an external force, acting against the electric force, to move unit charge from $\mathbf{x}_0$ to $\mathbf{x}$ along the path Γ, because $-\mathbf{E}$ is the force per unit charge that the external agent

[11] George Green (1793–1841) owned a grain mill in Nottingham, England, and was a self-taught amateur mathematician. In 1828 he published a paper on the mathematical theory of electricity and magnetism, introducing the concept of the *potential*, which was largely ignored. But the paper was rediscovered by Lord Kelvin in 1845, and became influential in the development of the theory. The Green's function is named after George Green.

must exert.[12] We will prove below that $V(\mathbf{x})$ does not depend on the particular choice of the path Γ from $\mathbf{x}_0$ to $\mathbf{x}$, because $\mathbf{E}$ is irrotational. The function $V(\mathbf{x})$ in (3.62) satisfies (3.59). Consider the change of $V(\mathbf{x})$ for an arbitrary infinitesimal displacement $\delta\mathbf{x}$,

$$V(\mathbf{x} + \delta\mathbf{x}) - V(\mathbf{x}) = \nabla V \cdot \delta\mathbf{x} = -\int_{\Gamma + \delta\mathbf{x}} \mathbf{E} \cdot d\mathbf{x} + \int_{\Gamma} \mathbf{E} \cdot d\mathbf{x} = -\mathbf{E}(\mathbf{x}) \cdot \delta\mathbf{x}.$$

The first equality is the definition of ∇V; the second equality follows from (3.62), because the path of integration for $V(\mathbf{x} + \delta\mathbf{x})$ is Γ from $\mathbf{x}_0$ to $\mathbf{x}$ followed by the extra displacement $\delta\mathbf{x}$ from $\mathbf{x}$ to $\mathbf{x} + \delta\mathbf{x}$. Thus $\mathbf{E}(\mathbf{x}) = -\nabla V$ as required.

Now, what about uniqueness? The line integral in (3.62) is set up with a reference point $\mathbf{x}_0$ and path Γ. These are not unique, so the question is whether $V(\mathbf{x})$ depends on $\mathbf{x}_0$ or Γ. In fact, $V(\mathbf{x})$ does not depend on Γ, but in general it does depend on $\mathbf{x}_0$. Let Γ_1 and Γ_2 be two different paths both from $\mathbf{x}_0$ to $\mathbf{x}$, and V_1 and V_2 the corresponding functions defined by (3.62). Then

$$V_1(\mathbf{x}) - V_2(\mathbf{x}) = -\int_{\Gamma_1} \mathbf{E} \cdot d\boldsymbol{\ell} + \int_{\Gamma_2} \mathbf{E} \cdot d\boldsymbol{\ell} = -\oint_C \mathbf{E} \cdot d\boldsymbol{\ell}, \qquad (3.63)$$

where C is the closed curve consisting of Γ_1 from $\mathbf{x}_0$ to $\mathbf{x}$, and the reverse of Γ_2 from $\mathbf{x}$ back to $\mathbf{x}_0$. But $\oint \mathbf{E} \cdot d\boldsymbol{\ell} = 0$ by Stokes's theorem, because $\nabla \times \mathbf{E} = 0$. Thus $V_1 = V_2$, proving that $V(\mathbf{x})$ does not depend on the path.

The path independence leads to an interpretation of the potential *difference* between two points. The difference $V(\mathbf{x}_2) - V(\mathbf{x}_1)$ is equal to the work that must be done by an external agent to move unit charge from $\mathbf{x}_1$ to $\mathbf{x}_2$ against the electric force, along any path. The work is independent of the path because the electrostatic force is conservative.

In general $V(\mathbf{x})$ does depend on the reference point $\mathbf{x}_0$. If we choose another reference point, say $\mathbf{x}_1$, then the new potential $V_1(\mathbf{x})$ may be written as the line integral from $\mathbf{x}_1$ to $\mathbf{x}_0$, and then from $\mathbf{x}_0$ to $\mathbf{x}$; the first part is $-V(\mathbf{x}_1)$, and the second part is $V(\mathbf{x})$, so

$$V_1(\mathbf{x}) = V(\mathbf{x}) - V(\mathbf{x}_1). \qquad (3.64)$$

Changing the reference point is equivalent to adding the constant $C = -V(\mathbf{x}_1)$ to the original potential. Thus the potential $V(\mathbf{x})$, as a function of $\mathbf{x}$, is determined only up to an additive constant. It is obvious that $V(\mathbf{x}) + C$ satisfies (3.59) for any constant C. The nontrivial point is that there is no other nonuniqueness of $V(\mathbf{x})$ besides an additive constant.

When solving problems, it often helps to make a judicious choice of the reference point. If the total charge in the system is finite, then the reference point is usully chosen to be at infinity. In that case $V(\mathbf{x})$ is the work required to bring unit charge from ∞ to $\mathbf{x}$, and $V(\mathbf{x})$ approaches 0 as $\mathbf{x}$ goes to ∞. If, however, the sys-

---

[12]Or, equivalently, $V(\mathbf{x})$ is the work per unit charge done by the electric force acting on a test charge that moves from $\mathbf{x}$ to $\mathbf{x}_0$.

tem has infinite charge, as in a charged line or plane, then choosing the reference point at ∞ is not valid; the potential diverges at ∞. In such cases, as we'll see in Example 8 following, one may choose any other convenient point, or leave the reference point unspecified.

We showed in Sec. 3.3 that the normal component of the electric field, at a surface, is discontinuous if there is a surface charge. The potential, however, is continuous, because it is equal to the line integral of $\mathbf{E}$ along a path. The variation of the integral, as the path crosses the surface, cannot be discontinuous unless the charge density is infinite, which would be unphysical. A discontinuity of $\mathbf{E}$ corresponds to a change in the gradient of V, rather than a discontinuity of V.

We recall from Chapter 2 that the gradient of a scalar function is everywhere perpendicular to any surface on which the function is constant. Applying this geometric relationship to $V(\mathbf{x})$ and $\mathbf{E}(\mathbf{x})$, the electric field vectors are everywhere normal to *equipotential surfaces*, pointing toward smaller V. We may visualize $\mathbf{E}(\mathbf{x})$ by drawing the *electric field lines*, which are the tangent curves of the vector field, i.e., the set of curves everywhere tangent to the field vectors. An electric field line crossing any equipotential surface is perpendicular to the surface.

3.6.2 ■ Poisson's Equation

The electrostatic field is determined throughout space by *local relations*, $\nabla \times \mathbf{E} = 0$ and $\nabla \cdot \mathbf{E} = \rho/\epsilon_0$. Introducing the potential, i.e., writing $\mathbf{E} = -\nabla V$, guarantees $\nabla \times \mathbf{E} = 0$, but that equation gives no information on $V(\mathbf{x})$. Gauss's Law becomes

$$-\nabla^2 V = \rho/\epsilon_0. \tag{3.65}$$

This equation, called *Poisson's equation*, together with the boundary conditions, must determine the potential function.

Green's Solution

The Green's function $G(\mathbf{x} - \mathbf{x}')$ is defined by (3.53). Comparing the definition with Poisson's equation shows that $G(\mathbf{x}-\mathbf{x}')/\epsilon_0$ is the solution of Poisson's equation for a point source of strength 1 at $\mathbf{x}'$, with the boundary condition $G \to 0$ at infinity. (The density function of a point source is proportional to the delta function—an infinitely sharp spike at the source point.) In other words, G/ϵ_0 *is the potential of a unit point source.*

We may write the general solution to Poisson's equation in terms of the Green's function, as

$$V(\mathbf{x}) = \frac{1}{\epsilon_0} \int G(\mathbf{x} - \mathbf{x}')\rho(\mathbf{x}')d^3x', \tag{3.66}$$

which is known as Green's solution to Poisson's equation for an unbounded space. (This equation is just the same as (3.60). For practice, the student should verify again, by the methods of Sec. 3.5, that this $V(\mathbf{x})$ satisfies (3.65).) Green's solution is an expression of the superposition principle: $G(\mathbf{x}-\mathbf{x}') \, dq/\epsilon_0$ is the potential for the field created by the point-like charge $dq = \rho(\mathbf{x}')d^3x'$, and the superposition of all elemental contributions is the total potential.

The Helmholtz Theorem

By writing $\mathbf{E} = -\nabla V$, and using Green's solution to Poisson's equation, we have solved the field equations $\nabla \times \mathbf{E} = 0$ and $\nabla \cdot \mathbf{E} = \rho/\epsilon_0$. This is a special case of a more general mathematical problem—to find a function $\mathbf{F}(\mathbf{x})$ with specified curl and divergence. (For static $\mathbf{E}$ the curl is always 0.) The Helmholtz theorem, which is discussed in Appendix B, supplies the solution to this general problem.

3.6.3 ▩ Example Calculations of $V(\mathbf{x})$

As we have seen, if an electrostatic system is highly symmetric then $\mathbf{E}$ may be calculated from the flux through an appropriate Gaussian surface. When there is not enough symmetry to apply this method, the simplest approach, if the charge density is known, is usually to calculate $V(\mathbf{x})$ from Green's solution, and then $\mathbf{E}$ is $-\nabla V$.

EXAMPLE 8   What is the electrostatic potential, and electric field, of a uniformly charged straight wire of length 2ℓ, for points on the midplane of the wire?

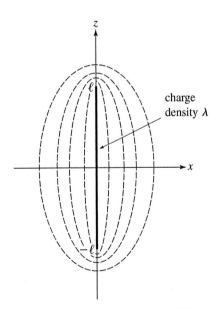

FIGURE 3.10   Examples 8 and 10: A uniformly charged line segment, with charge per unit length λ, extends from $z = -\ell$ to $z = +\ell$ on the z axis. In Example 8, V and $\mathbf{E}$ are found on the midplane ($z = 0$). In Example 10, $V(\mathbf{x})$ is found for all $\mathbf{x}$. The equipotentials in the xz plane are ellipses with foci at the ends of the line segment, shown as dashed curves. The equipotentials in three dimensions are ellipsoids of revolution.

The wire is shown in Fig. 3.10. It lies on the z axis from $z = -\ell$ to $z = +\ell$, the midplane is the xy plane, and λ denotes the charge per unit length. By (3.60)

$$V(r) = \frac{1}{4\pi\epsilon_0} \int_{-\ell}^{\ell} \frac{\lambda dz'}{\sqrt{r^2 + z'^2}} = \frac{\lambda}{4\pi\epsilon_0} \ln\left\{ \frac{\ell + \sqrt{r^2 + \ell^2}}{-\ell + \sqrt{r^2 + \ell^2}} \right\}, \qquad (3.67)$$

where r is the perpendicular distance to the wire, i.e., the radial coordinate of cylindrical coordinates. On the midplane the electric field is radial, by symmetry, so

$$\mathbf{E}(r) = -\hat{\mathbf{r}}\frac{\partial V}{\partial r} = \frac{\lambda \ell \hat{\mathbf{r}}}{2\pi\epsilon_0 r \sqrt{r^2 + \ell^2}}. \qquad (3.68)$$

The same result was obtained earlier in Example 2.

In the limit $\ell \to \infty$, the field approaches the result (3.44) of Example 6. In this limit the total charge is infinite. The midplane field is well-defined in the limit, but the potential (3.67) diverges as $\ell \to \infty$. For (3.67) the reference point (at which $V = 0$) is at $r = \infty$, which is fine for finite ℓ, but not valid for an infinite charged line. For an infinite wire the potential can be written as $V_{\text{line}}(r) = -(\lambda/2\pi\epsilon_0) \ln(r/r_0)$, where the reference point (where $V_{\text{line}} = 0$) is $r = r_0$, an arbitrary constant. This function $V_{\text{line}}(r)$ diverges as $r \to \infty$.

EXAMPLE 9   What is the electrostatic potential, and electric field, above a uniformly charged circular plate of radius a, on the axis of symmetry?

Let σ be the charge per unit area, and let the z axis be the axis of symmetry, as shown in Fig. 3.11. Then by (3.60), using cylindrical coordinates,

$$V(0, 0, z) = \frac{1}{4\pi\epsilon_0} \int_0^a \frac{\sigma 2\pi r' dr'}{\sqrt{r'^2 + z^2}} = \frac{\sigma}{2\epsilon_0}\left(\sqrt{a^2 + z^2} - |z| \right). \qquad (3.69)$$

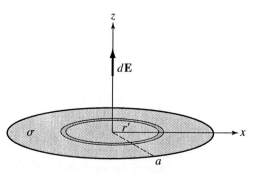

FIGURE 3.11   Example 9: A uniformly charged circular plate. $d\mathbf{E}$ is the field of the elemental ring.

On the positive z axis the electric field is in the direction of $\hat{\mathbf{k}}$, by symmetry; then

$$\mathbf{E}(0, 0, z) = -\hat{\mathbf{k}}\frac{\partial V}{\partial z} = \frac{\sigma\hat{\mathbf{k}}}{2\epsilon_0}\left[1 - \frac{z}{\sqrt{a^2 + z^2}}\right]. \tag{3.70}$$

In the limit $a \to \infty$ this field agrees with the result (3.46) of Example 7.

EXAMPLE 10   In the previous two examples we considered only points of symmetry of the system. In the current example we consider the full space. The calculation is more difficult but worth the effort. The system is a uniformly charged line segment of length 2ℓ with charge per unit length λ.

The potential function.   Set up a coordinate system with the line segment on the z axis from $z = -\ell$ to $z = \ell$; this is the same configuration as in Examples 2 and 8. Figure 3.10 shows the xz plane ($y = 0$). We shall first calculate the potential V in this plane, and then generalize the result to three dimensions by axial symmetry. At the point $(x, 0, z)$ the potential may be written

$$V(x, 0, z) = \frac{\lambda}{4\pi\epsilon_0} \int_{-\ell}^{+\ell} \frac{dz'}{\sqrt{x^2 + (z - z')^2}} \tag{3.71}$$

where $\lambda dz'$ is the element of charge on the line segment at $(0, 0, z')$. After changing the variable of integration to $\xi \equiv z - z'$ the integral is a standard form that can be found in integral tables, and the result is

$$V(x, 0, z) = \frac{\lambda}{4\pi\epsilon_0} \ln\left\{\frac{z + \ell + \sqrt{x^2 + (z + \ell)^2}}{z - \ell + \sqrt{x^2 + (z - \ell)^2}}\right\}. \tag{3.72}$$

To extend the result to three dimensions, simply replace x^2 by $x^2 + y^2$; because by rotational invariance about the z axis, V at a given z depends only on the perpendicular distance from the field point to the z axis. So, for all points

$$V(x, y, z) = \frac{\lambda}{4\pi\epsilon_0} \ln\left\{\frac{z + \ell + \sqrt{x^2 + y^2 + (z + \ell)^2}}{z - \ell + \sqrt{x^2 + y^2 + (z - \ell)^2}}\right\}. \tag{3.73}$$

The equipotential surfaces.   An equipotential surface, defined by the equation $V(x, y, z) = V_0$ where V_0 is a constant, is the locus of points where the argument of the logarithm in (3.73) is constant. Deducing the shape of the surface from this equation is not very easy, but we can guess the answer and then verify that our guess is correct. If the charge were restricted to a point, then the equipotentials would be spheres; such a symmetric system would necessarily have very symmetric equipotentials. Here the charge is spread out uniformly along a line segment, but that is still quite a symmetric system, so the equipotentials should again be quite symmetric, but now having two special points (the endpoints of the line). A natural guess is that the equipotentials are *ellipsoids of revolution*. An ellipsoid is the surface swept out when an ellipse in the xz plane is rotated about the z axis. We shall verify that an equipotential curve in the xz plane is an ellipse whose foci are at $(0, 0, \ell)$ and $(0, 0, -\ell)$.

The equation for an ellipse in the xz plane is

$$\frac{x^2}{a^2} + \frac{z^2}{c^2} = 1, \tag{3.74}$$

where c is called the semimajor axis and a the semiminor axis, assuming $c > a$. The ellipse intersects the z axis at $z = \pm c$, and the x axis at $x = \pm a$. Because the focal points are at $\pm \ell$ on the z axis, the minor axis may be written as $a = \sqrt{c^2 - \ell^2}$. [Note that the sum of the distances from (x, z) to the two foci is constant along the ellipse; on the x axis it is $2\sqrt{a^2 + \ell^2}$, and on the z axis $2c$.] Now we replace x^2 in the argument of the logarithm of (3.72) by $a^2(1 - z^2/c^2)$, and simplify the result in terms of c and ℓ. A little algebra shows that the numerator is $(c + \ell)(1 + z/c)$ and the denominator $(c - \ell)(1 + z/c)$, so for points on the ellipse the potential is a constant V_0 given by

$$V_0 = \frac{\lambda}{4\pi\epsilon_0} \ln \left\{ \frac{c + \ell}{c - \ell} \right\}. \tag{3.75}$$

Thus the equipotential curves in the xz plane are confocal ellipses, four of which are shown in Fig. 3.10. The potential V_0 depends on the semimajor axis c, and decreases as c increases.

By rotational invariance about the z axis the equipotential surfaces in three dimensions are confocal ellipsoids of revolution. These surfaces are *prolate spheroids*, characterized by the parameter $\varepsilon \equiv \ell/c$, which is the *eccentricity*. If ε approaches 0, i.e., far from the charged line segment, the surface becomes a sphere. If ε approaches 1, i.e., near the charged segment, the surface becomes needle-shaped with large curvature at the ends at $z = \pm c$. We'll make use of this example again in Chapter 4, to determine the field of a charged ellipsoid.

◼─────────

3.7 ◼ ENERGY OF THE ELECTRIC FIELD

The electrostatic potential is closely related to potential energy. What is the potential energy associated with the electrostatic force on a charge q? The work done by the electric field, as the charge moves along a curve Γ from the reference point $\mathbf{x}_0$ to $\mathbf{x}$, is

$$W = \int_\Gamma q\mathbf{E} \cdot d\boldsymbol{\ell} = -qV(\mathbf{x}). \tag{3.76}$$

The potential energy $U_q(\mathbf{x})$ with respect to the reference point is by definition $-W$; so

$$U_q(\mathbf{x}) = qV(\mathbf{x}). \tag{3.77}$$

Thus another way to define the electrostatic potential is that $V(\mathbf{x})$ is the *potential energy per unit charge* of the electric force on a test charge. [The potential energy of a force $\mathbf{F}$ is well-defined provided $\nabla \times \mathbf{F} = 0$, which is true for the electrostatic force because $\mathbf{E}(\mathbf{x})$ is irrotational.]

We may use this result to determine the *total electrostatic energy* of a charge distribution $\rho(\mathbf{x})$. First consider just two point charges q_1 and q_2. The work done by an external agent to bring q_2 from infinity to its position $\mathbf{x}_2$, against the field of q_1, is the potential energy of interaction between the charges

$$U_{12} = q_2 V_1 = \frac{q_1 q_2}{4\pi\epsilon_0 |\mathbf{x}_1 - \mathbf{x}_2|}. \tag{3.78}$$

For a set of charges $\{q_1, q_2, \ldots, q_N\}$ at positions $\{\mathbf{x}_1, \mathbf{x}_2, \ldots, \mathbf{x}_N\}$ each pair (ij) has potential energy U_{ij}, so the total energy is

$$U = \frac{1}{2} \sum_{i \neq j} \frac{q_i q_j}{4\pi\epsilon_0 |\mathbf{x}_i - \mathbf{x}_j|}; \tag{3.79}$$

the double sum over i and j excludes terms with $i = j$, and the factor $1/2$ accounts for the appearance of each pair twice in the double sum. U is the energy required to *assemble* the given charge distribution. For a continuum with charge density $\rho(\mathbf{x})$, all the charge in an infinitesimal volume d^3x is combined, and the sum over discrete particles becomes an integral over volume; then the energy is

$$U = \frac{1}{2} \int \frac{\rho(\mathbf{x})\rho(\mathbf{x}')}{4\pi\epsilon_0 |\mathbf{x} - \mathbf{x}'|} d^3x \, d^3x'. \tag{3.80}$$

Equivalently, we may write

$$U = \frac{1}{2} \int \rho(\mathbf{x}) V(\mathbf{x}) d^3x. \tag{3.81}$$

We regard the electric field as a real physical entity, possessing energy. This viewpoint becomes more important later, when we study light as an electromagnetic wave: The energy transported through vacuum by light must obviously belong to the fields. But even in electrostatics we adopt the view that the energy U resides in the field. Therefore, we will now rewrite the equation for U entirely in terms of the electric field $\mathbf{E}(\mathbf{x})$ produced by the charge density. The function $\rho(\mathbf{x}) V(\mathbf{x})$ is, by Poisson's equation,

$$\rho V = -\epsilon_0 (\nabla^2 V) V = -\epsilon_0 \nabla \cdot (V \nabla V) + \epsilon_0 (\nabla V)^2, \tag{3.82}$$

the second equality being an identity of vector calculus. The total volume integral of $\nabla \cdot (V \nabla V)$ is 0, because by Gauss's theorem it is equal to the flux of $V \nabla V$ through the surface at infinity, where $V = 0$. The final term in (3.82) is $\epsilon_0 E^2$.

Thus the total energy is

$$U = \frac{\epsilon_0}{2} \int E^2 d^3x. \tag{3.83}$$

The energy density (energy per unit volume) of the electric field is evidently

$$u_E(\mathbf{x}) = \frac{\epsilon_0}{2} E^2(\mathbf{x}). \tag{3.84}$$

The idea that energy belongs to the field is intriguing. For example, electrostatic energy stored in fields in the atmosphere, is, on a large scale, the origin of lightning; or, on a small scale, the origin of the spark between your hand and a metal doorknob.

EXAMPLE 11   The potential energy of two point charges q_1 and q_2 separated by d is $q_1 q_2/(4\pi\epsilon_0 d)$. How is this potential energy related to the total field energy (3.83)?

Instead of point charges, consider two charged spheres whose centers are separated by distance d. By the superposition principle, the electric field is $\mathbf{E}_1 + \mathbf{E}_2$, so the field energy is

$$U = \frac{\epsilon_0}{2} \int \left(E_1^2 + E_2^2 + 2\mathbf{E}_1 \cdot \mathbf{E}_2 \right) d^3x = U_1 + U_2 + U_{12}. \tag{3.85}$$

U_1 and U_2 are *self-energies*. U_1 is the energy of q_1 *alone*, and U_2 that of q_2. These are independent of the separation d. U_{12} is the interaction energy.

The self-energy U_1 is $Cq_1^2/(4\pi\epsilon_0 R_1)$, where R_1 is the radius of the first sphere, and C is a dimensionless constant that depends on how the charge is distributed in the sphere.[13] U_2 is analogous. Note that $U_1 \to \infty$ as $R_1 \to 0$; the self-energy of a *point charge* is infinite. This infinity is not relevant to the classical theory, which describes macroscopic systems, but it is an issue in quantum electrodynamics because the electron is a point particle.[14]

The interaction energy U_{12} remains finite in the limit of point charges. It can be shown[15] that for point charges, (3.85) gives $U_{12} = q_1 q_2/(4\pi\epsilon_0 d)$. Of course we obtained the same result earlier (3.78), but here it is derived from the field energy. The total field energy U (including self-energies) must be positive, because $u_E(\mathbf{x}) \geq 0$. But the interaction energy may be negative, implying attraction between the charges, if they have opposite signs.

---

[13] See Exercises 26 and 28.
[14] The procedures of *renormalization*, developed by Feynman and Schwinger, render the predictions of QED finite.
[15] See Exercise 27.

3.8 ■ THE MULTIPOLE EXPANSION

The potential and field due to a point charge at the origin are

$$V(\mathbf{x}) = \frac{q}{4\pi \epsilon_0 r} \quad \text{and} \quad \mathbf{E}(\mathbf{x}) = \frac{q\hat{\mathbf{r}}}{4\pi \epsilon_0 r^2}. \tag{3.86}$$

Here $r = |\mathbf{x}|$ and $\hat{\mathbf{r}} = \mathbf{x}/r$. Now consider a collection of charges in a region near the origin. The problem we shall solve in this section is to determine the asymptotic form of $V(\mathbf{x})$, i.e., far from the charge distribution. The complete treatment of this problem is to express $V(\mathbf{x})$ as an expansion in powers of $1/r$, called the *multipole expansion*. We won't derive the full multipole expansion, but rather consider the first few terms, which are the most important.

3.8.1 ■ Two Charges

As a first step, consider just two charges, q_1 at $\mathbf{x}_1$ and q_2 at $\mathbf{x}_2$. The exact potential function is the superposition

$$V(\mathbf{x}) = \frac{q_1}{4\pi \epsilon_0 |\mathbf{x} - \mathbf{x}_1|} + \frac{q_2}{4\pi \epsilon_0 |\mathbf{x} - \mathbf{x}_2|}. \tag{3.87}$$

Asymptotically, i.e., for $r \gg r_1$ and $r \gg r_2$, we may make the expansion (for $k = 1$ or 2)

$$\frac{1}{|\mathbf{x} - \mathbf{x}_k|} = \frac{1}{r} + \frac{\hat{\mathbf{r}} \cdot \mathbf{x}_k}{r^2} + \frac{3\left(\hat{\mathbf{r}} \cdot \mathbf{x}_k\right)^2 - r_k^2}{2r^3} + \mathcal{O}\left(\frac{1}{r^4}\right), \tag{3.88}$$

an expansion in powers of $1/r$. As $r \to \infty$, each term is smaller than the one before. The notation $\mathcal{O}(1/r^4)$ stands for terms proportional to powers of $1/r$ quartic or higher. To derive the expansion (3.88) write

$$\frac{1}{|\mathbf{x} - \mathbf{x}_k|} = \frac{1}{\sqrt{r^2 - 2rr_k \cos\theta + r_k^2}} = \frac{1}{r\sqrt{1 + \epsilon}} \tag{3.89}$$

where $\epsilon = \left(-2rr_k \cos\theta + r_k^2\right)/r^2$ and θ is the angle between $\mathbf{x}$ and $\mathbf{x}_k$. Expand the result in ϵ using the Taylor series

$$\frac{1}{\sqrt{1 + \epsilon}} = 1 - \frac{1}{2}\epsilon + \frac{3}{8}\epsilon^2 + \mathcal{O}(\epsilon^3); \tag{3.90}$$

and reexpress the result in powers of $1/r$, dropping terms of order r_k^3/r^4.

We obtain the multipole expansion for $V(\mathbf{x})$ by substituting (3.88) into (3.87). Neglecting terms higher than the quadrupole, the result is

$$V(\mathbf{x}) = \frac{1}{4\pi \epsilon_0} \left\{ \frac{Q}{r} + \frac{\hat{\mathbf{r}} \cdot \mathbf{p}}{r^2} + \frac{\hat{\mathbf{r}} \cdot \mathcal{Q}_2 \cdot \hat{\mathbf{r}}}{r^3} \right\}. \tag{3.91}$$

The three terms in (3.91) are called the monopole (Q), dipole ($\mathbf{p}$), and quadrupole ($\mathcal{Q}_2$) terms. Q is a scalar, equal to $q_1 + q_2$, the total charge. The monopole term is dominant as $r \to \infty$ unless $Q = 0$. The parameter $\mathbf{p}$ is a vector, given by

$$\mathbf{p} = q_1 \mathbf{x}_1 + q_2 \mathbf{x}_2, \tag{3.92}$$

called the *electric dipole moment* of the system. The dipole term is dominant at large r if $Q = 0$. The parameter $\mathcal{Q}_2$ is a tensor (or dyadic)[16], given by

$$\mathcal{Q}_2 = \frac{q_1}{2}\left[3\mathbf{x}_1\mathbf{x}_1 - r_1^2\mathbf{I}\right] + \frac{q_2}{2}\left[3\mathbf{x}_2\mathbf{x}_2 - r_2^2\mathbf{I}\right], \tag{3.93}$$

called the *quadrupole moment*. Here $\mathbf{I}$ denotes the unit tensor. As a vector has one index, a tensor has two indices; for example, $(\mathbf{x}_1\mathbf{x}_1)_{ij} = x_{1i}x_{1j}$ and $(\mathbf{I})_{ij} = \delta_{ij}$. The reader should not feel intimidated by the quadrupole. We shall often need the dipole potential—the second term in (3.91)—but rarely the higher multipoles.

Notice the dependence on r for the various multipoles: For a charge (monopole) the potential decreases as $1/r$ and the field as $1/r^2$; for a dipole the potential decreases as $1/r^2$ and the field as $1/r^3$; for a quadrupole the potential decreases as $1/r^3$ and the field as $1/r^4$. This pattern generalizes to higher multipoles.

EXAMPLE 12   To understand (3.91) it helps to work out a prototype case, with q_1 and q_2 on the z axis, at $z = d_1$ and $z = d_2$ respectively, illustrated in Fig. 3.12. By symmetry, $V(\mathbf{x})$ is $V(r, \theta)$ (independent of ϕ). The exact potential is

$$V(r, \theta) = \frac{1}{4\pi\epsilon_0}\left\{\frac{q_1}{\sqrt{r^2 - 2rd_1\cos\theta + d_1^2}} + \frac{q_2}{\sqrt{r^2 - 2rd_2\cos\theta + d_2^2}}\right\}.$$
$$\tag{3.94}$$

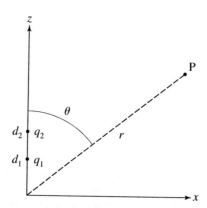

FIGURE 3.12   A system of 2 charges, q_1 and q_2. The potential at P is Eq. (3.94).

[16]The student may have encountered in mechanics, the *inertia tensor* of a mass distribution; the quadrupole tensor $\mathcal{Q}_2$ is analogous for a charge distribution.

The multipole expansion through order r^{-3}, valid for $r \gg d_1, d_2$, is

$$V(r, \theta) =$$

$$\frac{1}{4\pi\epsilon_0} \left\{ \frac{q_1 + q_2}{r} + \frac{(q_1 d_1 + q_2 d_2)\cos\theta}{r^2} + \frac{(q_1 d_1^2 + q_2 d_2^2)}{2r^3} \left(3\cos^2\theta - 1\right) \right\}.$$

$$(3.95)$$

This result agrees with the general equation (3.91). The total charge is $Q = q_1 + q_2$, and the dipole moment is $\mathbf{p} = (q_1 d_1 + q_2 d_2)\hat{\mathbf{k}}$. The quadrupole moment is

$$\mathcal{Q}_2 = \tfrac{1}{2}(q_1 d_1^2 + q_2 d_2^2)(3\hat{\mathbf{k}}\hat{\mathbf{k}} - \mathbf{I}). \qquad (3.96)$$

In general the dipole and quadrupole moments depend on the choice of origin. For example, (3.95) shows that the multipole expansion for a single charge displaced from the origin, e.g., q_1 at $(0, 0, d_1)$ with $q_2 = 0$, has nonzero dipole and quadrupole moments, $\mathbf{p} = q_1 d_1 \hat{\mathbf{k}}$ and $\mathcal{Q}_2 = q_1 d_1^2 (3\hat{\mathbf{k}}\hat{\mathbf{k}} - \mathbf{I})/2$; whereas if the origin were chosen to be the position of q_1 there would be only a monopole term. An interesting special case occurs if the total charge in the distribution is 0; then $\mathbf{p}$ does *not* depend on the choice of origin, as we shall prove generally later.

3.8.2 ▦ The Electric Dipole

For two equal but opposite charges, i.e., $q_1 = q$ and $q_2 = -q$, the total charge is $Q = 0$, and the dipole moment is

$$\mathbf{p} = q\mathbf{d} \qquad (3.97)$$

where $\mathbf{d} = \mathbf{x}_1 - \mathbf{x}_2$ is the vector from the negative charge ($q_2 = -q$) to the positive charge ($q_1 = q$). This system is called an electric dipole. An important limiting case is the limit $d \to 0$ with $\mathbf{p}$ fixed, called a *point-like electric dipole*. In this limit, and taking the origin to be the position of the dipole, all other multipole moments (quadrupole and higher) are 0. The potential of a point-like dipole is

$$V(\mathbf{x}) = \frac{\mathbf{p} \cdot \hat{\mathbf{r}}}{4\pi\epsilon_0 r^2} \qquad (3.98)$$

for all $\mathbf{x}$. If the dipole points in the z direction, i.e., $\mathbf{p} = p_0 \hat{\mathbf{k}}$, then in spherical coordinates the dipole potential is

$$V(r, \theta) = \frac{p_0 \cos\theta}{4\pi\epsilon_0 r^2}. \qquad (3.99)$$

The point-dipole potential is a good approximation for a neutral charge distribution that is small compared to all length scales in the problem. For example, a molecule, having a size of order 10^{-10} m, acts as a point-like dipole when placed

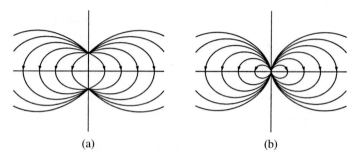

(a)   (b)

FIGURE 3.13   Field lines of (a) a finite dipole, and (b) a point dipole. For both cases the dipole moment points in the z direction.

in a laboratory-scale field. The permanent dipole moments of small molecules such as H_2O, NH_3, HCl, or CO, are of the order of 1 debye (D), where $1\,D = 3.33 \times 10^{-30}\,C\,m$. One debye is approximately the dipole moment of two charges $\pm 0.2e$ separated by 1 angstrom.

The electric field produced by a point-like electric dipole located at the origin is

$$E(x) = -\nabla V = \frac{3\hat{r}\,(p \cdot \hat{r}) - p}{4\pi\epsilon_0 r^3}. \tag{3.100}$$

Equation (3.100) is an important and interesting result, the derivation of which is left as an exercise in vector calculus.[17] The field lines for finite and point dipoles are plotted in Fig. 3.13. The reader should verify that (3.100) agrees with the figure. In particular, verify that E is antiparallel to p on the plane perpendicular to p, and parallel to p on the axis of p.

For the special case of a dipole $p = p_0\hat{k}$ located at the origin and pointing in the z direction, the field in spherical polar coordinates is

$$E(x) = -\nabla V = \frac{p_0}{4\pi\epsilon_0 r^3}\left(2\cos\theta\,\hat{r} + \sin\theta\,\hat{\theta}\right), \tag{3.101}$$

an expression that will prove useful for analyzing the field directions.

We will see in Chapter 6 that the atoms and molecules of a dielectric material behave as electric dipoles. Because a dipole has an associated electric field (3.100), the presence of a dielectric affects the total electric field in its neighborhood.

3.8.3 ■ Moments of a General Charge Distribution

For N point charges $q_1, q_2, \ldots, q_N$ at positions $x_1, x_2, \ldots, x_N$ the potential $V(x)$ can again be expanded as in (3.91). Generalizing from (3.92) and (3.93) the moments are

[17]See Exercise 37.

$$Q = \sum_{k=1}^{N} q_k$$

$$\mathbf{p} = \sum_{k=1}^{N} q_k \mathbf{x}_k \qquad (3.102)$$

$$\mathcal{Q}_2 = \sum_{k=1}^{N} \frac{q_k}{2} \left(3\mathbf{x}_k \mathbf{x}_k - r_k^2 \mathbf{I} \right).$$

For $N > 2$ it is possible to have both $Q = 0$ and $\mathbf{p} = 0$, in which case the asymptotic potential is the quadrupole potential, unless the quadrupole moment tensor $\mathcal{Q}_2$ happens to be 0.

If the charge distribution is a continuum with charge density $\rho(\mathbf{x}')$, which is assumed to have finite spatial extent, then for $|\mathbf{x}| \gg$ the size of the region of charge we may again use (3.88) for $1/|\mathbf{x} - \mathbf{x}'|$ in (3.60). In this case the multipole moments are

$$Q = \int \rho(\mathbf{x}') d^3 x' \qquad (3.103a)$$

$$\mathbf{p} = \int \mathbf{x}' \rho(\mathbf{x}') d^3 x' \qquad (3.103b)$$

$$\mathcal{Q}_2 = \int \frac{1}{2} \left(3\mathbf{x}' \mathbf{x}' - r'^2 \mathbf{I} \right) \rho(\mathbf{x}') d^3 x'. \qquad (3.103c)$$

Q is the total charge, $\mathbf{p}$ the electric dipole moment, and $\mathcal{Q}_2$ the quadrupole tensor, of the charge distribution. For example, a molecule is not simply a pair of opposite charges, but rather a collection of nuclei surrounded by a cloud of electrons. The dipole moment of a molecule is determined by (3.103b) for the charge density of the nuclei and electrons.

Using (3.103b) we may prove the theorem that the dipole moment is independent of the choice of origin if the total charge is 0. If the origin is moved to a point $\mathbf{x}_0$, then the dipole moment with respect to the new origin is

$$\mathbf{p}_0 = \int (\mathbf{x}' - \mathbf{x}_0) \rho(\mathbf{x}') d^3 x' = \mathbf{p} - Q\mathbf{x}_0,$$

which is the same as $\mathbf{p}$ if $Q = 0$.

3.8.4 ■ Equipotentials and Field Lines

The potential function of a point dipole is (3.98); or in polar coordinates, letting $\hat{\mathbf{k}}$ be the direction of $\mathbf{p}$, (3.99). Using the latter form, the equipotential surfaces are given by the equation $r^2 = C \cos \theta$ where $C = p_0/(4\pi\epsilon_0 V)$. Figure 3.14 shows the equipotentials (dashed curves) and electric field lines (solid curves) of the point dipole.

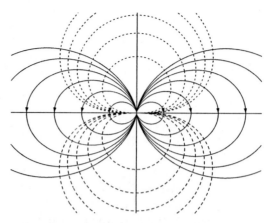

FIGURE 3.14   Equipotentials (dashed curves) and field lines (solid curves) of a point dipole. The dipole points in the z direction.

The potential function of a point-like *linear quadrupole*, lying along the z axis, is

$$V(r, \theta) = \frac{Q_2}{4\pi \epsilon_0} \frac{(3\cos^2 \theta - 1)}{2r^3} \tag{3.104}$$

where Q_2 is the (scalar) magnitude of the quadrupole moment tensor. For example, if there are charges $+q$ at $z = \pm d$ and $-2q$ at $z = 0$, then $Q_2 = 2qd^2$. The equipotential surfaces are given by the equation

$$r^3 = C(3\cos^2 \theta - 1), \tag{3.105}$$

where C is a constant depending on V. Figure 3.15 shows the equipotentials (dashed curves) and electric field lines (solid curves) of the point quadrupole.

3.8.5 ■ Torque and Potential Energy for a Dipole in an Electric Field

An electric dipole $\mathbf{p}$ in a field $\mathbf{E}$ experiences a torque that twists the dipole toward alignment with $\mathbf{E}$. An elementary way to determine the torque is to treat the dipole as a pair of opposite charges pulled in opposite directions along $\mathbf{E}$. But we will derive the result in a more general way. For a distribution with charge density $\rho(\mathbf{x})$, the torque (about the origin) is

$$\mathbf{N} = \int \mathbf{x} \times d\mathbf{F} = \int \mathbf{x} \times \mathbf{E}(\mathbf{x})\rho(\mathbf{x})d^3x, \tag{3.106}$$

because the force on the volume element d^3x is $\rho d^3x \mathbf{E}$. Assuming that $\mathbf{E}$ is constant on the length scale of $\rho(\mathbf{x})$, i.e., that we are dealing with a point-like dipole, $\times \mathbf{E}$ can be pulled out of the integral, and what remains is the dipole moment

FIGURE 3.15 Equipotentials (dashed curves) and field lines (solid curves) of a linear quadrupole.

(3.103b); that is,

$$\mathbf{N} = \mathbf{p} \times \mathbf{E}. \tag{3.107}$$

The torque is perpendicular to $\mathbf{E}$, so it twists $\mathbf{p}$ toward alignment with $\mathbf{E}$.

We may define a potential energy U for the dipole $\mathbf{p}$ in an external electric field. The interaction energy $\int dq\, V$ is

$$U = \int \rho(\mathbf{x}')V(\mathbf{x}')d^3x', \tag{3.108}$$

where $\rho(\mathbf{x}')$ is the charge density of the dipole. Assuming the dipole is pointlike, i.e., much smaller than the length scale for variation of $\mathbf{E}$, we may approximate $V(\mathbf{x}')$ using Taylor's theorem,

$$V(\mathbf{x}') \approx V(\mathbf{x}) + \left(\mathbf{x}' - \mathbf{x}\right) \cdot \nabla V(\mathbf{x}), \tag{3.109}$$

where $\mathbf{x}$ is at the center of the dipole. When this is inserted in (3.108) there are three terms. Two are 0 because $\int \rho(\mathbf{x}')d^3x' = 0$ for a neutral dipole. The remaining term is

$$U = \int \rho(\mathbf{x}')\mathbf{x}'d^3x' \cdot \nabla V(\mathbf{x})$$

$$= -\mathbf{p} \cdot \mathbf{E}(\mathbf{x}). \tag{3.110}$$

Considering all directions of $\mathbf{p}$, U is minimum when $\mathbf{p}$ is parallel to $\mathbf{E}$. Therefore $\mathbf{p} \parallel \mathbf{E}$ is the state of stable equilibrium for the dipole orientation. Note that the torque is 0 for this orientation.

The function $U(\mathbf{x})$ in (3.110) is more than the orientational potential energy; it is the full potential energy of the point-like dipole in the field $\mathbf{E}(\mathbf{x})$, taking the potential energy to be 0 at infinity. In other words, $U(\mathbf{x})$ is equal to the work required to bring the dipole to $\mathbf{x}$ from infinity. One way to prove this claim is to imagine that the dipole at infinity is compressed to an uncharged point particle; let $-w_0$ be the work supplied in this first step. No work is required to move the point particle through the electric field, to $\mathbf{x}$, because it is uncharged. When the particle is at $\mathbf{x}$, the charges $\pm q$ must be separated by $d\mathbf{x}$, say by moving $+q$ by $d\mathbf{x}$ but holding $-q$ fixed; the work done is $+w_0$ against the interaction between the charges, plus $-(q\mathbf{E}) \cdot d\mathbf{x} = -\mathbf{p} \cdot \mathbf{E}(\mathbf{x})$ against the electric field. Thus the total work is (3.110).

The net electric force on the dipole is, by (3.110),

$$\mathbf{F}(\mathbf{x}) = -\nabla U = \nabla \left(\mathbf{p} \cdot \mathbf{E}(\mathbf{x}) \right). \tag{3.111}$$

In a uniform field there is a torque on the dipole, but no net electric force. If the field varies with position, there is a net electric force as well as a torque. The force equation can be written another way. Substitute $\mathbf{E} = -\nabla V$ into (3.111); and, using the fact that $\mathbf{p}$ is constant, commute the ∇'s:

$$\mathbf{F}(\mathbf{x}) = -\nabla \left(\mathbf{p} \cdot \nabla V \right) = -\mathbf{p} \cdot \nabla \left(\nabla V \right)$$

$$= (\mathbf{p} \cdot \nabla) \mathbf{E}(\mathbf{x}). \tag{3.112}$$

Whichever of (3.111) or (3.112) is more convenient may be used to calculate $\mathbf{F}$.

We may explain one of the most basic demonstrations of static electricity by (3.111). A piece of amber, rubbed with a wool cloth, becomes charged, and will attract small pieces of paper.[18] But the paper is electrically neutral. Why does a charge Q attract a neutral object? The reason is that the object becomes *polarized* in the electric field of Q, and so has a dipole moment parallel to the field. The energy U decreases (becomes more negative) as the object approaches Q, because both $|\mathbf{p}|$ and $|\mathbf{E}|$ increase. The force $-\nabla U$ is in the direction of decreasing U, toward Q.

3.9 ■ APPLICATIONS

Applications of electrostatic fields range from large-scale ones using hundreds of kilowatts, e.g., precipitation of smokestack particles, to small-scale applications using about 1 watt, e.g., deposition of fibers on cloth or paper. The general principle in these applications is to put a charge on a particle in order to control its motion by an applied electrostatic field.

---

[18]Thales of Miletus (6th century BCE) is supposed to have discovered static electricity by observing that when amber is rubbed with animal fur, it attracts bits of feathers and the pith of plants. Thales is considered to be the first Western philosopher, and to him are attributed many early discoveries and ideas on science and mathematics.

Probably the most important industrial application is the removal of small solid particles from the efflux of smokestacks at large power plants. Cottrell electrostatic precipitators, which remove soot, smoke, fly ash, and other particle pollutants, are used. In the usual design the particles to be removed are first charged by passing them through ionized air; the ions attach themselves to the particles or transfer their charges upon collision. The charged solid particles are then collected on planes or coaxial charged conductors and thus prevented from leaving with the effluent gases. A typical large industrial process uses an electrostatic field with voltage of order 100 kV and collection current of about 2 A. Essentially the same process, but at lower voltage and current, is also used for smaller smokestacks, in which the purpose is to recover valuable dusts from the flue gases.

Another spectacular and commercially important application of electrostatic fields is in automobile painting. In practice liquid paint is fed onto a spinning, conducting disk or cone which is at a high voltage, say, 250 kV. The automobile body to be painted is nearby and kept grounded, i.e., at potential zero. Charged droplets spin off the disk by centrifugal force. The electric field between the charged disk and the automobile body exerts a force on the charged droplets toward the car, and deposits them there; a typical current is 5 mA. When operating properly, essentially all the paint is deposited on the car.

Some smaller-scale, but also commercially useful applications are: (a) Depositing the abrasive particles in sandpaper production; the electrostatic process has the desirable effect of tending to expose the sharp points of the particles; (b) Ink jet printers, which achieve remarkable control for depositing charged ink droplets on paper by using electrostatic fields; (c) Electrostatic fields used in depositing charged fiber-particles as flocking onto tee shirts, greeting cards, wallpaper, etc.

In most such applications the force is on a charged particle, as in Eq. (3.6). One can also control the motion of uncharged polarized particles by using nonuniform fields, as in (3.111). However, this is more difficult because the force is weaker. Finally, it is important to remember that although electrostatic control of charged particles is straightforward in principle, these applications require sophisticated engineering in practice.

3.10 ■ CHAPTER SUMMARY

Electrostatics is concerned with three functions, $\rho(\mathbf{x})$, $\mathbf{E}(\mathbf{x})$, and $V(\mathbf{x})$. To proceed to the next few chapters, it is necessary to comprehend the six relations among these three functions. For each pair of functions there is an integral formula and a partial differential equation.

1. Density and Field:

$$\mathbf{E} = \frac{1}{4\pi\epsilon_0} \int \frac{\hat{\mathbf{r}}}{r^2} \rho \, dV \quad \text{and} \quad \nabla \cdot \mathbf{E} = \rho/\epsilon_0.$$

The electric field is determined by the density by the integral formula (3.15), which is rewritten in a concise form above. The partial differential equation, which must hold at every point in space, is Gauss's Law. These are not independent principles. Either one may be derived from the other.

2. Field and Potential:

$$V(\mathbf{x}) = -\int_{\mathbf{x}_0}^{\mathbf{x}} \mathbf{E} \cdot d\boldsymbol{\ell} \quad \text{and} \quad \mathbf{E} = -\nabla V.$$

The reason a static $\mathbf{E}$ may always be written as a gradient, and that the line integral is independent of the path from $\mathbf{x}_0$ to $\mathbf{x}$, is that $\nabla \times \mathbf{E} = 0$.

3. Potential and Density:

$$V = \frac{1}{4\pi\epsilon_0} \int \frac{\rho \, dV}{r} \quad \text{and} \quad -\nabla^2 V = \rho/\epsilon_0.$$

The integral formula for $V(\mathbf{x})$ is (3.60), rewritten in a concise form above. The local equation is Poisson's equation. The integral is just Green's solution to Poisson's equation, discussed in Section 3.5.

FURTHER READING

The topics in this chapter, being basic, are covered in most textbooks of electricity and magnetism. For a student who wants to study other treatments of the subject, the books below are recommended.

D. J. Griffiths, *Introduction to Electrodynamics*, 3rd ed. (Prentice-Hall, Upper Saddle River NJ, 1999). [A popular textbook with many example calculations.]

J. R. Reitz, F. J. Milford, and R. W. Christy, *Foundations of Electromagnetic Theory*, 4th ed. (Addison Wesley, San Francisco CA, 1993).

J. C. Slater and N. H. Frank, *Electromagnetism* (McGraw-Hill, New York, 1947). [A classic text that emphasizes physical insight.]

R. H. Good, *Classical Electromagnetism* (Saunders, Fort Worth TX, 1999).

P. Lorrain, D. R. Corson, and F. Lorrain, *Fundamentals of Electromagnetic Phenomena* (Freeman, New York, 2000).

J. D. Jackson, *Classical Electrodynamics*, 3rd ed. (Wiley, New York, 1999). [The definitive graduate-level text.]

EXERCISES

Sec. 3.1. Coulomb's Law

3.1. Two protons interact, separated by distance d. Calculate the ratio of the electric force to the gravitational force for one of the protons.

3.2. The quantity $1/4\pi\epsilon_0$ is the constant of proportionality between the electric force and $q_1 q_2/r^2$. The parameter ϵ_0 is called the permittivity of the vacuum. Its value depends on the system of units being used.

(a) Use dimensional analysis to determine the SI units of ϵ_0.

(b) In SI units, $1/4\pi\epsilon_0$ is defined to be $10^{-7} c^2$ Ns^2/C^2 where c is the speed of light. From this, evaluate ϵ_0.

Sec. 3.2. The Electric Field

3.3. A proton moves in a uniform electric field $E_0 \hat{\mathbf{k}}$. The proton is released at the origin with initial velocity $v_0 \hat{\mathbf{i}}$. What is the position of the proton at time t?

3.4. For the charge configuration in Example 1: (a) find the electric field along the y axis; (b) find the electric field along the z axis.

3.5. For the line charge of Example 2, find the electric field along the z axis for $z > \ell$.

3.6. Starting from the result of Example 3, equation (3.20):

(a) Find **E** along the axis of a disk of radius R which has a constant surface charge density σ. (Build up the disk from elemental rings, and integrate the contributions from the rings.)

(b) From the result of (a) find $\mathbf{E}(\mathbf{x})$ due to an infinite plane with surface charge density σ.

3.7. Charges $+q$ and $-q$ are located on the z axis at $(0, 0, a)$ and $(0, 0, -a)$ respectively. Determine $\mathbf{E}(\mathbf{x})$ on the x axis, i.e., $\mathbf{x} = (x, 0, 0)$. Use computer graphics to plot $E_z(x)$ as a function of x/a.

3.8. A semi-infinite wire lies on the negative z axis, from $z = 0$ to $z = -\infty$, with constant linear charge density λ.

(a) Determine **E** at a point $(0, 0, z)$ on the positive z axis.

(b) Determine **E** at a point $(x, 0, 0)$ on the positive x axis.

3.9. A long, uniformly charged ribbon is located in the xz plane, parallel to the z axis, occupying the region $-\infty \le z \le \infty$ and $-a/2 \le x \le a/2$. The charge per unit area on the ribbon is σ.

(a) Determine **E** at $(x, 0, 0)$, where $x > a/2$. What is the asymptotic field on the x axis?

(b) Determine **E** at $(0, y, 0)$, where $y > 0$. What is the asymptotic field on the y axis?

Sec. 3.4. The Integral Form of Gauss's Law

3.10. There is a point charge Q at the origin.

(a) Calculate the flux of **E** through a cube centered at the origin and aligned with the Cartesian axes. Evaluate the surface integral directly, and verify that the result agrees with Gauss's Law.

(b) Calculate the flux of **E** through a cube with one vertex at the origin, and with the rest of the cube in the octant with $x, y, z > 0$. Evaluate the surface integral directly. How does the result square with Gauss's Law?

3.11. **(a)** Using Gauss's Law in integral form, prove that the electric field outside a uniformly charged spherical shell, with total charge Q and radius a, is $Q\hat{\mathbf{r}}/(4\pi\epsilon_0 r^2)$.

(b) Prove that the electric field inside the shell is 0. Note that these results are the same as obtained in Example 4 by integrating over the charge distribution.

3.12. A charge $+q$ is at $(0, 0, z_0)$.

(a) What is $\mathbf{E}(\mathbf{x})$?

(b) By explicitly integrating $E_n(\mathbf{x})$ over the surface of a sphere centered at the origin and with radius $R > z_0$, show that $\int \mathbf{E}\cdot\hat{\mathbf{n}}\,dA$ equals q_0/ϵ_0, as it must by Gauss's Law. (This is not an easy exercise.)

3.13. The integral and differential forms of Gauss's Law are equivalent, because either one implies the other. Derive the differential form of Gauss's Law from the integral form, by letting S be the surface of an infinitesimal cube.

3.14. Two large plates are parallel, have uniform surface charge density $+\sigma$ and $-\sigma$, and are separated by distance d.

(a) Determine the electric field between the plates far from the edges.

(b) Determine the force per unit area on either plate, neglecting edge effects. (Hint: A charged plate does not exert a force on itself.)

3.15. A hollow cylinder of radius R is uniformly charged with surface charge density σ. Use Gauss's Law to determine $\mathbf{E}$ for $r < R$ and $r > R$.

3.16. Two concentric thin spherical shells, with radii a and $b > a$, have uniformly distributed charges $+Q$ and $-Q$ respectively. What is $\mathbf{E}(\mathbf{x})$ in the region $a < r < b$? How does $\mathbf{E}(\mathbf{x})$ change if the outer shell is discharged?

Sec. 3.5. Green's Function and the Dirac delta Function

3.17. **(a)** Prove that $\int_{-\infty}^{\infty} \delta(ax + b)f(x)\,dx = f(-b/a)/|a|$.

(b) From (a) prove that $\int_{-\infty}^{\infty} \delta(x - y)f(y)\,dy = f(x)$.

(c) Prove that $\int \delta^3(A\mathbf{x} + \mathbf{b})f(\mathbf{x})d^3x = f(-A^{-1}\mathbf{b})/|\text{Det }A|$, where A is a 3×3 matrix. (Hint: When you change variables from $\mathbf{x}$ to $\mathbf{x}'$, remember the Jacobian factor $\mathcal{J}$ in $d^3x = \mathcal{J}\,d^3x'$.)

(d) From (c) or (b) prove that $\int \delta^3(\mathbf{x} - \mathbf{y})f(\mathbf{y})d^3y = f(\mathbf{x})$.

3.18. Consider $d(x, \epsilon) = \exp(-x^2/\epsilon)/\sqrt{\pi\epsilon}$.

(a) Use computer graphics to plot $d(x, \epsilon)$ versus x for $\epsilon = 0.3$, 1.0, and 3.0.

(b) Show that $\lim_{\epsilon\to 0} d(x, \epsilon) = 0$ for $x \neq 0$.

(c) Show that $\lim_{\epsilon\to 0} d(x, \epsilon) = \infty$ for $x = 0$.

(d) Determine $\int_{-\infty}^{\infty} d(x, \epsilon)dx$ for arbitrary ϵ.

(e) Explain the meaning of the statement $\lim_{\epsilon\to 0} d(x, \epsilon) = \delta(x)$.

3.19. **(a)** In one dimension (x), let $d_a(x)$ be $1/2a$ for $-a \leq x \leq a$, and 0 for $|x| > a$. Prove that $\lim_{a\to 0} d_a(x) = \delta(x)$.

(b) In three dimensions let $d_a(\mathbf{x})$ be $1/V$ for $r \leq a$, and 0 for $r > a$; here r denotes $|\mathbf{x}|$ and $V = \frac{4}{3}\pi a^3$. Prove that $\lim_{a\to 0} d_a(\mathbf{x}) = \delta^3(\mathbf{x})$.

(c) Consider a uniformly charged sphere with total charge Q and radius a; that is, the charge density is $\rho(\mathbf{x}) = Qd_a(\mathbf{x})$. Find the potential function $V(\mathbf{x})$ for $r \leq a$ and $r > a$.

(d) Verify by explicit calculation that $V(\mathbf{x})$ obeys Laplace's equation ($\nabla^2 V = 0$) for $r > a$ and Poisson's equation for $r < a$.

(e) Sketch plots of V versus r, for several values of a, showing how $V(\mathbf{x})$ approaches $Q/(4\pi\epsilon_0 r)$ as $a \to 0$.

Sec. 3.6. Electric Potential

3.20. (a) Consider the potential function $\mathbf{C} \cdot \mathbf{x}$ where $\mathbf{C}$ is a constant vector. What is $\mathbf{E}(\mathbf{x})$?

(b) What is the potential $V(\mathbf{x})$ for a uniform electric field of strength E_0 and direction $\hat{\mathbf{k}}$?

3.21. Show that the electrostatic potential of a charged circular ring with total charge Q and radius a, on the axis of symmetry of the ring, is $V(z) = Q/(4\pi\epsilon_0\sqrt{z^2 + a^2})$, where z is the distance from the center of the ring. From $V(z)$ determine the electric field on the axis of symmetry.

3.22. Use the integral (3.32) to determine the potential $V(\mathbf{x})$ both inside and outside a uniformly charged spherical surface, with total charge Q and radius R. (Hint: Divide the sphere into rings, and use the result given in Exercise 21.) From $V(\mathbf{x})$ calculate the electric field. Sketch plots of $V(r)$ and $E_r(r)$ versus r. Notice that the potential is continuous at $r = R$, but the electric field is discontinuous because of the surface charge.

3.23. Imagine a spherically symmetric charge density

$$\rho(\mathbf{x}) = \begin{cases} Cr & \text{for} \quad r \leq a \\ 0 & \text{for} \quad r > a. \end{cases}$$

(a) Determine the electric field $\mathbf{E}(\mathbf{x})$ and potential $V(r)$. Notice that $V(r)$ and $\mathbf{E}(\mathbf{x})$ are continuous at $r = a$.

(b) Now suppose additional charge is placed uniformly on the surface at $r = a$, with surface density σ_0. (σ_0 has units C/m^2.) Determine $\mathbf{E}(\mathbf{x})$ and $V(r)$ for this case. $\mathbf{E}(\mathbf{x})$ is discontinuous at $r = a$, but $V(r)$ is continuous. Explain why.

3.24. Consider the field

$$\mathbf{E} = \hat{\mathbf{i}}\left(2x^2 - 2xy - 2y^2\right) + \hat{\mathbf{j}}\left(-x^2 - 4xy + y^2\right).$$

Is it irrotational? If so, what is the potential function? Also, calculate $\nabla \cdot \mathbf{E}$.

3.25. Find the restrictions on C_1 and C_2 such that this function may be a potential function in a charge-free region of space: $V(r, \theta) = (C_1 \cos^2\theta + C_2)/r^3$.

Sec. 3.7. Energy of the Electric Field

3.26. Calculate the self-energy of a charged spherical surface, with total charge Q uniformly distributed on the surface, and radius R. What is the physical significance of this result? What is the limit of the self-energy as $R \to 0$?

3.27. Evaluate the interaction energy $U_{12} = \epsilon_0 \int \mathbf{E}_1 \cdot \mathbf{E}_2 d^3 x$ for two point charges separated by d. $\mathbf{E}_1$ is the field of q_1 and $\mathbf{E}_2$ is that of q_2. (Hint: U_{12} depends only on d, so without loss of generality let q_1 be at the origin, and q_2 on the z axis at $z = d$.)

3.28. (a) Show that the electrostatic energy of a uniformly charged solid sphere, with total charge Q and radius R, is $3Q^2/(20\pi\epsilon_0 R)$.

(b) Use (a) to compute the electrostatic energy of an atomic nucleus [charge $= Ze$, radius $= (1.2 \times 10^{-15}\,\text{m})A^{1/3}$] in MeV times $Z^2/A^{1/3}$.

(c) Calculate the change of electrostatic energy when a uranium nucleus ($Z = 92$, $A = 238$) fissions into two equal fragments.

3.29. Calculate the *energy per unit length* for two long coaxial cylindrical shells, neglecting end effects. The inner and outer cylinders have radii a and b, and linear charge densities λ and $-\lambda$, uniformly distributed on the surface, respectively.

Sec. 3.8. The Multipole Expansion

3.30. Let $V(z)$ be the potential of a ring of charge on the axis of symmetry at distance z from the center. (See Exercise 21.) Obtain the first two nonvanishing terms in $V(z)$ for $z \gg$ the radius of the ring. To which multipoles do they correspond? Can you see by symmetry that the dipole moment is 0?

3.31. Charges $\pm q$ separated by distance d make a dipole with dipole moment qd. For large r the potential approaches $qd\cos\theta/(4\pi\epsilon_0 r^2)$. For $\theta = 0$, how large must r be so that the asymptotic function is accurate to 1%?

3.32. A point dipole $\mathbf{p} = p\hat{\mathbf{k}}$ is at the origin.

(a) At point $P_1 = (x_1, 0, 0)$ there is a point charge q. What is the force on this charge due to the field of the dipole? What is the force on the dipole?
How much work is required to take the charge from P_1 to infinity if the dipole remains fixed at the origin? How much work is required to take the dipole from the origin to infinity if the charge remains fixed at P_1? What is the physics underlying the simple answers to these questions?

(b) The point charge is now moved from P_1 to another point $P_2 = (x_2, y_2, z_2)$ while the dipole remains fixed. How much work is required to move the charge? What is the force on q when it is at P_2?

3.33. The dipole $\mathbf{p} = p_1\hat{\mathbf{i}} + p_2\hat{\mathbf{j}} + p_3\hat{\mathbf{k}}$ is located at the origin.

(a) What is the potential $V(r, \theta, \phi)$ in spherical coordinates?

(b) What is the electric field $\mathbf{E}(r, \theta, \phi)$?

3.34. A point dipole $\mathbf{p} = p_0\hat{\mathbf{k}}$ is at the origin. A second dipole $\mathbf{p} = p_0\hat{\mathbf{k}}$ is at $(0, 0, z_0)$.

(a) What is the force on the second dipole?

(b) What is the interaction energy? [Hint: The energy is the work required to bring the second dipole to $(0, 0, z_0)$ with the first dipole fixed.]

(c) Show that the energy is $-\mathbf{p} \cdot \mathbf{E}(0, 0, z_0)$.

[Answer: (a) $F_z = -3p_0^2/(2\pi\epsilon_0 z_0^4)$; (b) $U = -p_0^2/(2\pi\epsilon_0 z_0^3)$.]

3.35. Figure 3.16 shows three charged line segments, each with linear charge density λ, extending from the origin O to $(a, 0, 0)$, from O to $(0, b, 0)$, and from O to $(0, 0, c)$.

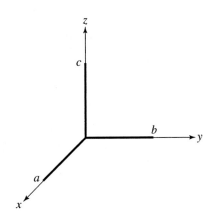

FIGURE 3.16 Three charged rods.

(a) Find the dipole moment of this charge distribution.
(b) Find the first two terms in the multipole expansion of the potential $V(0, 0, z)$ on the z axis, for $z \gg a, b, c$.
(c) What are the monopole and dipole contributions to $\mathbf{E}(0, 0, z)$ for $z \gg a, b, c$?

3.36. Repeat the previous exercise (35) if the line charges are placed symmetrically around the origin, i.e., from $(-a/2, 0, 0)$ to $(+a/2, 0, 0)$, from $(0, -b/2, 0)$ to $(0, +b/2, 0)$, and from $(0, 0, -c/2)$ to $(0, 0, +c/2)$.
[Answer: (b) $V(0, 0, z) = \lambda(a + b + c)/(4\pi\epsilon_0 z) - \lambda(a^3 + b^3 - 2c^3)/(96\pi\epsilon_0 z^3)$.]

3.37. Derive (3.100), the electric field of an electric dipole, from the dipole potential (3.98). Calculate $-\nabla(\mathbf{p} \cdot \mathbf{x}/r^3)$ either by using spherical coordinates or by the del identity for the gradient of a product. Sketch the electric field lines.

3.38. A linear quadrupole consists of three charges: q, $-2q$, and q, on the z axis. The positive charges are at $z = \pm d$ and the negative charge is at the origin.
(a) Show that this system is the same as *two dipoles*, with dipole moments $+qd\hat{\mathbf{k}}$ and $-qd\hat{\mathbf{k}}$, centered at $z = +d/2$ and $z = -d/2$, respectively.
(b) Calculate the potential $V(r, \theta)$ in spherical coordinates for $r \gg d$.
(c) Sketch the electric field lines in the xz plane.

3.39. Consider the electric quadrupole consisting of four charges in the xy plane: $+q$ at $(x, y) = (a, 0)$, $-q$ at $(0, a)$, $+q$ at $(-a, 0)$, and $-q$ at $(0, -a)$.
(a) Determine the electric potential on the x axis. What is the asymptotic form of $V(x, 0, 0)$? Use computer graphics to make a log-log plot of V versus x and verify the asymptotic behavior.
(b) Determine the electric field on the x axis for $x \gg a$.
(c) Determine the electric field on the z axis.

General Exercises

3.40. Imagine three concentric charged hollow spheres, with radii R, $2R$, and $3R$ and charges $+Q$, $-Q$, and $+Q$, respectively.

(a) Determine the electric field at four points a, b, c, d; these are all on a radial line at distances $0.5R$, $1.5R$, $2.5R$, and $3.5R$ from the center of the system, respectively.

(b) Determine the ratio E_d/E_b. [Answer: 9/49]

3.41. A large flat plate with constant surface charge density σ lies in the xy plane. What is the potential difference between the points $(0, 0, d)$ and $(d, d, 2d)$, which are far from the edges of the plate? [Answer: $-\sigma d/(2\epsilon_0)$]

3.42. A solid sphere of radius R centered at the origin is uniformly charged with charge density ρ. Then a small sphere is removed, making a spherical cavity centered at a point $\mathbf{a}$ within the sphere. Show that the electric field in the cavity is $\rho\mathbf{a}/3\epsilon_0$. (Surprisingly, it is uniform.)

3.43. A model for the electrostatic potential of an atom, due to the nucleus (charge $+Ze$) and electrons, is the so-called "screened Coulomb potential,"

$$V(\mathbf{x}) = \frac{Ze}{4\pi\epsilon_0 r}e^{-r/a}$$

in which a is an effective atomic radius and $r = |\mathbf{x}|$.

(a) What is the charge density $\rho(\mathbf{x})$? (Hint: If you calculate $-\nabla^2 V$ naively, e.g., using spherical polar coordinates, you will get the charge density of the electrons, but not the nuclear charge density, which is a delta function at $\mathbf{x} = 0$. Note that V diverges as $1/r$ as $r \to 0$, which by Gauss's Law implies a point charge at the origin.)

(b) Sketch a graph of ρ versus r.

(c) Show that the total charge is 0.

3.44. Determine the potential on the axis of symmetry of a uniformly charged disk. Expand the result in the region of z much larger than the disk's radius and show that the leading term is $Q/(4\pi\epsilon_0 z)$, as expected, where Q is the total charge on the disk.

3.45. For a uniformly charged line, determine $\mathbf{E}(\mathbf{x})$ in two ways: by integrating over the charge distribution, and by applying Gauss's Law.

Computer Exercises

3.46. We can use *numerical integration* to evaluate the potential $V(\mathbf{x})$ for a given charge distribution $\rho(\mathbf{x})$, when $\rho(\mathbf{x})$ is too complex for an easy analytic calculation. For example, consider a uniformly charged wire in the form of one turn of a spiral in the xy plane. The equation for the spiral is, in plane polar coordinates, $r(\theta) = a(1 - \theta/4\pi)$, for $0 \le \theta \le 2\pi$. At $\theta = 0$ the radius is a; at $\theta = 2\pi$ it is $a/2$. Find the potential at the origin. Use a computer program to evaluate the integral over the charge distribution numerically.

3.47. Use a computer program to make a contour plot, for points in the xy plane, of the electric potential of a system of three point charges. Let the charges be q, $2q$, and $3q$, located at the vertices of an equilateral triangle in the xy plane: $(a/2, 0, 0)$,

$(-a/2, 0, 0)$, and $(0, \sqrt{3}a/2, 0)$ respectively. The potential is

$$V(x, y, 0) = \frac{q}{4\pi\epsilon_0 a} f(x, y)$$

where $f(x, y)$ is a dimensionless function; set $a = 1$ to make a the unit of length. For example, some graphics programs have a *contour plotting* command, which can be used to plot the contours of $f(x, y)$. These are the equipotentials of the system.

3.48. Consider a dipole, consisting of charges $\pm 1\,\mu C$ at positions $(0, 0, \pm 1\,cm)$ respectively. The dipole moment is $2 \times 10^{-8}\,Cm$. Use a computer program to plot the electric field strength $E(x, y, 0)$ for points in the xy plane, as a function of $r = \sqrt{x^2 + y^2}$. Show on the same plot the electric dipole approximation for $E(x, y, 0)$. For what range of r is the dipole approximation accurate to 10%?

3.49. A wire is bent into a circular ring of radius R and given total charge Q. The cross section of the wire is negligible. Let the ring lie in the xy plane ($z = 0$) and use spherical polar coordinates (r, θ, ϕ) for points $\mathbf{x}$ off the ring. By rotational symmetry about the z axis, the potential $V(\mathbf{x})$ is independent of the azimuthal angle ϕ; that is, $V = V(r, \theta)$. The potential on the axis of the ring ($\theta = 0$) was calculated in Exercise 21. To calculate V off the axis requires a computer.

(a) Show that the potential function is

$$V(r, \theta) = \frac{Q}{4\pi\epsilon_0} \frac{1}{2\pi} \int_0^{2\pi} \frac{d\phi'}{\sqrt{r^2 + R^2 - 2rR\sin\theta\cos\phi'}}.$$

(b) The ϕ' integral can be evaluated numerically, e.g., with a computer program. A reasonable approximation is to replace the integral by a sum

$$V(r, \theta) \approx \frac{Q}{4\pi\epsilon_0} \frac{1}{N} \sum_{n=1}^{N} \frac{1}{\sqrt{r^2 + R^2 - 2rR\sin\theta\cos(2\pi n/N)}}$$

where N is a sufficiently large integer. For $r = 2R$ plot $V(r, \theta)$ as a function of θ, and explain the dependence on θ intuitively. (The domain of θ is $(0, \pi)$.) Use $Q/(4\pi\epsilon_0 R)$ as the unit of potential.

(c) For $r = R/2$ plot $V(r, \theta)$ as a function of θ, and explain the dependence on θ intuitively.

CHAPTER
4
Electrostatics and Conductors

The characteristic feature of a conductor is that some of its electrons are free to move macroscopic distances. These conduction electrons come from the atoms of the material, so there is also a lattice of positive ions, which move only by small vibrations about their equilibrium positions. In the absence of an electric field the free electrons are uniformly distributed and in rapid random motion, interacting with the lattice and each other. But if an electric field is applied to a conducting body, the free electrons inside will move in response to the electric force until a new equilibrium is reached, which happens when $\mathbf{E} = 0$ inside the conductor. The field outside the conductor is then different from the original field because it includes the Coulomb-field contributions of the displaced electrons. For this reason conductors play an important role in electrostatics.

To understand the physics of the approach to equilibrium, picture a conductor—to be specific let it be an uncharged sphere centered at the origin—to which is applied a uniform field $\mathbf{E} = E\hat{\mathbf{k}}$ in the $+z$ direction. Electrons in the conductor are subject to the force $\mathbf{F} = -e\mathbf{E}$, so the free electrons move in the $-z$ direction. They will accumulate somewhere on the conductor, displaced in the $-z$ direction from their original positions. A nonzero charge density develops, with a field that tends to cancel the applied field inside the conductor. The electrons stop moving, i.e., electrostatic equilibrium is reached, when the vector sum of all field contributions is zero everywhere in the conductor. We will solve this specific problem in Sec. 4.3, find that the displaced charges accumulate on the surface of the sphere, and calculate the surface charge density.

We usually think of conductors as metals. On a microscopic scale the conduction electrons inside a metal are in the periodic potential of a lattice of positive metallic ions. Their freedom to move, in other words, the high electrical conductivity of metals, can be explained by quantum mechanics. In a typical metal the conductivity is large, so the electrostatic equilibrium discussed in the previous paragraph is achieved in a very short time.[1] For example, in Cu the characteristic time for reaching equilibrium is about 10^{-19} s. Therefore, except for processes at very high frequencies, of order 10^{19} Hz and higher, equilibrium may be considered to be instantaneous. Semiconductors, e.g., graphite, also have conduction electrons, so their electrostatic behavior is similar to metals. There is active experimental and theoretical research on electrical properties of metals and semiconductors, about which much is understood, but also a lot is not yet understood.

[1] See Sec. 7.6.

4.1 ■ ELECTROSTATIC PROPERTIES OF CONDUCTORS

In this section we describe some general properties of conductors in equilibrium with static electric fields. These results will be used in the analysis of specific calculations later. In each case we'll state the property, and then explain the theory behind the statement.

The Electric Field Is Zero Inside a Conductor in Electrostatic Equilibrium. If **E** were nonzero then free charges would accelerate and the system would not be static.

The Electric Potential Is Constant Inside a Conductor in Electrostatic Equilibrium. Recall that the potential difference between two points is the work per unit charge required to move a test charge from one to the other against the electric force. But in a conductor **E** is zero, so there is no force on the test charge. No work is required to move the charge, the potential difference between any two internal points is zero, and we have $V(\mathbf{x}) = V_0$, a constant. In other words, the entire conductor is at potential V_0. The surface of the conductor, which is also at potential V_0, is an equipotential surface.

The Charge Density Is Zero Inside a Conductor in Electrostatic Equilibrium. Gauss's Law states that the flux of electric field through any closed surface is proportional to the charge enclosed by the surface. If a Gaussian surface lies entirely inside a conductor, then the flux is 0 because **E** = 0. Thus the net charge contained in any volume inside the conductor is 0. In other words, $\rho(\mathbf{x}) = 0$ inside the conductor. Any excess charge on a conductor must therefore reside on its surface, and must be described by a surface charge density σ.

Consider this question: If electrons (negative charge) appear on the surface of an uncharged conductor in an electric field, how can there then be zero charge density inside the conductor? Where do the electrons come from? The answer is that the electrons can be considered as coming from elsewhere on the surface of the conductor. The surface charges have been redistributed. An equivalent amount of positive charge must appear at surface points from which the electrons moved. We may therefore speak freely of flow of either negative or positive charge.

Either the Net Charge, or the Potential, of a Conductor in an Electrostatic System Can Be Specified. An isolated conductor may carry on its surface a net charge, negative or positive, or it may be uncharged. In an experiment the charge on a conductor is set by adding or removing electrons. If the conductor is then placed in an external field, or near other charges or conductors, its total charge stays the same, although the surface charge distribution changes so as to maintain **E** = 0 inside the conductor. In such an experiment we specify that the total charge on the conductor is fixed at some value Q_0. However, its potential and the surface charge distribution are determined by the other charges and conductors in the system.

Instead of specifying the charge on a conductor we may specify its potential. For example, a conductor connected to the Earth is said to be at zero potential, or *grounded*. The idea is that the Earth is a very large conductor, defined to be at zero potential itself, and it serves as a source or sink of electrons without its own potential being changed. In an experiment a conductor can be set to any specified potential V_0 by connecting one terminal of an appropriate dc-voltage source, say a battery, to the Earth and the other terminal to the conductor, as illustrated in Fig. 4.1. If the conductor is now placed in an external field, or near other charges or conductors, it will remain at the specified potential, although charge may be exchanged through the voltage source between the conductor and the Earth. The charge will distribute itself over the surface of the conductor so as to maintain $\mathbf{E} = 0$ in the conductor. In such an experiment we specify that the potential of the conductor is fixed at V_0. However, its net charge and surface charge distribution are determined by the other charges and conductors in the system.

At the Surface of a Conductor, $\mathbf{E}_{\text{tangential}} = 0$ *and* $E_{\text{normal}} = \sigma/\epsilon_0$. The surface of a conductor in electrostatic equilibrium is an equipotential, as we saw above, so no work is required to move a charge along the surface; i.e., the tangential electric force is 0. Recall from Sec. 3.3.2 that the tangential components of the electric field are continuous across *any* surface (conducting or insulating) because $\nabla \times \mathbf{E} = 0$. Therefore $\mathbf{E}_t$ is 0 just inside, just outside, and on the surface of a conductor. Indeed, if $\mathbf{E}_t$ were not zero on the surface, then there would be a flow of surface charge rather than static equilibrium.

However, in general there is a normal component E_n, i.e., perpendicular to the surface of a conductor. We can determine E_n by using the integral form of Gauss's Law. Draw a cylindrical Gaussian surface, with infinitesimal area dA and infinitesimal height, straddling the boundary surface as shown in Fig. 4.2. The electric flux across the flat surface inside the conductor is zero and that across the flat surface outside is $E_n dA$. The charge enclosed by the Gaussian surface is σdA, where σ is the local surface charge density. Therefore, Gauss's Law implies

$$E_n = \sigma/\epsilon_0. \qquad (4.1)$$

This is a special case of the boundary condition for E_n discussed in Sec. 3.3.2. Equation (4.1) is an important relation because we will use it to calculate the

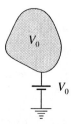

FIGURE 4.1   A conductor held at potential V_0.

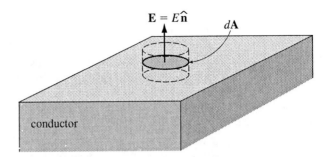

FIGURE 4.2 Gaussian surface at a conducting surface.

surface charge density on a conductor from the field. Alternatively, in terms of the potential function, $\sigma = -\epsilon_0 \hat{\mathbf{n}} \cdot \nabla V$ because $\mathbf{E} = -\nabla V$.

There is an outward force per unit area on the surface of a conductor, given by

$$\frac{F}{A} = \frac{\sigma^2}{2\epsilon_0}. \tag{4.2}$$

F/A may be called the *electrostatic pressure*. The ultimate origin of this force is the Coulomb repulsion between like charges. To calculate the electric force on charges at the surface is subtle. Naively the force on a small area A of the surface is $Q\mathbf{E} = \sigma A \mathbf{E}$. But $\mathbf{E}$ is discontinuous at the surface: It is 0 inside and $(\sigma/\epsilon_0)\hat{\mathbf{n}}$ outside. Which value should we use? One way to justify (4.2) is to realize that the field (4.1) is due partly to the local surface charges themselves, i.e., those in the immediate area, and partly to all the other charges on the conductor. But in calculating the force on the local surface charges in a small area, we should not include their own contribution to the field because that would be a self force. The contribution from local charges is $E_n(\text{local}) = \sigma/2\epsilon_0$ by (3.46), because near the surface the local charge density looks like an infinite plane of charge. The contribution to (4.1) from the other charges must be $E_n(\text{nonlocal}) = \sigma/2\epsilon_0$. Therefore $F/A = \sigma E_n(\text{nonlocal}) = \sigma^2/2\epsilon_0$.

A second way to understand (4.2) is by an energy argument. Suppose the small area A is moved outward by distance $\delta\ell$. The work done by the electrostatic force, $F_n\delta\ell$, would be $-\delta U$ where U is the electrostatic energy. The energy density is $\frac{1}{2}\epsilon_0 E^2$, and is nonzero only in the region *outside* the conductor, because $\mathbf{E} = 0$ inside. As A moves outward, the volume with nonzero energy density would *decrease* by $A\delta\ell$, so the energy would decrease, with

$$\delta U = -\frac{1}{2}\epsilon_0 E^2 A\delta\ell$$

where $E = \sigma/\epsilon_0$. Equating $F_n = -\delta U/\delta\ell$ leads to (4.2).

*The Potential V(**x**) throughout Space Is Determined (Uniquely up to an Additive Constant) When (a) The External Charge Densities, and (b) Either the Potentials or the Total Charges of All Conductors, Are Specified.* The physical idea here is that once all the available physical parameters of an electrostatic system have been set, then the fields are determined by the laws of nature. This is a statement of the *uniqueness theorem*. The practical importance of the uniqueness theorem is that when a potential function satisfying the equations and boundary conditions has been found, by any method, then we need look no further. Any other function satisfying these conditions can only differ by an additive constant. Potential differences and the electric field are unique; these experimental observables are not affected by an arbitrary constant in $V(\mathbf{x})$. We will frequently appeal to the uniqueness theorem in solving problems.

Laplace's Equation, Boundary Conditions, and the Uniqueness Theorem

The potential function $V(\mathbf{x})$ in a region of space where the charge density is zero, satisfies the partial differential equation $\nabla^2 V = 0$, which is called *Laplace's equation*. In a system of charged conductors, $V(\mathbf{x})$ satisfies Laplace's equation outside the conductors, and certain boundary conditions on the conductor surfaces. If the potential of a conductor is specified to be V_0, then the boundary condition is $V(\mathbf{x}) = V_0$, which is an example of a *Dirichlet boundary condition*. If the charge density on a surface is specified, then the boundary condition is $-\hat{\mathbf{n}} \cdot \nabla V = \sigma/\epsilon_0$, which is called a *Neumann boundary condition*. It is a classical theorem of mathematical analysis that Laplace's equation, with a complete set of Dirichlet and/or Neumann boundary conditions, has a unique solution.[2] If the net charge of a conductor is specified we may not know its distribution until the problem is solved. However, we do know that $V(\mathbf{x})$ is constant on the surface, and from the specified charge the value of the surface potential will be determined.

As an example of the use of the uniqueness theorem, we shall prove this interesting result: *The electric field is zero in any empty cavity in a conductor in electrostatic equilibrium, regardless of the electric field outside the conductor.* (See Fig. 4.3.) To prove the claim, we merely note that the potential function

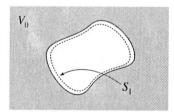

FIGURE 4.3   A cavity in a conductor. S_1 is an equipotential surface just inside the cavity.

---

[2]A mathematical proof of the uniqueness theorem for Laplace's equation with Dirichlet and Neumann boundary conditions is given in Chapter 5.

$V(\mathbf{x}) = V_0$, where V_0 is the constant potential of the conductor, does satisfy Laplace's equation in the cavity and does have the proper value on the surrounding surface. Hence, by the uniqueness theorem, V_0 must be the potential function throughout the cavity. The field in the cavity is then $\mathbf{E} = -\nabla V_0 = 0$. A conductor shields a cavity in its interior from any external electrostatic fields.

It is worthwhile to analyze this result in more detail. What happens if the conductor is charged? We will prove: *If a charged conductor has within it an empty cavity, the field in the cavity is* 0 *and the charge on the cavity wall is* 0. We already know that the net charge of a conductor must lie on its surface. The claim here is that it lies entirely on the *outer* surface.

Rather than just appealing to the uniqueness theorem, let's now analyze the potential physically. Let V_0 be the potential of the conductor, and V_1 the potential of an equipotential surface S_1 located just inside the cavity, illustrated in Fig. 4.3. If $V_1 > V_0$, then $\mathbf{E}$ points outward at all points of S_1. But that leads to a contradiction, because Gauss's Law implies there is a positive charge enclosed by S_1, contradicting the statement that the cavity is empty. So V_1 cannot be larger than V_0; and similarly it cannot be less than V_0 because that would imply a negative charge enclosed. Therefore $V_1 = V_0$, and by extension the same is true for any equipotential surface in the cavity. Hence $V(\mathbf{x}) = V_0$ and $\mathbf{E}(\mathbf{x}) = 0$ in the cavity. Now consider a Gaussian surface in the shape of a small pill box straddling the wall of the cavity. Since $\mathbf{E} = 0$ on both sides of the wall, the flux through the pill box is 0, and by Gauss's Law the charge on the wall must be 0.

This result, that $\mathbf{E} = 0$ inside the cavity, is ultimately a consequence of Coulomb's Law $F \propto 1/r^2$, i.e., that the force between charges varies inversely with the square of the distance. We derived the result from Gauss's Law, but Gauss's Law and Coulomb's Law are equivalent in that either one implies the other. If the electric force were proportional to r^{-p} with $p \neq 2$, then Gauss's Law would not be true, and $\mathbf{E}$ would be nonzero in the cavity.

The most sensitive experimental tests of the inverse square law ($p = 2$) have come from experiments that attempted to detect an electric force inside a cavity in a charged conductor. Cavendish[3] did one of the early experiments, in which a metal sphere was surrounded by a pair of hemispherical metal shells, and connected to the shells by a wire. Then charge was placed on the outer shells. If $\mathbf{E}$ were nonzero inside the conducting shell, some free charge would flow to the metal sphere inside. The hemispheres were then disconnected and removed, and the charge of the metal sphere was measured. This and all later experiments had a null result. Knowing the precision of the experiment, the null result sets an upper bound on $|p - 2|$, shown for several experiments in Table 4.1.[4]

Comment on General Electrostatics Problems
The most general electrostatics problem involving charges and conductors is to find $\mathbf{E}(\mathbf{x})$, or equivalently $V(\mathbf{x})$, for an arbitrary collection of charges and con-

[3] Henry Cavendish, b. 1731, d. 1810.
[4] In *quantum electrodynamics* the inverse square law $F \propto 1/r^2$ means that photons have no mass.

TABLE 4.1   Tests of Coulomb's Law

Experimenters	Date	$\lvert p - 2 \rvert$ bound
Cavendish	1773	$< 2 \times 10^{-2}$
Maxwell	1873	$< 5 \times 10^{-5}$
Plimpton and Lawton	1936	$< 2 \times 10^{-9}$
Bartlett, Goldhagen and Phillips	1970	$< 1.3 \times 10^{-16}$
Williams, Faller and Hill	1971	$< 1 \times 10^{-16}$

E.R. Williams, J.G. Faller, and H. Hill, *Phys. Rev. Lett.* **26**, 721 (1971).
A.S. Goldhaber and M.M. Nieto, "The mass of the photon", *Sci. Am.* (May, 1976).

ductors. One important source of difficulty in such problems is that we do not know *a priori* the charge distribution $\sigma(\mathbf{x})$ on the surfaces of the conductors; indeed part of the problem is to determine $\sigma(\mathbf{x})$ from the conditions of the system. If the charge distributions were known then the problem would reduce to calculating the integral in (3.15) or (3.60). What we do know is $\mathbf{E} = 0$ inside the conductors. That information together with the partial differential equations must somehow be used to find the complete field. An electrostatics problem is a *boundary value problem*. In the examples and exercises of this chapter we will consider some special, highly symmetric problems, that can be solved analytically. It should be understood, however, that problems of a more general form can be very difficult, and often require the use of numerical methods for their solution. In Chapter 5 we will develop some general methods for boundary value problems in electrostatics.

4.2 ■ ELECTROSTATIC PROBLEMS WITH RECTANGULAR SYMMETRY

General electrostatics problems, such as those that are important in applications, are often unsymmetrical; the conductors and charge distributions may have a complicated geometry. Such difficult problems must be solved by numerical methods, or by building and measuring analog models, or by other advanced techniques. In contrast, in the next sections we discuss some simple systems possessing symmetry, for which analytic solutions can be found. By studying such systems, which can be analyzed in detail, we will gain insight into the whole subject that will help to understand more complex systems. We'll consider rectangular symmetry in this section and spherical and cylindrical symmetry in later sections.

4.2.1 ■ Charged Plates

As a start, consider two plane conducting plates 1 and 2 separated by a vacuum gap, as shown in Fig. 4.4. The lower conductor is grounded, i.e., $V_1 = 0$, and the upper conductor is maintained at some positive potential V_0, i.e., $V_2 = V_0 > 0$. We ask: What are the potential, field, and surface charge densities? In Fig. 4.4 the plates have thicknesses d_1 and d_2, but for the present we are concerned mainly with the boundary planes $z = 0$ and $z = d$. We assume the plates each have a large

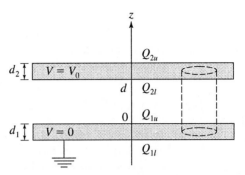

FIGURE 4.4 Parallel plate capacitor. The dashed cylinder is a Gaussian surface used in the analysis.

area A, and by large we mean that the lateral dimensions are much larger than the separation. Then there is a symmetry—translation invariance in directions parallel to the plates. The quantities we seek will not depend on the lateral coordinates, x or y. The potential and electric field will have the forms $V(\mathbf{x}) = V(z)$ and $\mathbf{E}(\mathbf{x}) = E_z(z)\hat{\mathbf{k}}$.

In the gap the potential must be a solution of Laplace's equation, which for this case is $\nabla^2 V = d^2V/dz^2 = 0$, satisfying the boundary conditions $V(0) = 0$ and $V(d) = V_0$. The solution, which is unique, is clearly $V(z) = V_0 z/d$. The corresponding field, obtained from $\mathbf{E} = -\nabla V$, is $E_z(z) = -V_0/d$. The surface charge densities are obtained from (4.1), in which $\hat{\mathbf{n}}$ is the outward normal at each surface. At the upper surface of Conductor 1 ($z = 0$) we have $\sigma_{1u} = -\epsilon_0 V_0/d$; and at the lower surface of Conductor 2 ($z = d$) we have $\sigma_{2l} = +\epsilon_0 V_0/d$. There are no surface charges on the other surfaces ($z = d+d_2$ or $z = -d_1$); this fact will become clear in Example 1 following. Thus the total charges on the conductors are equal and opposite, $Q_1 = -\epsilon_0 V_0 A/d$ and $Q_2 = +\epsilon_0 V_0 A/d$.

Notice that inside the conductors there is no electric field. Of course this is a condition for electrostatic equilibrium, but it is also a consequence of Equation (3.46), for the field of a planar charge. That is, the plane of surface charge at $z = 0$ contributes a field $E_z = -V_0/2d$ for all $z > 0$, and the plane of surface charge at $z = d$ contributes a field $E_z = +V_0/2d$ for all $z > d$. Thus $\mathbf{E} = 0$ for all $z > d$, in particular inside Conductor 2. Likewise $\mathbf{E} = 0$ for all $z < 0$, in particular inside Conductor 1. In the conductors the fields from the two inner surfaces cancel; in the gap they add.

The Parallel Plate Capacitor
Two conductors that carry equal but opposite charges $\pm Q$ form a capacitor. The capacitance of the system, which depends only on the geometry, i.e., on the sizes, shapes, and relative position of the conductors, is defined as

$$C = \frac{Q}{V}, \tag{4.3}$$

where V is the voltage difference between the conductors.[5] The unit of capacitance is the farad (F), defined by 1 farad $=$ 1 coulomb/volt; or, 1 F = 1 C/V. The conductors in Fig. 4.4 make up a parallel plate capacitor, whose capacitance is, by the previous results,

$$C = \frac{\epsilon_0 A}{d}. \tag{4.4}$$

In stating the problem at the start of this section we implied that the conductors were charged by connecting one to a battery and the other to ground. We could instead have considered the plates to be isolated and provided with charges of $\pm Q$. That system would be equivalent to Fig. 4.4, except for a constant added to the potential. The potential *difference* between the plates would be $V = Qd/(\epsilon_0 A) = V_0$, the same as before, and the electric field and surface charge densities would also be the same.

EXAMPLE 1   The previous discussion shows that if a potential difference V_0 is established between two, originally uncharged, conductors, by connecting them to opposite electrodes of a battery, then they obtain equal but opposite charges. What the battery does is transport electrons from one conductor to the other. In Fig. 4.4 the net charge accumulates on the surfaces facing the gap. However, if the conductors are isolated, from each other and from ground, then they can have arbitrary charges. In this example we ask the general question: How is the charge distributed for arbitrary charges Q' and Q'' on the two plates?

So, now suppose the plates in Fig. 4.4 are *isolated*, with total charge Q' on Conductor 1 and Q'' on Conductor 2. What will the charges be on each of the four surfaces? The four quantities that we seek are Q_{1u} and Q_{1l}, the charges on the upper and lower surfaces of Conductor 1, and Q_{2u} and Q_{2l}, the corresponding quantities for Conductor 2.

These charges may be determined from the following four equations:

$$Q_{1u} + Q_{1l} = Q', \tag{4.5}$$

$$Q_{2u} + Q_{2l} = Q'', \tag{4.6}$$

$$\frac{Q_{1u}}{A} + \frac{Q_{2l}}{A} = 0, \tag{4.7}$$

$$\frac{Q_{1l}}{A} - \frac{Q_{2u}}{A} = 0. \tag{4.8}$$

The first two equations express conservation of charge. The third equation is a consequence of Gauss's Law applied to the cylindrical surface shown dashed in Fig. 4.4. The electric flux is 0 so the charge enclosed is 0, which is (4.7). This equation implies $Q_{1u} = -Q_{2l}$, i.e., that the charges on the inner surfaces are

---

[5]A safer, but clumsy, notation would be to write the voltage difference as ΔV, so as not to confuse it with the absolute potential, whose value is only determined up to an additive constant.

equal but opposite. The fourth equation, which implies $Q_{1l} = Q_{2u}$, i.e., that the charges on the outer faces are equal, is true because E_z must be 0 inside the conductors: The field in either conductor is determined by the contributions of the four planes of charge, each of which gives $\mathbf{E}_{\text{plane}} = \hat{\mathbf{n}}\sigma/2\epsilon_0$. Since the contributions of the two inner surfaces are equal but opposite, the contributions of the outer two surfaces must also be equal but opposite to have $\mathbf{E} = 0$, which requires $Q_{1l} = Q_{2u}$.

The solution to the four equations is

$$Q_{1l} = Q_{2u} = (Q' + Q'')/2 \tag{4.9}$$

$$Q_{1u} = -Q_{2l} = (Q' - Q'')/2. \tag{4.10}$$

There is an interesting symmetry, in that the charges on the outer surfaces are equal, with each carrying half the total charge in the system. The field in the three regions outside the plates is

$$E_z(z) = \begin{cases} (Q' + Q'')/(2\epsilon_0 A) & \text{for } z > d + d_2 \\ (Q' - Q'')/(2\epsilon_0 A) & \text{for } 0 < z < d \\ -(Q' + Q'')/(2\epsilon_0 A) & \text{for } z < -d_1 \end{cases}. \tag{4.11}$$

These field strengths are independent of the thicknesses d, d_1, and d_2. If $Q'' = -Q'$ then we recover the previous results of a parallel plate capacitor.

Energy of a Capacitor

It is clear that energy is stored in the electric field of a capacitor, because electrical work is done by a battery in charging a capacitor, and mechanical work can be extracted from the capacitor by letting the two conductors, which attract each other, come together. The total electrostatic energy U in a system of charges and fields was determined in Chapter 3 to be

$$U = \frac{1}{2} \int \rho V d^3 x. \tag{4.12}$$

Using this general expression we may express the energy in a capacitor in terms of C, Q, and V.

Consider a capacitor with arbitrarily shaped conductors at potentials V_1 and V_2 and with respective total charges $-Q$ and $+Q$. Such a generalized capacitor is shown in Fig. 4.5. For clarity assume there is no external field or other charges anywhere. To find the energy we use (4.12) by substituting the surface charge element $\sigma(\mathbf{x})dA$, where $\sigma(\mathbf{x})$ is the surface charge density on the conducting surfaces of the capacitor, in place of the volume charge element $\rho(\mathbf{x})d^3x$. That is, $U = \frac{1}{2} \int \sigma V dA$. The integral is over the surfaces of both conductors. Since V is constant on either surface, the integral over the first conductor is $U_1 = -QV_1/2$, and that over the second conductor is $U_2 = +QV_2/2$. Thus the total electrostatic

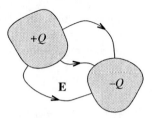

FIGURE 4.5 **A generalized capacitor**, consisting of two conductors with equal but opposite charge.

energy is $U = U_1 + U_2 = QV/2$, where the positive quantity $V = V_2 - V_1$ is the difference of potential between the conductors. For any capacitor,

$$U = \frac{1}{2}QV = \frac{1}{2}CV^2 = \frac{Q^2}{2C}. \tag{4.13}$$

Another way to derive the energy stored in a capacitor, from first principles, is to calculate the work done to separate the charge $+Q$ and $-Q$. To move a small amount of charge dQ' from the conductor with charge $-Q'$ to the conductor with charge $+Q'$ requires work $dW = V\,dQ'$, where $V = Q'/C$ is the potential difference. The total work to accumulate charges $\pm Q$ on the two conductors is

$$U = \int_0^Q \frac{Q'dQ'}{C} = \frac{Q^2}{2C}, \tag{4.14}$$

which is the same result as (4.13).

Equation (4.13) can be used to find the capacitance of two conductors whose shapes may be irregular. The method is: (a) Find the potential $V(\mathbf{x})$ between the conductors; (b) Calculate $\mathbf{E} = -\nabla V$; (c) Evaluate the total field energy with (3.84) and equate the result to one of the expressions in (4.13). This method is sometimes more convenient for calculating C than using the definition Q/V.

4.2.2 ■ Problems with Rectangular Symmetry and External Point Charges. The Method of Images

In the previous section we considered systems with charged plates, or with plates at given potentials. Next we consider problems with external point charges as well. In general these are more difficult because the external charges influence the charge distributions on the conductors and can affect the potentials. Furthermore, the system is no longer translation invariant in any direction, because the position of the charge is a preferred point in space.

The classic, simplest example of this kind is a point charge at distance z_0 from a grounded plane. The arrangement is shown in Fig. 4.6(a), in which the charge $+q$ is fixed at the point $(0, 0, z_0)$ and the grounded plane is $z = 0$. Formally the potential $V(x, y, z)$ we seek is the solution *in the region $z \geq 0$* to Poisson's

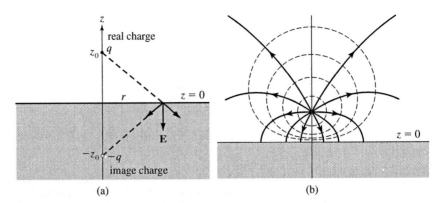

FIGURE 4.6   (a) Point charge over a conducting plane; q is the real charge and $-q$ is the fictitious image charge. (b) Electric field lines (solid curves) and equipotentials (dashed curves) for this system.

equation (3.66) with $\rho(\mathbf{x}) = q\delta(x)\delta(y)\delta(z - z_0)$, satisfying the boundary condition $V(x, y, 0) = 0$. For $z < 0$, i.e., inside the conductor, we know *a priori* that $V = 0$.

However, it is not necessary to proceed through a formal solution of the boundary value problem, because we can use a clever trick, called the *method of image charge*, to determine the potential. It is easy to see that if we construct $V(x, y, z)$ as the sum of the potential of the charge $+q$ plus the potential of a fictitious *image charge* $-q$ located at $(0, 0, -z_0)$, we will have the potential function that we seek. The potential for these two charges is

$$V(x, y, z) = \frac{q}{4\pi\epsilon_0}\left[\frac{1}{\sqrt{x^2 + y^2 + (z - z_0)^2}} - \frac{1}{\sqrt{x^2 + y^2 + (z + z_0)^2}}\right].$$
(4.15)

It is important to understand why this construction satisfies all the equations. The boundary condition is satisfied; that is, $V(x, y, 0) = 0$. This is obvious mathematically, from (4.15); physically, in the image system the plane $z = 0$ is equidistant from the charges $+q$ and $-q$ so its potential is 0. Also, Poisson's equation is satisfied in the region $z \geq 0$ because the second term in (4.15) is equivalent to a source in the region $z < 0$, so contributes nothing to $\nabla^2 V$ for $z \geq 0$. The corresponding field $\mathbf{E} = -\nabla V$ is

$$\mathbf{E}(x, y, z) = \frac{q}{4\pi\epsilon_0}\left[\frac{x\hat{\mathbf{i}} + y\hat{\mathbf{j}} + (z - z_0)\hat{\mathbf{k}}}{\left(x^2 + y^2 + (z - z_0)^2\right)^{3/2}} - \frac{x\hat{\mathbf{i}} + y\hat{\mathbf{j}} + (z + z_0)\hat{\mathbf{k}}}{\left(x^2 + y^2 + (z + z_0)^2\right)^{3/2}}\right].$$
(4.16)

We emphasize that (4.15) and (4.16) apply only in the region $z \geq 0$. In the region $z < 0$, V and $\mathbf{E}$ are 0.

Next we analyze the charge on the plate. We shall find that when q is placed at $(0, 0, z_0)$, an equal but opposite charge is drawn onto the conductor from ground.

The net charge of the conductor lies on the surface $z = 0$. We wish to determine the surface charge density σ. By symmetry, σ is a function only of $r = \sqrt{x^2 + y^2}$. According to (4.1) it is $\sigma(r) = \epsilon_0 E_n(r) = \epsilon_0 E_z(x, y, 0)$ where E_n is the field component along the *outward* normal of the surface. Using (4.16) to calculate the normal component on the surface, we see that the two terms contribute equally to $E_z(x, y, 0)$ because the real charge and image charge are equidistant from the surface. The result is

$$\sigma(r) = \frac{-qz_0}{2\pi \left(r^2 + z_0^2\right)^{3/2}}. \tag{4.17}$$

It is left as an exercise[6] to show that the total charge induced on the plane $\int_0^\infty \sigma(r) 2\pi r \, dr$ is $-q$. Figure 4.6(b) shows the electric field lines originating at $+q$ and terminating on the conductor, as solid curves. The equipotentials are shown as dashed curves.

A point charge near a grounded conducting plane experiences a force toward the plane. We leave it as an exercise[7] to calculate the force on the charge.

It is important to emphasize that there is no charge at $z = -z_0$. We know that the charge density inside a conductor in electrostatic equilibrium is 0. The electric field is actually the vector sum of the Coulomb fields for all real charges, namely, the point charge $+q$ and the surface charge of (4.17). For $z \geq 0$ these fields sum to give (4.16), and for $z < 0$ they sum to give $\mathbf{E} = 0$. Introducing the image charge at $z = -z_0$ is just a mathematical trick.

Figure 4.6 and the previous discussion can be extended to a technique for finding the electric field for *any* charge distribution above a grounded conducting plane. The underlying principle is the same as for the Green's function: If we know the potential for a point charge then we can construct the potential for a distributed charge by a superposition integral. This idea leads to the general rule that if $\rho(x, y, z)$ is the real charge distribution (for $z > 0$), then the appropriate image distribution in the region $z < 0$ is $-\rho(x, y, -z)$. For example, if a charged line or line segment lies above the grounded plane, then the image charge is its mirror image charged oppositely. An interesting case is a uniformly charged infinite line parallel to the surface. In that case, as we will see in Sec. 4.4 when we treat cylindrically symmetric systems, the equipotential surfaces are non-coaxial cylinders.

EXAMPLE 2   Figure 4.7 shows the case of an electric dipole $\mathbf{p} = p_0 \hat{\mathbf{j}}$ located at the point $(0, 0, z_0)$ above the grounded plane $z = 0$. What is the asymptotic electric field on the z axis for $z \gg z_0$?

The image charge configuration that makes $V = 0$ on the plane $z = 0$ is an image dipole $\mathbf{p}' = -p_0 \hat{\mathbf{j}}$ located at the point $(0, 0, -z_0)$. Equations (3.99) and (3.101) give the potential and field of a dipole. Proceeding from (3.101), note that

[6] See Exercise 6.
[7] See Exercise 7.

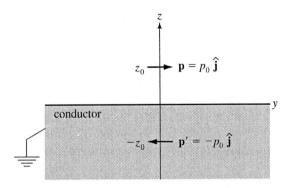

FIGURE 4.7 Example 2. An electric dipole over conducting plane. **p** is the real dipole, and **p'** is the fictitious image.

$\mathbf{p} \cdot \hat{\mathbf{r}} = 0$ and $\mathbf{p'} \cdot \hat{\mathbf{r}} = 0$ for points on the z axis, so

$$\mathbf{E}(0, 0, z) = E_y(z)\hat{\mathbf{j}} = \frac{p_0\hat{\mathbf{j}}}{4\pi\epsilon_0}\left[-\frac{1}{(z - z_0)^3} + \frac{1}{(z + z_0)^3}\right], \qquad (4.18)$$

where the contributions in the square brackets are from the real and image dipoles, respectively. The leading nonzero term in the expansion for $z \gg z_0$ is $E_y(z) = -3p_0z_0/(2\pi\epsilon_0 z^4)$. Not surprisingly, the far field is that of a quadrupole, because the total charge and total dipole moment are both zero for the combined system of real dipole plus image dipole.

Could the field have been obtained by calculating $V(\mathbf{x})$ and using $\mathbf{E} = -\nabla V$? The answer is that to find an expansion for $\mathbf{E}(\mathbf{x})$ at *arbitrary* $\mathbf{x}$ it would be easier to proceed by using the potential—a scalar—rather than the vector field. However, the potential is identically 0 at all points on the z axis, so to find $\mathbf{E}(0, 0, z)$ it is easier to compute the field itself as we have done.

In this example we could have considered the real dipole to be the limit of two individual charges $\pm q$ separated by a small distance. The image dipole would then be the combination of the images $\mp q$ of the real charges.

An interesting problem involving *multiple images* is shown in Fig. 4.8. A real charge $q_1 = +q$ is at $(0, y_0, z_0)$, and the mutually perpendicular planes $y = 0$ and $z = 0$ are both grounded. The image charge that is required to set the plane $y = 0$ to potential zero is $q_2 = -q$ at $(0, -y_0, z_0)$. The image charges then required to set the plane $z = 0$ to potential zero are $q_3 = -q$ at $(0, y_0, -z_0)$ and $q_4 = +q$ at $(0, -y_0, -z_0)$. The potential and field in the region $y, z \geq 0$, which contains the real charge, is the sum of the Coulomb potentials of q_1, q_2, q_3, q_4. The surface charge densities can be obtained with (4.1), noting that $\hat{\mathbf{n}} = \hat{\mathbf{k}}$ for the $z = 0$ plane, and $\hat{\mathbf{n}} = \hat{\mathbf{j}}$ for the $y = 0$ plane. In the regions behind and within the conductors, the potential and field vanish. In Fig. 4.8 the conducting planes are at an angle of

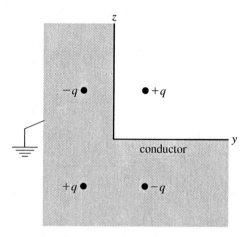

FIGURE 4.8　Multiple image charges.

90 degrees. Some problems with planes at other angles can be solved in a similar way.[8]

EXAMPLE 3　Figure 4.9 shows an electrostatics problem whose solution requires an infinite number of image charges. In this configuration there is a charge $+q$ at the origin, and the two planes $z = d$ and $z = -d$ are grounded.

We generate the image-charge solution by an iterative procedure. Start by putting an image charge $-q$ at $+2d$ on the z axis, in order to make $V = 0$ on the $z = +d$ plane. But now V is not 0 on the $z = -d$ plane, so put image charges $-q$ at $z = -2d$, and $+q$ at $z = -4d$ in order to make $V = 0$ on the the $z = -d$ plane. But now V is not 0 on the $z = +d$ plane, so put additional image charges in the region $z > +d$; and so on *ad infinitum*. In each iteration first set the upper conducting plane to zero potential, and then set the lower conducting plane to zero potential. The final result is that in the region $z > d$ there are image charges of $(-1)^k q$ at positions $z = 2dk$ with $k = 1, 2, 3 \ldots$; and in the region $z < -d$ there are image charges of $(-1)^k q$ at positions $z = -2dk$. The images are symmetric about the $z = 0$ plane. Finally, the potential on the $z = 0$ plane is the sum of Coulomb potentials of the real charge and all the images,

$$V(r) = \frac{q}{4\pi\epsilon_0}\left\{ \frac{1}{r} + \sum_{k=1}^{\infty} \frac{2(-1)^k}{\sqrt{r^2 + 4d^2 k^2}} \right\}, \tag{4.19}$$

where $r = \sqrt{x^2 + y^2}$ is the radial distance from the origin. Generally, the potential and field in the entire region $-d \leq z \leq d$ are superpositions of contributions

---

[8] See Exercises 11, 12.

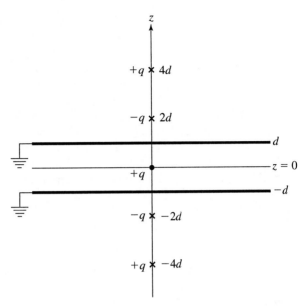

FIGURE 4.9 **Example 3.** A charge between grounded conducting planes has an infinite number of image charges, of which the first few are shown.

from $+q$ and all the image charges. Although the potential is a sum over an infinite number of image charges, the terms have alternating signs so convergence is rapid.

If there were less symmetry, the problem would still be tractable but harder. Suppose, in a configuration like Fig. 4.9, the charge were closer to one plate than the other. We could still solve the general problem, i.e., to find the potential and field between the plates, by orderly placing of image charges. However, for this less symmetric case, the charges will not be symmetric about the $z = 0$ plane.

4.3 ■ PROBLEMS WITH SPHERICAL SYMMETRY

4.3.1 ■ Charged Spheres

The simplest electrostatic system with spherical symmetry is a conducting sphere of radius a that carries a charge Q_0. What are the potential and field outside and inside the conductor? Because of the symmetry, they have the form $V = V(r)$ and $\mathbf{E}(\mathbf{x}) = E_r(r)\hat{\mathbf{r}}$; i.e., $V(r)$ and $E_r(r)$ do not depend on the polar coordinate θ or the azimuthal coordinate ϕ.

We can solve this problem from previous physical considerations. We know that all the charge must be uniformly distributed on the surface of the sphere, whether it is solid or a shell, i.e., $\sigma = Q_0/(4\pi a^2)$. Straightforward application of Gauss's Law gives $E_r(r) = Q_0/(4\pi\epsilon_0 r^2)$ for $r \geq a$, and $E_r(r) = 0$ for $r < a$.

The external field is the same as for a point charge at the origin, so the corresponding potential is $V(r) = Q_0/(4\pi\epsilon_0 r)$ for $r \geq a$, and $V(r) = Q_0/(4\pi\epsilon_0 a)$ for $r \leq a$. There is a discontinuity in E_r at $r = a$, because of the surface charge, equal to σ/ϵ_0, in agreement with (4.1). The potential, however, is continuous at $r = a$. We have described this system as an isolated conducting sphere with total charge Q_0. But we could instead specify the potential of the sphere to be, say, V_0. This case would be equivalent to the charged isolated sphere if the charge were taken to be $Q_0 = 4\pi\epsilon_0 V_0 a$. Then the potential for $r \geq a$ is $V(r) = V_0 a/r$.

We have obtained the potential function by physical arguments. A more formal approach would be to find the solution $V(r)$ to Laplace's equation, $\nabla^2 V = 0$, that satisfies the boundary conditions $V(a) = V_0$ and $V(\infty) = 0$. For a potential that depends only on r, Laplace's equation becomes

$$\frac{1}{r^2}\frac{d}{dr}\left(r^2\frac{dV}{dr}\right) = 0.$$

We may try to solve the equation with a power $V(r) \propto r^p$, and the equation requires $p = 0$ or $p = -1$. The general solution is

$$V(r) = \frac{A}{r} + B, \qquad (4.20)$$

where A and B are constants to be determined from the boundary conditions. The condition $V(\infty) = 0$ requires $B = 0$; and the condition $V(a) = V_0$ then requires $A = V_0 a$. The potential is thus $V(r) = V_0 a/r$ for $r \geq a$, as before.

Distribution of Charge on a Conductor

The surface charge distribution is uniform on a spherical conductor, but finding $\sigma(\mathbf{x})$ for an irregularly shaped conductor will be difficult. One useful principle is that charge on a conductor accumulates at those regions of the surface where the curvature is high. It follows that $\mathbf{E}$ is largest near sharp points.

As an illustration, Fig. 4.10 shows a schematic conductor consisting of a large sphere of radius a connected by a long wire of length d to a small sphere of

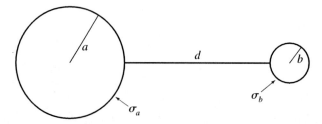

FIGURE 4.10   The system of two conducting spheres connected by a wire illustrates the fact that the charge density is greatest where the curvature is greatest, on the small sphere. The ratio σ_b/σ_a is a/b.

radius b; we have then $a \gg b$ and $d \gg a, b$. The large sphere represents a part of the conductor with small curvature, and the small sphere a region with large curvature. Because the entire conductor is at the same potential, $Q_a/(4\pi\epsilon_0 a) = Q_b/(4\pi\epsilon_0 b)$; from this we see that $Q_a = (a/b)Q_b$, where Q_a and Q_b are the charges on the spheres. The ratio of the *charge densities*, which by (4.1) is the same as the ratio of the normal surface electric fields, is

$$\frac{\sigma_b}{\sigma_a} = \frac{E_n(b)}{E_n(a)} = \frac{Q_b/b^2}{Q_a/a^2} = \frac{a}{b} \gg 1.$$

The phenomenon that $\mathbf{E}$ is largest near sharp points has important practical implications. The breakdown field for air is about 3×10^6 V/m. When the field reaches this value corona discharge can occur; the mechanism is that ions in the air are accelerated by the field and, by collisions, produce more ions and electrons. Corona discharges are important for xerography, ink-jet printers, field-ion microscopes, and other devices.

Two Concentric Conducting Spheres; Spherical Capacitor

It was hardly necessary to use the formal solution (4.20) for the problem of one conducting sphere, but it will be very useful for finding now the potential and field for the case of two concentric conducting spheres, shown in Fig. 4.11. Let the inner, solid sphere have radius a and be at potential V_0. Let the outer sphere, drawn as a spherical shell, have inner radius b and be grounded at potential zero. What is the potential $V(r)$ and the field $E_r(r)$ in the region $a \leq r \leq b$?

Stated as a boundary value problem, we seek the constants A and B in (4.20) such that $V(a) = V_0$ and $V(b) = 0$. The algebra reduces to solving two linear equations in the two unknowns A and B. The solutions for the potential and field in the region $a \leq r \leq b$ may be written in the form

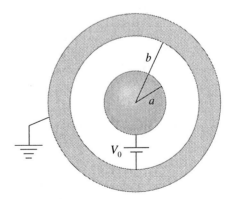

FIGURE 4.11   A spherical capacitor, consisting of an inner conducting sphere and a concentric conducting shell. The inner and outer conductors have equal but opposite charge.

$$V(r) = \frac{V_0 a(b-r)}{(b-a)r} \quad \text{and} \quad E_r(r) = \frac{V_0 ab}{(b-a)r^2}, \tag{4.21}$$

where $E_r(r) = -dV/dr$ from $\mathbf{E} = -\nabla V$. Inside the inner conductor, $r \le a$, the potential is V_0 and the field is zero. For $r > b$ the potential and field are both zero. It is left as an exercise[9] to sketch graphs of these functions.

The charges on the two spheres may be obtained from (4.1) with the field given in (4.21). Thus, on the surface of the inner sphere $\sigma = \epsilon_0 E_r(a) = \epsilon_0 V_0 b/(b-a)a$, and so the total charge is $Q(a) = 4\pi\epsilon_0 V_0 ab/(b-a)$. Similarly, on the inner surface of the outer sphere, $\sigma(b) = -\epsilon_0 E_r(b)$, and a short calculation shows that the total charge is $Q(b) = -Q(a)$. The charges are equal but opposite.

The capacitance of the two concentric spherical conductors shown in Fig. 4.11 may be determined from the definition (4.3). The potential difference is V_0. Using the charge found above, the capacitance of a spherical capacitor is

$$C = \frac{Q}{V_0} = \frac{4\pi\epsilon_0 ab}{b-a}. \tag{4.22}$$

An interesting limiting case occurs when $b \gg a$. That case, i.e., the capacitance with respect to infinity of a spherical conductor of radius a, has $C = 4\pi\epsilon_0 a$. For example, for the Earth, whose mean radius is $R = 6371$ km, the capacitance with respect to infinity is $C_{\text{Earth}} = 4\pi\epsilon_0 R = 7 \times 10^{-4}$ F. The actual capacitance of the Earth with respect to the atmosphere is larger, about 100 F [1].

Spherical Problems with Dependence on the Polar Angle θ
The solutions to Laplace's equation are called *harmonic functions* in mathematics, and the study of these functions is a large subject with a long history. There is an infinite set of solutions appropriate for spaces with spherical geometry, called *spherical harmonics*, and we will study the use of these general solutions in Chapter 5. But for now we will consider several classic systems for which the dependence on r and θ is quite simple.

We are already familiar with some solutions to Laplace's equation of the form $V(r, \theta)$. In particular consider

$$V(r, \theta) = \frac{A}{r} + B + \frac{C\cos\theta}{r^2} + Dr\cos\theta. \tag{4.23}$$

We know from past examples that each term in (4.23) satisfies Laplace's equation at all points, except for the singularities at $r = 0$ in the first and third terms.[10] Each term in (4.23) has an interesting meaning.

(a) The term A/r corresponds to a point charge at the origin or to the external potential of a spherically symmetric charge distribution. The total charge on the system is related to A by $A = Q_{\text{tot}}/4\pi\epsilon_0$.

---

[9] See Exercise 13.
[10] The reader can verify $\nabla^2 V = 0$ using polar coordinates.

(b) The term B is the arbitrary constant that can always be added to the potential without changing the field. These first two terms are the same as in (4.20).

(c) The third term is the potential of a dipole at the origin pointing in the direction $\hat{\mathbf{k}}$, as in (3.100). The dipole moment is related to C by $C = p/4\pi\epsilon_0$.

(d) The final term, which is new, corresponds to a field $\mathbf{E}(\mathbf{x}) = E_0\hat{\mathbf{k}}$ uniform throughout space. We may write this term as $V_4(r, \theta) = Dr\cos\theta = Dz = V_4(z)$, because $z = r\cos\theta$. The corresponding electric field is $\mathbf{E}(\mathbf{x}) = -\nabla V_4 = -D\hat{\mathbf{k}}$. Thus the field strength is related to D by $E_0 = -D$.

Knowing that (4.23) satisfies Laplace's equation in any region excluding the origin, for arbitrary constants A, B, C, D, we can use it to solve boundary value problems for which all the boundary conditions can be satisfied by this form.[11] We now apply (4.23) to an interesting and important problem.

EXAMPLE 4   Figure 4.12 shows a grounded conducting sphere, of radius a and centered at the origin, in an externally applied field $\mathbf{E}_{\text{appl}}(\mathbf{x}) = E_0\hat{\mathbf{k}}$ in the z direction. The presence of the sphere changes the field. What are the potential $V(r, \theta)$, field $\mathbf{E}(\mathbf{x})$, and surface charge density $\sigma(\theta)$?

We know that $V(r, \theta)$ must satisfy Laplace's equation outside the sphere, and (4.23) is a solution. Now we must satisfy the boundary conditions: (i) $V(a, \theta) = 0$, because the conductor is grounded; and (ii) $V \to -E_0 r \cos\theta$ as r approaches infinity, i.e., for $r \gg a$, which we have seen is the potential of the applied field. The conducting sphere affects the field near the sphere, but its effect must ap-

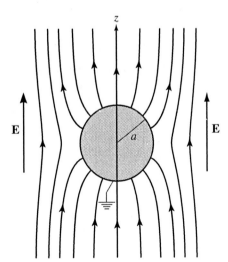

FIGURE 4.12   **Example 4.** A grounded conducting sphere in an applied electric field $E_0\hat{\mathbf{k}}$.

---

[11] In Chapter 5 we will develop a general method for solving Laplace's equation for $V(r, \theta)$. Equation (4.23) is a special case that suffices for several interesting problems.

proach 0 as $r \to \infty$; hence (ii). If we find values for A, B, C, and D, which satisfy these conditions, then the uniqueness theorem insures that they give the correct field.

First we argue that $A = 0$ because the applied field acts equally, but in opposite directions, on positive and negative charges. Thus whatever positive charge appears on the sphere above the $z = 0$ plane, an equal amount of negative charge will appear below it. Next we set the arbitrary constant B equal to 0. To satisfy condition (ii), notice that as $r \to \infty$ the term $C \cos\theta/r^2$ vanishes and leaves only the fourth term. Hence $D = -E_0$. Thus far by substituting these values into (4.23) we have $V(r, \theta) = -E_0 r \cos\theta + C \cos\theta/r^2$. To satisfy (i) requires $C = E_0 a^3$. Finally, then, for $r \geq a$,

$$V(r, \theta) = -E_0 r \cos\theta + \frac{E_0 a^3 \cos\theta}{r^2}. \tag{4.24}$$

Inside the sphere, $V = 0$. The term proportional to $\cos\theta/r^2$ is the potential of a dipole $\mathbf{p} = 4\pi\epsilon_0 a^3 E_0 \hat{\mathbf{k}}$ at the origin. The applied field polarizes the sphere, producing a surface charge distribution that creates a pure dipole field everywhere outside. The induced dipole moment $\mathbf{p}$ is proportional to the volume of the sphere.

The components of the electric field for this problem are, from $\mathbf{E} = -\nabla V$ in spherical coordinates,

$$E_r(r, \theta) = E_0 \left(1 + \frac{2a^3}{r^3}\right) \cos\theta \tag{4.25}$$

$$E_\theta(r, \theta) = -E_0 \left(1 - \frac{a^3}{r^3}\right) \sin\theta. \tag{4.26}$$

In these equations, the first terms are just the applied field. The second terms, which are proportional to a^3, are the contributions to the total field from the charge distribution on the sphere. As usual, the surface charge density can be obtained from (4.1) as $\sigma(\theta) = \epsilon_0 E_r(a, \theta)$. The result, which is remarkably simple, is

$$\sigma(\theta) = 3\epsilon_0 E_0 \cos\theta. \tag{4.27}$$

The total surface charge, which is given by $Q_{tot} = \int_0^\pi \sigma(\theta) 2\pi a^2 \sin\theta d\theta$, is 0 as we originally surmised. The result $Q_{tot} = 0$ implies that no charge is transferred between the sphere and ground. Therefore, if in the original statement of the problem we replace the grounded sphere by an uncharged isolated sphere, the same results (4.24)-(4.27) still hold.

An extension of this problem is for the sphere to be isolated and given a net charge Q_0. In that case, Q_0 must be distributed uniformly over the surface of the sphere, in order that the surface $r = a$ remain an equipotential. The potential of the sphere is then $Q_0/(4\pi\epsilon_0 a)$; a term $Q_0/(4\pi\epsilon_0 r)$ is added to the external potential; $\hat{\mathbf{r}} Q_0/(4\pi\epsilon_0 r^2)$ is added to the external field; and $Q_0/(4\pi a^2)$ is added to the surface charge density.

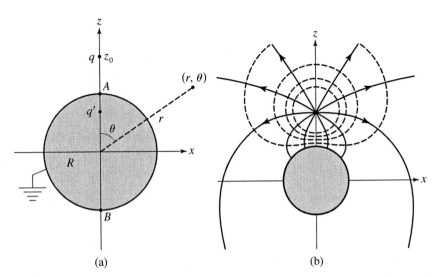

FIGURE 4.13   (a) A point charge outside a conducting sphere. q is the real charge and q' the fictitious image charge. (b) The electric field lines and equipotentials.

4.3.2 ■ Problems with Spherical Symmetry and External Charges

The principal problem that we will consider involving a spherical conductor and an external charge is shown in Fig. 4.13(a). A grounded conducting sphere of radius R has its center at the origin. A point charge q is at distance z_0 from the center of the sphere; we take q to be on the polar axis at $(0, 0, z_0)$. What are the potential $V(r, \theta)$ and field $\mathbf{E}(\mathbf{x})$ for $r \geq R$? What is the surface charge density $\sigma(\theta)$ on the sphere? Because of the axial symmetry, $V(r, \theta)$ and $\sigma(\theta)$ do not depend on the azimuthal angle ϕ.

We seek a function $V(r, \theta)$, satisfying Laplace's equation in the region $r \geq R$, except for a singularity at q, and equal to zero on the surface $r = R$. The sources of this potential will be q itself as well as the charge distributed on the surface of the sphere. A nice and rather surprising feature of this problem is that a single image charge can be found which, together with q, will give the correct potential. If we call that image charge q' and place it at $(0, 0, z_0')$ then the potential for $r \geq R$ is

$$V(r, \theta) = \frac{1}{4\pi\epsilon_0} \left\{ \frac{q}{\sqrt{r^2 + z_0^2 - 2rz_0 \cos\theta}} + \frac{q'}{\sqrt{r^2 + z_0'^2 - 2rz_0' \cos\theta}} \right\}. \quad (4.28)$$

Now we must determine q' and z_0' in terms of the experimentally selected parameters q, z_0, and R.

The image charge must lie inside the sphere, because there is only one singularity (q) for $r \geq R$. Ultimately q' and z_0' will give $V(R, \theta) = 0$ for all θ. But q' and z_0' can be determined from the simpler requirement that the potential be zero at the two points A and B on Fig. 4.13(a). The condition at A is $V(R, 0) = 0$; by (4.28) this implies $q(R-z_0')+q'(z_0-R) = 0$. The condition at B is $V(R, \pi) = 0$; by (4.28) this implies $q(R + z_0') + q'(R + z_0) = 0$. These two linear equations may be solved for q' and z_0' with the results

$$q' = \frac{-qR}{z_0} \quad \text{and} \quad z_0' = \frac{R^2}{z_0}. \tag{4.29}$$

Substituting the results back into (4.28), the potential function $V(r, \theta)$ may be written as

$$V(r, \theta) = \frac{q}{4\pi\epsilon_0} \left\{ \frac{1}{\sqrt{r^2 + z_0^2 - 2rz_0 \cos\theta}} - \frac{R}{\sqrt{r^2 z_0^2 + R^4 - 2rz_0 R^2 \cos\theta}} \right\}. \tag{4.30}$$

In (4.30) the first term is the contribution to the external potential due to the original charge q, and the second term is the contribution due to the surface charge on the sphere, which is equivalent to the potential of the fictitious image charge q' in the interior. It is obvious from (4.30) that $V(R, \theta) = 0$ for all θ. Thus the surface of the conductor, and indeed the entire conductor, is an equipotential at potential zero. The uniqueness theorem assures us that (4.30) is the correct potential for our problem. The two leading terms in the potential for the far-field, i.e., where $r \gg R$ and $r \gg z_0$, correspond to a point charge $q(1 - R/z_0)$ and a dipole $qz_0(1 - R^3/z_0^3)\hat{\mathbf{k}}$ at the origin.

The positions on the polar axis of the original charge and the image charge satisfy the relation $z_0 z_0' = R^2$. These two points are called *conjugate points*. More generally, for any point (r, θ, ϕ) outside the sphere there is a conjugate point (r', θ, ϕ) inside, at distance $r' = R^2/r$ from the center. If the exterior point is near the surface, then the conjugate point is approximately equidistant inside the surface, like the image in a plane. As the exterior point recedes from the sphere the conjugate point approaches the center.

The electric field $\mathbf{E}(\mathbf{x}) = -\nabla V$ can be found from (4.30) by taking the gradient in spherical coordinates. In Fig. 4.13(b) are shown some field lines and equipotentials for this problem. In general the field has both components $E_r(r, \theta)$ and $E_\theta(r, \theta)$; their calculation is left as an exercise.[12] Every point in space lies on some equipotential surface and the electric field vector at the point is perpendicular to the surface. In particular $E_\theta = 0$ at $r = R$, the surface of the sphere, in accord with the general result that there can be no component of electric field tangent to a conducting surface in electrostatic equilibrium.

Another interesting quantity is the surface charge density on the conductor. It is obtained as usual from (4.1). In this case, $\sigma(\theta) = \epsilon_0 E_r(R, \theta)$ where $E_r = -\partial V/\partial r$. The result is

$$\sigma(\theta) = \frac{q\left(R^2 - z_0^2\right)}{4\pi R\left(R^2 + z_0^2 - 2Rz_0 \cos\theta\right)^{3/2}}. \tag{4.31}$$

---

[12]See Exercise 15.

Notice that $\sigma(\theta)$ is negative (assuming q positive) for all θ. Its magnitude is largest at $\theta = 0$, the point on the sphere nearest q, and smallest at $\theta = \pi$, the point farthest from q. The physical explanation is that a positive charge q attracts from ground only negative charges. The total charge on the conductor may be calculated by integrating (4.31) over the area,

$$Q_{\text{tot}} = \int_0^\pi \sigma(\theta)2\pi R^2 \sin\theta d\theta = -qR/z_0, \tag{4.32}$$

which is just equal to the image charge q'. This result is obvious by applying Gauss's Law to a spherical surface lying just outside the conductor, because the field due to the surface charge is the same as for q'.

It is important to understand that q' is fictitious. The physical charge in the system consists of q and $\sigma(\theta)$. If the conductor itself could somehow be removed, leaving the original charge q at $(0, 0, z_0)$ and the surface charge $\sigma(\theta)$ at $r = R$, then the potential and field would be unchanged *for all r*, both inside and outside the conducting sphere. But $\mathbf{E}$ outside the sphere due to $\sigma(\theta)$ is the same as if there were a charge q' at the conjugate point. Because it is $\mathbf{E}$ that is detected, $\sigma(\theta)$ and q' are equivalent.

As an exercise[13] the student is asked to calculate the force on the charge q.

An Isolated Charged Conducting Sphere with an External Point Charge

It is natural to consider next a related problem. What would the potential and field be if the conducting sphere were not grounded but rather isolated with total charge Q_0? We solve this new problem, for the exterior region, by placing another fictitious charge $Q_0 - q'$ at the origin. That is, the potential in the region $r \geq R$, is that in (4.30) *plus* the term $(Q_0 - q')/(4\pi\epsilon_0 r)$. The entire conductor, including its surface, is again at constant potential, but the potential in this case is $(Q_0 - q')/(4\pi\epsilon_0 R)$. Mathematically, it is obvious that this two-image potential satsifies the equations. Physically, one can understand the result by starting with the problem of the grounded sphere, then isolating the sphere from ground and adding charge to it until the net charge on it is Q_0. The charge $(Q_0 - q')$ that must be added in this process distributes itself uniformly over the surface. But a uniform spherical charge has the same potential as a point charge at the origin for $r \geq R$.

The surface charge density of the isolated sphere is $\sigma(\theta)$ of (4.31) *plus* $(Q_0 - q')/(4\pi R^2)$. For the special case of an isolated *uncharged* sphere the value of the second fictitious charge at the center is just $-q'$.

Another related problem is for the conducting sphere to be held at a given potential V_0. The physical situation is that there is a battery connected between ground and the sphere in Fig. 4.13. For this situation the external potential is obtained by adding the term RV_0/r to (4.30), because this satisfies both Laplace's equation in the exterior region and the boundary condition. This result is equiva-

---

[13]See Exercise 15.

lent to placing a fictitious point charge $4\pi \epsilon_0 R V_0$ at the origin or, physically, to an additional surface charge density $\epsilon_0 V_0 / R$.

Spherical Problems with an Infinite Number of Image Charges
Example 3 showed how to find the potential for a point charge between two grounded planes by the image method. Its solution requires an infinite number of images, but the series converges because successive images become ever more remote from the region of interest, between the plates.

Similarly interesting problems involving conducting spheres can be solved with an infinite number of images. Three classic problems of this kind are:

(i)   To find the capacitance between a sphere and a plane.
(ii) To find the capacitance between two spheres of arbitrary radii; the spheres may be outside each other or one may be inside the other and off center.
(iii) To find the potential in space for a point charge between two conducting, grounded, spheres.

In each of these cases the strategy is basically the same idea as in Example 3. First, put a charge at the center of one spherical conductor, which makes its surface an equipotential but, in general, causes the second conductor not to be an equipotential. (In case (iii) the first image would be put at the conjugate point of one of the spheres.) Then put a suitable image within the second conductor that makes its surface an equipotential but causes the first conductor no longer to be an equipotential. Then a new image charge is put within the first conductor to restore its surface to an equipotential. This iteration is carried through until the resultant series for the potential converges as close to the ultimate value as required. In these problems the series converges because successive charges become smaller by a factor of order R/d where R is the radius of one of the conductors and d is the separation between conductors. If R/d is reduced the convergence is faster. These examples are treated in detail in the books listed at the end of the chapter (Refs. 2–5).

4.4 ■ PROBLEMS WITH CYLINDRICAL SYMMETRY

4.4.1 ■ Charged Lines and Cylinders

A charged line with constant linear charge density λ is the simplest cylindrically symmetric system. What are its field and potential? This problem was already considered in Chapter 3 (Examples 2 and 6) by using Gauss's Law, from which we derived the field. But we will start here by finding the potential in a more general way as a solution to Laplace's equation. There are two symmetries: translation invariance in z, the coordinate along the line; and rotational invariance in ϕ, the azimuthal angle. Therefore the potential can only depend on r, the perpendicular distance to the line. In cylindrical coordinates Laplace's equation becomes

$$\frac{1}{r}\frac{d}{dr}\left(r\frac{dV}{dr}\right) = 0,$$

which has the general solution

$$V(r) = A\ln(r/r_0) + B \tag{4.33}$$

where A and B are constants, and r_0 is an arbitrary scale of length (e.g., 1 m). (A change in r_0 can be absorbed into the constant B because

$$\ln r/r_0' = \ln r/r_0 + \ln r_0/r_0'$$

and the last term is a constant.) For example, recall for a line charge with charge density λ,

$$V(r) = -\frac{\lambda}{2\pi\epsilon_0}\ln\frac{r}{r_0}, \tag{4.34}$$

where r_0 is arbitrary. This result has the physically unrealistic property that it is infinite at $r = 0$ and negatively infinite at $r = \infty$, but that is because we have considered an ideal *line*, i.e., with infinite length and infinitesimal diameter. A real charged wire would be finite, and (4.34) would approximate the potential near the wire. The divergences at $r = 0$ and $r = \infty$ are tolerable because the expressions (4.33) and (4.34) give finite potential *differences*, which are the only physically measurable properties of $V(r)$ anyway. For example, the work required to take unit charge from r_1 to r_2 is $V(r_2) - V(r_1) = (\lambda/2\pi\epsilon_0)\ln(r_1/r_2)$.

Consider next a conducting cylinder of radius a which carries the constant line charge density λ spread uniformly on its surface.[14] For the potential in the region $r \geq a$ we may write

$$V(r) = -\frac{\lambda}{2\pi\epsilon_0}\ln\left(\frac{r}{a}\right). \tag{4.35}$$

Here r_0 and B have been chosen so that $V(a) = 0$. This choice of reference point is different from the spherical case, because here the potential blows up as $r \to \infty$ and the potential of the conductor is zero. This unusual situation occurs because the total charge on the cylinder is infinite.

Two Concentric Conducting Cylinders; Cylindrical Capacitor

The preceding results can be extended to find the potential between two concentric conducting cylinders and their capacitance per unit length. In Fig. 4.14 the potential of the inner cylinder, of radius a, is set at V_0; the outer cylinder, of radius b, is grounded. To find $V(r)$ for $a \leq r \leq b$ we choose the constants A and

---

[14]The surface charge density σ and linear charge density λ of a uniformly charged cylinder are related by $\lambda d\ell = \sigma dA$ where $dA = 2\pi a d\ell$; that is, $\lambda = 2\pi a\sigma$.

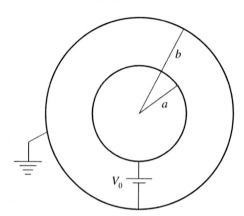

FIGURE 4.14   A cylindrical capacitor, shown in cross section.

B in (4.33) to satisfy the boundary conditions $V(a) = V_0$ and $V(b) = 0$. Solving these two linear equations we find

$$V(r) = V_0 \frac{\ln(b/r)}{\ln(b/a)} \quad \text{and} \quad E_r(r) = -\frac{\partial V}{\partial r} = \frac{V_0}{\ln(b/a)} \frac{1}{r}. \tag{4.36}$$

In order to find the capacitance per unit length C' of this configuration, we need λ in terms of V_0, a, and b. This relation can be obtained by applying Gauss's Law. The electric flux per unit length outward through a cylindrical Gaussian surface surrounding the inner conductor is $2\pi V_0 / \ln(b/a)$. The charge enclosed per unit length is, therefore, by Gauss's Law, $\lambda = 2\pi \epsilon_0 V_0 / \ln(b/a)$. Thus, for a cylindrical capacitor,

$$C' = \frac{\lambda}{V_0} = \frac{2\pi \epsilon_0}{\ln(b/a)}. \tag{4.37}$$

Potential $V(r, \phi)$ for Two Infinitely Long Line Charges $+\lambda$ and $-\lambda$
Figure 4.15 shows two infinitely long line charges with equal but opposite linear charge densities $+\lambda$ and $-\lambda$, parallel to the z axis, a distance $2d$ apart. We shall see later that the equipotentials for this system are circular cylinders, an interesting and useful result. For now we find the potential in plane polar coordinates (r, ϕ) defined by letting r be the perpendicular distance to the z axis, and letting ϕ be the angle around the z axis measured from the x axis, positive in the counterclockwise sense. (That is, ϕ is the azimuthal angle of cylindrical coordinates.) The potential is the superposition

$$V(r, \phi) = -\frac{\lambda}{2\pi \epsilon_0} \ln r_+ + \frac{\lambda}{2\pi \epsilon_0} \ln r_- = \frac{\lambda}{2\pi \epsilon_0} \ln \left(\frac{r_-}{r_+} \right) \tag{4.38}$$

where $r_\pm$ are the distances. The first term in (4.38) is the contribution of the charge $+\lambda$, the second term is that of $-\lambda$, with both in the form of (4.34). Using the law

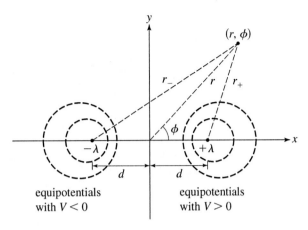

FIGURE 4.15 Two infinitely long line charges. The potential is (4.38).

of cosines to express r_+ and r_- in terms of (r, ϕ) gives

$$V(r, \phi) = \frac{\lambda}{4\pi\epsilon_0} \ln \frac{r^2 + d^2 + 2dr\cos\phi}{r^2 + d^2 - 2dr\cos\phi}. \tag{4.39}$$

The potential approaches zero as $r \to \infty$ because even though each line carries infinite charge, the net charge on the system is zero.

Problems with parallel lines and cylinders reduce to two dimensions with polar coordinates r and ϕ, because the system is translation invariant in z.

Potential of a Line Dipole

A special case of two infinite line charges is the dipole limit, where $d \to 0$, and $\lambda \to \infty$, with $2\lambda d$ remaining finite. In that case, again referring to Fig. 4.15, we have a *line dipole* described by the vector dipole moment per unit length $\mathbf{p}' = 2\lambda d\,\hat{\mathbf{i}}$. (The direction of $\mathbf{p}'$ is from the negative charge to the positive charge just as for a point dipole.) Note that the unit of $\mathbf{p}'$ is the coulomb.

To derive $V(r, \phi)$ in the line dipole limit, we rewrite (4.39) as

$$V(r, \phi) = \frac{\lambda}{4\pi\epsilon_0} \left[\ln\left(1 + 2\frac{d}{r}\cos\phi + \frac{d^2}{r^2}\right) - \ln\left(1 - 2\frac{d}{r}\cos\phi + \frac{d^2}{r^2}\right) \right],$$

and then expand the logarithms for $d/r \ll 1$ using $\ln(1+x) = x - x^2/2 + x^3/3 + \cdots$. Only the linear terms of this expansion need be retained, so the potential of a line dipole is

$$V(r, \phi) = \frac{2\lambda d\cos\phi}{2\pi\epsilon_0 r} = \frac{p_0'\cos\phi}{2\pi\epsilon_0 r} = \frac{\mathbf{p}'\cdot\hat{\mathbf{r}}}{2\pi\epsilon_0 r} \tag{4.40}$$

where $p_0' = 2\lambda d$. This result may be compared with the potential of a point dipole given in (3.100). Whereas for a point dipole the potential decreases with distance as $1/r^2$, for the line dipole the potential falls off more slowly, as $1/r$; that's because much more charge is on the infinite line than at a point. Both cases have a cosine angular dependence on the angle from the axis of the dipole.

Cylindrical Problems with Dependence on ϕ

Laplace's equation in cylindrical coordinates has many solutions, including the cylindrical harmonics, which we study later. But for now we are already familiar with solutions of the form

$$V(r, \phi) = A \ln r + B + \frac{C \cos \phi}{r} + Dr \cos \phi. \tag{4.41}$$

This expression may be compared to the analogous spherical solution (4.23), but it is important to bear in mind that the coordinates are different: In (4.41), r is the distance from the origin *in the plane*, and ϕ is measured from the x axis. The first two terms in (4.41) are the same as in (4.33). The third term is the potential of a line dipole $\mathbf{p}' = p'\hat{\mathbf{i}}$ at the origin for which $C = p'/2\pi\epsilon_0$. The fourth term, which can be written Dx, is the potential of the uniform field $\mathbf{E} = E_0\hat{\mathbf{i}}$ parallel to the x axis, with $D = -E_0$.

EXAMPLE 5   Figure 4.16 shows a grounded conducting cylinder, of radius a and centered at the origin, in an externally applied field $\mathbf{E}_{\text{app}}(\mathbf{x}) = E_0\hat{\mathbf{i}}$. The presence of the conductor changes the field. What are the potential $V(r, \phi)$, the field $\mathbf{E}(r, \phi)$, and the surface charge density $\sigma(\phi)$?

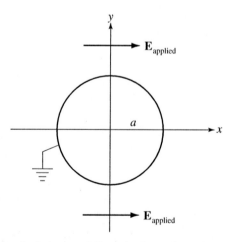

FIGURE 4.16   Example 5. A grounded conducting cylinder in an applied field perpendicular to the cylinder axis.

This problem is the cylindrical analogue to Example 4, and we'll solve it similarly, by evaluating the terms in (4.41). First we argue that $A = 0$ because there will be no net line charge drawn up from ground. For this case, whatever positive charge appears on the cylinder to the right of the $x = 0$ plane, an equal amount of negative charge will appear to the left. We set the arbitrary constant $B = 0$. To match the applied field far from the cylinder we must have $D = -E_0$. Thus far we have $V(r, \phi) = -E_0 r \cos \phi + C \cos \phi / r$. Finally, the boundary condition for the grounded cylinder is $V(a, \phi) = 0$, which requires $C = E_0 a^2$. The potential outside the cylinder is

$$V(r, \phi) = -E_0 r \cos \phi + \frac{E_0 a^2 \cos \phi}{r}. \tag{4.42}$$

The second term is the potential of a line dipole $\mathbf{p}' = 2\pi \epsilon_0 E_0 a^2 \, \hat{\mathbf{i}}$, whose magnitude is proportional to the cylinder's cross-sectional area.

The components of the electric field, obtained from $\mathbf{E} = -\nabla V$ in cylindrical coordinates, are

$$E_r(r, \phi) = E_0 \cos \phi \left(1 + \frac{a^2}{r^2} \right) \tag{4.43}$$

$$E_\phi(r, \phi) = E_0 \sin \phi \left(-1 + \frac{a^2}{r^2} \right). \tag{4.44}$$

In these equations the first terms are the applied field. The second terms are the field contributions from the surface charge on the cylinder, which are the same as for a line dipole. On the surface of the conductor the tangential field component vanishes, i.e., $E_\phi(a, \phi) = 0$. (Indeed, we could have determined C from this condition.) The surface charge density is obtained from the normal component of the field by $\sigma(\phi) = \epsilon_0 E_r(a, \phi)$, so

$$\sigma(\phi) = 2\epsilon_0 E_0 \cos \phi. \tag{4.45}$$

The total charge per unit length on the surface is zero, $\lambda_{\text{tot}} = \int_0^{2\pi} \sigma(\phi) a d\phi = 0$. We recall that in Example 4 there was likewise no net charge on the grounded sphere in an applied field.

St. Elmo's Fire

On ships and planes in electric storms a corona discharge, called St. Elmo's fire, may be observed in the vicinity of cylindrical conductors such as spars and wet ropes, or wing tips and antennas. If the external field were strong enough one would expect the discharge along all cylindrical conductors, of whatever diameter, to occur at the same time. This is because the electric field at the surface of a cylindrical conductor is, from (4.43), $E_r(a, \phi) = 2E_0 \cos \phi$, which doesn't depend on the curvature. In this context E_0 is the electric field of the storm. Although

E_0 itself may not be large enough to cause breakdown in air, the magnitude of the radial field at points on the conductor with $\phi = 0$ and π, which is $2E_0$, could be strong enough to cause coronal discharge. (For a sphere in an applied field, the field at the surface is $3E_0 \cos\theta\,\hat{\mathbf{r}}$.) In reality the electrical breakdown occurs first at sharp ends, where the field is largest, of spars, wires, and even, in literature,[15] a harpoon. The phenomenon had superstitious associations for sailors in the past.

Equipotentials for Two Infinite Line Charges

Earlier in this section we remarked that the equipotentials of two parallel line charges, with charge densities $\pm\lambda$ and separated by distance $2d$, are circular cylinders, and that this fact can be used in electrostatic analysis. The system is the one shown in Fig. 4.15, for which the potential $V(r, \phi)$ is (4.39).

To find the equipotentials of (4.39) it is convenient to use Cartesian coordinates, substituting $x = r\cos\phi$ and $y = r\sin\phi$, which gives

$$V(x, y) = \frac{\lambda}{4\pi\epsilon_0} \ln\frac{x^2 + y^2 + d^2 + 2dx}{x^2 + y^2 + d^2 - 2dx}. \tag{4.46}$$

The equation for an equipotential curve is $V(x, y) = V_0$; but then (4.46) implies a *quadratic relation* between coordinates x and y. In fact, the relation is the equation for a circle in the xy plane (i.e., a cylinder in three dimensions)

$$(x - x_0)^2 + y^2 = R^2 \tag{4.47}$$

where

$$x_0 = d\coth(2\pi\epsilon_0 V_0/\lambda) \tag{4.48}$$

$$R = d|\mathrm{csch}(2\pi\epsilon_0 V_0/\lambda)|. \tag{4.49}$$

The circle is centered at $(x_0, 0)$ and has radius R. The center of the equipotential circle is in the region $x > 0$ if V_0 and λ have the same sign, and in the region $x < 0$ if they have opposite signs. The equipotential surface with $V_0 = 0$ is the line $x = 0$ (i.e., the yz plane in three dimensions). Some equipotential curves— non-concentric circles—are shown in Fig. 4.15. We now apply these results to an interesting example.

EXAMPLE 6   Figure 4.17 shows two conducting cylinders each of radius R whose centers are $2x_0$ apart. Let the cylinder on the right be at potential $+V_0$ and that on the left at $-V_0$. What is $V(r, \phi)$ in the region between the cylinders? What is C', the capacitance per unit length?

We understand from the previous discussion that these cylinders are equipotentials for a system of two line charges $\pm\lambda$ separated by a distance $2d$, as shown in

---

[15]Herman Melville, *Moby Dick*, Chapter CXIX.

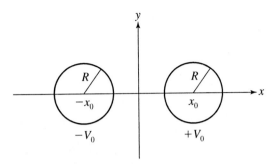

FIGURE 4.17   Example 6. Two conducting cylinders at equal but opposite potentials.

Fig. 4.15. Our task then reduces to finding λ and d in terms of the system parameters R, x_0, and V_0. This is straightforward, although the algebra is cumbersome.

From (4.46) and the discussion following it, we have the relations (4.48) and (4.49) for x_0 and R. To determine λ, divide (4.48) by (4.49) and solve for λ. The result is $\lambda = 2\pi\epsilon_0 V_0/\text{arccosh}(x_0/R)$. Now (4.48) and (4.49) imply

$$d = \sqrt{x_0^2 - R^2} = R\sinh(2\pi\epsilon_0 V_0/\lambda), \tag{4.50}$$

by the identity $\coth^2 x - \text{csch}^2 x = 1$. Substituting these results into (4.39) the potential is

$$V(r, \phi) = \frac{V_0}{2\text{arccosh}(x_0/R)} \ln \frac{r^2 + x_0^2 - R^2 + 2r\sqrt{x_0^2 - R^2}\cos\phi}{r^2 + x_0^2 - R^2 - 2r\sqrt{x_0^2 - R^2}\cos\phi}. \tag{4.51}$$

It isn't surprising that the result is somewhat cumbersome, because it contains all the information needed to find the field between the cylinders and their surface charge densities.

Now that we have done the hard work, it is simple to find the capacitance per unit length C'. We have, recalling that the potential difference between the cylinders is $2V_0$,

$$C' = \frac{\lambda}{2V_0} = \frac{\pi\epsilon_0}{\text{arccosh}(x_0/R)}. \tag{4.52}$$

Example 6 was set up to treat a symmetric case of two identical cylinders. But the technique we used to find $V(r, \phi)$ and C' can be generalized to the problem of two arbitrary infinite circular cylinders with parallel axes, i.e., any of the equipotential surfaces for Fig. 4.15, including cases in which one cylinder is inside the other and nonconcentric.

4.4.2 ◾ Problems with Cylindrical Symmetry and an External Line Charge

The prototype image problem for line charges and conductors is a single infinite line charge, with linear charge density $+\lambda$, above and parallel to an infinite conducting plane at perpendicular distance d. Picture the plane as the surface boundary of a conductor that fills the half-space below it. As might be guessed by analogy with the problem of a point charge above a conductor, the image is a line charge $-\lambda$, parallel to the physical line, and at distance d below the conducting plane. The field and potential above the plane are equivalent to those of the line charge and its image. The conductor is then at zero potential, and as usual there is no field inside it. Some implications are explored in the Exercises.[16]

There are also interesting problems involving cylindrical conductors and external line charges that can be solved with images. The image of a line charge in a cylindrical conductor is in one way simpler than for the spherical case, because the image of a line charge $+\lambda$ has linear charge density simply equal to $-\lambda$. However, these problems can have some difficult points associated with infinite amounts of charge.

Figure 4.18 shows the intersection of the xy plane and an infinite cylindrical conductor of radius R centered at the origin. An external line charge $+\lambda$ passes through the point $(x_0, 0)$, parallel to the z axis, which is the cylinder axis. We'll consider the case of an *isolated uncharged* conductor. The image required to make the conductor surface be at constant potential is a line charge $-\lambda$, also parallel to the z axis, which passes through the point $(x_0', 0)$, where $x_0 x_0' = R^2$. That is, $(x_0', 0)$ is the conjugate point to $(x_0, 0)$ in the circle of radius R, the analog in two dimensions of the conjugate point in the spherical case treated in Sec. 4.3.2. In addition, a fictitious line charge $+\lambda$, parallel to the other two, passes through the origin; this additional charge is needed to make the net line charge density of the cylinder zero.

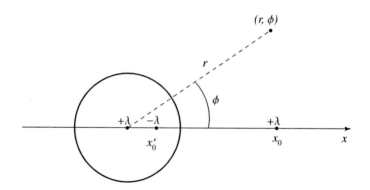

FIGURE 4.18   A line charge outside an uncharged cylinder.

[16] See Exercise 22.

Referring now to Fig. 4.18, the potential due to the three line charges is

$$V(r, \phi) = -\frac{\lambda}{2\pi\epsilon_0} \ln r_+ + \frac{\lambda}{2\pi\epsilon_0} \ln r_- - \frac{\lambda}{2\pi\epsilon_0} \ln r + \text{const.} \qquad (4.53)$$

The first three terms are, in order, the contributions of the external charge, of its image, and of the charge at the center. If we now express the distances in terms of r and ϕ by using the law of cosines, (4.53) becomes

$$V(r, \phi) = \frac{\lambda}{4\pi\epsilon_0} \ln \frac{r^2 + x_0'^2 - 2rx_0' \cos\phi}{r^2 + x_0^2 - 2rx_0 \cos\phi} - \frac{\lambda}{2\pi\epsilon_0} \ln \frac{r}{\alpha}. \qquad (4.54)$$

An arbitrary constant α in (4.54), with dimensions of length, has been introduced to make the last logarithm dimensionless. Finally we use the conjugate relation $x_0' = R^2/x_0$ to eliminate x_0' and combine the logs, to write

$$V(r, \phi) = \frac{\lambda}{4\pi\epsilon_0} \ln \left[\frac{R^2\alpha^2(R^2 + x_0^2r^2/R^2 - 2rx_0 \cos\phi)}{x_0^2r^2(r^2 + x_0^2 - 2rx_0 \cos\phi)} \right]. \qquad (4.55)$$

If (4.55) is now evaluated at the surface of the cylinder, the result is $V(R, \phi) = -(\lambda/2\pi\epsilon_0) \ln(x_0/\alpha)$, which is constant. By the uniqueness theorem we know that (4.55) is the potential for our problem. It can therefore be used to find the electric field outside the conductor and to find the surface charge density on the conductor. The constant α is still arbitrary and can be used to set the cylinder to any constant potential. We could, for example, choose $\alpha = x_0$ so that the potential on the cylinder would be zero. However, the choice of potential on the cylinder is not significant. In particular, it cannot be referred meaningfully to the potential at infinity. The reason is that $V \to \infty$ as $r \to \infty$, a consequence of the infinite amount of charge.

The case of an infinite line charge $+\lambda$ in the presence of an infinite *isolated charged* conducting cylinder with linear charge density $-\lambda$, is also interesting. In this case all flux lines from the external charge will end on the cylinder. The linear densities being equal but opposite implies that the asymptotic field decreases as $1/r^2$ for large r, so the field energy per unit length is finite. The potential for the region $r \geq R$ has then only two contributions, one from the real line charge and the other from its image line charge $-\lambda$ at the conjugate position. The potential for this case is then

$$V(r, \phi) = \frac{\lambda}{4\pi\epsilon_0} \ln \left[\frac{R^2(R^2 + x_0^2r^2/R^2 - 2rx_0 \cos\phi)}{x_0^2(r^2 + x_0^2 - 2rx_0 \cos\phi)} \right]. \qquad (4.56)$$

No constant term is added, so that the potential tends to 0 as $r \to \infty$. The potential on the conductor is finite,

$$V(R, \phi) = \frac{\lambda}{2\pi\epsilon_0} \ln \left(\frac{R}{x_0} \right). \qquad (4.57)$$

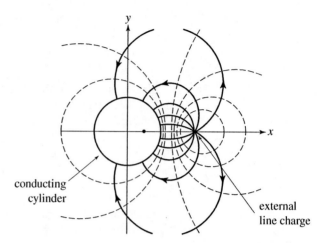

FIGURE 4.19   Field lines (solid curves) and equipotentials (dashed curves) for the problem of a charged conducting cylinder (with $dQ/dz = -\lambda$) and an external line charge (with $dQ/dz = +\lambda$). There is a fictitious image line inside the cylinder.

Figure 4.19 shows electric field lines and equipotentials for this problem.

The difficulties associated with infinite line charges are artificial in the sense that all real charge sources are finite. Near a long charged wire, i.e., at distances that are small compared to the length, the field is given accurately by the two-dimensional considerations with coordinates r and ϕ. However, at distances that are comparable or larger than the length, the problem becomes three-dimensional and the potential may be taken to be zero at infinity.

FURTHER READING

1. Joseph M. Crowley, *Fundamentals of Electrostatics* (Wiley, New York, 1986).
2. D. K. Cheng, *Field and Wave Electromagnetics*, 2nd ed. (Addison-Wesley, Reading, MA,1990).
3. E. Durand, *Électrostatique et Magnétostatique* (Masson et Cie., Paris, 1953).
4. L. Page and N. I. Adams, *Principles of Electricity*, 4th ed. (Van Nostrand, New York, 1969).
5. W. R. Smythe, *Static and Dynamic Electricity*, 3rd ed. (McGraw-Hill, New York, 1968).

EXERCISES

Sec. 4.1. Properties of Conductors

4.1. Suppose a conductor has a cavity inside it, and there is a point charge q somewhere in the cavity. Prove that the net charge on the wall of the cavity must be $-q$. If the net charge of the conductor is Q_0, what is the net charge on its outer surface? Why must every field line in the cavity begin on q and end on the cavity wall? (Hint: For the last statement, use the fact that $\oint \mathbf{E} \cdot d\boldsymbol{\ell} = 0$.)

Sec. 4.2. Charged Plates and Capacitors

4.2. Consider the generalization of Example 1, in which there are three isolated parallel conducting plates whose lateral dimensions are large compared to their separations. Let the total charges be $Q' = 1\,\mathrm{C}$, $Q'' = 4\,\mathrm{C}$, and $Q''' = 7\,\mathrm{C}$, on the bottom, middle, and top plate respectively. What are the charges on each of the six faces?
[Answer: 6,-5; 5,-1; 1,6 coulombs]

4.3. A parallel plate capacitor has plates of area A a distance x apart. The lateral dimensions of the plates are much larger than x.

   (a) The capacitor is charged to potential V_0 with a battery so that the plates carry charges $+Q$ and $-Q$. The battery is then disconnected. How much work must be done by an external force in order to increase the separation a small distance from x to $x + \Delta x$? What is $\Delta U = U_{\text{final}} - U_{\text{initial}}$, i.e., the change in energy of the capacitor?
   [Answer: $\Delta U = \left[Q^2/(2\epsilon_0 A)\right]\Delta x$]

   (b) Suppose the battery had remained connected as the external force increased the separation. How much work would then be done by the external force in increasing the separation from x to $x + \Delta x$? What is $\Delta U = U_{\text{final}} - U_{\text{initial}}$ for this case? Show that energy is conserved if all sources and sinks of energy, including the battery itself, are considered.
   [Answer: $\Delta U = -\epsilon_0 A V_0^2 \Delta x / (2x(x + \Delta x))$]

4.4. Prove $C = C_1 + C_2$ for two capacitors in parallel, and $C^{-1} = C_1^{-1} + C_2^{-1}$ for two capacitors in series.

4.5. (a) Derive the energy in a parallel plate capacitor, starting from the electric field energy density $u_E = \epsilon_0 E^2/2$.

   (b) Derive the energy in an arbitrary capacitor (two conductors with equal but opposite charges) by integrating the energy required to move charge dq from the conductor at lower potential to the conductor at higher potential.

Image Charges

4.6. A point charge $+q$ is fixed at the point $(0, 0, z_0)$ above the grounded, conducting plane $z = 0$.

   (a) What is the total charge induced on the plane?

   (b) Find the radius of the circle centered at the origin in the xy plane which encloses half of the total charge induced on the plane.

   [Answer: $R = z_0\sqrt{3}$]

4.7. The plane $z = 0$ is a grounded conductor. A point charge $+q$ is at the point $(0, 0, z_0)$.

   (a) What is the force on the charge?

(b) What is the energy in the field? How does this compare with the energy in the field of two charges $+q$ and $-q$ at a separation of $2z_0$?

[Answer: Energy in the field $= -q^2/(16\pi\epsilon_0 z_0)$]

4.8. Calculate the electric field on the z axis due to the planar surface charge density in Eq. (4.17), by integrating the Coulomb fields of elemental charges on the surface. Show that the result is the same as a point charge $-q$ at $(0, 0, -z_0)$ for field points in the region $z \geq 0$; and the same as a point charge $-q$ at $(0, 0, z_0)$ for field points in the region $z \leq 0$. Why does this justify the image charge method?

4.9. The planes $y = 0$ and $z = 0$ are grounded conducting surfaces. A point charge q is at $(0, a, a)$. Show that in the $x = 0$ plane, in the quadrant $y, z > 0$, the leading term in the potential for $r \gg a$ is

$$V(r, \theta) = \frac{3qa^2}{\pi\epsilon_0 r^3} \sin\theta \cos\theta$$

where r is the radial distance from the origin and θ is measured clockwise from the z axis. (Hints: First convince yourself that the monopole and dipole terms are zero for the charge configuration. Then, in the sum of the four inverse distances retain only terms of order a^2/r^3 for $r \gg a$. Take full advantage of the symmetry.)

4.10. **(a)** Consider a variation of Example 2, in which the $z = 0$ plane is grounded and there is a dipole $\mathbf{p} = p_0\hat{\mathbf{k}}$ at $(0, 0, z_0)$. What are the leading terms in the potential and electric field, on the z axis for $z \gg z_0$?

(b) Now consider the general case in which the $z = 0$ plane is grounded and there is a dipole $\mathbf{p} = p_0(\hat{\mathbf{j}} \sin\theta_0 + \hat{\mathbf{k}} \cos\theta_0)$ at $(0, 0, z_0)$. What are the leading terms in the potential and in each component of the electric field, on the z axis for $z \gg z_0$? Confirm that your result reduces to that of part (a) for $\theta_0 = 0$ and to that of Example 2 for $\theta_0 = \pi/2$.

[Answer (b):

$$E_z(0, 0, z) = p_0 \cos\theta_0/(\pi\epsilon_0 z^3) \quad \text{and} \quad E_y(0, 0, z) = -3p_0 z_0 \sin\theta_0/(2\pi\epsilon_0 z^4).]$$

4.11. Two grounded conducting planes meet at an angle $\theta = 60$ degrees at the origin. A point charge q is a distance r_0 from the origin along their angular bisector. What image charges are needed to satisfy the boundary conditions? What is the r dependence of the potential, on the angular bisector, for $r \gg r_0$? (Hint: Choose the most symmetric arrangement, with the real charge on the horizontal axis.)

[Answer: The leading term is $V(r) = 15qr_0^3/(16\pi\epsilon_0 r^4)$.]

4.12. Two grounded conducting planes meet at an angle θ at the origin, and a point charge q is a distance r_0 from the origin along their angular bisector. (See Exercise 11.) Show that if $\theta = 2\pi/n$, where n is any even integer, then $n - 1$ image charges are required to ground both sides of the wedge. Show that if, however, n is an odd integer then an image solution cannot be obtained. (Hint: Consider the case $n = 3$.)

This problem raises an interesting question about the connection between physics and mathematics. For the special cases where the angle between the plates is $2\pi/n$ with n even, it is not difficult to find the potential and field, exactly, by analytical mathematics. But for all other angles the mathematics is too hard for us. Why should this be? In nature there is nothing special about any angle, because in setting up an experiment we can set the plates as easily at one angle as at another.

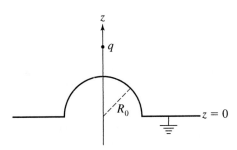

FIGURE 4.20   Exercise 16. A point charge above a grounded plane with a hemispherical boss.

Sec. 4.3. Spherical Symmetry

4.13. Sketch graphs of the potential $V(r)$ and electric field $E_r(r)$ for two concentric spheres held at potentials V_0 and 0, respectively.

4.14. Two concentric conducting spherical shells with radii a and b have charge $+Q$ and $-Q$ respectively. Calculate the total field energy U by integrating $\epsilon_0 E^2/2$ over all space. From the result, what is the capacitance?

4.15. A grounded conducting sphere of radius R has its center at the origin. Outside the sphere, at the point $(0, 0, z_0)$, is a point charge $+q$.

(a) What is the force on the charge?

(b) Plot the force as a function of z_0/R using a computer program.

(c) What is the potential energy as a function of z_0?
[Answer: $U = -q^2 R/[8\pi\epsilon_0(z_0^2 - R^2)]$.] (Hint: The potential energy is the external work done in bringing the charge from infinity to $z = z_0$.)

(d) What are the field components $E_r(r, \theta)$ and $E_\theta(r, \theta)$?

4.16. Figure 4.20 shows the grounded conducting plane $z = 0$, and on it a grounded conducting hemispherical boss of radius R_0 whose center is at the origin. A charge $+q$ is at the point $(0, 0, z_0)$.

(a) What is $V(r, \theta)$? (Hint: There will be three image charges.)

(b) What fraction of the total induced charge is induced on the hemispherical boss?

[Answer: $1 - (z_0^2 - R_0^2)/(z_0\sqrt{z_0^2 + R_0^2})$]

4.17. (a) Suppose that the inner conductor in Fig. 4.11 is a spherical shell instead of a solid sphere, and that both it and the outer spherical shell are isolated from each other and from ground. If charges Q' and Q'' are put on the inner and outer conductors, respectively, what will the charge be on each of the four spherical interfaces at equilibrium?

(b) Consider now the case of three concentric spherical, isolated, conducting shells. Charges Q', Q'', and Q''' are put on the innermost, intermediate, and outer conductors, respectively. What is the charge on each of the six spherical interfaces at equilibrium?

[Answer: From innermost to outermost the charges are: $0, Q'; -Q', Q'+Q''; -Q'-Q'', Q' + Q'' + Q'''$.] Notice how this distribution compares with that in Exercise 4.2, the analogous planar problem.

4.18. Any two conductors can be considered to be a capacitor. Suppose one conductor is a sphere of radius a and the other is a sphere of radius b. The centers of the conductors are a distance d apart, where $d \gg a, b$. What is the capacitance?
[Answer: $C = 4\pi\epsilon_0 ab/(a+b)$.]
  Notice that this interesting result is independent of the separation. The underlying physics is that the potential of each conductor is dominated by its own charge and shape, and only weakly dependent on the other conductor. This means that C for any two conductors with separation large compared to their dimensions will depend on d only weakly.

4.19. An arbitrarily shaped grounded conductor has within it a spherical cavity of radius a. Define the origin to be the center of the cavity. There is a point charge q in the cavity at distance b from the center. Define the z axis to be the direction from the origin to q.

(a) Use the method of images to find the potential $V(r)$ in the cavity.

(b) Determine the charge density $\sigma(\theta)$ on the cavity wall.

(c) Evaluate $\int_0^\pi \sigma(\theta)2\pi a^2 \sin\theta d\theta$, the total charge induced on the wall. Explain the result by Gauss's Law.

[Answer (b):

$$\sigma(\theta) = \frac{q(b^2 - a^2)}{4\pi a(a^2 + b^2 - 2ab\cos\theta)^{3/2}}]$$

4.20. An isolated spherical conductor of radius a and with charge Q_0 is placed in a uniform electric field whose magnitude is E_0.

(a) What must Q_0 be in order that $\sigma(\theta) \geq 0$ everywhere on the surface of the sphere?

(b) Sketch a graph of the function $\sigma(\theta)$.

4.21. A point charge q is at a distance z_0 from the center of an isolated spherical conductor of radius R. The charge q is outside the sphere ($z_0 > R$) and the conducting sphere has charge Q_0 on it.

(a) What must Q_0 be in order that $\sigma(\theta) \geq 0$ everywhere on the surface of the sphere?

(b) Plot the resulting $\sigma(\theta)$ using a computer program. For the plot assume $z_0 = 3R$.

Sec. 4.4. Cylindrical Symmetry

4.22. Consider $y = 0$ to be a grounded conducting plane. There is an infinitely long line charge above it. The line charge is parallel to the z axis, intersects the xy plane at the point $(0, y_0)$ and has constant linear charge density $+\lambda$.

(a) Find the surface charge density $\sigma(x, z)$ on the $y = 0$ plane.

(b) Show that the induced charge per unit length on the plane is $-\lambda$.

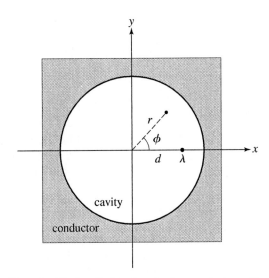

FIGURE 4.21 **Exercise 23.** A cylindrical cavity in a conductor, with a line charge λ at $(d, 0)$.

4.23. A cylindrical conductor with arbitrary cross-sectional shape has in it an infinitely long cavity with circular cross-section, radius R, centered at the origin, as shown in Fig. 4.21. The axis of the conductor and cavity is the z axis. An infinite line charge $+\lambda$ inside the cavity is parallel to the z axis and passes through the point $(d, 0)$.

(a) Find an expression for the potential $V(r, \theta)$ inside the cavity, in terms of plane polar coordinates. ($V(r, \theta)$ is determined up to an additive constant.)

(b) What is the charge density $\sigma(\theta)$ on the wall of the cavity?

[Answer (b):

$$\sigma(\theta) = \frac{\lambda(d^2 - R^2)}{2\pi R(R^2 + d^2 - 2dR\cos\theta)}]$$

(c) What is the total charge per unit length induced on the wall?

(d) What is the force per unit length on the line charge?

General Exercises

4.24. A point charge q (with $q > 0$) is located at distance $2a$ from the center of a grounded conducting sphere of radius a; let the center of the sphere be the origin and the position of q be on the z axis at $(0, 0, 2a)$. What is the electric field at $(0, 0, -2a)$? Is it directed toward or away from q?

4.25. An isolated conductor, in the shape of an ellipsoid of revolution with foci on the z axis at $z = \pm\ell$, has total charge Q.

(a) What is the potential function $V(\mathbf{x})$? Use cylindrical coordinates (r, ϕ, z) and the results from Chapter 3, Example 10. By the uniqueness theorem if you find a solution of Laplace's equation that satisfies the boundary conditions, it is the potential.

(b) Let c be one-half the length of the major axis of the ellipse, and a one-half the length of the minor axis, so that the equation for the ellipse in the xz plane is $(x/a)^2 + (z/c)^2 = 1$. Also, the eccentricity is $\varepsilon = \ell/c$, and $a = c\sqrt{1 - \varepsilon^2}$. What is the potential V_0 of the conductor, in terms of c and ε?

[Answer: $V_0 = (Q/8\pi\epsilon_0\varepsilon c)\ln\{(1 + \varepsilon)/(1 - \varepsilon)\}$]

(c) Determine the surface charge density at $(x, y, z) = (0, 0, c)$ and at $(x, y, z) = (a, 0, 0)$. Express the results in terms of c and ε. Note that for a needle ($\varepsilon \approx 1$) the surface charge at the end is much greater than at the middle.

Computer Exercises

4.26. Consider a charge q halfway between two grounded conducting plates. The electric field can be calculated by the image method, with an infinite number of images. On the midplane the field points away from q, by symmetry. Use a computer program to plot the contribution to the electric field from the image charges (i.e., the full field minus the Coulomb field of the real charge) on the midplane, as a function of distance from q. At what points on the midplane is the field due to the conducting planes maximum in magnitude?

4.27. The region $z \le 0$ is a conductor. On the positive z axis there is a point charge $+Q$ at $z = \ell$ and a point charge $-Q$ at $z = 2\ell$.

(a) Sketch a picture of the electric field lines.

(b) Use a computer to make a log-log plot of $E_z(0, 0, z)$ as a function of z, for $z > 2\ell$, and from the slope determine the asymptotic dependence on z.

(c) Use a computer to plot $E_z(x, 0, 0)$ versus x. On the xy plane, where is $\mathbf{E} = 0$?

CHAPTER

5

General Methods for Laplace's Equation

"The miracle of the appropriateness of the language of mathematics for the formulation of the laws of physics is a wonderful gift which we neither understand nor deserve."

– E. P. Wigner

"I never came across one of Laplace's 'Thus it plainly appears' *without feeling sure that I have hours of hard work before me to fill up the chasm and find out* how *it plainly appears."*

– Nathaniel Bowditch, American mathematician and astronomer

The most general problem in electrostatics is to determine the electric field $\mathbf{E}(\mathbf{x})$ for *any* static system of charges in the presence of solids, liquids, and gases. This kind of problem can be formidable; it frequently cannot be solved analytically. So far in our study we have limited ourselves to considering static systems of charges in the presence of conductors. Even those problems can be difficult, as we have seen. The approach used in physics to learn about a difficult general problem is to solve selected simple problems, which in electromagnetism means problems with a lot of symmetry. This chapter is another step for us in that approach.

We learned in Chapters 3 and 4 that the general problem can be reduced to finding a scalar function—the potential $V(\mathbf{x})$—which is usually easier to calculate than the vector function $\mathbf{E}(\mathbf{x})$. The field, charge density, and potential are connected by

$$\mathbf{E}(\mathbf{x}) = -\nabla V \quad \text{and} \quad \rho(\mathbf{x}) = \epsilon_0 \nabla \cdot \mathbf{E}. \tag{5.1}$$

The first of these equations permits us to find the field from the potential, and the second relates the field to the sources of charge.

Of particular interest to us in this chapter is the potential function in the *free space* of the system. Free space means the space outside the sources of charge. Those sources may be isolated charges, or charged surfaces, or charged volumes. But because $\rho(\mathbf{x}) = 0$ in free space, $V(\mathbf{x})$ satisfies *Laplace's equation*

$$\nabla^2 V = 0. \tag{5.2}$$

133

An important class of electrostatics problems are those in which the potential or charge distribution is given on conducting surfaces. In these problems we seek the solution to (5.2) in the region bounded by the surfaces, i.e., in the free space outside the conductor. Such problems are collectively called *boundary value problems*. We picture the boundary of the free space as a set of one or more surfaces at which $V(\mathbf{x})$ has a known behavior. On a conducting surface $V(\mathbf{x})$ is constant, independent of position $\mathbf{x}$ on the surface. At a charged conducting surface the normal ($\hat{\mathbf{n}}$) component of the electric field is σ/ϵ_0, where σ is the surface charge density. The general procedure to find $V(\mathbf{x})$ in the free space, i.e., to solve the boundary value problem, is to construct the solution of the partial differential equation (5.2) with known values of $V(\mathbf{x})$ or $\hat{\mathbf{n}} \cdot \nabla V$ on the boundary surfaces. Specifying $V(\mathbf{x})$ on a surface is called a Dirichlet boundary condition, and specifying $\hat{\mathbf{n}} \cdot \nabla V$ is called a Neumann boundary condition.

In Chapter 4 we studied a variety of solutions to Laplace's equation. However, all of those examples concerned highly symmetric systems. The solutions were generated by *mathematical tricks*, such as the method of images, or by writing the solution as a linear combination of a few known solutions and picking the coefficients to satisfy the boundary conditions. The purpose of the present chapter is to examine three general methods for constructing solutions to Laplace's equation. The method of *separation of variables* (Secs. 5.1–5.3) can be applied to problems with rectangular, spherical, or cylindrical boundaries; this material is the main subject of the chapter. The method of *conjugate functions* (Sec. 5.4) applies to two-dimensional problems. The method of *iterative relaxation* (Sec. 5.5) is a numerical method.

If we find a function $V(\mathbf{x})$ that satisfies Laplace's equation and the boundary conditions of a system, then $V(\mathbf{x})$ is the unique solution of the boundary value problem. No matter how we found the function—whether by one of the tricks in Chapter 4, by one of the classical techniques of the current chapter, or by a lucky guess—the problem is solved. General methods are important because they provide approaches to complicated cases. But solving Laplace's equation is never routine, and most solutions are found by a combination of educated guess-work and classical analytic techniques.

The Uniqueness Theorem

To justify the first statement in the previous paragraph, we shall prove the uniqueness theorem for Laplace's equation. Suppose $V_1(\mathbf{x})$ and $V_2(\mathbf{x})$ are both functions that satisfy Laplace's equation (5.2) in a volume $\mathcal{V}$ of space, and that satisfy the same boundary conditions on the set of boundary surfaces $\mathcal{S}$ of the volume. To prove the uniqueness theorem we must show that $V_1(\mathbf{x}) = V_2(\mathbf{x})$; i.e., there is only one solution.

Let $\mathbf{F}(\mathbf{x})$ be the function

$$\mathbf{F}(\mathbf{x}) = (V_1 - V_2) \, \nabla (V_1 - V_2), \tag{5.3}$$

and consider the integral of $\nabla \cdot \mathbf{F}$ over the volume $\mathcal{V}$ of interest. By Gauss's theorem,

$$\int_{\mathcal{V}} \nabla \cdot \mathbf{F} d^3 x = \int_{\mathcal{S}} (V_1 - V_2) \hat{\mathbf{n}} \cdot \nabla (V_1 - V_2) \, dA. \tag{5.4}$$

Again, $\mathcal{S}$ is the set of boundary surfaces, which may include surfaces at infinity. But both V_1 and V_2 satisfy the same boundary conditions on $\mathcal{S}$; so on $\mathcal{S}$, either $V_1 = V_2$ for the case of a Dirichlet boundary condition, or $\hat{\mathbf{n}} \cdot \nabla V_1 = \hat{\mathbf{n}} \cdot \nabla V_2$ for the case of a Neumann boundary condition. In either case the right-hand side of (5.4) is 0.

Now, $\mathbf{F}(\mathbf{x})$ has the form $f\mathbf{G}$, where f is the scalar $V_1 - V_2$ and $\mathbf{G}$ is the vector $\nabla(V_1 - V_2)$. Using the identity $\nabla \cdot (f\mathbf{G}) = \nabla f \cdot \mathbf{G} + f \nabla \cdot \mathbf{G}$, the divergence of $\mathbf{F}$ is

$$\nabla \cdot \mathbf{F} = \nabla(V_1 - V_2) \cdot \nabla(V_1 - V_2) + (V_1 - V_2)\nabla^2(V_1 - V_2). \tag{5.5}$$

The final term is 0 because both V_1 and V_2 satisfy Laplace's equation. Thus (5.4) becomes

$$\int_V [\nabla(V_1 - V_2)]^2 \, d^3 x = 0. \tag{5.6}$$

The integrand is never negative, so the only way the integral can be 0 is if $\nabla(V_1 - V_2) = 0$ for all $\mathbf{x}$; that is, $V_1(\mathbf{x}) - V_2(\mathbf{x})$ must be a constant C independent of $\mathbf{x}$. If there is any Dirichlet boundary condition in the system then C must be 0 because $V_1(\mathbf{x})$ and $V_2(\mathbf{x})$ would have the same value at the surface with a Dirichlet boundary condition. Hence,

$$V_1(\mathbf{x}) = V_2(\mathbf{x}), \tag{5.7}$$

and the uniqueness theorem is proven. If there are only Neumann boundary conditions (an uncommon case), then $V_1(\mathbf{x})$ and $V_2(\mathbf{x})$ may differ by an arbitrary constant, but the electric field $\mathbf{E}(\mathbf{x})$ is the same for $V_1(\mathbf{x})$ or $V_2(\mathbf{x})$.

5.1 ■ SEPARATION OF VARIABLES FOR CARTESIAN COORDINATES

An important general method for boundary value problems is *separation of variables*. The best way to learn the method is to apply it to some example problems, which we'll do below. But a general overview is also useful.

The first step in this method is to construct all solutions to Laplace's equation that have the form of a product of separate functions for each of the three coordinates. Then the solution of any boundary value problem can be written as a *superposition of separable functions*. The determination of $V(\mathbf{x})$ reduces to finding the coefficients of the superposition such that all the boundary conditions are satisfied. The method may be applied to Cartesian, cylindrical, or spherical coordinates, whichever choice is best suited to the system of interest.[1]

To analyze an electrostatic system with rectangular boundaries, it is natural to work in Cartesian coordinates x, y, z.

5.1.1 ■ Separable Solutions for Cartesian Coordinates

First, construct the solutions to $\nabla^2 V = 0$ that have the separable form

$$V(x, y, z) = X(x)Y(y)Z(z). \tag{5.8}$$

Not all solutions have this form, but we can use *linear combinations* of these separable solutions to solve any boundary value problem with rectangular boundaries. For (5.8) Laplace's equation becomes

$$\frac{d^2 X}{dx^2} YZ + \frac{d^2 Y}{dy^2} XZ + \frac{d^2 Z}{dz^2} XY = 0; \tag{5.9}$$

or, dividing by XYZ,

$$\frac{1}{X}\frac{d^2 X}{dx^2} + \frac{1}{Y}\frac{d^2 Y}{dy^2} + \frac{1}{Z}\frac{d^2 Z}{dz^2} = 0. \tag{5.10}$$

The first term depends only on x, the second on y, and the third on z. But these variables are independent, so the only way these three functions can sum to 0, for all values of x, y, and z, is if each term separately is a constant, and the three constants sum to 0. That is, Laplace's equation demands

$$\frac{1}{X}\frac{d^2 X}{dx^2} = \kappa_1^2 , \quad \frac{1}{Y}\frac{d^2 Y}{dy^2} = \kappa_2^2 , \quad \frac{1}{Z}\frac{d^2 Z}{dz^2} = \kappa_3^2 , \tag{5.11}$$

where

$$\kappa_1^2 + \kappa_2^2 + \kappa_3^2 = 0. \tag{5.12}$$

We denote the constants by $\kappa_1^2, \kappa_2^2, \kappa_3^2$ because then $X(x)$, $Y(y)$, $Z(z)$ have a convenient form. The partial differential equation $\nabla^2 V = 0$ has separated into three *ordinary* differential equations. The parameters $\kappa_1, \kappa_2, \kappa_3$ may be real or imaginary, but they cannot all be real, nor all imaginary, because then the sum of their squares would not be 0 as required by (5.12). At least one of them must be real and at least one must be imaginary. We will soon learn that the choice of real or imaginary κ_i is forced by the boundary conditions in an interesting way.

A special case of solutions to (5.11) is for $\kappa_1, \kappa_2, \kappa_3$ all equal to 0. In that case the solutions are $X(x) = ax + b$, $Y(y) = cy + d$, $Z(z) = ez + f$, where

---

[1] Separation of variables can be applied to other coordinate systems, such as ellipsoidal or toroidal coordinates. These are appropriate for boundary value problems where the boundary is a surface on which one of the coordinates is constant.

$a \dots f$ are constants. Combining these factors in (5.8) we get the special solution to Laplace's equation

$$V(x, y, z) = (ax + b)(cy + d)(ez + f), \qquad (5.13)$$

which is used if there is a uniform field or in other special cases.[2]

The general solution of the differential equation in (5.11) for $X(x)$ is a linear combination of exponentials $e^{\kappa_1 x}$ and $e^{-\kappa_1 x}$. The functions $Y(y)$ and $Z(z)$ are analogous. Thus

$$X(x) = C_1 e^{\kappa_1 x} + D_1 e^{-\kappa_1 x} \qquad (5.14)$$

$$Y(y) = C_2 e^{\kappa_2 y} + D_2 e^{-\kappa_2 y} \qquad (5.15)$$

$$Z(z) = C_3 e^{\kappa_3 z} + D_3 e^{-\kappa_3 z} \qquad (5.16)$$

where the C's and D's are constants. These functions deserve some discussion. If κ_1 is real, then as an alternative to (5.14) we may write $X(x) = C_1' \cosh \kappa_1 x + D_1' \sinh \kappa_1 x$. The choice of which expression to use is determined by the geometry of the system. For example, if $x = +\infty$ is included in the region where $V(\mathbf{x})$ is sought, then use $X(x) = D_1 e^{-\kappa_1 x}$, because any of the other candidate functions is infinite there. Similarly, if $x = -\infty$ is included in the region, use $X(x) = C_1 e^{\kappa_1 x}$. On the other hand, if the region of interest is bounded in x, then use a linear combination of cosh and sinh, which are even and odd functions of x, respectively. If from the symmetry of the system it can be seen that the potential is even in x, then use only $X(x) = C_1' \cosh \kappa_1 x$. Or, if the potential is odd in x, then use $X(x) = D_1' \sinh \kappa_1 x$. Use similar considerations for $Y(y)$ and $Z(z)$.

If κ_1 is imaginary, which is the case for κ_1^2 negative, then we may alternatively write (5.14) as $X(x) = C_1'' \cos k_1 x + D_1'' \sin k_1 x$ where $k_1 = |\kappa_1|$. This form has the advantage that it is manifestly real, and separated into functions with definite parity: $\cos k_1 x$ is even in x and $\sin k_1 x$ is odd. If from the symmetry of the problem the potential is even in x, then use only $X(x) = C_1'' \cos k_1 x$, and if the potential must be odd, then use only $X(x) = D_1'' \sin k_1 x$.

We have generated within the separable form (5.8) a large class of solutions to Laplace's equation. Furthermore, any linear combination of these solutions is also a solution, because Laplace's equation is linear. For a given boundary value problem we must find the linear combination that satisfies the boundary conditions. It is always possible to find such a linear combination, by the *completeness theorem*, which states that any solution of Laplace's equation can be expanded in functions of the form (5.8). Of course, as a practical matter, we would only use this class of separable solutions for problems with rectangular boundaries.

We now apply this technique for solving boundary value problems to some specific examples.

---

[2] See Exercise 1.

5.1.2 ■ Examples

EXAMPLE 1   Consider a square conducting pipe. The length L is much larger than the dimensions of the cross section, which is a square of size $a \times a$. The walls are insulated from each other. Two opposite walls are grounded ($V = 0$) and the other two are held at potential V_0. What is the potential function inside the pipe?

Set up coordinates as shown in Fig. 5.1. The z axis is the central axis of the pipe. The grounded walls are those at $x = \pm a/2$. Because the pipe is very long we may consider the potential to be independent of z, so $V = V(x, y)$. A separable solution for this effectively two-dimensional problem is

$$X(x)Y(y) = \cos(k_1 x)\, \cosh(\kappa_2 y) \tag{5.17}$$

with $-k_1^2 + \kappa_2^2 = 0$. Recall from the general discussion that one of the functions $X(x)$ and $Y(y)$ must be sinusoidal and the other hyperbolic. Note that $X(x)$ must be the sinusoidal one, to satisfy the boundary conditions $V(\pm a/2, y) = 0$; those conditions applied to the separable solution $X(x)Y(y)$ become

$$X(a/2) = X(-a/2) = 0. \tag{5.18}$$

A hyperbolic function (sinh or cosh) cannot have two zeroes,[3] so $X(x)$ must be sinusoidal. Then $Y(y)$ is hyperbolic, and it must be the even function $\cosh \kappa_2 y$

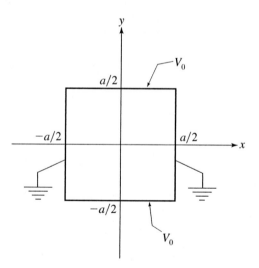

FIGURE 5.1   Example 1. A square conducting pipe shown in cross section. The axis of the pipe is the z axis, pointing out of the page. The walls at $x = \pm a/2$ are grounded, and the walls at $y = \pm a/2$ are held at the potential V_0.

[3] The definitions are $\cosh x = (e^x + e^{-x})/2$ and $\sinh x = (e^x - e^{-x})/2$. Note that $\cosh x$ is nowhere 0 and $\sinh x$ is 0 only at $x = 0$ (cf. Refs. 5–7).

by the symmetry of the system, because the system is invariant under the reflection $y \rightarrow -y$. The boundary condition (5.18) requires that $k_1 a$ is an odd integer multiple of π,

$$\cos (k_1 a/2) = 0 \quad \text{implies} \quad k_1 = (2n+1)\frac{\pi}{a}; \tag{5.19}$$

here n is a nonnegative integer. Also, for (5.17) to be a solution of Laplace's equation it is necessary that $\kappa_2 = k_1$.

The general solution of Laplace's equation for $V(x, y)$ in the square pipe, with $V = 0$ on the walls at $x = \pm a/2$ and $V(x, y)$ even in y, is a linear combination of the separable solutions (5.17)

$$V(x, y) = \sum_{n=0}^{\infty} C_n \cos\left[(2n+1)\frac{\pi x}{a}\right] \cosh\left[(2n+1)\frac{\pi y}{a}\right]. \tag{5.20}$$

Now we must determine the coefficients C_n to satisfy the other boundary conditions

$$V(x, a/2) = V(x, -a/2) = V_0, \tag{5.21}$$

which implies

$$V_0 = \sum_{n=0}^{\infty} C_n \cos\left[(2n+1)\frac{\pi x}{a}\right] \cosh\left[(2n+1)\frac{\pi}{2}\right]. \tag{5.22}$$

Equation (5.22) is quite remarkable: A sum of oscillatory functions (of x) on the right-hand side is a constant. What are the coefficients C_n that produce this result? Finding the C_n's is a problem in *Fourier analysis*.

Comments on Fourier Series
If $f(x)$ is a periodic function with period $2a$ then it can be expanded in sines and cosines in the form

$$f(x) = a_0 + \sum_{k=1}^{\infty} a_k \cos\left(\frac{k\pi x}{a}\right) + \sum_{k=1}^{\infty} b_k \sin\left(\frac{k\pi x}{a}\right). \tag{5.23}$$

The right-hand side is called the Fourier series or Fourier expansion of $f(x)$. (Note that $\cos(k\pi x/a)$ and $\sin(k\pi x/a)$ are periodic in x, with period $2a$, if k is an integer.) The set of sines and cosines are called orthogonal functions, and they satisfy the *orthogonality relations*

$$\int_{-a}^{a} \sin\frac{j\pi x}{a} \sin\frac{k\pi x}{a} \, dx = a\delta_{jk},$$

$$\int_{-a}^{a} \cos \frac{j\pi x}{a} \cos \frac{k\pi x}{a} \, dx = a\delta_{jk},$$

$$\int_{-a}^{a} \sin \frac{j\pi x}{a} \cos \frac{k\pi x}{a} \, dx = 0. \tag{5.24}$$

Here j and k are positive integers, and δ_{jk} is the Kronecker delta. The proof of (5.24) is a straightforward exercise.[4] If we multiply $f(x)$ by $\cos(j\pi x/a)$ or $\sin(j\pi x/a)$, and integrate over x, then only the cosine or sine term with $k = j$ in the expansion (5.23) is nonzero; so we *pick out* the coefficient a_j or b_j, respectively, by this procedure:

$$a_j = \frac{1}{a} \int_{-a}^{a} f(x) \cos \frac{j\pi x}{a} \, dx, \quad (\text{for } j \neq 0) \tag{5.25}$$

$$b_j = \frac{1}{a} \int_{-a}^{a} f(x) \sin \frac{j\pi x}{a} \, dx. \tag{5.26}$$

The case $j = 0$ is special:

$$a_0 = \frac{1}{2a} \int_{-a}^{a} f(x) \, dx. \tag{5.27}$$

As an example, let $f(x)$ be the square wave $S(x)$ plotted in Fig. 5.2. The function $S(x)$ is periodic in x with period $2a$. Because $S(x)$ is an even function the sine components in (5.23) are 0; that is, $b_k = 0$. Also, a_0 is 0 because the average of $S(x)$ over a cycle is 0. The Fourier coefficient a_k is

$$a_k = \frac{1}{a} \int_{-a}^{a} S(x) \cos \frac{k\pi x}{a} dx$$

$$= \frac{4}{k\pi} \sin \left(\frac{k\pi}{2} \right)$$

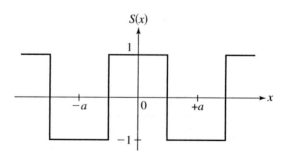

FIGURE 5.2   **Square wave.** $S(x)$ alternates between $+1$ and -1 with period $2a$, and is $+1$ in the domain $(-a/2, a/2)$.

[4] See Exercise 2.

$$= \frac{4}{k\pi} \begin{cases} (-1)^n & \text{for } k = 2n+1 \quad \text{(odd)} \\ 0 & \text{for } k = 2n \quad \text{(even)}. \end{cases} \tag{5.28}$$

Thus the Fourier series for the square wave $S(x)$ is

$$S(x) = \frac{4}{\pi} \sum_{n=0}^{\infty} \frac{(-1)^n}{(2n+1)} \cos\left[(2n+1)\frac{\pi x}{a}\right]. \tag{5.29}$$

In the region $-a/2 \le x \le a/2$ the square wave $S(x)$ is 1. For this domain equation (5.29) is like our earlier equation (5.22): a sum of cosines adding up to a constant. Of course the x domain $(-a/2, a/2)$ in our electrostatics problem is not one-half cycle of a periodic structure, as it would be for the square wave $S(x)$. But this difference doesn't matter: In (5.29) we have a linear combination of cosines that is constant in the domain $(-a/2, a/2)$. Comparing (5.22) and (5.29) we see that the coefficient in the former must be

$$C_n \cosh\left[(2n+1)\frac{\pi}{2}\right] = \frac{4V_0}{\pi} \frac{(-1)^n}{2n+1}, \tag{5.30}$$

which determines C_n. Finally then the potential $V(x, y)$ in the square pipe of Fig. 5.1 is

$$V(x, y) = \frac{4V_0}{\pi} \sum_{n=0}^{\infty} \frac{(-1)^n}{2n+1} \cos\left[(2n+1)\frac{\pi x}{a}\right] \frac{\cosh\left[(2n+1)\pi y/a\right]}{\cosh\left[(2n+1)\pi/2\right]}. \tag{5.31}$$

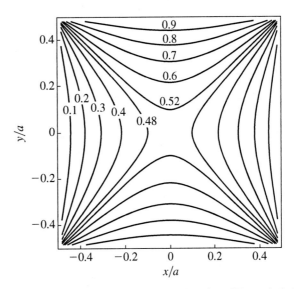

FIGURE 5.3   **Equipotentials** in the square conducting pipe of Example 1. At the top and bottom $V = V_0$ and at the sides $V = 0$. The potential values are shown in units of V_0.

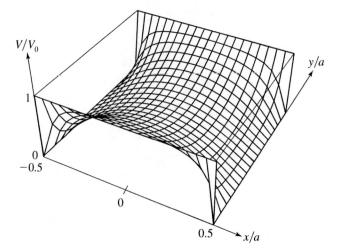

FIGURE 5.4   Surface plot of the potential function $V(x, y)$ in (5.31). The potential values are shown in units of V_0.

We have the solution to the boundary value problem in the form of an infinite series.

Figure 5.3 shows the equipotential curves of the potential $V(x, y)$. Figure 5.4 illustrates the potential in another way, as a three-dimensional surface plot of the equation $h = V(x, y)$. The surface has a saddle shape. Figure 5.5 has graphs of $V(x, 0)$ versus x, and $V(0, y)$ versus y, demonstrating again the saddle shape. In

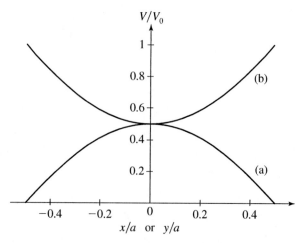

FIGURE 5.5   Plots of the potential function $V(x, y)$ in (5.31). (a) $V(x, 0)$ versus x, and (b) $V(0, y)$ versus y.

general, the solution of Laplace's equation $\nabla^2 V = 0$ is the function $V(\mathbf{x})$ that minimizes the integral of $(\nabla V)^2$ over the bounded space.[5] In this example the space is the two-dimensional square. The saddle shape minimizes the integral of $(\nabla V)^2$, subject to the boundary conditions. To make ∇V small, V makes itself as flat as possible with the boundary values.

We have a solution, but only in the form of an infinite series. To evaluate the potential $V(x, y)$ or field $\mathbf{E} = -\nabla V$ at a specific point, with the exception of points on the diagonal,[6] a computer must be used. For example, Figs. 5.3–5.5 were made by a computer. Or, the field at the point $(0, a/4)$ is $E_y(0, a/4) = -\partial V/\partial y = -1.07432\, V_0/a$, where a computer was used to get the final number. The potential at the origin is $V_0/2$ by symmetry.[6]

EXAMPLE 2   *A conducting pipe with specified end potential.*   In this example a rectangular region is bounded on five sides. Figure 5.6 shows a long metal pipe with an end cap at $x = 0$. The cross section is a rectangle of size $\ell_2 \times \ell_3$. The interior, where $\nabla^2 V = 0$, is the space

$$0 \le x < \infty, \quad 0 \le y \le \ell_2, \quad 0 \le z \le \ell_3. \tag{5.32}$$

The four sides (the bottom and top surfaces at $y = 0$ and $y = \ell_2$ and the side walls at $z = 0$ and $z = \ell_3$) are assumed to be grounded, so that $V = 0$. The

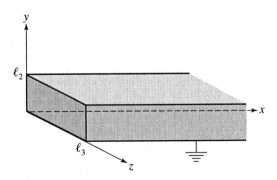

FIGURE 5.6   Example 2. A rectangular conducting pipe. The bottom and top surfaces ($y = 0$ and $y = \ell_2$) and the side walls ($z = 0$ and $z = \ell_3$) are grounded.

---

[5] See Exercise 24.

[6] See Exercise 27. Reflection symmetry about the diagonal implies that V is $V_0/2$ on the diagonals. A remarkable infinite series is revealed by this result.

end at $x = 0$ has some specified potential $V_0(y, z)$, which for the moment we leave arbitrary. The problem is to find $V(x, y, z)$. This is an example of Dirichlet boundary conditions, because $V(x, y, z)$ is specified on the surfaces. Unlike the previous example, which reduced to a two-dimensional problem, this problem will involve all three coordinates.

The separable solutions are again $X(x)Y(y)Z(z)$, where the separate functions are exponential or sinusoidal. They must satisfy the relevant boundary conditions. To have $V = 0$ on the grounded sides we require $Y(0) = Y(\ell_2) = 0$ and $Z(0) = Z(\ell_3) = 0$. A combination of real exponentials cannot have more than one zero, so $Y(y)$ and $Z(z)$ must be sinusoidal, with nodes at the boundaries; that is,

$$Y(y) = \sin\left(\frac{n\pi y}{\ell_2}\right) \quad \text{and} \quad Z(z) = \sin\left(\frac{m\pi z}{\ell_3}\right) \tag{5.33}$$

where n and m are integers ≥ 1. The constants Y''/Y and Z''/Z are

$$\kappa_2^2 = -(n\pi/\ell_2)^2 \quad \text{and} \quad \kappa_3^2 = -(m\pi/\ell_3)^2. \tag{5.34}$$

To satisfy Laplace's equation $X(x)$ must be a real exponential; to keep the potential finite as $x \to \infty$ it must be the decreasing exponential

$$X(x) = C_{nm}e^{-\kappa_{nm}x} \tag{5.35}$$

where the decay constant is

$$\kappa_{nm} = \sqrt{\left(\frac{n\pi}{\ell_2}\right)^2 + \left(\frac{m\pi}{\ell_3}\right)^2}. \tag{5.36}$$

The full potential of the system in Fig. 5.6 may be written as a superposition of these separable solutions,

$$V(x, y, z) = \sum_{n=1}^{\infty}\sum_{m=1}^{\infty} C_{nm}e^{-\kappa_{nm}x} \sin\left(\frac{n\pi y}{\ell_2}\right) \sin\left(\frac{m\pi z}{\ell_3}\right). \tag{5.37}$$

The remaining task is to determine the constants C_{nm}.

The final boundary condition, at the end at $x = 0$, is

$$V_0(y, z) = \sum_{n,m} C_{nm} \sin\left(\frac{n\pi y}{\ell_2}\right) \sin\left(\frac{m\pi z}{\ell_3}\right). \tag{5.38}$$

Equation (5.38) is a double Fourier series, in y and z. According to the *completeness theorem* of Fourier series, any function in a bounded domain can be expanded in trigonometric functions. The theorem guarantees that coefficients C_{nm} exist such that (5.38) is satisfied. The *orthogonality relation* of sine functions in this case is

$$\int_0^{\ell_2} \sin\left(\frac{j\pi y}{\ell_2}\right) \sin\left(\frac{n\pi y}{\ell_2}\right) dy = \frac{\ell_2}{2}\delta_{jn} \qquad (5.39)$$

where j and n are positive integers and δ_{jn} is the Kronecker delta.[7] It is important to appreciate (5.39) at an intuitive level. The functions $\sin(j\pi y/\ell_2)$ and $\sin(n\pi y/\ell_2)$ oscillate about 0 as functions of y. If $j = n$ they oscillate in phase, so the product is always ≥ 0 and the integral is nonzero, in fact $\ell_2/2$. If $j \neq n$ they are in phase at some positions and out of phase at others, so the product oscillates about 0 and the integral is 0. The orthogonality relation is used to calculate the coefficients C_{nm}. Orthogonality is applied to both variables y and z. Multiply both sides of (5.38) by $(2/\ell_2)\sin(j\pi y/\ell_2)(2/\ell_3)\sin(k\pi z/\ell_3)$, and integrate over y and z. This *projects out* on the right-hand side the term in (5.38) with $nm = jk$, and hence determines the coefficient C_{jk} to be

$$C_{jk} = \frac{2}{\ell_2}\frac{2}{\ell_3}\int_0^{\ell_3}\int_0^{\ell_2} V_0(y,z)\sin\left(\frac{j\pi y}{\ell_2}\right)\sin\left(\frac{k\pi z}{\ell_3}\right) dy dz. \qquad (5.40)$$

This holds for all j and k, so replacing jk by nm gives the coefficient in (5.37). We have solved the problem: Given $V_0(y,z)$ the Fourier coefficients are calculated by (5.40); and then $V(x,y,z)$ is given by (5.37). In practice the solution may require heavy numerical computation, e.g., to compute the values of the coefficients or the field $\mathbf{E}(x,y,z)$ at a specific point of interest, because $V(\mathbf{x})$ is expressed as an infinite series. With a computer one can evaluate many terms in the series, but it will be necessary to *truncate* the series, i.e., drop the terms after some maximum n and m, and the truncation produces some numerical error.

As a special case, suppose the end potential is just a constant V_0. Then the Fourier coefficients obtained from (5.40) are,[8] for n and m both odd integers,

$$C_{nm} = \frac{16V_0}{\pi^2 nm}. \qquad (5.41)$$

The coefficients with either n or m even are 0. Note that the coefficients C_{nm} factorize into the form $C_n C_m$. In the yz plane ($x = 0$), i.e., the end cap in Fig. 5.6, the double Fourier series reduces to the product of two single Fourier series, in y and z. The one-dimensional Fourier series with sine functions $\sin(n\pi y/\ell)$ and coefficients $4/(n\pi)$ for odd n, is a square wave with amplitude 1 and period 2ℓ.[8] Therefore $V(\mathbf{x})$ obeys the boundary condition $V(0,y,z) = V_0$ (a constant) for $0 \leq y \leq \ell_2$ and $0 \leq z \leq \ell_3$. For small x, i.e., near the end cap, $V(x,y,z)$ is approximately constant in y and z. For large x the Fourier series is dominated by a single term, because all the separable solutions decrease exponentially with x, with decay constants κ_{mn}. The dominant term is the one with the smallest decay

---

[7] The orthogonality relation (5.39) is analogous to those in (5.24), but has a slightly different form because here the domain is $(0, \ell_2)$ rather than $(-a, a)$. To prove (5.39), reexpress the integrand by using the trigonometric identity $2\sin A \sin B = \cos(A - B) - \cos(A + B)$.

[8] See Exercise 4.

constant, namely the term $n = m = 1$. Thus the asymptotic potential, for large x, is

$$V(x, y, z) \approx \frac{16V_0}{\pi^2} e^{-\kappa_{11}x} \sin\left(\frac{\pi y}{\ell_2}\right) \sin\left(\frac{\pi z}{\ell_3}\right), \tag{5.42}$$

decreasing exponentially with x.

In an *unbounded region* the asymptotic potential of a charge distribution decreases as $Q/(4\pi\epsilon_0 r)$, or some other power of $1/r$ if the total charge Q is 0. In the grounded pipe the potential decreases *exponentially* as $x \rightarrow \infty$. The reason for this marked difference is that in the latter case opposite charge is pulled from ground onto the inner walls of the pipe. The charge density on the walls is largest at small x. The potential due to the surface charge cancels the potential due to charge on the end cap in the limit of large x. This example illustrates the phenomenon that an electric field cannot penetrate far into a space with conducting walls. If the space is entirely enclosed, a configuration sometimes called a Faraday cage, then the electric field inside is 0. If the space has a small opening, or if the potential is fixed at one end as for the rectangular pipe in Fig. 5.6, then the field decreases exponentially with distance from the end.

Figures 5.7 and 5.8 illustrate the potential $V(\mathbf{x})$ in the pipe. In these figures ℓ_2 and ℓ_3 have been set equal to ℓ, and distances are measured in units of ℓ. Figure 5.7 is a graph of the potential $V(x, \ell_2/2, \ell_3/2)$ for points on the central axis of the pipe, as a function of distance x from the end. Note that the potential is plotted on a log scale. A straight line on a log plot implies exponential decay, so asymptotically the potential decreases exponentially. Figure 5.8 shows the equipotential curves on the vertical bisector plane $z = \ell_3/2$. It is left as an exercise[9] to sketch the corresponding electric field lines.

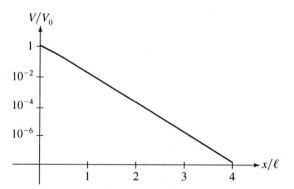

FIGURE 5.7   Potential on the central axis $V(x, \ell_2/2, \ell_3/2)$ of the rectangular pipe in Figure 5.6, as a function of x. The end potential ($x = 0$) is a constant V_0. Note that V/V_0 is plotted on a log scale. The dimensions ℓ_2 and ℓ_3 have been set equal to ℓ, and x is measured in units of ℓ.

[9]See Exercise 6.

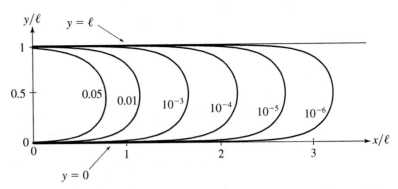

FIGURE 5.8 **Equipotential curves** on the vertical bisector plane $z = \ell_3/2$ of the rectangular pipe in Figure 5.6. The end potential $(x = 0)$ is a constant V_0. The dimensions ℓ_2 and ℓ_3 have been set equal to ℓ, and x is measured in units of ℓ. The potential values are shown in units of V_0.

These two examples show how the solution is expressed as a linear combination of separable solutions. But in each case we used symmetries, and our knowledge of sinusoidal and hyperbolic functions, to simplify the construction of the solution. As usual in physics the calculation is not just a routine process, but a combination of imagination and analysis.

5.2 ■ SEPARATION OF VARIABLES FOR SPHERICAL POLAR COORDINATES

For a system with spherical geometry, it is only natural that we work in polar coordinates r, θ, ϕ. That is, the potential function $V(\mathbf{x})$ is written as a function $V(r, \theta, \phi)$, and Laplace's equation is

$$\frac{1}{r^2} \frac{\partial}{\partial r} \left(r^2 \frac{\partial V}{\partial r} \right) + \frac{1}{r^2} \left[\frac{1}{\sin \theta} \frac{\partial}{\partial \theta} \left(\sin \theta \frac{\partial V}{\partial \theta} \right) + \frac{1}{\sin^2 \theta} \frac{\partial^2 V}{\partial \phi^2} \right] = 0. \quad (5.43)$$

To keep the discussion relatively simple, we restrict our attention to systems with *azimuthal symmetry*, i.e., invariant with respect to rotations around the z axis. This restriction still includes many interesting problems. To extend the method to problems that lack azimuthal symmetry is possible but more complicated. For a system with azimuthal symmetry the potential does not depend on ϕ; that is, $V(\mathbf{x}) = V(r, \theta)$. Then the last term on the left-hand side of (5.43) is 0.

5.2.1 ■ Separable Solutions for Spherical Coordinates

The key to separation of variables is to generate all solutions with the product form, as we did for Cartesian coordinates in Sec. 5.1. For the present case a product solution is

$$V(r, \theta) = R(r)\Theta(\theta). \tag{5.44}$$

Inserting this form into (5.43) and multiplying the result by $r^2/(R\Theta)$, we find

$$\frac{1}{R}\frac{d}{dr}\left(r^2\frac{dR}{dr}\right) + \frac{1}{\Theta\sin\theta}\frac{d}{d\theta}\left(\sin\theta\frac{d\Theta}{d\theta}\right) = 0. \tag{5.45}$$

The two terms depend on r and θ separately, and sum to 0 for all r and θ. Clearly this is only possible if both terms are constants that sum to 0, i.e., C and $-C$ respectively. It is convenient to denote the constant C by $\ell(\ell+1)$. The functions $R(r)$ and $\Theta(\theta)$ must satisfy the ordinary differential equations

$$\frac{d}{dr}\left(r^2\frac{dR}{dr}\right) - \ell(\ell+1)R = 0, \tag{5.46}$$

and

$$\frac{1}{\sin\theta}\frac{d}{d\theta}\left(\sin\theta\frac{d\Theta}{d\theta}\right) + \ell(\ell+1)\Theta = 0. \tag{5.47}$$

Later we shall see that ℓ must be an integer in order that the solution of (5.47) is finite in the whole domain of θ. The quantity $\ell(\ell+1)$ is the same for $\ell = n$ or $\ell = -n - 1$, so without loss of generality we may specify that ℓ is nonnegative.

Two independent solutions of the radial equation (5.46) are r^ℓ and $r^{-\ell-1}$, as can easily be verified. The most general solution is a linear combination

$$R(r) = Ar^\ell + \frac{B}{r^{\ell+1}} \tag{5.48}$$

where A and B are constants that will eventually be determined by boundary conditions. The two radial solutions have very different characters. We have specified that $\ell \geq 0$. Therefore the radial function r^ℓ is finite at $r = 0$ but diverges as $r \rightarrow \infty$. The function $r^{-\ell-1}$ diverges at $r = 0$ but approaches 0 as $r \rightarrow \infty$.[10] The behavior at $r = 0$ or $r = \infty$ will often dictate that either B or A must be 0. A finite solution for the potential in a region that includes the origin must have $B = 0$, and a finite solution in a region that extends to ∞ must have $A = 0$.

The angular equation (5.47) may be cast in the form of *Legendre's equation*. Rewrite the equation in terms of the variable $u \equiv \cos\theta$, and let $\Theta(\theta) = P(u)$. Note that

$$\frac{d\Theta}{d\theta} = \frac{dP}{du}\frac{du}{d\theta} = -\sin\theta\frac{dP}{du} = -\sqrt{1-u^2}\frac{dP}{du}.$$

Therefore the equation obeyed by $P(u)$ is

$$\frac{d}{du}\left[(1-u^2)\frac{dP}{du}\right] + \ell(\ell+1)P = 0, \tag{5.49}$$

---

[10]A solution of a differential equation that is finite in some domain of the variable is called a *regular solution*.

which is Legendre's equation. The domain of u is $(-1, 1)$. Equation (5.49) has two linearly independent solutions for any ℓ. However, we are only interested in *regular* solutions, i.e., that are finite for $-1 \le u \le 1$. As we shall see, the requirement that $P(u)$ be regular places severe restrictions on the solutions. Equation (5.49) has a regular solution only if ℓ is an integer, and the regular solution is simply a polynomial of degree ℓ called the *Legendre polynomial*. To get acquainted with the Legendre polynomials, a brief mathematical digression is in order.

5.2.2   ■ Legendre Polynomials

We may construct the solution of (5.49) by writing $P(u)$ as a power series

$$P(u) = \sum_{n=0}^{\infty} C_n u^n. \tag{5.50}$$

Inserting the series into (5.49) and combining the terms with equal powers of u, we find

$$\sum_{n=0}^{\infty} \left[C_{n+2}(n+2)(n+1) - C_n n(n+1) + C_n \ell(\ell+1) \right] u^n = 0, \tag{5.51}$$

which must hold for all u. The only way a convergent power series can be 0 for all values of the variable is if the coefficients are 0. The fact that the coefficient in square brackets must be 0, for all n, leads to the *recursion relation*

$$\frac{C_{n+2}}{C_n} = \frac{n(n+1) - \ell(\ell+1)}{(n+2)(n+1)}. \tag{5.52}$$

Given C_0 the recursion relation determines $C_2, C_4, \ldots$; or, given C_1 it determines $C_3, C_5, \ldots$. The solutions may be divided into two classes: even functions of u, for which $C_1, C_3, C_5, \ldots$ are 0; and odd functions, for which $C_2, C_4, C_6, \ldots$ are 0.

  To use the series (5.50) for the θ-dependence of $V(r, \theta)$, it must be finite in the entire domain $0 \le \theta \le \pi$. In other words, $P(u)$ must be a regular solution of Legendre's equation, finite for $-1 \le u \le 1$. We shall prove that *the series must terminate for the solution to be regular*. That is, all the C_n's must be 0 for n greater than some maximum value n_{max}. Suppose the series does not terminate.

TABLE 5.1
Legendre polynomials

$P_0(u) = 1$
$P_1(u) = u$
$P_2(u) = (3u^2 - 1)/2$
$P_3(u) = (5u^3 - 3u)/2$

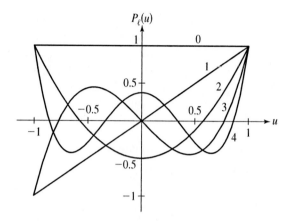

FIGURE 5.9   Legendre polynomials $P_\ell(u)$ for $\ell = 0, \ldots, 4$.

Then the recursion relation (5.52) requires $C_{n+2}/C_n \to 1$ as $n \to \infty$. But in that case the coefficients approach a constant as $n \to \infty$, and the power series will diverge at $u = \pm 1$, so the solution is not regular.

By the recursion relation (5.52) the series terminates at $n_{\max}$, giving a regular solution, if $\ell = n_{\max}$; because then the right-hand side of (5.52) vanishes for $n = n_{\max}$.

The result of the power series analysis is that Legendre's equation has a regular solution if and only if ℓ is an integer, in which case $P(u)$ is a polynomial of degree ℓ. The Legendre polynomials $P_\ell(u)$ are conventionally normalized to have $P_\ell(1) = 1$. We leave it as an exercise[11] to generate the first few Legendre polynomials from the recursion relation. Examples are listed in Table 5.1, and plotted in Fig. 5.9. Note that the *parity* of $P_\ell(u)$ (i.e., the evenness or oddness as a function of u) is $(-1)^\ell$. Some mathematical software packages have Legendre polynomials built in, so it is easy to look up higher polynomials.[12]

5.2.3 ■ Examples with Spherical Boundaries

From the previous section we know how to write a regular, azimuthally symmetric solution to Laplace's equation, in the form of a linear combination of separable solutions,

$$V(r, \theta) = \sum_{\ell=0}^{\infty} \left(A_\ell r^\ell + B_\ell r^{-\ell-1} \right) P_\ell(\cos \theta). \tag{5.53}$$

Indeed *any* regular solution may be expanded in this way, a fact known as the

---

[11] See Exercise 10.

[12] Also, there are mathematical handbooks, such as Refs. 5–7, that provide information on Legendre polynomials and other special functions.

completeness theorem of Legendre polynomials. The expansion coefficients A_ℓ and B_ℓ will be determined from the boundary conditions of the problem, as illustrated in the examples below.

The first few separable solutions are potential functions that we have encountered before. For $\ell = 0$ the two solutions are A_0 and B_0/r; the first of these is the constant that may always be added to the potential, and the second is the potential for a point charge at the origin. For $\ell = 1$ the two solutions are $A_1 r \cos\theta$ and $B_1 \cos\theta / r^2$; the first of these is the potential for a constant field in the z direction, and the second is the potential for a point dipole oriented in the z direction. Recall that in Sec. 4.3 we analyzed some simple examples using linear combinations of just these four solutions. By including all values of ℓ we can solve *any* problem with azimuthal symmetry.

EXAMPLE 3   Suppose a spherical surface of radius R is held at a specified potential $V_0(\theta)$, which is an azimuthally symmetric function of the polar angle θ. (The sphere cannot be a single conductor, because then it would be an equipotential, but it could be built out of sections of a thin conducting shell insulated from each other as in the special case considered below.) If there is no charge inside or outside the sphere, then the potential function $V(r, \theta)$ satisfies Laplace's equation $\nabla^2 V = 0$ for $r < R$ and $r > R$, with the boundary conditions $V(R, \theta) = V_0(\theta)$ and $V(\infty, \theta) = 0$. This example is a case of Dirichlet boundary conditions.

We expand $V(r, \theta)$ in separable functions, in both the interior and exterior regions, as

$$V_{\text{int}}(r, \theta) = \sum_{\ell=0}^{\infty} a_\ell \left(\frac{r}{R}\right)^{\ell} P_\ell(\cos\theta) \quad \text{for} \quad r \le R, \tag{5.54}$$

$$V_{\text{ext}}(r, \theta) = \sum_{\ell=0}^{\infty} a_\ell \left(\frac{R}{r}\right)^{\ell+1} P_\ell(\cos\theta) \quad \text{for} \quad r \ge R. \tag{5.55}$$

Writing the potential in this compact form, we have already used several ideas. In the interior $(r \le R)$ the radial function for angular mode ℓ must be r^ℓ so that $V(r, \theta)$ is finite as $r \to 0$. In the exterior $(r \ge R)$ the radial function for angular mode ℓ must be $r^{-\ell-1}$ so that $V(r, \theta) \to 0$ as $r \to \infty$. For convenience we have denoted the expansion coefficents by a_ℓ/R^ℓ and $a_\ell R^{\ell+1}$ in the interior and exterior regions, respectively. The constants a_ℓ are the same for the interior and exterior solutions, because the potential must be continuous at the boundary $(r = R)$. The a_ℓ's will be determined from the potential on the sphere $(r = R)$

$$V_0(\theta) = \sum_{\ell=0}^{\infty} a_\ell P_\ell(\cos\theta). \tag{5.56}$$

The set of Legendre polynomials is an example of *orthogonal polynomials*. With the conventional normalization the orthogonality relation is

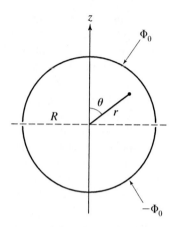

FIGURE 5.10   Example 3. The upper and lower hemispherical shells are held at constant equal but opposite potentials.

$$\int_{-1}^{1} P_k(u) P_\ell(u)\, du = \frac{2\delta_{k\ell}}{2k+1},$$
(5.57)

where $\delta_{k\ell}$ is the Kronecker delta. Or, in terms of the polar angle, replacing u by $\cos\theta$,

$$\int_{0}^{\pi} P_k(\cos\theta) P_\ell(\cos\theta) \sin\theta d\theta = \frac{2\delta_{k\ell}}{2k+1}.$$
(5.58)

To use the orthogonality to find a_k, multiply both sides of (5.56) by $P_k(\cos\theta)\sin\theta$ and integrate θ from 0 to π. Then the only term in the sum on the right-hand side that is not 0 is the one with $\ell = k$, and that term is $2a_k/(2k+1)$. Therefore,[13]

$$a_k = \frac{2k+1}{2} \int_{0}^{\pi} P_k(\cos\theta) V_0(\theta) \sin\theta d\theta.$$
(5.59)

This holds for all k, so we can replace k by ℓ for use in (5.56). In principle the problem is solved: Given $V_0(\theta)$ the a_ℓ's are known from this integral, and they determine $V(r, \theta)$ by the series in (5.54) and (5.55).

   As a specific case, suppose $V_0(\theta)$ is

$$V_0(\theta) = \begin{cases} +\Phi_0 & \text{for} \quad 0 \le \theta < \pi/2 \\ -\Phi_0 & \text{for} \quad \pi/2 < \theta \le \pi \end{cases}$$
(5.60)

as illustrated in Fig. 5.10. This electrostatic system consists of two thin hemispherical shells of a conducting material, glued together to make a sphere but insulated from one another by a thin insulating layer, and held at equal but oppo-

[13]Colloquially, we refer to this kind of calculation as "projecting out the kth coefficient."

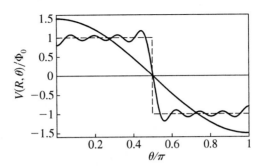

FIGURE 5.11   Legendre expansion for the potential (5.60). Two truncations of the series (5.56) are plotted, with maximum ℓ of 1 and 9. The *infinite* series is a discontinuous function equal to $+\Phi_0$ for θ in $(0, \pi/2)$ and $-\Phi_0$ for θ in $(\pi/2, \pi)$, shown as the dashed plot.

site potentials. The expansion coefficient for angular mode ℓ is

$$a_\ell = \frac{2\ell + 1}{2}\Phi_0 \left[\int_0^1 P_\ell(u)\, du - \int_{-1}^0 P_\ell(u)\, du \right]. \qquad (5.61)$$

If ℓ is an even integer, then $a_\ell = 0$ because $P_\ell(u)$ has even parity, so that the integrals over $(0, 1)$ and $(-1, 0)$ cancel. If ℓ is odd, then $P_\ell(u)$ has odd parity, so the two integrals add. To evaluate the integral $\int_0^1 P_\ell\, du$, integrate both sides of Legendre's equation (5.49) over the range $0 \le u \le 1$. The result is[14]

$$\int_0^1 P_\ell(u)\, du = \frac{P_\ell'(0)}{\ell(\ell + 1)}, \qquad (5.62)$$

where $P_\ell'(0)$ means the derivative of $P_\ell(u)$ evaluated at $u = 0$. So, for odd integer values of ℓ,

$$a_\ell = \frac{2\ell + 1}{\ell(\ell + 1)} P_\ell'(0)\Phi_0. \qquad (5.63)$$

The first three expansion coefficients in the series of (5.54) and (5.55) are

$$a_1, a_3, a_5 = \tfrac{3}{2}\Phi_0, -\tfrac{7}{8}\Phi_0, \tfrac{11}{16}\Phi_0. \qquad (5.64)$$

For r near R, $V(r, \theta)$ resembles the function $V_0(\theta)$, being $\approx +\Phi_0$ in the upper hemisphere and $\approx -\Phi_0$ in the lower hemisphere. Figure 5.11 shows how the Legendre expansion (5.56) produces this discontinuous function of θ. The functions plotted there are the first term in the expansion, which is just $1.5 \cos\theta$; and the sum of the first 10 terms, which resembles the step potential $V_0(\theta)$ except for some oscillations. To get rid of the oscillations requires the full infinite series. Terms with large values of ℓ are important if r is near R. But for $r \ll R$ or $r \gg R$

[14]See Exercise 10.

the series is dominated by the single term with $\ell = 1$. In particular, the asymptotic potential for large r is the first term in the series (5.55), which is the dipole potential

$$\frac{a_1 R^2 \cos\theta}{r^2} = \frac{p\cos\theta}{4\pi\epsilon_0 r^2};$$

the dipole moment of this bisected sphere is $p\hat{\mathbf{k}}$, with

$$p = 4\pi\epsilon_0 a_1 R^2 = 6\pi\epsilon_0 \Phi_0 R^2. \tag{5.65}$$

The terms with higher values of ℓ, which are proportional to $r^{-4}, r^{-6}, r^{-8}, \ldots$, are the higher multipole contributions to $V(r,\theta)$.

Now that we have found the exterior and interior potentials for the hemispheres, it is a natural next step to calculate their capacitance C, which has an interesting result. Recall that C is the ratio of the charge on one conductor to the potential difference between them. The potential difference here is $2\Phi_0$. We'll obtain the charge by integrating the charge density $\sigma(\theta)$ over the upper hemisphere.

Use the general boundary condition $\sigma = \epsilon_0(E_{1n} - E_{2n})$. In this case, the normal direction is radial, so

$$\sigma(\theta) = \epsilon_0 \left[-\frac{\partial V_{\text{ext}}}{\partial r} + \frac{\partial V_{\text{int}}}{\partial r} \right]\bigg|_{r=R}; \tag{5.66}$$

using the Legendre expansions (5.54) and (5.55) gives

$$\sigma(\theta) = \frac{\epsilon_0 \Phi_0}{R} \sum_{\ell,\text{odd}} \frac{(2\ell+1)^2}{\ell(\ell+1)} P'_\ell(0) P_\ell(\cos\theta). \tag{5.67}$$

The total charge on the upper hemisphere is then

$$Q = 2\pi R^2 \int_0^{\pi/2} \sigma(\theta) \sin\theta d\theta. \tag{5.68}$$

To do the integral, let $\cos\theta = u$ and refer back to (5.62). Finally the capacitance is

$$C = \frac{Q}{2\Phi_0} = \pi\epsilon_0 R \sum_{\ell,\text{odd}} \left[\frac{2\ell+1}{\ell(\ell+1)} P'_\ell(0) \right]^2. \tag{5.69}$$

Although the separation between the hemispheres is infinitesimal at the equator, C is finite. In contrast, a parallel plate capacitor with infinitesimal plate separation would have infinite C. The reason C is finite for the hemispheres is that we approximated the thickness of the shells by 0, so that the area of infinitesimal separation is 0.

Because of spherical symmetry, we have been able to do a lot with this problem. Indeed all of our analytic techniques require some degree of symmetry for the problem to be tractable. Even small changes in the system could make the problem much harder, or impossible, to solve analytically. For example, if the hemispheres were displaced horizontally, or tipped relative to one another, the Legendre expansion would not suffice to solve the problem. To find $V(\mathbf{x})$, $\mathbf{E}(\mathbf{x})$ and C for such unsymmetric problems would require numerical methods or direct measurement.

EXAMPLE 4 Consider again a sphere, but now instead of specifying the potential, specify the charge on the surface. Let the azimuthally symmetric surface density be $\sigma_0(\theta)$. What is the potential $V(r, \theta)$?

The same expansions in separable functions (5.54) and (5.55) can be used, because they are general solutions of Laplace's equation, but we must recalculate the coefficients a_ℓ. The boundary condition in this case is that the normal component of the electric field is discontinuous at the sphere, with discontinuity $\sigma_0(\theta)/\epsilon_0$; that is,

$$\left[-\frac{\partial V_{\text{ext}}}{\partial r} + \frac{\partial V_{\text{int}}}{\partial r}\right]\bigg|_{r=R} = \frac{\sigma_0(\theta)}{\epsilon_0}, \tag{5.70}$$

the same as (5.66) with surface density $\sigma_0(\theta)$. This case is an example with Neumann boundary conditions. Inserting the expansions (5.54) and (5.55) into the boundary condition gives

$$\sum_{\ell=0}^{\infty}(2\ell + 1)\frac{a_\ell}{R}P_\ell(\cos\theta) = \frac{\sigma_0(\theta)}{\epsilon_0}. \tag{5.71}$$

As before, we use the orthogonality relation (5.58) to project out the kth coefficient from the expansion, and here the result is

$$a_k = \frac{R}{2\epsilon_0}\int_0^\pi \sigma_0(\theta)P_k(\cos\theta)\sin\theta\,d\theta. \tag{5.72}$$

This expression is valid for any k so we can replace k by ℓ for use in (5.54) and (5.55).

Special case. Suppose the surface density is $\sigma_0(\theta) = \sigma\cos\theta$, where σ is a constant. The total charge for this distribution is 0, with positive charge above the equator and negative charge below. The surface density is large and positive at the north pole, large and negative on the south pole, and zero at the equator of the sphere. Because $\cos\theta = P_1(\cos\theta)$, the orthogonality of Legendre polynomials implies that $a_k = 0$ for $k \neq 1$, and $a_1 = \sigma R/(3\epsilon_0)$. The potential function is

$$V_{\text{int}}(r, \theta) = \frac{\sigma}{3\epsilon_0}r\cos\theta \quad \text{for} \quad r \leq R, \tag{5.73}$$

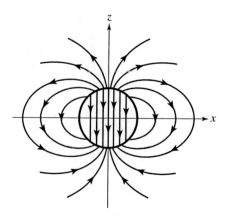

FIGURE 5.12   Example 4. The curves are electric field lines inside and outside a sphere with surface charge density $\sigma \cos \theta$. The field is constant inside the sphere and a pure dipole outside.

$$V_{\text{ext}}(r, \theta) = \frac{\sigma}{3\epsilon_0} \frac{R^3}{r^2} \cos \theta \quad \text{for } r \geq R. \tag{5.74}$$

The electric field inside the sphere is uniform, equal to $-\hat{\mathbf{k}}\sigma/(3\epsilon_0)$. The electric field outside the sphere is a pure dipole field, with dipole moment $\mathbf{p} = \frac{4}{3}\pi R^3 \sigma \hat{\mathbf{k}}$. The electric field lines for this system are shown in Fig. 5.12. Note that the normal component of the field is discontinuous at the surface.

In Example 4 of Chapter 4 we calculated the surface density of the charge induced on a conducting sphere placed in a uniform applied field $E_0\hat{\mathbf{k}}$. The result for that example is $\sigma(\theta) = 3\epsilon_0 E_0 \cos \theta$. If in our current example we set $\sigma = 3\epsilon_0 E_0$, then the field produced inside the sphere would be $-E_0\hat{\mathbf{k}}$. The electric field inside a conducting sphere in an applied field is the superposition of the applied field $E_0\hat{\mathbf{k}}$ and the field produced by the surface charge $\sigma \cos \theta$; the combined field is 0, as the field must be in a conductor.

EXAMPLE 5   The Coulomb potential.   Consider a point charge q displaced from the origin. We may set up a coordinate system with the charge on the z axis, as shown in Fig. 5.13. What is the potential due to this displaced charge in spherical polar coordinates?

The potential $V(\mathbf{x})$ will be independent of the azimuthal angle ϕ of the field point $\mathbf{x}$, and it will satisfy Laplace's equation everywhere except at the singular point $\mathbf{x}'$ where q is located. By the completeness of Legendre polynomials, we can write the potential in spherical coordinates as a Legendre series

$$V(r, \theta) = \sum_{\ell=0}^{\infty} R_\ell(r, r') P_\ell(\cos \theta), \tag{5.75}$$

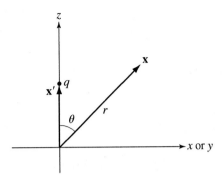

FIGURE 5.13   Example 5. The potential at **x** due to a point charge q at **x**$'$ on the z axis can be expanded in Legendre polynomials when expressed in spherical polar coordinates (r, θ). The distance from q to the field point is $|\mathbf{x} - \mathbf{x}'| = \sqrt{r^2 - 2rr'\cos\theta + r'^2}$.

where $r = |\mathbf{x}|$ and $r' = |\mathbf{x}'|$, and $R_\ell(r, r')$ is a function we will determine. Of course we know the potential; it is just the Coulomb potential

$$V(\mathbf{x}) = \frac{q}{4\pi\epsilon_0|\mathbf{x} - \mathbf{x}'|} = \frac{q}{4\pi\epsilon_0} \frac{1}{\sqrt{r^2 - 2rr'\cos\theta + r'^2}}. \qquad (5.76)$$

The question is, what are the radial functions $R_\ell(r, r')$ in the Legendre expansion? To answer the question we need a famous function in the theory of Legendre polynomials.

The Generating Function of Legendre Polynomials
We begin somewhat indirectly by defining, as a purely mathematical object, a function of two variables, by

$$g(u, s) = \sum_{\ell=0}^{\infty} P_\ell(u)s^\ell. \qquad (5.77)$$

In words, it is a power series in s whose coefficients are the Legendre polynomials. In general s may be complex, but that is not important here. The function $g(u, s)$ is called the generating function because the Legendre polynomials can be generated from it by differentiation; note that

$$\left.\frac{\partial^n g}{\partial s^n}\right|_{s=0} = n!\, P_n(u). \qquad (5.78)$$

If we find a simple expression for $g(u, s)$ then it determines in principle all the $P_\ell(u)$'s. Now, $g(u, s)$ satisfies a certain partial differential equation. By multiplying Legendre's equation (5.49) by s^ℓ and summing over ℓ, note that

$$\frac{\partial}{\partial u}\left[\left(1 - u^2\right)\frac{\partial g}{\partial u}\right] = -\sum_\ell \ell(\ell + 1)P_\ell(u)s^\ell; \qquad (5.79)$$

but the right-hand side may be reexpressed as

$$= -s\frac{\partial^2}{\partial s^2}\sum_\ell P_\ell(u)s^{\ell+1} = -s\frac{\partial^2}{\partial s^2}(sg). \tag{5.80}$$

Thus $g(u, s)$ obeys the partial differential equation

$$\frac{\partial}{\partial u}\left[(1 - u^2)\frac{\partial g}{\partial u}\right] + s\frac{\partial^2}{\partial s^2}(sg) = 0. \tag{5.81}$$

Also, there are some simple boundary values of $g(u, s)$. From (5.77) we see that

$$g(u, 0) = P_0(u) = 1,$$

$$\frac{\partial g}{\partial s}(u, 0) = P_1(u) = u,$$

$$g(1, s) = \sum_{\ell=0}^{\infty} s^\ell = \frac{1}{1 - s},$$

$$g(-1, s) = \sum_{\ell=0}^{\infty} (-1)^\ell s^\ell = \frac{1}{1 + s}. \tag{5.82}$$

For the last two equations $|s|$ must be < 1 in order that the series converge.

The solution to the equation (5.81) with the boundary values (5.82) is

$$g(u, s) = \frac{1}{\sqrt{1 - 2us + s^2}}. \tag{5.83}$$

To verify that this function satisfies (5.81) is straightforward. The domain of validity of the expansion (5.77) is $|s| < 1$, and $-1 \le u \le 1$ because we associate u with $\cos\theta$. The fact that $|s|$ must be < 1 will be important; if $|s| > 1$ then (5.77) is a divergent series.

The Legendre polynomials $P_\ell(u)$ and their generating function $g(u, s)$ are just one example in a general mathematics of orthogonal polynomials. Other examples, which a physics student will certainly encounter in quantum mechanics, are the Laguerre polynomials and the Hermite polynomials [5–7].

We are now ready to return to Example 5, and write the Coulomb potential in Legendre polynomials; that is, we will find the radial function $R_\ell(r, r')$ in (5.75). For $r < r'$,

$$\frac{1}{|\mathbf{x} - \mathbf{x}'|} = \frac{1}{r'}\frac{1}{\sqrt{1 - 2(r/r')\cos\theta + (r/r')^2}} = \frac{1}{r'}g(\cos\theta, r/r'). \tag{5.84}$$

We have factored out $1/r'$ so that the variable r/r' in the square root is < 1, which is the condition needed to write $g(\cos\theta, r/r')$ as the series in (5.77) with $s = r/r'$. Similarly, for $r > r'$, we have

$$\frac{1}{|\mathbf{x} - \mathbf{x}'|} = \frac{1}{r} \frac{1}{\sqrt{1 - 2(r'/r)\cos\theta + (r'/r)^2}} = \frac{1}{r} g(\cos\theta, r'/r). \qquad (5.85)$$

Now substitute the expansion (5.77) into (5.84) or (5.85). Comparing to (5.75) we see that the radial function $R_\ell(r, r')$ is, for $r < r'$,

$$R_\ell(r, r') = \frac{q}{4\pi\epsilon_0} \frac{r^\ell}{(r')^{\ell+1}}; \qquad (5.86)$$

and for $r > r'$,

$$R_\ell(r, r') = \frac{q}{4\pi\epsilon_0} \frac{(r')^\ell}{r^{\ell+1}}. \qquad (5.87)$$

Note that $R_\ell(r, r')$ is symmetric with respect to the interchange of r and r' because $|\mathbf{x} - \mathbf{x}'|$ has that symmetry. Our final expression for the Coulomb potential in terms of Legendre polynomials is

$$V(\mathbf{x}) = \frac{q}{4\pi\epsilon_0} \sum_{\ell=0}^{\infty} P_\ell(\cos\theta) \begin{cases} r^\ell/(r')^{\ell+1} & \text{for} \quad r \le r' \\ (r')^\ell/r^{\ell+1} & \text{for} \quad r \ge r'. \end{cases} \qquad (5.88)$$

The series is convergent for all points except $(r, \theta) = (r', 0)$, the singular point where q is located.

We derived (5.88) by taking $\mathbf{x}'$ to be on the z axis; then θ is the polar angle coordinate of $\mathbf{x}$ with respect to that axis. However, the equation holds for *arbitrary* $\mathbf{x}'$ and $\mathbf{x}$, i.e., independent of the choice of axes, if θ is the angle between the vectors $\mathbf{x}$ and $\mathbf{x}'$.

It was stated earlier that any azimuthally symmetric solution of Laplace's equation may be expanded in Legendre polynomials. Equation (5.88) is an example of that statement. Furthermore, $1/(4\pi|\mathbf{x} - \mathbf{x}'|)$ is the Green's function $G(\mathbf{x}, \mathbf{x}')$ of the Laplacian, which is used to construct solutions of Poisson's equation for a specified charge density.

5.3 ■ SEPARATION OF VARIABLES FOR CYLINDRICAL COORDINATES

When a problem has cylindrical symmetry the natural coordinate system to use is cylindrical coordinates r, ϕ, z.[15] This is appropriate if, for example, the equipotential surfaces or the surfaces on which the charge is known are cylinders. In these coordinates Laplace's equation, satisfied by the potential $V(\mathbf{x}) = V(r, \phi, z)$, is

$$\frac{1}{r} \frac{\partial}{\partial r} \left(r \frac{\partial V}{\partial r} \right) + \frac{1}{r^2} \frac{\partial^2 V}{\partial \phi^2} + \frac{\partial^2 V}{\partial z^2} = 0. \qquad (5.89)$$

[15]Whether to denote the angular variable by ϕ or by θ is a matter of convention. In this section we follow the notation of Chapter 2, and call it ϕ.

We will restrict our discussion to problems that are translationally invariant in the z direction; then V does does not depend on z.

5.3.1 ■ Separable Solutions for Cylindrical Coordinates

As before, we begin by considering the *separable* solutions. In cylindrical coordinates, with no z dependence, these take the form

$$V(r, \phi) = R(r)\Phi(\phi). \tag{5.90}$$

Substitute (5.90) into (5.89), and note that the term $\partial^2 V/\partial z^2$ is 0. Then multiply the result by $r^2/R(r)\Phi(\phi)$ in order to separate terms that depend only on r from those that depend only on ϕ. The result is

$$\frac{r}{R}\frac{d}{dr}\left(r\frac{dR}{dr}\right) + \frac{1}{\Phi}\frac{d^2\Phi}{d\phi^2} = 0. \tag{5.91}$$

This equation must hold for all r and ϕ, and these are independent variables, so each term must be constant. We also add the requirement that the potential be periodic in ϕ with period 2π, i.e., we require $V(r, \phi + 2\pi) = V(r, \phi)$. The periodicity implies we must set

$$\frac{1}{\Phi}\frac{d^2\Phi}{d\phi^2} = -n^2 \tag{5.92}$$

with n a nonnegative integer. The general solution for the ϕ dependence is

$$\Phi(\phi) = C_n \cos n\phi + D_n \sin n\phi, \tag{5.93}$$

where C_n and D_n are arbitrary constants.

We do not need to allow $n < 0$ because negative values are already included in the form (5.93). The case $n = 0$ is somewhat special. It would include $\Phi(\phi)$ a constant, which is relevant to problems with azimuthal invariance. A general solution with $n = 0$ is $\Phi = a\phi + b$, which is not periodic, but is relevant to some problems.[16]

Next, to find the form of $R(r)$ we replace the second term in (5.91) by $-n^2$. Then the radial equation may be rewritten in the form

$$r^2\frac{d^2R}{dr^2} + r\frac{dR}{dr} - n^2R = 0. \tag{5.94}$$

The general solution to this second-order ordinary differential equation is

$$R(r) = A_n r^n + B_n r^{-n} \tag{5.95}$$

[16]The potential $V(\phi) = V_0\phi/\phi_0$ is the potential in the wedge-shaped volume between the planes $\phi = 0$ and $\phi = \phi_0$ if the plane $\phi = 0$ is held at ground potential and the plane ϕ_0 is held at potential V_0. In this case the potential isn't required to be periodic because the range of ϕ is limited.

where A_n and B_n are additional arbitrary constants. To prove that this is the solution, just substitute it into (5.94). In a particular application, the constants A_n, B_n, C_n, and D_n will be determined by boundary conditions. The case $n = 0$ is again special. Its solution is $R = A \ln r + B$. We have met this important solution before as the potential of a uniform line charge.

Finally then we have the most general periodic solution for $V(r, \phi)$,

$$V(r, \phi) = A \ln r + B + \sum_{n=1}^{\infty} \left(A_n r^n + B_n r^{-n}\right) \left(C_n \cos n\phi + D_n \sin n\phi\right). \quad (5.96)$$

The ϕ dependent functions are called *zonal harmonics*, and they deserve some general discussion. For systems in which the potential is by some symmetry *even* in ϕ, we only use the functions $\cos n\phi$; that is, $D_n = 0$. Similarly, for systems in which the symmetry is *odd* in ϕ, we only use the functions $\sin n\phi$. If the system has neither symmetry, then use the linear combination $(C_n \cos n\phi + D_n \sin n\phi)$.

The terms with $n = 1$ and finite at the origin are proportional to $r \cos \phi$ and $r \sin \phi$, which describe uniform fields in the x direction and y direction, respectively. The field $\mathbf{E} = E_0 \hat{\mathbf{i}}$ has potential function $-E_0 r \cos \phi = -E_0 x$; similarly $\mathbf{E} = E_0 \hat{\mathbf{j}}$ has potential function $-E_0 r \sin \phi = -E_0 y$. Terms proportional to r^n are useful for interior regions, i.e., which include the origin, because r^n is nonsingular at $r = 0$. Terms proportional to r^{-n} are useful for exterior regions, i.e., which include $r = \infty$. In an annular region $R_1 \leq r \leq R_2$, which excludes $r = 0$ and $r = \infty$, the linear combination $\left(A_n r^n + B_n r^{-n}\right)$ must be used.

EXAMPLE 6 Figure 5.14(a) shows two half-cylinder shells that are the right and left halves of a cylinder with radius R. The shells' thickness is negligible compared to R, and they are made of a conducting material. They are separated from each other at $\phi = \pi/2$ and $\phi = 3\pi/2$ by small insulating gaps. The left half, for which $\pi/2 < \phi < 3\pi/2$, is held at potential $-V_0$, and the right half, which has $0 < \phi < \pi/2$ or $3\pi/2 < \phi < 2\pi$, is held at $+V_0$. We will find the potential function $V(r, \phi)$ inside and outside the cylinder.

We start with (5.96) but needn't use all the terms. Because of the left-right *antisymmetry*, the net charge on the cylinder is 0, and this means the term $A \ln r$ does not appear in $V(r, \phi)$. The system has even symmetry in ϕ, so the terms with $\sin n\phi$ do not appear. In the interior the terms $\propto r^{-n}$ do not appear, because they would be singular at $r = 0$, and in the exterior the terms $\propto r^n$ do not appear. Thus, for the interior $(r \leq R)$ and exterior $(r \geq R)$, respectively, the potential is

$$V_{\text{int}}(r, \phi) = \sum_{n=1}^{\infty} A_n r^n \cos n\phi \quad (5.97)$$

$$V_{\text{ext}}(r, \phi) = \sum_{n=1}^{\infty} B_n r^{-n} \cos n\phi. \quad (5.98)$$

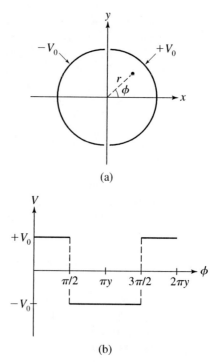

(a)

(b)

FIGURE 5.14   Example 6. (a) Two thin conducting half cylinders, the right held at potential $+V_0$ and the left at potential $-V_0$. The shells are separated by insulating gaps at $\phi = \pi/2$ and $\phi = 3\pi/2$. (b) The potential at the surface $r = R$ of the bisected cylinder.

Here is an often-used mathematical shortcut. (We used it in spherical coordinates in Example 3, and now we'll use it in cylindrical coordinates.) At $r = R$ we have the boundary condition $V_{\text{int}}(R, \theta) = V_{\text{ext}}(R, \theta)$, which implies $A_n R^n = B_n R^{-n} \equiv c_n$, a new constant for each n. Then the series take the more convenient form

$$V_{\text{int}}(r, \phi) = \sum_{n=1}^{\infty} c_n \left(\frac{r}{R}\right)^n \cos n\phi \qquad (5.99)$$

$$V_{\text{ext}}(r, \phi) = \sum_{n=1}^{\infty} c_n \left(\frac{R}{r}\right)^n \cos n\phi. \qquad (5.100)$$

On the surface of the cylinder,

$$V(R, \phi) = \sum_{n=1}^{\infty} c_n \cos n\phi. \qquad (5.101)$$

Equation (5.101) is a Fourier expansion of the periodic potential shown in Fig. 5.14(b). We obtain the coefficients c_n in the usual way from an orthogonality relation. Multiplying both sides of (5.101) by $\cos m\phi$ and integrating from $\phi = 0$ to $\phi = 2\pi$ gives

$$\int_0^{2\pi} V(R, \phi) \cos m\phi \, d\phi = c_m \pi \tag{5.102}$$

because the only term in the series that contributes to the integral is the term with $n = m$, by the orthogonality relation

$$\int_0^{2\pi} \cos n\phi \, \cos m\phi \, d\phi = \pi \delta_{nm}. \tag{5.103}$$

Now we substitute the function $V(R, \phi)$, which is piecewise constant ($-V_0$ for ϕ in $(\pi/2, 3\pi/2)$ and $+V_0$ in the rest of the domain of integration), and after a little algebra we find

$$c_m = \frac{4V_0}{\pi m} \sin \frac{m\pi}{2}. \tag{5.104}$$

Thus c_m is 0 if m is even; the sign is $+$ for $m = 1, 5, 9, \ldots$, or the sign is $-$ for $m = 3, 7, 11 \ldots$. To simplify the notation we will write $n = 2j + 1$ and sum over j in the series (5.99) and (5.100).

The final series for the complete potential are

$$V_{\text{int}}(r, \phi) = \frac{4V_0}{\pi} \sum_{j=0}^{\infty} \frac{(-1)^j}{2j + 1} \left(\frac{r}{R}\right)^{2j+1} \cos(2j + 1)\phi \tag{5.105}$$

$$V_{\text{ext}}(r, \phi) = \frac{4V_0}{\pi} \sum_{j=0}^{\infty} \frac{(-1)^j}{2j + 1} \left(\frac{R}{r}\right)^{2j+1} \cos(2j + 1)\phi. \tag{5.106}$$

5.4 ■ CONJUGATE FUNCTIONS IN 2 DIMENSIONS

An electrostatics problem in which the potential is translationally invariant in one direction, say the z direction, is called a two-dimensional (2D) problem. The potential is then $V(x, y)$ and Laplace's equation is

$$\frac{\partial^2 V}{\partial x^2} + \frac{\partial^2 V}{\partial y^2} = 0. \tag{5.107}$$

Although (5.107) is an idealization—it implies the system is infinite in the z direction—it is a good approximation for many real electrostatic systems in which the extent of one dimension (z) is much larger than the others (x, y).

Some 2D problems can be solved by separation of variables, e.g., in Cartesian or polar coordinates. However, there is another approach for 2D problems, called the method of *conjugate functions*, that solves Laplace's equation by combining x and y in a complex variable. This section describes the method and applies it to several problems in which we find $V(x, y)$ for charged 2D conductors. The mathematics is intricate, but it yields interesting results.

The central idea is to find two functions $g(x, y)$ and $h(x, y)$ that are conjugate to one another—a relationship we will explore. Once found, one function is the potential and the other describes the field lines, which are everywhere perpendicular to the equipotentials as always. The method is based on the mathematical theory of functions of a complex variable, and it brings to bear on electrostatics some of the mathematical elegance and power of this large subject [1,2].

The general 2D problem is to find $V(x, y)$ for specified boundary conditions. Picture the coordinate plane as the plane of complex numbers, by writing[17]

$$z = x + iy \tag{5.108}$$

where $i = \sqrt{-1}$. In words, z is a complex number with real part x and imaginary part y. The method of conjugate functions is based on the following theorem: *If $F(z)$ is an analytic function of $z = x + iy$, then Re $F(z)$ and Im $F(z)$ satisfy Laplace's equation* (5.107). The theorem provides a method for generating solutions of Laplace's equation. To solve a given boundary value problem it is "only" necessary to find the right function $F(z)$, namely one whose real part or imaginary part satisfies the boundary conditions. This may not be as easy as it sounds, and indeed the main difficulty is to find the appropriate $F(z)$.

The theorem is not difficult to prove. Write $F(z) = g(x, y) + ih(x, y)$, where g and h are real functions. We need to show that g and h are solutions of (5.107). By the chain rule of differentiation,

$$\frac{\partial F}{\partial x} = \frac{dF}{dz} \quad \text{and} \quad \frac{\partial F}{\partial y} = i\frac{dF}{dz}; \tag{5.109}$$

similarly,

$$\frac{\partial^2 F}{\partial x^2} = \frac{d^2 F}{dz^2} \quad \text{and} \quad \frac{\partial^2 F}{\partial y^2} = -\frac{d^2 F}{dz^2}. \tag{5.110}$$

Thus,

$$\nabla^2 F = \frac{\partial^2 F}{\partial x^2} + \frac{\partial^2 F}{\partial y^2} = \frac{d^2 F}{dz^2} - \frac{d^2 F}{dz^2} = 0; \tag{5.111}$$

that is, $F(z)$ satisfies Laplace's equation. Because ∇^2 is a real operator, both the real part $g(x, y)$ and the imaginary part $h(x, y)$ of $F(z)$ satisfy Laplace's equation.

We may use either $g(x, y)$ or $h(x, y)$ as the potential of an electrostatics problem, provided it satisfies the boundary conditions. Whichever we use, the *other*

---

[17]It requires some mental agility to not confuse the complex variable $z = x + iy$ with the Cartesian coordinate z. But the latter z never appears in a 2D problem.

function will tell us the electric field lines. Suppose $g(x, y)$ is the potential. Then the equipotentials are curves on which $g(x, y)$ is constant. We shall prove that the curves on which $h(x, y)$ is constant are the electric field lines.

Derivatives of $g(x, y)$ and $h(x, y)$ are intimately related because $F(z)$ is not a function of x and y independently, but of the special combination $x + iy$. Note that

$$\frac{\partial F}{\partial x} = \frac{\partial g}{\partial x} + i\frac{\partial h}{\partial x} = \frac{dF}{dz}, \tag{5.112}$$

$$\frac{\partial F}{\partial y} = \frac{\partial g}{\partial y} + i\frac{\partial h}{\partial y} = i\frac{dF}{dz}. \tag{5.113}$$

Multiplying the second equation by $-i$, and then equating the real and imaginary parts of the result to those of the first equation, we see that

$$\frac{\partial g}{\partial x} = \frac{\partial h}{\partial y} \quad \text{and} \quad \frac{\partial g}{\partial y} = -\frac{\partial h}{\partial x}. \tag{5.114}$$

These coupled partial differential equations are called the *Cauchy-Riemann equations*. Any pair of functions $g(x, y)$ and $h(x, y)$ that satisfy them are called *conjugate functions*. The real and imaginary parts of a complex function of z are always conjugates.

If $g(x, y)$ is the potential of an electrostatics problem, then the equation for an equipotential curve is $g(x, y) = g_0$, and so the displacements dx and dy along the curve are related by

$$dg = \frac{\partial g}{\partial x}dx + \frac{\partial g}{\partial y}dy = 0; \tag{5.115}$$

the slope of the curve in the xy plane is

$$\left(\frac{dy}{dx}\right)_g = -\frac{\partial g/\partial x}{\partial g/\partial y}. \tag{5.116}$$

Similarly, the slope of the curve $h(x, y) = h_0$, a constant, is

$$\left(\frac{dy}{dx}\right)_h = -\frac{\partial h/\partial x}{\partial h/\partial y} = +\frac{\partial g/\partial y}{\partial g/\partial x}, \tag{5.117}$$

where the second equality follows from the Cauchy-Riemann equations. Comparing the two slopes shows that the two curves are orthogonal where they cross, because the slope of one is the negative reciprocal of the slope of the other:

$$\left(\frac{dy}{dx}\right)_h = \frac{-1}{(dy/dx)_g}. \tag{5.118}$$

In conclusion, the family of curves with $h(x, y)$ constant are the electric field lines—curves everywhere orthogonal to the equipotentials.

We will now apply these ideas to three problems that are physically interesting, and are quite different from any problems studied so far in that they have less symmetry. The solutions are far from obvious, but can be found with conjugate functions.

■────────────

EXAMPLE 7   What is the potential function $V(x, y)$ for a region bounded by charged conducting plates that intersect at a right angle? The plates are in contact, so they have the same potential.

The geometry is shown in Fig. 5.15. The half of the xz plane ($y = 0$) with $x \geq 0$ is one boundary, and the half of the yz plane ($x = 0$) with $y \geq 0$ is the other boundary. Assume that the plates extend a large distance in z, so that the 2D approximation is valid. The boundary condition is that $V(x, y)$ is a constant on the two half planes, because the plates are conductors. We will specify $V = \Phi_0$ on the boundary, but the precise value of Φ_0 is not particularly relevant because we can always add a constant to the potential.

The complex function that solves this example is

$$F(z) = Az^2 + i\Phi_0 = A(x^2 - y^2) + 2iAxy + i\Phi_0, \qquad (5.119)$$

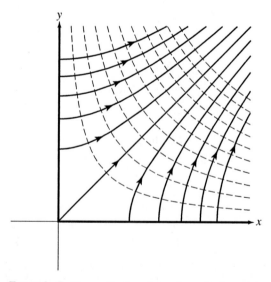

FIGURE 5.15   Example 7: Perpendicular charged conducting plates. The system is translation invariant in the z direction. Ideally the plates extend to ∞ in the positive x and y directions. The plates, at potential Φ_0, are half-planes shown on the positive x and y axes. The solid curves are electric field lines and the dashed curves are equipotentials.

where we use the imaginary part for the potential

$$V(x, y) = 2Axy + \Phi_0. \tag{5.120}$$

To motivate this choice of $F(z)$, note that z^2 is real on both the x and y axes (but nowhere else) so Im F is a constant, Φ_0, on the axes. But that's just the boundary condition we want for V. So by taking $V = $ Im F we have $V(0, y) = V(x, 0) = \Phi_0$ as required. The constant A will be interpreted later.

The equipotentials are curves with $xy = $ constant, which are hyperbolas with asymptotes along the axes (the dashed curves in Fig. 5.15). The electric field lines are curves with $x^2 - y^2 = $ constant, which are hyperbolas with asymptotes at $45°$ to the axes (the solid curves in Fig. 5.15).

The constant A is related to the surface charge density σ on the conductor. On the boundary σ is $\epsilon_0 E_n$. For example, on the horizontal conducting plate ($y = 0$) the surface charge density is

$$\sigma(x) = \epsilon_0 E_y(x, 0) = -2A\epsilon_0 x. \tag{5.121}$$

The density is 0 at the intersection of the plates, and increases in magnitude linearly with x. The charge on the vertical plate is analogous. Negative A corresponds to positive charge on the plates; in this case the electric field lines flow away from the conductors, terminating on negative charges.

Previously, we studied a charged plane. In that most symmetric problem **E** is uniform, in the normal direction, and σ is constant. Now imagine that the plane is bent in a right angle. The result is the problem we have just solved. The field and charge density are nonuniform.

We have taken the plates to be infinite in the positive x and y directions, which is why the total charge is infinite. In any real system, the extension of the plates in the x and y directions would be finite, the amount of charge would be finite, and the potential would approach 0 at infinity. The constant Φ_0 that we are using in this example is the potential difference between the plates and infinity. For finite plates, Fig. 5.15 would be accurate in the region near the intersection and far from the outer edges.[18]

EXAMPLE 8   What is the potential function $V(x, y)$ near the edge of a charged rectangular conducting plate?

The geometry is shown in Fig. 5.16. The plate lies on the half of the xz plane ($y = 0$) with $x \geq 0$. The potential must be a constant Φ_0 on this boundary, because it is a conductor. The complex function that solves this example is[19]

---

[18]See Exercise 19.
[19]For mavens of complex analysis, the branch cut of $z^{1/2}$ is along the positive real axis. See [1].

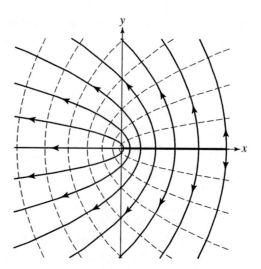

FIGURE 5.16   Example 8: Electric potential and field at the edge of a charged conducting plate. The plate is the half-plane $y = 0$, $x \geq 0$, shown on the positive x axis. The solid curves are electric field lines and the dashed curves are equipotentials.

$$F(z) = Az^{1/2} + i\Phi_0, \qquad (5.122)$$

where the imaginary part of $F(z)$ is the potential. To motivate this choice of $F(z)$, note that $z^{1/2}$ is purely real on the positive x axis, so Im F is a constant, Φ_0, there. That is exactly the boundary condition we want for V.

Some complex analysis is required to express $F(z)$ as a function of x and y. Write $z = re^{i\theta}$ where $\tan\theta = y/x$, and $r = \sqrt{x^2 + y^2}$; then

$$F(z) = Ar^{1/2}e^{i\theta/2} + i\Phi_0$$

$$= A(x^2 + y^2)^{1/4}[\cos\theta/2 + i\sin\theta/2] + i\Phi_0 \qquad (5.123)$$

$$= g(x, y) + iV(x, y). \qquad (5.124)$$

Using trigonometric identities it can be shown that

$$V(x, y) = \frac{A}{\sqrt{2}}\left(\sqrt{x^2 + y^2} - x\right)^{1/2} + \Phi_0 \qquad (5.125)$$

$$g(x, y) = \frac{A}{\sqrt{2}}\left(\sqrt{x^2 + y^2} + x\right)^{1/2}. \qquad (5.126)$$

Note that the potential $V(x, 0)$ on the xz plane ($y = 0$) is Φ_0 for $x \geq 0$, but $A\sqrt{|x|} + \Phi_0$ for $x \leq 0$. The equipotential curves are parabolas, shown in Fig. 5.16

as dashed curves. The electric field lines are also parabolas, shown as solid curves. Because we have taken the plate to be infinite in the $+x$ direction, y increases monotonically with x on an equipotential curve. For a plate that is finite in the x direction, i.e., a flat conducting strip along the z axis, the equipotential would curve back to the xz plane on the other side of the plate. The function $F(z)$ of (5.123) is for a semi-infinite plate. The problem of a finite strip can also be solved by conjugate functions, but that is more complicated.

It is interesting to calculate the surface charge density on the two sides of the conductor, from $\sigma = \epsilon_0 E_n$, where E_n is $-\partial V/\partial y$ on the upper surface and $+\partial V/\partial y$ on the lower. The charge density on the upper surface is

$$\sigma(x) = \epsilon_0 \lim_{y \to 0+} E_y(x, y) = -\frac{A\epsilon_0}{\sqrt{4x}}. \tag{5.127}$$

Some care is required in taking the limit. The charge density on the lower surface is the same. The calculated charge density approaches infinity at the edge of the plate $x = 0$, because we have neglected the thickness of the plate. For a plate of nonzero thickness the charge density would level off at a finite value as x decreases to 0. The relation $\sigma \propto x^{-1/2}$ would be a good approximation for a plate of finite length in the x direction (a strip) near the edge at $x = 0$.

For an infinite charged plane—a highly symmetric system we studied previously—there are no edge effects because there are no edges. Here we have *calculated the edge effects* for a semi-infinite plane!

EXAMPLE 9   *Fringing field of a parallel plate capacitor.*   Our final example is more complicated mathematically. Consider the function[20]

$$F(z) = \frac{V_0}{i\pi} \ln w = V(x, y) + ih(x, y), \tag{5.128}$$

where z is related to w by

$$z = \frac{d}{\pi} \left[\ln w + \frac{1}{2}(1 - w^2) \right]. \tag{5.129}$$

Both w and z are complex variables. As usual z is $x + iy$. The other complex plane corresponds to $w = u + iv$, with horizontal axis u and vertical axis v. The transformation $w \to z$ is an example of *conformal mapping* from the (u, v) plane to the (x, y) plane.

In this example the *real part* of $F(z)$ is the potential function $V(x, y)$. There is no way to solve for w as an explicit function of z, so the functional relationship between F and z, or between V and x, y, is parametric. Nevertheless we can

---

[20]The branch cut of $\ln w$ is along the positive real axis. The branch cut of a function of a complex variable is a curve along which the function has a discontinuity.

analyze the dependence and deduce what boundary value problem has $F(z)$ as the solution. In a way we are working backwards, from the solution to the problem.[21]

We need to connect the coordinates x and y to the potential $V(x, y)$ and the intensity function $h(x, y)$. Write $w = \rho e^{i\theta}$. Then using $\ln w = \ln \rho + i\theta$ and (5.129) it is straightforward to express x and y as

$$x = \frac{d}{\pi} \left[\ln \rho + \frac{1}{2} \left(1 - \rho^2 \cos 2\theta \right) \right], \tag{5.130}$$

$$y = \frac{d}{\pi} \left[\theta - \frac{1}{2} \rho^2 \sin 2\theta \right]. \tag{5.131}$$

The potential and intensity functions may also be written in terms of ρ and θ, as

$$V = V_0 \frac{\theta}{\pi} \quad \text{and} \quad h = -\frac{V_0}{\pi} \ln \rho. \tag{5.132}$$

Through these equations we have a parametric relation between (x, y) and V.

The range of θ is 0 to 2π, and the range of ρ is 0 to ∞. Curves in the xy plane with constant ρ have constant $h(x, y)$; these curves are the electric field lines, shown as solid curves in Fig. 5.17. These curves are generated by fixing ρ in (5.130) and (5.131) and varying θ from 0 to 2π; the point (x, y) traces out a

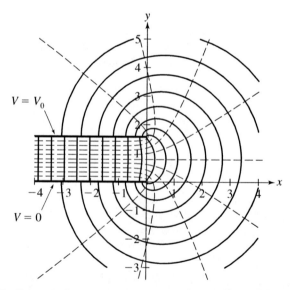

FIGURE 5.17   Example 9: Electric potential and field at the edge of a parallel plate capacitor. The solid curves are electric field lines and the dashed curves are equipotentials.

[21] There is no simple way to motivate the function $F(z)$ so this is an "exploratory" example. However, the conformal map (5.129) can be analyzed (Exercise 23) to see why it is related to parallel plates.

curve. Curves with constant θ are the equipotentials, shown as dashed curves in Fig. 5.17, generated in a similar way. The plots of curves in Fig. 5.17 are examples of parametric plots; y and x depend parametrically on the variables ρ and θ. As ρ varies with θ fixed, the point (x, y) follows an equipotential in the xy plane; as θ varies with ρ fixed, the point (x, y) follows an electric field line.

Points with $y = 0$ and $x \leq 0$ have parametric variables $\theta = 0$ and $0 \leq \rho < \infty$; i.e., w is real and positive. The potential $V(x, y)$ on this boundary is 0 by (5.132). Points with $y = d$ and $x \leq 0$ correspond to parameter values $\theta = \pi$ and $0 \leq \rho < \infty$; in this case w is real and negative. The potential $V(x, y)$ on this surface is V_0 by (5.132).

Now we can see what electrostatics problem is solved by the function $F(z)$ in (5.128). The system is a parallel plate capacitor with potential difference V_0, for which the plates are half-planes parallel to the xz plane, at $y = 0$ and $y = d$. The edge of the capacitor is at $x = 0$, and the plates extend to $-\infty$ in the negative x direction. The result in Fig. 5.17 is a good approximation for finite plates if their length is much larger than the separation d.

In our previous calculations involving parallel plate capacitors we neglected the edge effects. Here, however, we are calculating the edge effects! In the region deep between the plates ($x \ll -d$) the electric field is approximately constant $-\hat{\mathbf{j}} V_0/d$. Near the end ($x = 0$) however, the electric field bulges out into the unbounded region, as illustrated in Fig. 5.17. The field at the edge is called the *fringing field.* Figure 5.17 illustrates the directions of the fringing field but not the magnitudes. The field on the midplane, $y = d/2$, is

$$E_y = -\partial V/\partial y = +\partial h/\partial x. \tag{5.133}$$

It can be shown[22] that $E_y(x, d/2)$ is proportional to $1/x$ as $x \to \infty$.

The mathematics of conjugate functions is highly developed, both for its practical applications to 2D potential problems and for its mathematical beauty. Conformal maps such as the Schwarz-Christoffel transformations solve a great variety of interesting problems [2].

■────────

Examples 7–9 have presented, without much motivation, the functions that solve Laplace's equation for the appropriate 2D boundary conditions. The reader may well object that the solutions in Examples 7–9 were "pulled out of a hat like a magician's rabbit" rather than constructed logically. Of course the uniqueness theorem guarantees that the functions we have written are the correct, unique solutions. But where did they come from?

There are formal techniques for finding the $F(z)$ whose real or imaginary part gives the potential for a given 2D problem. These techniques have a certain kind of beauty, but they are beyond the scope of this book.[23] One trick is to construct a

---

[22] See Exercise 39.
[23] A famous discussion is Chapter IV of Smythe, Ref. 2.

conformal transformation. The idea is to start with a 2D problem whose solution is known and then transform it, using conformal mapping, into the arrangement of conducting boundaries in the problem to be solved. Dictionaries of conformal maps can be consulted.[24] For example, one 2D problem whose solution is well known is the potential of an infinite conducting plane. If one folds the plane to an angle of 90°, the new boundary is that of Example 7. The configuration of Example 8 results from folding the plane back on itself, i.e., by 180°.

Although formal methods for constructing solutions exist, perhaps it is worth repeating that in physics many results owe more to inspired guess-work than to routine calculation.

5.5 ■ ITERATIVE RELAXATION: A NUMERICAL METHOD

The use of computers is now ubiquitous in physics, so our discussion of Laplace's equation would not be complete if it did not include a numerical method. In fact there are many numerical techniques for solving boundary value problems on computers. We will consider one method, in which the potential is calculated approximately on a lattice of discrete points by solving a finite difference approximation of the partial differential equation. The method is not as powerful as other, more advanced methods, but it is simple and it illustrates the strengths and weaknesses of numerical calculations. An interested student should consult Refs. 3 and 4 to learn about other numerical methods.

For simplicity we limit the calculations to Laplace's equation in two dimensions, for $V(x, y)$. The generalization to three dimensions is obvious. We start by setting up a square lattice network of discrete points ("sites") and connecting line segments ("links") that covers the region of interest where $\nabla^2 V = 0$. The lattice is oriented with the x and y axes. The sites must be dense enough that the potential at any point (x, y) can be well approximated by interpolation between the values of V at nearby sites. Let ϵ be the distance between neighboring sites, in either the x or y direction. Then the sites are labeled by two integers m and n, with the position of the site (m, n) at

$$\mathbf{x}_{mn} = \hat{\mathbf{i}} m\epsilon + \hat{\mathbf{j}} n\epsilon. \tag{5.134}$$

The limits on m and n depend on the boundary of the region of interest. For example, for a square with $-a/2 \leq x \leq a/2$ and $-a/2 \leq y \leq a/2$ the integers m and n would run from $-N$ to N, where $N = \frac{1}{2}a/\epsilon$, as shown in Fig. 5.18. We denote by V_{mn} the potential at $\mathbf{x}_{mn}$.

The Laplacian ∇^2 is a differential operator. In the continuum it involves second derivatives, but on the lattice it will be approximated by a finite difference operation. The difference between values of V at points separated by $\epsilon \hat{\mathbf{i}}$ is approximately

[24]H. Kober, *Dictionary of Conformal Representations* (Dover, New York, 1952); V.I. Ivanov and M.K. Trubetskov, *Handbook of Conformal Mapping* (CRC Press, Boca Raton, FL, 1995).

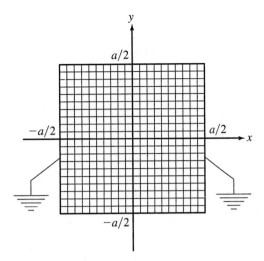

FIGURE 5.18   A square lattice. In Example 10 the sides at $x = \pm a/2$ are grounded ($V = 0$) and the sides at $y = \pm a/2$ are held at potential V_0.

$$V(\mathbf{x} + \epsilon\,\hat{\mathbf{i}}) - V(\mathbf{x}) \approx \epsilon\frac{\partial V}{\partial x}. \tag{5.135}$$

The right-hand side is the first term in a Taylor series. In this way the first derivative is related approximately to the finite difference. The second derivative may be approximated by a difference of differences, based on the approximate relation

$$\left[V(\mathbf{x} + \epsilon\,\hat{\mathbf{i}}) - V(\mathbf{x})\right] - \left[V(\mathbf{x}) - V(\mathbf{x} - \epsilon\,\hat{\mathbf{i}})\right] \approx \epsilon^2\frac{\partial^2 V}{\partial x^2}, \tag{5.136}$$

where the right-hand side is the first nonzero term in a Taylor series of the left-hand side. Taking $\mathbf{x}$ to be the position $\mathbf{x}_{mn}$ of site (m, n), Eq. (5.136) becomes

$$V_{m+1,n} - 2V_{mn} + V_{m-1,n} = \epsilon^2\frac{\partial^2 V}{\partial x^2}; \tag{5.137}$$

similarly, for differences in the y direction,

$$V_{m,n+1} - 2V_{mn} + V_{m,n-1} = \epsilon^2\frac{\partial^2 V}{\partial y^2}. \tag{5.138}$$

Laplace's equation is $\partial^2 V/\partial x^2 + \partial^2 V/\partial y^2 = 0$. If we substitute the finite difference approximation for the second derivatives then the partial differential equation becomes the *finite difference equation*

$$V_{m+1,n} + V_{m-1,n} + V_{m,n+1} + V_{m,n-1} = 4V_{mn}. \tag{5.139}$$

According to (5.139) *the potential at* $\mathbf{x}_{mn}$ *is equal to the average of the potential at the four nearest neighbor sites of the lattice.* The method of iterative relaxation makes use of this property of Laplace's equation to generate the solution. Start with a set of initial estimates for the potential values V_{mn} at the lattice sites $\mathbf{x}_{mn}$. Sites on the boundary must be assigned their specified values. Now move around the interior of the space, stopping at each lattice site exactly once; at each stop replace the current potential value V_{mn} by the average of the values at the neighboring sites. Readjusting the values V_{mn} in this way for all the interior sites is called one iteration. After many iterations the potential values converge to the solution of the difference equation (5.139).

It is clear that V_{mn} only approaches the solution in the limit of many iterations. The readjustment at one site (m, n) brings it into temporary agreement with the neighboring sites. But each of the neighboring sites will be changed later, during either the same iteration or the next one, after which V_{mn} will no longer agree with its neighbors. But the disagreement decreases as the number of iterations increases. When an iteration produces negligible changes in the potential values, then the process has converged to a satisfactory approximation of the solution to the difference equation (5.139).

The process is called *relaxation* because it is an approach to equilibrium. In the continuum, the state of static equilibrium occurs when $\nabla^2 V = 0$, with the boundary conditions satisfied. It can be shown[25] that the integral $\int (\nabla V)^2 d^3 x$ has its minimum value when $V(\mathbf{x})$ satisfies Laplace's equation, always keeping the boundary conditions satisfied. That is, equilibrium is the state with minimum electric field energy. Similarly, for the discrete lattice approximation the sum defined by

$$S = \sum_{m,n} \left[\left(V_{m+1,n} - V_{mn} \right)^2 + \left(V_{m,n+1} - V_{mn} \right)^2 \right] \qquad (5.140)$$

is minimum when V satisfies the difference equation (5.139). Each step in an iteration lowers the value of S, so during the iterative process the potential *relaxes* to the equilibrium state.

The iteration generates a solution to the finite difference approximation of Laplace's equation. How well that approximates the continuum potential depends on ϵ (the lattice spacing) that must be small compared to any length scale over which the potential varies by a significant amount. The smaller ϵ, the better the approximation; but also the longer the calculation, because each iteration must visit each site, and the number of sites is proportional to $1/\epsilon^2$ (in two dimensions). Numerical calculations require *compromises* between accuracy and speed.

The basic idea (5.139) that $V(\mathbf{x})$ equals the average of V at nearby sites, in the limit of small ϵ, can be extended to three dimensions. The analogous property in *one dimension* is particularly simple: The equation in one dimension (x) is $d^2 V/dx^2 = 0$, with general solution $V = ax + b$. This has the property that $V(x) = \frac{1}{2} [V(x - \delta) + V(x + \delta)]$ for any δ. In words, $V(x)$ equals the average

---

[25] See Exercise 24.

of V at the endpoints of any interval centered at x. There is a similar theorem for three dimensions: In a charge-free region of space, V at any point P equals the average of $V(\mathbf{x})$ over the surface of any sphere with P at the center.[26]

EXAMPLE 10   As an example of the relaxation method, we will calculate again the potential $V(x, y)$ in the infinitely long hollow square conducting pipe shown in Fig. 5.1. We constructed an analytic solution for this problem in Example 1 (the infinite series in (5.31)) so we shall be able to check the accuracy of the numerical method.

A lattice network for this problem is shown in Fig. 5.18. For this computer calculation the lattice spacing ϵ is taken to be $0.01a$ (much finer than the schematic mesh shown in Fig. 5.18) where a is the side of the square. The coordinates x and y vary from $-a/2$ to $a/2$, so the domains of the site labels m and n are the integers from -50 to 50. The boundary values in terms of site labels are

$$V_{-50,n} = V_{+50,n} = 0$$
$$V_{m,-50} = V_{m,+50} = V_0 = 1, \tag{5.141}$$

where we use V_0 as the unit of potential.

As a starting configuration we set $V_{mn} = 0.5$ at all interior points. The graph on the right of Figure 5.19 is the quantity S of (5.140) as a function of iteration number, for 1000 iterations. As the V_{mn} converge to the solution of (5.139), S decreases monotonically. The graph shows that it takes at least 1000 iterations to converge to a good approximation of the solution. We could follow other variables to double-check the convergence. Figure 5.20 shows $V_{m,0}$ as a function of m and $V_{0,n}$ as a function of n after 1000 iterations. These are the discrete approx-

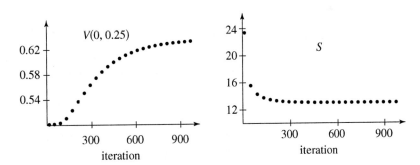

FIGURE 5.19   Convergence of the relaxation for Example 10. The graph on the left shows the variation of $V(0, 0.25)$ as the iteration proceeds for 1000 relaxation steps; the graph on the right shows the quantity S of (5.140), which decreases monotonically as the iteration proceeds.

[26]See Exercise 25.

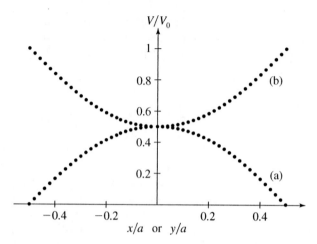

FIGURE 5.20   Potential in the square pipe calculated by the method of iterative relaxation. The potential values at every other lattice site (n even) along (a) the x axis, and (b) the y axis, are plotted.

imations of the potential on the x and y axes. Note that the lattice approximation agrees accurately with the saddle shape shown in Fig. 5.5, the result of the analytic calculation.

In some ways the numerical solution is superior to the analytic solution. The latter is still an infinite series, and it is nontrivial to evaluate $V(x, y)$ from this series, especially if x or y approaches the boundary. In a real application of electrostatics, having an array of numerical values of $V(\mathbf{x})$ on a dense set of points in the space, between which V may be estimated by interpolation, may be more useful than having a slowly convergent infinite series.

EXAMPLE 11   The real power of numerical methods is that, unlike analytic methods, they are not limited, practically, to highly symmetric systems. For example, we now relax the system shown in Fig. 5.21, the cross section of a conducting pipe crimped at one corner. Again we'll set $V = 0$ on the walls at $x = \pm a/2$, and $V = V_0 = 1$ on the other boundaries. Although it may be possible to solve Laplace's equation analytically with these boundary conditions, it would be very complicated. The numerical calculation is almost as simple as the earlier symmetric example. The only change is a slight modification in defining the boundary points around the crimp.

Figure 5.22 shows the potential values $V_{m,0}$ and $V_{0,n}$ where again m and n run from -50 to 50, after 1000 iterations. Having seen that the relaxation method performs well in the symmetric case, where we can compare it to analytic results, we may be confident that the relaxation method is accurate in the asymmetric case for which we have no analytic solution.

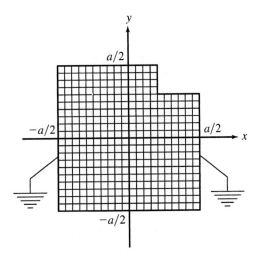

FIGURE 5.21 **An asymmetric boundary.** In Example 11 the sides at $x = \pm a/2$ are grounded ($V = 0$) and the other boundaries are held at potential V_0. The top boundary consists of three parts: $y = 0.5a$ for $-0.5a < x \le 0.2a$, $0.3a \le y \le 0.5a$ for $x = 0.2a$, and $y = 0.3a$ for $0.2a \le x < 0.5a$.

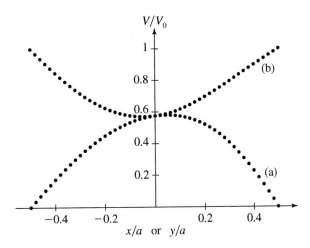

FIGURE 5.22 **Potential in the asymmetric pipe** calculated by the method of iterative relaxation. The potential values at every other lattice site (n even) along (a) the x axis, and (b) the y axis, are plotted.

Many subtle and difficult issues are associated with numerical methods. For example, any numerical method is an approximation, so it is important to estimate the *error* of the approximation. Despite such problems, the computer has replaced the classic analytic techniques as the best way to calculate complex fields for real applications encountered in engineering and experimental physics. This does not mean, though, that analytic methods are obsolete. Any computer code must be *verified* for correctness and accuracy. The best verification is to run the program for examples with known solutions, and obtain the right values. Computers extend our ability to do calculations, but that only *increases* the importance of analytic results.

■────────────

FURTHER READING

1. Tristan Needham, *Visual Complex Analysis* (Clarendon Press, Oxford, 1997).

2. W.R. Smythe, *Static and Dynamic Electricity*, 3rd ed. (McGraw-Hill, New York, 1968).

3. M.N.O. Sadiku, *Numerical Techniques in Electromagnetics* (CRC Press, Boca Raton, FL, 1992).

4. K.J. Binns, P.J. Lawrenson, and C.W. Trowbridge, *The Analytical and Numerical Solution of Electromagnetic Fields* (Wiley, Chichester, 1992).

5. M. Abramowitz and I.A. Stegun, *Handbook of Mathematical Functions* (U.S. Government Printing Office, Washington, 1972).

6. E. Jahnke, F. Emde, and F. Losch, *Tables of Higher Functions*, 6th ed. (McGraw-Hill, New York, 1960).

7. M.R. Spiegel, *Mathematical Handbook*, Schaum's Outline Series (McGraw-Hill, New York, 1968).

EXERCISES

Sec. 5.1. Separation of Variables for Cartesian Coordinates

5.1. (a) What choice of the parameters in (5.13) corresponds to the uniform field $\mathbf{E} = E_0\hat{\mathbf{k}}$?

   (b) Describe the equipotentials for (5.13) if $a = c$ and f are nonzero, and all other parameters are 0.

5.2. Prove the orthogonality relations in (5.24), which are important in Fourier analysis.

5.3. For the long square conducting pipe in Example 1, suppose that the potential on the sides at $x = \pm a/2$ is 0, and the potential on the sides at $y = \pm a/2$ is $V_0 \cos(\pi x/a)$. Then find $V(x, y)$ in the pipe, and sketch a graph of the equipotential curves in the xy plane.

5.4. Calculate the integral in (5.40) if $V_0(y, z)$ is a constant V_0, and verify that the result is (5.41).

5.5. In Example 2 we found that the potential inside the grounded conducting pipe in Fig. 5.6, with the end at $x = 0$ held at constant potential V_0, is

$$V(x, y, z) = \frac{16V_0}{\pi^2} \sum_{n \text{ odd}} \sum_{m \text{ odd}} \frac{e^{-\kappa_{nm}x}}{nm} \sin\left(\frac{n\pi y}{\ell_2}\right) \sin\left(\frac{m\pi z}{\ell_3}\right). \quad (5.142)$$

(a) Show that the Fourier series of a square wave $f(y)$ that has wavelength $2\ell_2$ and amplitude 1, with $f = 1$ for $0 < y < \ell_2$ and $f = -1$ for $-\ell_2 < y < 0$, is

$$f(y) = \frac{4}{\pi} \sum_{n \text{ odd}} \frac{1}{n} \sin\left(\frac{n\pi y}{\ell_2}\right).$$

(Hint: This square wave is like the one in Fig. 5.2, but with a phase shift of $90°$.)

(b) In particular, when the function in (a) is evaluated at $y = \ell_2/2$ (in the region where $f = 1$) the result is one of the most remarkable infinite series in mathematics

$$\frac{\pi}{4} = \frac{1}{1} - \frac{1}{3} + \frac{1}{5} - \frac{1}{7} + \frac{1}{9} - \frac{1}{11} + - \cdots.$$

This series, discovered by Leibniz in 1673, shows that π is actually a very simple number, because it can be expressed using every odd integer exactly once. Prove Leibniz's series from the Taylor series of $\arctan(x)$.

(c) Use a computer program to plot the series $f(y)$ truncated at $n = 1$ and at $n = 11$, as a function of y with $\ell_2 = 1$. (Superimpose the two curves, and show at least one full wavelength.)

(d) Use a computer program to plot $V(x, \frac{1}{2}, \frac{1}{2})$ as a function of x, for $\ell_2 = \ell_3 = 1$. The result shows that the electric field does not penetrate far into the pipe.

5.6. Sketch the electric field lines in a semi-infinite rectangular pipe (as in Fig. 5.6) if the end plate has constant potential V_0 and the walls are grounded. Also, describe the surface charge density.

5.7. Show that if in Example 1 the three sides of the square $x = \pm a/2$ and $y = -a/2$ are grounded and the fourth side $y = +a/2$ is held at potential V_0, then inside the square the potential is

$$V(x, y) = \frac{4V_0}{\pi} \sum_{n=0}^{\infty} \frac{(-1)^n}{2n+1} \cos\left[(2n+1)\frac{\pi x}{a}\right] \frac{\sinh\left[(2n+1)\pi(y/a + 1/2)\right]}{\sinh\left[(2n+1)\pi\right]}.$$

(Hint: In this configuration the potential is neither even nor odd in the variable y, so both $\sinh ky$ and $\cosh ky$ appear in the expression for $V(x, y)$.)

5.8. (a) From the result of the previous exercise and the superposition principle, show how to find $V(x, y)$ if the sides of the square in Example 1 are held at arbitrary constant potentials, e.g., V_1 at $x = a/2$, V_2 at $y = a/2$, V_3 at $x = -a/2$, and 0 at $y = -a/2$.

(b) What is the potential at the origin for this case? [Answer: $V(0, 0) = (V_1 + V_2 + V_3)/4$, which is just the average of the potential on the four sides of the square.]

Sec. 5.2. Separation of Variables for Spherical Coordinates

5.9. Two concentric spherical conducting shells of radii R_1 and R_2 have potentials V_1 and V_2, respectively. Find $V(r)$ between the shells and verify $\nabla^2 V = 0$ explicitly.

5.10. On Legendre polynomials.

    (a) Use the recursion relation (5.52), and the normalization condition $P_\ell(1) = 1$, to determine $P_0(u)$, $P_1(u)$, $P_2(u)$, $P_3(u)$. For the even polynomials, start with C_0 and use (5.52) to calculate $C_2, C_4, C_6 \ldots$ until the sequence terminates. Then determine C_0 such that $P_\ell(1) = 1$. For the odd polynomials start with C_1.

    (b) Use a mathematical software package to look up $P_{10}(u)$ and $P_{11}(u)$.

    (c) From Legendre's equation prove the orthogonality relation

$$\int_{-1}^{1} P_m(u) P_\ell(u)\, du = 0 \quad \text{if} \quad m \neq \ell.$$

    (d) Derive Eq. (5.62) from Legendre's equation.

5.11. **(a)** If an object in an applied electric field $\mathbf{E}_{\text{app}}$ acquires an induced dipole moment $\mathbf{p} = \alpha \mathbf{E}_{\text{app}}$, then α is called the *polarizability* of the object. Determine the polarizability of a conducting sphere of radius a.

    (b) Determine the surface charge density $\sigma(\theta)$ of the sphere, for $\mathbf{E}_{\text{app}} = E_0 \hat{\mathbf{k}}$.

    (c) Verify that $\mathbf{p} = \int \mathbf{r}\, dq$, which is the general definition of the dipole moment of a distribution of charge. (Here $dq = \sigma dA$.)

5.12. Derive the Taylor series in s for the function $g(u, s) = (1 - 2us + s^2)^{-1/2}$ through order s^3. Show that the coefficients are the first four Legendre polynomials.

5.13. Express the potential of a pointlike linear quadrupole, oriented parallel to the z axis, in Legendre polynomials. There are three charges: $+q$, $-2q$, and $+q$ on the z axis, with $-2q$ at the origin, and $+q$ at $z = -a$ and $+a$.

Sec. 5.3. Separation of Variables for Cylindrical Coordinates

5.14. Two concentric long circular cylinders of radii R_1 and R_2 have potentials V_1 and V_2, respectively. Find $V(r)$ between the cylinders and verify $\nabla^2 V = 0$ explicitly.

5.15. What is the potential $V(r, \phi)$ due to a cylinder of radius R, for which the upper-half cylinder $0 < \phi < \pi$ is a conductor at potential $+V_0$, and the lower-half cylinder $\pi < \phi < 2\pi$ is a conductor at potential $-V_0$? The two half cylinders are separated by small insulating gaps at $\phi = 0$ and $\phi = \pi$.

[Answer:

$$V_{\text{ext}}(r, \phi) = \frac{4V_0}{\pi} \sum_{k \text{ odd}} \frac{1}{k} \left(\frac{R}{r} \right)^k \sin k\phi,$$

$$V_{\text{int}}(r, \phi) = \frac{4V_0}{\pi} \sum_{k \text{ odd}} \frac{1}{k} \left(\frac{r}{R} \right)^k \sin k\phi.]$$

5.16. What is the potential $V(r, \phi)$ due to a cylindrical surface of radius R on which there is a surface charge density $\sigma(\phi) = \sigma_0 \cos \phi$? [Answer:

$$V_{\text{ext}}(r, \phi) = \frac{\sigma_0 R^2 \cos \phi}{2\epsilon_0 r} \quad \text{and} \quad V_{\text{int}}(r, \phi) = \frac{\sigma_0 r \cos \phi}{2\epsilon_0}.]$$

5.17. What is the potential $V(r, \phi)$ due to a cylindrical surface of radius R, on which there is a surface charge density $\sigma(\phi) = +\sigma_0$ on the right half $(3\pi/2 < \phi \le 2\pi$ and $0 \le \phi < \pi/2)$, and $\sigma(\phi) = -\sigma_0$ on the left half $(\pi/2 < \phi < 3\pi/2)$?

[Answer:

$$V_{\text{ext}}(r, \phi) = \frac{2\sigma_0 R}{\epsilon_0 \pi} \sum_{j=0}^{\infty} \frac{(-1)^j}{(2j+1)^2} \left(\frac{R}{r}\right)^{2j+1} \cos(2j+1)\phi,$$

$$V_{\text{int}}(r, \phi) = \frac{2\sigma_0 R}{\epsilon_0 \pi} \sum_{j=0}^{\infty} \frac{(-1)^j}{(2j+1)^2} \left(\frac{r}{R}\right)^{2j+1} \cos(2j+1)\phi.]$$

Sec. 5.4. Conjugate Functions

5.18. Show that the Cauchy-Riemann equations imply $\nabla^2 g = 0$ and $\nabla^2 h = 0$ in two dimensions.

5.19. Sketch, qualitatively, the equipotentials and electric field lines of two perpendicular charged conducting plates, similar to Fig. 5.15 but with finite extent in x and y. Include curves far from the plates.

5.20. Let $F(z) = Cz$ be the complex function in the method of conjugate functions, where C is a real constant. What boundary value problem is solved if Re F is taken to be the potential? What boundary value problem is solved if Im F is taken to be the potential?

5.21. Consider the function $F(z) = iC \ln z$, where $z = x + iy$ and C is a real constant.

(a) Let Im F be the potential function. What are the equipotentials and electric field lines? What is the charge distribution?

(b) What boundary value problem is solved if Re F is taken to be the potential?

5.22. In Example 7 we considered plates that intersect at a right angle. Find the potential $V(x, y)$ in the wedge-shaped region between the plates if they intersect at angle $\alpha < \pi/2$. (Hint: Try the complex function $F(z) = Az^p + i\Phi_0$ and determine the appropriate exponent p.)

5.23. For the conformal map $w \to z$ in (5.129), a curve in the complex w plane is mapped into a curve in the complex z plane; the curve in z is called the image of the curve in w.

(a) What is the image of the upper half of the unit circle of the w plane?

(b) What is the image of the upper side of the positive real w axis? What is the image of the lower side of the positive real w axis?

(c) What is the image of the negative real w axis? [Answer: The real w axis is mapped into the conductors of the semi-infinite parallel-plate capacitor of Fig. 5.17.]

General Exercises

5.24. Consider an electrostatic system that consists of a set of conductors $C_1, C_2, C_3, \ldots$ held at specified potentials $V_1, V_2, V_3, \ldots$. Prove that the electrostatic field is $-\nabla\Phi$, where $\Phi(\mathbf{x})$ is the function that minimizes $U = \frac{\epsilon_0}{2}\int_V (\nabla\Phi)^2 d^3x$, subject to the boundary conditions $\Phi = V_i$ on C_i. The volume of integration V is the free space—the space *outside* the conductors. Thus electrostatic equilibrium is the state with minimum field energy satisfying the boundary conditions.

5.25. Let P be an arbitrary point in a charge-free region of an electrostatic system, and set up a coordinate system with P at the origin. The average potential on a sphere S of radius r around P is

$$\overline{V}(r) = \frac{1}{4\pi r^2}\oint_S V(r, \theta, \phi)r^2 \sin\theta\, d\theta\, d\phi.$$

(Note, r is constant on S.)

(a) Prove that $d\overline{V}/dr = 0$, assuming there is no charge enclosed by S.

(b) Prove that $\overline{V}$ is equal to the potential at P.

5.26. From the result of the previous exercise, prove that $V(\mathbf{x})$ cannot be a maximum or a minimum at a charge-free point. Prove that a point charge cannot be at stable equilibrium with electrostatic forces only (Earnshaw's theorem).

5.27. Consider again the long square pipe shown in Fig. 5.1 with boundary potentials 0 at $x = \pm a/2$ and V_0 at $y = \pm a/2$.

(a) Show that the transformation $(x, y) \rightarrow (y, x)$, i.e., reflection about the diagonal $y = x$, produces an equivalent system. Show from this symmetry that the potential is constant on the diagonals of the square cross section. (Hint: Sketch qualitatively how the equipotentials must look.) Prove that this constant potential must be $V_0/2$.

(b) Verify numerically that $V(0, 0) = V_0/2$, by evaluating the series (5.31) by computer. Note, too, that we uncover from this exercise the beautiful and mysterious identity

$$\frac{\pi}{8} = \sum_{n=0}^{\infty} \frac{(-1)^n}{2n + 1}\frac{1}{\cosh\left[(2n + 1)\pi/2\right]}.$$

5.28. Consider the long conducting pipe shown in cross section in Fig. 5.1. But in this exercise let the walls at $x = \pm a/2$ be grounded ($V = 0$) and the walls at $y = \pm a/2$ be held at equal *but opposite* potentials $\pm V_0$.

(a) Determine $V(x, y)$. (Hint: Exploit the idea that it is now an odd function of y.)

(b) Use a computer to plot the potential $V(0, y)$ versus y. (Let V_0 be the unit of potential and a the unit of length.) A computer program is useful for this purpose, because the result of (a) is an infinite series which must be evaluated numerically.

(c) Calculate the electric field at the origin.

5.29. Solve Laplace's equation in *one dimension* (x) in the bounded domain $x_1 \le x \le x_2$ for boundary values $V(x_1) = V_1$ and $V(x_2) = V_2$.

5.30. Consider a cube made of 6 conducting plates of size $a \times a$, that encloses the space $-a/2 \leq x, y, z \leq a/2$. The plates at $z = \pm a/2$ are held at potentials $\pm V_0$, respectively. The other sides are grounded ($V = 0$). The separable solutions of Laplace's equation inside the cube have the form

$$\{\cosh(\kappa_{nm} z) \quad \text{or} \quad \sinh(\kappa_{nm} z)\} \cos\left[(2n + 1)\pi x/a\right] \cos\left[(2m + 1)\pi y/a\right]$$

where n and m are non-negative integers.

(a) Solve the boundary value problem for $V(x, y, z)$ inside the cube.

(b) The potential is obviously 0 at the center, but what is the electric field there? Express $E_z = -\left. \partial V/\partial z \right|_{(0,0,0)}$ as an infinite series, and use a computer to find the numerical value in terms of V_0/a.

5.31. Imagine a conductor consisting of three semi-infinite plates that intersect at right angles, making a corner like the floor and two walls of a room. Let the plates be the ++ quadrants of the xy, yz, zx planes.

(a) Show that the potential in the +++ octant is $V(\mathbf{x}) = C_1 xyz + C_2$. The constant C_1 depends on the charge on the plates.

(b) Describe the surface charge densities on the boundary planes.

5.32. Suppose a grounded spherical conducting shell of radius R surrounds a pointlike dipole at the center with $\mathbf{p} = p\hat{\mathbf{k}}$. Find the potential $V(r, \theta)$ for $r \leq R$. (Hint: Use spherical harmonics regular at $r = 0$ to satisfy the boundary condition.)

5.33. (a) Suppose on the surface of a sphere of radius R there is a surface charge density $\sigma(\theta) = \sigma_0 \cos^2 \theta$, where σ_0 is a constant. What is $V(r, \theta)$ inside and outside the sphere?

(b) Suppose now that $\sigma(\theta) = \sigma_0 \sin^2 \theta$. What is $V(r, \theta)$ inside and outside the sphere in this case?

[Answer to (b):

$$V_{\text{int}}(r, \theta) = \frac{2}{3} \frac{R\sigma_0}{\epsilon_0} - \frac{2R\sigma_0}{15\epsilon_0} \left(\frac{r}{R}\right)^2 P_2(\cos\theta)$$

$$V_{\text{ext}}(r, \theta) = \frac{2}{3} \frac{R\sigma_0}{\epsilon_0} - \frac{2R\sigma_0}{15\epsilon_0} \left(\frac{R}{r}\right)^3 P_2(\cos\theta).]$$

5.34. Consider a long circular cylinder of radius R bisected by a plane parallel to the axis of the cylinder, with the two halves insulated from one another. (Let the z axis be the central axis of the cylinder, and the bisecting plane the xz plane.) One hemicylinder $(0 < \phi < \pi)$ is grounded and the other $(\pi < \phi < 2\pi)$ is held at potential V_0.

(a) Use the method of separation of variables to determine the potential $V(r, \phi)$.

(b) Evaluate the potential at $r = 0$.

(c) Evaluate the electric field at $r = 0$.

(d) Use a computer to plot $V(R/2, \phi)$ as a function of ϕ. (It is most convenient to use a mathematical software package to calculate numerically the infinite series and plot the results.)

5.35. Suppose $V(\mathbf{x})$ satisfies Laplace's equation in three dimensions. Show that the potential at $\mathbf{x}$ is approximately equal to the average of the potentials at the six points

$\mathbf{x} \pm \epsilon \hat{\mathbf{i}}, \mathbf{x} \pm \epsilon \hat{\mathbf{j}}, \mathbf{x} \pm \epsilon \hat{\mathbf{k}}$, where ϵ is small, and that the error of the approximation is of order ϵ^3.

5.36. The real part of $F(z) = (-\lambda/2\pi\epsilon_0) \ln z$, where $z = x + iy$, is the potential function of a two-dimensional electrostatic system. What is the charge distribution? Sketch the equipotentials and electric field lines.

5.37. An array of long charged wires, parallel to the z axis and passing through the x axis at $x = 0, \pm\ell, \pm2\ell, \pm3\ell, \ldots$, is part of the design of a multiwire proportional counter (MWPC), invented by Charpak (Nobel prize, 1994). Neglecting end effects, this is a two-dimensional system.

(a) Show that the potential $V(x, y)$ is the real part of

$$F(z) = (-\lambda/2\pi\epsilon_0) \ln [\sin(\pi z/\ell)]$$

where $z = x + iy$. (Hints: We know that Re F satisfies Laplace's equation, so we only need to verify the boundary conditions at the wires. Show that Re F is a periodic function of x with period ℓ. What is $V(x, y)$ near the wire at $(x, y) = (0, 0)$?)

(b) Show that

$$V(x, y) = -\frac{\lambda}{4\pi\epsilon_0} \ln \left[\cosh\left(\frac{2\pi y}{\ell}\right) - \cos\left(\frac{2\pi x}{\ell}\right) \right].$$

(c) Sketch the equipotential lines and electric field lines, or, better, use computer graphics to make a contour plot of $V(x, y)$.

(d) Show that in the limit $y \to \infty$ the electric field approaches the field of a charged plane with charge per unit area λ/ℓ. Does this result make sense?

5.38. The sentence at the head of this chapter is part of a longer and rather serious quote from the essay: "The Unreasonable Effectiveness of Mathematics in the Natural Sciences," by E. P. Wigner. It is in the book *Symmetries and Reflections* (Indiana Univ. Press, Bloomington, 1967, W. J. Moore and M. Scriven, eds.)

> "The miracle of the appropriateness of the language of mathematics for the formulation of the laws of physics is a wonderful gift which we neither understand nor deserve. We should be grateful for it and hope that it will remain valid in future research and that it will extend, for better or for worse, to our pleasure, even though perhaps to our bafflement, to wide branches of learning."

Now that you've had experience solving many mathematical exercises, what do you think of this?

Computer Exercises

5.39. What is the electric field on the midplane of the plates of a semi-infinite parallel plate capacitor, both inside and outside the capacitor? The plates, shown in Fig. 5.17, are half-planes at $y = 0$ and $y = d$ with $x \leq 0$. The midplane is $y = d/2$. The problem is to calculate $E_y(x, d/2)$. In the notation of Example 9,

$$E_y(x, y) = -\left(\frac{\partial V}{\partial y}\right)_x = +\left(\frac{\partial h}{\partial x}\right)_y,$$

the second equality by the Cauchy-Riemann relations. In terms of the parameters ρ and θ of Example 9, the midplane has $\theta = \pi/2$, and

$$\frac{\partial h}{\partial x} = \frac{(\partial h/\partial \rho)_{\theta=\pi/2}}{(\partial x/\partial \rho)_{\theta=\pi/2}}.$$

(a) Compute $E_y(x, d/2)$ in units of V_0/d at $x = -2d, -d, 0, d$. (Neglecting edge effects E_y would be $-V_0/d$ between the plates.)

(b) Use computer graphics to plot $E_y(x, d/2)$ in units of V_0/d as a function of x/d. (Hint: Use parametric plot.)

(c) For what x on the midplane is the field strength $0.1\ V_0/d$?

5.40. Consider the system of Example 10, a long pipe with a square cross section of dimensions $a \times a$. Let the z axis be the central axis of the pipe, and suppose the sides at $x = \pm a/2$ are grounded ($V = 0$) while the sides at $y = \pm a/2$ are held at potential V_0. Write a computer program to solve Laplace's equation for $V(x, y)$ by iterative relaxation. Make plots of $V(x, 0)$ versus x and $V(0, y)$ versus y.

5.41. Modify the program in Exercise 40 to solve Laplace's equation for $V(x, y)$ with these boundary conditions:

$$V(\pm a/2, y) = 0$$

$$V(x, +a/2) = 0 \quad \text{and} \quad V(x, -a/2) = V_0;$$

that is, three sides are grounded, and the bottom side is at V_0. Make plots of $V(x, 0)$ versus x and $V(0, y)$ versus y. Show that $V(0, 0) = 0.25\ V_0$.

5.42. By iterative relaxation solve Laplace's equation in a square ($-a/2 \leq x \leq a/2$ and $-a/2 \leq y \leq a/2$) with these boundary conditions:

$$V(\pm a/2, y) = V(x, \pm a/2) = 0$$

$$V(x, a/10) = -V_0 \quad \text{for} \quad -a/4 \leq x \leq a/4$$

$$V(x, -a/10) = +V_0 \quad \text{for} \quad -a/4 \leq x \leq a/4.$$

That is, the four sides of the square are grounded ($V = 0$); and in the square there are parallel line segments (i.e., plates extending in the z direction) at $y = \pm 0.1a$ with x from $-0.25a$ to $0.25a$ held at potentials $\mp V_0$.

(a) Plot $E_y(x, 0)$ versus x. Like a capacitor, the electric field is approximately uniform between the line segments and decreases outside. In the lattice approximation E_y is computed from a finite difference

$$E_y(x, 0) = -\frac{V(x, +\epsilon) - V(x, -\epsilon)}{2\epsilon}$$

where ϵ is the lattice spacing.

(b) The energy density $u(x, y)$ is proportional to $E_x^2(x, y) + E_y^2(x, y)$. Calculate $E_x^2(x, y) + E_y^2(x, y)$ for representative points in the square: e.g., $(0, 0)$, $(0.25a, 0)$, $(0.4a, 0)$, and $(0, 0.3a)$.

CHAPTER

6

Electrostatics and Dielectrics

Up to now we have studied electrostatic systems of charges and conductors. Now we are ready to include insulators, also called dielectrics, in the systems we study.

Faraday coined the word "dielectric" to describe the effects he had discovered when an insulator is put into a capacitor.[1] He found that when a capacitor is filled with an insulating material—a dielectric—more charge and energy are stored at a given potential than without the dielectric. The capacitance of the system, the energy stored, and the charge on the conductors, are all greater by a factor κ, which is a characteristic property of the dielectric material called its *dielectric constant*. There is a wide range of κ among dielectrics. Capacitor design is an important topic in applied physics, and the properties of dielectrics are crucial, because capacitors are key elements in circuits.

In a metal, the atoms release one or more outer electrons to form a sea of electrons that are free to move throughout the conductor. When an external electric field is applied, as we have learned, free electrons become distributed on the surface in such a way that in electrostatic equilibrium there is zero field and zero net charge density inside the conductor.

In a dielectric, by contrast, the electrons are not free, but rather *bound* to their atoms or molecules. When an external electric field is applied to a dielectric, the electrons and nuclei become displaced by small distances, in the direction of **E** in the case of the nuclei, or in the opposite direction in the case of the electrons. But the electrons cannot escape from the Coulomb forces that bind them to the associated nuclei. In an atomic dielectric, solid argon for example, the electrons are bound to atoms. In a molecular dielectric, solid CO_2 for example, the electrons are bound to molecules. The opposite displacements of positive and negative charges result in a *polarized* atom or molecule, with a nonzero dipole moment. In cases such as these, the dipole is called an *induced dipole*. Figure 6.1 illustrates an induced atomic dipole. We also say that the bulk dielectric, made up of polarized atoms or molecules, has become polarized, because it has acquired a net dipole moment. The bulk material may also have acquired a net charge density inside, as we will see. So unlike a conductor, a dielectric may have a nonzero field and charge density inside.

Some dielectrics are composed of molecules whose electronic structure is asymmetric in space, so that the molecule has a permanent dipole moment, even in the absence of an external field. Water (H_2O), nitric oxide (NO), and car-

---

[1] Recall that the SI unit of capacitance, the farad (F), is named for Faraday because of his discoveries.

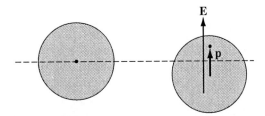

FIGURE 6.1   An induced atomic dipole. The atom on the left is unpolarized. When a field is applied, the nucleus is pulled in the direction of **E** and the electrons in the opposite direction, so the atom has a dipole moment proportional to **E**.

bon monoxide (CO) are examples. These are called *polar molecules*. In such dielectrics there is an additional mechanism of polarization, besides displacement of charges within the molecule, due to *orientation* of the permanent moment, which contributes to the polarization of the bulk material.

Because these polarization processes involve bound electrons, dielectric properties inevitably depend on the details of atomic and molecular structure. Therefore dielectric properties vary widely. Of course the properties of conductors also depend to some extent on atomic and molecular interactions, as we can see by comparing, say, lithium to lead. But the behavior of different metals in a static electric field is more or less the same. The behavior of different dielectrics can be quite different. Our presentation of this subject begins with a discussion of atoms and molecules as electric dipoles. That will prepare us to explore the connection between the dielectric properties of bulk matter and their microscopic origin.

6.1 ■ THE ATOM AS AN ELECTRIC DIPOLE

We can explain the magnitude of dielectric effects based on simple models of the atomic or molecular dipole moments.

6.1.1 ■ Induced Dipoles

In equilibrium, an isolated atom in an electric field **E** has a dipole moment **p** proportional to the field,

$$\mathbf{p} = \alpha \mathbf{E}, \tag{6.1}$$

because the field pulls the nucleus and the electrons in opposite directions, as illustrated in Fig. 6.1. The constant of proportionality α is called the *atomic polarizability*. Its units are rather clumsy, namely, Cm^2/V. Atomic polarizabilities must ultimately be measured by experiment, but we can estimate their order of magnitude by modeling the atom as a conducting sphere whose radius R is a typical atomic size, say $R \approx 10^{-10}$ m. We recall from Chapter 4 that the induced dipole moment of a conducting sphere is $\mathbf{p} = 4\pi\epsilon_0 R^3 \mathbf{E}$, which is pro-

TABLE 6.1   Atomic polarizabilities

$\alpha/(4\pi\epsilon_0)$ in units of 10^{-30} m^3

H	0.667	He	0.205
Li	24.3	Ne	0.396
Na	23.6	Ar	1.64
K	43.4	Kr	2.48
Rb	47.3	Xe	4.04
Cs	59.6		

portional to the sphere's volume. The model polarizability is $\alpha = 4\pi\epsilon_0 R^3$, so $\alpha/(4\pi\epsilon_0) = 10^{-30}$ m^3 for a typical atom. It is customary to give polarizabilities in the form $\alpha/(4\pi\epsilon_0)$ because this quantity has simply the units of volume.

Table 6.1 lists the polarizabilities of a few simple atoms. Note that in all cases $\alpha/(4\pi\epsilon_0)$ is of order 10^{-30} m^3, in accord with the above model, but there is marked variation. Atoms of the alkali metals, in the left column of Table 6.1, have large polarizabilities because they have a loosely bound outer electron. The values of α for Li, Na, K, Rb, and Cs were determined by measuring the deflections of atomic beams in a nonuniform electric field. Atoms of the noble gases, in the right column of Table 6.1, have small polarizabilities because their electron shells are all filled. The values of α for He, Ne, Ar, Kr, and Xe were determined by measuring the dielectric constant in a capacitor filled with the element. Of all the elements, He has the smallest polarizability. The largest atomic polarizability that has been measured is for Cs, but presumably that of Fr is still larger. As Table 6.1 shows, polarizabilities within each column do increase with atomic size. However, they are not simply proportional to R^3, as the naive model predicts.

A precise theory of atomic polarizability requires quantum mechanics. The simplest case is the hydrogen atom. In bare outline the procedure for calculating α of atomic hydrogen is as follows. The electron wavefunction $\psi(\mathbf{x})$ obeys the Schrödinger equation

$$-\frac{\hbar^2}{2m}\nabla^2\psi - \frac{e^2}{4\pi\epsilon_0 r}\psi + eE_0 z\psi = \mathcal{E}\psi \tag{6.2}$$

where $\mathcal{E}$ is the electron energy and $E_0\hat{\mathbf{k}}$ is the external electric field. The equation may be solved for small E_0 by perturbation theory [1,2]. The electric dipole moment is classically $-e\mathbf{x}$, because $\mathbf{x}$ is the vector from the proton to the electron. The mean dipole moment in quantum mechanics is the expectation value $\mathbf{p} = \langle\psi| - e\mathbf{x}|\psi\rangle$. From the solution of the Schrödinger equation, $\mathbf{p}$ turns out to be

$$\mathbf{p} = \tfrac{9}{2}\left(4\pi\epsilon_0 a_B^3\right)E_0\hat{\mathbf{k}} \tag{6.3}$$

where a_B is the Bohr radius. Comparing to (6.1) the quantum theory of the hydrogen atom gives

$$\frac{\alpha}{4\pi\epsilon_0} = \frac{9}{2}a_B^3 = 0.667 \times 10^{-30} \, \text{m}^3. \tag{6.4}$$

This calculated value is used for H in Table 6.1. The calculation for larger atoms is more difficult, but the results of quantum theory agree with the measurements in Table 6.1.

6.1.2 ■ Polar Molecules

Some molecules have a permanent dipole moment. For example, the electron density in an HCl molecule is not symmetric about the center, but rather skewed toward the Cl nucleus, so that there is a permanent dipole moment (in the body-fixed frame of reference) pointing from Cl to H. Another example is the water molecule H_2O, which is an asymmetric molecule with an electric dipole moment pointing from O toward the midpoint of the line connecting the H's.

Dipole moments of polar molecules are expressed in units of debyes (D) where $1 \, \text{D} = 3.34 \times 10^{-30} \, \text{C m}$. Some typical dipole moments of small polar molecules are 1.75 D for HF, 1.04 D for HCl, 0.80 D for HBr, 0.83 D for HI, 1.83 D for H_2O, and 1.48 D for the triangular pyramid molecule NH_3. For comparison, ea_B is 2.5 D. Large molecules can have large dipole moments; that of hemoglobin, for example, is hundreds of debyes.

If a material is composed of polar molecules, then the permanent molecular moments create a dielectric effect whose magnitude depends on temperature. When a polar molecule is placed in an electric field $\mathbf{E} = E_0\hat{\mathbf{k}}$, there is a torque $\mathbf{p} \times \mathbf{E}$ twisting the dipole toward alignment with $\mathbf{E}$; the potential energy is $U = -\mathbf{p} \cdot \mathbf{E}$. We may estimate the thermal average $\langle\mathbf{p}\rangle$ of the dipole moment, as a function of the temperature T, by treating the molecule as a classical rotor and applying statistical mechanics. The dipole orientations have a Boltzmann distribution, and the mean dipole moment is

$$\langle\mathbf{p}\rangle = \frac{\int e^{-U/kT}\, \mathbf{p}\, d\Omega_p}{\int e^{-U/kT}\, d\Omega_p}. \tag{6.5}$$

The dipole orientation can be specified by the polar angle θ (the angle between $\mathbf{p}$ and $\mathbf{E}$) and the azimuthal angle ϕ (the angle around the field line) as

$$\mathbf{p} = p\left(\hat{\mathbf{i}}\sin\theta\cos\phi + \hat{\mathbf{j}}\sin\theta\sin\phi + \hat{\mathbf{k}}\cos\theta\right). \tag{6.6}$$

Then the integration over the orientations of $\mathbf{p}$ is $d\Omega_p = \sin\theta d\theta d\phi$. The integral over ϕ is zero for the $\hat{\mathbf{i}}$ and $\hat{\mathbf{j}}$ components, because the energy $U = -pE_0\cos\theta$ is independent of ϕ. So the mean dipole moment points in the $\hat{\mathbf{k}}$ direction, $\langle\mathbf{p}\rangle = \langle p_z\rangle\hat{\mathbf{k}}$, with

$$\langle p_z\rangle = \frac{p\int_0^\pi e^{-U/kT}\cos\theta\sin\theta d\theta}{\int_0^\pi e^{-U/kT}\sin\theta d\theta} = \frac{p\int_{-1}^1 e^{au}u\,du}{\int_{-1}^1 e^{au}du}. \tag{6.7}$$

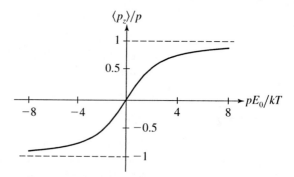

FIGURE 6.2   The Langevin formula. A molecule with a permanent dipole moment of magnitude p is placed in an electric field $E_0\hat{\mathbf{k}}$. The thermal average of the dipole moment vector at temperature T is $\langle p_z\rangle\hat{\mathbf{k}}$, and it depends on pE_0/kT.

In the second equality we have changed the variable of integration to $u = \cos\theta$ and defined the dimensionless parameter $a = pE_0/kT$. The integrals are elementary, and the result is

$$\langle p_z\rangle = p\left[-\frac{1}{a} + \coth a\right]. \tag{6.8}$$

This result, called Langevin's formula, is plotted in Fig. 6.2. At low temperature $(a \gg 1)$ the dipole aligns with $\mathbf{E}$ and so $\langle p_z\rangle$ approaches p. At high temperature $(a \ll 1)$ the thermal fluctuations of the dipole orientation are large, so the mean value $\langle p_z\rangle$ approaches 0. More precisely, in the case $pE_0 \ll kT$ we may approximate $\langle p_z\rangle$ by $pa/3$. (The Taylor series for $\coth a$ is $1/a + a/3 + a^3/45 + \cdots$.) Then the mean dipole moment is a linear function of $\mathbf{E}$, equal to $\alpha\mathbf{E}$ with

$$\alpha = \frac{p^2}{3kT}. \tag{6.9}$$

This α is the polarizability of a molecule with a permanent dipole moment p at temperature T. The temperature dependence $\alpha \propto T^{-1}$ is called a Curie law. In deriving (6.9) we ignored intermolecular interactions. Therefore the result holds only when interactions with other polar molecules are negligible, as in gases.

A polar molecule in an electric field also has an induced dipole moment due to distortion of the electron wave function, which contributes a term α_0 to the polarizability. The polarizability from both effects is the sum

$$\alpha = \alpha_0 + \frac{p^2}{3kT}. \tag{6.10}$$

The term α_0 from the induced moment is independent of T, so the two effects can be disentangled. Equation (6.10) is called the Langevin-Debye formula. Debye used experiments on dielectric effects to measure α, and thereby determine molecular size and structure.

In Sec. 6.3 we will calculate the response of a macroscopic dielectric sample to an applied electric field. The input to the calculation is the polarizability α of a single atom or molecule.

6.2 ■ POLARIZATION AND BOUND CHARGE

To account for the effects of atomic polarizability, it is necessary to define a new field (in this case a kind of *matter field*) called the *polarization* $\mathbf{P}(\mathbf{x})$. The polarization is a vector function of position in the dielectric. $\mathbf{P}(\mathbf{x})$ is defined to be the *mean dipole moment density*. This requires some explanation. The polarization at $\mathbf{x}$ is the dipole moment *per unit volume*, averaged over a subvolume that is small compared to the macroscopic scale of the system but large compared to a single atom,

$$\mathbf{P}(\mathbf{x}) = \frac{1}{\delta V} \sum_{i=1}^{\delta N} \mathbf{p}_i . \tag{6.11}$$

In the equation, $\mathbf{p}_i$ denotes the dipole moment of the ith atom in δV, and δN is the number of atoms. Because the number of atoms is very large, even in a small δV, $\mathbf{P}(\mathbf{x})$ is a smooth function of position. The *fluctuations* of individual atomic dipoles are washed out by averaging over many atoms.

The vector field $\mathbf{P}(\mathbf{x})$ is analogous to the scalar function $\rho(\mathbf{x})$, the macroscopic charge density, which is similarly defined as an average over subvolumes containing many atoms. In the case of ρ the quantity that is averaged is the charge. In the case of $\mathbf{P}$ it is the dipole moment. In fact, $\mathbf{P}(\mathbf{x})$ and $\rho(\mathbf{x})$ are related, and we now derive the relation.

Polarization is a shift of charge—positive charge being displaced down-field and negative charge up-field. In equilibrium *without* an applied field the mean charge density is 0 throughout the dielectric; i.e., the charge density due to electrons is equal but opposite to that due to the atomic nuclei. But when there is polarization, the displacement of charge may create a nonzero net charge density at some points in the material. The mean charge density resulting from polarization is called the *bound charge density*, and denoted $\rho_b(\mathbf{x})$. We shall prove that

$$\rho_b(\mathbf{x}) = -\nabla \cdot \mathbf{P}. \tag{6.12}$$

This result is so important that it is worthwhile to study two different derivations of it.

For the first derivation, recall the potential of an electric dipole $\mathbf{p}$,

$$V(\mathbf{x}) = \frac{\mathbf{p} \cdot (\mathbf{x} - \mathbf{x}')}{4\pi \epsilon_0 |\mathbf{x} - \mathbf{x}'|^3}, \tag{6.13}$$

where $\mathbf{x}'$ is the position of the dipole. The potential due to the superposition of all dipoles in the dielectric is

$$V(\mathbf{x}) = \int \frac{(\mathbf{x} - \mathbf{x}') \cdot \mathbf{P}(\mathbf{x}')d^3x'}{4\pi\epsilon_0|\mathbf{x} - \mathbf{x}'|^3} \tag{6.14}$$

because $\mathbf{P}(\mathbf{x}')d^3x'$ is the net dipole moment from all atoms in d^3x'. The integration region is the whole dielectric volume. Now we shall rewrite (6.14) in a familiar form. First, note that $(\mathbf{x} - \mathbf{x}')/|\mathbf{x} - \mathbf{x}'|^3$ is equal to $\nabla'(1/|\mathbf{x} - \mathbf{x}'|)$, where ∇' is the gradient operator for $\mathbf{x}'$. Next, convert the integrand by the identity

$$\mathbf{P}(\mathbf{x}') \cdot \left(\nabla'\frac{1}{|\mathbf{x} - \mathbf{x}'|}\right) = \nabla' \cdot \left(\frac{\mathbf{P}(\mathbf{x}')}{|\mathbf{x} - \mathbf{x}'|}\right) - \frac{\nabla' \cdot \mathbf{P}(\mathbf{x}')}{|\mathbf{x} - \mathbf{x}'|}. \tag{6.15}$$

Finally, apply Gauss's divergence theorem to the volume integral of the first term on the right side of (6.15)—a total divergence—obtaining a surface integral over the boundary of the dielectric. The result then becomes

$$V(\mathbf{x}) = \oint_S \frac{\mathbf{P}(\mathbf{x}') \cdot \hat{\mathbf{n}} dA'}{4\pi\epsilon_0|\mathbf{x} - \mathbf{x}'|} - \int \frac{\nabla' \cdot \mathbf{P}(\mathbf{x}')d^3x'}{4\pi\epsilon_0|\mathbf{x} - \mathbf{x}'|}, \tag{6.16}$$

where S is the surface of the dielectric. These integrals have forms that we have met previously in electrostatics, as potentials due to surface charge and volume charge, so we write

$$V(\mathbf{x}) = \oint_S \frac{\sigma_b dA'}{4\pi\epsilon_0|\mathbf{x} - \mathbf{x}'|} + \int \frac{\rho_b(\mathbf{x}')d^3x'}{4\pi\epsilon_0|\mathbf{x} - \mathbf{x}'|}. \tag{6.17}$$

The volume function $\rho_b(\mathbf{x}')$ is $-\nabla' \cdot \mathbf{P}(\mathbf{x}')$; the surface function $\sigma_b(\mathbf{x}')$ is

$$\sigma_b = \hat{\mathbf{n}} \cdot \mathbf{P} \tag{6.18}$$

where $\hat{\mathbf{n}}$ is the outward unit normal vector at the point $\mathbf{x}'$ on S. The interpretation of (6.17) is that the surface charge density on S is $\sigma_b(\mathbf{x}')$ and the volume charge density is $\rho_b(\mathbf{x}')$, hence proving (6.12).

The above derivation of (6.12) is rather formal, involving several mathematical steps whose physical meaning is obscure. Therefore, we should study another, more physical derivation of the result. Consider an arbitrary volume $\mathcal{V}$ entirely inside the dielectric. When the dielectric is unpolarized the total charge inside $\mathcal{V}$ is 0. As the material becomes polarized, in response to an applied field, some charge will cross the surface S of $\mathcal{V}$, either because atoms near S become distorted, or because polar molecules near S align with the field. Let Q_{across} be the net charge that moves across S outward from $\mathcal{V}$ as the material becomes polarized. By conservation of charge, the net charge in $\mathcal{V}$ after the dielectric is polarized is

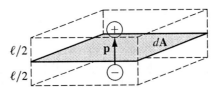

FIGURE 6.3   An ideal dipole—charges $\pm q$ separated by distance ℓ—straddling a surface element $d\mathbf{A}$, with $\mathbf{p}$ normal to the surface. If the center of the dipole is within distance $\ell/2$ above or below $d\mathbf{A}$, then the positive and negative charges are on opposite sides of $d\mathbf{A}$. The volume dV occupied by the atoms that straddle the surface is $dV = \ell\, d\mathbf{A}$.

$-Q_{\text{across}}$, so

$$\int_{\mathcal{V}} \rho_b d^3 x = -Q_{\text{across}} \tag{6.19}$$

where $\rho_b(\mathbf{x})$ is the bound charge density of the polarized material.

To calculate Q_{across}, consider a small patch dA on $\mathcal{S}$. If an atom is close enough to dA then it will straddle dA when polarized, as illustrated schematically in Fig. 6.3, and so contribute to Q_{across}. We'll calculate dQ_{across}, the charge crossing the small patch dA, by treating the atomic dipole as a pair of charges $+q$ and $-q$ with vector $\boldsymbol{\ell}$ from $-q$ to q. The dipole moment is $q\boldsymbol{\ell}$. If the center of a polarized atom lies within distance $\ell_n/2$ of the patch dA, inside or outside $\mathcal{V}$ (see Fig. 6.3) then net charge q will cross dA in the outward direction; note that $\ell_n \equiv \hat{\mathbf{n}} \cdot \boldsymbol{\ell}$ is the component of $\boldsymbol{\ell}$ normal to dA. If the center of the atom is inside $\mathcal{V}$ (within distance $\ell_n/2$ of dA) then $+q$ crosses dA in the $+\hat{\mathbf{n}}$ direction when the atom is polarized; if the center of the atom is outside $\mathcal{V}$ then $-q$ crosses dA in the $-\hat{\mathbf{n}}$ direction. In either case net charge q crosses dA in the outward direction. The *total charge* crossing dA is q times the number of atoms in the volume $\ell_n dA$; that is, $dQ_{\text{across}} = q\,(n_a dV)$, where n_a is the atomic density. But $q\ell_n n_a$ is just the normal component of the polarization field, so $dQ_{\text{across}} = P_n dA$. Integrating over the entire surface $\mathcal{S}$ yields

$$Q_{\text{across}} = \oint_{\mathcal{S}} P_n dA = \int_{\mathcal{V}} \nabla \cdot \mathbf{P} d^3 x, \tag{6.20}$$

the second equality by Gauss's theorem. Comparing (6.19) and (6.20), which must hold for any volume $\mathcal{V}$ in the dielectric, we see that $\rho_b = -\nabla \cdot \mathbf{P}$, as claimed.

If the polarization field is uniform, i.e., independent of $\mathbf{x}$, then $\rho_b(\mathbf{x})$ is 0. In this case the bound charge lies only on the surface of the dielectric, with density $\sigma_b = \hat{\mathbf{n}} \cdot \mathbf{P}$. If the polarization varies within the dielectric, then there is nonzero bound charge at points where $\mathbf{P}(\mathbf{x})$ diverges, with density $-\nabla \cdot \mathbf{P}$. *Divergence of polarization*, not merely polarization, produces bound charge.

There is a simple graphical demonstration that uniform polarization produces a surface charge on the boundary. On a sheet of paper draw a rectangle and fill it with small blue dots, which represent the positive atomic nuclei. On a transparency draw an identical rectangle but fill it with small red dots, which represent the negative electrons. First overlay the two rectangles, which together represent an unpolarized sample. The blue and red dots cancel throughout, so the charge density is 0. Now, to simulate uniform polarization, move the transparency a small distance. There will be a layer of blue dots, i.e., positive σ_b, on one side, and a layer of red dots, i.e., negative σ_b, on the other side. In the volume, the blue and red still cancel, so ρ_b is 0. (We could demonstrate volume bound charge if the transparency were stretchable!)

We have concentrated on the bound charge, a property of the polarized material. Now what about the field? The bound charge creates an electric field, so the dielectric becomes a field source. One way to calculate the field in the presence of a dielectric is first to determine the bound charge densities, ρ_b and σ_b, and then apply the standard methods for finding the field due to a known charge distribution.

EXAMPLE 1   Determine the bound charge density and the electric field of a uniformly polarized sphere.

Figure 6.4 shows the sphere of radius a. Let the z axis be in the direction of the polarization **P**. Then for $r \leq a$, $\mathbf{P(x)}$ is $P_0\hat{\mathbf{k}}$ where P_0 is a constant. The volume charge density ρ_b is 0, because $\nabla \cdot \mathbf{P} = 0$. The surface charge density on the spherical boundary $r = a$ is $\sigma_b(\theta) = \hat{\mathbf{r}} \cdot \mathbf{P} = P_0 \cos\theta$.

Figure 6.4(a) shows why polarization produces surface charge in this example. For $0 \leq \theta < \pi/2$ the surface density is positive, because there the positive ends of atomic dipoles are at the surface; for $\pi/2 < \theta \leq \pi$ the surface density is negative because the negative ends are at the surface.

We may construct the potential function for the polarized sphere by the methods of Chapters 4 and 5. The mathematical problem is identical to a spherical surface with charge density $\sigma(\theta) = P_0 \cos\theta$. The interior potential is $C_1 r \cos\theta$, and the exterior potential is $C_2 r^{-2} \cos\theta$, where C_1 and C_2 are constants. These functions satisfy Laplace's equation for $r < a$ and $r > a$. The constants are de-

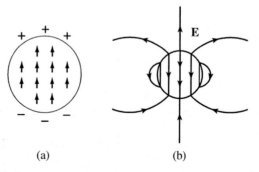

(a)   (b)

FIGURE 6.4 Example 1. A uniformly polarized dielectric sphere. (a) The atomic dipoles inside the sphere produce a bound charge on the surface, with density $\sigma(\theta) = P\cos\theta$. (b) Electric field lines of the sphere.

termined by the boundary conditions at $r = a$. The potential must be continuous at $r = a$,

$$V_{\text{int}}(a, \theta) = V_{\text{ext}}(a, \theta), \tag{6.21}$$

which implies $C_2 = C_1 a^3$. The normal component of $\mathbf{E}$ has a discontinuity at $r = a$ proportional to the surface charge density

$$E_{r,\text{ext}}(a, \theta) - E_{r,\text{int}}(a, \theta) = \frac{\sigma_b(\theta)}{\epsilon_0} = \frac{P_0 \cos \theta}{\epsilon_0} \tag{6.22}$$

and this implies $C_1 = P_0/3\epsilon_0$. Thus the potential function for the polarized sphere is

$$V_{\text{int}}(r, \theta) = \frac{P_0 r}{3\epsilon_0} \cos \theta \quad \text{for } r \le a, \tag{6.23}$$

$$V_{\text{ext}}(r, \theta) = \frac{P_0 a^3}{3\epsilon_0 r^2} \cos \theta \quad \text{for } r \ge a. \tag{6.24}$$

The electric field is $-\nabla V$. Inside the sphere $\mathbf{E}$ is $-P_0\hat{\mathbf{k}}/(3\epsilon_0)$, a uniform field in the $-\hat{\mathbf{k}}$ direction. Outside the sphere $\mathbf{E}$ is a pure dipole field. The dipole potential is $p \cos \theta/(4\pi\epsilon_0 r^2)$, so the dipole moment of the sphere is

$$\mathbf{p} = 4\pi\epsilon_0 \frac{P_0 a^3}{3\epsilon_0} \hat{\mathbf{k}} = \frac{4}{3}\pi a^3 \mathbf{P}. \tag{6.25}$$

The result makes sense: The dipole moment is the moment density ($\mathbf{P}$) times the volume of the sphere. Figure 6.4(b) shows the electric field lines for a uniformly polarized dielectric sphere.

6.3 ■ THE DISPLACEMENT FIELD

The fundamental equations for static electric fields are $\nabla \times \mathbf{E} = 0$ and $\nabla \cdot \mathbf{E} = \rho/\epsilon_0$. These equations are true in vacuum or in a dielectric. In either case ρ must include all the charge sources. In a dielectric there may be charge density from bound charge in addition to any other charge that is present. The bound charge depends on the behavior of the atoms that compose the dielectric. To determine the field and charge, we must know how the particular dielectric material responds to an electric field.

To analyze a dielectric system, it is useful to separate the charge density into *bound charge* $\rho_b(\mathbf{x})$, and *free charge* $\rho_f(\mathbf{x})$, by which we mean the charge placed in the system from the outside;

$$\rho(\mathbf{x}) = \rho_b(\mathbf{x}) + \rho_f(\mathbf{x}). \tag{6.26}$$

The total charge density is equal to $\epsilon_0 \nabla \cdot \mathbf{E}$, and the bound charge density is

$-\nabla \cdot \mathbf{P}$ by (6.12). So two of the terms in (6.26) are divergences, and we may combine these to write the equation as

$$\nabla \cdot (\epsilon_0 \mathbf{E} + \mathbf{P}) = \rho_f. \tag{6.27}$$

The combination $\epsilon_0 \mathbf{E} + \mathbf{P}$ is a new vector function denoted $\mathbf{D}(\mathbf{x})$, called the *displacement field*.

The basic equations for the displacement field are its definition

$$\mathbf{D} = \epsilon_0 \mathbf{E} + \mathbf{P} \tag{6.28}$$

and its field equation

$$\nabla \cdot \mathbf{D} = \rho_f. \tag{6.29}$$

By definition (6.28), $\mathbf{D}$ consists of a field term plus a matter term. The field equation (6.29) is another form of Gauss's Law. The corresponding integral relation is

$$\oint_S \mathbf{D} \cdot d\mathbf{A} = (Q_f)_{\text{enclosed}} . \tag{6.30}$$

The important point is that the field equations (6.29) or (6.30) relate $\mathbf{D}$ to the free charge density alone, i.e., without reference to the bound charge. In a highly symmetric system, for which the form of $\mathbf{D}(\mathbf{x})$ can be deduced from the symmetry alone, the "Gaussian surface trick" can be used to determine $\mathbf{D}(\mathbf{x})$ from the flux integral.

Note an important difference between $\mathbf{D}(\mathbf{x})$ and $\mathbf{E}(\mathbf{x})$. The electric field is irrotational, $\nabla \times \mathbf{E} = 0$, so $\mathbf{E}$ may be written as the gradient of a scalar. But in general $\nabla \times \mathbf{D}$ is not 0. Also, there are different *boundary conditions* for $\mathbf{E}$ and $\mathbf{D}$, which will be derived in Sec. 6.5.

A general problem in electrostatics with dielectrics is to determine $\mathbf{E}(\mathbf{x})$ and $\mathbf{P}(\mathbf{x})$ for a given set of *free charges* or *applied fields*. We may imagine an experiment. The observer can set up charges, conducting plates, insulators, etc., and then measure the electric field throughout the space. However, the observer does not get to specify the polarization in a dielectric, because that is determined by atomic or molecular properties of the material.

A common strategy for solving systems with dielectrics is to first find $\mathbf{D}$ for the specified free charge, using the symmetry and Gauss's Law (6.30). But once $\mathbf{D}$ is known, we still need more information to calculate $\mathbf{E}$ or, equivalently, $\mathbf{P}$.

To predict the behavior of the electrostatic system we must know how the dielectrics respond to an applied field. In principle $\mathbf{P}(\mathbf{x})$ could be calculated by solving the Schrödinger equation for all the bound electrons in the system, but that would not be practical. Rather, we need an empirical connection between $\mathbf{P}(\mathbf{x})$ and $\mathbf{E}(\mathbf{x})$, i.e., one more equation to close the set of simultaneous equations. The equation that relates $\mathbf{P}$ and $\mathbf{E}$, or equivalently $\mathbf{D}$ and $\mathbf{E}$, is called the *constitutive equation.* Here we will consider only the simplest case, in which the constitutive equation is an isotropic linear relationship.

6.3.1 ■ Linear Dielectrics

An insulating material for which $\mathbf{P}(\mathbf{x})$ is proportional to $\mathbf{E}(\mathbf{x})$, with a scalar constant of proportionality, is called an *isotropic linear dielectric.* All dilute gases are linear dielectrics, because the isolated molecules have dipole moments proportional to $\mathbf{E}$. Many other materials, especially liquids and amorphous solids like glass and plastics, are accurately described as linear dielectrics. Crystals are different. While there may be a linear relation between $\mathbf{P}$ and $\mathbf{E}$, in general it is a tensor equation because in a crystal $\mathbf{P}$ may have nonzero components perpendicular to $\mathbf{E}$.

Only one parameter is needed to relate $\mathbf{P}$ and $\mathbf{E}$ in an isotropic linear dielectric. However, three constants are conventionally used for the constitutive equation: the *susceptibility* χ_e, the *permittivity* ϵ, and the *dielectric constant* κ. They are defined in Table 6.2. These parameters are not independent; any one determines the other two. For example, from $\mathbf{D} = \epsilon_0 \mathbf{E} + \mathbf{P}$ and $\mathbf{D} = \epsilon \mathbf{E}$, we derive a relation between ϵ and χ_e,

$$\epsilon = \epsilon_0 (1 + \chi_e); \tag{6.31}$$

then by the definition of κ

$$\kappa = 1 + \chi_e. \tag{6.32}$$

Note that χ_e is always > 0, because the atomic dipoles align with $\mathbf{E}$ so that $\mathbf{P}$ points in the same direction as $\mathbf{E}$. Therefore $\epsilon > \epsilon_0$ and $\kappa > 1$ for any dielectric material.

We can now complete our strategy for solving symmetric problems with dielectrics. First find $\mathbf{D}$ from the free charge. Then calculate $\mathbf{E}$ and $\mathbf{P}$ from the linear constitutive equations. Finally, ρ and ρ_b can be determined from $\mathbf{E}$ and $\mathbf{P}$.

In vacuum, the susceptibility is obviously 0, so $\epsilon = \epsilon_0$ and $\kappa = 1$. In matter, these macroscopic parameters are ultimately determined by atomic properties, such as the atomic polarizability α and number density n. In a dilute gas the interactions between atoms are negligible, so each atom has dipole moment $\alpha \mathbf{E}$, and the polarization is $\mathbf{P} = n\alpha \mathbf{E}$; in that case, $\chi_e = n\alpha/\epsilon_0$ and $\kappa = 1 + n\alpha/\epsilon_0$. The

TABLE 6.2 Parameters for an isotropic linear dielectric with their definitions

Parameter	Symbol	Defining equation
susceptibility	χ_e	$\mathbf{P} = \chi_e \epsilon_0 \mathbf{E}$
permittivity	ϵ	$\mathbf{D} = \epsilon \mathbf{E}$
dielectric constant	κ	$\kappa = \epsilon/\epsilon_0$

TABLE 6.3   Dielectric properties of insulators

Material	Dielectric constant κ	Dielectric strength E_{max} in 10^6 V/m
air	1.00059	3
polystyrene	2.5	20
Lucite	2.8	20
Plexiglas	3.4	40
Teflon	2.1	60
Mylar	3.1	
paper	3.7	16
fused quartz	3.8 to 4.1	
Pyrex	4 to 6	14
water	80	
strontium titanate	332	8

dielectric parameters of a gas are not very different from vacuum because n is small. For example, the dielectric constant of air is 1.00059. In condensed matter the relation between κ and atomic parameters is more complicated, because each atom is influenced by nearby atoms. The dielectric parameters of condensed matter may be very different from vacuum. For example, the dielectric constant of water is 80. Table 6.3 lists the dielectric constant for some insulators.

Table 6.3 also lists the *dielectric strength* of the materials. The dielectric strength is the maximum electric field strength E_{max} for which the material remains an insulator. If the field strength exceeds E_{max}, then electrons are pulled away from the nuclei of atoms or molecules and become free to move through the material, making it a conductor. The process when current begins to flow is called *dielectric breakdown*. It is familiar in air as a spark. The dielectric strength is an important parameter if the material is used in a capacitor, because it sets a limit on the maximum potential that the capacitor can maintain without breakdown.

6.3.2 ▪ The Clausius-Mossotti Formula

We have seen that polarization of atoms or molecules by an electric field produces a macroscopic polarization field $\mathbf{P}(\mathbf{x})$, equal to $\chi_e \epsilon_0 \mathbf{E}(\mathbf{x})$ for a linear material. The purpose of this section is to derive a relation between the microscopic parameter α (atomic polarizability) and the macroscopic parameter χ_e (electric susceptibility).

In this derivation we approximate an atom as a small conducting sphere. Undoubtedly this model is an oversimplification. After all, an atom consists of a finite set of electrons in quantum states, rather than a continuous distribution of free charge in a sphere. Nevertheless, this simple model yields a reasonable theory of susceptibility.

The polarization in a dielectric is

$$\mathbf{P}(\mathbf{x}) = n\mathbf{p} \qquad (6.33)$$

where n is the atomic density and $\mathbf{p}$ the mean atomic dipole moment of atoms in the neighborhood of $\mathbf{x}$. We have previously defined the atomic polarizability α by the equation $\mathbf{p} = \alpha\mathbf{E}$, but now we need to be more precise in stating what field $\mathbf{E}$ contributes to the polarization. The total field $\mathbf{E}$ in the neighborhood of $\mathbf{x}$ is the sum of the field due to the atom itself $\mathbf{E}_{\text{self}}$ plus the field produced by all other sources $\mathbf{E}_{\text{other}}$. The field that polarizes the atom is $\mathbf{E}_{\text{other}}$, so

$$\mathbf{p} = \alpha\,(\mathbf{E}(\mathbf{x}) - \mathbf{E}_{\text{self}})\,; \tag{6.34}$$

the atom does not polarize itself, so we discount $\mathbf{E}_{\text{self}}$.

In a dilute gas the dielectric effect is small, so $\mathbf{E}_{\text{other}}$ and $\mathbf{E}(\mathbf{x})$ are both approximately equal to the externally applied field. Then $\mathbf{P} \approx n\alpha\mathbf{E}$, and $\chi_e \approx n\alpha/\epsilon_0$; note that $\chi_e \ll 1$ for a dilute gas. But in condensed matter $\mathbf{E}_{\text{other}}$ and $\mathbf{E}(\mathbf{x})$ differ significantly from the applied field, and from each other, so the calculation of χ_e is more intricate.

Treating the atom as a conducting sphere we can use results from Chapter 4 to estimate $\mathbf{E}_{\text{self}}$. If a conducting sphere of radius a is placed in a uniform field $\mathbf{E}_0$ then it acquires a dipole moment $\mathbf{p} = 4\pi a^3\epsilon_0\mathbf{E}_0$; the external field produced by the sphere itself is a pure dipole field. The *total field* inside the conducting sphere is, of course, 0, because a conductor is an equipotential. Therefore the *self-field* inside the sphere is $-\mathbf{E}_0$, which cancels the uniform applied field. So, in this model of the atom

$$\mathbf{E}_{\text{self}} = -\mathbf{E}_0 = \frac{-\mathbf{p}}{4\pi a^3\epsilon_0}. \tag{6.35}$$

The field produced by a conducting sphere in an applied field has the same form as the field of a uniformly polarized sphere (which we found in Example 1) so another way to characterize this model is that it approximates the atom by a polarized sphere.

To define the radius a of the model atomic sphere, we assume the atoms are densely packed, and set $\frac{4}{3}\pi a^3$ equal to the mean atomic volume in the material, $v = 1/n$. Then by the results of the previous paragraph, the self-field is

$$\mathbf{E}_{\text{self}} = -\frac{n\mathbf{p}}{3\epsilon_0} = -\frac{\mathbf{P}}{3\epsilon_0}. \tag{6.36}$$

Inserting this result into (6.34) the polarization is

$$\mathbf{P} = n\mathbf{p} = n\alpha\left(\mathbf{E} + \frac{\mathbf{P}}{3\epsilon_0}\right). \tag{6.37}$$

Thus there is indeed a linear relation between $\mathbf{P}$ and $\mathbf{E}$, and the susceptibility, defined in Table 6.2, is

$$\chi_e = \frac{n\alpha/\epsilon_0}{1 - n\alpha/(3\epsilon_0)}. \tag{6.38}$$

Or, we may solve for the atomic polarizability

$$\alpha = \frac{\epsilon_0}{n} \frac{\chi_e}{1 + \chi_e/3} = \frac{3\epsilon_0}{n} \frac{\kappa - 1}{\kappa + 2}, \tag{6.39}$$

where in the second equality κ is the dielectric constant, $\kappa = 1 + \chi_e$.

The relation (6.39) is called the Clausius-Mossotti formula. It was first obtained by Mossotti in 1850, by treating the atom as a conducting sphere. The derivation was refined by Clausius in 1879. Remarkably, this formula for the atomic parameter α was known before the structure of the atom was known. Of course a precise theory of the electric susceptibility must use quantum mechanics and modern condensed matter physics. But (6.39) does provide a reasonably accurate theory of the connection between microscopic and macroscopic parameters. For example, from the measured dielectric constant of water we may estimate the dipole moment of a water molecule.[2]

6.3.3 ■ Poisson's Equation in a Uniform Linear Dielectric

In a linear dielectric, $\mathbf{D} = \epsilon \mathbf{E} = -\epsilon \nabla V$. Now suppose the material is *uniform*, i.e., ϵ is constant. In a region with no free charge, $\nabla \cdot \mathbf{D}$ is 0 so $V(\mathbf{x})$ satisfies Laplace's equation. In a region with free charge density $\rho_f(\mathbf{x})$, $V(\mathbf{x})$ satisfies Poisson's equation

$$-\nabla^2 V = \rho_f/\epsilon. \tag{6.40}$$

The source is written as ρ_f/ϵ, rather than ρ/ϵ_0; these are in fact equal, but ρ_f is the quantity that can be specified in an experiment. For a given distribution of *free charge*, along with appropriate boundary conditions, (6.40) determines $V(\mathbf{x})$. From $V(\mathbf{x})$ any other electrostatic quantity can be calculated. So, in a uniform linear dielectric, electrostatics again reduces to a boundary value problem for $V(\mathbf{x})$. We should emphasize that (6.40) is valid if the dielectric is uniform.

EXAMPLE 2　Suppose two free charges are embedded in a uniform dielectric. What is the force on one of the charges?

Let the charges be q_1 located at $\mathbf{x}_1$, and q_2 at $\mathbf{x}_2$, as shown in Fig. 6.5. The free charge consists of q_1 and q_2. The force on q_1 is $\mathbf{F}_1 = q_1 \mathbf{E}_2(\mathbf{x}_1)$, where $\mathbf{E}_2(\mathbf{x})$ is the electric field due both to q_2 *and* to the polarization produced by q_2. The full electric field is $-\nabla V$, where $V(\mathbf{x})$, satisfying Poisson's equation with the source ρ_f/ϵ, is

$$V(\mathbf{x}) = \frac{q_1}{4\pi\epsilon|\mathbf{x} - \mathbf{x}_1|} + \frac{q_2}{4\pi\epsilon|\mathbf{x} - \mathbf{x}_2|}. \tag{6.41}$$

The potential differs from the potential in vacuum by the constant factor $\epsilon_0/\epsilon = 1/\kappa$. The second term in (6.41) is the contribution attributable to the charge q_2

[2] See Exercise 8.

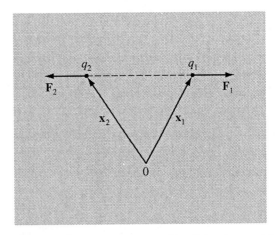

FIGURE 6.5   Charges q_1 and q_2 in a dielectric material.

including the polarization caused by q_2, and $-\nabla$ of that term is $\mathbf{E}_2$. Thus

$$\mathbf{F}_1 = \frac{q_1 q_2 \, (\mathbf{x}_1 - \mathbf{x}_2)}{4\pi\epsilon|\mathbf{x}_1 - \mathbf{x}_2|^3} = \frac{\mathbf{F}_{1\text{vacuum}}}{\kappa}. \tag{6.42}$$

$\mathbf{F}_{1\text{vacuum}}$ means the force on q_1 if no dielectric were present. Because $\kappa > 1$, the force on q_1 is *less than* it would be in vacuum. The charge q_2 is partially "screened" by bound charge of the opposite sign, which reduces its field strength relative to vacuum. Therefore $F_1 < F_{1\text{vacuum}}$.

■───────────────

In general the density of bound charge in a uniform dielectric is

$$\rho_b(\mathbf{x}) = \left(\frac{1}{\kappa} - 1\right)\rho_f(\mathbf{x}). \tag{6.43}$$

The bound charge has the same $\mathbf{x}$-dependence as the free charge, but opposite sign. For example, the bound charge associated with polarization by a point charge is itself a point charge of the opposite sign.[3] Equation (6.43) means that inside a uniform dielectric there can be bound charge only where there is free charge. Also, the total charge density $(\rho_b + \rho_f)$ is $\rho(\mathbf{x}) = \rho_f(\mathbf{x})/\kappa$.

6.4 ■ DIELECTRIC MATERIAL IN A CAPACITOR

Historically, Faraday conducted the first systematic study of dielectrics. He found that if a dielectric material is placed in a capacitor then the capacitance increases. The ratio of the capacitance C with the dielectric to the capacitance C_{vacuum} without the dielectric was the historical definition of the dielectric constant, and we

[3] See Exercise 9.

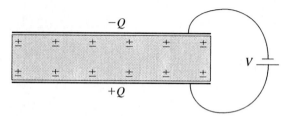

FIGURE 6.6 A parallel plate capacitor filled with a slab of dielectric material. The atomic dipoles produce positive bound charge on the upper surface of the slab, and negative on the lower. Let the z axis be perpendicular to the plates.

will show that the ratio is just κ, the parameter defined in Table 6.2. The relation $C/C_{\text{vacuum}} = \kappa$ provides a method for measuring κ.

Consider a parallel plate capacitor, as shown in Fig. 6.6, with plate area A and separation d, filled with an insulating material with dielectric constant κ. The potential difference across the plates is V; the plates have charge $\pm Q$, the *free charge* in this system. By Gauss's Law, the displacement field between the plates is[4]

$$\mathbf{D} = \sigma_f \hat{\mathbf{k}} = \frac{Q}{A} \hat{\mathbf{k}}. \tag{6.44}$$

To prove this result, apply (6.30) to a pill-box shaped Gaussian surface cutting through one of the plates. The electric field $\mathbf{E}$ between the plates is $\mathbf{D}/\epsilon$, so the potential difference is

$$V = \int_{-}^{+} \mathbf{E} \cdot d\boldsymbol{\ell} = \frac{Qd}{\epsilon A}. \tag{6.45}$$

Thus the capacitance is

$$C = \frac{Q}{V} = \frac{\epsilon A}{d}. \tag{6.46}$$

In terms of the capacitance without the dielectric ($C_{\text{vacuum}} = \epsilon_0 A/d$) the capacitance with the dielectric is

$$C = \kappa C_{\text{vacuum}} \tag{6.47}$$

where $\kappa = \epsilon/\epsilon_0$. The presence of the dielectric increases the capacitance by the factor κ. We have analyzed a parallel plate capacitor, but the same result is true for any generalized capacitor embedded in a dielectric.[5]

---

[4] The z axis is normal to the plates.
[5] See Exercise 12.

6.4.1 ■ Design of Capacitors

In the most general terms a capacitor consists of two conductors in close proximity. The capacitance is $C = Q/V$, where V is the potential difference when their charges are $+Q$ and $-Q$. But in practical terms a "capacitor" means a circuit element that stores charge and energy. A pair of metal plates or foil sheets separated by an insulator—air or a solid dielectric—is a common type of capacitor in circuits. The sandwich of two conductors and an insulator filling may be rolled or folded into a compact space. A coaxial cable is another kind of capacitor, in which charged regions travel along the cable as a wave.

There are several reasons for putting a dielectric material between the conductors of a capacitor. The charge that flows to a capacitor is supposed to pile up on the conductors, but not move through it. The solid dielectric holds the conductors apart, preventing charge from passing from one to the other. Also, the presence of the dielectric increases the capacitance, which is desirable if large capacitance is needed in a small volume. Finally, if the material has a large dielectric strength then breakdown will not occur at high voltage.

For a parallel plate capacitor, we have $C = \kappa \epsilon_0 A/d$. For typical circuit capacitors C is of the order of pF up to μF. These values can be achieved by making A large and d small, e.g., by folding or winding large thin sheets of metal foil separated by thin dielectric layers such as mica or paper. No special technique is needed to make capacitors in this range of C. But for large capacitance, of order 1 F, or even mF, clever designs are required. Two stategies are commonly used to make C large.[6]

In a *multilayer ceramic capacitor*, metal strips are folded into many layers and separated by a ceramic insulator such as barium titanate, which may have a dielectric constant as large as 2×10^4. The combination of multiple layers and large κ can achieve C of order mF.

In an *electrolytic capacitor*, the dielectric is a very thin layer of metal oxide. One conductor is a sheet of metal or metal foil on which the metal oxide is deposited. The other conductor is a conducting acid paste or liquid making good contact with the oxide layer. Aluminum, aluminum oxide, and hydrochloric acid are the classic materials in an electrolytic capacitor. The thickness d of the oxide layer may be as small as 10^{-8} m. Furthermore, the metal surface may be etched into sharp peaks and valleys before depositing the oxide, to increase the surface area A. The combination of small d and large A can achieve C of order 1 F or even larger. However, there are some limitations. Electrolytic capacitors are limited to low voltages, typically a few volts, because the breakdown voltage for such a thin insulating layer is small. Also, an electrolytic capacitor has a *definite directional polarity:* The metal foil must be positive with respect to the acid solution, because otherwise electrolytic action will destroy the oxide layer.

In some applications the capacitance must be kept small. For example, large C in a coaxial cable implies a long time constant for changes of current, and is therefore not suitable for fast switching.

---

[6]D. M. Trotter, Jr., Capacitors, *Sci. Am.,* July 1988.

Small Capacitance in Computer Design.   An actively pursued research problem with important commercial implications is to find insulating materials with very low κ, which can be used to separate the small wires and transistors in chips for the next generation of computers. Currently computers use fused quartz (SiO_2) for this purpose, but as the elements on computer chips get squeezed closer together, the capacitance between neighboring conductors increases. Also, as the conductors are made thinner their resistance increases. The characteristic time of RC circuits is[7] $\tau = RC$, so to have circuits with fast switching times, and to keep signal delays and crosstalk small, it is important for intra- and interlayer capacitive coupling to be small. The ideal is to have insulators with $\kappa \approx 1$, like air, so one approach is to use porous materials. As usual with engineering applications, there are a myriad of associated problems; e.g., the materials must be thermally stable, not chemically active with the Cu wires, and convenient to deposit.

All the capacitors described above would have fixed C. Variable capacitors, such as are used in radio tuners, have interleaved plates, with one set of plates fixed and the other free to rotate, so that the overlap area of the plates can be varied. The capacitance increases as the overlap area increases.

Historically, the first capacitor was the Leyden jar, invented in 1745 by van Musschenbroek in the Netherlands. The original Leyden jar consisted of a glass container partly filled with water, and a metal wire or chain with one end immersed in the water and the other end exposed. The Leyden jar would be charged by bringing the metal chain in contact with a machine that generated static electricity by friction. An improved version had separate metal foils coating the inner and outer surfaces of the glass container, making a capacitor capable of holding a large charge.

During the eighteenth century, demonstrations of static electricity with Leyden jars were a popular entertainment in Europe. Discharging a charged Leyden jar through human volunteers could produce startling, even fatal, effects! In 1752 Benjamin Franklin showed that lightning is an electric discharge in his famous kite experiment. He collected charge from a cloud by charging a Leyden jar from a metal key attached to the wet kite string. The Leyden jar charged by the kite behaved exactly the same as if it had been charged by an electrostatic generator.

6.4.2 ■ Microscopic Theory

Why, physically, does the dielectric in a capacitor increase the capacitance? To answer this question, let's consider again the parallel plates shown in Fig. 6.6. For a given potential V, a larger charge Q resides on the plates if the dielectric is present, i.e., if C is larger. As we have learned, there is bound charge of opposite sign on the dielectric surfaces, which partially screens the free charge on the plates.

Figure 6.6 illustrates the atomic dipoles aligned with the electric field in the dielectric slab. At the upper surface of the slab there is a positive bound charge

---

[7] See Section 7.5.

density $\sigma_{b,u}$ due to the positive ends of dipoles at the surface; at the lower surface the charge density is $\sigma_{b,l} = -\sigma_{b,u}$. The net bound charge is 0. The surface density is $\sigma_{b,u} = P = \chi_e \epsilon_0 E$, by (6.18). Thus the electric field E_z between the plates is a superposition of two terms: $Q/(A\epsilon_0)$ due to the free charge, which is the same as for vacuum between the plates; and $-\sigma_{b,u}/\epsilon_0$ due to the bound charge, which equals $-\chi_e E_z$. That is,

$$E_z = \frac{Q}{A\epsilon_0} - \chi_e E_z.$$

Solving for Q we find that the free charge for a given field strength $E_z = V/d$ is

$$Q = A\epsilon_0(1 + \chi_e)E_z = \frac{\epsilon A V}{d}. \tag{6.48}$$

Q is larger for a dielectric-filled capacitor than for a vacuum-filled capacitor by the factor $\kappa = \epsilon/\epsilon_0$.

6.4.3 ■ Energy in a Capacitor

In Chapter 3 we showed that the *field energy* in an electrostatic system is $(\epsilon_0/2) \int E^2 d^3x$. But this is not the *total energy* stored in a charged capacitor with a dielectric, because there is also some *strain energy* associated with the distortion of the polarized atoms. We can derive the total energy U_{cap} of the capacitor, by calculating the work done to separate the free charge $\pm Q$. If charge dQ is moved from the negative plate to the positive, then the work done is $dW = VdQ$, where $V = Q/C$. The total work that must be done to transfer charge Q, which is equal to U_{cap}, is

$$U_{\text{cap}} = \int_0^Q \frac{Q}{C} dQ = \frac{Q^2}{2C}. \tag{6.49}$$

But compare U_{cap} to the field energy U_{field}. The electric field between the plates is $Q/(\epsilon A)$, and the volume is Ad, so

$$U_{\text{field}} = \frac{\epsilon_0}{2} \left(\frac{Q}{\epsilon A}\right)^2 Ad = \frac{1}{\kappa}\left(\frac{Q^2}{2C}\right), \tag{6.50}$$

where we used (6.46) in the second step. The field energy is smaller than the total energy by the factor $1/\kappa$. The additional energy in U_{cap} is equal to the work that must be supplied to pull the atoms into induced dipoles, against the internal forces of the atom.

The result (6.49) can be generalized. We shall show that the total energy of a linear dielectric is

$$U = \frac{1}{2} \int \mathbf{D} \cdot \mathbf{E} d^3x. \tag{6.51}$$

This agrees with (6.49) because it implies the result $U = U_{cap} = \kappa U_{field}$ that we found before.

To justify (6.51) in general, consider the work done to change the free charge density by a small amount $\delta \rho_f$. The energy added is $\delta U = \int \delta \rho_f V d^3 x$. Now, $\delta \rho_f = \nabla \cdot (\delta \mathbf{D})$ by Gauss's Law, so

$$\delta U = \int \nabla \cdot (\delta \mathbf{D}) \, V d^3 x = \int \delta \mathbf{D} \cdot (-\nabla V) \, d^3 x. \tag{6.52}$$

To obtain the second equality in (6.52) we have integrated by parts and discarded the surface term at infinity. But $-\nabla V$ is $\mathbf{E}$, and for a linear dielectric

$$\delta \mathbf{D} \cdot \mathbf{E} = \epsilon \delta \mathbf{E} \cdot \mathbf{E} = \frac{\epsilon}{2} \delta(E^2) = \frac{1}{2} \delta(\mathbf{D} \cdot \mathbf{E}). \tag{6.53}$$

Thus δU is equal to $\delta \left(\frac{1}{2} \int \mathbf{D} \cdot \mathbf{E} d^3 x \right)$ for an arbitrary change, proving (6.51). We may interpret $\frac{1}{2}\mathbf{D} \cdot \mathbf{E}$ as the total energy density in a linear dielectric, including both field energy and strain energy of the atoms.[8]

EXAMPLE 3 A parallel plate capacitor has plate area A and separation d. Denote by U the total energy if the space between the plates is filled with a dielectric with permittivity ϵ, and by U_0 the energy if the space is vacuum.

Case 1. Calculate U/U_0 if the charge on the plates is specified to be $\pm Q$.

The displacement field is $\mathbf{D} = \sigma_f \hat{\mathbf{n}}$, where $\sigma_f = Q/A$ and $\hat{\mathbf{n}}$ is normal to the plates. (Prove this from Gauss's Law (6.30).) It is important to understand that in this case $\mathbf{D}$ is the same for dielectric or vacuum, because the free charge is specified to be the same. The energy U for the dielectric-filled capacitor is

$$U = \frac{1}{2} \int \mathbf{D} \cdot \mathbf{E} d^3 x = \frac{D^2}{2\epsilon} Ad = \frac{Q^2 d}{2\epsilon A}, \tag{6.54}$$

which is just $Q^2/2C$. The energy for the vacuum-filled capacitor is

$$U_0 = \frac{Q^2 d}{2\epsilon_0 A}. \tag{6.55}$$

The ratio of the energies is

$$\frac{U}{U_0} = \frac{\epsilon_0}{\epsilon} = \frac{1}{\kappa}. \tag{6.56}$$

For specified charge, the dielectric-filled capacitor has *less energy* than the vacuum-filled capacitor.

----

[8]Note that (6.51) is not true for nonlinear dielectrics.

Case 2. Calculate U/U_0 if the potential difference between the plates is specified to be V.

The electric field is $\mathbf{E} = \hat{\mathbf{n}} V/d$ because $\int \mathbf{E} \cdot d\boldsymbol{\ell} = V$. It is important to understand that in this case $\mathbf{E}$ is the same for dielectric or vacuum, because V is specified to be the same. The energy U for the dielectric-filled capacitor is

$$U = \frac{1}{2} \int \mathbf{D} \cdot \mathbf{E} d^3 x = \frac{\epsilon E^2}{2} Ad = \frac{V^2 \epsilon A}{2d}, \tag{6.57}$$

which is just $\frac{1}{2} C V^2$. The energy for the vacuum-filled capacitor is

$$U_0 = \frac{V^2 \epsilon_0 A}{2d}. \tag{6.58}$$

Thus the ratio of the energies is

$$\frac{U}{U_0} = \frac{\epsilon}{\epsilon_0} = \kappa. \tag{6.59}$$

For specified potential difference, the dielectric-filled capacitor has *more energy* than the vacuum-filled capacitor.

6.4.4 ■ A Concrete Model of a Dielectric

By formal considerations we found that the energy density in a linear dielectric is $\frac{1}{2} \mathbf{D} \cdot \mathbf{E}$, where $\mathbf{D} = \epsilon \mathbf{E}$ is the displacement field. This energy includes both field energy and strain energy of the polarized atoms. Can we understand this result from more physical considerations? A complete theory of a polarized atom would require quantum mechanics, but we can gain some insight from a classical model. Treat the atomic dipole as a pair of charges $+q$ and $-q$ bound together by a spring force, with potential energy $\frac{1}{2} K x^2$ where K is the spring constant and x is the separation of the charges in the direction of the applied electric field. We now analyze a dilute gas of these model atoms, to learn something about a dielectric in a field $\mathbf{E} = E_0 \hat{\mathbf{i}}$.

The model atom is in static equilibrium, with charge separation x_{eq}, when the electric force $q E_0$ balances the spring force $K x_{\mathrm{eq}}$ on q (or $-q$); that is, $x_{\mathrm{eq}} = q E_0/K$. Then the dipole moment is $q x_{\mathrm{eq}} = q^2 E_0/K$, and therefore the polarization field is $P_x = n p_x = n q^2 E_0/K$, where n is the atomic density. The displacement field in the material is

$$D_x = \epsilon_0 E_x + P_x = \epsilon_0 \left(1 + \frac{nq^2}{\epsilon_0 K} \right) E_0. \tag{6.60}$$

Now, what is the energy density, u? According to our formal result (6.51),

$$u = \frac{1}{2} \mathbf{D} \cdot \mathbf{E} = \frac{\epsilon_0}{2} \left(1 + \frac{nq^2}{\epsilon_0 K} \right) E_0^2. \tag{6.61}$$

The first term on the right-hand side of (6.61) is the energy density of the electric field, $u_{\text{field}} = \epsilon_0 E_0^2/2$. The second term is the potential energy per unit volume stored in the atomic springs, because

$$\frac{1}{2}\frac{nq^2}{K}E_0^2 = \frac{1}{2}Kx_{\text{eq}}^2 n = u_{\text{spring}}. \tag{6.62}$$

So this classical model shows that

$$\tfrac{1}{2}\mathbf{D}\cdot\mathbf{E} = u_{\text{field}} + u_{\text{spring}},$$

and verifies the interpretation of $\tfrac{1}{2}\mathbf{D}\cdot\mathbf{E}$ as the *total* energy density in the system of atoms and field.

6.5 ■ BOUNDARY VALUE PROBLEMS WITH DIELECTRICS

Something important that we need to learn about dielectrics, and remember for later applications, are the boundary conditions for **E** and **D** at dielectric surfaces. In this section we derive them first, and then use them in some interesting examples.

6.5.1 ■ The Boundary Conditions

Because $\nabla \times \mathbf{E} = 0$, *the tangential components of* **E** *are continuous across any surface in space.* To prove this statement, consider Figure 6.7(a) which shows a surface S cut by a small loop Γ. The loop integral $\oint_\Gamma \mathbf{E}\cdot d\boldsymbol{\ell}$ is 0 for any Γ, by Stokes's theorem. Now let Γ shrink, until its width across S approaches 0; the opposite sides of Γ then coincide with a curve C on S. The loop integral

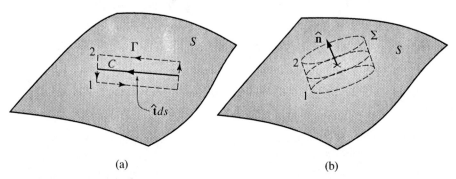

(a)   (b)

FIGURE 6.7   Derivation of the boundary conditions at a surface. (a) The tangential component of **E** is continuous, by applying Stokes's theorem to Γ. (b) The normal component of **D** jumps by σ_{f}, by applying Gauss's theorem to Σ. These conditions might be applied, for example, at the interface between two dielectrics.

approaches $\int_C (\mathbf{E}_2 - \mathbf{E}_1) \cdot \hat{\mathbf{t}} ds$, where $\mathbf{E}_2$ and $\mathbf{E}_1$ are the field vectors just above and below S, respectively, and $\hat{\mathbf{t}}$ is the tangent to C. The fact that this integral is 0 for any curve C implies that the tangential components of $\mathbf{E}$ are continuous across S, as claimed. This is the same argument as in Sec. 3.3.2 in slightly different terms.

Gauss's Law, $\nabla \cdot \mathbf{D} = \rho_f$, implies that the normal component of $\mathbf{D}$ is continuous across S, unless there is a free surface charge, in which case $D_{2n} - D_{1n} = \sigma_f$. To prove this statement, consider Fig. 6.7(b) which shows a Gaussian pillbox Σ cutting through S. In the limit that the height of the pillbox approaches 0, the flux of $\mathbf{D}$ through Σ approaches $\int (D_{2n} - D_{1n}) dA$, where dA is the area of an infinitesimal patch on S. By Gauss's theorem the flux is $\int \sigma_f dA$. The fact that these surface integrals are equal for any area on S implies that $D_{2n} - D_{1n} = \sigma_f$, as claimed. At an interface between two dielectrics, the normal component of $\mathbf{E}$ is usually discontinuous because of bound charge on the surface. If there is no free charge then D_n is continuous, and $E_{2n} - E_{1n} = D_n (\epsilon_2^{-1} - \epsilon_1^{-1})$.

The methods of boundary value problems may be used to calculate the fields in an electrostatic system with dielectrics, by satisfying the above boundary conditions. The next three subsections illustrate the methods.

6.5.2 ■ A Dielectric Sphere in an Applied Field

Figure 6.8 shows a dielectric sphere of radius a in an applied field. If the sphere were not present then the electric field would be $E_0 \hat{\mathbf{k}}$. The sphere distorts the field nearby, but asymptotically $\mathbf{E}$ approaches $E_0 \hat{\mathbf{k}}$.

Because $\nabla \times \mathbf{E} = 0$, we may write $\mathbf{E}$ in terms of a potential,[9] as $\mathbf{E} = -\nabla V$. There is no free charge so $\nabla \cdot \mathbf{D} = 0$. Assuming the interior is a uniform lin-

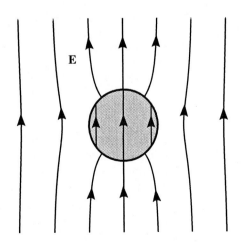

FIGURE 6.8 A dielectric sphere in an applied field.

[9] In general we cannot write $\mathbf{D}$ as a gradient. However, for the special case of a *uniform* linear dielectric, $\mathbf{D} = -\epsilon \nabla V = -\nabla (\epsilon V)$.

ear dielectric ($D = \epsilon E$), and the exterior is vacuum ($D = \epsilon_0 E$), $\nabla \cdot E$ is 0 in both regions, and $\nabla^2 V = 0$. So, the problem is to solve Laplace's equation with appropriate boundary conditions.

Applying methods from Chapters 4 and 5, we shall try interior and exterior solutions of the form

$$V_{\text{int}}(r, \theta) = -C_1 r \cos\theta \quad \text{for} \quad r \leq a, \tag{6.63}$$

$$V_{\text{ext}}(r, \theta) = -E_0 r \cos\theta + \frac{C_2 a^3}{r^2} \cos\theta \quad \text{for} \quad r \geq a. \tag{6.64}$$

These satisfy Laplace's equation in both regions. V_{int} is finite at $r = 0$; and V_{ext} has the correct asymptotic limit as $r \to \infty$, such that $E \to E_0\hat{k}$. The constants C_1 and C_2 must be chosen to satisfy the boundary conditions at $r = a$.

For E_t to be continuous at the surface $r = a$, V must be continuous; therefore $C_1 = E_0 - C_2$. The condition for D_n to be continuous is

$$-\epsilon \left.\frac{\partial V_{\text{int}}}{\partial r}\right|_a = -\epsilon_0 \left.\frac{\partial V_{\text{ext}}}{\partial r}\right|_a ;$$

therefore $\kappa C_1 = E_0 + 2C_2$. The solution to these simultaneous equations is

$$C_1 = \frac{3}{\kappa + 2} E_0 \quad \text{and} \quad C_2 = \frac{\kappa - 1}{\kappa + 2} E_0. \tag{6.65}$$

Some electric field lines are shown in Fig. 6.8. The electric field inside the sphere is $C_1\hat{k}$, uniform in the $\hat{k}$ direction but with strength reduced by the factor $3/(\kappa + 2)$ compared to the applied field. The field outside is the superposition of the applied field and a pure dipole field. Inspecting (6.64) we see that the dipole moment is $p = 4\pi\epsilon_0 C_2 a^3 \hat{k}$. Because $C_2 \propto E_0$, the dipole moment is proportional to the applied field. The *polarizabilty* of the dielectric sphere is the constant of proportionality α in the linear relation $p = \alpha E_0$. The previous result (6.65) implies

$$\alpha = 4\pi\epsilon_0 a^3 \left(\frac{\kappa - 1}{\kappa + 2}\right). \tag{6.66}$$

For comparison, recall that the polarizability of a conducting sphere is $4\pi\epsilon_0 a^3$. If $\kappa = 1$ there is no dielectric and the polarizability is 0. As κ increases α increases, and when κ is large α approaches the value for a conductor.

Finally, it is useful to stand back and compare the results just obtained for a dielectric sphere in an applied field $E_0\hat{k}$, to the results of Example 1 for a sphere with uniform polarization $P_0\hat{k}$. The external field for the polarized sphere is a pure dipole field, while that for the dielectric sphere is a dipole field superimposed on the uniform applied field. These results are perhaps expected. The internal fields are more surprising. Although in both cases the internal field is uniform, the directions are different: E_{int} is in the $-\hat{k}$ direction for the polarized sphere, but in the $+\hat{\mathbf{k}}$ direction for the dielectric sphere. The two problems are closely related, because there is uniform polarization of the sphere in both cases.

6.5.3 ▧ A Point Charge above a Dielectric with a Planar Boundary Surface

Figure 6.9 shows a charge q at distance d from a dielectric. The xy plane is the surface of the dielectric. Again, the potential $V(\mathbf{x})$ must satisfy Laplace's equation in both regions $z \geq 0$ (except at the position of q) and $z \leq 0$. We can construct $V(\mathbf{x})$ by the method of images. However, the image charges for a dielectric are more complicated than the image charge for a conductor.

There are *different* image charges for the regions $z \geq 0$ and $z \leq 0$. In each case the image charge lies in the *other region*, so that Laplace's equation holds for the contribution to V from the image charge. In the region $z \geq 0$, i.e., outside the dielectric, the potential is

$$V_{\text{above}}(x, y, z) = \frac{1}{4\pi\epsilon_0} \left\{ \frac{q}{\sqrt{x^2 + y^2 + (z-d)^2}} + \frac{q'}{\sqrt{x^2 + y^2 + (z+d)^2}} \right\};$$

$$(6.67)$$

q is the real charge at $(0, 0, d)$ and q' is an image charge at $(0, 0, -d)$. In the region $z \leq 0$, i.e., inside the dielectric, the potential is

$$V_{\text{below}}(x, y, z) = \frac{1}{4\pi\epsilon_0} \frac{q''}{\sqrt{x^2 + y^2 + (z-d)^2}};$$
$$(6.68)$$

q'' is an effective charge at $(0, 0, d)$ in the region above the dielectric. Note that the field *inside* the dielectric ($z \leq 0$) is the same as for a point charge at the position of q, but with a different strength q''.

The image charges q' and q'' must be chosen so that the boundary conditions are satisfied. The condition that V is continuous at $z = 0$ implies $q + q' = q''$.

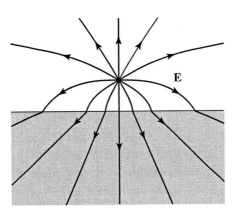

FIGURE 6.9   A point charge above a planar dielectric surface.

A brief calculation[10] shows that the condition for D_n to be continuous at $z = 0$ implies $q - q' = \kappa q''$. The solution to these simultaneous equations is

$$q' = -\frac{\kappa - 1}{\kappa + 1}q \quad \text{and} \quad q'' = \frac{2}{\kappa + 1}q. \qquad (6.69)$$

If $\kappa = 1$ then there is no dielectric, and the image charges are $q' = 0$ and $q'' = q$; in other words, $V(\mathbf{x})$ is just the Coulomb potential of q. If $\kappa \gg 1$ then the image charges are approximately $q' = -q$ and $q'' = 0$, the same as for a conducting plane.

Of course q' and q'' are fictitious. The real charge consists of the free charge q and the bound charge σ_b on the dielectric surface. In the region $z > 0$ the electric field due to σ_b is the same as if there were a point charge q' at $z = -d$. By reflection symmetry, in the region $z < 0$ the electric field due to σ_b is the same as if there were a point charge q' at $z = +d$; adding this image charge to the real charge q gives the effective charge $q'' = q + q'$ at $z = d$.

6.5.4 ■ A Capacitor Partially Filled with Dielectric

Figure 6.10 shows parallel plates of size $\ell \times w$ and separation d, partially filled with a dielectric slab. The dimensions of the slab are $x \times w \times d$, with $x < \ell$. What is the capacitance? Assume the separation d is small compared to x and $\ell - x$, so that end effects are negligible.

Let the potential difference between the plates be V. The key to this example is that *the electric field $\mathbf{E}$ is the same in the two regions (dielectric and vacuum) between the plates.* By planar symmetry, the field is uniform in both regions and normal to the plates.[11] But $\mathbf{E}$, being normal to the plates, is *tangent* to the boundary surface of the dielectric. Since the tangential field must be continuous, the field is the same on either side of the boundary. Thus the electric field is $\mathbf{E} = \hat{\mathbf{n}}V/d$ in both regions, where $\hat{\mathbf{n}}$ is normal to the plates.

To determine the capacitance we must relate V to the charge $\pm Q$ on the plates. Apply Gauss's Law to a surface surrounding one plate, e.g., the lower plate: The flux of $\mathbf{D}$ is equal to Q. That is, since $\mathbf{D}$ is $\epsilon\mathbf{E}$ in the dielectric (of area xw) and

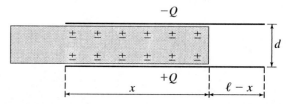

FIGURE 6.10   A dielectric slab inserted part-way between charged plates. A force is pulling the slab into the space between the plates.

[10]See Exercise 18.

[11]Approximating $\mathbf{E}$ as uniform neglects the end effects at the end of the dielectric.

$\epsilon_0 \mathbf{E}$ in the vacuum region [of area $(\ell - x)w$],

$$\epsilon Exw + \epsilon_0 E(\ell - x)w = Q. \tag{6.70}$$

Substituting $E = V/d$ we find that the capacitance is

$$C = \frac{Q}{V} = [\epsilon x + \epsilon_0(\ell - x)]\frac{w}{d}. \tag{6.71}$$

We might have anticipated this result by noting that the system is equivalent to two capacitors in parallel, with $C_1 = \epsilon x w/d$ and $C_2 = \epsilon_0(\ell - x)w/d$. Then the combined capacitance is $C_1 + C_2$.

There is a related example, in which the dielectric slab has the full area of the plates but a thickness less than the separation between the plates. In this case it is the displacement field $\mathbf{D}$ rather than $\mathbf{E}$ that is the same in the slab and the vacuum, because the normal component of $\mathbf{D}$ is continuous.[12]

The Force on the Dielectric Slab

The energy formula (6.51) can be used to evaluate the force on a dielectric slab inserted part way between oppositely charged plates, as shown in Fig. 6.10. The plates have dimensions $\ell \times w$, and are separated by d. A length x of the dielectric lies between the plates. The slab will experience a force pulling it into the capacitor. Qualitatively it is easy to see why the slab is pulled in. The atoms in the material are polarized by the electric field, and therefore attracted to the charged plates. But the problem is to calculate the force F. We'll determine F in two different ways in the following examples.

EXAMPLE 4 Plates with fixed charge. Suppose the plates are isolated so that their charges $\pm Q$ are constant. The work done by the electrostatic force if the slab moves a distance dx into the capacitor is $F\,dx$. Because the system is isolated, conservation of energy says that the work is equal to $-dU$ where U is the total energy of the capacitor. Therefore the inward force on the slab is $F = -dU/dx$.

To calculate U for this example it is natural to use the equation $U = Q^2/2C$, because Q is constant. The capacitance is given by (6.71).[13] Thus the energy is

$$U(x) = \frac{Q^2 d}{2w\,[\epsilon x + \epsilon_0(\ell - x)]} \tag{6.72}$$

and the force on the slab is

$$F = \frac{Q^2 d\,(\epsilon - \epsilon_0)}{2w\,[\epsilon x + \epsilon_0(\ell - x)]^2}. \tag{6.73}$$

---

[12]See Exercise 21.
[13]The capacitance in (6.71) was derived under the assumption that end effects are negligible, which is valid if $d \ll x$ and $d \ll \ell - x$.

For fixed charge the force decreases as x increases.[14] We may also express F in terms of the potential difference between the plates, $V = Q/C$, as

$$F = \frac{V^2 w \, (\epsilon - \epsilon_0)}{2d}. \tag{6.74}$$

EXAMPLE 5 Plates with fixed potential difference.   Now suppose the plates in Fig. 6.10 are connected to a battery so that the potential difference V is fixed. In this case the work done by the electrostatic force F is not equal to $-dU$, because additional energy is supplied by the battery. If the slab moves by distance dx, then charge dQ is transferred to the plates from the battery, and so the battery supplies energy $(dQ)V$. The conservation of energy in this case is

$$dU = (dQ)V - F \, dx. \tag{6.75}$$

Here it is natural to use the equation $U = CV^2/2$, because V is constant. Also, $Q = CV$ so $dQ = (dC)V$. Inserting these relations into the conservation law gives

$$F \, dx = -\tfrac{1}{2}(dC)V^2 + (dC)V^2 = \tfrac{1}{2}(dC)V^2. \tag{6.76}$$

Thus the force on the slab is

$$F = \frac{1}{2}\frac{dC}{dx}V^2 = \frac{V^2 w \, (\epsilon - \epsilon_0)}{2d}. \tag{6.77}$$

This result is the same as (6.74) obtained earlier by considering the plates to be isolated.

If V is held constant then the force is independent of x, according to (6.77). This is true if the fraction of the slab between the plates is not too close to 0 or 1. An approximation has been made in deriving (6.77), that end effects can be neglected at the end of the dielectric. This is valid if $x \gg d$ and $\ell - x \gg d$, so for that range of x the force for constant V is approximately independent of x. But if $x = \ell$, i.e., the slab is completely inserted, then it is obvious by symmetry that the force must be 0, not (6.77).

FURTHER READING

1. S. Gasiorowicz, *Quantum Physics*, 2nd ed. (Wiley, New York, 1996). Chapter 16 describes the H atom in an electric field.

2. H. C. Ohanian, *Principles of Quantum Mechanics* (Prentice Hall, Englewood Cliffs, NJ, 1990). Section 10.1 describes the H atom in an electric field.

[14]See Exercise 22.

EXERCISES

Sec. 6.1. Atomic and Molecular Dipoles

6.1. Consider two electric dipoles $\mathbf{p}_1$ and $\mathbf{p}_2$ located, respectively, at the origin and on the x axis at $(a, 0, 0)$. There are no other charges or fields present.

  (a) Let $\mathbf{p}_1 = p\,\hat{\mathbf{i}}$ and $\mathbf{p}_2 = p\,\hat{\mathbf{k}}$. If $\mathbf{p}_1$ is held fixed, what is the direction of the torque on $\mathbf{p}_2$? If $\mathbf{p}_2$ is held fixed, what is the direction of the torque on $\mathbf{p}_1$? If both vectors are free to rotate, what are the orientations at static equilibrium?

  (b) Rank the following four configurations in order of the interaction energy, from smallest energy to largest: (i) $\mathbf{p}_1 = p\,\hat{\mathbf{k}}$ and $\mathbf{p}_2 = p\,\hat{\mathbf{k}}$, (ii) $\mathbf{p}_1 = p\,\hat{\mathbf{k}}$ and $\mathbf{p}_2 = -p\,\hat{\mathbf{k}}$, (iii) $\mathbf{p}_1 = p\,\hat{\mathbf{i}}$ and $\mathbf{p}_2 = p\,\hat{\mathbf{i}}$, (iv) $\mathbf{p}_1 = p\,\hat{\mathbf{i}}$ and $\mathbf{p}_2 = -p\,\hat{\mathbf{i}}$. (This exercise can be solved analytically, but you might find it easier to use a computer to calculate the energies, treating each dipole as 2 charges $+q$ and $-q$ separated by a small distance, say $0.01a$.)

6.2. Figure 6.11 shows a dipole $\mathbf{p}$ at distance z_0 from a grounded conducting plane, taken to be the xy plane. The direction of $\mathbf{p}$ is at an angle θ to the normal of the plane. Find the torque on $\mathbf{p}$. What are the equilibrium values of θ? [Answer: $\mathbf{N} = -p^2 \sin\theta \cos\theta\,\hat{\mathbf{j}}/(32\pi\epsilon_0 z_0^3)$]

6.3. Suppose the magnitude of the permanent dipole moment of a polar molecule is 1 D (Debye). If the molecule is placed in an electric field $E_0\hat{\mathbf{k}}$, then the thermal average $\langle p_z \rangle$ is given by the Langevin formula (6.8).

  (a) Plot $\langle p_z \rangle$ versus temperature T in Kelvin, for $E_0 = 10^6$ V/m. Be sure to label the axes. Notice that the interesting temperature range is at low T.

  (b) Plot the polarizability, defined by $\alpha = d\langle p_z \rangle/dE_0$, versus T.

6.4. (a) A dipole $\mathbf{p}$ with fixed magnitude $|\mathbf{p}| = p_0$ is placed in an external electric field $\mathbf{E}(\mathbf{x})$ that varies with position. Show that there is a force on the dipole, and find how this force depends on the orientation of the dipole and the variation of the field. [Answer: $\mathbf{F} = (\mathbf{p} \cdot \nabla)\,\mathbf{E} = \nabla(\mathbf{p} \cdot \mathbf{E})$]

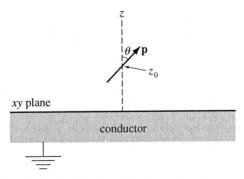

FIGURE 6.11   **Exercise 2.** An electric dipole interacting with a grounded conducting plane.

(b) Find the force and torque on a dipole in the field of a point charge. Let the charge q be at the origin, and the dipole

$$\mathbf{p} = p_0 \left(\sin \xi_0 \,\hat{\mathbf{i}} + \cos \xi_0 \hat{\mathbf{k}} \right)$$

be at the point $(0, 0, z_0)$. Also, find the force on q, and verify Newton's third law. [Answer: The force on q is

$$\frac{p_0 q}{4\pi \epsilon_0 z_0^3} \left(- \sin \xi_0 \,\hat{\mathbf{i}} + 2 \cos \xi_0 \hat{\mathbf{k}} \right) .]$$

Sec. 6.2. Polarization and Bound Charge

6.5. A dielectric object that has a quasi-permanent polarization when the applied field is 0 is called an *electret*. Consider a uniformly polarized electret in the shape of a cylinder of height h and radius $10h$. The polarization in the dielectric is $P\hat{\mathbf{k}}$, where $\hat{\mathbf{k}}$ is parallel to the cylinder axis.

(a) Sketch the electric field lines.

(b) Calculate the electric field $\mathbf{E}$ at the center of the cylinder. Because the radius is large compared to the height, you may neglect edge effects.

(c) Calculate the electric field $\mathbf{E}$ on the midplane of the cylinder, at distance $100h$ from the center. Because the distance is large compared to the radius, the dipole dominates the multipole expansion.

6.6. Many microphones manufactured today are based on the "foil electret" design. Obtain information on the foil electret from a library or the internet. What is a foil electret, and how is it used in a microphone?

Sec. 6.3. The Displacement Field

6.7. Two concentric conducting spherical shells, with radii a and $2a$, have charge $+Q$ and $-Q$ respectively. The space between the shells is filled with a linear dielectric with permittivity

$$\epsilon(r) = \frac{\epsilon_0 a}{1.5a - 0.5r} \, ,$$

which varies with radial distance r from ϵ_0 at $r = a$ to $2\epsilon_0$ at $r = 2a$.

(a) Use Gauss's Law to determine the displacement field between the spherical shells.

(b) Determine the bound charge density between the spherical shells.

(c) Determine the total energy U of this system, from the relation $U = \frac{1}{2} \int \mathbf{D} \cdot \mathbf{E} d^3 x$.

(d) Determine the capacitance. [Answer: $C = 4\pi \epsilon_0 a / (0.75 - 0.5 \ln 2)$]

6.8. On the Clausius-Mossotti formula

(a) Sketch a plot of the dielectric constant κ versus atomic density n.

(b) According to Langevin's formula, the polarizability of a polar molecule at temperature T, for $pE \ll kT$, is $\alpha = p^2/(3kT)$, where p is the permanent dipole moment. From the measurement $\kappa = 80$ for water, calculate the dipole moment of a water molecule. Express the result in units of ea_B.

6.9. A point charge Q is embedded in a dielectric medium with dielectric constant κ.

  (a) What is the free charge enclosed by a sphere of radius R centered at Q?

  (b) What is the bound charge enclosed by the sphere? How does the result vary with R?

  (c) What is the total charge enclosed by the sphere?

  (d) Explain the result of (b) microscopically, assuming an atom to be a tiny pair of equal but opposite charges $\pm e$ with dipole moment $\mathbf{p} = \alpha \mathbf{E}$. (Hint: For small atomic density n, approximate $\chi_e = n\alpha/\epsilon_0$.)

6.10. The dielectric constant of air is 1.00059. From this determine the mean polarizability of atmospheric molecules. Compare the result to the atomic polarizabilities in Table 6.1.

Sec. 6.4. Dielectric in a Capacitor

6.11. Consider a parallel plate capacitor, with plate area A, plate separation d, and dielectric constant κ of the material between the plates.

  (a) Evaluate the capacitance if $A = 100\,\text{cm}^2$, $d = 1\,\text{mm}$, and $\kappa = 120$.

  (b) If the potential difference between the plates is 6 V, calculate the energy of the capacitor. [Answer: $U = 1.9 \times 10^{-7}\,\text{J}$]

6.12. Two conductors of arbitrary geometry are embedded in a uniform linear dielectric with dielectric constant κ. Show that the capacitance is equal to κ times the capacitance for the same conductors in vacuum.

6.13. A multi-plate capacitor consists of 6 parallel plates, as shown in Fig. 6.12. Alternating plates are connected to opposite terminals. Calculate the capacitance if the plates have area A, separation d, and are separated by a material with permittivity ϵ. [Answer: $5\epsilon A/d$]

6.14. Two isolated square parallel conducting plates, of side L and separation d, are charged with surface densities $+\sigma$ on the upper plate and $-\sigma$ on the lower. Two dielectric slabs, each with thickness $d/2$ and area $L \times L$, are inserted between the plates, one slab above the other as shown in Fig. 6.13(a). The dielectric constants are κ_1 and κ_2. Assume $d \ll L$. Determine:

  (a) $\mathbf{D}$ everywhere between the plates.

  (b) $\mathbf{E}$ everywhere between the plates.

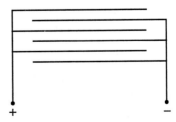

FIGURE 6.12   **Exercise 13.** A multi-plate capacitor with six plates.

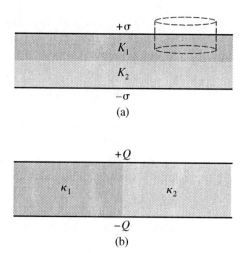

FIGURE 6.13   (a) Exercise 14. A cylindrical Gaussian surface is indicated by the dashed outline. (b) Exercise 15.

   (c) The bound surface charge densities σ_b on the three dielectric surfaces.

   (d) The capacitance. [Answer: $C = (\epsilon_0 L^2/d)\{2\kappa_1\kappa_2/(\kappa_1 + \kappa_2)\}]$

   Finally, verify Gauss's Law in the form $\oint \mathbf{D} \cdot d\mathbf{A} = Q_f$ and in the form $\oint \mathbf{E} \cdot d\mathbf{A} = Q/\epsilon_0$.

6.15. Suppose the volume between the two plates in the previous exercise is filled with two dielectric slabs, each with thickness d and area $L \times L/2$, inserted between the plates side by side as shown in Fig. 6.13(b). The dielectric constants are κ_1 and κ_2. Charge $+Q$ is placed on the upper plate and charge $-Q$ on the lower plate. Determine:

   (a) $\mathbf{D}$ everywhere between the plates. (Hint: σ is not constant on the plates.)

   (b) $\mathbf{E}$ everywhere between the plates.

   (c) The bound surface charge densities σ_b at the four areas where the dielectric slabs touch the plates.

   (d) The capacitance. [Answer: $C = (\epsilon_0 L^2/d)\{(\kappa_1 + \kappa_2)/2\}]$

   (e) Compare the capacitance for this configuration to that of the configuration in the previous exercise, for the special case $\kappa_2 = 1$. (Note that the capacitances of the two configurations are the same if $\kappa_1 = \kappa_2$.)

6.16. Consider a spherical capacitor consisting of concentric conducting spheres of radii a and b. If the region between the spheres is vacuum, then the capacitance is $4\pi\epsilon_0 ab/(b - a)$.

   (a) What is the capacitance if the volume between the conductors is partly filled with dielectric as shown in Fig. 6.14(a)? (Hint: D_n is continuous at the dielectric boundary.) [Answer: $C = 4\pi\kappa\epsilon_0 ab(a + b)/((\kappa a + b)(b - a))]$

   (b) What is the capacitance if the volume between the conductors is half-filled with dielectric as shown in Fig. 6.14(b)? (Hint: E_t is continuous at the dielectric boundary.) [Answer: $C = 2\pi\epsilon_0(\kappa + 1)ab/(b - a)]$

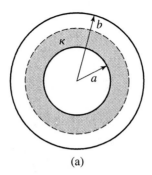
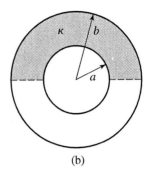

(a)   (b)

FIGURE 6.14   Exercise 16. The concentric conducting spherical shells have radii a and b. A dielectric material occupies part of the volume between the spheres. In (a) the dielectric is a spherical shell with inner radius a and outer radius $(a + b)/2$. In (b) the dielectric is a hemispherical shell of thickness $b - a$.

Sec. 6.5. Boundary Value Problems with Dielectrics

6.17. Consider an electric field line passing through a planar interface between two insulating media with dielectric constants κ_1 and κ_2. Assume there is no free charge on the interface. Let θ_1 and θ_2 be the angles between the field line and the normal to the interface in the two regions. Prove that $\kappa_1 \cot \theta_1 = \kappa_2 \cot \theta_2$.

6.18. For a point charge q above a planar dielectric surface, shown in Fig. 6.9, show that the boundary conditions on $\mathbf{E}_t$ and D_n imply $q + q' = q''$ and $q - q' = \kappa q''$. Here q' and q'' are the image charges; see Eqs. (6.67) and (6.68).

6.19. A point charge q is located in vacuum at distance z from a dielectric with a planar boundary surface.

(a) Determine the force on the charge.

(b) Determine the surface charge density on the dielectric boundary. [Answer: $\sigma = -(2dq/4\pi r^3)\{(\kappa - 1)/(\kappa + 1)\}$]

6.20. (a) Sketch a plot of the electric field lines for a dielectric sphere in a uniform applied field. Show the directions, and explain the directions from the distribution of bound charge.

(b) Sketch a plot of the electric field lines for an electrostatic system consisting of a point charge above a planar dielectric surface.

6.21. A slab of dielectric with area A and thickness t is placed in a parallel plate capacitor having plate area A and separation d, with $d > t$. Show that the capacitance is $\epsilon_0 \epsilon A/[\epsilon(d - t) + \epsilon_0 t]$.

6.22. Plot a graph of the force pulling a dielectric slab into the region between parallel plates with fixed charges $\pm Q$ (the system shown in Fig. 6.10) as a function of the length x of slab between the plates. Let $Q = 1\,\mu C$, $\epsilon = 5\epsilon_0$, $d = 1\,\text{cm}$, $w = 5\,\text{cm}$, and $\ell = 20\,\text{cm}$. Note that Eq. (6.73) was derived assuming $d \ll x$ and $d \ll \ell - x$. What is the force on the slab if $x = \ell$? Why is (6.73) not correct in the latter case?

6.23. A hollow dielectric sphere, with dielectric constant κ, inner radius a and outer radius b, is placed in a uniform applied field $E_0\hat{\mathbf{k}}$. The presence of the sphere changes the field. Find the field inside and outside the sphere, and in the dielectric material. What is the field at the center of the sphere? What is the dipole moment of the sphere? [Answer:

$$p_z = \frac{4\pi\epsilon_0 b^3(a^3 - b^3)(2\kappa + 1)(\kappa - 1)E_0}{2a^3(\kappa - 1)^2 - b^3(2\kappa + 1)(\kappa + 2)}]$$

General Exercises

6.24. A high voltage coaxial cable has an inner wire with radius 0.3 cm and an outer sheath with radius 0.8 cm. The insulating material is a plastic with dielectric strength 40 MV/m and dielectric constant $\kappa = 3$. What is the maximum voltage difference V_{max} that can be supported without dielectric breakdown? Does V_{max} depend on κ?

6.25. (a) Two parallel conducting plates in air are connected to the terminals of a 12 V battery. The capacitance in air is $1.0\,\mu$F. Then the battery is disconnected and a plexiglas sheet that just fits the gap is inserted between the plates. What is the final potential difference, in V? What is the stored energy, in J?

(b) Again the two plates in air are connected to the 12 V battery. While the battery is still connected, the plexiglas sheet is inserted between the plates. What is the stored energy, in J?

6.26. The space between two parallel conducting disks, which have radii R and separation $d \ll R$, is filled with a nonuniform dielectric whose permittivity varies linearly as a function of the distance from the center,

$$\epsilon(r) = \epsilon_1 + (\epsilon_2 - \epsilon_1)r/R.$$

Calculate the capacitance. [Answer: $(\pi R^2/d)(\epsilon_1/3 + 2\epsilon_2/3)$]

6.27. Two identical parallel-plate capacitors, each with capacitance C, are connected in series across a fixed total potential V. Then a slab of dielectric, with dielectric constant κ, is inserted into one of the capacitors. Calculate the change of total energy stored in the two capacitors, the work done by the electrostatic force on the slab as it is inserted, and the energy supplied by the voltage source.

6.28. Consider two insulating media with dielectric constants κ_1 and κ_2, placed together with a planar interface between them. In the region of κ_1 there is a line charge with charge per unit length λ, parallel to the interface at perpendicular distance a. Use the method of images to find the electric field in both media. Show that the force per unit length on the line charge is

$$\frac{dF}{dL} = \frac{\lambda^2(\kappa_1 - \kappa_2)}{4\pi\epsilon_1 a(\kappa_1 + \kappa_2)}.$$

Explain physically why the force is away from the interface for $\kappa_1 > \kappa_2$ and toward the interface for $\kappa_1 < \kappa_2$.

6.29. A spherical capacitor consists of an inner metal sphere at radius a, an insulating shell with dielectric constant κ_1 for $a < r < b$, another insulating shell with dielectric constant κ_2 for $b < r < c$, and an outer metal sphere at radius c. Determine the capacitance.

6.30. Two coaxial thin-walled conducting tubes with radii a and b are dipped vertically into a dielectric liquid of suspectibility χ_e and mass density ρ. If a voltage difference V_0 is applied to the tubes, the liquid rises to a height h in the space between the tube walls.

   (a) What is h in terms of the other parameters of the problem? [Answer: $h = (\epsilon_0 \chi_e V_0^2)/(\rho g (b^2 - a^2) \ln(b/a))$.]

   (b) Note that this is generally a small effect. For example, calculate h if the dielectric is water, for $a = 1.0$ cm, $b = 1.2$ cm, and $V_0 = 500$ volts. [Answer: 2.2 mm]

6.31. A very long dielectric cylinder of radius a and dielectric constant κ is placed in a field $\mathbf{E}_0$ perpendicular to its axis.

   (a) What is the electric field inside the dielectric cylinder?

   (b) What is the induced dipole moment per unit length?

   (Hint: Take the cylinder axis to be the z axis, and $\mathbf{E}_0 = E_0 \hat{\mathbf{i}}$. The potential for a line dipole, with dipole moment per unit length $\mathbf{p}' = C \hat{\mathbf{i}}$, is $C \cos \phi / (2 \pi \epsilon_0 r)$ in cylindrical coordinates.)

CHAPTER

7

Electric Currents

So far we have only considered electrostatics, the physics of charges at rest. In this chapter we begin the study of *electrodynamics*, the physics of charges in motion. An electric current is a net flow of charge, due to motion of charged particles, as illustrated in Fig. 7.1. The mathematical description of current is a necessary preparation for the theory of magnetism, because electric current is a basic source of the magnetic field.

The most familiar electric currents are those in metal wires, such as incandescent bulb filaments, power cords of appliances, and windings of electric motors. But there can also be currents in regions where there are no metal conductors. There are currents in fluorescent lamps and neon signs, in which charges move through a gas at low pressure in a tube. Lightning is flow of charge through the open air. There are electric currents in nerve cells, due to motion of positive and negative ions through plasma membranes.

A simple, prototype current is steady-state, or direct current, i.e., one in which the mean velocity of the charges is constant. Ordinary household and industrial applications involve alternating currents, which oscillate at a frequency of 60 Hz. The essential physics of steady-state currents applies also to ac currents because the time scales of the microscopic processes that determine electrical conductivity are very short. For example, Ohm's law, and the resistance of a wire, are essentially the same for direct current or 60 Hz alternating current.

"Flow" is a continuum concept. Think of water flowing in a river. But charge resides in discrete atomic particles, which move individually in very irregular motions. Because current is the net flow of charge, averaged over many particles, it is a smooth function of **x** and t.

Some results in this chapter, particularly the continuity equation, are true for currents with arbitrary time dependence, as we will see. The continuity equation is a fundamental result, relating current and charge densities, and expressing the conservation of charge.

7.1 ■ ELECTRIC CURRENT IN A WIRE

The conceptually simplest example of an electric current is the current in a thin conducting wire. In an ideal one-dimensional wire the current I is defined as the net charge passing a point P per unit time. In a real wire I is defined as the charge

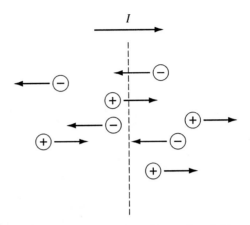

FIGURE 7.1   Current. Current may be produced by motion of either positive or negative
charged particles, or both.

per unit time passing through a cross section of the wire at P

$$I = \frac{dQ}{dt},$$ (7.1)

where dQ denotes the *net charge* passing the position P during the infinitesimal
time interval dt. For a current to the right (see Fig. 7.1),

$$dQ = dQ_+(R) - dQ_-(L),$$

where $dQ_+(R)$ is the charge of positive carriers moving to the right past P, and
$dQ_-(L)$ is the charge of negative carriers moving to the left, as illustrated in
Fig. 7.1. The unit of current is the ampere (A), which is the basic electric unit in
the SI system. The definition of the ampere is discussed in Chapter 8, in terms of
the magnetic force on a current carrying wire.[1] A coulomb of charge is defined
by $1\,C = 1\,A\,s$. If the wire carries a current of 1 A, then 1 C of net charge passes
P each second.

   If the current in a wire is due to charges q moving with mean velocity v, and
the charges have linear density n_L (= number of charge carriers per unit length),
then

$$I = qn_L v.$$ (7.2)

Note that the units of $qn_L v$ are C/s $=$ A. To appreciate the meaning of (7.2),
consider the wire shown in Fig. 7.2. Assuming the mobile charges are positive,
the amount of charge that will pass the cross section at P, during dt, is equal to the

[1] By definition, the force per unit length on very long parallel wires carrying 1 A and separated by 1 m
is exactly 2×10^{-7} N/m.

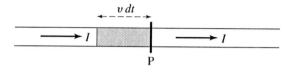

FIGURE 7.2   Current in a wire. For positive charge carriers, the charge in the length $v dt$ passes a cross section at P during the time dt. Therefore $I = q n_L v$.

charge of the mobile particles contained in the length $v\, dt$ upstream of P, because those are the particles that pass P during dt. That charge is $dQ = q n_L (v\, dt)$; hence (7.2).

In a metal wire the charge carriers are the conduction electrons, and each electron has *negative* charge $-e$ where $e = 1.602 \times 10^{-19}$ C. A current in the wire corresponds to electrons moving on average in the direction opposite to the current. dQ in (7.1) is the *net charge* passing P: negative charge moving in the negative direction is a positive current, as is positive charge moving in the positive direction. If there are both positive and negative charge carriers $\pm q$ with equal linear density n_L and moving with velocities $v_\pm$ in the one-dimensional space, then $I = q n_L (v_+ - v_-)$.

7.2 ■ CURRENT DENSITY AND THE CONTINUITY EQUATION

Equation (7.1) defines the current in a wire, i.e., along a curve. More generally we need to define current throughout a volume of space. For example, the space might be occupied by a metal or a plasma. How should we describe the current *at a point* in the material? We denote by $\mathbf{J}(\mathbf{x})$ the *volume current density* at the point $\mathbf{x}$. The definition of $\mathbf{J}(\mathbf{x})$ is that if $d\mathbf{A}$ is an infinitesimal area at $\mathbf{x}$ in the volume, then $\mathbf{J}(\mathbf{x}) \cdot d\mathbf{A}$ is the net charge per unit time passing through $d\mathbf{A}$; that is,

$$dI = \mathbf{J} \cdot d\mathbf{A}, \tag{7.3}$$

as illustrated in Fig. 7.3. In other words, $J_i(\mathbf{x})$ is the current per unit area at $\mathbf{x}$ in the direction of $\hat{\mathbf{e}}_i$.[2] The unit of current density is A/m^2.

If the current in a volume of space is due to charges q with volume number density n (= number of charge carriers per unit volume) moving with mean velocity $\mathbf{v}$, then the current density is

$$\mathbf{J} = q n \mathbf{v}. \tag{7.4}$$

Note that the units of $q n \mathbf{v}$ are A/m^2. Current density is the flux of electric charge. In general flux is equal to density times velocity. For example, in hydrodynamics the mass flux is the mass density times the fluid velocity. Or, another example will

---

[2] The suffix $i = 1, 2, 3$ corresponds to the x, y, z direction, respectively.

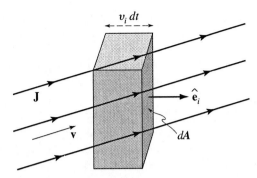

FIGURE 7.3 **Volume current density. J** is the current per unit area, so the charge passing the area $d\mathbf{A}$ during dt is $\mathbf{J} \cdot \hat{\mathbf{e}}_i \, dA \, dt = J_i \, dA \, dt$, which is equal to $qn v_i \, dA \, dt$ if the charge carriers have velocity **v**.

be discussed later in this book when we will learn that the energy flux in an electromagnetic wave is the field energy density times the velocity of propagation. In (7.4) qn is the charge density of the current carriers. To understand (7.4) consider Fig. 7.3, which demonstrates that the flowing charge contained in a volume of size $d\mathbf{A} \cdot \mathbf{v} dt$ will pass through $d\mathbf{A}$ during the time interval dt. That charge, $qn(d\mathbf{A}) \cdot \mathbf{v} dt$, is by definition $\mathbf{J} \cdot (d\mathbf{A}) dt$; hence (7.4).

Besides one-dimensional current (I) in a wire, and three-dimensional current density (**J**) in a volume, we also encounter two-dimensional current density (**K**) for charge flow on a surface. The definition of surface current density is that the charge per unit time flowing across a line segment $d\ell$ on the surface is

$$dI = \mathbf{K} \cdot \hat{\mathbf{e}}_\perp d\ell, \tag{7.5}$$

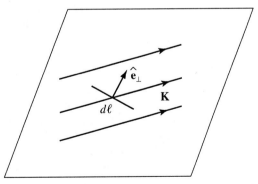

FIGURE 7.4 **Surface current density. K** is the current per unit length on the surface. The charge per unit time passing through a line segment $d\ell$ is $\mathbf{K} \cdot \hat{\mathbf{e}}_\perp d\ell$, where $\hat{\mathbf{e}}_\perp$ is perpendicular to $d\ell$, and lies on the surface. In the figure the direction of $\hat{\mathbf{e}}_\perp$ has been chosen such that $\mathbf{K} \cdot \hat{\mathbf{e}}_\perp$ is positive.

where $\hat{\mathbf{e}}_\perp$ is a unit vector lying in the surface and perpendicular to $d\ell$. The surface current is illustrated in Fig. 7.4. If the current in the surface is perpendicular to $d\ell$ then $dI = K\,d\ell$; if the current is parallel to $d\ell$ then dI is 0. The units of $\mathbf{K}$ are A/m. If σ is the charge per unit area of charges moving on the surface with velocity $\mathbf{v}$, then $\mathbf{K} = \sigma\mathbf{v}$.

7.2.1 ■ Local Conservation of Charge

The net charge of an isolated system is constant. But *furthermore*, charge is not only conserved overall in a system, it is also conserved point by point throughout the system. This *local conservation of charge* is described mathematically by the *continuity equation*

$$\nabla \cdot \mathbf{J} = -\frac{\partial \rho}{\partial t}. \tag{7.6}$$

Here $\rho(\mathbf{x}, t)$ is the volume charge density ($=$ charge per unit volume), and $\mathbf{J}(\mathbf{x}, t)$ is the volume current density ($=$ current per unit area). Equation (7.6) is universally true, for arbitrary time dependence.

The continuity equation (7.6) states that the rate of charge flow away from the point $\mathbf{x}$ is equal to the rate of decrease of charge at $\mathbf{x}$. Recall the meaning of divergence from Chapter 2. By the definition (2-34), $\nabla\cdot\mathbf{J}d^3x$ is the flux of $\mathbf{J}$ through the boundary of the infinitesimal volume d^3x. Because charge is conserved, that flux must equal the rate of decrease of charge inside d^3x, i.e., $-(\partial\rho/\partial t)d^3x$; hence (7.6). If we integrate Eq. (7.6) over an arbitrary volume V, and use Gauss's theorem to convert the volume integral of $\nabla \cdot \mathbf{J}$ to the surface integral of $\mathbf{J}$, then we obtain the *integral form* of the continuity equation,

$$\oint_S \mathbf{J} \cdot d\mathbf{A} = -\frac{d}{dt}\int_V \rho d^3x. \tag{7.7}$$

Equation (7.7) states that the rate of charge passing outward through the closed surface S is equal to the rate of decrease of charge in the enclosed volume V. Equation (7.6), or equivalently (7.7), is a basic equation of electrodynamics, expressing local conservation of charge.

Conservation of charge is one of the most profound laws of nature. From an experimental viewpoint, charge is conserved because the elementary charged particles—electrons and protons in ordinary matter—are themselves conserved, never appearing from, nor disappearing into, nothingness. In modern theories of the fundamental interactions, conservation of charge is a consequence of *gauge symmetry*, i.e., the fact that the dynamics is invariant under certain transformations of the basic fields called gauge transformations. This abstract idea has far-reaching theoretical implications.

7.2.2 ■ Boundary Condition on J(x, t)

In field theory the primary equations are partial differential equations, which specify how the fields vary in space and time. If there are boundaries in the system,

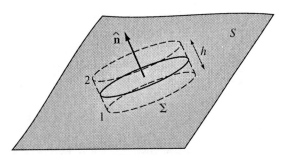

FIGURE 7.5   The boundary condition for J. For a steady current, $\nabla \cdot \mathbf{J} = 0$. Then $J_{2n} = J_{1n}$.

such as interfaces between different materials, then the differential equations must be augmented by *boundary conditions*. The appropriate boundary conditions are themselves determined by the partial differential equations. We encounter numerous examples of boundary conditions in field theory. For example, we have already seen that the tangential components of $\mathbf{E}(\mathbf{x})$ are continuous across a boundary surface. Also, the normal component of $\mathbf{D}(\mathbf{x})$ is continuous across a surface without free charge; or, D_n jumps by σ_{free} across a surface with free charge density.

In this section we derive the boundary condition satisfied by the current density $\mathbf{J}(\mathbf{x}, t)$. Let S be a surface in space. S may be any surface—even an imaginary mathematical surface—but the interesting case is where S separates two different materials. Now consider a cylindrical Gaussian surface Σ, in the shape of a pill box of height h, cutting across S, as shown in Fig. 7.5. According to the continuity equation, the flux of $\mathbf{J}$ out through Σ is equal to $-dQ/dt$, where Q is the charge enclosed by Σ. To derive the boundary condition, let $h \to 0$. In the limit, the cylindrical side of Σ approaches 0 in area, so the flux through it approaches 0; and the top and bottom faces of Σ each approach a patch of area A on S, from opposite sides of S. Thus the flux integral approaches $\int_A (J_{2n} - J_{1n}) dA$, where J_{1n} and J_{2n} denote the component of $\mathbf{J}(\mathbf{x}, t)$ normal to the surface (in the direction from 1 to 2) on the opposite sides of A. As $h \to 0$, the enclosed charge Q approaches the surface charge $\int_A \sigma dA$. Using the fact that these surface integrals must be equal for *any* area A on S, we see that the normal component of $\mathbf{J}(\mathbf{x}, t)$ must satisfy

$$J_{2n} - J_{1n} = -\frac{\partial \sigma}{\partial t}. \tag{7.8}$$

If the surface charge density σ on S is constant in time, for example 0, then the normal component of $\mathbf{J}(\mathbf{x}, t)$ is continuous across S. That is a consequence of charge conservation: If charge flows into the surface on one side, then an equal amount must flow away from the surface on the other side if σ remains constant. Conversely, if J_n is discontinuous, then the difference between the current away from the surface on one side and that into the other side must equal the rate of decrease of σ.

The argument leading to (7.8) is a good example of a derivation of the boundary condition for a vector function. The starting point was the partial differential equation $\nabla \cdot \mathbf{J} = -\partial\rho/\partial t$. Because this involves the divergence, it implies a condition on the normal component at any surface.

7.3 ■ CURRENT AND RESISTANCE

We learned in Chapter 4 that a conductor in electrostatic equilibrium is an equipotential, with $\mathbf{E} = 0$ inside it. However, if there is current in the conductor then it is not an equipotential. The current is driven by the potential gradient, i.e., by nonzero $\mathbf{E}$ in the conductor. At the end of this section we will see that for a steady current, as for electrostatics, the interior of the conductor is electrically neutral; any *excess charge* must reside on the surface. Charge moves around inside the conductor in response to the electric field but $\rho = 0$ if the current is constant in time.

7.3.1 ■ Ohm's Law

How is the current related to the potential gradient in a conductor? If two terminal points on a conductor are held at a constant potential difference V, e.g., by connecting them to the electrodes of a battery, then in equilibrium a steady current flows through the conductor. Let I be the total current at either point, i.e., the integrated flux $\int \mathbf{J} \cdot d\mathbf{A}$ through a surface inside the conductor surrounding the point.[3] It is found empirically that for many cases the current and potential difference are proportional,

$$V = IR. \tag{7.9}$$

Equation (7.9) is *Ohm's law*.[4] The constant of proportionality R is called the *resistance* of the conductor. The SI unit of resistance is the ohm (Ω), defined by $1\,\Omega = 1$ V/A. The reciprocal $1/R$ is called the *conductance*, and its unit is Ω^{-1}, or siemens (S).[5,6] Ohm's law holds to a very good approximation for many conductors. However, it is not a universal principle, and in Sec. 7.7 we'll see an example where it does not hold.

The resistance R of a sample of matter is a function of the geometry (size and shape) of the sample, and of the material composition. For example, the resistance of a uniform cylinder, of length L and cross section A, is proportional to L and inversely proportional to A,

---

[3] The surface integral does not include the wire from the battery. For a *closed* surface integral, which would include the cross section of the wire, $\oint \mathbf{J} \cdot d\mathbf{A} = 0$ by conservation of charge.

[4] Georg Simon Ohm, b. Erlangen in 1789, d. Munich in 1854.

[5] William Siemens, b. Hannover in 1823, d. London in 1883.

[6] In some places the unit of conductance is called by the cute name "mho" for 1/ohm. Honoring Herr Siemens is preferred.

TABLE 7.1 Resistivity ρ for some materials at 0° C or at room temperature

Material	Resistivity ρ ($\Omega\,m$)
pure metals[a] (0° C)	
Ag	1.47×10^{-8}
Cu	1.54×10^{-8}
Al	2.43×10^{-8}
Be	2.71×10^{-8}
W	4.82×10^{-8}
Zn	5.59×10^{-8}
Fe	8.64×10^{-8}
Pt	9.8×10^{-8}
Cs	18.0×10^{-8}
Pb	19.2×10^{-8}
Ti	45×10^{-8}
Hg[b] (liq., 20° C)	95.8×10^{-8}
Bi	127×10^{-8}
α-Mn	137×10^{-8}
sea water	0.21
semiconductors (20°C)	
Ge	~ 0.5
Si	~ 2300
insulators	
wood	10^8 to 10^{11}
glass	10^{10} to 10^{15}

[a] J. Bass, Landolt-Börnstein Volume 15. *Metals: Electronic Transport Phenomena* (Springer-Verlag, Berlin, 1982).

[b] *American Institute of Physics Handbook*, 2nd ed. (McGraw-Hill, New York, 1963).

$$R = \rho L / A. \qquad (7.10)$$

The parameter ρ (not to be confused with charge density $\rho(\mathbf{x})$!) is an intrinsic property of the material called the *resistivity*. Table 7.1 lists the approximate resistivities of some materials. Metals, semiconductors, and insulators have vastly different values of resistivity; the mechanisms of charge transport are different for these different classes of materials. Typical ranges of resistivity are: ρ(pure metals) $\approx 10^{-8}\,\Omega\,m$, ρ(alloys) $\approx 10^{-6}\,\Omega\,m$, ρ(semiconductors) $\approx 10^{-5} - 10^3\,\Omega\,m$, and ρ(insulators) $\approx 10^8 - 10^{17}\,\Omega\,m$. We also define the *conductivity* σ of the material by $\sigma = 1/\rho$; the units of σ are $(\Omega\,m)^{-1}$, or S/m.

Table 7.1 illustrates the full range of ρ for pure metals, which is only two orders of magnitude. The table also shows that resistivities of insulators are much higher than those of metals; indeed the common definition of an insulator is that the material does not conduct electricity. The microscopic reason for large ρ is that the energy required to move electrons from bound states in the insulator's atoms, into states in which they can move through the insulator, is very large, so that essentially no electrons are available to transport charge.

Local Form of Ohm's Law

We may also write a *local form* of Ohm's law, which is a more basic equation than (7.9):

$$\mathbf{J(x)} = \sigma \mathbf{E(x)}. \tag{7.11}$$

Again, σ is the conductivity. We will often use σ rather than ρ to express the material property; to use Table 7.1, $\sigma = 1/\rho$. This proportionality between the current density $\mathbf{J}$ and the electric field $\mathbf{E}$, point by point in the material, holds to a good approximation throughout uniform isotropic conductors. Equation (7.11) is not universally true, but rather an empirical fact for many isotropic materials. A physical model that suggests why $\mathbf{J}$ should be proportional to $\mathbf{E}$ in simple materials will be explored in the next section. In *crystals* it is found experimentally that an electric field in one direction can cause current in another direction; in this case the conductivity σ must be generalized to a tensor quantity.

Resistance Calculations

We can use the local form of Ohm's law (7.11) to calculate the resistance of a sample of matter with a given geometry. For example, consider a uniform cylinder of length L and cross section A, with a potential difference V between the ends. The electric field in the cylinder is $E = V/L$, and the current is $I = JA = \sigma EA$. Thus the resistance is

$$R = \frac{V}{I} = \frac{L}{\sigma A}, \tag{7.12}$$

which is the relation (7.10) stated earlier.

EXAMPLE 1   Consider a cylindrical resistor with a resistivity $\rho(z)$ that varies along the length as shown in Fig. 7.6. The resistant material occupies the region $0 \le z \le h$ and the cross section is A. Above and below the resistor is a perfect conductor ($\rho = 0$). We'll calculate the total resistance, and the *net charge distribution* of the resistor if there is a steady current I.

The resistor can be subdivided into slices of height dz, and the resistance of each slice is $dR = \rho(z)dz/A$. Because the current flows uniformly through the slices in series, the total resistance is the sum of the resistances of the slices,

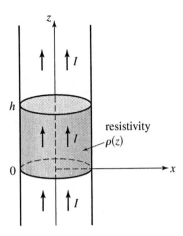

FIGURE 7.6   Example 1. A cylindrical resistor with resistivity $\rho(z)$ that varies along its length.

$$R = \int dR = \frac{1}{A} \int_0^h \rho(z)dz. \tag{7.13}$$

If the resistivity is a constant ρ_0, then $R = \rho_0 h/A$.

Inside the resistor, and on the top and bottom faces, we can determine the interesting charge distribution using the methods of Chapter 6. The current is uniform so $\mathbf{J}(\mathbf{x}) = I\hat{\mathbf{k}}/A$. Then the electric field is[7]

$$\mathbf{E}(\mathbf{x}) = \frac{\rho(z)I}{A}\hat{\mathbf{k}}. \tag{7.14}$$

Assuming the permittivity ϵ is constant, the free charge density in the volume of the resistor is

$$\rho_{\text{free}}(\mathbf{x}) = \epsilon \nabla \cdot \mathbf{E} = \frac{\epsilon I}{A}\left(\frac{d\rho}{dz}\right). \tag{7.15}$$

The surface charge densities on the boundaries at $z = 0$ and $z = h$ are

$$\sigma_{\text{free}} = \begin{cases} +\epsilon E_z(0) = \dfrac{\epsilon I}{A}\rho(0) & \text{for } z = 0 \\[2mm] -\epsilon E_z(h) = -\dfrac{\epsilon I}{A}\rho(h) & \text{for } z = h \end{cases}. \tag{7.16}$$

Thus a positive charge exists on the upstream boundary surface, e.g., at $z = 0$ if $I > 0$, and a negative charge exists on the downstream surface. However, the total charge on the resistor is 0 because the integrated volume charge, which is $\epsilon I\,[\rho(h) - \rho(0)]$, is equal but opposite to the net charge on the surfaces.

---

[7]Note that all field equations and boundary conditions are satisfied by this solution: The curl of $\mathbf{E}$ is 0, $\mathbf{E}$ is normal to the equipotential surfaces at $z = 0$ and $z = h$, and $\mathbf{J}$ is tangential on the cylinder.

EXAMPLE 2   What is the resistance for a radial current between concentric conducting spheres with radii a and b, if there is a material with conductivity σ between the spheres?

We'll solve this problem in two different ways. First, using the geometrical method of the previous example, divide the conducting volume into spherical shells of radius r and thickness dr, and treat the shells as resistors in series. The resistance of the shell at r is $dR = dr/(\sigma\, 4\pi r^2)$. The total resistance is $\int dR$; that is,

$$R = \frac{1}{4\pi\sigma}\left(\frac{1}{a} - \frac{1}{b}\right). \tag{7.17}$$

Second, calculate V/I for a given potential difference V between the spheres. To have the potential difference V we know from electrostatics (Gauss's Law) there must be a charge Q on the inner sphere, with

$$V = \frac{Q}{4\pi\epsilon_0}\left(\frac{1}{a} - \frac{1}{b}\right). \tag{7.18}$$

The current is obtained by integrating $\mathbf{J}$, which is $\sigma\mathbf{E}$, over any concentric sphere in the conducting region

$$I = \oint \sigma\mathbf{E}\cdot d\mathbf{A} = \frac{\sigma Q}{\epsilon_0}, \tag{7.19}$$

where the second equality follows from Gauss's Law. The resistance $R = V/I$ is again (7.17).

Now is a good time to derive an interesting *general* relation connecting the capacitance C of two conductors and the resistance R between them. Consider Conductors 1 and 2 shown in Fig. 7.7. A potential difference V has been set up between them, say by connecting and then disconnecting a battery, so that they have charges $\pm Q$, where $C = Q/V$. The space around the conductors is filled with a material of permittivity ϵ and conductivity σ. Some lines of $\mathbf{E}$ are also shown on the figure. Current $I = V/R$ will flow from $1 \to 2$ along the lines of $\mathbf{E}$. Suppose σ is small so that I is small.

Now consider the Gaussian surface S, shown dashed in Fig. 7.7, which completely surrounds Conductor 1. Just after the battery is disconnected we may write, using $\mathbf{J} = \sigma\mathbf{E}$ and $\nabla\cdot\mathbf{E} = \rho_f/\epsilon$,

$$I = \oint_S \mathbf{J}\cdot\hat{\mathbf{n}}\,dA = \sigma\oint_S \mathbf{E}\cdot\hat{\mathbf{n}}\,dA = \sigma\int \nabla\cdot\mathbf{E}\,d^3x = \frac{\sigma}{\epsilon}\int \rho_f d^3x = \frac{\sigma}{\epsilon}Q. \tag{7.20}$$

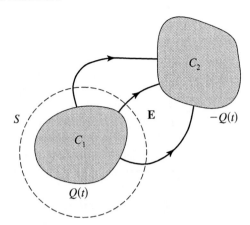

FIGURE 7.7   Two conductors in a conducting medium. The capacitance and resistance are related by (7.21). S is a Gaussian surface.

Going from the third expression to the fourth we've used Gauss's theorem. If we now substitute $I = V/R$ and $Q = CV$ in the first and last terms of (7.20) we find a relation between C and R,

$$RC = \frac{\epsilon}{\sigma}. \tag{7.21}$$

Equation (7.21) is very interesting because it connects two quantities R and C that depend in a complicated way on the configuration (shape, size, separation) of the two conductors. The physics underlying this connection is that the same $\mathbf{E}(\mathbf{x})$ determines both the current and the potential difference.

Notice that both sides of (7.21) have dimensions of time. We will see in Section 7.6 that the ratio ϵ/σ is the time constant for decay of a charge concentration in a conductor, and the product RC is the decay time for a series RC-circuit. Also, R varies inversely with σ, and C is proportional to ϵ. Equation (7.21) has practical implications too. It shows that any *real* capacitor has a finite leakage resistance because $\sigma \neq 0$, and any *real* resistor has a finite capacitance because $\epsilon \neq \infty$. These nonideal properties of real circuit devices obey (7.21).

7.3.2 ▪ Fabrication of Resistors

A *resistor* is a circuit element with a specified value of resistance. Resistors are found in all practical electric circuits, with resistances ranging from a few Ω to $k\Omega$, or even $M\Omega$. These resistances are much larger than those in the wires or contacts in the circuit.

Three common types of resistors are wire-wound, thin film, and composition. A *wire-wound resistor* is a long, very fine wire wound on an insulating support. The resistance, $\rho L/A$, can be predicted precisely, so this type is useful where an accurate resistance value is important. The wire is usually a metal alloy, such as manganin or constantin. These materials have resistivities almost independent of temperature, so such resistors can be used in applications where the Joule heat is so large that it raises the temperature significantly. In a *thin-film resistor*, the conducting path is a thin film of metal on a cylindrical insulating support. The resistance is inversely proportional to the thickness of the film. The advantage of this design is that the inductance is small, whereas a long wire, wound in coils, would have a large inductance. Thin-film resistors are used in high-frequency applications, where inductance would produce undesired effects. A *composition resistor* consists of graphite (C) granules pressed together and held in an insulating glue. Graphite is a semimetal, i.e., a conductor with high resistivity. Composition resistors are often used in applications requiring large resistance.

7.3.3 ▒ The Surface Charge on a Current Carrying Wire

Picture a wire whose ends are connected to the electrodes of a battery. The wire will carry a constant current because its ends are held at different potentials maintained by the chemical emf of the battery. Now consider a local view: The current is distributed over the cross section of the wire and is related to the field by the local version of Ohm's law, $\mathbf{J} = \sigma \mathbf{E}$. There is a local field $\mathbf{E}$ everywhere in the wire where current exists, and it is fair to say that this $\mathbf{E}$ drives the charges in the current. Now, what is the source of $\mathbf{E}$? This may seem like a simple question, but the answer is surprisingly difficult.

Because there is no time dependence, the source of $\mathbf{E}$ can only be a charge distribution. But where can the charges be? Inside the wire the charge density must be 0. To see this, note that in the steady state the rate of charge entering one end of the wire equals the rate of charge flowing out the other end, so that no charge is accumulating in the wire. This means that $\partial \rho / \partial t = 0$ in the wire, and so $\nabla \cdot \mathbf{J} = 0$ by the continuity equation. Then by the local form of Ohm's law, $\nabla \cdot \mathbf{E} = 0$ in the wire, and so $\rho = 0$ by Gauss's law, as claimed. Thus the charge producing $\mathbf{E}$ is not inside the wire. Outside the wire could be vacuum, so the charge isn't there. All that remains is the surface. The answer to the questions we have posed is that there is in general a *surface charge distribution* on the wire. This surface charge determines $\mathbf{E}$, by Gauss's Law, both outside and inside the wire, and hence the current in the wire.[8]

Calculating the surface charge distribution is generally difficult. The reason is that it depends sensitively on the geometry of the *entire* circuit including the battery. A few ideal models can be solved analytically, involving for example coaxial conductors for the current path and parallel plates or a "spherical battery" for the emf.[9]

Although the calculation is difficult for us, nature has no apparent difficulty in solving even complicated boundary value problems of this kind. After all, the

---

[8] J. D. Jackson, *Am. J. Phys.* **64**, 855 (1996). Jackson describes three roles of the surface charge: "(1) to maintain the potential around the circuit, (2) to provide the electric field in the space around the circuit, and (3) to assure the confined flow of current."

[9] N. W. Preyer, *Am. J. Phys.* **68**, 1002 (2000). Preyer reviews the history of this interesting subject.

wire can be bent into a hairpin, or even tied in a bow, and immediately nature arranges the surface charge such that its $\mathbf{E}$ maintains the current $\mathbf{J} = \sigma\mathbf{E}$ confined in the wire.[10] Although the net charge on the surface of a current-carrying wire is small, because the capacitance is small, e.g., $C' \approx 1\,\text{pF/m}$ for household lamp wire, it is important to the phenomenon.

The field inside a wire.   To find the field inside a current-carrying wire, it is necessary in general to solve a difficult boundary-value problem. However, an ideal straight wire with circular cross section can be analyzed as follows. Let the z axis be the location of the wire, so that by translation invariance the current density has the form $\mathbf{J} = J_z(r)\hat{\mathbf{k}}$. The field is electrostatic so $\nabla \times \mathbf{E} = 0$; then because $\mathbf{J} = \sigma\mathbf{E}$, also $\nabla \times \mathbf{J} = 0$ in a uniform conductor. Evaluating the curl in cylindrical coordinates gives $\nabla \times \mathbf{J} = -(\partial J_z/\partial r)\,\hat{\boldsymbol{\phi}} = 0$. Therefore J_z is independent of r, i.e., the current must be uniformly distributed on any plane orthogonal to the z axis. It follows that $\mathbf{E} = E_z\hat{\mathbf{k}}$ is uniform in the wire, which is the solution to the problem. But this ideal straight wire is only a prototype model. For a curved wire, $\mathbf{J}$ and $\mathbf{E}$ will not be strictly uniform, although they will be approximately uniform if the radius of curvature is large compared to the radius of the wire.

The Capacitance of a Wire

In Chapters 3 and 4 we discussed the properties of conducting ellipsoids, and we can use the results obtained there to derive the capacitance per unit length of a wire. The idea is to approximate the wire by a long thin ellipsoid. The result of Exercise 24(b) of Chapter 4 gives the potential V_0 of a conducting ellipsoid whose total charge is Q. From the result, the capacitance with respect to potential zero at infinity is $C = Q/V_0$, given by

$$C = \frac{8\pi\epsilon_0\sqrt{c^2 - a^2}}{\ln\left\{\left(c + \sqrt{c^2 - a^2}\right)\big/\left(c - \sqrt{c^2 - a^2}\right)\right\}}, \tag{7.22}$$

in terms of a and c, the semiminor and semimajor axes.

Now, for a long thin ellipsoid, $a \ll c$; so the argument of the logarithm may be expanded as $\{\ldots\} \approx (2c/a)^2$. We may then write

$$C' = \frac{C}{L} \approx \frac{2\pi\epsilon_0}{\ln(L/a)}, \tag{7.23}$$

where L, the length of the wire, is approximated by $2c$. For a wire of length $L = 20\,\text{cm}$ with radius $a = 0.5\,\text{mm}$, (7.23) gives $C' = 6\,\text{pF/m}$, in agreement with the estimate used earlier.

----

[10]"Nature laughs at the difficulties of integration." —P.-S. Laplace

7.4 ■ A CLASSICAL MODEL OF CONDUCTIVITY

Ohm's law is not a universal principle but it is, for many materials and for a wide range of currents, a very good approximation. Why is Ohm's law valid? We can gain some insight into electrical resistance by examining a simple theoretical model of conduction, developed by Drude in 1900, based on classical mechanics. Let n be the density of mobile charge carriers in a material, q the charge of each carrier, and $\langle \mathbf{v} \rangle$ the mean velocity of the carriers, so that the current density is $\mathbf{J} = qn\langle \mathbf{v} \rangle$, as in (7.4).

In a metal the charge carriers are conduction electrons moving at a characteristic speed and undergoing frequent collisions with microscopic scatterers. (This speed is called the thermal speed for historical reasons. We will see below that it is characteristic of the metal rather than the temperature.) The velocity of any specific electron changes frequently because of the collisions. If there is no current, then the electrons undergo random motion with no net drift. If, however, an electric field is applied, there will be a current because the electrons will then move by a combination of random walk and a slow drift with the mean velocity $\langle \mathbf{v} \rangle$ in the direction of $-\mathbf{E}$. The motion of a conduction electron is illustrated in Fig. 7.8.

Let λ be the mean free path between collisions of the electrons. Then the mean time τ between collisions is λ/v_{th}, where v_{th} is the thermal speed, which is much larger than the drift speed $|\langle \mathbf{v} \rangle|$.[11] Now consider a certain electron. Just after a collision, the electron has velocity $\mathbf{v}_0$. The average of $\mathbf{v}_0$ over many collisions is 0, because the collisions randomize the direction of motion. Between collisions the electron has acceleration $\mathbf{a} = q\mathbf{E}/m$. Therefore its average velocity between collisions is of order $\mathbf{a}\tau$, and we do not need to be more precise than this for our estimate of $\langle \mathbf{v} \rangle$. Substituting this drift velocity into (7.4), we find the proportionality $\mathbf{J} = \sigma \mathbf{E}$, where the conductivity is

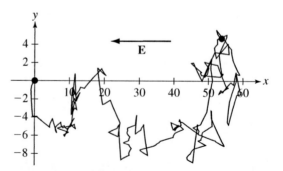

FIGURE 7.8   The Drude model. An electron in a current moves by a combination of random walk, and drift in the direction opposite to **E**. The figure shows a random path in two dimensions, but in a real metal the motion is a random path in three dimensions.

[11] See Exercise 1.

$$\sigma = \frac{nq^2\lambda}{mv_{\text{th}}}. \tag{7.24}$$

Although this model of electric conductivity is only heuristic, it does give a reasonable picture of resistance. For example, it predicts that conductivity σ decreases as temperature increases, which is true for most metals at room temperature, although not for semiconductors. For most simple metals the temperature coefficient of resistivity, defined by

$$\alpha = \frac{1}{\rho}\left(\frac{d\rho}{dT}\right),$$

is approximately constant and equal to $0.004\ \text{K}^{-1}$.

A characteristic property of semiconductors is that, in contrast to metals, their conductivity *increases* with increasing temperature. The microscopic reason is that in semiconductors the energy gap between the bound electronic states in the atoms and the conducting states is small enough that as temperature increases, electrons are freed from the atoms and thus the density of charge carriers increases; this is in contrast to insulators, as we learned earlier. In condensed matter physics one says that electrons in semiconductors are promoted across the energy gap between the top of the valence band and the bottom of the conduction band by increasing temperature. In the commonly used semiconductors, Ge, Si, InSb, and GaAs, that energy gap ranges from 0.2 to 1.4 eV.

It is appropriate here to discuss our current knowledge of the parameters in (7.24). First, the density n of conduction electrons ranges for common metals from about $0.9 \times 10^{28}\ \text{m}^{-3}$ for Cs to $25 \times 10^{28}\ \text{m}^{-3}$ for Be; for Cu at room temperature $n = 8.47 \times 10^{28}\ \text{m}^{-3}$. Second, there is an interesting history and physics associated with v_{th}. Drude's theory (1900) predated quantum mechanics. Classically v_{th} would be calculated from the equipartition theorem

$$\tfrac{1}{2}mv_{\text{th}}^2 = \tfrac{3}{2}kT \tag{7.25}$$

where k is Boltzmann's constant and T is, say, room temperature. But we now know that electrons in a metal do not behave classically, but must be treated by quantum mechanics. Because of the Pauli exclusion principle, conduction electrons occupy states that correspond to velocities from 0 up to the Fermi velocity v_F, defined by

$$\tfrac{1}{2}mv_F^2 = kT_F \tag{7.26}$$

where T_F is the *Fermi temperature*. The relevant electron speed in (7.24) is v_F rather than the classical thermal speed v_{th}. T_F is much higher than room temperature; for example, T_F is 81,600 K for Cu. Therefore the electron speed is ten times larger than the classical value, and so the corresponding mean-free path deduced from (7.24) is ten times longer than in the classical model.[12] Finally, we note that precise calculations of the mean-free time τ, a parameter implicit in (7.24), depend on details of the scattering mechanisms for the conduction electrons. Scattering occurs from phonons (quantized lattice vibrations), impurities and other lattice imperfections, polycrystalline boundaries, etc. Such calculations play an important role in modern theoretical studies of conduction in metals.

7.5 ■ JOULE'S LAW

Resistance is a kind of friction. The current carriers in a resistor, e.g., conduction electrons in a metal, lose energy by collision with atomic-scale scatterers. The energy is transferred to lattice vibrations as heat. We may determine the power dissipated in the resistor, as a function of the current I and resistance R. The work per unit time done by $\mathbf{E}$ as charge passes through the potential difference V, is IV. By energy conservation this must equal the power P dissipated as heat

$$P = IV = I^2R, \tag{7.27}$$

a result known as *Joule's law*. The relation $P = I^2R$ was discovered by Joule in 1841, by experiment. Table 7.2 gives some typical examples of current and power for everyday devices.

We may also write a *local expression* for the power dissipated in resistance, in terms of the current density $\mathbf{J(x)}$ in the conducting material. The work per unit time done on an isolated charged particle q is $\mathbf{F} \cdot \mathbf{v} = q\mathbf{E} \cdot \mathbf{v}$, where $\mathbf{v}$ is the particle velocity. This is the power supplied to q. Similarly, the power supplied to all mobile charge dQ in a volume element d^3x is

$$dP = dQ\mathbf{v} \cdot \mathbf{E} = \rho(\mathbf{x})\mathbf{v} \cdot \mathbf{E}d^3x \tag{7.28}$$

where $\mathbf{v}$ is the mean velocity of the charge carriers, and $\rho(\mathbf{x})$ their charge density. But $\rho(\mathbf{x})\mathbf{v}$ is the current density $\mathbf{J(x)}$, so the power supplied to dQ is $dP = \mathbf{J} \cdot \mathbf{E}d^3x$. That is, the *local* power per unit volume is $\mathbf{J(x)} \cdot \mathbf{E(x)}$, and the *total*

TABLE 7.2   Currents and Powers

Device	Current (A)	Power (W)
light emitting diode	2×10^{-2}	10^{-2}
electric shaver	8×10^{-2}	10
table lamp	0.8	100
hand iron	8	1000
starter motor of a car	180	2200
central air conditioner	20	5000

[12]See Exercise 12.

power is

$$P = \int \mathbf{J} \cdot \mathbf{E} d^3 x = \int \rho J^2 d^3 x. \qquad (7.29)$$

(The ρ in (7.29) is the resistivity, not the charge density!) Equation (7.29) is the generalization of (7.27). Another form of this equation that we will use later, is that the power density is σE^2.

Because energy is dissipated in resistance, a source of energy is necessary to maintain a current in a resistor. For example, a battery is a chemical reactor that maintains a constant potential difference V between its electrodes. The chemical reaction is the energy source. If the battery is used in a circuit with resistance R, then the ideal lifetime of the battery is $U_0 R / V^2$ where U_0 is the initial stored chemical energy.

If there is no energy source in a circuit with resistance, then the current must decay. What would happen to the current in a battery-driven circuit if the battery were replaced by a capacitor with the same voltage? The current would decay to 0.

7.6 ■ DECAY OF A CHARGE DENSITY FLUCTUATION

In electrostatics the net charge density inside a conductor is 0; in equilibrium any net charge of the conductor resides on the surface. But suppose there is at some initial time a nonzero charge density $\rho_0(\mathbf{x})$ in an isolated conductor with conductivity σ. How long does it take to reach equilibrium, i.e., for the charge density to decay? We will answer this question by finding $\rho(\mathbf{x}, t)$. To begin with, we assume that the polarization is negligible in the conductor, so $\epsilon = \epsilon_0$.

The charge density decays as charge flows to the surface. According to the continuity equation, $\partial \rho / \partial t = -\nabla \cdot \mathbf{J}$. In an ohmic conductor, $\mathbf{J}$ is $\sigma \mathbf{E}$. Also, $\nabla \cdot \mathbf{E}$ is ρ / ϵ_0 by Gauss's Law. Taken together these equations imply

$$\frac{\partial \rho}{\partial t} = -\frac{\sigma}{\epsilon_0} \rho. \qquad (7.30)$$

The solution to this differential equation is

$$\rho(\mathbf{x}, t) = \rho_0(\mathbf{x}) e^{-t/\tau}. \qquad (7.31)$$

The charge density decays exponentially with time constant $\tau \equiv \epsilon_0 / \sigma$. For an ideal conductor, i.e., $\sigma = \infty$, the decay time would be 0. For metals τ is very short, and for insulators it is very long. We leave it as an exercise for the reader to substitute physical values, and to calculate τ for various materials.[13]

---

[13] See Exercise 14.

Energy is conserved, so the total energy dissipated as heat and emitted as radiation while the charge moves to equilibrium must equal the decrease in the total electrostatic energy initially associated with the charge density $\rho_0(\mathbf{x})$. At equilibrium the excess charge resides on the surface of the conductor. The energy dissipated thermally is the integrated Joule power,

$$\int_0^\infty P \, dt = \int_0^\infty \int \sigma E^2 d^3x \, dt. \tag{7.32}$$

The spatial integral is over the volume of the conductor. By Gauss's Law, $\mathbf{E}(\mathbf{x}, t)$ must decay with the same time dependence as $\rho(\mathbf{x}, t)$, so $\mathbf{E} = \mathbf{E}_0 e^{-t/\tau}$ where again $\tau = \epsilon_0/\sigma$ and $\mathbf{E}_0$ is the initial field. Integrating over the time in (7.32) gives

$$\int_0^\infty P \, dt = \int_0^\infty \int \sigma E_0^2 e^{-2t/\tau} d^3x \, dt = \frac{\epsilon_0}{2} \int E_0^2 d^3x \tag{7.33}$$

which is indeed the initial field energy *inside* the conductor. At equilibrium there is no field energy inside the conductor because the internal field is zero. But we must also consider the electrostatic energy *outside* the conductor. In general the external field changes because the charge distribution changes, but (7.33) shows that this change does not contribute to the dissipation by heat. Rather the external field energy decreases because of radiation. In Chapter 15 we will study radiation from time-dependent currents, of which this is an example.

In our calculations we neglected polarization in the conductor. This is a good approximation for metals at low frequencies. For a *static* system the polarization in a conductor is exactly 0 because $\mathbf{E} = 0$. But in the presence of $\mathbf{E}$ and $\mathbf{J}$, the ions in the crystal lattice may be polarized, leading to bound charge. How are the results different if the permittivity is ϵ? In this case Gauss's Law is $\nabla \cdot \mathbf{E} = \rho_f/\epsilon$, where ρ_f is the density of free charge, i.e., conduction electrons. Again $\mathbf{J} = \sigma \mathbf{E}$, and here $\mathbf{J}$ is the current of the conduction electrons. The only change is to replace ϵ_0 by ϵ, so the decay time constant is $\tau = \epsilon/\sigma$.

EXAMPLE 3   As another example of the decay of a charge distribution, consider the RC circuit shown in Fig. 7.9. With the switch as shown, the charge on the capacitor is $Q_0 = CV_0$. Suppose the switch is flipped at $t = 0$. Then the charge at time t is[14]

$$Q(t) = Q_0 e^{-t/RC}; \tag{7.34}$$

so Q decays exponentially with time constant RC. The power dissipated in the resistor is $I^2 R$, where by charge conservation $I = dQ/dt$. The total energy dissipated is

[14]See Exercise 15.

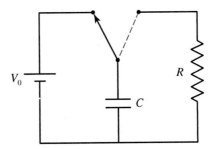

FIGURE 7.9   An RC circuit.

$$\int_0^\infty I^2 R\,dt = \int_0^\infty \left(\frac{Q_0}{RC}\right)^2 e^{-2t/\tau} R\,dt = \frac{Q_0^2}{2C} \tag{7.35}$$

which is the initial energy stored in the capacitor.

■─────────

7.7 ■ *I–V* CHARACTERISTIC OF A VACUUM-TUBE DIODE

A vacuum-tube diode consists of a heated cathode at ground potential and an anode at positive potential V_0. Electrons are emitted by the cathode and accelerate to the anode. The potential difference V_0 controls the net current. This seems like a very elementary example of electric current, but we will see that it has some interesting complexity. We will analyze the *steady state* of this simple system, with charge and current densities independent of time.

The electrodes are shown in Fig. 7.10. The plates have area A and separation d, and are perpendicular to the x axis at $x = 0$ and $x = d$. To make the calculation tractable we assume $d \ll \sqrt{A}$ so that edge effects are negligible. Then the variables of the system depend only on the coordinate x. The charge density and current density are $\rho(x)$ and $\mathbf{J} = \hat{\mathbf{i}}\,J(x)$; the potential is $V(x)$.

The current is due to motion of the electrons, so we need to determine the electron velocity as a function of position x in the gap. At the cathode the electrons have small velocity, which we take to be 0. They accelerate through the gap to the anode. At position x the electron velocity is $\mathbf{v} = \hat{\mathbf{i}}\,v(x)$. By conservation of energy, $\frac{1}{2}mv^2(x) = eV(x)$, so the velocity at x is

$$v(x) = \left[\frac{2eV(x)}{m}\right]^{1/2}. \tag{7.36}$$

We don't know *a priori* the potential function $V(x)$. If there were no charge between the plates, then $V(x)$ would just be $V_0 x/d$. But the electron charge also contributes to $V(x)$. We need some other equations to determine the potential as a function of x.

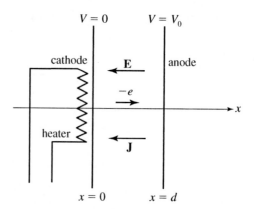

FIGURE 7.10   Cathode and anode of a vacuum tube diode. Electrons boil off the hot cathode and accelerate in **E** to the anode.

Before we start writing the equations, let's try to understand the principles in physical terms. First, the density $\rho(x)$ and electron velocity $v(x)$ determine the current in the gap. Second, in the steady state the current must be the same through any cross section of the gap; otherwise the amount of charge between two cross sections would be changing, which is not the steady state. Finally, the potential $V(x)$ throughout the gap is determined by the charge density $\rho(x)$, by Poisson's equation. These principles, together with the relation (7.36) between the speed and potential, imply a unique solution for $\rho(x)$, $V(x)$, and $J(x)$.

We are now ready to analyze the equations. Note first that the current density $J(x)$ is $\rho(x)v(x)$. Conservation of charge is expressed by the continuity equation (7.6). In the steady state $\partial\rho/\partial t$ is 0; and for this one-dimensional current $\nabla \cdot \mathbf{J}$ is dJ/dx. Thus the continuity equation reduces to the statement

$$\frac{dJ}{dx} = 0, \tag{7.37}$$

i.e., $J(x)$ is a constant J in the gap. (Note that J is negative: The current in Fig. 7.10 is due to electrons moving to the right, i.e., the positive x direction, so $J = J_x$ is negative.) Combining (7.36) and the equation $J = \rho(x)v(x)$ gives one relation between $\rho(x)$ and $V(x)$,

$$\rho(x) = \frac{J}{v(x)} = -C\,[V(x)]^{-1/2} \tag{7.38}$$

where C is a positive constant, defined by $C = -J[m/(2e)]^{1/2}$.

Another relation between $V(x)$ and $\rho(x)$ is Poisson's equation, $-\nabla^2 V = \rho/\epsilon_0$. For this one-dimensional system, $\nabla^2 V$ is just d^2V/dx^2. Then using (7.38) the potential function must satisfy

$$\frac{d^2V}{dx^2} = \frac{C}{\epsilon_0}V^{-1/2}. \tag{7.39}$$

This ordinary differential equation can be solved by conventional methods. We'll solve it by guessing that the solution is a power of x. We must also satisfy the boundary conditions, $V(0) = 0$ and $V(d) = V_0$, so we write the trial solution as

$$V(x) = V_0 \left(\frac{x}{d}\right)^p \tag{7.40}$$

where the power p remains to be determined. Substituting this form into (7.39) we find

$$\frac{V_0}{d^p} p(p-1) x^{p-2} = \frac{C}{\epsilon_0} \frac{d^{p/2}}{V_0^{1/2}} x^{-p/2}. \tag{7.41}$$

The equation is satisfied for

$$p = \frac{4}{3} \quad \text{and} \quad C = \frac{4\epsilon_0 V_0^{3/2}}{9d^2}. \tag{7.42}$$

Our first result is that $V(x)$ is proportional to $x^{4/3}$. The charge density in the gap is now determined by (7.38)

$$\rho(x) = -\frac{4\epsilon_0 V_0}{9d^2} \left(\frac{d}{x}\right)^{2/3}. \tag{7.43}$$

Note that $\rho(x)$ is proportional to $x^{-2/3}$.

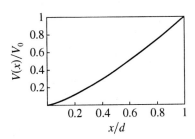

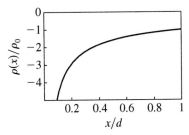

FIGURE 7.11 $V(x)$ **and** $\rho(x)$ **in a vacuum tube diode.** ρ_0 is the coefficient of $(d/x)^{2/3}$ in (7.43).

Figure 7.11 shows plots of $V(x)$ and $\rho(x)$ as functions of x. The charge in the gap is called *space charge*. If the space charge is initially 0, then as electrons are emitted by the cathode a charge density will build up in the gap until (7.38) is true. This final state—the steady state—is a state of dynamic equilibrium.

The total current through any cross section of the gap is $I = JA$. According to the solution (7.42), remembering that $J = -C(2e/m)^{1/2}$, the current is

$$I = -\left(\frac{2e}{m}\right)^{1/2} \frac{4\epsilon_0 A}{9d^2} V_0^{3/2}. \tag{7.44}$$

This current is said to be "limited by the space charge," because the potential gradient near the cathode is less than it would be if there were no space charge. In fact, the electric field approaches 0 at the cathode.

For this elementary device, Ohm's law is not valid; i.e., I is not proportional to V_0. That is not surprising. Ohm's law is associated with dissipation by collisions, but in this vacuum tube there are no electron collisions. The characteristic IV dependence $I \propto V_0^{3/2}$ is called the *Child-Langmuir law*. It is common for currents limited by space charge.

Ion rockets. An interesting application of charged particle emission with acceleration by an electrostatic field is the ion rocket. Ion rockets have been used for three decades as small thrusters in satellites, e.g., to control orientation of a satellite. The basic principle of an ion rocket is that electrostatic acceleration of heavy positive ions produces a reaction force on the satellite. If a positive space charge were to build up, then the thrust would be reduced. Therefore electrons are also emitted to neutralize the space charge.

Diodes. A diode is a circuit element in which current can flow in only one direction. Diodes are used in rectifiers and filters. A simple diode acts as a half-wave rectifier: If an alternating current is the input to the diode, then the output current is just the positive half of the input current.

Before the development of solid-state electronics, *vacuum-tube diodes* of the kind described above were common. In such devices a hot cathode emits electrons that accelerate to the anode. The electrons can only flow in one direction—from cathode to anode—and the current is in the opposite direction because the electron charge is negative. Figure 7.12 shows the IV curve for a vacuum-tube diode, plotting the magnitude of I. I is 0 for $V < 0$, and $|I| \propto V^{3/2}$ for $V > 0$.

The system in Fig. 7.10 is an idealized model to explain the physics. In practice, the cathode is a small cylindrical metal surface with a (red-hot) heating filament enclosed. The anode is typically a concentric cylindrical surface. The reason for heating the cathode is so that electrons come off the metal even though the field is zero at its surface, as we found. To facilitate electron emission the cathode is coated with an oxide (BaO or SrO). A typical potential for the anode (often called the plate) in the conducting mode is $V_0 = +10$ to $+100$ volts with respect to the cathode. The anode is cold and uncoated so that it does not emit electrons even if the sign of V_0 is reversed.

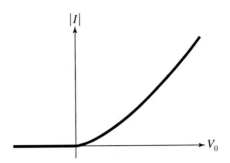

FIGURE 7.12   *I V* **curve of a vacuum-tube diode.**

The pressure in a vacuum-tube diode is 10^{-6} Torr or less. Under these conditions the buildup of negative space charge decreases the field at the cathode, as we have seen, and inhibits electron flow. The electron space charge is reduced in a *gas-filled thermionic diode*, in which a low pressure of Hg vapor is present between the cathode and anode. Then if V_0 is greater than the ionization potential of Hg (10.4 V), energetic electrons ionize Hg atoms, and the resulting Hg^+ ions move to the cathode, tending to neutralize the negative space charge and enhance the electron flow.

Vacuum-tube diodes have long ago been replaced in most applications by *semiconductor pn-junction diodes*. These solid-state devices are smaller, cheaper, more rugged, and require less power (there is no heating filament) than vacuum-tube devices. The physics of semiconductors is essentially quantum mechanical. The idea is that pure semiconductors, Group IV elements such as Si and Ge, have a low density of conduction electrons and therefore do not conduct well. Their conductivity can be increased by adding a small concentration (parts per million) of a Group V element such as As or Sb. The added *donor* atoms donate electrons to the semiconductor lattice, increasing the conductivity; such a material is called an n-type doped semiconductor.[15] The conductivity of a pure semiconductor can also be increased by adding a small concentration of a Group III element such as Al or In. In the resulting p-type doped semiconductor the added *acceptor* atoms accept electrons from the lattice and the conductivity is increased because there is an increase in the concentration of holes, states from which electrons have been removed.

At a pn-junction, i.e., the boundary between an n-type region and a p-type region, the electrons and holes diffuse across the boundary into, respectively, the p-type and n-type region. The reason a pn-junction makes a good diode, so that such diodes are widely used as rectifiers, comes from the following properties of the junction: If an external voltage is applied across the pn-junction in the forward bias direction, i.e., with the p-region positive and the n-region negative, then high current exists across the junction, in the direction from p to n, with a

---

[15]n-type and p-type refer to negative or positive charge carriers, respectively.

low voltage drop. This is analogous to the situation in a vacuum-tube diode, in which the anode is positive with respect to the cathode. If, however, the external voltage is applied in the reverse bias direction, i.e., with the p-region negative and the n-region positive, charge cannot flow across the junction, as in a vacuum-tube diode with the anode negative with respect to the cathode. Semiconductor pn-junction diodes in use today range from those in integrated circuits in consumer electronics, which carry currents of a few μA and are a few μm in diameter, to those in large industrial power rectification applications, which carry currents of thousands of amperes and have diameters of order 10 cm.

7.8 ■ CHAPTER SUMMARY

Three concepts introduced in this chapter will be very important in later chapters. The reader should know and understand these equations:

1. The definition of current density, $dI = \mathbf{J} \cdot d\mathbf{A}$.
2. The continuity equation, $\nabla \cdot \mathbf{J} = -\partial \rho / \partial t$.
3. The two versions of Ohm's law, $\mathbf{J}(\mathbf{x}) = \sigma \mathbf{E}(\mathbf{x})$ and $V = IR$.

It is a useful exercise to sketch drawings that explain the meanings of each of these equations.

FURTHER READING

The physics of resistance is an important topic in solid-state physics. To learn more, see C. Kittel, *Introduction to Solid State Physics*, 7th ed. (Wiley, New York, 1996).

Lightning is a dramatic example of electric current that everyone has observed. It is described in *The Feynman Lectures*, Volume II, Chapter 9.

EXERCISES

Sec. 7.1. 1D Currents

7.1. Estimate the mean drift velocity of a conduction electron in copper, assuming that there is one conduction electron per atom. Consider No. 14 wire (diameter 0.16 cm), common in houses, carrying 5 A. The result should be very small. Why then does an electric light turn on almost instantly when you flip the switch? Also, compare the drift velocity to the Fermi velocity $v_F = 1.57 \times 10^8$ cm/s for Cu, which is a characteristic speed of the electrons' random motion.

7.2. A charged ring rotates about its axis. What is the current? Let the ring have radius a, linear charge density λ, and angular speed ω.

Sec. 7.2. Volume Current

7.3. A solid sphere of radius a has total charge Q uniformly distributed throughout its volume. The sphere rotates with angular velocity $\omega = \omega \hat{\mathbf{k}}$. Find the current density $\mathbf{J}(\mathbf{x})$. Use spherical polar coordinates.

Sec. 7.3. Ohm's Law

7.4. Explain why the resistance of two resistors in series is $R_1 + R_2$, and why the resistance of two resistors in parallel is $R_1 R_2 / (R_1 + R_2)$.

7.5. A wire has resistance R. Determine the resistance of a wire of twice the length and half the diameter, made from the same material.

7.6. What is the resistance of a copper wire with length 10 m and diameter 1 mm?

7.7. A conducting sphere of radius a is imbedded in a large volume filled with a material with electric conductivity σ. The sphere is held at potential V_0 with respect to the boundary, which may be approximated as spatial infinity. Determine the current density $\mathbf{J}(\mathbf{x})$, the total current I, and the resistance. Verify that the continuity equation is satisfied.

7.8. Two small conducting spheres with radii a and b are imbedded in a medium of resistivity ρ and permittivity ϵ with their centers separated by a distance $d \gg a, b$.

    (a) What is the resistance R between them? Notice that the result is independent of ϵ.

    (b) If $a = b = 1$ cm and $d = 10$ m, and the medium is sea water, what is the value of R?

7.9. Consider two coaxial conducting cylinders with radii a and $3a$ and length L. The region $a < r < 2a$ is filled with a material with conductivity σ_1, and the region $2a < r < 3a$ has conductivity σ_2. (Assume $\epsilon_1 = \epsilon_2 = \epsilon_0$.) The inner cylinder is held at potential V_0 and the outer cylinder at $V = 0$, so there is a radial current I.

    (a) Determine the resistance.

    (b) Determine the surface charge density on the boundary at $r = 2a$.

7.10. A parallel-plate capacitor with plates of area A and separation d is imbedded in a large volume filled with a material with electric conductivity σ. Determine the total current I if the two capacitor plates are maintained at a potential difference V_0. Explain how such a device could be used to measure the salinity of salt water.

7.11. A solid cylindrical conductor of length L and cross section A has resistivity ρ_1 in half its length and ρ_2 in the other half. There exists a current I in the cylinder parallel to the axis and uniformly distributed over the cross section. Find the surface charge density σ on the interface between ρ_1 and ρ_2. Assume the permittivity of both materials is ϵ_0.

Sec. 7.4. The Drude Model

7.12. From Eq. (7.24), which is based on the classical model of conductivity, estimate the mean free path λ of a conduction electron in copper at room temperature, assuming there is one conduction electron per atom. Do the calculation for both the classical thermal velocity and the Fermi velocity. Discuss the resulting values of λ.

7.13. As a classical model of conductivity, assume an electron in a conductor moves subject to two forces: $-e\mathbf{E}$ and $-\gamma\mathbf{v}$, where $\mathbf{v}$ is its velocity. (The damping force is a model for the energy loss by collisions.) Determine the conductivity in terms of e, γ, and the electron density.

Sec. 7.6. Decay Time

7.14. Calculate the time constant, in seconds, for decay of a charge density fluctuation in copper and in germanium.

7.15. **(a)** Derive (7.34) for the charge on a capacitor as a function of time in an RC circuit. The *time constant* of the decay is $\tau = RC$. By what factor does Q decrease between $t = 0$ and $t = \tau$? Between $t = \tau$ and $t = 2\tau$? Between $t = 5\tau$ and $t = 6\tau$? At what time is the charge one-tenth of its original value?

(b) Verify that the total energy is conserved as the charge comes to equilibrium.

General Exercises

7.16. *On atmospheric electricity.* At the surface of the Earth (radius 6400 km) there is an average electric field $E_r \approx -100 \,\text{V/m}$, and a corresponding current density $J_r \approx -3.5 \times 10^{-12} \,\text{A/m}^2$ that carries electrons away from the Earth to the upper atmosphere. The potential difference between the Earth and the upper atmosphere is 400 kV. What is the total current that reaches the Earth's surface? What power is required to maintain this current? How does it compare to the output of a power plant? Earth's surface charge is continually restored by lightning, which carries down negative charge from the upper atmosphere. A typical lightning stroke has a peak current of 10^4 A and carries 20 C. How many lightning flashes occur each second, on average, worldwide? (See *The Feynman Lectures*, Vol. II, Chap. 9.)

7.17. You are given three resistors, of $1 \,\Omega$, $2 \,\Omega$, and $3 \,\Omega$. What are all the resistances you can make, using either one, two, or all three resistors? (Hint: There are 17 arrangements.)

7.18. Consider the resistor network shown in Fig. 7.13. There are n resistors crossing from the top line to the bottom line, and $n - 1$ resistors along the top or bottom line. The total number of resistors is $3n - 2$. All the resistors have equal resistance R. Let R_n be the resistance of the network. Calculate R_1, R_2, R_3, and R_∞. [Answer: $R_\infty = R(\sqrt{3} - 1)$.]

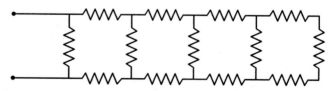

FIGURE 7.13 **A resistor array.** The resistance is R_n where n is the number of rungs in the ladder. In the illustration, $n = 5$.

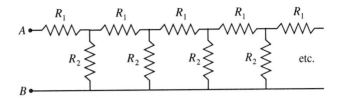

FIGURE 7.14 A resistor array. The network continues *ad infinitum*.

7.19. Consider the resistor array shown in Fig. 7.14. The total resistance between A and B is R_0. The ladder of resistors extends to the right *ad infinitum*. (This model resembles the resistance in an axon nerve; R_1 represents the resistance of unit length along the cell wall, and R_2 the resistance of unit length across the membrane.)

   (a) Note that the resistance from the part of the array to the right of the left-most rung of the ladder is also R_0. From this show that $R_0 = R_1 + R_2 R_0/(R_2 + R_0)$.

   (b) Determine R_0.

   (c) Now suppose a current flows into A and away from B. Let V_n be the potential difference across the nth rung to the right of A and B. Show that the ratio V_{n+1}/V_n is $R_2/(R_0 + R_2)$.

   (d) Show that V_n decreases exponentially with n.

7.20. A wire of length L and cross section A lies along the positive x axis. The resistivity varies with x, as

$$\rho(x) = \rho_0 + \rho_1 e^{-x/d}.$$

(For example, the end at $x = 0$ may have become oxidized, and therefore have a larger resistivity than the pure metal.) The end at $x = 0$ is held at potential V_0 and the other end is grounded. Determine the current I and power per unit length dP/dx. Make a sketch of dP/dx as a function of x.

7.21. The resistivity of a wire increases with temperature, according to $\rho = \rho_0 e^{\alpha(T-T_0)}$, where T_0 is room temperature. (The parameter α is called the temperature coefficient of resistivity.) Let C be the heat capacity of the wire, in units of Joules/K. The length of the wire is ℓ and its cross sectional area is A. At time $t = 0$ the wire is connected to a battery with constant emf V_0. Assuming all the Joule heat goes into raising the temperature, determine T as a function of t. Show that $T = T_0 + \alpha^{-1} \ln(1 + \alpha\beta t)$, where β is a constant that depends on the parameters of the problem, and find β.

7.22. Show that the time constant for discharge of a leaky parallel plate capacitor, from current through the dielectric, is $\tau = \epsilon_0\kappa\rho$, where κ is the dielectric constant and ρ the resistivity. Use the equations for R and C of parallel plates to show that $\tau = RC$.

7.23. The region between concentric conducting spheres is filled with a material with conductivity σ and dielectric constant κ. The inner sphere (radius a) is connected to the positive terminal of a battery with emf V_0, and the outer sphere (radius b) is connected to the negative terminal. At $t = 0$ the battery is disconnected.

   (a) Calculate $Q(t)$, the charge on the inner conductor, and sketch a plot of this function. Show that the decay constant is $\kappa\epsilon_0/\sigma$.

   (b) Calculate $I(t)$.

Computer Exercises

7.24. The current in a thin film may be described by a surface current density **K**, with units A/m. Consider a sample of the form shown in Fig. 7.15. The film lies on the xy plane and has a small thickness δ in the z direction. Then

$$\mathbf{K} = \mathbf{J}\delta = K_x\,\hat{\mathbf{i}} + K_y\,\hat{\mathbf{j}}.$$

The continuity equation for a steady current is $\nabla \cdot \mathbf{K} = 0$. Also, because $\mathbf{J} = \sigma\mathbf{E}$, **K** is irrotational, $\nabla \times \mathbf{K} = 0$. Together these imply that $\mathbf{K} = -\nabla\Phi$, where Φ satisfies Laplace's equation, $\nabla^2\Phi = 0$.

Use the method of iterative relaxation (Sec. 5.5) to solve Laplace's equation in the thin film with the boundaries shown in Fig. 7.15. There are mixed Dirichlet and Neumann boundary conditions:

$$\begin{aligned}
\Phi(-1, y) &= 1 \quad \text{for} \quad -1 \le y \le 1 \\
\Phi(1, y) &= 0 \quad \text{for} \quad -1 \le y \le 0; \qquad\qquad (7.45) \\
\partial\Phi/\partial y &= 0 \quad \text{at} \quad y = 1 \text{ for } -1 \le x \le 0, \\
\partial\Phi/\partial y &= 0 \quad \text{at} \quad y = 0 \text{ for } 0 \le x \le 1, \\
\partial\Phi/\partial y &= 0 \quad \text{at} \quad y = -1 \text{ for } -1 \le x \le 1, \\
\partial\Phi/\partial x &= 0 \quad \text{at} \quad x = 0 \text{ for } 0 \le y \le 1. \qquad\qquad (7.46)
\end{aligned}$$

The conditions (7.45) state that the current enters and leaves the film in the $\hat{\mathbf{i}}$ direction at the two ends; the conditions (7.46) state that the current is tangential on the sides.

(a) Make a contour plot of $\Phi(x, y)$ and on it sketch the current vectors.

(b) Calculate **K** at some representative points. For example, where is $|\mathbf{K}|$ greatest?

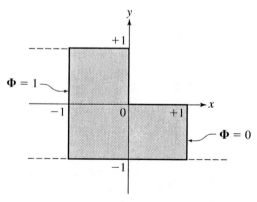

FIGURE 7.15   A current-carrying thin film. A current $\mathbf{K} = -\nabla\Phi$ enters at $x = -1$ and exits at $x = +1$.

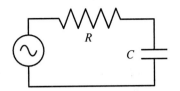

FIGURE 7.16   A driven RC circuit.

7.25. Figure 7.16 shows an RC circuit driven by a harmonic EMF. The differential equation for the charge on the capacitor is

$$\mathcal{E}_0 \sin \omega t = IR + Q/C,$$

where $I = dQ/dt$. The solution has the form

$$Q(t) = Q_1 e^{-t/\tau} + A \sin(\omega t - \phi),$$

where $\tau = RC$ is the time constant. Q_1, A, and ϕ are constants, which must be determined from the differential equation and the initial conditions. The first term $Q_1 e^{-t/\tau}$ is the *transient*.

Rather than solve the equation analytically, use a computer program to solve it numerically. (It is just an ordinary differential equation.) Assume at $t = 0$ the charge on the capacitor is 0. Take RC to be the unit of time, and to be specific let $\omega = 4/RC$ and $\mathcal{E}_0/R = 1$. Plot Q as a function of time. Also, make a plot showing both $Q(t)$ and $\sin \omega t$ on the same graph, and estimate the phase shift of the charge. The calculations can also be done analytically.

CHAPTER

8

Magnetostatics

*"To those who do not know mathematics it is difficult to get across a real
feeling as to the beauty, the deepest beauty, of nature.... If you want to
learn about nature, to appreciate nature, it is necessary to understand the
language that she speaks in."*

—R. P. Feynman, *The Character of Physical Law*

Everyone has played with magnets, and felt the mystery of their forces. What
causes the magnetic force? In 1820 Oersted discovered that *magnetic forces are
produced by electric currents*, by demonstrating that a compass needle is deflected
if placed near a current-carrying wire. Conversely, a magnet creates a force on
a current. The effect was further studied by Biot and Savart, and by Ampère.
Ampère in particular explored in detail the magnetic forces between electric currents. Later, Faraday and Maxwell developed the field theory of electric and magnetic forces, in which currents create magnetic fields.

The subject of this chapter is magnetostatics—*time-independent magnetic
fields*, and their interaction with *constant currents*. The principles we will study
are exact for magnetic systems with constant currents, and apply approximately
to systems in which the currents change slowly in time, called the quasistatic
approximation.

Some magnetic phenomena were known long before experiments with currents and wires. Naturally occuring lodestones, which attract iron, were known
to the people of Ancient Greece. The magnetic compass was invented in China,
probably over 2000 years ago, and came to Europe in the 12th century. Gerhard
Mercator, in 1546, introduced the idea that the point a magnetic needle seeks is
terrestrial, and not in the stars. In 1600, William Gilbert concluded from his experiments on spherical lodestones that the Earth itself is a great magnet; in his
words, "*magnus magnes ipse est globus terrestris*." In modern terms we'd say he
explored the external magnetic field of a spherical permanent magnet. (We will
calculate the same field in Chapter 9.) It is interesting to note that Gilbert's experiments predate by 90 years Newton's understanding of the terrestrial origin of
gravity.[1]

---

[1] As an historical aside, before Gilbert's time people had the idea that the behavior of a magnetic
compass is governed by the stars, because the north pole of a compass points to the North Star. We
owe a lot to Mercator and Gilbert for looking for a natural cause in the Earth, nearer at hand than the
stars.

In Chapter 9 we will see how the magnetic field of a permanent ferromagnet can be attributed to electric currents on the atomic scale. But first we need to comprehend the basic principles of magnetostatics.

8.1   ■ THE MAGNETIC FORCE AND THE MAGNETIC FIELD

8.1.1   ■ Force on a Moving Charge

The most basic form of the magnetic force is the force on a charge q moving with velocity $\mathbf{v}$, exerted by a magnetic field $\mathbf{B}$, which is

$$\mathbf{F} = q\mathbf{v} \times \mathbf{B}. \tag{8.1}$$

The SI unit of magnetic field is the tesla (T). One tesla is defined as one newton per ampere-meter. By (8.1) the force on 1 C moving 1 m/s perpendicular to a field of 1 T is 1 N. A magnetic field of 1 T is a rather strong field. The largest fields that can be produced by conventional electromagnets are about 2 T. For high-field research in laboratories, fields of 10–12 T, produced with superconducting magnets, are used. The record for steady-state fields is about 30 T. At the surface of the Earth the magnetic field is about 0.5×10^{-4} T. Very large and very small fields are important in astronomy; e.g., 10^8 T near pulsars, and 10^{-10} T in interstellar space in the galaxy.

The direction of the magnetic force on q is *sideways*, i.e., perpendicular to $\mathbf{v}$ and to $\mathbf{B}$, as shown in Fig. 8.1 for a positive charge. Therefore, the magnetic force does no work on q:

$$dW = \mathbf{F} \cdot d\mathbf{x} = \mathbf{F} \cdot \mathbf{v}dt = 0. \tag{8.2}$$

The magnetic force affects the direction of motion of q, but not its kinetic energy. Because the magnetic force does no work, it is not possible to define a potential energy function for the magnetic force. The magnetic force is *velocity-dependent*, which is quite different from the other fundamental forces we encounter in physics.

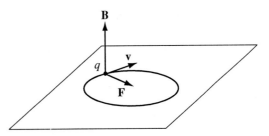

FIGURE 8.1   The magnetic force. The force on a moving charge is sideways, perpendicular to both $\mathbf{v}$ and $\mathbf{B}$. The figure shows $\mathbf{F}$ and the circular trajectory for a positive charge in a uniform field. What would the trajectory be for a negative charge $-q$ projected with the same initial velocity $\mathbf{v}$?

We can demonstrate the magnetic force on a moving charge with a Crookes tube and a bar magnet. A Crookes tube[2] is essentially a glass cylinder filled with gas at low pressure, with a sealed-in electrode at each of its closed ends. A high voltage is applied between the electrodes, creating a beam of electrons traveling from the cathode to the anode. The path of the electrons is visible because gas atoms along the path are excited by the energetic electrons and emit light. If there is no magnetic field the beam is a straight line. But if a magnet is brought near then the beam is *curved* because of the sideways magnetic force (8.1) on each electron.

Equation (8.1) may be taken as the *definition* of the magnetic field. From the measurement of the force on a test charge q, e.g., by observing its deflection for a known velocity, the defined quantity **B** could in principle be deduced from (8.1).

In using (8.1) we picture a laboratory setup in which there is a **B** field, e.g., due to a large magnet, and **v** is the velocity of the charge q in the laboratory—physicists say "in the reference frame of the laboratory." The particle is deflected by the force calculated by (8.1). But with this velocity-dependent force, it's natural to ask questions like these: What is the force on q in a reference frame that is *moving* in the laboratory? In that frame **v** will be different, but what will **B** be? Does (8.1) hold in the moving frame? What is the force on the particle in the moving frame? In particular, suppose the reference frame is moving along with the particle at some instant; in that reference frame the particle has zero velocity. Is the force on the particle then 0 as (8.1) implies? Surely not. Is the particle *undeflected* in that reference frame? Surely not.

The full resolution to these difficult questions will come in Chapter 12. The basic point is that (8.1) is only part of a more general force equation, called the Lorentz force

$$\mathbf{F} = q\left(\mathbf{E} + \mathbf{v} \times \mathbf{B}\right). \tag{8.3}$$

In the reference frame in which $\mathbf{v} = 0$, the particle still experiences a force and is deflected. The force on it is, however, not magnetic, but rather electric, given by $q\mathbf{E}$, even though there is no electric field in the laboratory! The idea, which we will explore fully in Chapter 12, is that the fields **E** and **B** are inextricably intertwined in the sense that they transform together if the reference frame changes. For this reason we properly speak of an *electromagnetic field*, and the subject is called *electromagnetism*.

In principle, **B** can be measured by observing moving test charges, but in practice, **B** is measured with solid-state magnetometers, of which there are many kinds. Some examples are Hall-effect probes,[3] and magnetoresistors, which are materials whose resistance varies with the applied magnetic field. For measurements of very small magnetic fields, as in magnetoencephalography where the neuromagnetic fields are of order 10^{-12} T, or in magnetocardiography where

---

[2] Named for William Crookes, one of the discoverers of cathode rays.
[3] See Exercise 1.

the field is of order 2×10^{-11} T, superconducting quantum interference devices (SQUIDs) are used.

8.1.2 ■ Force on a Current-Carrying Wire

A current-carrying wire in a magnetic field experiences a force, from the magnetic force on the individual moving charges that make the current. Electric current is motion of charge, and (8.1) is the force on a moving charge. Suppose the current consists of particles with charge q and linear density n_1, moving with mean velocity $\mathbf{v}$. Then the net force $d\mathbf{F}$ on a small segment $d\boldsymbol{\ell}$ of the wire is

$$d\mathbf{F} = (n_1 d\ell) q \mathbf{v} \times \mathbf{B} \qquad (8.4)$$

because $n_1 d\ell$ is the number of moving charges in $d\ell$. The current I in the wire is $q n_1 v$, and $\mathbf{v}$ is parallel to $d\boldsymbol{\ell}$, so the *total force* on the wire is

$$\mathbf{F} = \int_{\text{wire}} I d\boldsymbol{\ell} \times \mathbf{B}. \qquad (8.5)$$

Biot and Savart, and Ampère, measured magnetic forces on current-carrying wires in their studies of the basic nature of magnetism.

8.2 ■ APPLICATIONS OF THE MAGNETIC FORCE

Many devices in science and technology are based on magnetic deflection of charged particles. In this section we will study the equations of motion, and describe some applications.

8.2.1 ■ Helical or Circular Motion of q in Uniform B

What is the general motion of a charge q in a uniform magnetic field? We choose the z axis to be the direction of $\mathbf{B}$, so $\mathbf{B} = B\hat{\mathbf{k}}$. The equation of motion of the charge is $d\mathbf{p}/dt = q\mathbf{v} \times \mathbf{B}$, where $\mathbf{p}$ is its momentum. In general $\mathbf{F}$ is perpendicular to $\mathbf{B}$, so in this case $F_z = 0$. Therefore the z component of the motion has constant momentum, and $v_z = $ constant. In the special case $v_z = 0$, the motion of the charge is parallel to the xy plane. If v_z is not zero, then superimposed on the xy motion there is a constant drift parallel to the z axis. To obtain the xy motion we note that the kinetic energy and the magnitude $|\mathbf{p}|$ of momentum are constant because $\mathbf{B}$ does no work. But a constant force perpendicular to the motion, and constant speed, are characteristics of uniform circular motion. Thus the general trajectory is a *helix*, around $\hat{\mathbf{k}}$.

   If $v_z = 0$ then the orbit is a circle, as illustrated in Fig. 8.1. For nonrelativistic mechanics, i.e., for v much less than the speed of light c, the equation for circular motion with radius R is

$$\frac{mv^2}{R} = -F_r = qvB. \tag{8.6}$$

The magnetic force is *centripetal*. The magnitude is qvB because $\mathbf{v}$ is perpendicular to $\mathbf{B}$, so that $|\mathbf{v} \times \mathbf{B}| = vB$. The angular speed ω of the particle is

$$\omega = \frac{v}{R} = \frac{qB}{m}. \tag{8.7}$$

This parameter ω is called the *cyclotron frequency*; the period of revolution of q is $2\pi/\omega$. For nonrelativistic motion ω is independent of both the orbit radius and the speed.[4]

The fact that a charged particle moves on a circle in a uniform field $\mathbf{B}$ (for $\mathbf{v}$ perpendicular to $\mathbf{B}$) is the basis for several important devices, of which we will describe two.

Cyclotron

A cyclotron is a charged-particle accelerator used for scattering experiments in nuclear physics, and for production of radioactive isotopes for medical applications. In the classic design, shown schematically in Fig. 8.2, particles in a "bunch" move on semicircular orbits in the two D-shaped regions, which are called the "dees." In each dee there is a uniform $\mathbf{B}$ perpendicular to the orbit to make the particle trajectory circular. The magnetic force does no work, so to raise the kinetic energy of the circling particles an electric field is necessary, which resides in the gap between the dees. The direction of $\mathbf{E}$ reverses for each half cycle, so that the particles in a bunch are accelerated to higher energy each time the bunch passes through the gap. The frequency of oscillation of $\mathbf{E}$ in Hz is $\omega/(2\pi)$, where ω is the cyclotron frequency. The cyclotron frequency remains constant ($= qB/m$) as the particles gain energy, as long as $v \ll c$. Because ω is independent of radius and energy, multi-bunch acceleration of charges is possible in a cyclotron, by injecting particles at appropriate times in the cycle.

For higher energies, i.e., v a significant fraction of c, relativistic effects become important. In this case the cyclotron frequency does vary with v, so the frequency of $\mathbf{E}$ must change to remain synchronized with the particle motion. This is the principle of the synchrotron. In a synchrotron only one bunch of charges at a time can be accelerated in the machine.

By (8.6) the orbit radius increases as $\sqrt{E_K}$, where E_K is the particle kinetic energy, so the bunch spirals out as the particles gain energy. When the particle energy reaches the desired energy, the beam is extracted by a deflecting magnet, and directed to a target where the scattering experiment occurs. Protons, for example, may be accelerated to energies of order $100 \, \text{MeV}$.

A different design is necessary for very high energies. In a high-energy synchrotron the orbit radius is fixed, and the strength of $\mathbf{B}$ increases as the particle

---

[4]By (8.7) the momentum magnitude is $p = mv = qBR$. The solution for *relativistic dynamics* is also $p = qBR$, but p is the magnitude of the relativistic momentum $\mathbf{p} = m\mathbf{v}/\sqrt{1 - v^2/c^2}$.

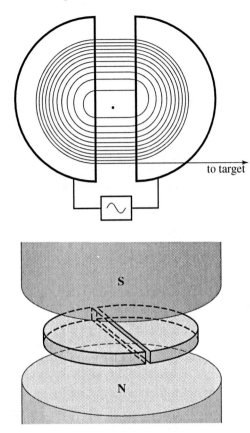

FIGURE 8.2   Cyclotron. There is a magnetic field perpendicular to the plane of the dees, and an alternating electric field across the gap. The gap between the dees is greatly exaggerated for illustration.

energy increases. The accelerator ring has alternating regions of **B** and **E**, the first to keep the particles in a circle and the second to raise their energy. The Tevatron accelerator at Fermilab is a synchrotron, 2 km in diameter, that accelerates protons and antiprotons to 1 TeV.

Mass Spectrometer
A mass spectrometer, shown schematically in Fig. 8.3, is a device invented by J. J. Thomson to separate charged particles by mass. The design in Fig. 8.3 has three parts, velocity selector, magnetic deflector, and particle detector. A beam composed of several species of ion passes into the velocity selector. In the velocity selector there are crossed **E** and **B** fields, $\mathbf{E} = E\,\hat{\mathbf{j}}$ and $\mathbf{B} = B\hat{\mathbf{k}}$. If a charge has velocity $\mathbf{v} = (E/B)\,\hat{\mathbf{i}}$ then it passes through the fields undeflected,[5] because the

---
[5] See Exercise 2.

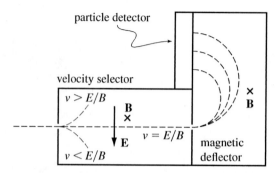

FIGURE 8.3   Mass spectrometer. Positive ions moving with a definite velocity separate in the magnetic field according to the ratio q/m. Only those particles with $\mathbf{v} = (E/B)\hat{\mathbf{i}}$ (the x axis is to the right) make it through the velocity selector and into the magnetic deflector.

net force (electric plus magnetic) is 0. Other particles will be deflected out of the beam. The particles that emerge from the velocity selector have this known velocity.

Particles from the velocity selector then enter a region of uniform magnetic field, where they travel on circular paths. By (8.6) the orbit radius is proportional to the ratio m/q, so the different particle species travel along circular arcs with different radii, and separate in space, as shown in Fig. 8.3. By observing the final particle positions with a charged-particle detector, the experimenter can deduce what species of particles were present in the initial beam. Mass spectrometers are used to identify the products of chemical reactions, or of particles produced in nuclear collisions.

The mass-spectrometer principle is used also in residual gas analysers, laboratory devices that are commonly employed to identify, and locate the source of, gas-borne impurities in high-vacuum systems. An interesting aspect of a residual gas analyzer is that molecules that have the same mass, and therefore the same ratio of m/q when singly ionized, such as CO^+ and N_2^+, are difficult to tell apart because their orbits are the same.

8.2.2 ■ Cycloidal Motion of q in Crossed E and B

As another example of the magnetic force, we will determine the motion of a positive charge q that is released from rest at the origin in the orthogonal uniform fields $\mathbf{E} = E\hat{\mathbf{j}}$ and $\mathbf{B} = B\hat{\mathbf{k}}$, as shown in Fig. 8.4. The motion is quite different from the previous case because of the electric field. When released, q initially experiences an electric force $qE\hat{\mathbf{j}}$ and accelerates in the $+y$ direction. The magnetic force for $\mathbf{v}$ in the y direction is in the x direction, so q begins to curve away from the y axis toward positive x. With a component of velocity in the $+x$ direction there will be a component of acceleration in the $-y$ direction. Now, what is the full trajectory?

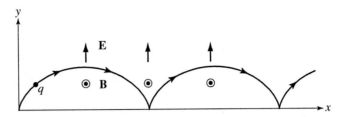

FIGURE 8.4 Motion in crossed E and B fields. A positive charge is released from rest from the origin. **E** and **B** point in the y and z directions, respectively. The trajectory is a cycloid curve.

The equation of motion that we must solve is

$$m\frac{d\mathbf{v}}{dt} = q(\mathbf{E} + \mathbf{v} \times \mathbf{B});$$ (8.8)

or, in component form, where ω is the cyclotron frequency,

$$dv_x/dt = \omega v_y$$ (8.9)

$$dv_y/dt = -\omega v_x + \frac{\omega E}{B}$$ (8.10)

$$dv_z/dt = 0.$$ (8.11)

We see that v_z remains constant, equal to the initial value 0. Differentiating (8.9) and substituting for dv_y/dt from (8.10),

$$\frac{d^2 v_x}{dt^2} + \omega^2 v_x = \frac{\omega^2 E}{B}.$$ (8.12)

The solution of this inhomogeneous linear equation is constructed as the sum of a particular solution ($v_{xp} = E/B$) plus a general solution of the homogeneous equation. That is,

$$v_x(t) = \frac{E}{B} + c_1 \cos \omega t + c_2 \sin \omega t,$$ (8.13)

where c_1 and c_2 are constants to be determined from the initial conditions. Then by (8.9)

$$v_y(t) = -c_1 \sin \omega t + c_2 \cos \omega t.$$ (8.14)

The initial values $v_x(0) = v_y(0) = 0$ determine that the constants c_1 and c_2 are

$$c_1 = -E/B \quad \text{and} \quad c_2 = 0.$$ (8.15)

It is easy to integrate $v_x(t)$ and $v_y(t)$ with respect to t to find the trajectory of q

in space; the result is

$$x(t) = \frac{E}{\omega B}(\omega t - \sin \omega t), \qquad (8.16)$$

$$y(t) = \frac{E}{\omega B}(1 - \cos \omega t). \qquad (8.17)$$

These are parametric equations for a *cycloid curve*, shown in Fig. 8.4.

The motion of q is a combination of harmonic oscillations in x and y, and a constant *drift* in the x direction. The *drift velocity*, i.e., the average velocity over one or more periods of oscillation, is

$$\mathbf{v}_D = \frac{E}{B}\,\hat{\mathbf{i}}, \qquad (8.18)$$

because the averages of the oscillating terms are 0. Equation (8.18) is equivalent to

$$\mathbf{v}_D = \frac{\mathbf{E} \times \mathbf{B}}{B^2}. \qquad (8.19)$$

The charge drifts in a direction perpendicular to **E**.

The result (8.19) has a surprising consequence in plasma physics. The drift velocity is the same for positive and negative charges. This fact is obvious because (8.19) does not depend on q. (The reader should analyze the forces *qualitatively* and figure out why $\mathbf{v}_D$ has the same direction for + or − charges!) Therefore a constant electric field, in the presence of an orthogonal magnetic field, does not produce a net electric current in a dilute (i.e., collisionless) neutral plasma, because the positive and negative charges drift together.

Some interesting variations of the cycloidal motion are explored in a computer exercise.[6]

8.2.3 ■ Electric Motors

An everyday application of the magnetic force on a current is in electric motors. The basic design principle of any electric motor is that an electric current in a magnetic field experiences a force perpendicular to the current. In a simple DC motor the current is driven by a constant EMF, e.g., a battery. The current exists in a coil of wire, which is free to rotate on an axle attached to the coil. The motor has a magnet—either a permanent ferromagnet or an electromagnet—that exerts a magnetic force on the current. The force acts in opposite directions on opposite sides of the coil, creating a torque on the coil, as shown schematically in Fig. 8.5. To keep the direction of the torque constant as the coil rotates, the current in the coil must reverse every half cycle. The switch that reverses the current direction is called a commutator. In a motor with only two poles, as in Fig. 8.5, the commutator is a split ring consisting of two semicircular conductors insulated from

---

[6]See Exercise 36.

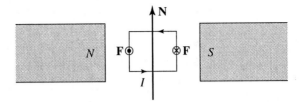

FIGURE 8.5 Electric motor. Because of the magnetic force on the current around the coil, there is a torque **N** on the coil. In the figure the forces on the left and right vertical segments point out of and into the page, respectively.

one another. In a multipole motor the commutator ring has several conducting segments. If a machine is connected to the axle of the motor, then the torque on the coil drives the machine.

Another type of electric motor, that uses an alternating source of EMF, is an induction motor. An induction motor uses a time-dependent field $\mathbf{B}(t)$ to create current in the rotor coil by electromagnetic induction (Chapter 10). Other applied fields exert forces on the current to drive the motor.

Historically, the first demonstration model of an electric motor, a device using a magnet to produce rotation of a current-carrying wire, was built by Faraday. The model, illustrated in Fig. 8.6, was not a practical design, but it illustrated the principle.

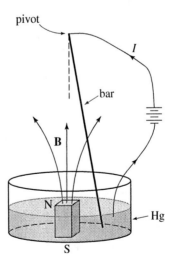

FIGURE 8.6 Faraday's motor. The conducting bar is pivoted at the top at a fixed angle to the vertical, and its lower end is free to move through the liquid mercury. The wires connected to the battery are held fixed, and there is a field **B** due to a magnet below the pivot. In what direction will the bar spin?

8.3 ■ ELECTRIC CURRENT AS A SOURCE OF MAGNETIC FIELD

Where does a magnetic field come from? What is its *source*? The most familiar source is a *permanent magnet*, i.e., a piece of magnetized iron or other ferromagnetic material. But at a more basic level the source of **B** is to be found in electric current. In Chapter 9 we will study magnetic materials, and see that the magnetic field of a ferromagnet comes from properties of atomic electrons—their spin and orbital motion—which, although not classical currents, do involve dynamics of charged particles. However, in this chapter we are concerned with **B** from macroscopic steady currents.[7]

8.3.1 ■ The Biot-Savart Law

Oersted discovered that an electric current creates a force on a magnetized compass needle. In the language of field theory, electric current is a source of magnetic field. Biot and Savart[8] measured the forces between currents and magnets.

The experiments on which the Biot-Savart law is based were done with currents in *macroscopic* wires in circuits. But to start, we consider a more elementary formula: The field $d\mathbf{B}$ at a point P, due to an *infinitesimal* current element $I d\boldsymbol{\ell}$ at a point P'. Figure 8.7 shows the geometry. For this elemental case the Biot-Savart law is

$$d\mathbf{B} = \frac{\mu_0}{4\pi} \frac{I d\boldsymbol{\ell} \times \hat{\mathbf{r}}}{r^2}. \tag{8.20}$$

Here r is the distance between P' (the source point) and P (the field point), $\hat{\mathbf{r}}$ is the unit vector in the direction from P' to P, and $\mathbf{r} = r\hat{\mathbf{r}}$. Note that the Biot-Savart law is an inverse-square law, like Coulomb's law, but the direction of the magnetic field is azimuthal, around the axis of $I d\boldsymbol{\ell}$.

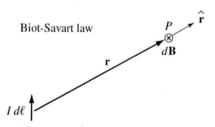

FIGURE 8.7   Elemental form of the Biot-Savart law. $I d\boldsymbol{\ell}$ is the source of magnetic field, and $d\mathbf{B}$ is the resulting field at P.

---

[7] In Chapter 11 we'll study another source of **B**—the Maxwell displacement current.
[8] J. B. Biot (b. 1774, d. 1862 Paris), F. Savart (b. 1791 Mézières, d. 1841 Paris), French physicists. Besides their work on magnetism, Biot had a particular interest in polarized light and Savart had a particular interest in the physics of violins.

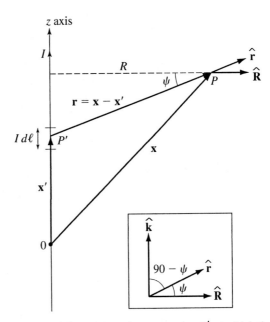

FIGURE 8.8   A long straight current. A source point P', at which the current element $I\,d\boldsymbol{\ell}$ is located, is on the z axis at $(0, 0, z')$. The vector from the origin to P' is $\mathbf{x}'$. The field point P, at which $\mathbf{B}(\mathbf{x})$ is calculated in Example 1, is at perpendicular distance R from the z axis. The cylindrical coordinates of P are (R, ϕ, z). The vector from the origin to P is $\mathbf{x}$ and the vector from P' to P is $\mathbf{r} = \mathbf{x} - \mathbf{x}'$. The inset shows the relationships among $\hat{\mathbf{R}}$ and $\hat{\mathbf{k}}$, which are the unit vectors of cylindrical coordinates, and $\hat{\mathbf{r}} = \mathbf{r}/r$.

To get the field due to a macroscopic wire we must integrate (8.20). Then, letting $\mathbf{x}$ denote the position of P, the magnetic field $\mathbf{B}(\mathbf{x})$ is

$$\mathbf{B}(\mathbf{x}) = \frac{\mu_0}{4\pi} \int_{\text{wire}} \frac{I\,d\boldsymbol{\ell} \times \hat{\mathbf{r}}}{r^2} = \frac{\mu_0}{4\pi} \int_{\text{wire}} \frac{I\,d\boldsymbol{\ell} \times (\mathbf{x} - \mathbf{x}')}{|\mathbf{x} - \mathbf{x}'|^3}. \tag{8.21}$$

Here $\mathbf{r} = \mathbf{x} - \mathbf{x}'$ is the vector from the source point $\mathbf{x}'$, i.e., the position of the wire element $d\boldsymbol{\ell}$, to the field point $\mathbf{x}$, as illustrated for the special case of a straight wire in Fig. 8.8. Again, $r = |\mathbf{r}|$ and $\hat{\mathbf{r}} = \mathbf{r}/r$. We take (8.21) as an empirical law, deduced from experiments on the forces between currents and magnets. The parameter μ_0, called the *permeability of the vacuum*, depends on the system of units. In SI units it is assigned the value $\mu_0 = 4\pi \times 10^{-7}$ Tm/A, exactly; that is, the unit of current (ampere) is defined such that μ_0 has this value.

Examples 1 and 2 below show that calculating $\mathbf{B}$ from the Biot-Savart integral is not simple, because of the vector character of the integral. We will later develop simpler methods for calculating $\mathbf{B}$, especially by Ampère's Law. Nevertheless, it is important to see how (8.21) works, by studying these examples.

EXAMPLE 1   Determine the magnetic field due to a current I in a long straight wire.

Figure 8.8 shows the geometry. The direction of I is $\hat{\mathbf{k}}$. The field point $\mathbf{x}$ has *cylindrical coordinates* (R, ϕ, z). In (8.21) the length element $d\boldsymbol{\ell}$ is $\hat{\mathbf{k}}dz'$ at the source point $\mathbf{x}' = \hat{\mathbf{k}}z'$, and

$$r = |\mathbf{x} - \mathbf{x}'| = \sqrt{(z - z')^2 + R^2} \qquad (8.22)$$

so

$$\mathbf{B}(\mathbf{x}) = \frac{\mu_0 I}{4\pi} \int_{-\infty}^{\infty} \frac{\hat{\mathbf{k}} \times \hat{\mathbf{r}}dz'}{(z - z')^2 + R^2}. \qquad (8.23)$$

Note that we have approximated the wire by an infinite line—an idealization but a valid approximation for points $\mathbf{x}$ near the wire and far from the ends. We assume the return path for I is so far away that it makes a negligible contribution to $\mathbf{B}$ at $\mathbf{x}$. In this approximation $\mathbf{B}$ is independent of z so without loss of generality we can let $z = 0$.

The unit vector $\hat{\mathbf{r}}$ points in the direction from the location of the current element $Id\boldsymbol{\ell}$ to $\mathbf{x}$. Examining Fig. 8.8 we see that

$$\hat{\mathbf{r}} = \hat{\mathbf{R}}\cos\psi + \hat{\mathbf{k}}\sin\psi, \qquad (8.24)$$

where $\hat{\mathbf{R}}$ is the cylindrical radial unit vector at $\mathbf{x}$ and ψ is the angle between $\hat{\mathbf{R}}$ and $\hat{\mathbf{r}}$. The ability to expand one unit vector as a linear combination of other unit vectors is a useful skill in vector analysis. First note that $\hat{\mathbf{r}}$ is in the plane spanned by $\hat{\mathbf{R}}$ and $\hat{\mathbf{k}}$, and write $\hat{\mathbf{r}} = c_1\hat{\mathbf{R}} + c_2\hat{\mathbf{k}}$. Then determine the coefficients by dotting $\hat{\mathbf{r}}$ with $\hat{\mathbf{R}}$ and $\hat{\mathbf{k}}$: $c_1 = \hat{\mathbf{r}} \cdot \hat{\mathbf{R}} = \cos\psi$ and $c_2 = \hat{\mathbf{r}} \cdot \hat{\mathbf{k}} = \sin\psi$. (See the inset in Fig. 8.8.) Hence (8.24).

Now, in (8.23) $\hat{\mathbf{k}} \times \hat{\mathbf{r}} = \hat{\boldsymbol{\phi}}\cos\psi$ and $\cos\psi = R/r$. Making these substitutions we have

$$\mathbf{B}(\mathbf{x}) = \frac{\mu_0 I}{4\pi}\hat{\boldsymbol{\phi}}R \int_{-\infty}^{\infty} \frac{dz'}{(z'^2 + R^2)^{3/2}}. \qquad (8.25)$$

The evaluation of the z' integral is an exercise in calculus,

$$\int_{-\infty}^{\infty} \frac{dz'}{(z'^2 + R^2)^{3/2}} = \frac{z'}{R^2(z'^2 + R^2)^{1/2}}\Bigg|_{-\infty}^{\infty} = \frac{2}{R^2}.$$

Thus the magnetic field of a long straight wire is

$$\mathbf{B}(\mathbf{x}) = \frac{\mu_0 I}{2\pi R}\hat{\boldsymbol{\phi}}. \qquad (8.26)$$

The direction of **B** is azimuthal. That is, **B** "curls around the current," which is a useful mnemonic for the direction of **B**. With the thumb of your right hand pointing in the direction of $I\hat{\mathbf{k}}$, the fingers curl in the direction of **B**.

EXAMPLE 2   Determine the magnetic field due to a circular current loop, at an arbitrary point on the axis of symmetry.

Let the z axis be the axis of symmetry, with the current loop in the xy plane, as shown in Fig. 8.9. The loop has radius a and current I. The line element at angular position ϕ on the loop is $d\ell = \hat{\phi}a d\phi$, so by (8.21)

$$\mathbf{B}(z) = \frac{\mu_0 I}{4\pi} \int_0^{2\pi} \frac{a\hat{\phi} \times \hat{\mathbf{r}} d\phi}{a^2 + z^2}. \tag{8.27}$$

The field point P is on the z axis at $(0, 0, z)$. The unit vector $\hat{\mathbf{r}}$ points in the direction from $d\ell$ to the field point $z\hat{\mathbf{k}}$. The inset in Fig. 8.9 shows that

$$\hat{\mathbf{r}} = -\hat{\mathbf{R}}\cos\psi + \hat{\mathbf{k}}\sin\psi \tag{8.28}$$

where $\hat{\mathbf{R}}$ is the cylindrical radial vector and ψ is the angle between $\hat{\mathbf{r}}$ and $-\hat{\mathbf{R}}$. Therefore in (8.27) $\hat{\phi} \times \hat{\mathbf{r}}$ is a sum of two terms,

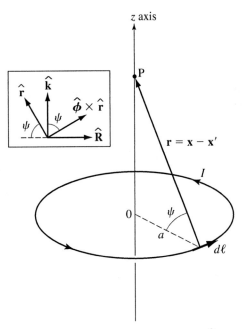

FIGURE 8.9   A circular current. The field at the point $\mathbf{x} = z\hat{\mathbf{k}}$ is calculated in Example 2 from the Biot-Savart law. The element $d\ell$ of the wire is located at the point $\mathbf{x}' = a\hat{\mathbf{R}}$.

$$\hat{\boldsymbol{\phi}} \times \hat{\mathbf{r}} = \hat{\mathbf{k}} \cos \psi + \hat{\mathbf{R}} \sin \psi. \tag{8.29}$$

In (8.27) $\hat{\boldsymbol{\phi}} \times \hat{\mathbf{r}}$ is integrated from 0 to 2π in ϕ. The second term in (8.29) integrates to 0; this can be seen by symmetry, or more directly by noting that $\hat{\mathbf{R}} = \hat{\mathbf{i}} \cos \phi + \hat{\mathbf{j}} \sin \phi$ and both $\cos \phi$ and $\sin \phi$ have integral 0. Therefore the field at $z\hat{\mathbf{k}}$ is in the $\hat{\mathbf{k}}$ direction. Note, too, from Fig. 8.9 that $\cos \psi = a/r$. The evaluation of the remaining integral is trivial because the integrand is independent of ϕ, and the result is

$$\mathbf{B}(z) = \frac{\mu_0 I a^2}{2 \left(a^2 + z^2\right)^{3/2}} \hat{\mathbf{k}}. \tag{8.30}$$

Asymptotically, as $z \to \infty$, the magnetic field decreases as $1/z^3$.

On the z axis the field points in the $\hat{\mathbf{k}}$ direction. This is again consistent with the mnemonic that $\mathbf{B}$ curls around the current. Anywhere on the loop, if the thumb of your right hand is tangent to the circle then the fingers curl up through the loop in the $\hat{\mathbf{k}}$ direction. (Conversely, if the fingers curl with the current then the thumb points along $\mathbf{B}$.)

8.3.2 ■ Forces on Parallel Wires

Consider two long parallel wires, with currents I_1 and I_2, having length ℓ and separation d. The situation is shown in Fig. 8.10. The force on wire 2 is

$$\mathbf{F}_2 = \int I_2 d\boldsymbol{\ell}_2 \times \mathbf{B}_1 \tag{8.31}$$

where $\mathbf{B}_1$ is the field due to wire 1 at the position of $d\boldsymbol{\ell}_2$. The field $\mathbf{B}_1$ may be calculated from (8.26), neglecting end effects, i.e., treating $\mathbf{B}_1$ as constant along

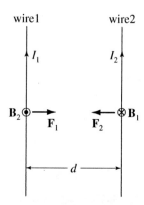

FIGURE 8.10   The forces on parallel wires.

wire 2, which gives

$$\mathbf{B}_1 = \frac{\mu_0 I_1}{2\pi d}\hat{\boldsymbol{\phi}}_1; \tag{8.32}$$

this is a valid approximation for $\mathbf{B}_1$ if $\ell \gg d$. The vector $d\boldsymbol{\ell}_2 \times \hat{\boldsymbol{\phi}}_1$ is $\hat{\mathbf{r}}_{21}d\ell_2$, where $\hat{\mathbf{r}}_{21}$ lies in the plane of the wires and points from wire 2 toward wire 1. Therefore the force on wire 2 is

$$\mathbf{F}_2 = \frac{\mu_0 I_1 I_2 \ell}{2\pi d}\hat{\mathbf{r}}_{21}. \tag{8.33}$$

The force on wire 1 is equal but opposite to $\mathbf{F}_2$.

We see from (8.33) that if the currents I_1 and I_2 are in the same direction then the force is attractive; or, if they are in opposite directions the force is repulsive. For static *charges*, opposites attract; but for *currents*, opposites repel. The reader should study Fig. 8.10, figure out the current, field and force vectors, and verify that the direction of the forces is as described.

Ampère studied the force between currents, and determined by experiment the result (8.33). However, Ampère's work predated the development of field theory. Ampère attributed the force to "action at a distance" between currents. Field theory is different: Wire 1 produces a field $\mathbf{B}_1$ throughout the space, and that field *at wire 2* exerts the force $\mathbf{F}_2$.

The SI unit of current, which is the basic SI electrical unit, is the ampere (A), or "amp" for short. The ampere is defined by the statement that the force per unit length on long parallel wires is exactly 2×10^{-7} N/m if $I_1 = I_2 = 1$ A and $d = 1$ m. It follows that μ_0 has exactly the value $4\pi \times 10^{-7}$ Tm/A.

8.3.3 ■ General Field Equations for B(x)

Equation (8.21) gives the field produced by a one-dimensional current, e.g., I in a thin wire. We must generalize the equation to the case of an arbitrary *volume current density* $\mathbf{J}(\mathbf{x})$. The generalization follows from the superposition principle. Recall that the dimensions of $\mathbf{J}$ are A/m^2, so $\mathbf{J} \cdot d\mathbf{A}$ is the current through a small area element $d\mathbf{A}$. Consider an infinitesmal *volume* d^3x' at $\mathbf{x}'$, consisting of *area* $d\mathbf{A}$ perpendicular to $\mathbf{J}(\mathbf{x}')$ *times length* $d\ell$ parallel to $\mathbf{J}(\mathbf{x}')$. The current through $d\mathbf{A}$ is $\mathbf{J} \cdot d\mathbf{A} = J\, dA$, and the contribution to $\mathbf{B}(\mathbf{x})$ from this small current element is, according to the Biot-Savart law,

$$d\mathbf{B}(\mathbf{x}) = \frac{\mu_0}{4\pi}\frac{J\,dA\,d\ell \times \hat{\mathbf{r}}}{r^2}, \tag{8.34}$$

where again $\mathbf{r} = \mathbf{x} - \mathbf{x}'$. Now, $dA\,d\ell$ is just the volume d^3x', and $d\ell$ is parallel to $\mathbf{J}(\mathbf{x}')$. Thus the complete field at $\mathbf{x}$ is

$$\mathbf{B}(\mathbf{x}) = \frac{\mu_0}{4\pi}\int \frac{\mathbf{J}(\mathbf{x}') \times (\mathbf{x} - \mathbf{x}')}{|\mathbf{x} - \mathbf{x}'|^3}d^3x'. \tag{8.35}$$

(Note that $r = |\mathbf{x} - \mathbf{x}'|$ and $\hat{\mathbf{r}} = (\mathbf{x} - \mathbf{x}')/|\mathbf{x} - \mathbf{x}'|$.) This is the generalization of the Biot-Savart law (8.21) to an arbitrary current distribution throughout a volume of space.

The general equation (8.35), obtained by replacing the infinitesimal *linear* current element $I d\ell$ by the infinitesimal *volume* current element $\mathbf{J}(\mathbf{x}')d^3x'$, can also be extended to *surface* currents. Thus if $\mathbf{K}$ is a current per unit length flowing on a surface S' then the "current times length" $dI \, d\ell$ is $\mathbf{K}(\mathbf{x}')dA'$. We picture $\mathbf{K}$ as the current crossing unit length perpendicular to the flow direction. As a rule the substitutions $I d\ell \rightarrow \mathbf{K}(\mathbf{x}')dA' \rightarrow \mathbf{J}(\mathbf{x}')d^3x'$ apply in the Biot-Savart law (8.21), and in the magnetic force law (8.5). These substitutions are analogous to the electrostatics cases (Chapter 3) where we have $dq \rightarrow \sigma(\mathbf{x}')dA' \rightarrow \rho(\mathbf{x}')d^3x'$ connecting, respectively, infinitesimal charges, surface charge densities, and volume charge densities.

Equation (8.35) is one of the basic equations of magnetism. It determines the magnetic field created by any steady current source. Equation (8.35) is analogous to the Coulomb integral (3-9) for a static electric field. In electrostatics the charge density $\rho(\mathbf{x}')$ creates the electric field $\mathbf{E}(\mathbf{x})$; in magnetostatics the current density $\mathbf{J}(\mathbf{x}')$ creates the magnetic field $\mathbf{B}(\mathbf{x})$.

From (8.35) we will derive two partial differential equations satisfied by the field $\mathbf{B}(\mathbf{x})$. We will eventually prove that

$$\nabla \cdot \mathbf{B} = 0, \qquad (8.36)$$

and

$$\nabla \times \mathbf{B} = \mu_0 \mathbf{J}. \qquad (8.37)$$

Equation (8.36) is universal, i.e., *always true*, assuming that magnetic monopoles do not exist.[9] Equation (8.37), called Ampère's Law, is only valid for magnetostatics, i.e., for the field from a current distribution that is constant in time.

The last point in the previous paragraph will be very important when we consider time-dependent fields, so it deserves some more explanation here, although that will take us away from magnetostatics briefly. Recall from Chapter 2 that the divergence of the curl of any vector function is identically 0. In particular, $\nabla \cdot (\nabla \times \mathbf{B}) = 0$. Therefore (8.37) implies

$$\nabla \cdot \mathbf{J} = 0. \qquad (8.38)$$

This equation is satisfied in magnetostatics. Recall the continuity equation, $\nabla \cdot \mathbf{J} = -\partial\rho/\partial t$, which was discussed in Chapter 7 and is always true. For a static system $\partial\rho/\partial t$ is 0, so then $\nabla \cdot \mathbf{J} = 0$. However, for time-dependent systems (8.38) is not true, and (8.37) must be changed by including the *displacement current*. That is,

---

[9] See Sec. 8.5.

generally $\mathbf{J}$ in (8.37) must be replaced by $\mathbf{J} + \epsilon_0 \partial \mathbf{E}/\partial t$, so that

$$\nabla \times \mathbf{B} = \mu_0 \left(\mathbf{J} + \epsilon_0 \partial \mathbf{E}/\partial t \right),$$

one of the Maxwell equations. The divergence of the right-hand side is always 0:

$$\nabla \cdot \left(\mathbf{J} + \epsilon_0 \frac{\partial \mathbf{E}}{\partial t} \right) = \nabla \cdot \mathbf{J} + \frac{\partial \rho}{\partial t} = 0,$$

because $\epsilon_0 \nabla \cdot \mathbf{E}$ is equal to ρ by Gauss's Law. We shall study the time-dependent Maxwell equation in Chapter 11, but we restrict ourselves in the present chapter to static systems, with (8.37).

Returning to magnetostatics, we must prove (8.36) and (8.37). The starting point is the integral representation of $\mathbf{B}$ in (8.35). Now, note that

$$\frac{(\mathbf{x} - \mathbf{x}')}{|\mathbf{x} - \mathbf{x}'|^3} = -\nabla \frac{1}{|\mathbf{x} - \mathbf{x}'|}. \tag{8.39}$$

Make this replacement in (8.35), and reverse the order of terms in the cross product, remembering to change the sign. Next, rewrite the integrand using the vector identity

$$\left(\nabla \frac{1}{|\mathbf{x} - \mathbf{x}'|} \right) \times \mathbf{J}(\mathbf{x}') = \nabla \times \left(\frac{\mathbf{J}(\mathbf{x}')}{|\mathbf{x} - \mathbf{x}'|} \right). \tag{8.40}$$

Finally, pull the curl operator out of the integral in (8.35), which is allowed because $\nabla \times$ is the curl with respect to $\mathbf{x}$, whereas the integral is over $\mathbf{x}'$. (It may help to write the curl in Cartesian coordinates to see that this step is correct.) Hence

$$\mathbf{B}(\mathbf{x}) = \nabla \times \left[\frac{\mu_0}{4\pi} \int \frac{\mathbf{J}(\mathbf{x}') d^3 x'}{|\mathbf{x} - \mathbf{x}'|} \right]. \tag{8.41}$$

Equation (8.41) is important. It says that $\mathbf{B}(\mathbf{x})$ is equal to the curl of a vector function, $\mathbf{B} = \nabla \times \mathbf{A}$. The function $\mathbf{A}(\mathbf{x})$ is called the *vector potential* and will be discussed further in Sec. 8.5. The first field equation (8.36) follows immediately from (8.41), because the divergence of the curl of a vector function is identically 0.

Equation (8.41) is a special case of the Helmholtz theorem, which is discussed and proven in Appendix B. Any vector function that approaches 0 at ∞ can be written as the sum of a curl and a gradient, as in equations (B.3) to (B.5). Applying the theorem to $\mathbf{B}$, the gradient term (B.4) is 0 because $\nabla \cdot \mathbf{B} = 0$. The curl term is $\nabla \times \mathbf{A}$. Furthermore, comparing (8.41) and (B.5) we see that $\nabla \times \mathbf{B} = \mu_0 \mathbf{J}$. This proves, by a rather roundabout argument, the second field equation (8.37). Because Ampère's law (8.37) is very important, a more direct proof will be given in Sec. 8.4.

8.4 ■ AMPÈRE'S LAW

Equation (8.37) is the *differential form of Ampère's law*: $\nabla \times \mathbf{B} = \mu_0 \mathbf{J}$. We may write an equivalent integral form. Let C be an arbitrary closed curve (loop) in space, and consider the line integral of $\mathbf{B}$ around C. Also, let S be any surface bounded by C, as illustrated in Fig. 8.11. By Stokes's theorem, the loop integral of $\mathbf{B}$ is equal to the surface integral of $\nabla \times \mathbf{B}$ on S. That is, using (8.37),

$$\oint_C \mathbf{B} \cdot d\boldsymbol{\ell} = \mu_0 \int_S \mathbf{J} \cdot d\mathbf{A} = \mu_0 I \text{ (enclosed)}. \tag{8.42}$$

This is the *integral form of Ampère's law*, also called the *circuital law*. I is the current *flowing through* S, or equivalently enclosed by C. The flux of $\mathbf{J}$ is independent of S, for a given boundary C, because $\nabla \cdot \mathbf{J} = 0$. Recall from Stokes's theorem that the directions of $d\boldsymbol{\ell}$ and $d\mathbf{A}$ are correlated by the right-hand rule: With the thumb pointing in the direction of $d\mathbf{A}$, the fingers curl in the direction of $d\boldsymbol{\ell}$.

We derived the integral (8.42) from the differential equation (8.37). We can reverse the derivation, because the integral must hold for *any* closed curve C. The differential and integral relations are equivalent.

Although we are postponing the proof of Ampère's law until the end of this section, we can easily check that the integral relation (8.42) is true for some simple cases. Let I be the current in a long straight wire, for which the magnetic field is (8.26). First, suppose C is a circle perpendicular to I with I at the center, shown in Fig. 8.12(a). Then

$$\oint_C \mathbf{B} \cdot d\boldsymbol{\ell} = (\mu_0 I / 2\pi r)(2\pi r) = \mu_0 I,$$

consistent with (8.42). Second, suppose C is the boundary of an angular section cut out of a circular annulus centered at I, shown in Fig. 8.12(b). Along the radial sides of C, $\mathbf{B} \cdot d\boldsymbol{\ell} = 0$; along the azimuthal arcs,

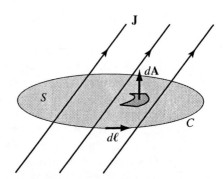

FIGURE 8.11   Geometry for Ampère's Law.

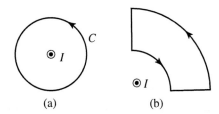

FIGURE 8.12   Two examples of Ampère's Law. (a) A circle enclosing a wire; (b) an annular section not enclosing the wire.

$$\int_{\text{arc}} \mathbf{B} \cdot d\boldsymbol{\ell} = \pm(\mu_0 I/2\pi r)(r\Delta\phi) = \pm\mu_0 I(\Delta\phi/2\pi),$$

where $\Delta\phi$ is the angle of the section. The arcs cancel in the loop integral, so $\oint \mathbf{B} \cdot d\boldsymbol{\ell} = 0$, consistent with (8.42) because there is no current enclosed by C.

8.4.1 ■ Ampère Law Calculations

For current distributions with a high degree of symmetry it is often possible to derive the magnetic field by direct application of the circuital law, taking into account the symmetry. The next four examples illustrate the method. When studying these examples, the reader should draw a picture showing the current, field, and Amperian loop.

EXAMPLE 3   What is the field due to a current I in a long straight wire? We previously calculated $\mathbf{B}(\mathbf{x})$ from the Biot-Savart law. Here we will determine the field from (8.42).

Approximate the wire as an infinite line. By the cylindrical symmetry of the problem, and the fact that $\mathbf{B}$ "curls around" the current, $\mathbf{B}$ must be azimuthally directed around the wire. The magnitude of $\mathbf{B}$ cannot depend on either the position z along the wire or the angle ϕ around the wire. Therefore in cylindrical coordinates, $\mathbf{B}(\mathbf{x})$ must have the form

$$\mathbf{B}(\mathbf{x}) = B(r)\hat{\phi}. \tag{8.43}$$

Now, apply (8.42) to a circular loop centered at the wire with radius r. The loop integral is $\oint \mathbf{B} \cdot d\boldsymbol{\ell} = 2\pi r B(r)$, and this must equal $\mu_0 I$ by (8.42), so

$$B(r) = \frac{\mu_0 I}{2\pi r}. \tag{8.44}$$

The result is, of course, the same as (8.26), but the derivation here is much easier.

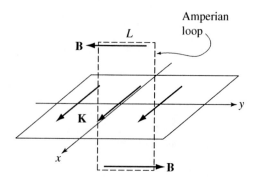

FIGURE 8.13   Example 4. The magnetic field produced by a current sheet can be determined from Ampère's Law.

<hr/>

EXAMPLE 4   What is the magnetic field due to a planar current sheet?

Referring to Fig. 8.13, let $\mathbf{K} = K\,\hat{\mathbf{i}}$ be the *surface current density* (current per unit of length perpendicular to the current) on the xy plane. That is, the current is in the x direction, and $K\,dy$ is the charge per unit time passing across the perpendicular length dy. By symmetry, and the fact that $\mathbf{B}$ curls around the current, $\mathbf{B}(\mathbf{x})$ must have the form

$$\mathbf{B}(\mathbf{x}) = \begin{cases} -B(|z|)\,\hat{\mathbf{j}} & \text{for} \quad z > 0 \\ +B(|z|)\,\hat{\mathbf{j}} & \text{for} \quad z < 0, \end{cases} \tag{8.45}$$

as shown in Fig. 8.13. In this equation we have allowed the field to have z dependence, although we shall soon see that it is independent of z. Now, apply (8.42) to a rectangular loop of size $L \times 2z$ parallel to the yz plane and bisected by the xy plane, also shown in Fig. 8.13. Let the top side of the loop be at $+z$ and the bottom side at $-z$. The loop integral is $\oint \mathbf{B} \cdot d\boldsymbol{\ell} = 2BL$, and the current enclosed by the loop is KL; so by (8.42)

$$B(z) = \mu_0 K / 2. \tag{8.46}$$

Note that the field is independent of the distance z from the plane, because we have considered an infinite plane of current. Recall that $\mathbf{E}$ due to an infinite charged plane is similarly independent of the distance from the plane. We say in problems involving only infinite planes *"there is no scale of length."* Equation (8.46) is a valid approximation for the magnetic field near a current sheet far from the edges.

<hr/>

An extension of Example 4 is to find $\mathbf{B}$ everywhere in space for a configuration with surface current $\mathbf{K} = +K\,\hat{\mathbf{i}}$ on the plane $z = +a$, and $\mathbf{K} = -K\,\hat{\mathbf{i}}$ on the plane

$z = -a$. This configuration, called a bus bar, with two parallel conducting plates carrying large currents in opposite directions, is used in power stations. By the superposition principle, i.e., adding the fields of the two plates, the magnetic field is $\mu_0 K \hat{\mathbf{j}}$ between the plates, and 0 above or below the plates. In practice, the two parallel conductors of a bus bar are parallel conducting ribbons, not infinite planes. But because the separation between the ribbons is small compared to their width, the simplifying assumptions we've used hold to a good approximation.

EXAMPLE 5   As an extension of Example 4, we now find the magnetic field due to the constant volume current density $\mathbf{J} = J_0 \hat{\mathbf{i}}$ flowing in the conducting slab shown in Fig. 8.14. The slab extends between $-a \leq z \leq a$ and is infinite in the x and y directions.

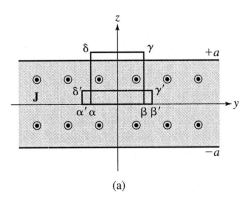

(a)

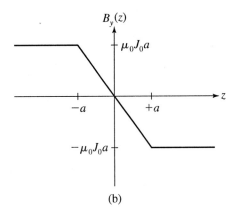

(b)

FIGURE 8.14   A slab of current. (a) The slab geometry. The current density in the slab is uniform, $\mathbf{J} = J_0 \hat{\mathbf{i}}$. The magnetic field due to this current is in the $\pm \hat{\mathbf{j}}$ direction. (b) The magnetic field as a function of z.

This problem has enough symmetry that it can be solved by Ampère's law, but first the symmetry must be exploited to gain insights as to what form **B** may have.

Notice first that **B** cannot depend on the coordinates x and y because there is no variation in the geometry as these coordinates are changed; thus, $\mathbf{B} = \mathbf{B}(z)$. Notice next that the finite slab can be considered as a stack of infinitesimal slabs, or current sheets, of thickness dz, each carrying the current per unit length $dK\,\hat{\mathbf{i}} = J_0 dz\,\hat{\mathbf{i}}$. The idea is that each section of slab, with unit length in the y direction and dz thick, carries current $J_0 dz$ in the x direction. But, as we know from Example 4, the field due to such a current sheet is constant in magnitude and parallel to the y axis, in the $-y$ direction above the sheet, and in the $+y$ direction below the sheet. The field due to the stack of such sheets must also be parallel to the y axis, so we may write $\mathbf{B} = B_y(z)\,\hat{\mathbf{j}}$. Notice finally that $\mathbf{B} = 0$ on the plane $z = 0$ because the field due to the current sheet at $+z$ cancels the field due to the symmetrically opposite one at $-z$.

We will apply Ampère's law (8.42) to the two rectangular loops shown in Fig. 8.14. The direction of traversal of all the loops is counterclockwise, consistent with the right-hand rule and current out of the page.

First apply (8.42) to the loop $\alpha\beta\gamma\delta\alpha$, which has dimensions $\ell \times z$ with $z > a$. The only nonzero contribution to $\oint \mathbf{B} \cdot d\boldsymbol{\ell}$ comes from the segment $\gamma\delta$, because $\mathbf{B} = 0$ along $\alpha\beta$, and $\mathbf{B}$ is perpendicular to $d\boldsymbol{\ell}$ along $\beta\gamma$ and $\delta\alpha$. Using (8.42), $\oint \mathbf{B} \cdot d\boldsymbol{\ell} = B_y(z)(-\ell) = \mu_0 J_0 \ell a$. The reason for $(-\ell)$ in the line integral is that the integral from γ to δ is in the $-\hat{\mathbf{j}}$ direction. The total current passing through $\alpha\beta\gamma\delta\alpha$ is $J_0 \ell a$; there is no current for $z > a$. The result is $B_y(z) = -\mu_0 J_0 a$.

The field is constant above the slab: $\mathbf{B} = -\mu_0 J_0 a\,\hat{\mathbf{j}}$, no matter how large z is, as long as $z \geq a$. We can also obtain this result by integrating the contributions of infinitesimal current sheets making up the finite slab. By (8.46), $d\mathbf{B} = -\frac{1}{2}\mu_0 dK\,\hat{\mathbf{j}}$, where $dK = J_0 dz$; the integral over $-a \leq z \leq a$ gives again $\mathbf{B} = -\mu_0 J_0 a\,\hat{\mathbf{j}}$. Similarly, considering the region $z < -a$ below the slab, one finds there $\mathbf{B} = +\mu_0 J_0 a\,\hat{\mathbf{j}}$.

To find **B** in the region $0 \leq z \leq a$, inside the slab and above the midplane, apply (8.42) to the loop $\alpha'\beta'\gamma'\delta'\alpha'$, of dimensions $\ell \times z$ with $z < a$. The only nonzero contribution to $\oint \mathbf{B} \cdot d\boldsymbol{\ell}$ is from the segment $\gamma'\delta'$, so $\oint \mathbf{B} \cdot d\boldsymbol{\ell} = B_y(z)(-\ell) = \mu_0 J_0 \ell z$. Because the loop is traversed counterclockwise, the right-hand convention requires that current in the $+x$ direction (i.e., out of the page) be positive. Hence for $z > 0$ we wrote the current as $J_0 \ell z$, which is proportional to the height z. The result is $B_y(z) = -\mu_0 J_0 z$. Similarly, for the region $-a \leq z \leq 0$, the field is $B_y(z) = -\mu_0 J_0 z$; notice that z is negative in this region so **B** points in the positive y direction.

Figure 8.14(b) is a plot of $B_y(z)$ versus z. Note that B_y is continuous at the interfaces $z = \pm a$ because there is no surface current there. Recall from Sec. 7.2 what is meant by surface current: There must be *finite* current on the surface S through a rectangle of *infinitesimal* height dz perpendicular to S. (The current through a unit length of the rectangle due to the volume current is only the infinitesimal $J_0 dz$.) For a surface with zero surface current, the tangential components of **B** are continuous across the surface. On the other hand, $B_y(z)$ is discontinuous at $z = 0$ in Example 4; in general the tangential component of **B** has a discontinuity at a current-carrying surface, proportional to K.

■──────────

EXAMPLE 6 What is the field due to a long, densely wound cylindrical solenoid with circular cross section?

A solenoid is a coil of wire carrying current I wound around a cylinder. If the coil is densely wound then the current is approximately the same as a current sheet rolled up into a cylinder, as illustrated in Fig. 8.15. (The surface current density is $\mathbf{K} = nI\hat{\phi}$, where n is the number of turns per unit length.) We consider this ideal case. It would be difficult to find **B** directly from the Biot-Savart integral, but it is easy by Ampère's law.

By symmetry, treating the solenoid as infinitely long as well as densely wound, $\mathbf{B}(\mathbf{x})$ must point in the direction $\hat{\mathbf{k}}$ of the solenoid axis, and cannot depend on the position z along the axis, nor on the azimuthal angle ϕ around the z axis. That is, in cylindrical coordinates, $\mathbf{B}(\mathbf{x}) = B(r)\hat{\mathbf{k}}$. First consider a rectangular Amperian loop C_1 outside the solenoid; see Fig. 8.15. The loop integral taken in the counterclockwise direction around C_1 is $[B(r_1) - B(r_2)]\ell$, and this is 0 by the circuital law because there is no current enclosed by C_1. Thus $B(r)$ is constant outside the solenoid. Furthermore, the constant must be 0 because **B** is 0 at $r = \infty$. **B** must be 0 at ∞ because it is the field due to a finite current I. We conclude that

$$\mathbf{B} = 0 \text{ outside.} \tag{8.47}$$

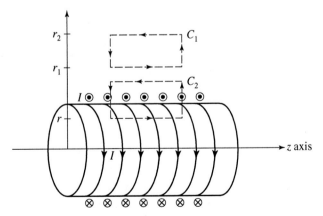

FIGURE 8.15 **Example 6.** The magnetic field of a long solenoid can be calculated from Ampère's Law. C_1 and C_2 are Amperian loops used in the analysis. The figure shows a finite solenoid for convenience but Example 6 considers the ideal case of a closely wound solenoid of infinite length.

This result is for an ideal solenoid. For a real solenoid with one layer of wire wound on a cylinder, there is a component of current in the axial direction, so that an Amperian loop surrounding the solenoid (coaxial with the cylinder) encloses the current I. Then **B** has a small azimuthal component outside the solenoid, $B_\phi(r) \approx \mu_0 I/(2\pi r)$ for a circular solenoid. This external field would be very weak compared to the field inside the solenoid.

Next consider a rectangular Amperian loop C_2 cutting through the solenoid surface. The loop integral around C_2 is $B(r)\ell$; this is equal to $\mu_0 n\ell I$ by the circuital law, where n is the number of turns per unit length, because $n\ell I$ is the total current enclosed by C_2. Thus we have

$$\mathbf{B} = \mu_0 n I \hat{\mathbf{k}} \text{ inside.} \tag{8.48}$$

Note that $\mathbf{K} = n I \hat{\phi}$ is the surface current density, so $B = \mu_0 K$ inside the solenoid. Remarkably, **B** *is constant in the solenoid*—independent of r.

We derived (8.47) and (8.48) for a solenoid with circular cross section, and can show that the same equations hold for a square cross section, or indeed any cross sectional shape, assuming the shape is constant in the z direction.[10] To prove the latter result, build up the generalized cylinder from infinitesimal square solenoids inside it in such a way that all currents cancel except those on the outer surface.

■────────────────

Example 6 is an infinite solenoid—an idealization that approximates a long solenoid (length $\gg$ radius) far from the ends. But real solenoids do not all satisfy these simplifying assumptions. The field on the axis of a densely wound solenoid of *finite* length can be calculated[11] by integrating $d\mathbf{B}$ of the circular coils that compose the solenoid; we know $d\mathbf{B}$ from Example 2. Figure 8.16 is a plot of

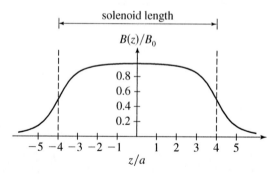

FIGURE 8.16   **Magnetic field of a finite solenoid.** The field is plotted along the axis of a circular cylindrical solenoid for which the ratio of length to diameter is 4. $B(z)$ is plotted in units of B_0, the field of an infinite solenoid with the same surface current.

[10]Equations (8.47) and (8.48) are correct for any cross sectional shape because they satisfy the field equations and the boundary conditions. See Chapter 9.
[11]See Exercise 13.

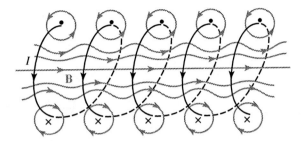

FIGURE 8.17 A loosely wound solenoid.

$B_z(0, 0, z)/B_0$ for a solenoid of radius a centered on the z axis, extending from $z = -4a$ to $z = 4a$; the reference field B_0 is the field $\mu_0 n I$ of an infinite solenoid.

Figure 8.17 shows the magnetic field for a solenoid that is loosely wound and finite in length, i.e., a helical wire. The reader should imagine how the ideal solenoid emerges as the limit of very closely wound coils.

Solenoids are commonly used for electromechanical relay switches. Current produces a magnetic field that pulls a ferromagnetic plunger into the solenoid, closing the switch. The starter solenoid of a car works this way; a small current supplied through the ignition switch to a solenoidal coil closes a circuit through which the battery supplies a much larger current to the starter motor.

Examples 3, 4, and 6 are summarized in Fig. 8.18. Note in each case how **B** curls around I .

8.4.2 ■ Formal Proof of Ampère's Law

We have seen how useful Ampère's law can be. But how is it related to the Biot-Savart law? We will construct a mathematical proof of Ampère's law, starting from the Biot-Savart law. The differential and integral forms of Ampère's law are

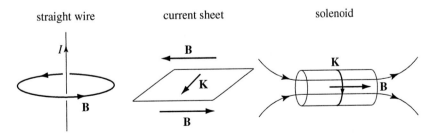

FIGURE 8.18 Fields and currents of Examples 3, 4, and 6. Note in each case how the field curls around the current. The ideal analytical solutions require translation invariance in each case.

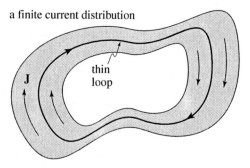

FIGURE 8.19   Any finite current distribution may be subdivided into thin loops.

equivalent, so we may prove either one. We'll prove the integral relation (8.42). The mathematics of the proof is difficult but beautiful.[12]

The starting point is the Biot-Savart law (8.21), which is an experimental fact. Now, it is sufficient to consider as the source of **B** a current I on a closed curve, because any finite current distribution can be subdivided into thin loops, as illustrated in Fig. 8.19. By the superposition principle, if Ampère's law is true for one loop then it is true for any superposition of loops.

Now consider Fig. 8.20(a). Let C' be the closed curve carrying current I. For simplicity we assume that C' is planar, but the proof can be generalized to a non-planar current loop. For an Amperian loop C (see Fig. 8.20(a)) we will evaluate formally the circulation integral of **B**, defined as

$$\gamma \equiv \oint_C \mathbf{B} \cdot d\mathbf{x} = \sum_{\delta\mathbf{x}} \mathbf{B} \cdot \delta\mathbf{x} \qquad (8.49)$$

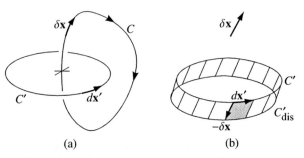

(a)   (b)

FIGURE 8.20   Proof of Ampère's Law. (a) C' carries current I and C is an Amperian loop. (b) $-\delta\mathbf{x} \times d\mathbf{x}'$ is an area element of the surface that would be swept out if C' were to be displaced by $-\delta\mathbf{x}$.

[12]J. C. Slater and N. H. Frank, *Electromagnetism* (Dover, New York, 1969), Chapter 5.

where the limit $|\delta \mathbf{x}| \to 0$ is understood. We must prove that $\gamma = \mu_0 I$ if C and C' are linked, but $\gamma = 0$ if C and C' are not linked.

By the Biot-Savart law,

$$\mathbf{B}(\mathbf{x}) = \frac{\mu_0 I}{4\pi} \oint_{C'} \frac{d\mathbf{x}' \times \hat{\mathbf{r}}}{r^2}, \tag{8.50}$$

where $\mathbf{r} = \mathbf{x} - \mathbf{x}'$ is the vector from the source point $\mathbf{x}'$ on C' to the field point $\mathbf{x}$ on C. The contribution to γ from the small segment $\delta \mathbf{x}$ of C is

$$\delta \gamma = \frac{\mu_0 I}{4\pi} \oint_{C'} \frac{\left(d\mathbf{x}' \times \hat{\mathbf{r}}\right) \cdot \delta \mathbf{x}}{r^2} = \frac{\mu_0 I}{4\pi} \oint_{C'} \frac{\left(\delta \mathbf{x} \times d\mathbf{x}'\right) \cdot \hat{\mathbf{r}}}{r^2}, \tag{8.51}$$

where the second form follows from a vector cross product identity. Now, $-\delta \mathbf{x} \times d\mathbf{x}'$ is a certain area element, and we will evaluate the integral by analytic geometry.

Let C'_{dis} denote the loop that results from displacing C' by $-\delta \mathbf{x}$, as shown in Fig. 8.20(b). Also, let $\delta \mathcal{R}$ denote the surface between C' and C'_{dis} (a loop of ribbon). $\delta \mathcal{R}$ is the surface that is swept out by C' as it is translated by $-\delta \mathbf{x}$. Figure 8.20(b) demonstrates that $-\delta \mathbf{x} \times d\mathbf{x}'$ is an area element $d\mathbf{A}$ of the ribbon $\delta \mathcal{R}$, shown as a shaded region in the figure. The key to this proof of Ampère's law is to see that $-\hat{\mathbf{r}} \cdot d\mathbf{A}/r^2$ is the *solid angle* subtended by $d\mathbf{A}$ viewed from $\mathbf{x}$ ($-\hat{\mathbf{r}}$ is the unit vector pointing from $\mathbf{x}$ to $\mathbf{x}'$). Therefore the integral around C' in (8.51) gives the net solid angle $\delta \Omega$ subtended by the ribbon loop $\delta \mathcal{R}$,

$$\oint_{C'} \frac{\left(\delta \mathbf{x} \times d\mathbf{x}'\right) \cdot \hat{\mathbf{r}}}{r^2} = \delta \Omega. \tag{8.52}$$

But $\delta \Omega$ is just the difference between the solid angles Ω' and Ω'_{dis} subtended, respectively, by the planar surfaces bounded by C' and C'_{dis}; that is, $\delta \Omega = \Omega' - \Omega'_{\text{dis}}$. As C' moves away from $\mathbf{x}$ (by $-\delta \mathbf{x}$) the solid angle decreases; that is, $\Omega'_{\text{dis}} < \Omega'$. In other words, letting S' denote the planar surface bounded by C', $\delta \Omega$ is the decrease of the solid angle subtended by S' either when C' moves by $-\delta \mathbf{x}$ with $\mathbf{x}$ fixed, or when $\mathbf{x}$ moves by $+\delta \mathbf{x}$ with C' fixed. The sign convention here is that Ω' is positive viewed from a point above S', as in Fig. 8.20(b), and negative when viewed from a point below S'; equivalently, the direction of an area element $d\mathbf{A}'$ on S' points downward.

Now we can evaluate the circulation integral γ in (8.49). Combining (8.51) and (8.52) we have

$$\gamma = \sum_{\delta \mathbf{x}} \delta \gamma = \sum_{\delta \mathbf{x}} \frac{\mu_0 I}{4\pi} \left[\Omega'(\mathbf{x}) - \Omega'(\mathbf{x} + \delta \mathbf{x}) \right], \tag{8.53}$$

where $\Omega'(\mathbf{x})$ and $\Omega'(\mathbf{x} + \delta \mathbf{x})$ are the solid angles subtended by S' viewed from $\mathbf{x}$ and $\mathbf{x} + \delta \mathbf{x}$, respectively. Adding all the small differences of solid angle as $\mathbf{x}$ steps around the loop C yields the net change in the solid angle subtended by S', so

$$\gamma = \frac{\mu_0 I}{4\pi} \left(-\Delta\Omega'\right), \tag{8.54}$$

where $\Delta\Omega'$ is the net change $\Omega'_{final} - \Omega'_{initial}$ from an initial point on C infinitesimally above S' to a final point on C infinitesimally below S'.

If C threads through C', as in Fig. 8.20(a), then $\Delta\Omega' = -4\pi$: Imagine traversing C starting from a point infinitesimally above the planar surface S' bounded by C'. The initial solid angle subtended by S' is 2π. As $\mathbf{x}$ moves around C, Ω' decreases, becomes 0 at some point, and then negative. As $\mathbf{x}$ approaches the final point infinitesimally below S', Ω' approaches -2π. The net change is $\Delta\Omega' = -4\pi$. Hence Ampère's law is proven, because $\gamma = \mu_0 I$ and the current I is enclosed by C.

If C does not thread through C' then the initial and final solid angles, subtended by C' at points infinitesimally above and below the plane of C' respectively, are both 0. In this case $\Delta\Omega' = 0$ so the circulation γ is 0, in accord with Ampère's law because zero current is enclosed by C.

The integral form of Ampère's law is thus proven. The proof is formal rather than practical. It shows that Ampère's law and the Biot-Savart law are not independent principles, but equivalent.

8.5 ■ THE VECTOR POTENTIAL

At this point it is useful to compare the principles of magnetostatics and electrostatics, given in the table below.

Magnetostatics	Electrostatics
$\nabla \times \mathbf{B} = \mu_0 \mathbf{J}$	$\nabla \cdot \mathbf{E} = \rho/\epsilon_0$
$\nabla \cdot \mathbf{B} = 0$	$\nabla \times \mathbf{E} = 0$
$\mathbf{B}$ curls around I	$\mathbf{E}$ diverges from q

Note this difference: The electric field has scalar sources, but the magnetic field has vector sources. Electric charge, i.e., a point source of $\mathbf{E}$ from which $\mathbf{E}$ diverges, is a common property of elementary particles. Motion of charged particles—current—is the source of $\mathbf{B}$, around which $\mathbf{B}$ curls.

Magnetic charge, i.e., a point source from which $\mathbf{B}$ diverges in the rest frame of the particle, apparently does not exist in nature, or at least it has not been observed. Such a hypothetical charge is called a *magnetic monopole*. Dirac showed that it is possible to construct a consistent quantum theory with both electric charges and magnetic monopoles. However, the fundamental magnetic charge g and electric charge e would necessarily be quantized, and satisfy the relation $eg = n/2$, where n is an integer. Many experimenters have searched for magnetic monopoles, but so far the results are negative. Some speculative theories of high-energy physics, such as grand unified field theories, predict the existence of very massive magnetic monopoles, too massive to be produced at current high-energy accelerators, but which might have been produced in the big bang. Searching for magnetic monopoles continues to be an interesting experimental challenge.

If magnetic monopoles do not exist, then the equation $\nabla \cdot \mathbf{B} = 0$ is a universal equation of magnetism. Whether magnetic monopoles exist or not, the source of the magnetic fields we encounter in physics are not point magnetic charges but rather currents of electric charge, corresponding to the source equation $\nabla \times \mathbf{B} = \mu_0 \mathbf{J}$.

We found in electrostatics that it is useful to introduce a scalar potential $V(\mathbf{x})$ for the electrostatic field, such that $\mathbf{E} = -\nabla V$. This guarantees that $\nabla \times \mathbf{E} = 0$. In an analogous way we may introduce a *vector potential* $\mathbf{A}(\mathbf{x})$ for the magnetic field, such that

$$\mathbf{B} = \nabla \times \mathbf{A}. \tag{8.55}$$

This guarantees that $\nabla \cdot \mathbf{B} = 0$. Indeed, we have already seen $\mathbf{B}(\mathbf{x})$ written as a curl in (8.41).

However, (8.55) does not uniquely determine $\mathbf{A}$ for a given magnetic field $\mathbf{B}$: If $f(\mathbf{x})$ is an arbitrary scalar function, then $\mathbf{A} + \nabla f$ has the same curl as $\mathbf{A}$ (namely $\mathbf{B}$) because $\nabla \times \nabla f$ is identically 0. Therefore we may impose a condition on $\mathbf{A}$, called a gauge condition, to remove this ambiguity. The *Coulomb gauge condition*, which we will use in magnetostatics, is

$$\nabla \cdot \mathbf{A} = 0. \tag{8.56}$$

Taking (8.55) and (8.56) together is still not enough to make $\mathbf{A}(\mathbf{x})$ unique, because adding a constant does not change either $\nabla \times \mathbf{A}$ or $\nabla \cdot \mathbf{A}$. But imposing an appropriate boundary condition, such as requiring $\mathbf{A} \to 0$ at infinity, makes $\mathbf{A}$ unique.

In Sec. 8.5.1 we will find an integral formula for the potential $\mathbf{A}(\mathbf{x})$ of localized current sources. But first, as a simple example of the vector potential, consider the *uniform field* $\mathbf{B} = B_0 \hat{\mathbf{k}}$. A vector potential function, satisfying (8.55) for the uniform field and the Coulomb gauge condition (8.56), is

$$\mathbf{A}(\mathbf{x}) = \tfrac{1}{2} \mathbf{B} \times \mathbf{x} = \tfrac{1}{2} B_0 \left(-y\hat{\mathbf{i}} + x\hat{\mathbf{j}} \right). \tag{8.57}$$

An example of uniform $\mathbf{B}$ is the field inside a solenoid, so for $\mathbf{B} = B_0 \hat{\mathbf{k}}$ we may picture a long, tightly and uniformly wound solenoid, whose axis is the z axis. Note for this case that $\mathbf{A}(\mathbf{x})$ is parallel to the surface currents (azimuthal) and that $\mathbf{A}(\mathbf{x})$ curls around the $\mathbf{B}$ field.

8.5.1 ■ General Solution for $\mathbf{A}(\mathbf{x})$

By Ampère's law we have

$$\nabla \times (\nabla \times \mathbf{A}) = \mu_0 \mathbf{J}. \tag{8.58}$$

The double cross product can be reduced to

$$\nabla \times (\nabla \times \mathbf{A}) = \nabla(\nabla \cdot \mathbf{A}) - \nabla^2 \mathbf{A} = -\nabla^2 \mathbf{A}, \qquad (8.59)$$

the last equality by the Coulomb gauge condition (8.56). Now (8.58) becomes $-\nabla^2 \mathbf{A} = \mu_0 \mathbf{J}$, which is *Poisson's equation* for each Cartesian component of $\mathbf{A}(\mathbf{x})$. Recall from Chapter 3 how Poisson's equation can be solved with the Green's function of $-\nabla^2$, which is $1/4\pi |\mathbf{x} - \mathbf{x}'|$. The solution is[13]

$$\mathbf{A}(\mathbf{x}) = \frac{\mu_0}{4\pi} \int \frac{\mathbf{J}(\mathbf{x}')d^3x'}{|\mathbf{x} - \mathbf{x}'|}. \qquad (8.60)$$

For self-consistency we must have $\nabla \cdot \mathbf{A} = 0$, which is true because $\nabla \cdot \mathbf{J} = 0$.[14] (Show $\nabla \cdot \mathbf{A} = 0$ by a trick: ∇ acts on the $\mathbf{x}$ in $|\mathbf{x} - \mathbf{x}'|$. But ∇ on a function of $\mathbf{x} - \mathbf{x}'$ is the same as $-\nabla'$. In the latter form integrate by parts; the surface term is 0 because $\mathbf{J} = 0$ at infinity, and the other term is proportional to $\nabla' \cdot \mathbf{J}(\mathbf{x}')$, also 0. Exercise 21 supplies the details.)

The vector potential in (8.60) is the same as in (8.41), the result obtained earlier from the Biot-Savart law. In other words, we obtain the same field from the Biot-Savart law and from Ampère's law. This provides another proof that Ampère's law is equivalent to the Biot-Savart law. We have written $\mathbf{A}$ in the Coulomb gauge, but the equivalence of Ampère's law and the Biot-Savart law is independent of gauge choice because these laws involve only the gauge-invariant quantities $\mathbf{B}$ and $\mathbf{J}$.

Equation (8.60) is an example of the superposition principle. The small current source $\mathbf{J}(\mathbf{x}')d^3x'$ contributes its part to the potential, $d\mathbf{A}(\mathbf{x})$; the full potential is $\mathbf{A}(\mathbf{x}) = \int d\mathbf{A}(\mathbf{x})$. Note from (8.60) that the contribution $d\mathbf{A}(\mathbf{x})$ from any current element $\mathbf{J}(\mathbf{x}')d^3x'$ is parallel to the current element.

EXAMPLE 7   Determine the vector potential for the magnetic field produced by a hollow charged sphere that rotates with constant angular velocity $\boldsymbol{\omega} = \omega\hat{\mathbf{k}}$. The sphere is shown in Fig. 8.21.

Let σ be the surface charge density, and a the radius of the sphere. The surface current density at a point $\mathbf{x}'$ on the sphere is $\mathbf{K}(\mathbf{x}') = \sigma\mathbf{v}(\mathbf{x}')$, where $\mathbf{v}(\mathbf{x}')$ is the velocity at the point $\mathbf{x}'$. The direction of $\mathbf{v}(\mathbf{x}')$ is azimuthal and the magnitude is $\omega a \sin\theta$, where θ is the angle between $\hat{\mathbf{k}}$ and $\mathbf{x}'$. Therefore $\mathbf{v}(\mathbf{x}') = \omega\hat{\mathbf{k}} \times \mathbf{x}'$, and the surface current density is $\mathbf{K}(\mathbf{x}') = \sigma\boldsymbol{\omega} \times \mathbf{x}'$.

Now calculate $\mathbf{A}(\mathbf{x})$ using (8.60). Divide the surface into small patches $dA' = a^2d\Omega'$, and replace the current element $\mathbf{J}(\mathbf{x}')d^3x'$ appropriate for a volume by $\mathbf{K}(\mathbf{x}')dA'$ for the surface; that is,

---

[13] Equation (8.60) is safe in a Cartesian coordinate system, where the direction vectors are constants. In spherical or cylindrical coordinates (8.60) may be misunderstood, because the directions of the basis vectors in the integrand depend on the point $\mathbf{x}'$, which varies over the integral, and not on $\mathbf{x}$. For example, to find $A_r(\mathbf{x}) = \hat{\mathbf{r}} \cdot \mathbf{A}(\mathbf{x})$ we need more than $J_r(\mathbf{x}')$.

[14] See Exercise 21.

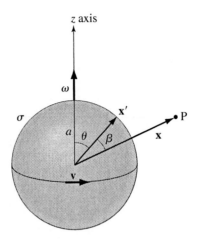

z axis

FIGURE 8.21 Example 7. A hollow sphere with surface charge density σ rotates with angular velocity $\boldsymbol{\omega} = \omega\hat{\mathbf{k}}$. What is the magnetic field? The point $\mathbf{x}'$ is a source point and the point $\mathbf{x}$ is a field point.

$$\mathbf{A}(\mathbf{x}) = \frac{\mu_0}{4\pi} \int \frac{\mathbf{K}(\mathbf{x}')dA'}{|\mathbf{x} - \mathbf{x}'|} = \frac{\mu_0\sigma a^2}{4\pi}\boldsymbol{\omega} \times \mathbf{f}(\mathbf{x}), \qquad (8.61)$$

where

$$\mathbf{f}(\mathbf{x}) = \int \frac{\mathbf{x}'\,d\Omega'}{|\mathbf{x} - \mathbf{x}'|}. \qquad (8.62)$$

Again, $\mathbf{x}'$ is a source point on the sphere with $|\mathbf{x}'| = a$, and $\mathbf{x}$ is an arbitrary field point with $|\mathbf{x}| = r$. The integral is over all points on the sphere, and we have used $dA' = a^2 d\Omega'$ to express it as an integral over all solid angles.

To evaluate the integral for $\mathbf{f}(\mathbf{x})$ requires some clever tricks. First, $\mathbf{f}(\mathbf{x})$ is a vector, depending only on $\mathbf{x}$, so it must be $C\mathbf{x}$ where C is a scalar. (Since $\mathbf{f}$ depends only on $\mathbf{x}$, and there is no preferred direction in (8.62), what other vector could it be?) But then $\mathbf{x} \cdot \mathbf{f} = \mathbf{x} \cdot C\mathbf{x} = Cr^2$, so the scalar C may be evaluated as $\mathbf{x} \cdot \mathbf{f}/r^2$, i.e.,

$$C = \frac{\mathbf{x} \cdot \mathbf{f}}{r^2} = \frac{1}{r^2} \int \frac{\mathbf{x} \cdot \mathbf{x}'\,d\Omega'}{|\mathbf{x} - \mathbf{x}'|}. \qquad (8.63)$$

Next, let β be the angle between $\mathbf{x}$ and $\mathbf{x}'$, as in Fig. 8.21, so that $\mathbf{x} \cdot \mathbf{x}' = ra\cos\beta$; and write C as an integral over β

$$C = \frac{a}{r} \int_0^\pi \frac{\cos\beta\, 2\pi\sin\beta d\beta}{\sqrt{r^2 + a^2 - 2ra\cos\beta}} \qquad (8.64)$$

using the fact that $2\pi\sin\beta d\beta$ is the solid angle $d\Omega'$ of a strip of the unit sphere around the axis defined by $\mathbf{x}$. Finally, change the variable of integration from β

to $u \equiv \cos \beta$, and evaluate the integral over u with an integral table, or with an analytic computer program. But *be careful*; the result is different for $r < a$ and $r > a$:

$$\int_{-1}^{1} \frac{u\,du}{\sqrt{r^2 + a^2 - 2rau}} = \frac{2r_<}{3r_>^2}, \tag{8.65}$$

where $r_<$ is the smaller of r and a, and $r_>$ is the larger of r and a. The final result is

$$\mathbf{A}(\mathbf{x}) = \frac{\mu_0 \sigma a}{3} \, \boldsymbol{\omega} \times \mathbf{x} \quad \text{if} \quad r < a \tag{8.66}$$

$$\mathbf{A}(\mathbf{x}) = \frac{\mu_0 \sigma a^4}{3r^3} \, \boldsymbol{\omega} \times \mathbf{x} \quad \text{if} \quad r > a. \tag{8.67}$$

Now calculate $\mathbf{B}(\mathbf{x}) = \nabla \times \mathbf{A}$, as an exercise in vector calculus, and sketch a graph of the field lines.[15] The field has a very simple form, as one might expect from such a simple current source: Inside the sphere $\mathbf{B}$ is uniform and in the direction of $\boldsymbol{\omega}$, because $\mathbf{A}(\mathbf{x})$ has the same form as (8.57). Outside the sphere $\mathbf{B}$ is a pure dipole field, i.e., the same as if there were a point magnetic dipole at the origin—a field configuration that we study next.

◼━━━━━━━━━━━━━━

8.6 ◼ THE MAGNETIC DIPOLE

What is the asymptotic magnetic field far from a static current distribution of finite extent? In Chapter 3 we addressed the analogous question for electrostatics, and derived the multipole expansion for the electric potential. We now construct the similar expansion for the vector potential. For an arbitrary current distribution the multipole expansion of $\mathbf{A}(\mathbf{x})$ has in order dipole, quadrupole, octopole terms, etc.; there is no monopole term. We will derive only the dipole term, which is the dominant term at large distances unless the dipole moment happens to be 0.

8.6.1 ◼ Asymptotic Analysis

The vector potential $\mathbf{A}(\mathbf{x})$ is given by (8.60). The source $\mathbf{J}(\mathbf{x}')$ is assumed to be localized in a limited region of space. Far from the source, i.e., for $r \gg r'$ where $r = |\mathbf{x}|$ and $r' = |\mathbf{x}'|$, the following approximation is valid:

$$\frac{1}{|\mathbf{x} - \mathbf{x}'|} = \frac{1}{r} + \frac{\hat{\mathbf{r}} \cdot \mathbf{x}'}{r^2} + \mathcal{O}\left(\frac{r'^2}{r^3}\right). \tag{8.68}$$

We used the same expansion in Eq. (3.61) of Chapter 3, when deriving the asymptotic $V(\mathbf{x})$ of a static charge distribution. In (8.68) $\hat{\mathbf{r}}$ is $\mathbf{x}/r$, the unit vector parallel to $\mathbf{x}$. We will neglect the last term in (8.68), i.e., terms that decrease as r^{-3} or faster as r increases; these are quadrupole and higher multipole terms.

---

[15]See Exercise 22.

We will first show that the first term on the right-hand side of (8.68) gives no contribution to the vector potential. If we substitute that term into (8.60), the result is proportional to the integral of $\mathbf{J}$ over all space, which we will prove is 0,

$$\int \mathbf{J}(\mathbf{x}')d^3x' = 0. \tag{8.69}$$

How to prove this simple equation is not so obvious! Note that

$$J_i(\mathbf{x}) = \mathbf{\nabla} \cdot (x_i \mathbf{J}) - x_i \mathbf{\nabla} \cdot \mathbf{J}. \tag{8.70}$$

(We temporarily drop the prime on the source point to simplify the notation.) But $\mathbf{\nabla} \cdot \mathbf{J} = 0$ for magnetostatics, by the continuity equation, so the second term on the right side of (8.70) vanishes. And the integral of the first term $\mathbf{\nabla} \cdot (x_i \mathbf{J})$ is 0 by Gauss's theorem, because $x_i \mathbf{J}$ is 0 on the sphere at infinity; $\mathbf{J}$ has finite extent. Hence (8.69).

The asymptotic potential is obtained by substituting the *second term* of (8.68) into (8.60)

$$\mathbf{A}(\mathbf{x}) = \frac{\mu_0}{4\pi r^2} \int \mathbf{J}(\mathbf{x}')\hat{\mathbf{r}} \cdot \mathbf{x}' d^3x'. \tag{8.71}$$

We will show that the integral in (8.71) can be reexpressed as

$$\int \mathbf{J}(\mathbf{x}')\hat{\mathbf{r}} \cdot \mathbf{x}' d^3x' = \mathbf{m} \times \hat{\mathbf{r}} \tag{8.72}$$

where $\mathbf{m}$, called the *magnetic dipole moment* of the current distribution, is defined by

$$\mathbf{m} = \frac{1}{2} \int \mathbf{x}' \times \mathbf{J}(\mathbf{x}')d^3x'. \tag{8.73}$$

The dipole moment $\mathbf{m}$ is an intrinsic property of the current distribution. That is, $\mathbf{m}$ does not depend on the observation point $\mathbf{x}$.

To derive (8.72) requires some real ingenuity! First note that[16]

$$\int \mathbf{\nabla} \cdot \left(x_i x_j \mathbf{J}\right) d^3x = \int \left(x_i J_j + x_j J_i\right) d^3x = 0. \tag{8.74}$$

The first equality is a vector identity, using the fact that $\mathbf{\nabla} \cdot \mathbf{J} = 0$ for magnetostatics. The second equality is true because the far left-hand side is 0 by Gauss's theorem, since $x_i x_j \mathbf{J}$ is 0 on the surface at infinity; $\mathbf{J}$ has limited extent.[17] Now, (8.74) says that $\int x_i J_j d^3x$ is antisymmetric in ij; therefore we have

---

[16] As usual the indices i, j, k take the values $1, 2, 3$ corresponding to Cartesian coordinates x, y, z. Also, we temporarily drop the prime on the source point to simplify the notation.

[17] A mathematician would say "$\mathbf{J}$ has compact support." In other words, the region of space where $\mathbf{J} \neq 0$ does not extend to infinity.

$$\int \left(x_i J_j - x_j J_i \right) d^3 x = 2 \int x_i J_j d^3 x = \epsilon_{ijk} \int (\mathbf{x} \times \mathbf{J})_k \, d^3 x$$

where k is summed from 1 to 3, and ϵ_{ijk} is the Levi-Civita tensor. We may then write

$$\int x_i J_j d^3 x = \epsilon_{ijk} m_k. \tag{8.75}$$

Equation (8.72) now follows: Restoring the prime on the integration variable, i.e., the source point, multiplying both sides of the equation by $(\hat{\mathbf{r}})_i$, and summing over i,

$$\int J_j(\mathbf{x}')\mathbf{x}' \cdot \hat{\mathbf{r}} d^3 x' = \epsilon_{ijk} m_k \hat{\mathbf{r}}_i = \left(\mathbf{m} \times \hat{\mathbf{r}} \right)_j. \tag{8.76}$$

Equation (8.76) is (8.72).

The dipole vector potential is, from (8.71) and (8.72),

$$\mathbf{A(x)} = \frac{\mu_0}{4\pi} \frac{\mathbf{m} \times \hat{\mathbf{r}}}{r^2} = \frac{\mu_0}{4\pi} \frac{\mathbf{m} \times \mathbf{x}}{|\mathbf{x}|^3}. \tag{8.77}$$

This is the asymptotic vector potential of a finite current distribution, assuming $\mathbf{m}$ is not zero. Equation (8.77) is also the vector potential of a *point-like magnetic dipole*. What we mean by point-like is that the spatial extent of the current distribution is much smaller than any other relevant lengths in the system. The magnetic field due to the dipole is

$$\mathbf{B(x)} = \nabla \times \mathbf{A} = \frac{\mu_0}{4\pi} \frac{\left[3\hat{\mathbf{r}} \left(\mathbf{m} \cdot \hat{\mathbf{r}} \right) - \mathbf{m} \right]}{r^3}. \tag{8.78}$$

Note that $\mathbf{B(x)}$ for a point magnetic dipole has exactly the same pattern of field lines as $\mathbf{E(x)}$ for a point electric dipole (3.72).

In spherical polar coordinates the vector potential and magnetic field of a point-like magnetic dipole $\mathbf{m} = m\hat{\mathbf{k}}$, located at the origin and pointing in the z direction, are

$$\mathbf{A(x)} = \frac{\mu_0 m \sin \theta}{4\pi r^2} \hat{\phi}, \tag{8.79}$$

$$\mathbf{B(x)} = \frac{\mu_0 m}{4\pi r^3} \left[2\hat{\mathbf{r}} \cos \theta + \hat{\theta} \sin \theta \right]. \tag{8.80}$$

8.6.2 ■ Dipole Moment of a Planar Loop

As an example of a finite current distribution, consider a planar current loop C with current I. We may picture a wire loop carrying the current, as in Fig. 8.22. What is the dipole moment? The shape of the loop is irrelevant as long as it is planar. Recall that $\mathbf{J(x')}d^3 x'$ is the same as $I \, d\boldsymbol{\ell}$ for current in a wire, where $d\boldsymbol{\ell}$ is a loop segment at $\mathbf{x'}$ as shown in Fig. 8.22. With this replacement in (8.73), the

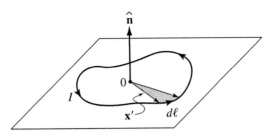

FIGURE 8.22   A planar current loop. The magnetic moment is $\mathbf{m} = I A \hat{\mathbf{n}}$ because $\int \frac{1}{2} \mathbf{x}' \times d\boldsymbol{\ell} = A \hat{\mathbf{n}}$. The origin 0 is arbitrary.

dipole moment of a wire loop is

$$\mathbf{m} = I \oint_C \frac{1}{2} \mathbf{x}' \times d\boldsymbol{\ell}. \tag{8.81}$$

The vector $\frac{1}{2} \mathbf{x}' \times d\boldsymbol{\ell}$ is perpendicular to the plane of the loop, and its magnitude is equal to the area of the infinitesimal triangular sector bounded by $\mathbf{x}'$, $\mathbf{x}' + d\boldsymbol{\ell}$, and $d\boldsymbol{\ell}$; that is, $\frac{1}{2} \mathbf{x}' \times d\boldsymbol{\ell}$ is an infinitesimal area element $d\mathbf{A}$ of the loop. Thus

$$\mathbf{m} = I A \hat{\mathbf{n}} \tag{8.82}$$

where $\hat{\mathbf{n}}$ is the normal unit vector, and A is the area of the loop. The dipole moment points perpendicular to the loop, with magnitude $I A$.

8.6.3 ■ Torque and Potential Energy of a Magnetic Dipole

Up to now we have considered the field created by a magnetic dipole. Now we turn to the dynamics of a dipole. What is the force or torque on a dipole in an applied field $\mathbf{B}$? (It is important to understand that $\mathbf{B}$ is an external field applied to the dipole, not the field produced by the dipole.) We'll answer this question by considering the special case of a planar current loop, although the result we will obtain holds in general.

The *torque* on a planar current loop, due to the magnetic force on the current distribution, is

$$\mathbf{N} = \oint_C \mathbf{x} \times d\mathbf{F} = \oint_C \mathbf{x} \times (I d\mathbf{x} \times \mathbf{B}), \tag{8.83}$$

where $d\mathbf{x}$ is an infinitesimal segment of the curve at $\mathbf{x}$. The equation simplifies if $\mathbf{B}$ is uniform in the region of the current, by some cross product identities. First note that

$$d\left[\mathbf{x} \times (\mathbf{x} \times \mathbf{B})\right] = \mathbf{x} \times (d\mathbf{x} \times \mathbf{B}) + d\mathbf{x} \times (\mathbf{x} \times \mathbf{B}) \tag{8.84}$$

assuming $\mathbf{B}$ is constant. The loop integral of a perfect differential is 0, because for a loop the endpoints of integration are the same point, so the loop integrals of the two terms on the right side are equal but opposite. Therefore, we may write **N** as

$$\mathbf{N} = \frac{I}{2} \oint_C [\mathbf{x} \times (d\mathbf{x} \times \mathbf{B}) - d\mathbf{x} \times (\mathbf{x} \times \mathbf{B})] \tag{8.85}$$

$$= \frac{I}{2} \oint_C [(\mathbf{x} \times d\mathbf{x}) \times \mathbf{B}] \tag{8.86}$$

where the second equality is an example of the cross-product identity

$$\mathbf{a} \times (\mathbf{b} \times \mathbf{c}) + \mathbf{b} \times (\mathbf{c} \times \mathbf{a}) + \mathbf{c} \times (\mathbf{a} \times \mathbf{b}) = 0. \tag{8.87}$$

Again assuming that **B** is constant, we may pull $\times \mathbf{B}$ out of the integral. (To justify this step it may be helpful to express (8.86) in Cartesian coordinates.) The remaining integral is **m** by (8.81); hence

$$\mathbf{N} = \mathbf{m} \times \mathbf{B}. \tag{8.88}$$

The result (8.88) may be applied generally to any point-like magnetic dipole. The torque on a magnetic dipole is in the direction twisting the dipole moment **m** toward alignment with the field **B**; that is, the torque is perpendicular to **B**.

Because we have derived (8.88) for an *arbitrary* planar loop the calculation is rather intricate. To gain some intuition, study Fig. 8.23, which shows specifically a square loop.

The *orientational potential energy* associated with the torque (8.88) is given by

$$U = -\mathbf{m} \cdot \mathbf{B}. \tag{8.89}$$

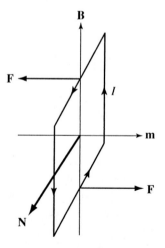

FIGURE 8.23   A square current loop. The magnetic forces on currents in the top and bottom sides are equal but opposite, making a torque **m** × **B** on the loop, pointing out of the page.

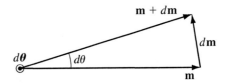

FIGURE 8.24   Rotation of m. The change of **m** due to the rotation $d\theta$ is $d\mathbf{m} = d\theta \times \mathbf{m}$.

This result can be derived by a purely mechanical argument. Assume the internal currents that make **m** are not changed by rotation of **m**, so that the magnitude $|\mathbf{m}|$ is constant. The potential energy U is defined by the statement that the work done by the torque, during an infinitesimal rotation of the dipole, is $-dU$. The work is $\mathbf{N} \cdot d\theta$, where $d\theta$ is rotation; the direction of $d\theta$ is the axis of rotation and the magnitude is the angle. The change of **m** is $d\mathbf{m} = d\theta \times \mathbf{m}$, by Fig. 8.24, so

$$dU = -\mathbf{N} \cdot d\theta = -(\mathbf{m} \times \mathbf{B}) \cdot d\theta = -(d\theta \times \mathbf{m}) \cdot \mathbf{B} = -d\mathbf{m} \cdot \mathbf{B};$$

hence (8.89).

If the magnetic field varies with **x**, then there is also a *translational force* on a point-like dipole, given by

$$\mathbf{F} = -\nabla U = \nabla \left(\mathbf{m} \cdot \mathbf{B}(\mathbf{x}) \right). \tag{8.90}$$

The force is 0 if **B** is uniform, just as the force on an electric dipole is 0 if **E** is uniform.

If a magnetized needle is put through a cork floating on the surface of water, the needle becomes aligned with the Earth's magnetic field, pointing north, because of the torque (8.88). However, the cork does not move toward the north, because for that to happen there would have to be a gradient in the Earth's field, as shown by (8.90). In fact the Earth's field does have a gradient, but it is much too small to affect the cork.

A famous example of a translational force on a magnetic dipole due to a nonuniform **B(x)** is the Stern-Gerlach experiment. A beam of neutral silver atoms is sent through a region between the poles of a magnet. The pole shapes are asymmetric, designed to create a gradient of **B**. The beam splits into two parts, because some Ag atoms have their dipole moments parallel to **B** while the others are antiparallel. The force (8.90) deflects the two groups of atoms in opposite directions. This experiment led to the discovery of electron spin, which contributes to the magnetic moment of an atom.

It is interesting to compare the torque on a magnetic dipole **m** in **B**, to the torque on an electric dipole **p** in **E**. The comparison is illustrated in Fig. 8.25. The electric dipole consists of equal but opposite charges, which are forced in opposite directions along **E**. The magnetic dipole may be visualized as a small square loop of current, opposite sides of which are forced in opposite directions perpendicular to **B**. In either case the torque is perpendicular to the field, twisting the dipole toward alignment with the field.

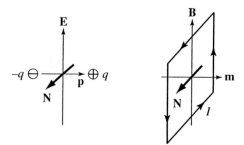

FIGURE 8.25   Torques on electric and magnetic dipoles. Where, and in what directions, are the forces in each case?

A notable difference between **p** *and* **m.**   It is interesting to compare (8.89) with the analogous potential energy $U = -\mathbf{p} \cdot \mathbf{E}$, equation (3.78), of an electric dipole **p** in an electric field **E**. The results are analogous because simply interchanging **m** with **p** and **B** with **E** interchanges the magnetic and electric expressions. But there is a fundamental difference between the two cases which is interesting to explore. Equation (8.89) is an orientational potential energy, equal to the work that must be done to turn **m** from the orientation with zero energy (**m** perpendicular to **B**) to any other orientation, keeping $|\mathbf{m}|$, $|\mathbf{B}|$, and **x** fixed. Equation (3.78) is something more general, because it is equal to the total work that must be done to bring the dipole from infinity, where the potential energy is taken to be 0, to the position **x**. The derivation of (3.78) depended on the existence of monopole charges $+q$ and $-q$. The magnetic case cannot be treated in the same way because there are no magnetic monopoles.

What then is the total potential energy U_{total} required to bring **m** from infinity to **x**? To consider this, it is helpful to think of **m** as a current loop, and of the source of **B** as current-carrying coils. A proper calculation of U_{total} must take into account the work done to keep the loop and coil currents constant against the forces associated with moving the coil through the field.[18, 19] The rather surprising result of this analysis is that $U_{\text{total}} = +\mathbf{m} \cdot \mathbf{B}$. To analyze the complete energetics of the system it is U_{total} that is the relevant potential energy. However, to analyze the motion of the dipole itself, assuming the currents that create **m** and **B** are constant in magnitude, it is correct to calculate forces and torques with (8.88), (8.89), and (8.90).

Magnetic dipole dynamics.   The rotational motion of a dipole under the influence of the magnetic torque is determined by the equation of motion $d\mathbf{L}/dt = \mathbf{m} \times \mathbf{B}$, where **L** is the angular momentum of the object. Rotational dynamics is diverse, and sometimes nonintuitive. The rotational motion of the dipole depends on the relation between **m** and **L**. If **m** is constant along a body-fixed axis, as in

---

[18] *The Feynman Lectures*, Vol. II, Secs. 15-1 and 15-2.
[19] G. H. Goedecke and R. C. Wood, Magnetic dipole orientation energies, *Am. J. Phys.* **67**, 45 (1999).

the case of a magnetized needle, then one class of rotational motion is oscillation about the equilibrium. The equilibrium is **m** aligned with **B**. An example of this case is the oscillation of a compass needle about the north direction.[20] For the oscillating needle, **m** and **L** are perpendicular; **m** is in the plane of the compass and **L** is along the pivot axis. On the other hand, if **m** is proportional to **L**, so that these vectors are parallel or antiparallel (as is the case for electrons, protons, neutrons, and other elementary particles) then the motion is *precession about* **B**, analogous to precession of a top.

The interaction energy (8.89) is important in nuclear magnetic resonance (NMR). The dipole moment of an atomic nucleus is proportional to the angular momentum, which is quantized. Therefore the energy of the nucleus in an applied magnetic field is quantized. NMR is the absorption of radiation by quantum transitions between the magnetic energy levels. NMR is the basis for magnetic resonance imaging (MRI) widely used in medical applications.

8.6.4 ▦ The Magnetic Field of the Earth

Understanding the Earth's magnetic field is one of the oldest problems in science, and it is still not completely solved. The sources of the field are self-generated electric currents in the Earth's outer core. The outer core is a conducting liquid with resistivity $\rho \approx 1.7 \times 10^{-6}$ Ωm, made of molten Fe and Ni under high pressure, extending in radius from about $0.2 R_{\text{Earth}}$ to $0.55 R_{\text{Earth}}$. Ganymede, a moon of Jupiter and the largest moon in the solar system, also has a magnetic field due to currents in a molten core. From radiometric dating of magnetic rocks there is evidence that the Earth's magnetic field is changing slowly and has been changing for the last 160 My or longer. The magnetic poles at the surface wander around the Earth's rotation axis on a long time scale and the polarity has reversed many times, on a time scale of about 200 ky, over millions of years.

At the surface of the Earth the magnetic field is approximately equal to a dipole field whose magnetic moment is $m_{\text{Earth}} = 7.79 \times 10^{22}$ Am^2. The field at the surface is predominantly the asymptotic dipole field, but accurate measurements reveal the small multipole components. We can deduce the field inside the Earth from field measurements at the Earth's surface, at least to the depth of the source currents, from the source-free field equations $\nabla \cdot \mathbf{B} = 0$ and $\nabla \times \mathbf{B} = 0$. The form of the magnetic field near its source is very different from a dipole field, and is quite complex.[21]

8.7 ▰ THE FULL FIELD OF A CURRENT LOOP

In this chapter we have considered a circular loop of current several times. In Example 2 we found the magnetic field on the axis of the loop. In Sec. 8.6.1 we

---

[20]See Exercise 33.

[21]D. R. Stump and G. L. Pollack, A current sheet model for the Earth's magnetic field, *Am. J. Phys.* **66**, 802 (1998).

found the vector potential of a magnetic dipole, which may be thought of as a "point" current loop. In Sec. 8.6.2 we found the dipole moment of a loop. It is natural to discuss the circular loop, because it is highly symmetric and, in some sense, the most elementary source of a magnetic field.

It is also natural to ask: Can we find the *full field, everywhere,* for a circular loop of current, by using the analytical techniques developed in this chapter? The answer is yes, but the analysis is somewhat more challenging than the problems we have done so far, and the result is left in terms of elliptic integrals—tabulated functions less familiar than algebraic or trigonometric functions. In this section we find that magnetic field, as an illustration of the power and limitations of analytical methods, and their connection to numerical methods.

Let the loop lie in the xy plane, have radius R, and carry current I, as shown in Fig. 8.26. We know from the symmetry of the problem that the magnitude of $\mathbf{B}(\mathbf{x})$ cannot depend on the aximuthal angle ϕ. For convenience we choose to consider a field point P in the yz plane, with coordinates $(0, y, z)$. The position vector of P is

$$\mathbf{x} = y\hat{\mathbf{j}} + z\hat{\mathbf{k}}. \tag{8.91}$$

An arbitrary source point has coordinates $(x', y', z') = (R\cos\phi', R\sin\phi', 0)$, so that its position vector is

$$\mathbf{x}' = R\cos\phi'\hat{\mathbf{i}} + R\sin\phi'\hat{\mathbf{j}}. \tag{8.92}$$

The current element $I d\ell$, shown in Fig. 8.26, is tangent to the loop at $\mathbf{x}'$, counterclockwise in the xy plane, so the length element is

$$d\ell = R d\phi' \left(-\sin\phi'\hat{\mathbf{i}} + \cos\phi'\hat{\mathbf{j}}\right). \tag{8.93}$$

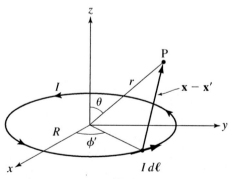

FIGURE 8.26   Geometry of a current ring. The current element $I d\ell$ is at $\mathbf{x}'$ in the xy plane. The field point P is at $\mathbf{x}$ in the yz plane.

The Magnetic Field

We'll first find $\mathbf{B}(\mathbf{x})$ from the Biot-Savart law (8.21),

$$\mathbf{B}(\mathbf{x}) = \frac{\mu_0}{4\pi} \int \frac{I d\boldsymbol{\ell} \times (\mathbf{x} - \mathbf{x}')}{|\mathbf{x} - \mathbf{x}'|^3}. \tag{8.94}$$

The relative vector, and its magnitude, are

$$\mathbf{x} - \mathbf{x}' = -R \cos \phi' \hat{\mathbf{i}} + (y - R \sin \phi') \hat{\mathbf{j}} + z \hat{\mathbf{k}} \tag{8.95}$$

$$|\mathbf{x} - \mathbf{x}'| = \sqrt{R^2 + y^2 + z^2 - 2 y R \sin \phi'}. \tag{8.96}$$

The cross product in the numerator of (8.94) can be evaluated by expanding in the Cartesian unit vectors, leading to the result

$$\mathbf{B}(0, y, z) = \frac{\mu_0 I R}{4\pi} \int_0^{2\pi} \frac{z \cos \phi' \hat{\mathbf{i}} + z \sin \phi' \hat{\mathbf{j}} + (R - y \sin \phi') \hat{\mathbf{k}}}{\left(R^2 + y^2 + z^2 - 2 y R \sin \phi'\right)^{3/2}} d\phi'. \tag{8.97}$$

The x component of $\mathbf{B}$ in (8.97) can be evaluated explicitly. But before doing so, we can see that $B_x = 0$ from symmetry, by examining Fig. 8.26: The contribution to B_x at P due to the current element at ϕ' is canceled by that due to the current element at $\pi - \phi'$. To evaluate B_x explicitly, change the variable of integration from ϕ' to $u = R^2 + y^2 + z^2 - 2 y R \sin \phi'$. That is,

$$B_x = \frac{\mu_0 I z}{4\pi y} \int_{\phi'=0}^{\phi'=2\pi} d\left(u^{-1/2}\right) = 0. \tag{8.98}$$

The integration is over an entire cycle of ϕ', so u is the same at the two endpoints, and the integral is 0. The integrals for B_y and B_z in (8.97) can be expressed in terms of elliptic integrals, an analysis that can be found in advanced books,[22] but we won't pursue that approach here. Alternatively, the integral over ϕ' can be evaluated numerically by computer.[23]

The symmetry of this problem suggests that it is natural to rewrite the field in terms of spherical components B_r and B_θ, and the spherical coordinates r and θ of P, rather than Cartesian components and coordinates. After all, we chose P to be in the yz plane only for convenience in calculation. We already know that $B_\phi = 0$, because in the yz plane the ϕ-component of $\mathbf{B}$ is parallel to the x axis. Thus by symmetry $\mathbf{B}$ has the form $\hat{\mathbf{r}} B_r + \hat{\boldsymbol{\theta}} B_\theta$, where the components are independent of ϕ. To get $B_r(r, \theta)$ and $B_\theta(r, \theta)$ from the Cartesian components $B_y(y, z)$ and $B_z(y, z)$ at the point P, we make in (8.97) the substitutions $y = r \sin \theta$ and $z = r \cos \theta$. Also, the connections between the spherical and Cartesian components at P are

$$B_r = B_y \sin \theta + B_z \cos \theta, \tag{8.99}$$

[22] Jackson; Smythe.
[23] See Exercise 41.

$$B_\theta = B_y \cos\theta - B_z \sin\theta; \qquad (8.100)$$

because in the yz plane, $\hat{\mathbf{r}} = \hat{\mathbf{j}}\sin\theta + \hat{\mathbf{k}}\cos\theta$ and $\hat{\boldsymbol{\theta}} = \hat{\mathbf{j}}\cos\theta - \hat{\mathbf{k}}\sin\theta$. With (8.97) to (8.100) we have determined $\mathbf{B}$ in terms of integrals over ϕ'. The results in elliptic integrals are

$$B_r(r,\theta) = \frac{\mu_0 I R^2 \cos\theta}{4\pi} \int_0^{2\pi} \frac{d\phi'}{\left(R^2 + r^2 - 2rR\sin\theta\sin\phi'\right)^{3/2}} \qquad (8.101)$$

$$B_\theta(r,\theta) = \frac{\mu_0 I R}{4\pi} \int_0^{2\pi} \frac{(r\sin\phi' - R\sin\theta)\,d\phi'}{\left(R^2 + r^2 - 2rR\sin\theta\sin\phi'\right)^{3/2}}. \qquad (8.102)$$

The Vector Potential

Another way to analyze the current loop of Fig. 8.26 is to find the vector potential $\mathbf{A}(\mathbf{x})$; then $\mathbf{B} = \nabla \times \mathbf{A}$. By (8.60)

$$\mathbf{A}(\mathbf{x}) = \frac{\mu_0 I}{4\pi} \int \frac{d\boldsymbol{\ell}}{|\mathbf{x} - \mathbf{x}'|}. \qquad (8.103)$$

Considering again the point P in the yz plane, the Cartesian representation of $\mathbf{A}(\mathbf{x})$ is

$$\mathbf{A}(\mathbf{x}) = A_x \hat{\mathbf{i}} + A_y \hat{\mathbf{j}} = \frac{\mu_0 I R}{4\pi} \int_0^{2\pi} \frac{-\sin\phi'\,\hat{\mathbf{i}} + \cos\phi'\,\hat{\mathbf{j}}}{\sqrt{R^2 + y^2 + z^2 - 2yR\sin\phi'}} d\phi'. \qquad (8.104)$$

Note that $A_y = 0$, because the integral for A_y has essentially the same form as the integral for B_x that we found to be 0 in (8.98). We could have anticipated this result, because (8.103) shows that the contribution to the vector potential from a small current element is parallel to the current. The current in Fig. 8.26 runs in the $\hat{\boldsymbol{\phi}}$ direction, and contributions to $A_y(0, y, z)$ from pairs of points on the ring cancel.

At this point it is natural to use spherical coordinates. We have seen that the vector potential is azimuthal, so $\mathbf{A}(\mathbf{x}) = A_\phi(r,\theta)\hat{\boldsymbol{\phi}}$; and by converting from y and z in the yz plane to r and θ,

$$A_\phi(r,\theta) = \frac{+\mu_0 I R}{4\pi} \int_0^{2\pi} \frac{\sin\phi'\,d\phi'}{\sqrt{R^2 + r^2 - 2rR\sin\theta\sin\phi'}}. \qquad (8.105)$$

This reduces $\mathbf{A}(\mathbf{x})$ to a finite integral.

By calculating $\mathbf{A}(\mathbf{x})$, we gain something and lose something. The gain is that instead of the two integrals (8.101) for B_r and (8.102) for B_θ, we now have only the one integral (8.105) for A_ϕ. However, any of these integrals would have to be evaluated numerically. In the case of (8.97) we are getting the components of $\mathbf{B}$ itself, i.e., what we are interested in. In the case of (8.105) we must calculate

derivatives of A_ϕ to obtain **B**. Furthermore, to calculate the derivatives numerically, it would be necessary to evaluate **A(x)** on a fine mesh of points. The loss is
that in a numerical computation it would be necessary to compute **A(x)** precisely.

Expansion in Legendre Polynomials

It is interesting that a further simplification can be applied to the integrals for B_r,
B_θ, and A_ϕ. They can be expanded in Legendre polynomials. We shall apply this
technique to (8.105).

In Sec. 5.2.3 we learned that $1/|\mathbf{x} - \mathbf{x}'|$ is the generating function of Legen-
dre polynomials. Equations (5.84a) and (5.84b) express this inverse distance as a
series. Applying then (5.84a) to the integrand in (8.105) we may write

$$A_\phi(r, \theta) = \frac{\mu_0 I R}{4\pi r} \sum_{n=0}^{\infty} \left(\frac{R}{r}\right)^n \int_0^{2\pi} P_n(\sin\theta \sin\phi') \sin\phi' d\phi'. \qquad (8.106)$$

Equation (8.106) holds for the region $r > R$, i.e., beyond one ring radius from
the center; for the region $r < R$ one must use (5.84b). Equation (8.106) is the
multipole expansion for a circular ring. The terms in the sum with n even are 0:
$P_n(\sin\theta \sin\phi')$ is a polynomial in $\sin\phi'$ with only even powers if n is even; when
multiplied by the factor $\sin\phi'$ they integrate to 0 over the range from $\phi' = 0$ to
2π. [The integral of any odd power of $\sin\phi'$ over the cycle $(0, 2\pi)$ is zero.] Thus
the first nonzero term in the series is the $n = 1$ term, the dipole term; the series
has only odd n's.

The series in (8.106) converges rapidly unless R/r is close to 1. To compute **B**
for $R/r < 1$ one can evaluate $\nabla \times \mathbf{A}$ analytically, keeping only the first few terms
of the series for sufficient accuracy, from

$$B_r = \frac{1}{r\sin\theta} \frac{\partial}{\partial\theta} \left(\sin\theta\, A_\phi\right) \quad \text{and} \quad B_\theta = -\frac{1}{r}\frac{\partial}{\partial r}\left(r\, A_\phi\right). \qquad (8.107)$$

Figure 8.27 shows the magnetic field of a loop of current, in a half-plane perpendicular to the wire. Very close to the wire the lines of **B** circle around the
current, like the field of a straight wire. Far from the loop, the term in (8.106)
with $n = 1$ dominates the series, and **A** approaches the dipole vector potential

$$A_\phi \approx \frac{\mu_0 m \sin\theta}{4\pi r^2}, \qquad (8.108)$$

where $m = I\pi R^2$. It is the intermediate fields, i.e., in the region where r is
comparable to R, that are hard to calculate. This foreshadows what we will learn
about the fields radiated by oscillating dipoles and antennas in Chapter 15. For that
case, too, the near and intermediate fields are complicated, while the far fields are
(relatively) simple.

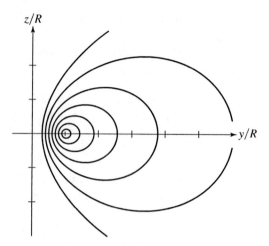

FIGURE 8.27   Magnetic field lines of a current ring. The ring lies in the xy plane centered at the origin. Field lines in the yz plane are shown.

EXERCISES

Sec. 8.1. The Magnetic Field

8.1. A Hall probe is a device for measuring magnetic fields. In a typical device a current I flows in a ribbon of n-type doped semiconductor located in the magnetic field. In an n-type doped semiconductor the charge carriers are electrons. In the steady state there is a voltage difference V_H between the edges of the ribbon, such that the net perpendicular force on a charge carrier $q\mathbf{E}_\perp + q(\mathbf{v} \times \mathbf{B})_\perp$ is 0. From V_H one can determine $\mathbf{B}$.

(a) Assume the geometry shown in Fig. 8.28, with dimensions 1.0 cm by 0.2 cm by 0.005 cm. The material is As-doped Si, with 2×10^{15} conduction electrons per

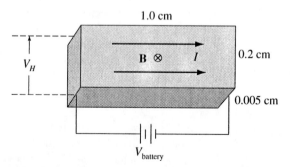

FIGURE 8.28   Exercise 1. The Hall effect. The sign of V_H depends on the sign of the mobile charges. The battery supplies the voltage difference that produces the current.

cm^3, and resistivity 1.6 Ω cm. The current is driven by a 3 V battery across the 1.0 cm length. What Hall voltage will be measured across the 0.2 cm width if the field strength is 0.1 T? [Answer: $V_H = 11.7$ mV]

(b) Show that the sign of V_H depends on the sign of the charge carriers. Edwin H. Hall measured V_H in metals in 1879, and found that the charge carriers are negative.

Sec. 8.2. Applications of the Magnetic Force

8.2. The experiment by which Thomson discovered the electron, consisted of a cathode ray passing between parallel capacitor plates in a uniform magnetic field. The electrons travel parallel to the plates and **B** is perpendicular to both **E** and **v**. Derive the condition relating the potential difference V_0 between the plates and the magnetic field strength, along with any other relevant parameters, such that the cathode ray is undeflected, assuming the cathode ray is a beam of electrons. (This is the principle of the *velocity selector* in a mass spectrometer.)

8.3. Consider a proton with kinetic energy 10 MeV in a cyclotron with magnetic field 1 T. Calculate the cyclotron frequency and the radius of the orbit.

8.4. *Quadrupole focusing.* A magnetic quadrupole field can be used as a focusing field for a charged particle beam. The cross section of the pole faces is shown in Fig. 8.29. The pole faces are hyperbolas of the form $xy = $ constant. There are two north poles and two south poles, marked on the figure. The dimension of the magnet perpendicular to the cross section is ℓ. The magnetic field in the region $0 \leq z \leq \ell$ is

$$\mathbf{B}(x, y, z) = b\left(y\hat{\mathbf{i}} + x\hat{\mathbf{j}}\right)$$

in which $b > 0$; the field is 0 for $z < 0$ and $z > \ell$. Particles enter from negative z with velocity $\mathbf{v}_0 = v_0\hat{\mathbf{k}}$ and are deflected by the force $\mathbf{F} = q\mathbf{v}_0 \times \mathbf{B}$. (Neglect the small components v_x and v_y in calculating the force.)

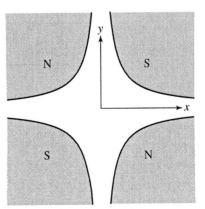

FIGURE 8.29   **Exercise 4.** A quadrupole focusing magnet. The hyperbolic curves are the boundaries of the pole faces. A charged particle beam moves out of the page.

(a) Sketch the **B** field lines in the xy plane.

(b) Explain qualitatively why **B** produces focusing in the x direction and defocusing in the y direction, assuming the beam particles are positively charged.

(c) Write the equations of motion for a beam particle with charge q and mass m, using the approximate force given above. Solve for x as a function of z for $z > 0$, assuming $x = x_0$ and $v_x = 0$ at $z = 0$. Sketch a graph of $x(z)$.

8.5. (a) Use a computer program to plot the cycloid curve in the xy plane, given by the parametric equations

$$x(\theta) = \theta - \sin\theta,$$

$$y(\theta) = 1 - \cos\theta.$$

(The cycloid is the curve traced out by a point on a circle of radius 1 as the circle rolls on the x axis.)

(b) For a charge q moving in crossed (i.e., orthogonal) **E** and **B** fields, starting from rest at the origin, plot the kinetic energy as a function of time.

8.6. Figure 8.30 shows the essential features of an early mass spectrograph of A. Dempster. Singly positive ions enter the vacuum chamber vertically through the slit, after having been accelerated through a voltage of 20.0 kV. Their paths are bent by the magnetic field **B** and they are deposited a distance s from the slit on a photographic plate.

(a) If $s = 0.250$ m for ions of Samarium with mass number 150, i.e., $^{150}\mathrm{Sm}_{62}$, what is **B**?

(b) What is the range of s for the stable isotopes of Sm, whose mass numbers range from 144 to 154?

8.7. Figure 8.31 shows a schematic version of a mass spectrograph of K. T. Bainbridge. It is a modification of the design of Fig. 8.30 because it includes a velocity selector in the vacuum chamber through which the positive ions pass. The selector uses a horizontal **E** field and the same **B** field that bends the path of the ions in the spectrograph proper.

What is the mass of the ions that impinge on the photographic plate a distance s from the slit? [Answer: $M = eB^2s/2E$]

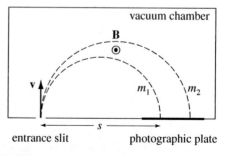

FIGURE 8.30   Exercise 6. The Dempster mass spectrograph.

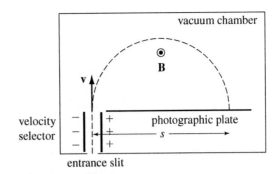

FIGURE 8.31 Exercise 7. The Bainbridge mass spectrograph.

Sec. 8.3. Biot-Savart Law

8.8. By pictures showing the directions of the currents, fields, and forces, prove that parallel but opposite currents repel each other.

8.9. Consider two very long parallel wires a distance $d = 1$ m apart, carrying equal currents I in the same direction.

    (a) What is the force per unit length f if $I = 1$ A? (Notice that if we define $\mu_0 = 4\pi \times 10^{-7}$ N/A^2, then a specified value of f can be used to define the unit of current. This is in fact the way the ampere is defined.)

    (b) What must I be to have $f = 1$ N/m? [Answer: About 2 kA]

8.10. *Helmholtz coils.* Two circular coils of radius a, each carrying current I in the same sense, are parallel with the xy plane with their centers at $(0, 0, \pm s/2)$. On the z axis the magnetic field is $\mathbf{B} = B(z)\hat{\mathbf{k}}$, in the z direction; and at $z = 0$, i.e., halfway between the coils, $\partial B/\partial z$ is 0.

    (a) Determine s such that $\partial^2 B/\partial z^2$ is 0 at $z = 0$ on the z axis. This configuration is called Helmholtz coils, and it produces a very uniform magnetic field in a neighborhood of the origin. Show that for this configuration the *third* derivative with respect to z is also 0 at $z = 0$.

    (b) Use a computer program to plot the magnetic field $B(z)$ as a function of position z along the axis of the Helmholtz coils.

        Helmholtz coils are often used in laboratories to cancel out the Earth's magnetic field in a small region of space. However, if a large magnetic-field-free region is needed, say room sized, the necessary coils would be impractically large, as you can see from your answer to (b). In practice, to produce a field-free volume one surrounds it with a material with high magnetic susceptibility, e.g., Mumetal, which concentrates the magnetic field in the walls.

8.11. Find the magnetic field at the center of a square loop of size $2a \times 2a$ carrying current I. Repeat the calculation for a regular polygon with n sides, letting the perpendicular distance from the center to any side be a. Show that the result approaches the field at the center of a circular loop of radius a in the limit $n \to \infty$.

8.12. A square wire loop of size $2a \times 2a$ lies in the xy plane with its center at the origin and sides parallel to the x and y axes. A counterclockwise current I runs around the loop.

(a) Find the magnetic field on the z axis. [Answer: $B_z(z) = 2\mu_0 I a^2/[\pi(a^2 + z^2)(2a^2 + z^2)^{1/2}]$]

(b) Show that for $z/a \gg 1$ the field becomes that of a magnetic dipole, and find the magnetic moment. (See (8.78) and (8.80).)

(c) Compare the field at the center of this square loop with that at the center of a circular loop of diameter $2a$.

8.13. Consider a circular cylindrical solenoid of finite length L, radius a, with N turns of wire carrying current I_0. The current may be approximated by a surface current density K (= azimuthal current per unit length along the cylinder) equal to $N I_0/L$.

(a) Calculate the magnetic field on the axis of the cylinder halfway between the ends. (Hint: Subdivide the solenoid into infinitesimal current rings $dI = K dz'$ and use (8.30).)

(b) Calculate the magnetic field on the axis of the cylinder at either end.

(c) Show that B_{end}/B_{center} approaches $1/2$ as $L/a \to \infty$.

8.14. Consider the magnetic field $\mathbf{B}(\mathbf{x}) = axy\,\hat{\mathbf{i}} + by^2\,\hat{\mathbf{j}}$.

(a) What relation must connect the constants a and b?

(b) What current density $\mathbf{J}(\mathbf{x})$ produces this field? Describe the current distribution in words and pictures.

Sec. 8.4. Ampère's Law

8.15. Consider a circular current loop in the xy plane, with radius a, centered at the origin, and carrying current I. On the z axis the magnetic field has the form $B_z(z)\hat{\mathbf{k}}$. Determine $B_z(z)$ and evaluate $\int_{-\infty}^{\infty} B_z dz$. By Ampère's law the integral must be $\mu_0 I$. Why?

8.16. Use Ampère's Law to determine the magnetic field in a densely wound toroidal solenoid with inner radius a, outer radius b, and N total turns of wire. The cross section of the toroid is a rectangle of width $b - a$ and height h. Show that in the limit $a, b \to \infty$ with $b - a$ fixed, and with the linear density of turns fixed, the result agrees with a straight solenoid.

8.17. Let the current distribution in the slab of Example 5 be $\mathbf{J}(z) = (j_0|z|/a)\,\hat{\mathbf{i}}$, where j_0 is a constant with dimensions A/m².

(a) What is $\mathbf{B}$ inside the slab, and above and below it?

(b) Sketch a plot of $B_y(z)$.

8.18. Figure 8.32 shows the cross section of a long coaxial cable. The center conductor ($r \le a$) carries a current I_0 in the direction out of the page. The outer conductor ($b \le r \le c$) carries the return current I_0 into the page. The currents are uniformly distributed in the conductors. Between the conductors there is an insulating material with permittivity ϵ_0 and permeability μ_0.

Find $\mathbf{B}$ for all r and sketch a plot of $\mathbf{B}$ versus r.

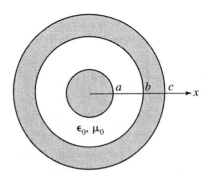

FIGURE 8.32   Exercise 18. Cross section of a coaxial cable.

8.19. Let **B** be the magnetic field due to a long thin wire carrying current I. Let the wire be parallel to the z axis and passing through the point $(x, y) = (a, 0)$ in the xy plane. Explicitly evaluate $\oint \mathbf{B} \cdot d\boldsymbol{\ell}$ for a circle of radius R, with $R > a$, *centered at the origin*. (The answer must be $\mu_0 I$ by Ampère's Law.)

(Hint: This calculation is not as easy as it looks! On the circle, $d\boldsymbol{\ell}$ is $R d\phi \hat{\boldsymbol{\phi}}$, but the field **B** is not in the direction $\hat{\boldsymbol{\phi}}$. Rather, **B** is in the direction $\hat{\mathbf{t}}$, tangent to a circle centered at $(a, 0)$. Use Cartesian coordinates to show that

$$\hat{\boldsymbol{\phi}} \cdot \hat{\mathbf{t}} = \frac{R - a \cos \phi}{(R^2 + a^2 - 2aR \cos \phi)^{1/2}} .)$$

Sec. 8.5. Vector Potential

8.20. Consider the vector potential $\mathbf{A}(\mathbf{x}) = \frac{1}{2}\mathbf{c} \times \mathbf{x}$, where **c** is a constant vector. Does this potential satisfy the Coulomb gauge condition? What is the magnetic field?

8.21. Equation (8.60) is a general formula for the vector potential of a localized static current density. Prove that $\nabla \cdot \mathbf{A} = 0$ and $-\nabla^2 \mathbf{A} = \mu_0 \mathbf{J}$.

8.22. The vector potential $\mathbf{A}(\mathbf{x})$ of a spinning sphere (radius a, angular velocity vector ω) with surface charge density σ is given in Example 7. Calculate $\mathbf{B}(\mathbf{x})$, for points **x** both inside and outside the sphere, sketch the magnetic field lines, and describe the results in words.

Sec. 8.6. Magnetic Dipole

8.23. (a) Derive (8.78).

(b) A small circular loop carrying current I is in the xy plane with its center fixed at the origin. The dipole moment is in the same direction as $+\hat{\mathbf{k}}$. Sketch the magnetic field lines.

(c) A second identical current loop is located on the x axis at some fixed distance d from the origin. Assume the axis of the first loop is fixed in the z direction, but the second loop is free to rotate. What is the equilibrium orientation of the second loop? What is the direction of the second dipole moment in equilibrium?

8.24. At the surface of the Earth the magnetic field of the Earth is approximately the same as the field from a point dipole $\mathbf{m}_E$ at the center of the Earth. The dipole moment is

$$\mathbf{m}_E = m_E \left(\hat{\mathbf{i}} \sin\theta_0 \cos\phi_0 + \hat{\mathbf{j}} \sin\theta_0 \sin\phi_0 + \hat{\mathbf{k}} \cos\theta_0 \right)$$

where $m_E = 7.79 \times 10^{22}$ A m^2, and (θ_0, ϕ_0)=(169 degrees, 109 degrees). The z axis is the Earth's rotation axis and the x axis passes through the Prime Meridian, on which Greenwich lies; positive ϕ is to the east.

(a) Calculate the magnetic field $\mathbf{B}$ at a point on the Earth with colatitude θ and longitude ϕ. Give the components of $\mathbf{B}$ to the north $(-\hat{\boldsymbol{\theta}})$, to the east $(\hat{\boldsymbol{\phi}})$, and vertical $(\hat{\mathbf{r}})$.

(b) Calculate these components for your home town, in gauss (1 gauss = 10^{-4} T).

8.25. Two identical point-like dipoles are located, respectively, at the origin and at $\mathbf{x} = z\hat{\mathbf{k}}$ on the z axis. Both dipole moments point in the $+z$ direction. Determine the force on the dipole at $z\hat{\mathbf{k}}$. What happens if the dipoles point in opposite directions?

8.26. Let the dipole $\mathbf{m} = m\hat{\mathbf{k}}$ be at the origin, and call a certain horizontal axis the y axis.

(a) On the z axis, what is the angle between the z axis and $\mathbf{B}$?

(b) On the y axis, what is the angle between the z axis and $\mathbf{B}$?

(c) On the cone $\theta = 45$ degrees, what is the angle between the z axis and $\mathbf{B}$?

(d) What is the angle of the cone on which $\mathbf{B}$ is horizontal? [Answer: $\arccos(1/\sqrt{3})$]

8.27. Figure 8.33 shows three parallel, identical magnetic dipoles. Each has dipole moment $m_0\hat{\mathbf{j}}$. They are situated at the corners of a 45-90-45 right triangle whose sides are s, $\sqrt{2}s$, and s.

(a) How much work is required to reverse the direction of $\mathbf{m}_2$, if the others are held fixed?

(b) What is the torque on $\mathbf{m}_2$, in the position shown?

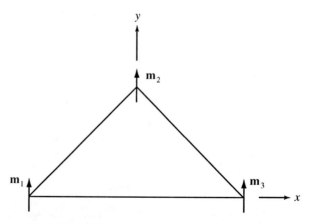

FIGURE 8.33   Exercise 27. Three magnetic dipoles.

(c) What is the torque on $\mathbf{m}_1$, in the position shown? [Answers: (a) $\mu_0 m_0^2/(2\pi s^3)$; (b) 0; (c) $3\mu_0 m_0^2/(8\pi s^3)\hat{\mathbf{k}}$]

General Exercises

8.28. (a) A long cylindrical conductor of radius R_0, whose axis coincides with the z axis, carries a uniformly distributed current I_0 in the $+z$ direction. Find the magnetic field $\mathbf{B}(\mathbf{x})$ inside and outside the cylinder.

   (b) A cylindrical hole is now drilled out of the conductor, parallel to the axis, so that the cross section is as shown in Fig. 8.34. The center of the hole is at $x = a$, and the radius is b. The conductor carries the same current I_0 as in (a). Determine the magnetic field in the hole. You will find that the magnetic field in the hole is uniform.

8.29. The *magnetron* is a vacuum-tube device that is used to generate ultra-high frequency currents in microwave sources, like microwave ovens or radar transmitters. The frequency range is 10^9 Hz to 10^{11} Hz. A schematic design for a magnetron is shown in Fig. 8.35. An electron bunch circulates in a constant magnetic field $\mathbf{B}$, passing electrodes at opposite ends of a diameter of the orbit. The potential V at either electrode oscillates with the distance from the electron bunch.

   (a) Determine the frequency of the alternating potential.

   (b) Determine B for a microwave frequency of 10^{10} Hz.

8.30. A beam of hydrogen isotopes enters a mass spectrometer. The protons and deuterons have been accelerated from rest by a potential drop V_0. The radius of the proton orbit is 10 cm. Calculate the radius of the deuteron orbit.

8.31. A thin disk of radius R carries a surface charge σ. It rotates with angular frequency ω about the z axis, which is perpendicular to the disk and through its center. What is $\mathbf{B}$ along the z axis? What is the magnetic moment of the spinning disk? [Answer: $m_z = \pi\sigma\omega R^4/4$]

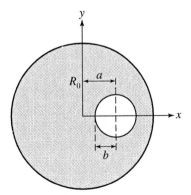

FIGURE 8.34   **Exercise 28.** Cylinder with a cylindrical cavity.

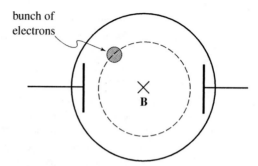

FIGURE 8.35   Exercise 29. Magnetron.

8.32. A charged annular disk with inner radius a and outer radius b, and surface charge density σ, lies in the xy plane with its center at the origin. It rotates about the z axis with angular velocity ω.

(a) What is the magnetic moment **m**?

(b) What is **B** in the xy plane at distance ρ from the origin, where $\rho \gg a, b$?

(c) Find $B_z(0, 0, z)$ for all values of z.

8.33. A circular coil (radius R and mass M) has N turns of wire carrying current I_0. The coil is free to rotate about the z axis, which lies in the plane of the coil and passes through its center, as shown in Fig. 8.36. There is a constant magnetic field $\mathbf{B} = B_0 \hat{\mathbf{i}}$. Initially the coil is in stable equilibrium with its magnetic moment parallel to the field.

(a) What is the frequency for small angle oscillations in ϕ about the equilibrium position?

(b) Now let $M = 0.10\,\text{kg}$, $N = 100$ turns, $I_0 = 0.1\,\text{A}$, and $B_0 = 0.05\,\text{T}$. Evaluate the frequency of small angle oscillations.

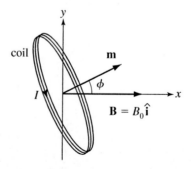

FIGURE 8.36   Exercise 33. Coil in a magnetic field.

(c) If the coil is released from rest at the small angle $\phi_0 = 15$ degrees, what is its angular velocity as it swings through the equilibrium position?

8.34. Show that the magnetic field on the axis of a finite solenoid of radius a centered on the z axis, extending from $z = -\ell/2$ to $z = \ell/2$, and with n turns per unit length of wire carrying current I, is

$$\mathbf{B}(0, 0, z) = \frac{\mu_0 n I}{2} \left[\frac{\ell/2 - z}{\sqrt{a^2 + (\ell/2 - z)^2}} + \frac{\ell/2 + z}{\sqrt{a^2 + (\ell/2 + z)^2}} \right] \hat{\mathbf{k}}.$$

Computer Exercises

8.35. The equations of motion of a charge q in a magnetic field $B_0 \hat{\mathbf{k}}$ are

$$dx/dt = v_x, \qquad dy/dt = v_y$$
$$dv_x/dt = \omega v_y, \qquad dv_y/dt = -\omega v_x$$

where $\omega = q B_0/m$. Solve the equations numerically on a computer. Set $\omega = 1$ and take initial values $(x_0, y_0, v_{0x}, v_{0y}) = (1, 0, 0, 1)$.

You might integrate the equations stepwise for a small time step. Or, more simply, use an analytic computer program with a built-in differential equation solver.

Plot the trajectory, i.e., $(x(t), y(t))$ as a function of t. It should be a circle.

8.36. The equations of motion of a charge q in crossed electric and magnetic fields $E_0 \hat{\mathbf{j}}$ and $B_0 \hat{\mathbf{k}}$ for motion with $v_z = 0$ are

$$dx/dt = v_x, \qquad dy/dt = v_y$$
$$dv_x/dt = \omega v_y, \qquad dv_y/dt = a - \omega v_x$$

where $\omega = q B_0/m$ and $a = q E_0/m$. Solve the equations numerically on a computer. For illustration purposes choose units with $\omega = 1$ and $a = 1$. If the particle starts at rest at the origin the trajectory is a cycloid. Explore what happens for different initial values of v_{0x}, both positive and negative, keeping $v_{0y} = 0$. Explain why if $v_{0x} = 1$ (in these units) the particle moves on a straight line.

8.37. A familiar type of problem in magnetostatics is to determine the magnetic field for a specified current distribution. Here we consider the inverse problem: Given the field $\mathbf{B}(\mathbf{x})$, what current density $\mathbf{J}(\mathbf{x})$ would produce that field?

(a) Suppose the vector potential is

$$\mathbf{A}(\mathbf{x}) = \frac{A_0 a}{r} e^{-r^2/a^2} \hat{\boldsymbol{\phi}}.$$

(We specify $\mathbf{A}$ because that guarantees that $\nabla \cdot \mathbf{B} = 0$.) r is the *cylindrical* radial coordinate, i.e., the perpendicular distance from the z axis to $\mathbf{x}$. Determine $\mathbf{B}(\mathbf{x})$ and $\mathbf{J}(\mathbf{x})$. Describe in words and pictures the field and current. Use an analytical and graphical computer program to carry out the rather intricate calculations, and to plot the results.

(b) Suppose instead the vector potential is

$$\mathbf{A}(\mathbf{x}) = A_0 e^{-r^2/a^2} \hat{\mathbf{k}},$$

where again r is the cylindrical radial coordinate. Analyze $\mathbf{B}(\mathbf{x})$ and $\mathbf{J}(\mathbf{x})$ as in (a).

8.38. The magnetic field of a magnetic dipole at the origin is given in (8.78). Let the dipole moment point in the z direction. Use computer graphics to make a figure showing the vector field: At each point of a dense grid in the xz plane, have the computer draw an arrow in the direction of $\mathbf{B}$.

8.39. The vector potential for a pointlike magnetic dipole $m_0\hat{\mathbf{k}}$ at the origin is (8.77). It is infinite at $r = 0$. Consider now the *nonsingular* potential

$$\mathbf{A}(\mathbf{x}) = \frac{\mu_0 m_0}{4\pi} \frac{\hat{\mathbf{k}} \times \hat{\mathbf{r}}}{r^2 + a^2} = \frac{\mu_0 m_0}{4\pi} \frac{\sin\theta}{r^2 + a^2}\hat{\boldsymbol{\phi}},$$

which approaches (8.77) for $r \gg a$. Determine the current density $\mathbf{J}(\mathbf{x})$ that would create this vector potential. Use computer graphics to make a *contour plot* of J_ϕ in the yz plane.

8.40. Use a computer program to plot the magnetic field of a ring of current, on the axis of the ring, as a function of distance from the center of the ring.

8.41. Consider the current loop of Sec. 8.7—a circle in the xy plane centered at the origin and with radius R. For points on the xy plane the magnetic field has the form $\mathbf{B} = B_z(r)\hat{\mathbf{k}}$, where $r = \sqrt{x^2 + y^2}$ is the distance from the origin. By computing $B_z(r)$ numerically, plot a graph of $B_z(r)$ as a function of r. (Use R as the unit of length and $\mu_0 I/R$ as the unit of magnetic field.) Be careful: The field has a singularity at the wire. On the same graph show the field of a pointlike dipole with the same dipole moment $m = \pi R^2 I$.

CHAPTER

9

Magnetic Fields and Matter

The most general problem in magnetostatics is to calculate the magnetic field everywhere in a system of specified current loops and magnetic materials. We saw in Chapter 8 how the Biot-Savart law determines the magnetic field produced by a steady current, but we assumed that any matter that might affect the field was far away and negligible. In other words, we only described a magnetic field in a vacuum.

But the presence of matter—solids, liquids, or gases—changes the magnetic field from what it would be in vacuum. In this sense all materials are magnetic materials. They change the field because they contain microscopic, atomic-scale currents and magnets which are themselves sources of **B**. In this chapter we study why the field changes and how to calculate the new field. We shall acquire a qualitative and semiquantitative understanding of magnetic interactions in matter, learn how and why magnetic properties vary among materials, and look at some applications.

Most magnetic materials—*diamagnetic* or *paramagnetic* materials—interact with externally applied fields in the manner just described. These interactions are usually weak, so although such materials have an important role in physics for understanding magnetism, their role is not familiar from everyday use. But some magnetic materials—*ferromagnetic* materials—interact strongly with external fields. Furthermore, ferromagnetic materials can be *permanent magnets*; that is, they can be sources of strong magnetic fields by themselves. Some applications of ferromagnets are large permanent magnets used industrially, small refrigerator magnets or those in toys, large solenoids in magnetic resonance imaging and other medical applications, and small solenoids in relay switches on cars and electrical devices.

9.1 ■ THE ATOM AS A MAGNETIC DIPOLE

We learned in Chapter 6 that when an electric field **E** is applied to an isotropic insulator the material becomes electrically polarized, i.e., it develops an electric dipole moment per unit volume **P(x)**, which is called the *polarization*, parallel to **E**. Similarly, when a magnetic field **B** is applied to a magnetic material the material becomes magnetically polarized, i.e., it develops a *magnetic dipole moment per unit volume* **M(x)**, which is called the *magnetization*. Magnetization is a density—the density of magnetic moment. The dimensions of **M** are A m^2/m^3

or A/m. The phenomena associated with magnetic polarization are rather more complex than those of electric polarization. For example, in some materials **M** points in the direction of the applied field **B**, in which case the material is called paramagnetic; in others **M** points opposite to the direction of **B**, in which case the material is called diamagnetic. In ferromagnetic materials, whose consideration we leave for Section 9.6, **M** is parallel to **B** and very large. It is important to realize that **M**(**x**) may vary with position **x** in the material.

Electrons are the microscopic origin of the macroscopic magnetic properties of matter. Every electron is a small magnet, because it has a magnetic moment $\mathbf{m}_{\text{spin}}$ associated with its *spin*. The spin is the electron's intrinsic angular momentum. For electrons, the direction of $\mathbf{m}_{\text{spin}}$ is opposite to the spin direction, and the magnitude, called the *Bohr magneton*, is

$$\mu_B = \frac{e\hbar}{2m_e} = 9.27 \times 10^{-24} \, \text{A m}^2 \tag{9.1}$$

where m_e is the electron mass. The sum of electron spin moments in the neighborhood of **x** is one contribution to **M**(**x**).

Another electronic contribution to **M**(**x**) comes from the atomic orbital motion of electrons. Quantum mechanics is necessary for an accurate theory of the electronic structure of matter, but we can get some understanding of magnetic materials from simpler semiclassical models. An electron (charge $-e$) moving with velocity v in a circular orbit of radius r is equivalent to a current $I = ev/(2\pi r)$ circulating around a loop of area $A = \pi r^2$. Figure 9.1 is a schematic picture of an electron revolving around a nucleus whose charge is Ze. The orbital magnetic moment is $m_{\text{orbital}} = IA = evr/2$. Thus m_{orbital} is directly related to the orbital angular momentum $L = m_e vr$, by

$$m_{\text{orbital}} = \frac{eL}{2m_e}. \tag{9.2}$$

(The direction of **m** is opposite to **L** because the electron is negative.) The angular momentum and magnetic moment are both quantized; L_z must be $m_\ell \hbar$, where

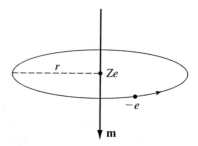

FIGURE 9.1   Classical electron orbit. In a classical model the electron revolves around the nucleus, creating a magnetic moment.

$m_\ell = 0, \pm1, \pm2, \ldots, \pm\ell$ is the angular momentum quantum number. It follows that

$$\hat{\mathbf{k}} \cdot \mathbf{m}_{\text{orbital}} = -\mu_B m_\ell,$$

so m_{orbital} is also of order μ_B. The sum of orbital moments in a neighborhood of $\mathbf{x}$ contributes to $\mathbf{M}(\mathbf{x})$.

If no magnetic field is present, then usually the $\mathbf{m}$'s of different atoms point in random directions, and so $\mathbf{M}(\mathbf{x}) = 0$. However, if there is a net dipole moment $\delta\mathbf{m}$ in a small volume δV at $\mathbf{x}$ then $\mathbf{M}(\mathbf{x}) = \delta\mathbf{m}/\delta V$. This is the usual averaging procedure: We obtain the macroscopic $\mathbf{M}(\mathbf{x})$ by averaging over small volumes that yet contain many atoms.

Table 9.1 lists some magnetic moments of fundamental interest, and it is clear from the table that the range of magnetic moments is very large. (There are astronomical objects—pulsars—with much larger magnetic moments than the Earth's.) From Table 9.1 we see that the magnetic moments of the proton and neutron are much smaller than that of the electron; the proton moment is of order $e\hbar/2m_p$, smaller than μ_B by the ratio $m_e/m_p \approx 0.5 \times 10^{-3}$. Therefore the nuclear particles do not contribute significantly to the macroscopic magnetic properties of matter. However, the proton magnetic moment and the magnetic moments of heavier nuclei are the subjects of important research and applications in nuclear magnetic resonance (NMR) and magnetic resonance imaging (MRI). In this context it is interesting to note the importance of the magnetic nuclei Fluorine-19 and Phosphorus-31, which, because they are the only stable isotopes of these elements, find wide use in magnetic resonance studies.

It might seem surprising that the neutron, a neutral particle, has a magnetic moment, but if we recall that the neutron is a bound state of quarks (which are

TABLE 9.1   Some magnetic dipole moments

Magnetic system	Symbol	Value in A-m^2 [a]
Electron spin	$\mu_e = -\mu_B$	-9.27×10^{-24}
Proton spin	μ_p	1.41×10^{-26}
Neutron spin	μ_n	-0.966×10^{-26} [b]
Typical compass needle		10^{-2} [c]
Typical superconducting current loop		20 [c]
Earth		7.79×10^{22} [d]

[a] The unit A-m^2 is equivalent to J/T. The Bohr magneton μ_B is 9.27×10^{-24} J/T, or 5.79×10^{-5} eV/T.

[b] The minus sign means that the direction of the neutron's magnetic moment is opposite that of its angular momentum.

[c] D. R. Stump and G. L. Pollack, *Am. J. Phys.* **65**, 81 (1997).

[d] IGRF model for 2000. The Earth's magnetic moment is directed approximately opposite to its angular momentum.

charged spin-1/2 particles) it is not surprising at all. Neutron beams are used to probe the microscopic magnetic structure of matter, for example, the magnetic structure of superconductors. This technique exploits the interaction between the neutron's magnetic moment, small though it is, and the atomic magnetic moments in the superconductors.

9.1.1 ▦ Diamagnetism

Many everyday materials, e.g., water, wood, glass, and polyethylene, as well as many elements, e.g., H_2, N_2, Ar, Cu, Ag, Pb, are diamagnetic. When a diamagnetic material is placed in an external **B** field, it becomes magnetized: It acquires a magnetization **M** proportional to |**B**|, but **M** points in the direction opposite to the direction of **B**. If the field is removed the magnetization of a diamagnetic substance vanishes. Because **M** is opposite to **B**, a diamagnetic material will be repelled, although weakly, by a bar magnet; like poles repel. Diamagnetism is essentially independent of temperature.

Diamagnetism was first observed by Brugmans in 1778, in Bi and Sb. A needle of Bi, or any diamagnetic material, suspended on a pivot in a magnetic field, has its equilibrium position *perpendicular* to the field. In contrast, when a needle of a paramagnetic material is thus suspended it comes to equilibrium aligned *parallel* to the field; this is like the familiar behavior of a compass needle in Earth's magnetic field. It is natural to ask: Why does a diamagnetic needle align cross-wise to the field? This question requires some consideration. The principle is that in a diamagnetic material the atomic dipoles, which are induced by **B**, always point opposite to **B**, whatever the orientation of the needle, so the lowest energy state is the perpendicular orientation. Figure 9.2 shows the equilibrium positions of diamagnetic and paramagnetic needles in a magnetic field.

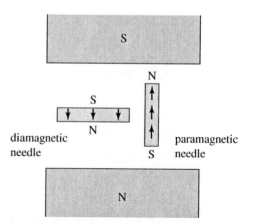

FIGURE 9.2   A needle in a magnetic field. Equilibrium for a diamagnetic needle is to be perpendicular to **B**; or, for a paramagnetic needle to be parallel with **B**.

The names "diamagnetism" and "paramagnetism" were invented by Faraday, who studied these phenomena extensively beginning in 1845. The names are based on the criteria that equilibrium is parallel to **B** for a paramagnetic needle and perpendicular ("diametrical") for a diamagnetic needle. In Faraday's time there was no knowledge of atomic structure (nor even of the existence of atoms!); the fact that some materials align parallel to **B** and others perpendicular must have seemed a real puzzle.

Diamagnetism is a rather anti-intuitive phenomenon because one would expect, from $\mathbf{N} = \mathbf{m} \times \mathbf{B}$, that the torque on microscopic dipoles in the material would tend to line them up parallel to the applied field. But this model is too simplistic, because diamagnetism, and indeed every magnetic phenomenon, is really a quantum mechanical phenomenon. It is natural to use quantum theory to explain magnetism, because the magnetic properties of matter originate in the electronic structure of the atom—a quantum system Unfortunately a full quantum discussion is rather intricate, and beyond the scope of this book. So instead it is customary[1] to give a classical demonstration of the plausibility of diamagnetism and we'll study that now. This explanation, proposed by J. J. Larmor in the early 1900's, is based on Faraday's Law of electromagnetic induction.

The classical picture of diamagnetism is based on a planetary model of the atom, reminiscent of the Bohr model but without quantization. An atomic electron is assumed to revolve around the nucleus on a circular orbit. For simplicity we'll imagine a single electron. This model is only heuristic, and of course differs from the real hydrogen atom, which has zero angular momentum in the ground state.

Figure 9.3 represents the model atom in a magnetic field. The electron revolves counterclockwise around the proton at radius r. The magnetic field $\mathbf{B} = B\hat{\mathbf{k}}$ is

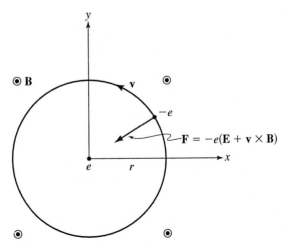

FIGURE 9.3   Induced dipole moment. In a classical model, applying a magnetic field changes the speed of the electron, and so the magnetic moment.

[1] *The Feynman Lectures*, Vol. II, Chap. 34.

perpendicular to the electron orbit, out of the page. When no magnetic field is present we can write the equation of motion $e^2/(4\pi\epsilon_0 r^2) = m_e v^2/r$, calling v the electron speed; i.e., the Coulomb attraction is the centripetal force on the electron. However, when **B** is present there is an additional inward force on the electron, due to the Lorentz force, as shown in Fig. 9.3. If the electron orbit remains the same while the magnetic field is turned on, i.e., r is the same with or without **B**, then the electron must speed up as **B** increases, reaching a new speed v' when the field strength is B. Adding the magnetic force the equation of motion then becomes

$$\frac{e^2}{4\pi\epsilon_0 r^2} + ev'B = m_e \frac{v'^2}{r}. \tag{9.3}$$

We defer until Chapter 10 the proof that r remains constant when **B** is applied, because that proof relies on Faraday's Law.

Now, if $m_e v^2/r$ is substituted for $e^2/(4\pi\epsilon_0 r^2)$ in (9.3) then, after rearranging, (9.3) becomes

$$\frac{m_e}{r}(v'^2 - v^2) = \frac{m_e}{r}(v' + v)(v' - v) = ev'B. \tag{9.4}$$

For actual atomic sizes and realistic fields the velocity increase $(v' - v) = \delta v$ is very small compared to v. We may therefore say $(v' + v) \approx 2v'$ so that (9.4) implies $\delta v = eBr/2m_e$. Then from the argument leading to (9.2) the change in $\mathbf{m}_{\text{orbital}}$ associated with Fig. 9.3 is

$$\delta\mathbf{m} = -\frac{er}{2}(\delta v)\hat{\mathbf{k}} = -\frac{e^2 r^2}{4m_e}B\hat{\mathbf{k}}. \tag{9.5}$$

The minus sign in (9.5) tells us that the change in the magnetic moment of this classical atom, brought about by the application of **B**, is *opposite* to the direction of **B**. Figure 9.3 shows counterclockwise motion of the electron, but we would get the same result if we considered clockwise motion, because in that case the additional force on the electron is radially outward. Then the electron slows down as **B** is applied, so that again the change in the magnetic moment is in the direction opposite to the applied field. If the net moment in a neighborhood of $\mathbf{x}$ is 0 for $\mathbf{B} = 0$, it will be opposite to **B** after **B** is applied. The material is therefore diamagnetic according to this classical picture, as claimed.

We have seen how the direction of **m** ends up opposite to **B**. The final result (9.5) is even reasonable quantitatively.[2] However, the crucial step in the argument was passed over lightly: The orbital radius r remains constant while **B** is changing. That follows from Faraday's Law of electromagnetic induction; we'll see that in Sec. 10.3. In qualitative terms the changing **B** induces a current opposing the change of **B**, by Lenz's law, i.e., an induced moment opposite to **B**.

[2] See Exercise 5.

The result (9.5), although obtained from classical physics, is the same as the quantum mechanical result if r^2 is properly interpreted. We shouldn't read too much into that; some similarity in the quantum and classical results is inevitable by dimensional analysis and the correspondence principle. Roughly, r^2 is the mean square radius of the electron quantum state. The quantum mechanical explanation for diamagnetism is based on the momentum term in the electron Hamiltonian. If there is no magnetic field then the momentum term is the familiar $\mathbf{p}^2/2m_e$, where $\mathbf{p}$ is the momentum. If there is a magnetic field then this term becomes $(\mathbf{p} - e\mathbf{A})^2/2m_e$, where $\mathbf{A}$ is the vector potential. It is the small term $e^2\mathbf{A}^2/2m_e$ that explains diamagnetism.

The principles of quantum mechanics apply, of course, to all materials so all materials have a diamagnetic response. However, in paramagnetic and ferromagnetic materials the diamagnetism is masked by larger effects. Materials that have complete electron shells, and therefore no net orbital or spin angular momentum in zero field, have only diamagnetic behavior. Examples of such purely diamagnetic materials are the inert gases, He, Ne, Ar, Kr, Xe, and Rn. Some diamagnetic ions are singly positive ions of alkali metals, Li^+, Na^+, K^+, Rb^+, Cs^+, and singly negative halide ions, F^-, Cl^-, Br^-, I^-, and At^-. We will see other examples later.

We have considered just the response of one electron. The response of the whole atom will be the sum of the effects of all the electrons. Because diamagnetism depends on the internal dynamics of the atom in its ground state, it does not depend on the temperature.

9.1.2 ■ Paramagnetism

When a paramagnetic substance is placed in an external $\mathbf{B}$ field it becomes magnetized with $\mathbf{M}$ proportional to $\mathbf{B}$ and in the same direction as $\mathbf{B}$. The magnetization vanishes when the field is removed. Because $\mathbf{M}$ is parallel to $\mathbf{B}$, a paramagnetic material will be drawn into a magnetic field; e.g., it will be attracted by a bar magnet.

The cause of paramagnetism is that the atom or molecule of the material has a *permanent dipole moment*. In other words, the electronic structure of the ground state of the atom has a nonzero $\mathbf{m}$, even if $\mathbf{B} = 0$. In the presence of $\mathbf{B}$ these permanent moments align partially with the field, because of the magnetic torque, and make up the magnetization. Not all materials have a permanent atomic or molecular moment; those with $\mathbf{m} = 0$ are diamagnetic.

The typical order of magnitude of a permanent atomic dipole moment is μ_B, coming from unpaired electron spins or orbital angular momenta in the atom. This is large enough that the paramagnetic response dominates the diamagnetic response. Any atom with an odd number of electrons will have a magnetic moment and will therefore be paramagnetic. Some atoms with an even number of electrons have a permanent magnetic moment because inner electron shells are unfilled; examples of such atoms are Cr, Fe, Ni, Pd, and Pt. Also, almost all the rare earth elements have permanent magnetic moments because they have electrons in an unfilled 4f-shell. The 4f states have large orbital angular momentum

($\ell = 3$) and degeneracy ($2\ell + 1 = 7$). Some rare earth elements have very large atomic moments because they have many electrons with their spins aligned, and also with large orbital angular momentum. Molecules in most cases are not paramagnetic because the electron spins are paired. However, some molecules, for example NO which has 15 electrons, are paramagnetic. A particularly interesting and important case is O_2 which, although it has 16 electrons, has two unpaired electrons and is therefore strongly paramagnetic.

The preferential alignment of the permanent moments parallel to **B** is favored by the interaction energy $U = -\mathbf{m} \cdot \mathbf{B}$, which is minimum if **m** and **B** point in the same direction; but it is opposed by thermal fluctuations, which are characterized by the energy kT at temperature T. The angle θ between **m** of a particular atom and **B** fluctuates between 0 and π; but it is a little more likely to have $\theta < \pi/2$ than $\theta > \pi/2$, so the *mean* orientation is parallel to **B**.

In classical statistical mechanics the thermal average of $\cos\theta$ at high temperatures is

$$\langle \cos \theta \rangle = \frac{mB}{3kT}.$$

As $T \to \infty$ the direction of **m** becomes random, equally likely to have $\theta < \pi/2$ or $\theta > \pi/2$, so $\langle \cos\theta \rangle \to 0$. Paramagnetism is a *small effect* at room temperature because $mB \ll kT$. For example, suppose $m = \mu_B$, a typical value, and $B = 1\,T$ and $T = 300\,\mathrm{K}$; then $\langle \cos\theta \rangle = 7.5 \times 10^{-4}$, which corresponds to an angle of 89.96 degrees, an alignment of only 0.04 degrees in the direction of **B**.

If the atomic density is n and each atom has a magnetic moment with magnitude m_0 then the magnetization in the presence of **B**, neglecting interactions between the atoms, is

$$\mathbf{M} = nm_0 \langle \cos\theta \rangle \hat{\mathbf{k}} = \frac{nm_0^2}{3kT}\mathbf{B}. \tag{9.6}$$

The result shows that the paramagnetic response varies with temperature as $1/T$. The linear relationship between **M** and **B** holds down to low temperatures and high fields. At very low T and high $|\mathbf{B}|$, the atomic dipoles approach a state of complete alignment with the field. In this limit **M** approaches a limiting, or saturation, value, and thus remains constant if $|\mathbf{B}|$ increases still further.

9.2 ■ MAGNETIZATION AND BOUND CURRENTS

If a magnetic body has a magnetization $\mathbf{M}(\mathbf{x}')$ as a function of position $\mathbf{x}'$ in the material, what magnetic field $\mathbf{B}(\mathbf{x})$ does it produce? We'll restate this question by asking for the vector potential $\mathbf{A}(\mathbf{x})$, from which $\mathbf{B}(\mathbf{x})$ can be calculated by $\mathbf{B} = \nabla \times \mathbf{A}$. The magnetization $\mathbf{M}(\mathbf{x}')$ can be either permanent magnetization in the absence of an applied field, as in ferromagnets, or the result of magnetic polarization by an applied field, as in diamagnets or paramagnets. In the latter case

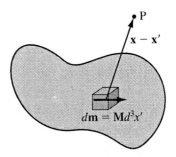

FIGURE 9.4   A magnetic object. $d\mathbf{m}$ is the dipole moment of a small volume d^3x' at $\mathbf{x}'$. The field at P is the integral of fields produced by elemental moments.

the resultant field is the superposition of the applied field and the field produced by the magnetized material.

Because individual electrons are the ultimate source of $\mathbf{M}$, the magnetic moment density fluctuates widely on a *microscopic scale* depending on how many electrons or magnetic atoms are included in a microscopic volume. For our *macroscopic* purposes we can ignore this fine-scale variation by considering volumes d^3x' that contain many atoms but are small on the scale of the macroscopic system. On the macroscopic scale $\mathbf{M}(\mathbf{x}')$ is a smooth function and we can use it in integrals.

Figure 9.4 shows such a magnetic object. The magnetic moment of the volume element d^3x' is $d\mathbf{m} = \mathbf{M}(\mathbf{x}')d^3x'$, density times volume. We seek $\mathbf{A}(\mathbf{x})$, then, at the point P. Recall Eq. (8.71), which gives $\mathbf{A}(\mathbf{x})$ for a point-like magnetic moment $\mathbf{m}$, apply it to the elemental moments $d\mathbf{m}$, and integrate over the volume V of the body; the result is

$$\mathbf{A}(\mathbf{x}) = \frac{\mu_0}{4\pi}\int_V \frac{d\mathbf{m}\times(\mathbf{x}-\mathbf{x}')}{|\mathbf{x}-\mathbf{x}'|^3} = \frac{\mu_0}{4\pi}\int_V \frac{\mathbf{M}(\mathbf{x}')\times(\mathbf{x}-\mathbf{x}')}{|\mathbf{x}-\mathbf{x}'|^3}d^3x'. \qquad (9.7)$$

There are many problems in which $\mathbf{M}(\mathbf{x}')$ is given and the field is sought, but (9.7) is seldom used, as written, to solve such problems. One reason is that the integrand can be unwieldy. But a more important reason is that (9.7) can be recast into a form that is easier to use.

First note that the integrand in (9.7) may be written as

$$\text{integrand} = \frac{\mathbf{M}(\mathbf{x}')\times(\mathbf{x}-\mathbf{x}')}{|\mathbf{x}-\mathbf{x}'|^3} = \mathbf{M}(\mathbf{x}')\times\left(\nabla'\frac{1}{|\mathbf{x}-\mathbf{x}'|}\right) \qquad (9.8)$$

where ∇' is the gradient with respect to $\mathbf{x}'$. Now integrate by parts. That is, swing the del operator over to the other term in the integrand by writing

$$\text{integrand} = \frac{\nabla'\times\mathbf{M}(\mathbf{x}')}{|\mathbf{x}-\mathbf{x}'|} - \nabla'\times\left(\frac{\mathbf{M}(\mathbf{x}')}{|\mathbf{x}-\mathbf{x}'|}\right). \qquad (9.9)$$

(The latter result is an application of the vector identity

$$\nabla \times (f\mathbf{F}) = (\nabla f) \times \mathbf{F} + f\nabla \times \mathbf{F},$$

applied to the variable $\mathbf{x}'$ with $\mathbf{F} = \mathbf{M}(\mathbf{x}')$ and $f = |\mathbf{x} - \mathbf{x}'|^{-1}$.) The volume integral of the second term in (9.9), which is a total derivative, may be converted to a surface integral by the identity[3]

$$\int_V \nabla' \times \mathbf{G}d^3x' = \int_S \hat{\mathbf{n}} \times \mathbf{G}dA'; \tag{9.10}$$

S is the boundary surface of V, and $\hat{\mathbf{n}}$ is the outward normal at dA'. Combining these results, we write the vector potential as

$$\mathbf{A}(\mathbf{x}) = \frac{\mu_0}{4\pi} \int_V \frac{\nabla' \times \mathbf{M}(\mathbf{x}')}{|\mathbf{x} - \mathbf{x}'|}d^3x' + \frac{\mu_0}{4\pi} \int_S \frac{\mathbf{M}(\mathbf{x}') \times \hat{\mathbf{n}}dA'}{|\mathbf{x} - \mathbf{x}'|}. \tag{9.11}$$

Equation (9.11) should look familiar. It has the same form as equations (8.54) and (8.55) for the vector potential of specified current distributions, volume and surface respectively. Comparing the first term on the right of (9.11) with (8.54) shows that $\nabla' \times \mathbf{M}(\mathbf{x}')$ occupies the place of a volume current density; we'll call it $\mathbf{J}_b(\mathbf{x}')$, the density of *bound current* at $\mathbf{x}'$ in the magnetized body. Likewise, the second term on the right of (9.11) may be compared with (8.55), showing that $\mathbf{M}(\mathbf{x}') \times \hat{\mathbf{n}}$ may be considered to be a surface current density; we'll call that $\mathbf{K}_b(\mathbf{x}')$, the density of *bound current* at $\mathbf{x}'$ on the surface of the magnetized body. Equation (9.11) thus tells us something interesting about magnetism and gives us an important tool for calculating magnetic fields: The magnetic field produced by a magnetized object is the same as that of a volume current $\mathbf{J}_b$ and a surface current $\mathbf{K}_b$, where

$$\mathbf{J}_b(\mathbf{x}') = \nabla' \times \mathbf{M}(\mathbf{x}') \tag{9.12}$$

and

$$\mathbf{K}_b(\mathbf{x}') = \mathbf{M}(\mathbf{x}') \times \hat{\mathbf{n}}. \tag{9.13}$$

To calculate the field we may replace the actual magnetization $\mathbf{M}(\mathbf{x}')$ in the object by the effective currents $\mathbf{J}_b(\mathbf{x}')$ and $\mathbf{K}_b(\mathbf{x}')$. Finally, to express this explicitly, we can write (9.11) as

$$\mathbf{A}(\mathbf{x}) = \frac{\mu_0}{4\pi} \int_V \frac{\mathbf{J}_b(\mathbf{x}')}{|\mathbf{x} - \mathbf{x}'|}d^3x' + \frac{\mu_0}{4\pi} \int_S \frac{\mathbf{K}_b(\mathbf{x}')}{|\mathbf{x} - \mathbf{x}'|}dA'. \tag{9.14}$$

9.2.1 ■ Examples

We will now apply this approach to two examples. The examples are relatively simple—limited to highly symmetric bodies in which $\mathbf{M}$ is constant. Even these

---

[3] This integration identity may be derived by applying Gauss's divergence theorem to $\mathbf{C} \times \mathbf{G}$, where $\mathbf{C}$ is an arbitrary constant vector, and using the identity $\hat{\mathbf{n}} \cdot (\mathbf{C} \times \mathbf{G}) = -\mathbf{C} \cdot (\hat{\mathbf{n}} \times \mathbf{G})$.

problems are challenging. Clearly for more general problems—consider for example a geological ore with inclusions of variously magnetized material mixed with nonmagnetic substances—the integrals over volume and surface in (9.14) would be very difficult.

EXAMPLE 1   What is the field of a uniformly magnetized sphere?

Let the magnetization in the sphere be $\mathbf{M} = M\hat{\mathbf{k}}$ and its radius be a, as shown in Fig. 9.5. Because $\mathbf{M}$ is constant there is no volume current density: $\mathbf{J_b} = \nabla \times \mathbf{M} = 0$. There is, however, a surface current density $\mathbf{K_b} = \mathbf{M} \times \hat{\mathbf{r}} = M \sin\theta\hat{\boldsymbol{\phi}}$.

It will not be necessary to redo the integral over $\mathbf{K_b}(\mathbf{x}')$ in (9.14) because we have evaluated an integral of the same form in Example 7 of Chapter 8 (the rotating charged sphere). Both of these examples have a surface current proportional to $\hat{\boldsymbol{\phi}} \sin\theta$ on a sphere. Notice the similarity between Figs. 9.5 and 8.17; in particular $\mathbf{K_b}$ in Fig. 9.5 is parallel to $\mathbf{v}$ of Fig. 8.17. If we replace $\mathbf{M}$ by $\sigma a\omega\hat{\mathbf{k}}$, then the field of this example is the same as that of a sphere of radius a with surface charge density σ rotating with $\boldsymbol{\omega} = \omega\hat{\mathbf{k}}$.

Carrying over the result of Example 7 of Chapter 8, and replacing $\sigma a\omega$ by M, we obtain the vector potential of the magnetized sphere. In spherical coordinates the result is

$$\mathbf{A}(r,\theta) = A_\phi(r,\theta)\hat{\boldsymbol{\phi}} = \frac{\mu_0 M}{3} r \sin\theta\hat{\boldsymbol{\phi}} \quad \text{for} \quad r < a, \tag{9.15}$$

$$\mathbf{A}(r,\theta) = A_\phi(r,\theta)\hat{\boldsymbol{\phi}} = \frac{\mu_0 M}{3} \frac{a^3}{r^2} \sin\theta\hat{\boldsymbol{\phi}} \quad \text{for} \quad r > a. \tag{9.16}$$

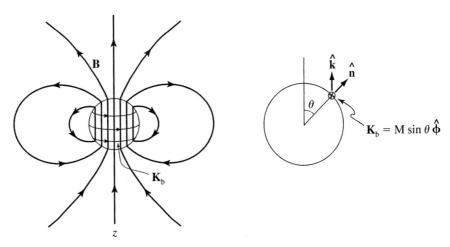

FIGURE 9.5   A magnetized sphere. The magnetization is uniform inside the sphere. There is a corresponding bound surface current density $\mathbf{K_b}$ on the sphere.

The corresponding magnetic field $\nabla \times \mathbf{A}$ is

$$\mathbf{B}(r, \theta) = \frac{2}{3}\mu_0 \mathbf{M} = \frac{2}{3}\mu_0 M \left(\hat{\mathbf{r}} \cos\theta - \hat{\boldsymbol{\theta}} \sin\theta\right) \quad \text{for} \quad r < a, \quad (9.17)$$

$$\mathbf{B}(r, \theta) = \frac{\mu_0 M}{3} \frac{a^3}{r^3}(2\hat{\mathbf{r}} \cos\theta + \hat{\boldsymbol{\theta}} \sin\theta) \quad \text{for} \quad r > a. \quad (9.18)$$

In writing the second equality of (9.17) we used the unit vector identity $\hat{\mathbf{k}} = \hat{\mathbf{r}} \cos\theta - \hat{\boldsymbol{\theta}} \sin\theta$.

In magnetostatics the field $\mathbf{B}$ satisfies certain boundary conditions. We will develop these conditions in detail later[4] but it is worthwhile to anticipate that development by noting the behavior of $\mathbf{B}$ at the boundary of the sphere in this example. From (9.17) and (9.18) at $r = a$ we have:

(i) The normal component of $\mathbf{B}$ is continuous,

$$B_{\text{int},r}(a, \theta) = B_{\text{ext},r}(a, \theta) = \tfrac{2}{3}\mu_0 M \cos\theta;$$

(ii) The tangential components of $\mathbf{B}$ are discontinuous and the discontinuity is related to the surface current density,

$$B_{\text{ext},\theta}(a, \theta) - B_{\text{int},\theta}(a, \theta) = \mu_0 M \sin\theta = \mu_0 K_{\text{b},\phi}.$$

Equation (9.17) shows that *inside* the sphere $\mathbf{B}$ is uniform and parallel to $\mathbf{M}$. Equation (9.18) shows that *outside* the sphere $\mathbf{B}$ is the same as the field of a point magnetic dipole at the origin (see (8.73)) with dipole moment $\mathbf{m} = (4/3)\pi a^3 \mathbf{M}$. We know that the far field ($r \gg a$) must have this form because $\mathbf{M}$ is the dipole moment per unit volume and $(4/3)\pi a^3$ is the volume of the sphere. But it is somewhat surprising here, as it was in Example 7 of Chapter 8, that the external field is a pure dipole field. It is pleasing, but maybe now not so surprising, to know that there is a theorem generalizing this result.[5] The theorem states that if a sphere has any spherically symmetric multipole density, i.e., a multipole moment per unit volume that depends only on r, then the whole external field is the same as if the total multipole moment were at the origin.

EXAMPLE 2   What is the magnetic field *on the axis* of a uniformly magnetized cylinder? A physical realization of this example could be a cylindrical bar magnet.

Let the magnetization of the cylinder be $\mathbf{M} = M\hat{\mathbf{k}}$, its radius a, and its height h, as shown in Fig. 9.6. Again, as in Example 1, there is no volume current because $\mathbf{J}_b = \nabla \times \mathbf{M} = 0$ for uniform magnetization. As for surface current, there is none on the flat surfaces, $z = 0$ and $z = h$, because $\mathbf{M}$ is, respectively, antipar-

---

[4]See Section 9.3.4.

[5]M. J. Harrison and R. D. Spence, *Am. J. Phys.* **62**, 828 (1994).

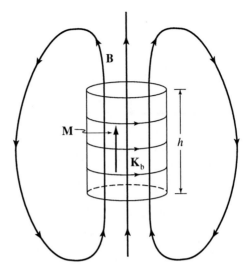

FIGURE 9.6 **A magnetized cylinder.** The magnetization is uniform inside the cylinder. There is a corresponding bound surface current density $\mathbf{K_b}$ on the cylinder.

allel and parallel to $\hat{\mathbf{n}}$ there. There is bound surface current on the curved surface $r = a$, because there $\mathbf{K_b} = \mathbf{M} \times \hat{\mathbf{r}} = M\hat{\boldsymbol{\phi}}$. Thus for purposes of calculating $\mathbf{B}(\mathbf{x})$, a uniformly magnetized finite cylinder is equivalent to a stack of elemental *rings* each with radius a and current $dI = Mdz'$. Note too that this problem is equivalent to finding $\mathbf{B}$ for a finite solenoid.

To calculate the magnetic field we first note, from symmetry, that on the z axis $\mathbf{B} = B_z(z)\hat{\mathbf{k}}$. Using (8.26) for the field on the axis of a current ring and integrating elemental rings over the length of the cylinder yields

$$B_z(z) = \frac{\mu_0 M a^2}{2} \int_0^h \frac{dz'}{[a^2 + (z - z')^2]^{3/2}}$$

$$= \frac{\mu_0 M}{2} \left(\frac{h - z}{\sqrt{a^2 + (h - z)^2}} + \frac{z}{\sqrt{a^2 + z^2}} \right), \qquad (9.19)$$

which answers the question we posed. We leave it as an exercise[6] to plot $B_z(z)$.

In Example 1 we found $\mathbf{B}(\mathbf{x})$ *everywhere* in space for the magnetized sphere, but in Example 2 we limited ourselves to finding $\mathbf{B}$ on the cylinder axis. Can we find $\mathbf{B}(\mathbf{x})$ everywhere in space for the magnetized finite cylinder of Fig. 9.6? How would we proceed?

The answer is that we can find $\mathbf{B}(\mathbf{x})$ everywhere for the cylinder but it is analytically difficult. One approach is to obtain $\mathbf{A}(\mathbf{x})$ for the finite cylinder by integrating $d\mathbf{A}(\mathbf{x})$ for elemental current loops making up the bound surface current. (We know

---

[6]See Exercise 2.

the vector potential of a ring from Sec. 8.7.) In practice it would be necessary to use numerical techniques to do the calculation. It seems somewhat perverse that a magnetostatics problem so easily stated and with so much symmetry as this one should be analytically so difficult. After all, cylindrical bar magnets and solenoids are in many everyday devices. We learn from this that, in electromagnetism, even easy problems can be hard.

On the other hand, it is straightforward to find the field of the magnetized cylinder far away. For $r \gg a$ and h, $\mathbf{B}$ approaches a dipole field (8.73) with dipole moment $\mathbf{m} = \pi a^2 h M \hat{\mathbf{k}}$, the total dipole moment of the cylinder. On the far z axis, i.e., $z \gg a$ and h, we have $B_z(z) = \mu_0 M a^2 h / 2 z^3$; on the far $z = 0$ plane, i.e., $r \gg a$ and h, we have $B_z(r) = -\mu_0 M a^2 h / 4 r^3$. Another case for which we can easily write the answer is an infinitely long cylinder. That case is equivalent to an infinite solenoid with current per unit length $M \hat{\boldsymbol{\phi}}$, so $\mathbf{B}_{\text{outside}} = 0$ and $\mathbf{B}_{\text{inside}} = \mu_0 M \hat{\mathbf{k}}$.

9.2.2 ■ A Geometric Derivation of the Bound Currents

The equations (9.12) and (9.13) for bound current densities $\mathbf{J}_b$ and $\mathbf{K}_b$ are important to our understanding of magnetism in matter. Their derivation has been rigorous, but mathematical and not intuitive. It is worthwhile to derive the equations again, by geometric arguments, to better appreciate their meaning. In particular, we shall see that $\nabla \times \mathbf{M}$ is indeed a *current density*.

Imagine a magnetized body subdivided into small rectangular cells. Each cell contains many atoms, although it is small on the overall scale of the body. For simplicity we'll assume the magnetization is everywhere parallel to the z axis $\mathbf{M} = M \hat{\mathbf{k}}$ but the magnitude may vary. $\mathbf{M}$ is the magnetic dipole moment density, so the total dipole moment of a cell of size $\delta x \times \delta y \times \delta z$ is $M \hat{\mathbf{k}} \delta x \delta y \delta z$. We will attribute this total moment—the sum of the moments of many atoms—to a surface current $\mathbf{K}$ on the four faces of the cell parallel to the z axis, as shown in Fig. 9.7(a). The dipole moment due to the surface current is $\hat{\mathbf{k}} I A$, where I is the integrated current density $I = K \delta z$ and A is the area $A = \delta x \delta y$; comparing with the formula for M,

$$M \delta x \delta y \delta z = (K \delta z)(\delta x \delta y),$$

so the effective surface current on the faces has $K = M$. The direction of $\mathbf{K}$ on any of the four boundary surfaces of the cell is tangent to the surface, and we see from Fig. 9.7(a) that $\mathbf{K} = \mathbf{M} \times \hat{\mathbf{n}}$, where $\hat{\mathbf{n}}$ is the outward normal of that surface.

If $\mathbf{M}$ is constant throughout the sample, then the surface currents on cell-faces inside the sample cancel; each interior face is part of the boundary of two neighboring cells, with surface currents in opposite directions. In this case the only bound current is $\mathbf{K}_b = \mathbf{M} \times \hat{\mathbf{n}}$ on the exterior surface of the sample.

If $\mathbf{M}(\mathbf{x})$ varies with position in the material, then the internal bound currents do not cancel, so there is a volume current density $\mathbf{J}_b(\mathbf{x})$. Again for simplicity

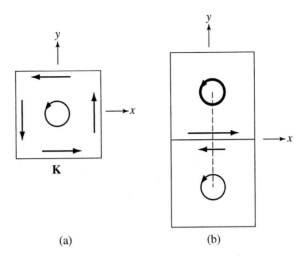

(a)   (b)

FIGURE 9.7   Geometric derivation of bound current. (a) A small rectangular cell with magnetization **M** out of the page and equivalent surface current **K** counterclockwise. (b) Two neighboring cells; the magnetization in the upper cell is greater than that in the lower cell, so there is an effective current to the right at the interface.

taking $\mathbf{M} = M\hat{\mathbf{k}}$, consider neighboring cells centered at $y + \delta y/2$ and $y - \delta y/2$, as shown in Fig. 9.7(b). The net current in the x direction through the area $\delta y \delta z$ between the cell centers (indicated by a dashed line) is

$$J_{bx}\delta y \delta z = M_z(y + \delta y/2)\delta z - M_z(y - \delta y/2)\delta z,$$

because $K_x = \pm M_z$ is the surface current density at either side of the interface between the cells. In this expression, $M_z(y \pm \delta y/2)$ means M_z evaluated at $y \pm \delta y/2$. Now, treating δy as infinitesimal, $J_{bx} = \partial M_z/\partial y$. Similarly, by considering neighboring cells centered at $x + \delta x/2$ and $x - \delta x/2$, there is a volume current density in the y direction of $J_{by} = -\partial M_z/\partial x$. These results are precisely the x and y components of the vector equation $\mathbf{J}_b = \nabla \times \mathbf{M}$ for the case when $\mathbf{M} = M\hat{\mathbf{k}}$.

In a magnetized object there are no charged particles moving macroscopic distances through the material, but there is electric current. The current comes from electrons bound in atoms, and its macroscopic effects are described by $\mathbf{J}_b(\mathbf{x})$ and $\mathbf{K}_b(\mathbf{x})$. These *effective* currents result from adding the atomic currents of *many atoms* in a neighborhood of **x**. Both the geometric arguments and the rigorous calculus proof leading to (9.11) show that the bound *volume current density* is $\mathbf{J}_b = \nabla \times \mathbf{M}$ and the bound *surface current density* is $\mathbf{K}_b = \mathbf{M} \times \hat{\mathbf{n}}$.

Pictorial Description of Bound Current
Some simple pictures can help to visualize the relation $\mathbf{J}_b = \nabla \times \mathbf{M}$, which is the main principle for understanding the magnetic effects of matter.

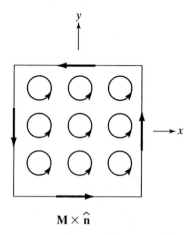

FIGURE 9.8   Magnetization and bound current. A small cell containing many aligned dipoles, and the equivalent effective surface current $\mathbf{K} = \mathbf{M} \times \hat{\mathbf{n}}$. The magnetization points out of the page.

Figure 9.8 shows a small cell in the material with many aligned dipoles, represented as small current loops. On an atomic scale the magnetization fluctuates wildly with position, but the *macroscopic* magnetization, i.e., due to the combined effects of many atoms, is uniform $\mathbf{M} = nm\hat{\mathbf{k}}$. The same magnetization would be produced by a surface current density $\mathbf{M} \times \hat{\mathbf{n}}$ around the four faces of the cell.

Figure 9.9 shows 5 contiguous cells with magnetization $M_z(x)\hat{\mathbf{k}}$ pointing in the z direction with M_z increasing in the $+x$ direction. Just by looking at the picture we see that $\mathbf{M}$ curls counterclockwise around $-\hat{\mathbf{j}}$. Therefore there must be a bound current out of the page ($-\hat{\mathbf{j}}$ direction) with

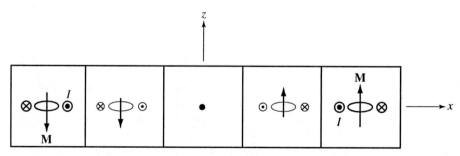

FIGURE 9.9   Curl of M is $\mathbf{J}_b$. The magnetization points in the z direction and increases in the $+x$ direction. At the center of each cell a typical atom is represented as a small current loop. There is a bound current out of the page.

$$\mathbf{J}_b = \nabla \times \mathbf{M} = -\frac{\partial M_z}{\partial x} \hat{\mathbf{j}}. \tag{9.20}$$

The figure shows at the center of each cell a representative atomic dipole (there are many dipoles in each cell) as a small current loop. (Of course the electrons revolve in the opposite direction.) The picture shows why there is a bound current out of the page. For example, consider the region near the interface between the two right-most cells. The current out of the page on the left side of the cell to the right of the interface is greater in magnitude than the current into the page on the right side of the cell to the left of the interface—making a net current out of the page near the interface. The picture is consistent with (9.20). By a similar argument the student can verify that there is a net current out of the page near any of the interfaces between neighboring cells.

9.3 ■ AMPÈRE'S LAW FOR FREE CURRENTS, AND H

We have seen that there are two kinds of current in matter, free current $\mathbf{J}_f$ and bound current $\mathbf{J}_b$. Free current is familiar; it's the mean translational motion of charges that are free to move throughout the material. Any single charge moves in an irregular random walk, but summing over the many particles in a neighborhood of $\mathbf{x}$ creates the current density $\mathbf{J}_f(\mathbf{x})$. Bound current is different, and somewhat difficult to visualize. It is also due to motion of charge,[7] but the charge is bound within atoms or molecules. Orbital motion and spin of electrons contribute to the magnetic dipole moment of the atom; and those atomic currents, summed over the many atoms in a neighborhood of $\mathbf{x}$, produce a current density $\mathbf{J}_b(\mathbf{x}) = \nabla \times \mathbf{M}$, where $\mathbf{M}(\mathbf{x})$ is the magnetization. A bound current exists at $\mathbf{x}$ if the magnetization curls there. The bound current is just as much a source of **B** as the free current. Therefore Ampère's law in a magnetic material is

$$\nabla \times \mathbf{B} = \mu_0 \left(\mathbf{J}_f + \mathbf{J}_b \right). \tag{9.21}$$

Now, $\mathbf{J}_b$ is the curl $\nabla \times \mathbf{M}$, and it is convenient to combine the curls in (9.21) by writing

$$\nabla \times \mathbf{H} = \mathbf{J}_f, \tag{9.22}$$

where

$$\mathbf{H} = \frac{\mathbf{B}}{\mu_0} - \mathbf{M}. \tag{9.23}$$

Equation (9.22) is another form of *Ampère's law*. Equation (9.23) defines a new field $\mathbf{H}(\mathbf{x})$ that is a combination of the magnetic field $\mathbf{B}(\mathbf{x})$ and the material field $\mathbf{M}(\mathbf{x})$.

---

[7]We're using poetic license here. It is not rigorously correct to think of electron spin—a quantum phenomenon—as motion.

The nomenclature for the related fields $\mathbf{B}(\mathbf{x})$ and $\mathbf{H}(\mathbf{x})$ is not as standardized as one might wish.[8] One convention, used for example in the famous graduate textbook by J. D. Jackson, is to name $\mathbf{B}$ the "magnetic induction" and $\mathbf{H}$ the "magnetic field." However, $\mathbf{B}$ is the fundamental field that exerts the magnetic force on charged particles, whereas $\mathbf{H}$ is a derived *auxiliary* quantity that we introduce for convenience; so it is preferable to name $\mathbf{B}$ the magnetic field, in analogy with $\mathbf{E}$ the electric field. (This is the term used for $\mathbf{B}$ in this book.) An informal language is to call the two fields the $\mathbf{B}$-field and the $\mathbf{H}$-field, which is precise but not elegant. A common usage, especially in conversations, is just to call either one the magnetic field!

Terminology aside, $\mathbf{B}$ and $\mathbf{H}$ are distinct fields. Note that $\mathbf{B}$ and $\mathbf{H}$ have different units. $\mathbf{B}$ is measured in teslas, where 1 T=1 N/(Am). $\mathbf{H}$ is measured in A/m, which doesn't have a special name. In many cases $\mathbf{B}$ and $\mathbf{H}$ are proportional. For example, in a vacuum $\mathbf{B} = \mu_0\mathbf{H}$. In common materials (diamagnetic or paramagnetic) they are also proportional, $\mathbf{B} = \mu\mathbf{H}$, in which the parameter μ is called the *permeability* of the material. In ferromagnets $\mathbf{B}$ is a nonlinear function of $\mathbf{H}$, as we shall see in Sec. 9.6.

In certain systems the forms of $\mathbf{B}$ and $\mathbf{H}$ are quite different, as we shall see in some examples. A fundamental principle of magnetism is that $\nabla \cdot \mathbf{B} = 0$, expressing that magnetic monopoles—point sources from which $\mathbf{B}$ diverges—do not exist. In contrast $\nabla \cdot \mathbf{H}$ is not necessarily 0. This difference implies that $\mathbf{B}$ and $\mathbf{H}$ satisfy different boundary conditions at an interface between materials: The normal component of $\mathbf{B}$ is continuous; and, if there is no free surface current, the tangential component of $\mathbf{H}$ is continuous. We shall derive these boundary conditions presently.[9] Example 3 shows how $\mathbf{B}$ and $\mathbf{H}$ could be measured in a magnetic material, in principle.

EXAMPLE 3   Consider a large piece of material that has a constant magnetic field and magnetization, respectively, $\mathbf{B}_1$ and $\mathbf{M}_1$, which are parallel to each other; (9.23) holds, as always, so $\mathbf{H}_1 = \mathbf{B}_1/\mu_0 - \mathbf{M}_1$.

(a) If a thin disk-shaped cavity is hollowed out of the material with the disk axis parallel to $\mathbf{B}_1$, what are the fields $\mathbf{B}_0$ and $\mathbf{H}_0$ in the cavity near the middle of it?

   By symmetry $\mathbf{B}_0$ near the middle of the cavity is perpendicular to the faces of the disk. The normal component of $\mathbf{B}$ is continuous across a boundary so $\mathbf{B}_0 = \mathbf{B}_1$ and, from (9.23), $\mathbf{H}_0 = \mathbf{H}_1 + \mathbf{M}_1$. (We have used $\mathbf{B}_0 = \mu_0\mathbf{H}_0$ in the cavity.)

------

[8]The long and interesting history of the magnetic vectors $\mathbf{B}$ and $\mathbf{H}$, as well as the subtle concepts associated with these quantities, are discussed in Ref. [1].

[9]See Section 9.3.4. The reader may wish to read Sec. 9.3.4 before Example 3, because the boundary conditions must be used in the example.

(b) If a thin needle-shaped cavity is hollowed out of the material with the needle axis parallel to $\mathbf{B}_1$, what are the fields $\mathbf{B}_0$ and $\mathbf{H}_0$ in the cavity near the middle?

By symmetry $\mathbf{B}_0$ and $\mathbf{H}_0$ near the middle of the cavity are parallel to $\mathbf{B}_1$ and $\mathbf{H}_1$, and tangent to the cavity walls. The tangential component of $\mathbf{H}$ is continuous across the boundary so in this case $\mathbf{H}_0 = \mathbf{H}_1$ and, from (9.23) $\mathbf{B}_0 = \mathbf{B}_1 - \mu_0\mathbf{M}_1$.

(c) If a small spherical cavity is hollowed out, what are $\mathbf{B}_0$ and $\mathbf{H}_0$ in the cavity?

By the superposition principle the field inside the cavity is the sum of the field in the bulk material plus the field of a superposed sphere with magnetization $-\mathbf{M}_1$. Example 1 shows that the magnetic field in the superposed sphere is $-\frac{2}{3}\mu_0\mathbf{M}_1$. Therefore inside the cavity we have

$$\mathbf{B}_0 = \mathbf{B}_1 - \tfrac{2}{3}\mu_0\mathbf{M}_1 \quad \text{and} \quad \mathbf{H}_0 = \mathbf{H}_1 + \tfrac{1}{3}\mathbf{M}_1.$$

The results of these three cases are summarized in a table:

Cavity shape	$\mathbf{B}_0$	$\mathbf{H}_0$
Disk	$\mathbf{B}_1$	$\mathbf{H}_1 + \mathbf{M}_1$
Sphere	$\mathbf{B}_1 - \frac{2}{3}\mu_0\mathbf{M}_1$	$\mathbf{H}_1 + \frac{1}{3}\mathbf{M}_1$
Needle	$\mathbf{B}_1 - \mu_0\mathbf{M}_1$	$\mathbf{H}_1$

Notice that the fields in the spherical cavity are intermediate between the fields in the other cavities. This is expected, because all three cavities are similar to special cases of an ellipsoidal cavity, with the sphere intermediate between the extremes of disk and needle. (An extremely oblate ellipsoid is disklike, but with a smooth surface, and an extremely prolate ellipsoid is needlelike.) The general problem of an ellipsoid of magnetic material in an external magnetic field can be solved analytically in terms of elliptic integrals. Therefore it is convenient to use ellipsoidal samples for measuring magnetic properties. The corresponding electrostatics problem of an ellipsoid of dielectric material in an external **E** can be solved similarly.

Because the field $\mathbf{B}_0$ in a disk-shaped cavity is the same as the field $\mathbf{B}_1$ in the bulk material, and the field $\mathbf{H}_0$ in a needle-shaped cavity is the same as the field $\mathbf{H}_1$ in the bulk, Lord Kelvin introduced these cavities as operational ways of measuring the fields inside a magnetic material. An analogous procedure can be used to measure **E** and **D** in a dielectric material.[10] This emphasis on the importance of direct measurement of physical quantities dates from a time, a hundred years ago, when a quantity was only considered meaningful if it could be measured directly. But we don't follow that ideal so closely anymore in physics.

---

[10]See Exercise 8.

9.3.1 ■ The Integral Form of Ampère's Law

The differential equation (9.22) is equivalent to an integral relation for the circulation of **H** around a closed curve, i.e., an Amperian loop. By Stokes's theorem

$$\oint_C \mathbf{H} \cdot d\boldsymbol{\ell} = \int_S \mathbf{J}_f \cdot d\mathbf{A} = I_f(\text{enclosed}). \tag{9.24}$$

Here C is any closed curve, S is any open surface bounded by C, and $I_f(\text{enclosed})$ is the free current through S. The directions of $d\boldsymbol{\ell}$ and $d\mathbf{A}$ are correlated by the right-hand rule, in the usual way for Stokes's theorem. We have derived (9.24) from (9.22) but the derivation can be reversed because (9.24) must hold for arbitrary C and S.

9.3.2 ■ The Constitutive Equation

The reason (9.22), or (9.24), is so important and useful is that the source term is only the free current—the familiar current due to motion of free charges. The bound current is hidden in **H**. But we cannot just ignore the bound current; it has physical effects. To use the field **H** in a calculation of the magnetic properties of a system, it is necessary to know *from some other information* the relation between **B** and **H**, or equivalently between **M** and **H**. The required information is a question of materials science. It is determined by the electronic structure of the atom. The equation relating **B** and **H** is called the *constitutive equation*.

In most materials the magnetization $\mathbf{M}(\mathbf{x})$ is proportional to the field $\mathbf{H}(\mathbf{x})$. Such a material is called a *linear magnetic material*. The constant of proportionality χ_m defined by

$$\mathbf{M} = \chi_m \mathbf{H}, \tag{9.25}$$

is called the *magnetic susceptibility*. Then **B** is also a linear function of **H**

$$\mathbf{B} = \mu \mathbf{H}, \tag{9.26}$$

where the permeability μ is

$$\mu = \mu_0 (1 + \chi_m) \tag{9.27}$$

by the definition (9.23) of **H**.

9.3.3 ■ Magnetic Susceptibilities

Linear materials are classified as diamagnetic or paramagnetic. The definition of a diamagnetic material is that $\chi_m < 0$. In this class of materials the magnetization points in the direction opposite to the **H** field, $\mathbf{M} \propto -\mathbf{H}$. In other words, the atomic dipoles are aligned opposite to **H** and **B**. Recall from Sec. 9.1 that putting an atom in a magnetic field creates a contribution to the magnetic moment, in the direction opposite to the field. The classical picture of diamagnetism, described in Sec. 9.1, attributes it to electromagnetic induction, and the direction of the induced moment is opposite to the field by Lenz's law. In diamagnetic materials this contribution to the atomic dipole moment is dominant, so $\mathbf{M} \propto -\mathbf{H}$ and $\chi_m < 0$. As Table 9.2 shows, $|\chi_m| \ll 1$; in this sense diamagnetism is a small magnetic effect.

TABLE 9.2 Magnetic Susceptibilities of Some Elements and Compounds

Diamagnetic elements	χ_m[a]
H_2 (STP)	-2.2×10^{-9}
He (STP)	-1.1×10^{-9}
N_2 (STP)	-6.7×10^{-9}
Si	-3.3×10^{-6}
Ar (STP)	-1.1×10^{-8}
Cu	-9.6×10^{-6}
Xe (STP)	-2.6×10^{-8}
Au	-3.4×10^{-5}
Pb	-1.6×10^{-5}

Paramagnetic elements	χ_m
O_2 (STP)	$+1.9 \times 10^{-6}$
Na	$+8.5 \times 10^{-6}$
Al	$+2.1 \times 10^{-5}$
K	$+5.7 \times 10^{-6}$
Cr	$+2.9 \times 10^{-4}$
Rb	$+3.7 \times 10^{-6}$
W	$+7.0 \times 10^{-5}$
Nd	$+2.8 \times 10^{-4}$
Gd	$+8.7 \times 10^{-3}$

Compounds	χ_m
$H_2O(\ell, 293\,K)$	-9.0×10^{-6}
CO (STP)	-5.5×10^{-9}
NO (STP)	$+8.2 \times 10^{-7}$
CO_2 (STP)	-1.2×10^{-8}
SiO_2	-1.4×10^{-5}

[a] We use SI units, in which χ_m is dimensionless. There will come a time when comprehensive tables of χ_m are available in SI units. Today, however, the most comprehensive tables (*Handbook of Chemistry and Physics*, CRC Press, Boca Raton, 1999) list χ_m in Gaussian units of cm^3/mol. To convert to SI units, multiply by $4\pi/v_m$, where v_m is the molar volume in cm^3/mol.

The definition of a paramagnetic material is that $\chi_m > 0$. In this class of materials the magnetization is in the same direction as the magnetizing field. Recall from Sec. 9.1 that an atom with a permanent magnetic moment will tend to align with **B** or **H** because of the torque $\mathbf{m} \times \mathbf{B}$ on the dipole moment. In paramagnetic materials the permanent dipole moment is greater than the induced dipole moment, so $\mathbf{M} \propto +\mathbf{H}$ and $\chi_m > 0$.

Table 9.2 shows values of susceptibility χ_m for some interesting elements and compounds; for gases the data are for STP, and for solids the data are for room temperature. The elements have been chosen so that there are at least two from each row of the periodic table. For each row, except the first, there is shown χ_m for at least one diamagnetic and one paramagnetic element. (The first row of the periodic table contains only the elements hydrogen and helium; both H_2 and He are diamagnetic.) Notice that $|\chi_m|$ for paramagnetic elements is generally larger than for diamagnetic elements. But even for paramagnetic materials, $|\chi_m| \ll 1$.

The inert-gas elements are all diamagnetic; notice in Table 9.2 that for He, Ar, and Xe, the magnitude of χ_m increases as the size of the atom increases. This reflects the increase in the factor r^2 in (9.5). The same effect shows up for CO and CO_2; the triatomic CO_2 molecule is larger than diatomic CO so that, in some sense, the average r^2 in (9.5) is larger. In general if an atom has a permanent dipole moment then the paramagnetic effect dominates the diamagnetic effect and the material is paramagnetic. Table 9.2 shows examples with a large range of χ_m for both diamagnetic and paramagnetic elements.

The magnetic susceptibilities of the elements vary with position in the periodic table in a way that can be explained in terms of elementary atomic structure.[11]

EXAMPLE 4 (a) Sample calculation of diamagnetic χ_m for H_2. Now that we have a quantitative measure of magnetic susceptibility, it is instructive to investigate how well (9.5), the result of our model calculation of diamagnetism, compares to an experimental value. The simplest example is hydrogen.

Equation (9.5) gives the contribution to the magnetic moment of an atom from the change of the orbital motion of one electron in response to an applied field **H**. For an H atom, using $r = a_0 = 5.29 \times 10^{-11}$ m (the Bohr radius) and substituting[12] $\mathbf{B} = \mu_0\mathbf{H}$ we obtain $\delta\mathbf{m} = -(e^2 a_0{}^2 \mu_0/4m_e)\mathbf{H}$. Hydrogen gas is diatomic, so there are no data on χ_m for a gas of H atoms, but as a rough approximation we can consider that 1 mole of H_2 molecules consists of 2 moles of H atoms. Then the number density of H atoms in H_2 at STP is $n = 2N_A/2.24 \times 10^{-2}$ atoms/m^3. The magnetic moment per unit volume—the magnetization—is then

$$\mathbf{M} = -\frac{ne^2 a_0^2 \mu_0}{4m_e}\mathbf{H} = \chi_m\mathbf{H}, \tag{9.28}$$

---

[11] K. Krane, *Modern Physics*, 2nd ed. (Wiley, 1996).

[12] In a magnetic material $B = \mu H$, but it is a good approximation to use $B = \mu_0 H$ in (9.5) because $\chi_m \ll 1$.

which implies $\chi_m = -1.3 \times 10^{-9}$ for H_2 at STP. (There is no net contribution from the electron spins because the total spin is $S = 0$ in an H_2 molecule.) Considering the approximations we have made this estimate is gratifyingly close to the experimental value -2.2×10^{-9} shown in Table 9.2.

(b) Sample calculations of paramagnetic χ_m.   It is also instructive to apply (9.6), the result of our model calculation for the magnetization of a paramagnetic substance, to experimental values.

We first apply (9.6), or rather a quantum modification,[13] to NO, which has one unpaired electron spin per molecule. The molecular density is $n = N_A/2.24 \times 10^{-2}\,\text{m}^{-3}$, and again we may approximate **B** by $\mu_0\mathbf{H}$, so

$$\mathbf{M} = \frac{n\mu_B^2\mu_0}{kT}\mathbf{H} = \chi_m\mathbf{H}, \tag{9.29}$$

which gives $\chi_m = 7.7 \times 10^{-7}$ for NO at STP. This agrees quite well with the experimental value 8.2×10^{-7} in Table 9.2. If we do the same calculation for O_2 gas, considering that the O_2 molecule has 2 unpaired electrons, we find $\chi_m = 1.5 \times 10^{-6}$, which compares favorably to the experimental value for O_2.

If we carry out a similar calculation for a metal, say Al, we obtain a value for χ_m that is about 10^2 too large. The reason for this discrepancy is that in a metal the temperature that characterizes the electrons whose spins can line up paramagnetically, is not room temperature but a much higher quantum-mechanical temperature called the Fermi temperature. This is another indication that magnetism is a quantum phenomenon. There is even a theorem[14] that according to classical physics there can be no magnetism.[15]

■———————

The classic technique for measuring susceptibility is to measure the force—attractive or repulsive, respectively, on a paramagnetic or diamagnetic sample—produced by an external magnet, using a sensitive force balance. From the force, the magnetization and susceptibility can be calculated.

9.3.4 ▨ Boundary Conditions for Magnetic Fields

A standard problem in magnetostatics is to calculate the fields for a specified system of free currents and magnetic materials. To solve such problems it is nec-

[13]Equation (9.6) is the classical formula for the magnetization. The quantum formula for a spin 1/2 atom, which we use here, is the classical result multiplied by 3.

[14]The theorem is due to J. H. van Leeuwen. An accessible explanation is in *The Feynman Lectures*, Vol. II, Sec. 34-6.

[15]C. Kittel once put it this way: "...and were the value of $\hbar$ to go to zero, the loss of the science of magnetism is one of the catastrophes that would overwhelm the universe." C. Kittel, *Introduction to Solid State Physics*, 4th ed., Chap. 15 (Wiley, New York, 1971).

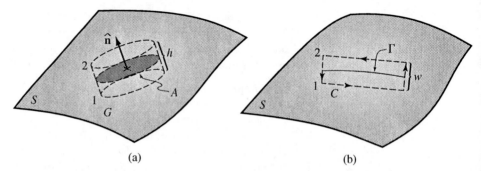

FIGURE 9.10   Boundary conditions for magnetic fields. S is a surface separating regions with different permeabilities. (a) A Gaussian pill box G cuts through S, with area A on opposite sides of S and height h. (b) An Amperian loop C cuts through S, with sections Γ on opposite sides of S and width w.

essary to know the boundary conditions for the fields. We will derive the general boundary conditions from the field equations.

The field $\mathbf{B}(\mathbf{x})$ satisfies $\nabla \cdot \mathbf{B} = 0$ in any magnetic system, because there are no magnetic monopoles. This equation implies that *the normal component of* $\mathbf{B}$ *is continuous across any two-dimensional surface S.* Figure 9.10(a) shows a Gaussian surface G shaped like a pill box, with similar faces A on opposite sides of S, and height h. The flux of $\mathbf{B}$ outward through G is 0, by Gauss's theorem. In the limit $h \to 0$ the flux through the edge approaches 0, so the flux through the two faces must be equal but opposite. Thus, letting $\mathbf{B}_1$ and $\mathbf{B}_2$ be the limits of $\mathbf{B}$ approaching opposite sides of S, and $\hat{\mathbf{n}}$ the unit normal vector in the direction $1 \to 2$, Gauss's law implies

$$\int_A \mathbf{B}_2 \cdot \hat{\mathbf{n}} dA = \int_A \mathbf{B}_1 \cdot \hat{\mathbf{n}} dA. \tag{9.30}$$

As this must be true for any area A on S,

$$\mathbf{B}_2 \cdot \hat{\mathbf{n}} = \mathbf{B}_1 \cdot \hat{\mathbf{n}}. \tag{9.31}$$

B_n is continuous.

The field $\mathbf{H}(\mathbf{x})$ satisfies $\nabla \times \mathbf{H} = \mathbf{J}_f$. This relation implies that *the tangential components of* $\mathbf{H}$ *are continuous across S if there is no free surface current.* Otherwise, if there is a free surface current density $\mathbf{K}_f$, then $\mathbf{H}$ has the discontinuity $\mathbf{K}_f \times \hat{\mathbf{n}}$. Figure 9.10(b) shows an Amperian loop with width w, whose sides on opposite sides of S approach a curve Γ on S as $w \to 0$. By Stokes's theorem, the circulation of $\mathbf{H}$ around the loop equals the free current enclosed by the loop. In the limit $w \to 0$, the circulation approaches the difference of the line integrals of $\mathbf{H}_1$ and $\mathbf{H}_2$ along Γ, and the enclosed current is the free surface current integrated across Γ (out of the page in Fig. 9.10(b)). Thus by Ampère's law

$$\int_{\Gamma} (\mathbf{H}_2 - \mathbf{H}_1) \cdot d\boldsymbol{\ell} = \int_{\Gamma} (\mathbf{K}_f \times \hat{\mathbf{n}}) \cdot d\boldsymbol{\ell}. \tag{9.32}$$

To see that the right-hand side of (9.32) is the surface current passing through the loop, consider that the vector $\mathbf{K}_f \times \hat{\mathbf{n}}$ lies on S and is perpendicular to $\mathbf{K}_f$. Only the component of $\mathbf{K}_f$ perpendicular to Γ contributes to the line integral on the right-hand side. As (9.32) must hold for any Γ, but only depends on the components of $\mathbf{H}$ tangent to S, we may conclude

$$\mathbf{H}_{2t} - \mathbf{H}_{1t} = \mathbf{K}_f \times \hat{\mathbf{n}} \tag{9.33}$$

where $\mathbf{H}_t$ denotes the tangential part of $\mathbf{H}$, i.e., $\mathbf{H} - H_n \hat{\mathbf{n}}$. If there is no free surface current, then $\mathbf{H}_t$ is continuous. The conditions (9.31) and (9.33) are true for any surface. The most important applications are where S is an interface between different materials.

We will use these boundary conditions in some interesting magnetic field calculations.

Boundary Conditions of Magnetic Potentials

In magnetic field calculations it is often useful to introduce potentials, $\mathbf{A}(\mathbf{x})$ or $\phi_m(\mathbf{x})$. For any magnetic field we may write $\mathbf{B} = \nabla \times \mathbf{A}$, because $\nabla \cdot \mathbf{B} = 0$. The vector potential $\mathbf{A}(\mathbf{x})$ was introduced in Sec. 8.5. The boundary condition on $\mathbf{A}(\mathbf{x})$ is that *the tangential components of* $\mathbf{A}$ *are continuous across* S. Consider again the Amperian loop in Fig. 9.10(b). By Stokes's theorem, the circulation of $\mathbf{A}$ around the loop equals the flux of $\mathbf{B}$ through the loop. As $w \to 0$ the flux approaches 0 with the area of the loop, and the circulation approaches $\int_{\Gamma} (\mathbf{A}_2 - \mathbf{A}_1) \cdot d\boldsymbol{\ell}$; thus the boundary condition on the vector potential is

$$\mathbf{A}_{2t} = \mathbf{A}_{1t}. \tag{9.34}$$

For certain problems a scalar potential $\phi_m(\mathbf{x})$ may be employed. In a region where there is *no free current* we may write $\mathbf{H} = -\nabla \phi_m$, because $\nabla \times \mathbf{H} = 0$. (This technique will be used in Sec. 9.5.) The boundary condition on $\phi_m(\mathbf{x})$ is that ϕ_m *is continuous across S,*

$$\phi_{m2} = \phi_{m1}. \tag{9.35}$$

It can be shown that the continuity of $\phi_m(\mathbf{x})$ implies that the tangential components of $\nabla \phi_m$ are continuous, as required by (9.33) with $\mathbf{K}_f = 0$.[16]

9.4 ■ PROBLEMS INVOLVING FREE CURRENTS AND MAGNETIC MATERIALS

In Examples 1 and 2 we found the $\mathbf{B}$ field due to magnetized objects, but with no free currents. A sphere or cylinder made of magnetized iron is equivalent to

[16]See Exercise 10.

Example 1 or 2, respectively. Now that we have developed the mathematical principles and physical insight required to find the **B** and **H** fields due to free currents, we can solve problems involving both currents and linear magnetic materials. Examples 5 and 6, which have planar and cylindrical symmetry respectively, are problems of this kind.

■─────

EXAMPLE 5   In this example we consider an infinite slab of a conducting material with magnetic susceptibility χ_m, carrying a certain current distribution. Figure 9.11 shows the system. The slab is parallel to the xy plane, between $z = -a$ and $z = a$. It carries a free volume current density $\mathbf{J}_f(z) = (J_0 z/a)\,\hat{\mathbf{i}}$, which is plotted in Fig. 9.12. Above the xy plane the current is out of the page, below it is into the page, and the integrated current density is 0. Outside the slab is vacuum. What are **H**, **M**, and **B**?

We'll solve this problem in two ways: first using Ampère's law and second by the superposition principle. In order to use (9.24) effectively we take advantage of two symmetry considerations. Notice first that the conducting slab may be considered as made up of infinitesimal sheets of free current parallel to the xy plane. In Example 4 of Chapter 8 we found the magnetic field **B** due to a current sheet with surface current density $K\,\hat{\mathbf{i}}$; from Ampère's law the field is $-(\mu_0 K/2)\,\hat{\mathbf{j}}$ above the sheet and $+(\mu_0 K/2)\,\hat{\mathbf{j}}$ below. We are now interested in **H** for a current sheet with *free current* per unit length $\mathbf{K} = K\,\hat{\mathbf{i}}$; by (9.24)

$$\mathbf{H}(\mathbf{x}) = -(K/2)\,\hat{\mathbf{j}} \quad \text{above the current sheet,} \qquad (9.36)$$

$$\mathbf{H}(\mathbf{x}) = +(K/2)\,\hat{\mathbf{j}} \quad \text{below the current sheet.} \qquad (9.37)$$

Similarly, the planar symmetry implies for the slab of current of Fig. 9.11 that the

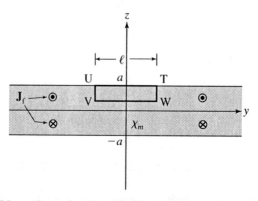

FIGURE 9.11   Magnetic conducting slab. The slab of material has free current density $(J_0 z/a)\,\hat{\mathbf{i}}$, out of the page for $z > 0$ and into the page for $z < 0$. The rectangular loop TUVWT is used to determine the field **H** from Ampère's law.

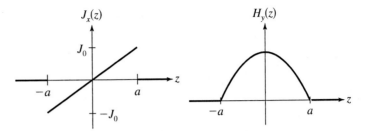

FIGURE 9.12   **Current density J and field H** of the magnetic conducting slab.

field cannot depend on x or y, and **H** must have the form $\mathbf{H}(\mathbf{x}) = H_y(z)\,\hat{\mathbf{j}}$; i.e., **H** has only a y component but it may vary with z.

The second symmetry consideration is that $\mathbf{H} = 0$ outside the slab, i.e., for $z \geq a$ or $z \leq -a$. The reason is that the *net current* flowing in the slab is zero, as is clear from the graph on the left of Fig. 9.12. $\mathbf{H}(\mathbf{x})$ from a sheet of current is independent of the distance from the sheet, so the field above or below the slab, due to the superposition of thin sheets with net current 0, is 0.

Now apply Ampère's law to the closed Amperian path TUVWT in Fig. 9.11. The only contribution to the line integral comes from VW, whose length we call ℓ; **H** is 0 along TU, and **H** is perpendicular to UV and WT. Thus Ampère's law states

$$H_y(z)\ell = \int_z^a \frac{J_0 z'}{a}\ell\,dz' \tag{9.38}$$

from which

$$H_y(z) = \frac{J_0}{2a}(a^2 - z^2) \quad \text{for} \quad -a \leq z \leq a. \tag{9.39}$$

Equation (9.39) holds throughout the slab because (9.38) would be the same even if the Amperian path were extended into the region $z < 0$. Figure 9.12 shows a plot of $H_y(z)$ on the right.

Let us now calculate $H_y(z)$ in another way—by integrating the contributions of the infinitesimal sheets that make up the entire current flow. The idea here is that $d\mathbf{H}(z) = \mp (dK/2)\,\hat{\mathbf{j}}$ above or below the sheet, where the sheet current density is $dK = (J_0 z'/a)dz'$. For a field point z between $-a$ and a

$$H_y(z) = +\frac{1}{2}\int_z^a \frac{J_0 z'}{a}dz' - \frac{1}{2}\int_{-a}^z \frac{J_0 z'}{a}dz', \tag{9.40}$$

which gives again (9.39). Notice that the first term in (9.40), which comes from the current above the field point, equals the second term, which comes from the current below the field point.

Finally we calculate the magnetization, magnetic field, and bound currents. Inside the slab, $\mathbf{M} = \chi_m \mathbf{H}$ and $\mathbf{B} = \mu_0(1 + \chi_m)\mathbf{H}$. Outside the slab both $\mathbf{M}$ and $\mathbf{B}$ are zero. The bound surface current density $\mathbf{K_b} = \mathbf{M} \times \hat{\mathbf{n}}$ is 0 at both upper and lower slab surfaces because $\mathbf{H} = 0$ there, and therefore $\mathbf{M} = 0$ too. The bound volume current density is

$$\mathbf{J_b} = \nabla \times \mathbf{M} = \frac{\chi_m J_0 z}{a} \hat{\mathbf{i}}. \tag{9.41}$$

Note that $\mathbf{J_b}$ is proportional to $\mathbf{J_f}$, so it too is an odd function of z. Therefore the net bound current in the slab is zero, like the net free current.

EXAMPLE 6   What are $\mathbf{H}$, $\mathbf{M}$, and $\mathbf{B}$ for the coaxial cable shown in cross section in Fig. 9.13(a)? The inner and outer conductors carry equal currents in opposite directions. The currents are distributed uniformly in the conductors. The conductors are separated by an insulator with susceptibility χ_m.

Let I be the current, and let the z axis point out of the page. The current densities are

$$\mathbf{J}_{\text{inner}} = \frac{I\hat{\mathbf{k}}}{\pi a^2} \tag{9.42}$$

$$\mathbf{J}_{\text{outer}} = \frac{-I\hat{\mathbf{k}}}{\pi(c^2 - b^2)}. \tag{9.43}$$

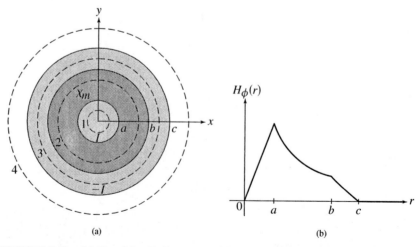

(a)   (b)

FIGURE 9.13   (a) Coaxial cylinders separated by a magnetic material. There is current out of the page in the inner cylinder ($0 < r < a$) and into the page in the outer cylinder ($b < r < c$) both distributed uniformly. Between the cylinders ($a < r < b$) the magnetic susceptibility is χ_m. **(b)** Magnetic field as a function of radius for the coaxial cable.

I is a free current, and by symmetry $\mathbf{H}$ must have the form $\mathbf{H} = H_\phi(r)\hat{\phi}$. Now we apply Ampère's law. Taking the line integrals counterclockwise around each of the four circular paths shown in Fig. 9.13(a) gives

$$H_\phi = \frac{Ir}{2\pi a^2} \quad \text{for} \quad r \le a \quad \text{(Path 1)} \tag{9.44}$$

$$H_\phi = \frac{I}{2\pi r} \quad \text{for} \quad a \le r \le b \quad \text{(Path 2)} \tag{9.45}$$

$$H_\phi = \frac{I(c^2 - r^2)}{2\pi r(c^2 - b^2)} \quad \text{for} \quad b \le r \le c \quad \text{(Path 3)} \tag{9.46}$$

$$H_\phi = 0 \quad \text{for} \quad r \ge c \quad \text{(Path 4)}. \tag{9.47}$$

In deriving these results notice that Path 1 encloses only that part of $+I$ inside radius r, Path 2 encloses all the current $+I$, Path 3 encloses the current $+I$ and also that part of $-I$ between b and r, and, finally, the total current enclosed by Path 4 is 0. Fig. 9.13(b) shows $H_\phi(r)$.

As usual, in the region of the magnetic material $a \le r \le b$, the field $\mathbf{B}$ is $\mu_0(1+\chi_m)\mathbf{H}$; elsewhere it is just $\mu_0\mathbf{H}$. In the magnetic material the magnetization is $\mathbf{M} = \chi_m(I/2\pi r)\hat{\phi}$; this function is curl free so there is no bound current in the volume of the material. There are, however, bound surface currents: At $r = a$ the normal vector out of the magnetic material is $\hat{\mathbf{n}} = -\hat{\mathbf{r}}$, so $\mathbf{K_b} = \mathbf{M} \times (-\hat{\mathbf{r}}) = \chi_m(I/2\pi a)\hat{\mathbf{k}}$; the total bound current at this interface is $\chi_m I$ in the $+\hat{\mathbf{k}}$ direction. At $r = b$, where $\hat{\mathbf{n}} = \hat{\mathbf{r}}$, we have $\mathbf{K_b} = -\chi_m(I/2\pi b)\hat{\mathbf{k}}$; the total bound current at this interface is $\chi_m I$ in the $-\hat{\mathbf{k}}$ direction. The sum of the bound currents is zero.

■————————————

Example 5 is a situation in which there is $\mathbf{J}_b$ but no $\mathbf{K}_b$. Example 6 is a situation in which there is $\mathbf{K}_b$ but no $\mathbf{J}_b$. The general rule is that in a uniform linear magnetic material there can be bound *volume* current only in a region where free current flows. Consider the curl of both sides of the relation $\mathbf{M} = \chi_m\mathbf{H}$; the result is

$$\nabla \times \mathbf{M} = \chi_m \nabla \times \mathbf{H} \quad \text{or} \quad \mathbf{J_b} = \chi_m\mathbf{J_f}. \tag{9.48}$$

9.5 ■ A MAGNETIC BODY IN AN EXTERNAL FIELD: THE MAGNETIC SCALAR POTENTIAL $\phi_m(\mathbf{x})$

We consider next a class of magnetostatics problems that can be conveniently solved by using a magnetic scalar potential $\phi_m(\mathbf{x})$ which is a solution to Laplace's equation. This method has the advantage that we'll be able to bring to bear on magnetostatics the considerable mathematical armamentarium we developed for electrostatics.

The situation in magnetostatics is analogous to the electrostatic case so a quick review is in order. In electrostatics we always have $\mathbf{\nabla} \times \mathbf{E} = 0$, which permits us to write $\mathbf{E} = -\mathbf{\nabla}V$, where V is the electrostatic potential. For uniform linear dielectrics we also have $\mathbf{\nabla} \cdot \mathbf{D} = \epsilon \mathbf{\nabla} \cdot \mathbf{E} = \rho_f$, so that if there is no free charge $\mathbf{\nabla} \cdot \mathbf{E} = 0$, and then V satisfies Laplace's equation.

In magnetostatics we have $\mathbf{\nabla} \times \mathbf{H} = \mathbf{J}_f$, so that if there are no free currents $\mathbf{\nabla} \times \mathbf{H} = 0$. This permits us to introduce the magnetic scalar potential $\phi_m(\mathbf{x})$ where

$$\mathbf{H}(\mathbf{x}) = -\mathbf{\nabla}\phi_m. \tag{9.49}$$

We emphasize that the scalar potential (9.49) can *only* be used in regions of $\mathbf{x}$ where there is no free current. If the magnetic medium is linear then we also have $\mathbf{\nabla} \cdot \mathbf{B} = \mathbf{\nabla} \cdot (\mu\mathbf{H}) = 0$; and if, in addition, μ is constant in the region of interest, then $\mathbf{\nabla} \cdot \mathbf{H} = 0$. Combining this with (9.49) gives

$$\nabla^2\phi_m = 0. \tag{9.50}$$

Equations (9.49) and (9.50) taken together with the boundary conditions (9.31) and (9.33) will enable us to find $\phi_m(\mathbf{x})$ and the resulting fields. We will now use this method to solve two interesting problems.

EXAMPLE 7   What are the fields inside and outside a sphere of permeability μ in a uniform applied field $\mathbf{B} = B_0\hat{\mathbf{k}}$?

The sphere, with radius b, and the magnetic field are shown in Fig. 9.14. We will proceed in the now familiar way by judiciously guessing the form of $\phi_m(\mathbf{x})$ from the symmetry of the problem. If $\phi_m(\mathbf{x})$ satisfies (9.50) and the proper boundary conditions then it will give the correct fields, by the uniqueness theorem. But if our guess for the form of $\phi_m(\mathbf{x})$ is wrong, we will not be able to satisfy the boundary conditions.

Notice first that the scalar potential corresponding to the external field alone is

$$\phi_m^0(\mathbf{x}) = -\frac{B_0}{\mu_0}z = -\frac{B_0}{\mu_0}r\cos\theta. \tag{9.51}$$

This suggests that $\phi_m(\mathbf{x})$ may need no other angular dependence than $\cos\theta$. The only two solutions to Laplace's equation in spherical coordinates that are proportional to $\cos\theta$ are $r\cos\theta$ and $\cos\theta/r^2$. The first corresponds to a uniform field, and the second to a point dipole at the origin. Thus we are led to try

$$\phi_m^{\text{int}}(r, \theta) = \alpha r\cos\theta \quad \text{for} \quad r < b \tag{9.52}$$

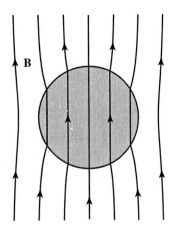

FIGURE 9.14   Example 7. A solid sphere is placed in a uniform applied magnetic field, but the total magnetic field is not uniform because of the effect of the material. The figure shows the **B**-field lines for the case of a paramagnetic sphere.

$$\phi_m^{\text{ext}}(r, \theta) = -\frac{B_0}{\mu_0} r \cos\theta + \beta \frac{\cos\theta}{r^2} \quad \text{for} \quad r > b. \tag{9.53}$$

The scalar potential for a dipole **m** at the origin is

$$\phi_m^{\text{dipole}}(\mathbf{x}) = \frac{1}{4\pi} \frac{\mathbf{m} \cdot \hat{\mathbf{r}}}{r^2}, \tag{9.54}$$

so the second term in (9.53) corresponds to a dipole $\mathbf{m} = 4\pi\beta\hat{\mathbf{k}}$. A dipole term cannot be included in (9.52) because it would be singular at the origin.

The two constants α and β will now be determined from the two boundary conditions. One boundary condition is the continuity of the tangential component H_θ of **H** at $r = b$. In terms of the scalar potential this condition is

$$-\frac{1}{b}\left(\frac{\partial \phi_m^{\text{int}}}{\partial \theta}\right)_{r=b} = -\frac{1}{b}\left(\frac{\partial \phi_m^{\text{ext}}}{\partial \theta}\right)_{r=b}, \tag{9.55}$$

which gives

$$b^3\alpha - \beta = -b^3 B_0/\mu_0. \tag{9.56}$$

(The same equation could be obtained from the condition that ϕ_m is continuous.) The other boundary condition is the continuity of the normal component B_r of **B** at $r = b$. Because **B** is $\mu\mathbf{H}$ inside the sphere and $\mu_0\mathbf{H}$ outside, the condition is

$$-\mu\left(\frac{\partial \phi_m^{\text{int}}}{\partial r}\right)_{r=b} = -\mu_0\left(\frac{\partial \phi_m^{\text{ext}}}{\partial r}\right)_{r=b}, \tag{9.57}$$

which gives

$$\kappa_m b^3 \alpha + 2\beta = -b^3 B_0/\mu_0. \tag{9.58}$$

In (9.58), $\kappa_m \equiv \mu/\mu_0$ is the relative permeability.

The solutions to (9.56) and (9.58) are

$$\alpha = \frac{-3B_0}{(\kappa_m + 2)\mu_0} \quad \text{and} \quad \beta = \frac{(\kappa_m - 1)B_0 b^3}{(\kappa_m + 2)\mu_0}. \tag{9.59}$$

Finally, substituting these into (9.52) and (9.53) gives the potentials

$$\phi_m^{\text{int}}(r, \theta) = \frac{-3B_0}{(\kappa_m + 2)\mu_0} r \cos\theta, \tag{9.60}$$

$$\phi_m^{\text{ext}}(r, \theta) = -\frac{B_0}{\mu_0} r \cos\theta + \frac{(\kappa_m - 1)B_0 b^3}{(\kappa_m + 2)\mu_0} \frac{\cos\theta}{r^2}. \tag{9.61}$$

We note again that the magnetic scalar potential is continuous at $r = b$ in accord with the boundary condition (9.35).

The internal fields.   The calculations of the fields in the region $r < b$ are

$$\mathbf{H}^{\text{int}} = -\nabla \phi_m^{\text{int}} = \frac{3B_0}{(\kappa_m + 2)\mu_0} \hat{\mathbf{k}}, \tag{9.62}$$

$$\mathbf{B}^{\text{int}} = \mu \mathbf{H}^{\text{int}} = \frac{3\kappa_m B_0}{(\kappa_m + 2)} \hat{\mathbf{k}}, \tag{9.63}$$

$$\mathbf{M} = \chi_m \mathbf{H}^{\text{int}} = \frac{3(\kappa_m - 1)B_0}{(\kappa_m + 2)\mu_0} \hat{\mathbf{k}}. \tag{9.64}$$

The interior fields are uniform, i.e., independent of $\mathbf{x}$, and point in the $+\hat{\mathbf{k}}$ direction. The magnetization is also uniform. $\mathbf{M}$ points in the $+\hat{\mathbf{k}}$ direction if $\kappa_m > 1$, a paramagnet; or, $\mathbf{M}$ points in the $-\hat{\mathbf{k}}$ direction if $\kappa_m < 1$, a diamagnet.

If the sphere were nonmagnetic, i.e., $\kappa_m = 1$, then $\mathbf{B}^{\text{int}} = \mathbf{B}_0$ as expected. A paramagnetic sphere has $\kappa_m > 1$ and so $B^{\text{int}} > B_0$; a paramagnet concentrates the magnetic field lines. (Figure 9.14 illustrates this case.) As κ_m increases $\mathbf{B}^{\text{int}}$ increases until, in the limit of large κ_m, $\mathbf{B}^{\text{int}} \to 3\mathbf{B}_0$. This limit is easy to reach because for many ferromagnetic substances κ_m is larger than 10^3. A diamagnetic sphere has $\kappa_m < 1$ and so $B^{\text{int}} < B_0$; a diamagnet expels the magnetic field lines. The extreme limit $\kappa_m \to 0$ is called a *perfect diamagnet*. In this case the field $\mathbf{B}^{\text{int}}$ is 0, i.e., the magnetic field does not penetrate into the material. A superconductor is essentially a perfect diamagnet.

The external fields.   In the external region $r > b$ we have

$$\mathbf{B}^{\text{ext}} = \mu_0 \mathbf{H}^{\text{ext}} = -\mu_0 \nabla \phi_m^{\text{ext}} = B_0 \hat{\mathbf{k}} + \frac{\kappa_m - 1}{\kappa_m + 2} \left(\frac{b^3}{r^3} \right) B_0 (2\hat{\mathbf{r}} \cos\theta + \hat{\boldsymbol{\theta}} \sin\theta). \tag{9.65}$$

The first term in (9.65) is the external field that we specified. The second term is the field of a point dipole at the origin with $\mathbf{m}_{total} = (4/3)\pi b^3 \mathbf{M}$, the total magnetic moment of the sphere. This shows, as we also found in Example 1, that the external field produced by a sphere with constant magnetization is the same as that of a point dipole at the origin.

We now ask a natural question about this problem. All the fields, **H**, **B**, and **M**, have zero divergence and zero curl both inside and outside the sphere. Yet the fields themselves are nonzero and interesting. What and where are the sources of the fields? The answer is that the sources of the applied field are essentially at infinity—picture large north and south magnetic poles far away—and the other sources are bound surface currents at $r = b$.

Magnetic Shielding

Now we turn to an example with practical importance. It is sometimes desirable to produce a magnetic-field-free region, say a laboratory room or a box for electronic circuit components. *Magnetic shielding* means surrounding the volume of interest with a layer of material with high permeability μ. A material used for this purpose is Mumetal, the alloy $Ni_{0.77}Fe_{0.16}Cu_{0.05}Cr_{0.02}$ with $\mu/\mu_0 = 10^5$. If μ is large and the layer is thick then any external **B** concentrates in the magnetic layer so that the field inside the cavity can, in principle, be made arbitrarily small. The next example illustrates magnetic shielding for spherical geometry.

EXAMPLE 8   What is the field inside a spherical cavity of radius R_i that is surrounded by a concentric spherical shell of permeability μ, if there is an external applied field $\mathbf{B} = B_0 \hat{\mathbf{k}}$?

Figure 9.15 shows the arrangement. The magnetic layer—the shield—extends from radius R_i to R_o. The problem is to find the field inside radius R_i as a function of the relative permeability $\kappa_m = \mu/\mu_0$.

This problem is similar to Example 7 because it has the same applied field and a spherical body. The key difference is that here we must find the magnetic scalar potential in *three* regions. Because the only angular dependence is proportional to $\cos\theta$ we are now led to try

$$\phi_m^{int}(r, \theta) = \alpha r \cos\theta \quad \text{for} \quad r < R_i \tag{9.66}$$

$$\phi_m^{betw}(r, \theta) = \gamma r \cos\theta + \delta \frac{\cos\theta}{r^2} \quad \text{for} \quad R_i < r < R_o \tag{9.67}$$

$$\phi_m^{ext}(r, \theta) = -\frac{B_0}{\mu_0} r \cos\theta + \beta \frac{\cos\theta}{r^2} \quad \text{for} \quad r > R_o. \tag{9.68}$$

The four coefficients α, β, γ, and δ are determined as before from the boundary conditions. First, B_r is continuous at the interfaces, $r = R_i$ and $r = R_o$,

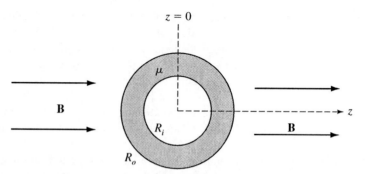

FIGURE 9.15 Example 8: Magnetic shielding. A magnetic material fills a spherical shell with inner radius R_i and outer radius R_o. To what extent does the external field penetrate into the interior?

$$-\mu_0 \frac{\partial \phi_m^{int}}{\partial r} = -\mu \frac{\partial \phi_m^{betw}}{\partial r} \quad \text{at} \quad r = R_i$$

$$-\mu \frac{\partial \phi_m^{betw}}{\partial r} = -\mu_0 \frac{\partial \phi_m^{ext}}{\partial r} \quad \text{at} \quad r = R_o.$$

Also, H_θ is continuous at the interfaces, or equivalently, and more simply, ϕ_m is continuous there,

$$\phi_m^{int} = \phi_m^{betw} \text{ at } r = R_i$$

$$\phi_m^{betw} = \phi_m^{ext} \text{ at } r = R_o.$$

These conditions give the following four linear equations in the unknowns

$$R_i{}^3\alpha - \kappa_m R_i{}^3\gamma + 2\kappa_m\delta = 0 \tag{9.69}$$

$$R_i{}^3\alpha - R_i{}^3\gamma - \delta = 0 \tag{9.70}$$

$$2\beta + \kappa_m R_o{}^3\gamma - 2\kappa_m\delta = -R_o{}^3 B_0/\mu_0 \tag{9.71}$$

$$\beta - R_o{}^3\gamma - \delta = R_o{}^3 B_0/\mu_0. \tag{9.72}$$

The solutions are

$$\alpha = \frac{-9\kappa_m}{(2\kappa_m + 1)(\kappa_m + 2) - 2(R_i/R_o)^3(\kappa_m - 1)^2} \left(\frac{B_0}{\mu_0}\right) \tag{9.73}$$

$$\gamma = \left(\frac{2\kappa_m + 1}{3\kappa_m}\right)\alpha \tag{9.74}$$

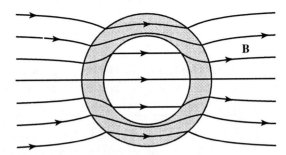

FIGURE 9.16 **Field lines for magnetic shielding.**

$$\delta = \left(\frac{\kappa_m - 1}{3\kappa_m}\right) R_i{}^3 \alpha \tag{9.75}$$

$$\beta = \frac{(2\kappa_m + 1)(\kappa_m - 1)(R_o{}^3 - R_i{}^3)}{-9\kappa_m}\alpha. \tag{9.76}$$

In taking ϕ_m to be solutions of Laplace's equation we assumed that $\nabla \times \mathbf{H} = 0$ and $\nabla \cdot \mathbf{H} = 0$. Because there is no free current it is clear that $\nabla \times \mathbf{H} = 0$. Straightforward calculation from the fields in the three regions shows that $\nabla \cdot \mathbf{H} = 0$ everywhere, as well.

For the sake of completeness all the coefficients have been given. But we are particularly interested in α, because the field in the cavity is $\mathbf{B}^{\text{int}} = -\mu_0\alpha\hat{\mathbf{k}} = B_z^{\text{int}}\hat{\mathbf{k}}$. In the limit of large κ_m we have, from (9.73),

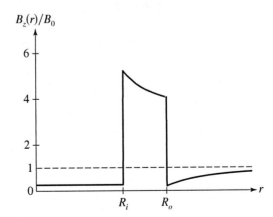

FIGURE 9.17 **Field in a shield.** The plot shows $B_z(r)/B_0$ on the xy plane for Example 8, with $\mu/\mu_0 = 20$. Here B_0 is the asymptotic field strength. Note that the field inside the cavity is smaller than the asymptotic field. In the limit $\mu \to \infty$ the interior field $\to 0$.

$$B_z^{\text{int}} \approx \frac{9 B_0}{2\kappa_m[1 - (R_i/R_o)^3]} \propto \frac{B_0}{\kappa_m} \quad (\text{for } \kappa_m \gg 1). \tag{9.77}$$

Equation (9.77) means that if κ_m is large then the interior region $r < R_i$ is *magnetically shielded* from external fields. Figure 9.16 illustrates how the field lines of the external field concentrate in the magnetic shield. As a numerical example of the effectiveness of this shielding: The magnetic field inside a room whose sides are about 4 m, and which is shielded by a 1 cm thick layer of material whose $\kappa_m = 10^4$, would be about 0.03 of the field outside.

Figure 9.17 shows $B_z(r)$ on the midplane ($z = 0$) of the sphere, for $\mu/\mu_0 = 20$, showing quantitatively how **B** concentrates in the shield.

■────────

9.6 ■ FERROMAGNETISM

Ferromagnetism is the magnetic phenomenon most familiar from everyday experience. Horseshoe magnets, bar magnets, refrigerator magnets, and compass needles are all ferromagnets. Ferromagnets are also key components in motors and generators, transformers, loudspeakers, microwave ovens, and other common household and workplace devices.

Ferromagnetic materials have a wide range of physical properties, but one key characteristic is that they are easy to magnetize. Even a small magnetic field applied to a ferromagnet can result in its developing a large magnetization. A ferromagnet can have large magnetization even in the absence of an externally applied field—that's what a permanent magnet is.

Only three elements, Fe, Co, and Ni, are ferromagnetic at room temperature, although many alloys are ferromagnetic at room temperature and several elements become ferromagnetic at lower temperatures. Indeed, one of the important characteristics of ferromagnetism is its temperature dependence. Each ferromagnet has a characteristic temperature, called the Curie temperature (T_C), above which it is no longer ferromagnetic but becomes paramagnetic. For Fe, $T_C = 1043$ K; for Co, $T_C = 1388$ K, and for Ni, $T_C = 627$ K. Two elements that are ferromagnetic below room temperature are Gd ($T_C = 292$ K) and Dy ($T_C = 88$ K). A ferromagnetic material undergoes a magnetic phase transformation—a phenomenon of great interest in statistical mechanics—at the Curie temperature; i.e., its magnetization behaves in a characteristic way as T approaches T_C from below and its susceptibility behaves in a characteristic way as T approaches T_C from above. This contrasts with the temperature dependences of diamagnetism and paramagnetism, which were discussed previously.

What is the microscopic source of the large magnetization of ferromagnets? Why is the magnetization permanently retained even in the absence of an externally applied field? Why does ferromagnetism disappear when the temperature is raised?

The microscopic sources of magnetization in ferromagnets are the individual atoms, which have permanent magnetic moments due to unpaired spins. An iron atom, for example, has two unpaired electron spins, in the 3d shell, so that the atom has a net magnetic moment of $2\mu_B$. If the magnetic moments in neighboring atoms interacted only via the classical dipole-dipole interaction, then they would align themselves antiparallel, like this, $\uparrow\downarrow$, because that is the lowest energy state for two classical dipoles. However, in a ferromagnetic material there is a much stronger interaction between neighboring atoms, called the *exchange interaction*, and of quantum mechanical origin, with the property that the lowest energy state is for the atomic magnetic moments to be parallel, like this, $\uparrow\uparrow$.

A macroscopic piece of unmagnetized Fe consists of many magnetic domains, each containing from 10^{17} to 10^{21} atoms. In each domain essentially all the Fe atoms have their moments aligned, even at room temperature. Even a single crystal of Fe has many domains of this kind. If the surface of the material is suitably etched the domains, which have volumes of 10^{-8} to 10^{-12} m^3, can be seen under a low power microscope. An unmagnetized piece of Fe has no net magnetic moment because the domains are randomly oriented. It is interesting to calculate the magnetization M in a domain. The number density of Fe atoms in the solid is $n = 8.5 \times 10^{28}$ m^{-3} so that $M = 2\mu_B n = 1.6 \times 10^6$ A/m. This corresponds to a large internal magnetic field as we can see by finding the field inside an Fe sphere uniformly magnetized to the saturation magnetization. That field, according to (9.17) of Example 1, is $B = \frac{2}{3}\mu_0 M = 1.3$ T.

When an external H field is applied to unmagnetized Fe, the domains gradually become aligned in the direction of the field. (If the Fe sample is placed in a detection coil with an audio amplifier, the flipping of domains is audible, a demonstration known as the Barkhausen effect.) More favorably oriented domains grow at the expense of less favorably oriented ones as the boundaries between domains move, and unaligned domains can rotate in the direction of the applied field. In single crystals the alignment occurs easily, but generally domain growth can be slow, depending on the grain structure of the sample. As a result of the orientation of domains, the piece of Fe becomes magnetized in the direction of the applied field. This magnetized state is energetically favored because of the exchange interaction, so the Fe *remains magnetized* when the external field is removed. Because ferromagnetism is a result of magnetic ordering it is not surprising that raising the temperature sufficiently destroys ferromagnetism. If kT is large compared to the energy of the exchange interaction then thermal fluctuations eliminate the alignment. The precise calculation of Curie temperatures and magnetic behavior near T_C is a large and complex subject.

9.6.1 ■ Measuring Magnetization Curves: The Rowland Ring

To characterize the properties of a ferromagnetic material, we measure its magnetization curve, which is the curve of M versus H or, more commonly, B versus $\mu_0 H$. These are equivalent ways to express the properties of the magnetic material because the quantities are related by $\mathbf{B} = \mu_0(\mathbf{H} + \mathbf{M})$, according to (9.23).

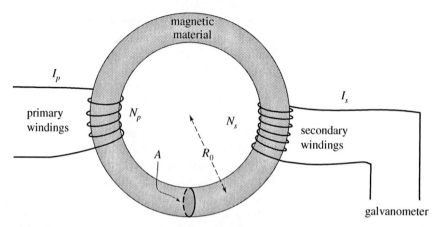

FIGURE 9.18 **A Rowland ring** for measuring magnetization curves. When the current in the primary windings is increased incrementally there are corresponding incremental increases ΔH and ΔB in the toroid of magnetic material. The amount of charge that flows through a galvanometer is proportional to ΔB.

    Figure 9.18 shows a Rowland ring, a standard device for measuring magnetization curves. The toroid of magnetic material is wrapped with N_p turns of primary windings. A current I_p is sent through them. That produces in the material the field $\mathbf{H} = H\hat{\theta} = (\mu_0 N_p I_p / 2\pi R)\hat{\theta}$, by Ampère's law. N_s turns of secondary windings are also wrapped around the toroidal core, which are connected to a galvanometer.

    The principle involved in measuring magnetization curves with a Rowland ring is that H is applied by the experimenter—H is the independent variable—and in response to H the material develops a magnetization M and an internal magnetic field B. Assume the material is isotropic so that all these fields are parallel, in the azimuthal direction. In practice H is incremented in known steps ΔH, by changing I_p in steps ΔI_p. This produces changes ΔB of the magnetic field inside the material. What is actually measured is the amount of charge ΔQ that is driven through the galvanometer by electromagnetic induction as a result of the change ΔB of magnetic field. We defer explanation of electromagnetic induction until Chapter 10 but anticipate the result (Exercise 30 of Chap. 10), which is that ΔB is proportional to ΔQ and can therefore be determined from the measurement of ΔQ. Thus, one starts with $H = 0$ and an unmagnetized sample. The magnetization curve is then built up by measuring $\Delta B / \Delta H$ for the incremental changes described. Magnetization curves are usually plotted as B versus $\mu_0 H$, because then both ordinate and abscissa have the same units. This choice of axes has the advantage that the slope of a straight line, or the derivative of the magnetization curve, is dimensionless. Notice that if the core material were removed from the Rowland ring then the magnetic field would be $\mu_0 H$ in the empty toroid.

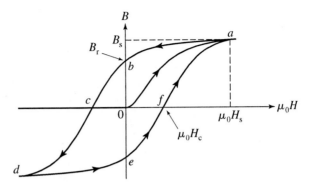

FIGURE 9.19   Ferromagnetic behavior. The magnetic field B is plotted versus $\mu_0 H$ in the material. Some important parameters are the saturation magnetic field B_s, the remanent field B_r at $\mu_0 H = 0$, and the coercive force H_c.

9.6.2 ■ Magnetization Curves of Ferromagnetic Materials

For diamagnetic and paramagnetic materials the magnetization curves are straight lines, with slope of B versus $\mu_0 H$ equal to $\kappa_m = \mu/\mu_0 = 1 + \chi_m$, where κ_m is the *relative permeability*, a dimensionless quantity. For paramagnets κ_m is slightly larger than 1 and for diamagnets it is slightly less than 1, as we saw in Table 9.2. The magnetization curves are straight lines because in these linear materials B is directly proportional to H. In other words, the susceptibility χ_m and therefore the permeability μ are independent of H.

Ferromagnetic materials are nonlinear so their magnetization curves are complicated. Figure 9.19 shows a schematic plot of B versus $\mu_0 H$ for a typical ferromagnetic material. Starting at the origin with an unmagnetized sample, the magnetization curve follows the path $0a$ as the experimenter increases H. The initial increase in B is the result of domain growth. Favorably oriented magnetic domains, i.e., those with **M** pointing in or near the direction of **H**, grow at the expense of less favorably oriented ones, and larger domains grow at the expense of smaller ones. The curve eventually flattens out as it approaches a, where the material reaches its saturation magnetization B_s; the corresponding abscissa at saturation is $\mu_0 H_s$. Further increase in H does not result in increasing magnetization. The reason the curve flattens out approaching saturation is because the dominant mechanism in this region is rotation of domains to line up more closely with H, a difficult process.

If the experimenter then decreases H, the magnetization decreases but the dependence follows a different path—the path ab—to the point where $H = 0$. At $H = 0$ the material is left with the *remanent field* B_r. This B_r is the field of the magnetization that remains in the sample with zero applied field. For a good permanent magnet B_r is large, almost as large as B_s. Continuing from b to c, H is becoming more negative, i.e., it is increasing in the direction opposite to its initial direction, so that the sample is becoming demagnetized. When H reaches the value H_c, B is zero. H_c is called the "coercive force," although its units are not those of force but rather A/m. If H is then further increased in the negative direction the curve eventually reaches the point d, at which the sample is again magnetized to saturation but this time in the opposite direction from that at a.

The entire cycle $abcdefa$ is called the *hysteresis loop* for the material. If H is varied slowly enough then the path $defa$ is symmetric with the path $abcd$, and the loop can be reproducibly retraced many times. The area inside the loop, $\oint \mathbf{B} \cdot d\mathbf{H}$, is the work that must be supplied to bring unit volume (1 m³) of the material through a cycle. This work goes into heating the magnet.

Figure 9.19 is a qualitative schematic drawing. But it is important to know that the magnetic properties of real ferromagnets vary over a wide range. This subject is large, of course, but three examples in the table below give an illustration of the range.

Material	B_s	H_c	Energy loss per hysteresis cycle
Commercial Fe (99.8% pure)	2.2 T	80 A/m	250 J/m³
Fe (99.95% pure)	2.2 T	4 A/m	30 J/m³
Supermalloy (Ni, Fe, Mo)	0.8 T	0.3 A/m	2 J/m³

Ref.: *AIP Handbook,* Table 5f-11, 3rd ed., 1982.

It is also important to know that the range of the horizontal axis ($\mu_0 H$) is much smaller than the range of the vertical axis (B), because in a ferromagnet, a little H goes a long way toward producing magnetization.

The quantitative properties of ferromagnetic materials are important in the design of applications. For the cores of transformers, ferromagnetic materials are used for which H_c is small (1–10^2 A/m) and for which the energy loss per cycle is small. The idea is that the internal field can be easily reversed and the transformer can be run at, say, 60 Hz, without overheating the core. Hysteresis curves of such materials are very narrow. For the core of relay solenoids, ferromagnetic materials are chosen that have a small B_r so that when the current in the energizing coil is turned off the relay will open easily; materials with this property are called "soft" magnets. For permanent magnets the desirable properties are large B_r and large H_c (10^3–10^5 A/m). Materials with large B_r are called "hard" magnets. A large H_c prevents the magnetization from being disturbed by stray fields, a property that is important in magnetic tapes, for example, on which information is stored in small ferromagnets with large H_c.

9.6.3 ■ The Permeability of a Ferromagnetic Material

A ferromagnetic material is often characterized by its permeability μ or its relative permeability $\kappa_m = \mu/\mu_0$. For ferromagnetic materials, μ and κ_m are functions of H. This is in contrast to diamagnetic and paramagnetic materials, in which they are constants independent of H.

As we've learned from Fig. 9.19 and the associated discussion, the B versus $\mu_0 H$ magnetization curves of ferromagnets have complicated shapes, of the form $B = B(H)$. For ferromagnets we define $\mu(H)$ and $\kappa_m(H)$ by

$$\mu(H) = \frac{B(H)}{H} \quad \text{and} \quad \kappa_m(H) = \frac{B(H)}{\mu_0 H}. \tag{9.78}$$

Therefore, in Fig. 9.19, κ_m is the slope of the straight line drawn from the origin to the point $(\mu_0 H, B)$. Correspondingly, if we had plotted the magnetization curve as B versus H, the slope of the straight line drawn from the origin to the point (H, B) would be μ. To characterize fully the permeability of a ferromagnet would require the whole permeability function, but in practice the initial and maximum values of κ_m are given for the path Oa, over which the ferromagnet is initially magnetized. For the three materials in the table above these quantities are:

Material	κ_m(initial)	κ_m(max)
Commercial iron	200	5,000
Pure iron	10,000	200,000
Supermalloy	10^5	10^6

In some contexts μ is defined as $\Delta B / \Delta H$, where ΔH is a small change in H. As we saw previously, this ratio is the quantity measured with a Rowland ring. For diamagnetic and paramagnetic materials this definition is equivalent to that in (9.78). But for ferromagnets the two definitions are very different things. For example, at saturation $B(H)/H$ is usually very large, but $\Delta B / \Delta H$, which is essentially the slope at saturation, is near zero.

Figure 9.20 shows actual magnetization curves for some ferromagnetic materials. This plot shows $\log B$ versus $\log H$, so that κ_m is constant, as labeled,

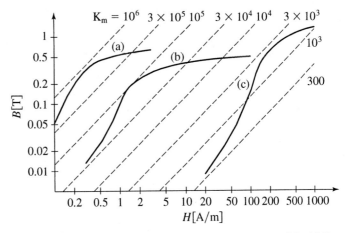

FIGURE 9.20 **Magnetization curves** of some ferromagnetic materials. (a) Supermalloy, (b) Mumetal, (c) Pure Iron Sheet. The diagonal lines correspond to constant values of the relative permeability $\kappa_m = \mu/\mu_0$.

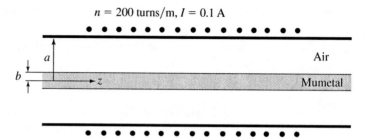

FIGURE 9.21   Example 9. A pencil of Mumetal in a solenoid.

on the dashed straight lines. This plot also shows clearly how κ_m varies as the field changes. For example, for Mumetal (curve b), κ_m starts at 30,000, then increases to 100,000, and then decreases to below 1,000. Notice also the difference in ranges on the abscissa and ordinate, mentioned previously; the value $H = 10$ A/m corresponds to the value $\mu_0 H = 4\pi \times 10^{-6}$ T.

We now give two examples which use these magnetization curves on prototype problems of magnet design.

EXAMPLE 9   A long cylindrical solenoid of radius $a = 5$ cm is wound with $n = 200$ turns/m of wire carrying current 0.1 A. There is a cylindrical core of Mumetal with radius $b = 1$ cm on the axis of the solenoid. What are **H**, **B**, and **M** everywhere? Also, what is the magnetic flux through a cross section of the core, and through a cross section of the solenoid outside the core?

The system is shown in Fig. 9.21. All the fields point in the z direction, by symmetry. Outside the solenoid H, B, and M are 0.

Everywhere inside the solenoid ($0 \leq r < a$) we have $H = nI = 20$ A/m, by Ampère's law. In the air-filled region ($b < r < a$) $B_1 = \mu_0 H = 2.5 \times 10^{-5}$ T, and $M_1 = 0$.

Inside the Mumetal ($0 \leq r < b$) we read from the magnetization curve in Fig. 9.20 that $B_2 = 0.45$ T. Note that the solenoid is operating near the saturation region of Mumetal. Then $M_2 = B_2/\mu_0 - H = 3.6 \times 10^5$ A/m.

The magnetic flux through a cross section of the Mumetal is $\Phi_2 = B_2 \pi b^2 = 1.4 \times 10^{-4}$ Wb. The flux through a cross section of the rest of the solenoid is $\Phi_1 = B_1 \pi (a^2 - b^2) = 1.9 \times 10^{-7}$ Wb. We see that the magnetic flux is overwhelmingly concentrated in the magnetic material, even though the Mumetal is only a thin stick of a core. This same effect occurs in magnetic shielding.

EXAMPLE 10 Electromagnet design.   Figure 9.22 shows a symmetric electromagnet with an iron yoke wound with two identical coils, each with $N = 500$ turns carrying current I. What I is needed to produce a field in the gap of $B_g =$

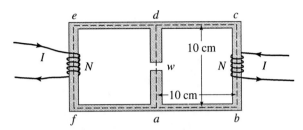

FIGURE 9.22 **Example 10.** A magnetic circuit.

1.0 T if the gap width is $w = 1.0$ cm? What I would be needed for the same B_g if $w = 0.5$ cm?

Our strategy will be to apply Ampère's law to the path $abcda$. Let H be denoted by H_F along the segments ab, bc, and cd; by H_g in the gap; and by H'_F along the part of da not in the gap. Then

$$\oint_{abcda} \mathbf{H} \cdot d\boldsymbol{\ell} = (0.30 \,\text{m})H_F + (0.10 \,\text{m} - w)H'_F + H_g w = 500 \, I. \qquad (9.79)$$

The integral on the left is called the magnetomotive force (mmf) although it is not a force because its unit is the ampere. In this expression w is known; H_F, H'_F, and H_g will be found from magnetic properties and symmetry; we seek I.

In the gap $H_g = B_g/\mu_0 = 8.0 \times 10^5$ A/m. Because the normal component of **B** is continuous at the gap surfaces, $B'_F = B_g = 1.0$ T. Reading from the curve for pure iron in Fig. 9.20 we find that $H'_F = 500$ A/m. Because of the bilateral symmetry we reason that half the magnetic flux in da comes from each side of the yoke, so $B_F = \frac{1}{2}B'_F = 0.50$ T. Reading again from Fig. 9.20 we find that $H_F = 185$ A/m. Substituting these values into (9.79) gives

$$(185)(0.30) + (500)(0.10 - w) + 8.0 \times 10^5 w = 500 \, I$$

with w in m and I in A. The dominant term on the left is the one from the path through the gap; we say that "the gap has much the largest contribution to the mmf."

For $w = 1.0$ cm, I must be 16 A. For $w = 0.5$ cm, half as wide, I must be 8.2 A.

In this calculation we tacitly made several simplifications, e.g., neglecting leakage of magnetic flux at the gap, and assuming that the field is uniform across cross sections of the yoke. It is not hard to imagine that without such simplifications, or without the bilateral symmetry, this would be a much harder calculation.

The general magnet-design problem is to construct a magnet whose field has a given configuration. Such magnets are important for many applications, e.g., industrial, medical, particle accelerators and detectors, laboratory experiments, and magnetic confinement of plasmas. The principles involved are the ones we have discussed, but the details required for an application can make the problem very difficult.

FURTHER READING

1. J. J. Roche, *Am. J. Phys.* **68**, 438–449 (2000).
2. *The Feynman Lectures*, Vol. II, Chapter 34.
3. Magnetism in matter is a major topic in solid-state physics, discussed in textbooks of that subject, such as C. Kittel, *Introduction to Solid State Physics*, 7th ed. (Wiley, New York, 1996).

EXERCISES

Sec. 9.1. Atomic Dipoles

9.1. In the Bohr model the radius of the first Bohr orbit is 0.529×10^{-10} m. Calculate the orbital frequency if the orbital magnetic moment is 1 Bohr magneton. Compare the corresponding electron speed to the speed of light. [Answer: 1/137]

Sec. 9.2. Magnetization and Bound Current

9.2. Use a computer to plot the field $B(z)$ on the axis of a uniformly magnetized cylinder with $\mathbf{M} = M_0 \hat{\mathbf{k}}$, as a function of z/a, where the cylinder has radius a and length $3a$.

Sec. 9.3. Ampère's Law and **H**; Susceptibilities

9.3. Estimate the maximum magnetization (**M**) of iron. Assume the atomic dipole moment of an iron atom is due to two aligned (unpaired) electron spins. The mass density of iron is $7.87 \times 10^3 \, \text{kg/m}^3$, and the atomic mass is 55.85 u.

9.4. A long cylindrical bar magnet has a uniform magnetization 5×10^5 A/m. What is **B** in the middle of the bar? (The result is the same for any cross-sectional shape.) What is **B** at the end of the bar if the cross section of the cylinder is a circle? (Hint: See Chapter 8, Exercise 13.)

9.5. A simple theoretical model of the susceptibility χ_m of a diamagnetic material, based on (9.5), is

$$\chi_m = -\frac{ne^2 r_0^2}{6m_e} \mu_0 Z,$$

where r_0 is the RMS atomic radius, n is the atomic density of the material, and Z is the number of electrons in the atom. (The factor of 2/3 relative to (9.5) is because not all electron orbits are orthogonal to the field.) For this model calculate r_0 for copper, which has mass density $8.92 \times 10^3 \, \text{kg/m}^3$ and atomic mass 63.55 u. Does your result make sense?

9.6. A simple theoretical model of the susceptibility χ_m of a paramagnetic material, based on (9.6), is

$$\chi_m = \mu_0 n \left(\frac{m^2}{3kT} - \frac{e^2 r_0^2}{6m_e} Z \right),$$

where m is the permanent dipole moment of the atom. Other parameters are as in the previous exercise. From this relation calculate m for aluminum, which has mass density 2.70×10^3 kg/m^3 and atomic mass 26.98 u. Assume r_0 is 0.5×10^{-10} m. Express the result in Bohr magnetons. Does your result make sense? What should be used for the temperature T?

9.7. The magnetic susceptibilities of the first three noble gases are $\chi_m(\text{He}) = -1.1 \times 10^{-9}$, $\chi_m(\text{Ne}) = -3.9 \times 10^{-9}$, and $\chi_m(\text{Ar}) = -1.1 \times 10^{-8}$. The magnetic susceptibilities of the first three rare earth elements are $\chi_m(\text{La}) = 5.33 \times 10^{-5}$, $\chi_m(\text{Ce}) = 1.50 \times 10^{-3}$, and $\chi_m(\text{Pr}) = 3.34 \times 10^{-3}$. Explain these results.

9.8. Consider a large piece of dielectric material that has a constant electric field $\mathbf{E}_1$ and polarization $\mathbf{P}_1$, which are parallel to each other; as always (6.24) holds, $\mathbf{D}_1 = \epsilon_0 \mathbf{E}_1 + \mathbf{P}_1$.

   (a) If a thin disk-shaped cavity is hollowed out of the material with the disk axis parallel to $\mathbf{E}_1$, what are the fields $\mathbf{E}_0$ and $\mathbf{D}_0$ in the cavity near the middle of it?

   (b) If a thin needle-shaped cavity is hollowed out of the material with the needle axis parallel to $\mathbf{E}_1$, what are the fields $\mathbf{E}_0$ and $\mathbf{D}_0$ in the cavity near the middle?

   (c) If a small spherical cavity is hollowed out, what are $\mathbf{E}_0$ and $\mathbf{D}_0$ in the cavity?

   Summarize your results by completing the table:

Cavity shape	$\mathbf{E}_0$	$\mathbf{D}_0$
Disk		
Sphere		$\mathbf{D}_1 - \frac{2}{3}\mathbf{P}_1$
Needle		

This is a combination of Example 1 of Chapter 6 and Example 3 of Chapter 9. It is interesting to compare the results of this exercise to the table in Example 3. Why are the results symmetric?

9.9. The magnetic field of a long solenoid, far from the ends, is 0 outside and $\mu_0 n I \hat{\mathbf{k}}$ inside. Prove that this field satisfies the appropriate field equations and boundary conditions.

9.10. Show that if $\phi_m(\mathbf{x})$ is continuous at a surface S, where ϕ_m is some scalar function, then the tangential components of $\nabla\phi_m$ are continuous at S.

Sec. 9.4. Free Current and Materials

9.11. Calculate $\nabla \times \mathbf{H}$ for the field in Example 5, and sketch the field lines.

9.12. A conducting slab, parallel to the xy plane and extending from $z = -a$ to $z = a$, carries a uniform free current density $\mathbf{J}_f = J_0 \hat{\mathbf{i}}$. The magnetic susceptibility is 0 in the slab and χ_m outside. Determine the magnetic field and the bound current distribution.

9.13. Consider an iron ring (torus) with the radius of the ring $R = 0.2$ m and the radius of the cross section of the ring $r = 1$ cm. Around the ring there are 1000 turns of copper wire carrying 1 A of current. Determine the magnetic fields $\mathbf{H}$ and $\mathbf{B}$, and the magnetization $\mathbf{M}$. Take the magnetic susceptibility of the iron to be 5.5×10^3.

9.14. A long straight wire of circular cross section with radius r_0 carries current I and is immersed in a large volume of water. The wire is nonmagnetic. Calculate the magnetization $\mathbf{M}(\mathbf{x})$ in the water. (The susceptibility is -9.0×10^{-6}.) What is the bound current density $\mathbf{J}_b(\mathbf{x})$ for $r > r_0$? Calculate the *total* current (free plus bound). [Answer: $I(1 + \chi_m)$]

9.15. Two thin-walled long coaxial cylinders with radii a and b carry equal but opposite currents $\pm I$ parallel to the axis. The surface current densities are $I/2\pi a$ and $-I/2\pi b$. Between the cylinders is a material with magnetic susceptibility χ_m. Determine the field $\mathbf{H}(r, \phi)$ and magnetization $\mathbf{M}(r, \phi)$. Plot $H_\phi(r)$. Describe in words and pictures the bound currents.

Sec. 9.5. Body in a Magnetic Field

9.16. Suppose in Example 7 the region inside the sphere has susceptibility χ_1 and the region outside the sphere has susceptibility χ_2. Then solve the boundary value problem.

9.17. **(a)** A long copper cylinder is placed in an applied magnetic field $B_0\hat{\mathbf{k}}$ parallel to the axis, with $B_0 = 1$ T. Find the magnetic field $\mathbf{B}$ inside and outside the cylinder, and the bound surface current density $\mathbf{K}_b$.

(b) Now suppose the applied field is perpendicular to the axis of the cylinder in the x direction, $B_0\hat{\mathbf{i}}$ with $B_0 = 1$ T. Find the magnetic field $\mathbf{B}$ inside and outside the cylinder if the radius of the cylinder is $a = 1$ m. (Hint: Use the magnetic scalar potential in cylindrical coordinates.)

Sec. 9.6. Ferromagnetism

9.18. Figure 9.23 shows a toroidal solenoid with N turns and current I, with an iron core. For the magnetic condition of the iron the relative permeability is $\kappa_m = \mu/\mu_0$. The cross section of the torus is small compared to the radius r. The core has length

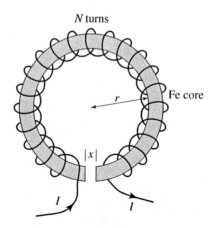

FIGURE 9.23   Iron toroid with an air gap.

$2\pi r$ minus a small air gap of length x. Because x is small, we may approximate the magnetic field in the torus to be azimuthal around the axis of symmetry with no fringing at the gap. The problem is to find the magnetic field B_g in the gap.

(a) Show from Ampère's law that $L H_i + x H_g = N I$, where $L = 2\pi r - x$ and the subscripts i and g refer to the iron and the gap.

(b) Show that because B_n is continuous, $H_g = \kappa_m H_i$.

(c) Show that

$$B_g = \frac{\kappa_m (L + x)}{L + \kappa_m x} B_0,$$

where B_0 is the field if there is no iron core.

(d) Suppose $\kappa_m = 10^3$, $r = 10\,\text{cm}$, and $NI = 10^3$ A. Evaluate B_g for $x = 1\,\text{mm}$, and for $x = 1\,\text{cm}$. What is B in the toroid if the core is iron and there is no gap? What is B in the air if there is no core?

General Exercises

9.19. (a) Using classical statistical mechanics for a gas of magnetic dipoles (moments m_0 and density n) free to rotate in any direction in a magnetic field $B_0 \hat{\mathbf{k}}$, show that the magnetization is

$$\mathbf{M} = n m_0 \left(\coth b - 1/b \right) \hat{\mathbf{k}},$$

where $b = m_0 B_0 / kT$ and T is the temperature. This result is called the Langevin formula. Plot the magnetization M_z as a function of b, in units of $n m_0$. Show that for $kT \gg m_0 B_0$ the susceptibility approaches $n m_0^2 \mu_0 / 3kT$.

(b) Derive the magnetization assuming instead that the atoms have spin 1/2, i.e., two states ↑ and ↓ with magnetic moments $\pm \mu_B$.

9.20. (a) If an atom with one unpaired electron spin is placed in a magnetic field $B_0 \hat{\mathbf{k}}$ then there are two energy levels, corresponding to the two values of the quantum number M_S. Calculate the energy difference ΔE between the states.

(b) If the atom is irradiated it will absorb energy if the frequency ν is $\Delta E / h$. This phenomenon is called *electron spin resonance*. Calculate the frequency if B_0 is 2 T. In what part of the electromagnetic spectrum is this frequency?

9.21. In the Bohr model of the atom an electron circles the nucleus with radius r, speed v, and angular momentum $L = mvr = h/2\pi$, where h is Planck's constant. Find the magnetic moment in SI units. As a simple model of a ferromagnetic domain, assume the atomic moments (with the value just calculated) are aligned and located at the vertices of a cubic lattice with lattice spacing 3 Angstroms. Calculate the magnetization.

9.22. Is the following statement true or false? In a uniform material with magnetic susceptibility χ_m and electric conductivity 0, any bound current distribution can only be a surface current (assume no time dependence).

9.23. A long straight wire carrying current I is parallel to the z axis and passes through the point $(a, 0, 0)$. The region $x > 0$ is vacuum, and the region $x < 0$ is a material with magnetic susceptibility χ_m.

Show that the **H**-field in the region $x < 0$ is the same as **H** that would be produced by a current $2I/(2 + \chi_m)$ in the wire with the material everywhere; and the **H**-field in the region $x > 0$ is the same as **H** that would be produced by the combination of the current I in the wire and a current $\chi_m I/(2 + \chi_m)$ along the line parallel to the z axis through $(-a, 0, 0)$, with vacuum everywhere. (Hint: Appeal to the uniqueness theorem. What are the boundary conditions?)

Calculate the force per unit length on the wire.

Computer Exercise

9.24. Use a computer to plot the magnetic field strength at the center of the sphere in Example 8—magnetically shielded by a shell of material—as a function of μ/μ_0. Let $B_0 = 1\,T$ and $R_o/R_i = 1.5$.

CHAPTER

10

Electromagnetic Induction

"I am busy just now again on electro-magnetism, and think I have got hold of a good thing, but can't say. It may be a weed instead of a fish that, after all my labor, I may at last pull up."

Michael Faraday, prior to his discovery of electromagnetic induction

So far we have studied only fields that are independent of time. Now we will be concerned with electric and magnetic fields, $\mathbf{E}(\mathbf{x}, t)$ and $\mathbf{B}(\mathbf{x}, t)$, that vary with time. The time dependence produces physical phenomena that are not present for static fields—*electromagnetic induction* and the *displacement current*.

Electromagnetic induction is the production of an electric field by a magnetic field that varies with time. Its discovery is generally credited to Faraday, in 1831; but there is evidence that it was discovered independently in 1830 by Joseph Henry. Before the experiments of Faraday and Henry it was known, from the work of Oersted in 1820, that a magnetic field is produced by a steady electric current, i.e., a current that is constant in time. Therefore Faraday first attempted to detect an electric field from a strong steady magnetic field. This idea proved to be wrong. But Faraday, who was a skilled experimenter, did notice a different effect: When the magnetic field *changes* in the region of a loop of wire, an electric current occurs, as indicated by deflection of a galvanometer needle. The essential idea of electromagnetic induction is that an electric field is produced by nonzero $\partial \mathbf{B}/\partial t$.

It is natural to ask whether the situation is symmetric: Does nonzero $\partial \mathbf{E}/\partial t$ produce a magnetic field? The answer, whose consequences will be discussed in Chapter 11 in connection with the displacement current, is yes.

From the scientific discoveries of Faraday and Henry—small, esoteric effects at the time of their discovery[1]—technology has developed that has changed the human condition. Electromagnetic induction is the basis for the generation of electricity that we all use in daily life. The goal of this chapter is to describe the mathematical theory of electromagnetic induction and some of its applications.

---

[1] At a public lecture, Faraday was asked: "Mr. Faraday, of what use are your discoveries?" His answer: "Madam, of what use is a baby?"

10.1 ■ MOTIONAL EMF

As a preliminary step it is useful to study *motional EMF*—the EMF due to motion of a conductor in a magnetic field.

10.1.1 ▓ Electromotive Force

The most general definition of electromotive force (EMF) is that the EMF $\mathcal{E}$ along a curve Γ due to a force $\mathbf{F}$ acting on charges is the work per unit charge that would be done by $\mathbf{F}$ on a test charge q that moves along Γ; i.e., $\mathcal{E} = \int \mathbf{F} \cdot d\boldsymbol{\ell}/q$. The term "electromotive force" is a misnomer, because $\mathcal{E}$ is not a force; it is the work per unit charge, with units of volts. So, for example, the "EMF of a battery" is the energy per unit charge supplied by the battery to transport positive ions from the cathode to the anode within the battery, or equivalently the work per unit charge done by the electric field on an electron that moves around the wire from the cathode to the anode.

In this chapter we are concerned more specifically with *circuital EMF*, i.e., EMF for a *closed curve*, or loop. From now on, EMF will mean electromotive force around a circuit. Again, the definition is the work per unit charge on a test charge that is carried around the circuit,

$$\mathcal{E} = \oint_C \frac{\mathbf{F} \cdot d\boldsymbol{\ell}}{q}, \tag{10.1}$$

where $\mathbf{F}$ is the force on q. Here C may be any closed curve in space, not necessarily occupied by a conducting material.

The general electromagnetic force on a test charge q is the Lorentz force $\mathbf{F} = q(\mathbf{E} + \mathbf{v} \times \mathbf{B})$. In this first section we are mainly concerned with the contribution to the EMF (10.1) that comes from the magnetic term $q\mathbf{v} \times \mathbf{B}$ in $\mathbf{F}$. This contribution is called the *motional EMF*, which may be nonzero if the circuit moves in a magnetic field.

The electric term $q\mathbf{E}$ in $\mathbf{F}$ needs a little explanation now, in anticipation of a much fuller discussion later. There are two sources for $\mathbf{E}$, and they play different roles with respect to EMF. The two sources are *electrostatic*, due to charges at rest or at least moving slowly enough to make the quasistatic approximation, and *inductive*, due to $\partial \mathbf{B}/\partial t$. There is no EMF *around a closed loop* associated with the electrostatic field because $\nabla \times \mathbf{E} = 0$ for static charges; by Stokes's theorem,

$$\oint_C \mathbf{E} \cdot d\boldsymbol{\ell} = \int_S \nabla \times \mathbf{E} \cdot d\mathbf{A} = 0. \tag{10.2}$$

Indeed, generally, the EMF from any conservative force is 0, because a conservative force has, by definition, $\oint \mathbf{F} \cdot d\boldsymbol{\ell} = 0$. There is however an EMF associated with the inductive part of $\mathbf{E}$, which, as we shall see, has $\nabla \times \mathbf{E} \neq 0$.

But first we ask: What is the EMF from the magnetic force when the circuit moves in a magnetic field?

10.1.2 ■ EMF from Motion in B

Consider a rigid wire loop C moving in a magnetic field. What is the EMF $\mathcal{E}$? Suppose the magnetic field $\mathbf{B}(\mathbf{x})$ is static in time but nonuniform in space. The magnetic force on a test charge q in the wire is $q\mathbf{v} \times \mathbf{B}$, where $\mathbf{v}$ is the macroscopic velocity of the point in the wire where q is located. The EMF around the loop is therefore, by the definition (10.1),

$$\mathcal{E} = \oint_C (\mathbf{v} \times \mathbf{B}) \cdot d\boldsymbol{\ell} = \oint_C \mathbf{B} \cdot (d\boldsymbol{\ell} \times \mathbf{v}), \tag{10.3}$$

using a vector product identity in the second equality. The subsequent analysis is based on Fig. 10.1, which shows the loop C at times t and $t + \delta t$. (The limit $\delta t \rightarrow 0$ is understood.) The surface swept out as C moves, denoted $\delta\Sigma$, is shaped like a loop of ribbon. Note that $d\boldsymbol{\ell} \times \mathbf{v}$ is $d\boldsymbol{\ell} \times \delta\mathbf{x}/\delta t$, where $\delta\mathbf{x}$ is the displacement of $d\boldsymbol{\ell}$ during δt; from Fig. 10.1, $d\boldsymbol{\ell} \times \delta\mathbf{x}$ is the outward directed area element of a section of the ribbon. Thus the right-hand side of (10.3) is $\delta\Phi_r/\delta t$, where $\delta\Phi_r$ is the flux of $\mathbf{B}$ through the ribbon surface $\delta\Sigma$ in the direction defined by $d\boldsymbol{\ell} \times \delta\mathbf{x}$. Now consider the *closed surface* composed of $\delta\Sigma$ and surfaces $S(t)$ and $S(t+\delta t)$ bounded by $C(t)$ and $C(t + \delta t)$ respectively: By Gauss's theorem the flux of $\mathbf{B}$ outward through the closed surface is 0, so

$$\delta\Phi_r + \int_{S(t+dt)} \mathbf{B} \cdot d\mathbf{A} - \int_{S(t)} \mathbf{B} \cdot d\mathbf{A} = 0. \tag{10.4}$$

The sense of $d\mathbf{A}$ on $S(t)$ is correlated with that of $d\boldsymbol{\ell}$ by the right-hand rule. With the fingers curling in the direction of $d\boldsymbol{\ell}$ the thumb points in the direction of $d\mathbf{A}$. This direction is outward for the surface $S(t + dt)$ but inward for $S(t)$; hence the minus sign in the third term of (10.4). Finally, then, since $\delta\Phi_r = \mathcal{E}\delta t$ by (10.3),

$$\mathcal{E} = -\frac{d}{dt}\int_{S(t)} \mathbf{B} \cdot d\mathbf{A} = -\frac{d\Phi}{dt}, \tag{10.5}$$

where Φ is the magnetic flux through the surface S bounded by C. The dimensions of magnetic flux Φ are $\mathrm{T\,m^2}$, or webers (Wb). Equation (10.5) is the formula for

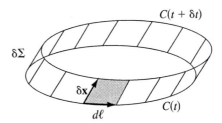

FIGURE 10.1 Motional EMF. The surface $\delta\Sigma$ (a ribbon loop) is swept out as the circuit C moves between times t and $t+\delta t$. Surfaces $S(t)$ and $S(t+\delta t)$ are bounded by the curves $C(t)$ and $C(t + \delta t)$, and the union of $S(t)$, $S(t + \delta t)$ and $\delta\Sigma$ is a closed surface. For the section of $\delta\Sigma$ shown shaded, the directed area element is $d\boldsymbol{\ell} \times \delta\mathbf{x}$.

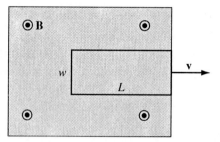

FIGURE 10.2   Motional EMF. B points out of the page in the shaded region. As the conducting loop moves out of **B** there is a counterclockwise induced EMF equal to vBw, which is $-d\Phi/dt$ where Φ is the flux of **B** through the loop.

motional EMF around a circuit moving in a magnetic field. The flux $\Phi(t)$ varies in time if the wire loop $C(t)$ moves through a nonuniform field.

We derived (10.5) in a general but abstract way. We'll understand the result better if we verify it for a simple specific example. Figure 10.2 shows a rectangular loop of size $L \times w$ moving with velocity $v\,\hat{\mathbf{i}}$ out of the region between the rectangular pole faces of a magnet. We may approximate the **B** field as uniform in the region between the pole faces, and zero outside. Figure 10.2 shows the configuration at $t = 0$. Assume the linear size of the magnet is larger than w. The motional EMF is $\mathcal{E} = vBw$ by (10.3), because only the left end of the loop (which is in **B**) contributes to the line integral. As a function of time t, the length of the loop between the pole faces is $L - vt$, so the magnetic field flux through the loop is $\Phi = B(L - vt)w$. The rate of change of Φ is $d\Phi/dt = -Bvw$. Hence $\mathcal{E} = -d\Phi/dt$, in accord with the general result.

Equation (10.5) was derived for motion of a rigid loop. But the same equation would hold if the conducting loop were to change in ways other than by translation, such as expansion or contraction, or a change in shape.

10.1.3 ■ The Faraday Disk Generator

As another example of motional EMF, Fig. 10.3 shows a copper disk spinning in a uniform external magnetic field $\mathbf{B} = B\hat{\mathbf{k}}$. It is natural to describe the system in cylindrical coordinates (r, ϕ, z) with the z axis orthogonal to the disk. Let a be the radius of the disk, and let $\boldsymbol{\omega} = \omega\hat{\mathbf{k}}$ be the angular velocity. There is a conducting wire, fixed in space, with sliding contacts at the rim and center of the disk. A current I will flow, in the sense shown, through the circuit composed of the disk and the wire, driven by the motional EMF. The EMF from the center to the rim is, by (10.3),

$$\mathcal{E} = \int_0^a \omega r B\, dr = \frac{1}{2}\omega B a^2. \tag{10.6}$$

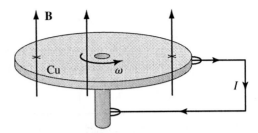

FIGURE 10.3 Faraday disk generator. A conducting disk rotates in a magnetic field. Because of the Lorentz force on mobile charges, a charge separation (+ at the rim, − at the center) would develop if there were no wire. The conducting wire connecting the rim and the center completes the circuit, and a current flows.

By Ohm's law the current is

$$I = \frac{\omega B a^2}{2R},$$ (10.7)

where R is the total resistance in the circuit. We can produce large currents with such a device, if the radius and angular velocity are large. The power dissipated by resistance must be replaced by the agent that keeps the disk spinning. If the disk is not driven then it will slow down and stop, a phenomenon called *magnetic braking*. The magnetic force on the current in an area element dA of the disk is $d\mathbf{F} = (\mathbf{K}dA) \times \mathbf{B}$, where $\mathbf{K}$ is the surface current density and $dA = r\,dr\,d\phi$. $\mathbf{K}$ is essentially radial from the center to the rim. Thus the direction of $d\mathbf{F}$ is $\hat{\mathbf{r}} \times \hat{\mathbf{k}} = -\hat{\boldsymbol{\phi}}$, opposite to the velocity of dA, and so the external field acts as a brake. The torque on the disk is

$$\mathbf{N} = \int_{\text{disk}} \mathbf{r} \times d\mathbf{F} = -\mathbf{B} \int \mathbf{r} \cdot \mathbf{K} r\,dr\,d\phi.$$ (10.8)

(The double cross product has been reduced in the second equality.) The azimuthal integral $\int_0^{2\pi} K_r r\,d\phi$ is the total current I through a circle of radius r on the disk, and I is independent of r by charge conservation. The radial integral gives

$$\mathbf{N} = -B\hat{\mathbf{k}} \int_0^a I r\,dr = -\frac{I B a^2}{2} \hat{\mathbf{k}}.$$ (10.9)

The direction of $\mathbf{N}$ is opposite to the angular velocity $\omega\hat{\mathbf{k}}$, so $\mathbf{N}$ is a braking torque.

Faraday used this device as a very basic demonstration that when a conductor moves through magnetic lines of force, a current is induced.

Self-Excited Dynamo

We consider next an interesting modification of Fig. 10.3, in which the stationary wire from the rim of the disk back to the axle is replaced by a stationary solenoidal coil through which the induced current I flows. The coil is placed near the rotating disk, and has the same axis as the disk. Then the magnetic field due to I is orthogonal to the disk's plane. The coil is wound in the sense such that its current I produces a magnetic self-field $\mathbf{B}_{self}$ that is *in the same direction as the external field* $\mathbf{B}$. Then, once current starts to flow, the external field may be turned off, and if the generator is driven it will continue to produce an EMF.

In principle, even if there is no external $\mathbf{B}$, a generator driven at a large enough angular frequency $\omega > \omega_{critical}$, which depends on the geometry and the electrical resistivity of the system, will produce a $\mathbf{B}_{self}$ large enough to produce an EMF and current flow. Such a system may be called, even in the strictest sense, a self-excited dc-generator, or a dynamo. In practice, for a system of the kind just described, made only of copper, $\omega_{critical}$ would be very large. If, however, the solenoidal coil is wound around a ferromagnetic core, then a self-excited generator becomes practical. Generators in common use with self-generated fields have a ferromagnetic core that retains some magnetization if the system stops. Then when the system is restarted the magnetic field of the core serves the same purpose as an external field.

To calculate $\omega_{critical}$ is difficult. However, dimensional analysis suggests that $\omega_{critical}$ is of order $K\left[\rho/(\ell^2\mu_0)\right]$, where K is a dimensionless constant, ρ is the resistivity of copper, and ℓ is a characteristic linear dimension of the system. For laboratory-size systems, ℓ is the size of the apparatus, and $\omega_{critical}$ is very large, practically unattainable. For the Earth, however, where ℓ is much larger, $\omega_{critical}$ is small enough that Earth's magnetic field is due to a self-excited dynamo.

The terrestrial self-excited dynamo that is generally accepted to be the origin of the Earth's magnetic field, is due to currents in the Earth's outer core—a conducting liquid—which extends from about $0.2\,R_{Earth}$ to $0.55\,R_{Earth}$. This core is molten Fe, with a few percent Ni, at about $4000°$ C, far too hot for any ferromagnetism. The precise details of the current distribution are unknown, although model calculations can be made.[2]

Even starting with a small applied external magnetic field, it is very difficult to construct a laboratory-scale fluid dynamo. The first successful demonstration, which used molten sodium as the conducting liquid and took 25 years to achieve, has been reported.[3]

10.2 ■ FARADAY'S LAW OF ELECTROMAGNETIC INDUCTION

We have seen that a conducting loop moving in a static but nonuniform magnetic field experiences an EMF. For example, in Fig. 10.2 the nonuniformity is that $\mathbf{B}$ is confined to a finite region, out of which the conducting loop is being pulled. In such a situation, the conductor moves and the field is at rest. By the principle of relativity we should expect the converse phenomenon to occur: If the conductor is at rest, and the field moves, there should again be an induced EMF. After all,

---

[2]D. R. Stump and G. L. Pollack, *Am. J. Phys.* **66**, 802–810 (1998).
[3]A. Gailitis, et al., *Phys. Rev. Lett.* **84**, 4365 (2000).

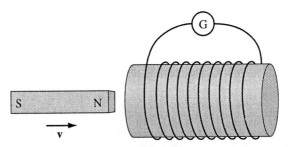

FIGURE 10.4   Demonstration of electromagnetic induction. When the magnet moves with respect to the solenoid, the galvanometer needle deflects.

the principle of relativity implies that one cannot determine an absolute state of motion. Who is to say which is at rest—the conductor or the magnet? Indeed a moving magnet *does* produce an EMF around a nearby circuit. An even stronger statement is true. If the magnetic field changes in time *for any reason* then there is an induced EMF from an electric field. This phenomenon is called *electromagnetic induction.*

A simple demonstration of electromagnetic induction is illustrated in Fig. 10.4. A bar magnet is moved into or out of a coil of wire, which is connected to a galvanometer. When the magnet moves, a current flows in the solenoid, driven by the induced EMF, causing deflection of the galvanometer. This demonstration is not exactly the way Faraday first observed electromagnetic induction, but Faraday did similar demonstrations.

Note the similarity and difference between electromagnetic induction and motional EMF. We obtain a current either by moving the coil with the magnet at rest (motional EMF), in which case the force on the charges in the current is magnetic; or, we obtain a current by moving the magnet with the solenoid at rest (electromagnetic induction), in which case the force is electric. (In the latter case the force cannot be magnetic because the conductor is at rest; a magnetic force, which would be proportional to $\mathbf{v} \times \mathbf{B}$, would be 0.) These are two aspects of the same phenomenon. In Chapter 12, on relativity, we will study in more detail the close relation between electric and magnetic fields in different frames of reference.

10.2.1 ■ Mathematical Statement

A qualitative statement of Faraday's Law is that *a changing magnetic field creates an electric field.* A quantitative statement, found in introductory textbooks, is the relation

$$\mathcal{E} = -\frac{d\Phi}{dt}, \tag{10.10}$$

where $\mathcal{E}$ is the EMF around a closed curve C due to the induced electric field, and Φ is the flux of magnetic field through any surface with boundary C. Equation

(10.10) is the same as the equation (10.5) for motional EMF, as we have anticipated by considerations of relativity. The flux Φ can change for various reasons: The field may change with the loop fixed; the position, orientation, or shape of the loop may change with **B** fixed; or some combination of these changes may occur. In any case, (10.10) holds.

Equation (10.10) is an *integral relation* between the line integral $\mathcal{E}$ and the area integral Φ. An equivalent, but more powerful statement of Faraday's law, is a *local relation* between partial derivatives of the fields $\mathbf{E}(\mathbf{x}, t)$ and $\mathbf{B}(\mathbf{x}, t)$, which may now depend on time. We can derive the differential relation from (10.10) by Stokes's theorem. The EMF around C due to the *induced electric field* is

$$\mathcal{E} = \oint_C \mathbf{E} \cdot d\boldsymbol{\ell} = \int_S \boldsymbol{\nabla} \times \mathbf{E} \cdot d\mathbf{A}. \tag{10.11}$$

Again, C is an arbitrary closed curve fixed in space, not necessarily occupied by a conductor. Faraday's Law is really a statement about the fields, not about current, although it was first discovered by observing the current produced in a conductor by the induced **E**. The second equality in (10.11) is just Stokes's theorem, and S may be any fixed surface bounded by C. In electrostatics $\mathcal{E}$ would be 0, because an electrostatic field is irrotational, i.e., has curl equal to 0. But the electric field produced by electromagnetic induction is not irrotational.

Now, the right-hand side of (10.10) is

$$-\frac{d\Phi}{dt} = -\int_S \frac{\partial \mathbf{B}}{\partial t} \cdot d\mathbf{A}. \tag{10.12}$$

We are justified in pulling the time derivative into the integral because the surface S is fixed in space. The surface integrals in (10.11) and (10.12) must be equal *for any surface S*, because the original loop C is arbitrary. The only way these two integrals can be equal for all surfaces is if the integrands are equal,

$$\boldsymbol{\nabla} \times \mathbf{E} = -\frac{\partial \mathbf{B}}{\partial t}. \tag{10.13}$$

Equation (10.13) is Faraday's Law. It is one of the universal equations of electromagnetism. It relates the variation of **E** in space to the variation of **B** in time. Or, there is another way to regard the meaning of this equation. It says that there exists another source of electric field besides electric charge: A magnetic field changing in time is a source of electric field. This source is not like charge, from which **E** diverges, but rather like a "current" around which **E** curls. Of course it is not a charge current, but a magnetic-field current. Static charge creates an irrotational field. Nonzero $\partial \mathbf{B}/\partial t$ creates a curling field.

The Helmholtz theorem (Appendix B) states that a vector function is determined by its divergence and curl with suitable boundary conditions. In the case of **E** the divergence is $\boldsymbol{\nabla} \cdot \mathbf{E} = \rho/\epsilon_0$, which involves charges; the curl is $\boldsymbol{\nabla} \times \mathbf{E} = -\partial \mathbf{B}/\partial t$, which involves electromagnetic induction.

The reader may feel that while (10.10) and (10.13) may seem plausible, based on the similarity to motional EMF and the relativity principle, the relation has not been derived with logical necessity. But of course it cannot be derived from more fundamental principles, because no principle of electromagnetism is more fundamental than Faraday's Law. Faraday's Law, in either of the equivalent formulations (10.10) or (10.13), is an empirical observation, based on Faraday's experiments, and is the most basic relation between time-dependent fields.

10.2.2 ▩ Lenz's Law

Lenz's law is a very useful statement about the direction of the induced **E** in electromagnetic induction. It is a corollary of Faraday's Law.

Faraday's Law is a statement about the fields. The direction of the induced **E** follows from (10.13) or (10.10): If the thumb of your right hand points *opposite* to the direction of the change of **B**, then the fingers curl in the direction of **E**.

Lenz's law is a statement about the direction of the induced current if a conductor is present: *If a conducting loop forms a closed curve C in a changing magnetic field, then the direction of the induced current opposes the change of* **B**. Let's check that this statement is right. Let the thumb of your right hand point *opposite* to the change of **B**. The fingers then curl in the direction of the current driven by the induced **E**; the magnetic field produced by *that current* is in the direction of your thumb (by Ampère's Law) opposite to the change of **B**. Thus the induced current *opposes* the change of **B**. In other words, the current induced by a changing **B** field flows in the direction that tends to keep the flux from changing. The principle is illustrated in Fig. 10.5. But note that the field of the induced current does not cancel the change of **B**; the flux *is changing*.

Lenz's law is helpful when we need to figure out the direction of the induced electric field. Of course we can always derive the direction by a very careful application of Stokes's theorem, remembering that the directions of $d\mathbf{A}$ of S and $d\boldsymbol{\ell}$

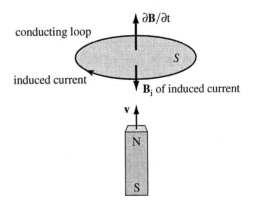

FIGURE 10.5 Lenz's law. C is a conducting loop and S is a surface bounded by C. As the magnet moves toward C, the flux through S increases, and the field at a point on S changes at the rate $\partial \mathbf{B}/\partial t$. While the flux is changing the induced current opposes the change by making a field $\mathbf{B_i}$ opposite to $\partial \mathbf{B}/\partial t$.

of C are related by the right-hand rule. But it is usually easier to apply Lenz's law, by imagining what would happen if a conductor were present.

EXAMPLE 1   Suppose the circular pole faces of an electromagnet are above and below the xy plane, and parallel to the xy plane, so that the magnetic field is in the $+\hat{\mathbf{k}}$ direction. (The north pole is below and the south pole above the plane.) What is the electric field in the xy plane if the magnitude of $\mathbf{B}$ is increasing in time? In particular, what is the direction of $\mathbf{E}$?

We can solve this problem by the "Amperian loop trick." The problem has cylindrical symmetry. The direction of the electric field must be azimuthal (curling around $-\partial\mathbf{B}/\partial t$) and its magnitude cannot depend on ϕ; i.e., the field $\mathbf{E}(x, y, 0)$ has the form $E(r)\hat{\phi}$. Applying (10.10) to a circular loop of radius r, we find

$$2\pi r E = -\frac{d\overline{B}}{dt}\,\pi r^2, \tag{10.14}$$

where $\overline{B}$ is the average of the field over the area of the circle. (The definition of the average field is $\overline{B} = \left(\int \mathbf{B}\cdot d\mathbf{A}\right)/\pi r^2$; if the field is uniform then $\overline{B} = B$.) Thus the electric field is

$$\mathbf{E} = -\frac{r}{2}\frac{d\overline{B}}{dt}\,\hat{\phi}. \tag{10.15}$$

According to (10.15), $\mathbf{E}(\mathbf{x})$ is in the direction $-\hat{\phi}$ if B_z is increasing in time. This direction is in accord with Lenz's law: The change of $\mathbf{B}$ is in the direction of $+\hat{\mathbf{k}}$. If C were occupied by a conductor, then current in the direction $-\hat{\phi}$ would produce a magnetic field in the $-\hat{\mathbf{k}}$ direction by Ampère's Law, i.e., opposing the change of $\mathbf{B}$.

10.2.3 ■ Eddy Currents

Currents that are induced in conductors, especially conducting plates, are called eddy currents. Figure 10.6 illustrates the eddy currents associated with magnetic braking of a Cu plate swinging as the bob of a pendulum in a magnetic field, a common classroom demonstration.

EXAMPLE 2 Magnetic braking.   If there is no magnetic field, the plate, swinging as a pendulum from a suspension point high on the z axis, oscillates freely in the $x = 0$ plane. But if a magnetic field $\mathbf{B}_0$ is applied in the shaded region of Fig. 10.6, perpendicular to the plate, the motion is damped. Suppose $\mathbf{B}_0$ points in the $+x$ direction, as shown. This may be arranged by letting the plate swing between the poles of an electromagnet, with the N-pole at $x = -\Delta x_0$ and the

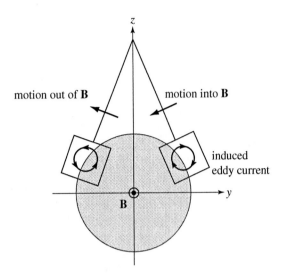

FIGURE 10.6   Eddy current demonstration. B points out of the page in the shaded region. As the Cu plate swings into or out of the magnetic field, the induced eddy currents oppose the change of flux.

S-pole at $x = +\Delta x_0$. As the plate swings into the magnetic field region from the right, eddy currents are induced in the direction shown, clockwise by Lenz's law. The induced magnetic moment **m** of the plate then points in the $-x$ direction. At this stage of the motion the plate is *repelled* from the field region because of the force on the induced current exerted by the magnet. If the plate swings all the way through the field region, then counterclockwise eddy currents are induced as it leaves the region. At that stage the induced **m** is in the $+x$ direction. The plate is *attracted* to the field region because of the force on the induced current exerted by the magnet.

Both while entering and while leaving the field, the induced eddy currents experience a force braking the plate's motion. This is the principle of magnetic brakes, which are used to stop circular motion of conducting disks, and are applied in, e.g., circular saws, lawn mowers, and high-speed trains.

In magnetic-braking applications it is the conducting disk whose motion in a stationary magnetic field is damped. As we know, however, only the *relative* motion between conductor and field is relevant. Therefore, if one rotates a permanent magnet in front of a stationary conducting disk—we imagine a bar magnet rotating in a plane parallel to the disk—then the interaction of the bar magnet and the induced eddy currents will result in a drag torque that depends on the rotation frequency. This effect is applied in torque transmission equipment and in tachometers.

EXAMPLE 3   As a quantitative example of magnetic braking due to eddy currents, consider a thin conducting rectangular frame released from rest that falls from a region of constant magnetic field normal to the frame into a region of zero field. We'll show that its fall is held back by eddy currents.

The frame and magnetic field are shown in Fig. 10.7. Call the long, vertical dimension of the frame ℓ, and its width w. The horizontal line is the boundary of the magnetic field. As long as the entire frame is within the magnetic field, the flux through it is constant, there is no net EMF around it, no current flows, and there is no magnetic braking, so the frame behaves like any other falling body.

But what is the subsequent motion if the frame is released from rest, at $t = 0$, in the position shown in Fig. 10.7? Let downward be the positive direction. Counteracting the gravitational force Mg, there will be a magnetic braking force opposing the loss of flux through the frame as it moves downward out of the field region. If v is the downward velocity then there is a magnetic force on the electrons along the side ab, creating a current. According to (10.3) the EMF is $\mathcal{E} = vBw$. If we take **B** to point out of page then the induced current is $I = vBw/R$, where R is the resistance of the entire conducting path $abcda$, flowing in the counterclockwise direction around the frame. The magnetic force acting on the current in ab, the braking force, is upwards, and is given by

$$F_{\text{braking}} = -IwB = -\frac{B^2 w^2 v}{R},$$

negative implying upwards. Thus the equation of motion of the frame is

$$M\frac{dv}{dt} = Mg - \frac{B^2 w^2}{R}v, \tag{10.16}$$

where M is its mass. To simplify the solution to (10.16) we use the notation $gMR/(B^2w^2) = v_\infty$ and rewrite (10.16) as

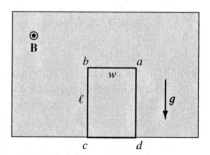

FIGURE 10.7   Falling frame. B points out of the page in the shaded region. The conducting frame (*abcd*) falls in Earth's gravity. The induced current and the braking force are calculated in Example 3.

$$\frac{dv}{dt} + \frac{gv}{v_\infty} = g. \tag{10.17}$$

In this form we see that v_∞ is the *terminal velocity*, i.e., the constant downward velocity the frame would reach if it could fall until it reached the steady state $dv/dt = 0$.

Equation (10.17) is a first-order linear differential equation with constant coefficients. Its solution can therefore be written as a sum of a particular solution and a solution to the homogeneous equation: $v(t) = v_p + v_h = v_\infty + C \exp(-gt/v_\infty)$, where C is determined from the initial condition $v(0) = 0$. Finally then

$$v(t) = v_\infty \left(1 - e^{-gt/v_\infty} \right). \tag{10.18}$$

$v(t)$ starts from 0, and approaches v_∞ exponentially. In a real experiment the frame might not reach terminal velocity. After it has fallen entirely out of the field, it will again fall freely. If a piece were cut out of the frame, i.e., if there were an insulating break in the path *abcda*, then there would be no current around the loop so the braking effect wouldn't occur.

In this example we used a rectangular frame because the eddy current for that case is easy to calculate. But it's natural to ask also what would happen if instead of a frame the conductor were a thin sheet, as used in lecture demonstrations and described in Example 2. In that case there would also be induced eddy currents and a concomitant magnetic braking effect. However, the current flow is more complicated, and it can only be calculated by advanced analytical methods.

EXAMPLE 4 Magnetic levitation. A potential future application of eddy currents is in mass transport. The idea is to use the repulsion between induced currents and magnets to levitate vehicles, with the resulting advantages of lower friction and less noise. Magnetic levitation has been used for trains in Germany, Japan, and England, but the cost is so high the future of this technology is in doubt.

We already saw in connection with Fig. 10.6 that eddy currents produce repulsive and drag forces when there is a relative sideways motion between magnets and conductors. Figure 10.8 will help to illustrate further some principles related to magnetic levitation. Imagine that the bar magnet in Fig. 10.8 is lowered from above into the position shown. If the plate is a good conductor then the eddy currents induced in it will repel the magnet and tend to keep it suspended. If the magnet is now moved sideways, say to the right as indicated by the arrow, then new eddy currents produced at the front of the moving magnet will tend to keep it suspended. There will also be drag forces due to eddy currents in back of the magnet, so that a levitated vehicle, here represented by the magnet, must have a source of propulsion if it is to move.

The development of high-temperature superconducting materials has increased the interest in magnetic levitation, because they may offer the possibility of pro-

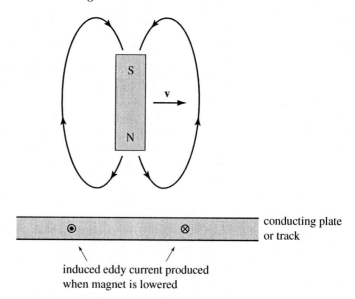

conducting plate
or track

induced eddy current produced
when magnet is lowered

FIGURE 10.8   Example 4. Magnetic levitation.

ducing zero-resistivity materials with much lower refrigeration requirements compared to low-temperature superconductors. Interactions between superconductors and magnetic fields are fascinating. Superconductors can act as perfect diamagnets, in the sense shown in Fig. 10.8, as long as the magnetic field is less than a critical value. In these materials, surface currents are induced that prevent the magnetic field from penetrating the bulk. In fact, if there is a magnetic field inside a metal in its normal state, then it will be *expelled* if the metal becomes superconducting as the temperature is lowered. This is a quantum-mechanical phenomenon called the *Meissner–Ochsenfeld effect*. This behavior does not occur in non-superconducting metals. In a non-superconductor, any magnetic field flux that has penetrated the bulk will remain in it as the temperature is lowered and the resistivity decreases. If one tries to remove the field from inside the metal at low temperature, say by turning off the source, then induced currents will flow according to Lenz's law, in the sense so as to *maintain* the field inside. The study of magnetic properties of superconductors is closely bound up with their atomic structure and electric properties, and it is likely to remain an active field of research for a long time.

10.3 ■ APPLICATIONS OF FARADAY'S LAW

Faraday's Law is not just an abstract field equation. It describes a physical phenomenon—electromagnetic induction—that is one of the most important in technology. Here we will look at three practical applications.

10.3.1 ■ The Electric Generator and Induction Motor

Faraday's Law is the principle at work in an electric generator. The essential design is a conducting coil rotating in the magnetic field of a fixed magnet, as illustrated in Fig. 10.9(a). For constant angular velocity the magnetic flux through the coil area A is

$$\Phi = \int \mathbf{B} \cdot d\mathbf{A} = BA\cos\omega t \qquad (10.19)$$

because the angle between $\mathbf{B}$ and $d\mathbf{A}$ is ωt as the coil spins. Thus the induced EMF around the coil is, by (10.10),

$$\mathcal{E} = BA\omega\sin\omega t. \qquad (10.20)$$

Figure 10.9(a) illustrates a basic generator. The magnetic field is supplied by a stationary permanent magnet, the stator. The coil, or rotor, shown as a single loop of wire, is rotationally driven with angular velocity ω by an external source of power. An alternating EMF (10.20) of angular frequency ω is produced. The EMF is led to the outside via conductors on the rotation shaft, and eventually supplied to, say, an electric socket. In the position shown, the induced EMFs across the upper and lower segments of the coil drive currents in the indicated directions,

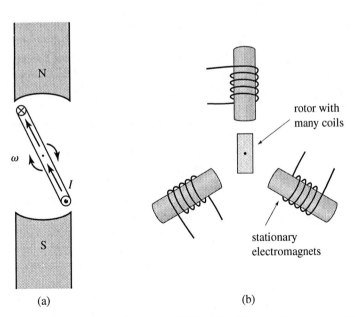

(a)   (b)

FIGURE 10.9   Generators and motors. (a) Principle of the electric generator; (b) three-phase generator or induction motor.

in agreement with Lenz's law.[4] In Fig. 10.9(a), when the plane of the loop is horizontal (i.e., perpendicular to **B**) the induced EMF is zero, because at that instant $\partial\Phi/\partial t = 0$. When the coil is vertical (i.e., parallel to **B**) the induced EMF is maximum. After the plane of the coil has turned through 180 degrees, the current direction in each segment is reversed. All of this behavior follows from (10.13) and Lenz's law. The magnetic moment **m** induced in the loop is in the direction such that the torque on the loop $\mathbf{N} = \mathbf{m} \times \mathbf{B}$ is in the direction opposite to ω. (In Fig. 10.9(a), ω is into the page and $\mathbf{m} \times \mathbf{B}$ is out of the page.) Power $-\mathbf{N} \cdot \omega$ is required to drive the generator at constant ω; that power is expended as I^2R and mechanical losses in the load connected across the electric socket.

Motors and Generators: A Symmetry.   Figure 10.9(a) can also be used to illustrate a basic ac-electric motor. There is a general equivalence—a symmetry—between motors and generators, related to conservation of energy. In running the system as an electric motor, alternating current at frequency ω is supplied to the coil from outside via conductors along the rotation shaft. When the motor coil is in the position shown in Fig. 10.9(a) the current flows *opposite* to the directions shown for the generator mode, so that the torque is now *parallel* to ω and the coil is driven in the direction shown. When the coil passes through horizontal, the direction of the supplied current must reverse for the torque to remain parallel to ω. In the motor, power $P = IV$ is supplied as current I at voltage V; that power is expended as mechanical work done by the motor. In general only small motors use permanent magnets as the source of the magnetic field.

Induction Motors.   Most large electric motors are induction motors, in which the source of the magnetic field is alternating current in a stator. This current is the means by which the external power to run the motor is supplied. The stator consists of several stationary electromagnets whose poles are arrayed at axially symmetric positions around the rotation axis of the motor. The spatially and temporally varying magnetic field of the electromagnets—a *rotating field*—induces time-dependent EMFs in the coils of the rotor in the sense to drive the rotor at a frequency equal to an integer multiple of the frequency of the supplied current in the electromagnets. Figure 10.9(b) shows a schematic diagram of a three-pole induction motor. There are iron cores in the electromagnets, which increase the magnetic field by a factor of order 10^4 compared to what copper coils alone would provide. The rotor is made up of many coils of wire. The current in the rotor coils is an induced current; there is no contact between the coil wire and the outside! In a real motor the gap between the outer surface of the rotor and the stator poles would be small and the number of poles would range from 2 to 100.

Most industrial distribution systems supply electricity to run induction motors in the form of three-phase power. To obtain three-phase power one uses, for example, a generator like that in Fig. 10.9(a), but with three loops, or three coils,

---

[4]Lenz's law gives the correct current direction in the loop; however, it is easier in this case to figure out the current direction from the direction of the magnetic force $q\mathbf{v} \times \mathbf{B}$.

each with its own feedout, making angles of 120 degrees with each other, on the same rotation shaft. The output of such a generator is the superposition of three EMFs with the harmonic form of (10.20); but each is shifted in phase by one-third of a cycle, 120 degrees, with respect to the others. This output can be applied to energize the three electromagnets of Fig. 10.9(b) so that the magnetic field rotates and the coil rotates with it. The principles of polyphase ac-induction motors and rotating magnetic fields were invented by Nikola Tesla in the 1880s and first applied commercially by the Westinghouse Electric Company. Common motors of this kind range from very small ones in servomechanisms, whose power is a fraction of a watt (10^{-4} hp), to large motors used in pumps and wind tunnels, whose power is several thousand hp.

10.3.2 ■ The Betatron

A betatron is an electron accelerator that uses the electric field induced by a changing magnetic field to accelerate electrons to relativistic energies. (In nuclear physics an electron is called a beta particle, because electrons are products of some beta decays.) The first betatron was built by Kerst in 1940, for electron scattering experiments in nuclear physics. Betatrons are still used today to produce hard X rays for studies of solid-state structure, and for medical purposes.

A schematic diagram of a betatron is shown in Fig. 10.10. Electrons circulate in a toroidal vacuum chamber of radius r_0 between the poles of an electromagnet. The magnetic field in the plane of the torus is $B(r, t)\hat{\mathbf{k}}$. What is the condition on $B(r, t)$ such that the radius r_0 of the electron orbit remains constant as the electrons gain kinetic energy?

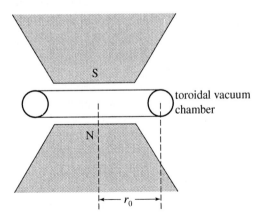

FIGURE 10.10   Betatron. The electrons travel in the toroidal vacuum chamber at radius r_0. The magnetic field strength increases from 0 to a large value B_{max} during the acceleration phase. As B_z increases an azimuthal induced electric field raises the electron kinetic energy. The pole shape is designed such that the radius of the electron orbit remains constant as the electrons gain kinetic energy. The betatron has been likened to a slingshot: B ramps up to B_{max} and then the electrons are released to the target.

For circular motion with radius r_0 the momentum is $\mathbf{p}(t) = p(t)\hat{\phi}$, where the magnitude of the momentum is $p(t) = er_0 B(r_0, t)$, just as in a cyclotron discussed in Chapter 8. But in the betatron, the magnitude of the momentum increases because of the force exerted by the induced electric field. By Newton's second law and (10.15),

$$\frac{dp}{dt} = -eE_\phi = \frac{er_0}{2}\frac{d\overline{B}}{dt}, \tag{10.21}$$

where $\overline{B}$ is the average of B_z over the area inside the orbit. Comparing the two relations between p and B shows that the condition for r_0 to be constant is that

$$B(r_0, t) = \tfrac{1}{2}\overline{B}(t). \tag{10.22}$$

That is, the pole shape must be designed in such a way that the vacuum ring is at the radius where the field strength is $\overline{B}/2$.

A betatron can be used to accelerate electrons up to energies of a few hundred MeV. For higher energies it is difficult to get the large average magnetic fields that are required; also, high-energy electrons radiate a lot of electromagnetic energy, so that (10.21) no longer holds. Therefore to reach higher energies, synchrotrons or linear accelerators are necessary.

10.3.3 ■ Self-Inductance

An inductor is a circuit device that stores energy in a magnetic field. In this section we will consider *self-inductance* of a conducting loop.

If a conducting loop carries current I then the magnetic flux Φ through the area bounded by the loop is, by the Biot-Savart law, proportional to I:

$$\Phi = LI. \tag{10.23}$$

(This assumes any magnetic materials in the system are linear.) The constant of proportionality L, called the self-inductance, depends on the geometry (size and shape) of the loop, but not on the current. The SI unit of inductance is the henry (H), which is equal to $T\,m^2/A$, or Wb/A.[5]

If I changes then there is an induced EMF $\mathcal{E}$ around the loop by Faraday's Law,

$$\mathcal{E} = -\frac{d\Phi}{dt} = -L\frac{dI}{dt}. \tag{10.24}$$

This EMF is called a *back EMF* because by Lenz's law it opposes any change of I.

---

[5]Wb, for weber, is a unit of magnetic flux, equal to $T\,m^2$.

EXAMPLE 5   What is the self-inductance of a long densely wound cylindrical solenoid?

Suppose the solenoid has N turns of wire, length ℓ, and cross section A. From magnetostatics the magnetic field in the solenoid is $B = \mu_0 N I / \ell$ in the direction parallel to the axis. The magnetic flux through the area enclosed by one loop of wire is BA, and the total flux, summed over all loops of the solenoid, is $\Phi = BAN$. Therefore the self-inductance is

$$L = \frac{\Phi}{I} = \frac{\mu_0 N^2 A}{\ell}. \tag{10.25}$$

Equation (10.25) is the inductance for a cylindrical solenoid with any cross sectional shape—it need not be circular. The *self-inductance per unit length L/ℓ* of a long solenoid, denoted L', is $\mu_0 n^2 A$, where $n = N/\ell$ is the number of turns per unit length.

EXAMPLE 6   What is the self-inductance of a closely wound toroidal coil with rectangular cross section?

Figure 10.11 shows a top view and a cutoff view of the toroidal coil. Each loop has height b, and extends from the inner radius R to the outer radius $R + a$.

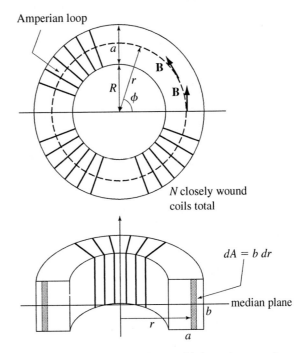

FIGURE 10.11   Example 6. The toroidal solenoid shown in a top view and a cut-away side view. The dashed circle is an Amperian loop used in the analysis.

For convenience, take the xy plane to be the median plane of the toroid, passing horizontally through the centers of the rectangular windings. The z axis goes vertically through the center of the torus. N is the total number of windings.

Recall that in Example 6 of Chapter 8 we learned that the magnetic field inside a long straight solenoid is uniform $\mathbf{B} = \mu_0 n I \hat{\mathbf{k}}$, and outside it $\mathbf{B} = 0$, for any cross sectional shape as long as it is constant along the entire length of the solenoid. A similar symmetry, which is, however, more difficult to analyze, holds for a toroidal solenoid.

For a toroidal solenoid, symmetry requires that inside the solenoid the magnetic field have the form $\mathbf{B} = B_\phi(r)\hat{\phi}$, and outside it $\mathbf{B} = 0$, for any cross sectional shape as long as it is the same all the way around the solenoid. In this case we can obtain the field by applying Ampère's Law to the circular path of radius r shown in Fig. 10.11(a). The result is

$$\oint \mathbf{B} \cdot d\boldsymbol{\ell} = B_\phi(r) 2\pi r = \mu_0 N I, \tag{10.26}$$

from which

$$B_\phi(r) = \frac{\mu_0 N I}{2\pi r}. \tag{10.27}$$

(Note that the current enclosed by the circular path is the same for any height z inside the toroid, so B_ϕ is independent of z.) Then the total flux through all N rectangular loops of wire is

$$\Phi = N \int B_\phi(r) \, dA = \frac{\mu_0 N^2 I}{2\pi} \int \frac{dA}{r} = \frac{\mu_0 N^2 I}{2\pi} \int_R^{R+a} \frac{b \, dr}{r}. \tag{10.28}$$

Finally, the self-inductance is

$$L = \frac{\Phi}{I} = \frac{\mu_0 N^2 b}{2\pi} \ln\left(\frac{R+a}{R}\right). \tag{10.29}$$

EXAMPLE 7   An important application of self-inductance is in *resonant oscillator circuits*. How does the response of a harmonically driven LC circuit, shown in Figure 10.12, depend on the driving frequency ω? Realistically we should also include resistance, but to understand the principle we may neglect resistance. The potential difference across the capacitor is

$$\frac{Q}{C} = \mathcal{E}_0 \cos \omega t - L \frac{dI}{dt}, \tag{10.30}$$

where Q is the capacitor charge and $I = dQ/dt$ the current. In (10.30) the term $\mathcal{E}_0 \cos \omega t$ is the alternating EMF of the power supply, and $-L dI/dt$ is the back EMF of the inductor. The equation for the time dependence of Q is

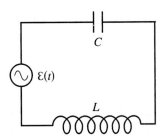

FIGURE 10.12   Example 7. A harmonically driven LC circuit is an example of resonance. The applied EMF is $\mathcal{E}(t) = \mathcal{E}_0 \cos \omega t$.

$$\frac{d^2 Q}{dt^2} + \omega_0^2 Q = \frac{\mathcal{E}_0}{L} \cos \omega t, \tag{10.31}$$

where the resonant frequency ω_0 is

$$\omega_0 = 1/\sqrt{LC}. \tag{10.32}$$

The steady-state solution of this inhomogeneous linear differential equation has $Q(t)$ oscillating at the driving frequency ω, i.e., $Q(t) = Q_0 \cos \omega t$. Substituting this form into (10.31) we find that the steady-state solution is

$$Q(t) = \frac{\mathcal{E}_0}{L} \frac{\cos \omega t}{\omega_0^2 - \omega^2}. \tag{10.33}$$

For $\omega < \omega_0$, Q is in phase with $\mathcal{E}$; for $\omega > \omega_0$, it is 180 degrees out of phase. The amplitude of oscillation diverges as the driving frequency approaches the "natural frequency" ω_0. (Resistance, which we have neglected, would make the amplitude finite and introduce a phase shift.) An oscillator circuit has maximum energy when L and C are tuned such that $\omega = \omega_0$, an example of *resonance*. Resonant LC circuits are used in radio tuners. A variable capacitor makes it possible to tune to a continuous range of frequencies.

■──────────────

10.3.4 ▓ Classical Model of Diamagnetism

In Sec. 9.1 we studied a classical model of the response of an electron in an atom to an applied field **B**. The result (9.5) is an equation for the resulting change of the orbital magnetic moment, which, because it is opposite in direction to **B**, provides a model of diamagnetism. The crucial step in the calculation was to state that the orbital radius remains constant as the magnetic field increases from 0 to **B**. That step is justified by Faraday's Law.

Refer to Fig. 9.1. For $\mathbf{B} = 0$ the electron $(-e)$ revolves around the nucleus (Ze) in a circular orbit satisfying $mv_0^2/r = Ze^2/(4\pi\epsilon_0 r^2)$. If the field is $B\hat{\mathbf{k}}$ then the equation of motion for the same radius r is

$$\frac{mv^2}{r} = evB + \frac{Ze^2}{4\pi\epsilon_0 r^2}.$$

The difference between the speeds $\delta v = v - v_0$ for the two cases, is small; neglecting terms of order $(\delta v)^2$ we may write

$$\delta v = \frac{erB}{2m}. \tag{10.34}$$

If the speed changes in this way as B is applied, then the orbit remains a circle with constant radius r.

The cause of the increase of the electron speed is not the magnetic force, which does no work, but the electric force exerted by the induced electric field of Faraday's Law. We'll now calculate this increased v and show that it's the same as required by (10.34). The kinetic energy changes by $-e\mathcal{E}$ for each revolution, where the EMF around the orbit is

$$\mathcal{E} = -\frac{d\Phi}{dt} = -\pi r^2 \frac{dB}{dt}.$$

That is, the rate of change of kinetic energy is

$$\frac{d}{dt}\left(\frac{1}{2}mv^2\right) = \frac{-e\mathcal{E}}{T},$$

where $T = 2\pi r/v$ is the period of revolution. Substituting Faraday's Law for $\mathcal{E}$ we find

$$\frac{dv}{dt} = \frac{er}{2m}\frac{dB}{dt}. \tag{10.35}$$

This equation for the change of the electron speed is identical to the condition (10.34) for the radius to be constant.

We shouldn't read too much into this model, because electrons do not obey classical mechanics in atoms. But it does give a simple explanation, and about the right magnitude, for the diamagnetic response of an atom.

10.4 ■ MUTUAL INDUCTANCE

We have earlier (Sec. 10.3.3) considered self-inductance of a single current loop. In this section we consider a general system of interacting current loops $C_1, C_2, C_3, \ldots$. For steady currents $I_1, I_2, I_3, \ldots$ the magnetic field is the superposition of the fields from the individual loops, and is linear in each of the currents.[6] The magnetic flux Φ_i through loop C_i is a linear combination of the

---

[6]For simplicity we ignore magnetic susceptibility of any materials in the system.

currents

$$\Phi_i = L_i I_i + \sum_{j \neq i} M_{ij} I_j. \tag{10.36}$$

Here L_i is the self-inductance of C_i, and M_{ij} is the *mutual inductance* of C_i and C_j. The inductances depend on the geometry of the system, but not on the currents. The unit of mutual inductance is the henry (H), the same as self-inductance.

We define $M_{ij} \equiv \Phi_{ij}/I_j$, where Φ_{ij} is the flux through loop C_i of the magnetic field produced by I_j. Similarly, $M_{ji} \equiv \Phi_{ji}/I_i$. The simplest case is two loops. In that case $M_{21} = \Phi_{21}/I_1$ is the flux through loop 2 due to unit current in loop 1; $M_{12} = \Phi_{12}/I_2$ is the flux through loop 1 due to unit current in loop 2. We shall prove that these mutual inductances are in fact equal. In general, $M_{ij} = M_{ji}$.

Let $\mathbf{B}_j$ denote the field produced by I_j, and $\mathbf{A}_j$ the corresponding vector potential, so that $\mathbf{B}_j = \nabla \times \mathbf{A}_j$. The flux Φ_{ij} is

$$M_{ij} I_j = \int_{S_i} \mathbf{B}_j \cdot d\mathbf{S}_i = \oint_{C_i} \mathbf{A}_j(\mathbf{x}_i) \cdot d\boldsymbol{\ell}_i, \tag{10.37}$$

where S_i is a surface bounded by C_i; the second equality is by *Stokes's theorem*.[7] In Chapter 8 we obtained a general formula (8.54) for the vector potential, which is for this case

$$\mathbf{A}_j(\mathbf{x}_i) = \frac{\mu_0}{4\pi} \oint_{C_j} \frac{I_j d\boldsymbol{\ell}_j}{|\mathbf{x}_i - \mathbf{x}_j|}. \tag{10.38}$$

Substituting this result into (10.37), and canceling the common factor I_j, we find an equation for the mutual inductance

$$M_{ij} = \frac{\mu_0}{4\pi} \oint_{C_i} \oint_{C_j} \frac{d\boldsymbol{\ell}_i \cdot d\boldsymbol{\ell}_j}{|\mathbf{x}_i - \mathbf{x}_j|}, \tag{10.39}$$

a result known as *Neumann's equation*.[8] This result shows explicitly that M_{ij} is entirely determined by the geometry of the system. Also, the double integral is unchanged if we interchange the subscripts i and j, proving that $M_{ij} = M_{ji}$. For the simplest case—two loops—this becomes $M_{12} = M_{21}$. That means that for *any* two loops, the flux through loop 1 due to unit current in loop 2 is the same as the flux through loop 2 due to unit current in loop 1, a remarkable result!

There is no expression analogous to (10.39) for self-inductance. In fact, (10.39) is an approximation: We have treated the currents as one-dimensional curves, which is a good approximation if the wire diameter is small compared to all other distances. But this approximation would break down for self-inductance. In fact,

---

[7] Be sure to understand the notation in (10.37): $d\mathbf{S}_i$ is the area element of the surface S_i; $\mathbf{A}_j$ is the vector potential of the field. We usually use $d\mathbf{A}$ for the area element, to emphasize that it has units of area, but here the symbol $\mathbf{A}$ is reserved for the vector potential!

[8] The integral in (10.39) could have either sign, depending on the relative directions of the two loops. It is conventional to make M_{ij} positive.

we shall see in Example 12 that the *self-inductance* of a current loop diverges as the wire diameter approaches 0.

The general expression (10.39) looks pretty unwieldy (a double line integral) but it can be evaluated analytically in simple cases. More importantly the result $M_{ij} = M_{ji}$ is handy for finding mutual inductances, because one of M_{ij} or M_{ji} might be easier to figure out than the other, as is the case in the following examples.

EXAMPLE 8   Find the mutual inductance for the circuits 1 and 2 shown in Fig. 10.13. C_1 is a rectangle in the xy plane, of size $b \times c$. C_2 is a larger rectangle in the xy plane, infinite in the y direction, and at distance a from the parallel sides of C_1 in the x direction. It is simple to calculate Φ_{12}, because $\mathbf{B}_2$ is just the field due to a pair of long wires. The mutual inductance is

$$M_{12} = \frac{\Phi_{12}}{I_2} = \frac{1}{I_2} \int_{S_1} \mathbf{B}_2 \cdot d\mathbf{A}_1$$

$$= \frac{1}{I_2} \int_{-b/2}^{b/2} \left[\frac{\mu_0 I_2}{2\pi(a + b/2 - x)} + \frac{\mu_0 I_2}{2\pi(a + b/2 + x)} \right] c\, dx$$

$$= \frac{\mu_0 c}{\pi} \ln\left(\frac{a+b}{a}\right). \tag{10.40}$$

The first term of the integrand is the flux through C_1 of the magnetic field due to the current in the wire on the right of C_2, and the second term is the equal flux

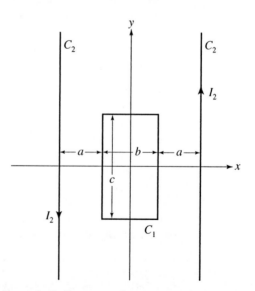

FIGURE 10.13   Example 8. Geometry of two rectangular loops for which the mutual inductance is to be calculated.

through C_1 due to the wire on the left of C_2. It would be harder to calculate the other flux Φ_{21}, because $\mathbf{B}_1$—the field of a rectangular loop—is not so simple. But that calculation is unnecessary because we know that $M_{21} = M_{12}$.

EXAMPLE 9   Find the mutual inductance for the circuits 1 and 2 shown in Fig. 10.14. Imagine that C_1 is a very long, tightly wound, solenoid, with N_1 turns and length ℓ_1. C_2 is a loosely wound solenoid, with N_2 turns and length ℓ_2. Both have cross section S.

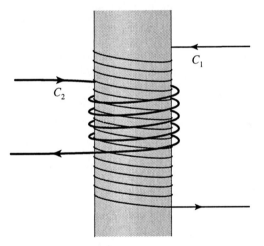

FIGURE 10.14   Example 9. Geometry of two concentric coils for which the mutual inductance is to be calculated.

In this example it is easy to calculate Φ_{21}, because $\mathbf{B}_1$ is just the field of a tightly wound solenoid $(\mu_0 N_1 I_1 / \ell_1)$ and $\Phi_{21} = N_2 B_1 S$. Thus the mutual inductance is

$$M_{21} = \frac{\Phi_{21}}{I_1} = \frac{\mu_0 N_1 N_2 S}{\ell_1}. \tag{10.41}$$

The result is not invariant under the interchange of 1 and 2. The inductance M_{12} is *not* obtained by interchanging 1 and 2 in the result, because the two solenoids are different; rather, M_{12} has the same value as M_{21} by the general theorem (10.39).

Mutual inductance is the phenomenon at work in a *transformer*. The EMF around a secondary coil C_2 is related to the changing current in a primary coil C_1

by Faraday's Law:

$$\mathcal{E}_2 = -\frac{d\Phi_2}{dt} = -M_{21}\frac{dI_1}{dt}. \tag{10.42}$$

An AC transformer may increase (step up) or decrease (step down) the voltage from the primary to the secondary, depending on the mutual inductance. For example, the ignition coil in a car takes input voltage alternating between 0 and 12 V—alternating because of the opening and closing of contact points—and steps it up to a high output voltage. Then the electric field at the spark plug exceeds the breakdown field for the air-fuel mixture in the cylinder, a spark is produced, and the mixture ignites.

The development of transformers was an important step in the history of AC electric power, because long-distance transmission lines use a much higher voltage than household appliances or electric lights. Throughout cities and towns transformers which step down the voltage for household use, can be seen on telephone poles.

There is a subtle approximation in (10.42), or indeed in any circuit calculation involving inductors. We use magnetostatics to evaluate M_{21}, but in (10.42) the magnetic field is not static! This is an example of the *quasistatic approximation*: If the current changes slowly, then the flux calculated from magnetostatics is a good enough approximation. For very high-frequency alternating currents there would be radiation fields, so the quasistatic approximation would break down. But in common applications of electrical engineering, the quasistatic approximation is very accurate.

In Section 8.7 we found the vector potential of a current loop; the result (8.106) has the form $\mathbf{A}(\mathbf{x}) = A_\phi(r, \theta)\hat{\phi}$. We now apply that result to a sophisticated example.

EXAMPLE 10   Find the mutual inductance between two coaxial current loops. The arrangement is shown in Fig. 10.15.

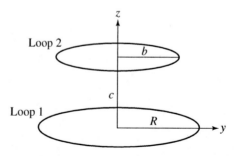

FIGURE 10.15   Example 10. Geometry of two coaxial circular loops for which the mutual inductance is to be calculated.

   To do this problem starting from first principles would be difficult, but by having (8.106) and (10.37) we've already prepared the way. Notice that to use (10.37) to find M_{ij}, we need $\mathbf{A}_j(\mathbf{x}_i)$, which is the vector potential at the ith loop due to the current in the jth loop. In our case the vector potential at Loop 2 due to the current in Loop 1 is *constant* because the spherical coordinates of points on Loop 2 are $r = \sqrt{b^2 + c^2}$ and $\theta = \arcsin(b/\sqrt{b^2 + c^2})$, which are both constant. Thus (10.37) when applied to our problem becomes, since $\mathbf{A}_1(\mathbf{x}_2) = A_\phi(r, \theta)\hat{\phi}$,

$$M_{21} I_1 = \int_{\phi_2=0}^{2\pi} A_\phi \left(\sqrt{b^2 + c^2}, \arcsin(b/\sqrt{b^2 + c^2}) \right) \hat{\phi} \cdot \hat{\phi}\, b\, d\phi_2. \quad (10.43)$$

Substituting the vector potential of Loop 1 from (8.106) into (10.43), and evaluating it with the values of r and θ at Loop 2 gives, after simplification

$$M(\text{far}) = M_{21}$$

$$= \frac{\mu_0 R b}{2\sqrt{b^2 + c^2}} \sum_{\text{odd } n=1}^{\infty} \left(\frac{R}{\sqrt{b^2 + c^2}} \right)^n \int_0^{2\pi} P_n \left(\frac{b \sin \phi'}{\sqrt{b^2 + c^2}} \right) \sin \phi'\, d\phi',$$

$$(10.44)$$

for the mutual inductance of the loops. (Recall that P_n is the Legendre polynomial of degree n.)

   It is important to note that (10.44) applies only when $\sqrt{b^2 + c^2} > R$, the condition for which (8.106) was derived; that is the significance of the notation $M(\text{far})$. Although (10.44) is an infinite series, containing only terms for odd n as discussed in Section 8.7, it converges rapidly if $R/\sqrt{b^2 + c^2}$ is small compared to unity, i.e., if Loop 2 is far away compared to R. The series starts off with the leading term ($n = 1$) as $M = \mu_0 R^2 b^2 \pi / 2(b^2 + c^2)^{3/2} + \cdots$; the higher terms can be evaluated numerically.

   If the two loops are near enough together so that $\sqrt{b^2 + c^2} < R$, then the mutual inductance becomes

$$M(\text{near}) = M_{21}$$

$$= \frac{\mu_0 b}{2} \sum_{\text{odd } n=1}^{\infty} \left(\frac{\sqrt{b^2 + c^2}}{R} \right)^n \int_0^{2\pi} P_n \left(\frac{b \sin \phi'}{\sqrt{b^2 + c^2}} \right) \sin \phi'\, d\phi'.$$

$$(10.45)$$

The mutual inductance for the interesting basic problem of two coaxial and coplanar loops is obtained by setting $c = 0$ in (10.45).

10.5 ■ MAGNETIC FIELD ENERGY

By analyzing the energy balance for an isolated inductor, we can deduce the formula for the energy content of a magnetic field. The power that must be supplied to maintain a current I in an inductor while the current is changing, in the presence of the back EMF $\mathcal{E} = -L\,dI/dt$, is

$$P = -I\mathcal{E} = +IL\frac{dI}{dt}. \tag{10.46}$$

Thus the total energy that must be supplied to establish a current I, starting from 0, is $\int P\,dt = \int L I\,dI = \frac{1}{2}LI^2$. By conservation of energy, that amount of energy is stored in the inductor

$$U = \frac{1}{2}LI^2. \tag{10.47}$$

We attribute the stored energy to the magnetic field **B** produced by the current I. Therefore, we must be able to rewrite U directly in terms of **B**.

As a first step toward relating U and **B**, we express U in terms of the current and vector potential **A**, where $\mathbf{B} = \nabla \times \mathbf{A}$. The quantity LI is, by definition, the flux of **B** through a surface S bounded by the current loop C, so

$$LI = \int_S \mathbf{B} \cdot d\mathbf{S} = \oint_C \mathbf{A} \cdot d\boldsymbol{\ell}, \tag{10.48}$$

the second equality by Stokes's theorem. Thus the magnetic energy can be written as

$$U_M = \frac{1}{2} \oint_C I\mathbf{A} \cdot d\boldsymbol{\ell}. \tag{10.49}$$

Although we have derived (10.49) for a single current loop, the same formula holds for multiple loops if we sum the right-hand side over all the loops.[9] Also, we may generalize the result to a continuous distribution of current, with volume current density $\mathbf{J}(\mathbf{x})$, by replacing $I\,d\boldsymbol{\ell}$ by $\mathbf{J}d^3x$, because $\mathbf{J}dA$ is the current dI in a tube of area dA. Thus, in general,

$$U_M = \frac{1}{2} \int \mathbf{J} \cdot \mathbf{A}\, d^3x. \tag{10.50}$$

Note that (10.50) is analogous to the equation for electrostatic energy $U_E = \frac{1}{2} \int \rho V d^3x$ that we derived in Chapter 3; in both cases, one-half of source density times potential is energy density.

Finally, we may write U_M entirely in terms of the magnetic field. For magnetostatics, Ampère's Law is that $\mathbf{J} = \nabla \times \mathbf{H}$. The integrand in (10.50) is

$$(\nabla \times \mathbf{H}) \cdot \mathbf{A} = \nabla \cdot (\mathbf{H} \times \mathbf{A}) + \mathbf{H} \cdot (\nabla \times \mathbf{A}), \tag{10.51}$$

---

[9] See Exercise 26.

where the right-hand side follows from a vector identity. The volume integral of $\nabla \cdot (\mathbf{H} \times \mathbf{A})$ is 0 for a localized field, i.e., a field that approaches 0 at ∞, by Gauss's theorem: It equals the flux of $\mathbf{H} \times \mathbf{A}$ through the surface at ∞, where the field is 0. The final term in (10.51) is $\mathbf{H} \cdot \mathbf{B}$. The total energy is

$$U_M = \frac{1}{2} \int \mathbf{H} \cdot \mathbf{B} d^3 x. \tag{10.52}$$

Attributing this energy to the magnetic field, we deduce that the local energy density is

$$u_M(\mathbf{x}) = \tfrac{1}{2}\mathbf{H}(\mathbf{x}) \cdot \mathbf{B}(\mathbf{x}). \tag{10.53}$$

In the absence of magnetic materials, $\mathbf{H}$ is $\mathbf{B}/\mu_0$. Then the magnetic energy density, associated only with the field $\mathbf{B}$, is $u_M = \tfrac{1}{2}B^2/\mu_0$. Recall for comparison that the energy density of the electric field is $u_E = \tfrac{1}{2}\epsilon_0 E^2$ in the absence of dielectric materials. In the presence of matter, the electric energy density is $\tfrac{1}{2}\mathbf{D} \cdot \mathbf{E}$ and the magnetic energy density is $\tfrac{1}{2}\mathbf{H} \cdot \mathbf{B}$; these energies include both field energy and polarization energy of the matter.

The reader has the right to feel dissatisfied with the above derivation of the important fundamental result (10.53), which was only heuristic and had some gaps. A rigorous derivation will be given in Chapter 11, based on local conservation of energy.

EXAMPLE 11   What is the magnetic energy per unit length for a long cylindrical solenoid, tightly wound with n turns per unit length of wire carrying current I? This is a good example, because it is the simplest form of magnetic field—a uniform field inside the cylinder.

Figure 10.16 shows two views of the solenoid. We will use three methods to find its energy per unit length U'_M with units J/m.

Method 1.   We first find U'_M by using (10.53), which says that the energy density is $u_M = B^2/2\mu_0$. Inside the solenoid $\mathbf{B} = \mu_0 n I \hat{\mathbf{k}}$, and outside $\mathbf{B} = 0$. Therefore the energy density inside ($r < R$) is $u_M = \mu_0 n^2 I^2/2$, and outside ($r > R$) it is zero. The volume enclosed by a length ℓ of the solenoid is $\pi R^2 \ell$. Finally then

$$U'_M = \frac{u_M \pi R^2 \ell}{\ell} = \frac{\pi}{2}\mu_0 n^2 I^2 R^2. \tag{10.54}$$

Method 2.   We next find U'_M by using (10.47), in the form $U' = L'I^2/2$. In (10.25) we found the total inductance L of a solenoid of length ℓ with N turns. Dividing by ℓ gives the inductance per unit length $L' = L/\ell = \mu_0 n^2 A$, where $n = N/\ell$ and the area of the cylinder is $A = \pi R^2$ for our case. If we substitute (10.25) into (10.47) the result for U'_M is the same as (10.54), as we expected.

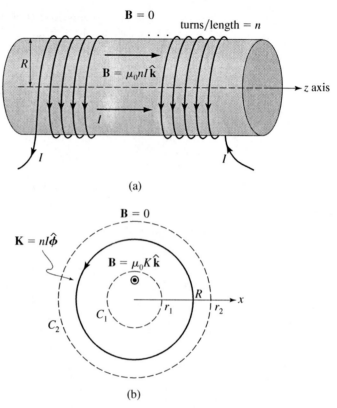

FIGURE 10.16 **Example 11.** The magnetic field energy in a solenoid. The view in (b) is from the $+z$ axis looking in the direction of $-\hat{\mathbf{k}}$.

Method 3. In the two previous calculations the magnetic energy was taken to be inside the cylindrical volume of the solenoid—that assumption was explicit in Method 1 and implicit in Method 2. However, we know from (10.49) and (10.50) that U_M may be associated with the current itself. It will therefore be interesting to find U'_M from that principle. We proceed by writing (10.50) in the equivalent form, appropriate for surface currents,

$$U_M = \frac{1}{2} \oint \mathbf{K} \cdot \mathbf{A} \, dS. \tag{10.55}$$

In applying (10.55) to our problem we have, for the current per unit length on the solenoid's surface, $\mathbf{K} = nI\hat{\phi}$ as shown in Fig. 10.16(b). Also, the surface area of a length ℓ of the solenoid is $S = 2\pi R\ell$. Now we need $\mathbf{A}(\mathbf{x})$.

A Nice Amperian Trick. To evaluate (10.55) requires the vector potential $\mathbf{A}(\mathbf{x})$ for a solenoid. We'll determine this interesting quantity by an Amperian loop trick! By Stokes's theorem, the flux of **B** through a cross section of the solenoid is

$$\Phi = \oint_C \mathbf{A} \cdot d\ell, \tag{10.56}$$

where C is the boundary curve. Equation (10.56) is reminiscent of Ampère's Law. If there's enough symmetry, we can use (10.56) to find **A** if Φ is known, just as if there's enough symmetry we can use Ampère's Law to find **B** if I is known. Figure 10.16(b) shows an end view of the solenoid with two closed curves, C_1 and C_2, along which we can apply (10.56). The current around the solenoid is constant and flows in the $\hat{\phi}$ direction, so from symmetry we take the form $\mathbf{A}(\mathbf{x}) = A_\phi(r)\hat{\phi}$ along both C_1 and C_2. Now for C_2 the enclosed flux is within R, so that $\Phi_2 = B_z \pi R^2 = \mu_0 n I \pi R^2$. For C_1 the enclosed flux is $\Phi_1 = B_z \pi r^2 = \mu_0 n I \pi r^2$. In both cases the right-hand side of (10.56) is $A_\phi(r) 2\pi r$. Therefore

$$\mathbf{A}_{ext} = A_\phi(r)\hat{\phi} = \frac{\mu_0 n I R^2}{2r}\hat{\phi} \quad \text{for} \quad r > R, \tag{10.57}$$

and

$$\mathbf{A}_{int} = A_\phi(r)\hat{\phi} = \frac{\mu_0 n I r}{2}\hat{\phi} \quad \text{for} \quad r < R. \tag{10.58}$$

Notice that **A** is continuous at $r = R$ and has there the value $\mathbf{A}(R) = (\mu_0 n I R/2)\hat{\phi}$.

Now we return to the calculation of the magnetic energy (10.55). Substituting **K**, **A** and the surface area S, we find

$$U'_M = \frac{U_M}{\ell} = \frac{1}{2}(nI)\hat{\phi} \cdot \left(\frac{1}{2}\mu_0 n I R\right)\hat{\phi} \, \frac{2\pi R\ell}{\ell} = \frac{1}{2}\mu_0 \pi n^2 I^2 R^2, \tag{10.59}$$

in agreement with the previous methods.

The vector potentials in the exterior and interior regions (10.57) and (10.58) are related to the respective magnetic fields by $\mathbf{B} = \nabla \times \mathbf{A}$. It's interesting to note that although $\mathbf{B} = 0$ outside the solenoid, **A** is not zero there and does have physical significance. For example, if the solenoid's current were to vary with time, **A** would also become a function of time, of the form $\mathbf{A} = A_\phi(r, t)\hat{\phi}$. Whenever **A** changes with time there is an associated electric field $\mathbf{E} = -\partial\mathbf{A}/\partial t$, because Faraday's law written in terms of **A** is $\nabla \times \mathbf{E} = \nabla \times (-\partial\mathbf{A}/\partial t)$.[10] For a solenoid with a varying current there is an azimuthal electric field and so nonzero EMF around C_1 and C_2.

The fact that the energy in an inductor is $\frac{1}{2}LI^2$ provides another basis for calculating the self-inductance L of an inductor: Calculate the field energy (10.52),

[10]We'll learn more about the connection between **E**, **A**, and the scalar potential V in Chapter 11.

and set it equal to $\frac{1}{2}LI^2$. This technique is sometimes easier than using the definition $L = \Phi/I$, because calculating Φ can be difficult if it involves self-flux through wires of finite size. The next example illustrates the method.

EXAMPLE 12   Find the self-inductance per unit length of a long coaxial cable for which the inner conductor is a solid cylinder of radius a and the outer conductor has radius b. Let ℓ be the length of the cable. Assume that the current in the inner cable is uniformly distributed over the cross section.

Current I flows in the inner conductor, and $-I$ in the outer conductor. By Ampère's Law the magnetic field is $B(r)\hat{\phi}$, and

$$\mathbf{B}(r) = \begin{cases} \mu_0 Ir/(2\pi a^2) & \text{for} \quad r \leq a \\ \mu_0 I/(2\pi r) & \text{for} \quad a \leq r \leq b \\ 0 & \text{for} \quad r > b. \end{cases} \tag{10.60}$$

Therefore, using (10.52), the total energy is

$$\frac{1}{2}LI^2 = \frac{\ell}{2\mu_0}\left[\int_a^b \left(\frac{\mu_0 I}{2\pi r}\right)^2 2\pi r\, dr + \int_0^a \left(\frac{\mu_0 Ir}{2\pi a^2}\right)^2 2\pi r\, dr\right]. \tag{10.61}$$

The integrals over r are elementary, and it is straightforward to show that the self-inductance per unit length is

$$L' = \frac{L}{\ell} = \frac{\mu_0}{2\pi}\ln\frac{b}{a} + \frac{\mu_0}{8\pi}. \tag{10.62}$$

Note that the self-inductance of the cable diverges logarithmically in the limit $a \to 0$. If the inner conductor were a thin cylindrical sheet of radius a, then L' would be just the first term on the right of (10.62). The second term, which arises from the field energy inside the inner conductor, is the internal self-inductance per unit length of a wire of *finite* radius.[11]

The case of an *isolated* wire is similar. For a wire of finite radius a the internal self-inductance per unit length is again $\mu_0/8\pi$. However, in the limit $a \to 0$ the external magnetic field energy density near the wire, which is $B^2/2\mu_0 = \mu_0 I^2/(8\pi^2 r^2)$, goes to infinity as $r \to 0$ so fast that the external contribution to the inductance becomes infinite.[12]

10.5.1 ■ Energy in a Ferromagnet

Equation (10.52) for the magnetic energy is correct if the only magnetic materials present are linear materials—diamagnetic or paramagnetic. Its derivation relies

---

[11] Smythe, Sec. 8.09.

[12] Recall (7.23) that the capacitance per unit length of a wire with length L and radius a is $C' = 2\pi\epsilon_0/\ln(L/a)$.

on the assumption that **B** is proportional to I. In the presence of ferromagnetic materials, that assumption is not true. I is the free current, and although **H** is proportional to I, **B** is not. In ferromagnetic systems, the definition of magnetic energy is problematic, because of the phenomenon of *hysteresis*. The energy required to prepare the system in a specified state depends on the path by which the state is created.

To analyze the energy for a system with a ferromagnet, consider a current-carrying coil of N turns, wound around a solid ferromagnetic ring, as shown in Fig. 10.17. If the flux per turn changes by $\delta\Phi$ in time δt, then the EMF induced in each turn is $\mathcal{E} = -\delta\Phi/\delta t$. The work that must be done against this back EMF to maintain the current I in all N turns of wire is $\delta W = -\mathcal{E}NI\delta t = NI\delta\Phi$. Now, $NI = \oint_C \mathbf{H} \cdot d\ell$ by Ampère's Law, where C is an Amperian loop linked with the coil (i.e., C is a circle inside the iron ring). Thus the work is

$$\delta W = \oint_C \delta\Phi\,\mathbf{H} \cdot d\ell. \tag{10.63}$$

For a process in which the magnetic field changes by $\delta\mathbf{B}$ the change in flux is $\delta\Phi = S\delta B$, where S is the area enclosed by one turn of the coil. Inserting this result in (10.63) we encounter $Sd\ell$, which is a volume element d^3x in the field. That is, the work done to change the magnetic field by $\delta\mathbf{B}$ is

$$\delta W = \int \delta\mathbf{B} \cdot \mathbf{H} d^3x. \tag{10.64}$$

For a linear material ($\mathbf{B} = \mu\mathbf{H}$), δW is equal to the change of magnetic energy δU_M, with U_M given by (10.52). In that case the total work done as the system undergoes a cyclic process, starting and ending in the same state, is $U_M^{\text{final}} - U_M^{\text{initial}} = 0$. But for a ferromagnet the total work done in a cyclic process is not 0. In other words, the process of changing **B** is not a conservative process. There is heating of the system as **B** changes. We cannot define the magnetic energy merely by integrating the work done on the system.

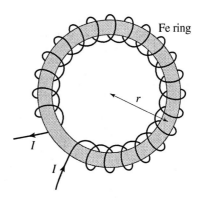

FIGURE 10.17   Example 13. A toroidal solenoid filled by a ferromagnetic material.

EXAMPLE 13   Consider an iron ring wound with N turns of wire carrying current I. For simplicity assume the radius r of the ring is much larger than the dimensions of its cross-sectional area A. How much energy must be supplied to carry the iron through one hysteresis cycle?

By Ampère's Law, **H** inside the torus, at distance r from the symmetry axis, is

$$\mathbf{H} = \frac{NI}{2\pi r}\hat{\phi}. \tag{10.65}$$

We may treat r as a constant. Now let I vary to change **H** and **B**. The power that must be supplied by the power supply is $P = -I\mathcal{E}$, where $\mathcal{E}$ is the back EMF, which is $-NAdB/dt$ by Faraday's Law. So,

$$P = \frac{2\pi r H}{N} NA\frac{dB}{dt} = VH\frac{dB}{dt}, \tag{10.66}$$

where $V = 2\pi r A$ is the volume of iron. The total energy supplied as **H** and **B** vary around one hysteresis cycle is

$$W = \int P\,dt = V \oint H\,dB. \tag{10.67}$$

In other words, $\oint H\,dB$ is the energy loss in the iron, per unit volume per hysteresis cycle.

If H is plotted as a function of B, as illustrated previously in Fig. 9.19, then the area enclosed by the hysteresis loop in the BH plane is $\oint H\,dB$. The energy loss per unit volume equals the area enclosed by the hysteresis loop.

EXERCISES

Sec. 10.1. Motional EMF

10.1. A metal bar (mass m, length ℓ) slides on horizontal conducting rails in a vertical uniform magnetic field, as shown in Fig. 10.18. The electrical resistance of the closed circuit consisting of the rails and the bar is R, independent of the position x of the bar.

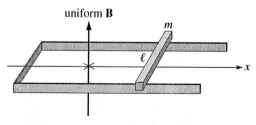

FIGURE 10.18   Exercise 1. A conducting bar of mass m slides on conducting rails.

(a) If the bar moves with velocity v in the x direction, what is the current in the circuit?

(b) What is the force on the bar? Note that this is a braking force.

(c) If the initial velocity of the bar is v_0, how far does the bar slide before stopping? (Neglect friction with the rails.)

(d) Show that the initial kinetic energy of the bar has all been expended as Joule heat when the bar stops.

10.2. In 1996 astronauts on the space shuttle Columbia did an interesting experiment, to test an idea for generating electric power in satellites in orbit around the Earth, using a *conducting tether*. Columbia deployed a satellite attached to the space shuttle by a 20-km long tether. Motion of the tether through the Earth's magnetic field would generate a voltage drop across the tether. (In vacuum, charge would build up on the ends, but in the ionosphere current would flow, the return current being in the plasma.)

Assume the tether is a straight wire, held taut by the shuttle and satellite at its ends. The whole system revolves around the Earth with speed 8 km/s. Assume for simplicity that the wire direction $\boldsymbol{\ell}$, velocity $\mathbf{v}$, and Earth's magnetic field $\mathbf{B}$ are all mutually orthogonal. The strength of the magnetic field is 0.3×10^{-4} T at the height of the satellite. Calculate the voltage difference between the ends of the wire. (In the NASA experiment the tether broke or melted through before the end of the experiment. But data collected before this catastrophe showed the potential for generating power by this method.)

10.3. A copper disk of radius 5 cm rotates at 20 revolutions per second, in a magnetic field $\mathbf{B} = 0.5\,\mathrm{T}\,\hat{\mathbf{n}}$ perpendicular to the disk. The rim and center are connected electrically by a fixed wire with sliding contacts. The total resistance is $10\,\Omega$. Calculate the induced current.

10.4. A *magnetohydrodynamic (MHD) generator* is a device that has been proposed for generating power from flow of ionized plasma, e.g., in nuclear fusion reactors. The plasma flows in the z direction through a rectangular pipe, whose cross section is parallel to the xy plane, and there is a magnetic field $\mathbf{B} = B\,\hat{\mathbf{i}}$ in the plasma. The x, y, z dimensions of the pipe are w, h, ℓ, respectively. The walls at $x = \pm w/2$ are insulating, and the walls at $y = \pm h/2$ are conducting.

(a) Show that the potential between the conducting walls is $V = vBh$, where v is the fluid velocity.

(b) Suppose the conducting walls are connected by a wire with resistance R. Determine the current in the wire, if ρ is the resistivity of the plasma. (Hint: There are currents in series in the wire and in the plasma.)

10.5. A method for measuring the Earth's magnetic field is to use a *flip coil* and *ballistic galvanometer*. The coil, of radius a, turns quickly through 180 degrees. The galvanometer measures the total charge that flows through the coil as it flips.

(a) Explain how the vector $\mathbf{B}$ can be measured.

(b) Suppose the coil axis is initially parallel to a magnetic field of strength 0.5×10^{-4} T, and $a = 5$ cm and the resistance of the coil is $R = 0.1\,\Omega$. What total charge flows through the coil as it flips over?

Sec. 10.2. Faraday's Law

10.6. There is a time-dependent current $I_s(t)$ in a long and densely wound solenoid.

    (a) Determine the electric field at radius r on the midplane of the solenoid, both inside and outside the solenoid. (Hint: The direction of **E** is azimuthal; use the Amperian loop trick.) This problem is another example of the quasistatic approximation. Use magnetostatics to derive a formula for **B** due to a steady current, and use the same formula for the varying current.

    (b) From your result of (a) calculate the curl of **E** at radius r.

10.7. How can the purely azimuthal electric field $\mathbf{E} = E_\phi \hat{\phi} = -(E_0 r/a)\hat{\phi}$ be produced in a cylindrical region of radius a?

10.8. A long straight wire carries an alternating current $I(t) = I_0 \cos \omega t$. Nearby is a square loop. The wire lies in the plane of the loop, parallel to two sides of the square, which are at distances a and b from the wire. (The side of the square is $b - a$.) Determine the current induced in the square loop if its resistance is R.

10.9. A circular loop of wire with radius a and electrical resistance R lies in the xy plane. A uniform magnetic field is turned on at time $t = 0$; for $t > 0$ the field is

$$\mathbf{B}(t) = \frac{B_0}{\sqrt{2}} \left(\hat{\mathbf{j}} + \hat{\mathbf{k}}\right)\left[1 - e^{-\lambda t}\right].$$

    (a) Determine the current $I(t)$ induced in the loop.

    (b) Sketch a graph of $I(t)$ versus t.

10.10. A metal disk of radius a, thickness d, and conductivity σ is located in the xy plane, centered at the origin. There is a time-dependent uniform magnetic field $\mathbf{B}(t) = B(t)\hat{\mathbf{k}}$. Determine the induced current density $\mathbf{J}(\mathbf{x}, t)$ in the disk.

10.11. A classroom demonstration of eddy currents and magnetic braking is to let a small cylindrical magnet, e.g., a cow magnet, fall through a vertical copper tube with a slightly larger diameter than the magnet. Analyze the induced eddy currents, and the force on the magnet, and explain why the magnet falls at a slow terminal velocity.

10.12. Consider the circuit shown in Fig. 10.19, in the center of which there is a long solenoid with a changing magnetic field $\mathbf{B}(t)$. Two voltmeters are connected as shown across the two resistors R_1 and R_2.

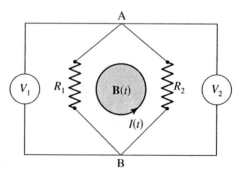

FIGURE 10.19   **Exercise 12.** The two voltmeters V_1 and V_2 have different readings.

(a) Determine the readings on the two voltmeters.

(b) How is it possible that the readings are different? The readings are different even if $R_1 = R_2$. Should not there be a unique voltage drop between points A and B? (Reference: R. H. Romer, *Am. J. Phys.* **50**, 1089 (1982).)

Sec. 10.3. Applications

10.13. The alternator in a car consists of a rectangular coil, with 250 turns of wire and area $0.01\,\mathrm{m}^2$, rotating in a 0.1 T magnetic field. (The field is produced by a direct-current electromagnet.) If the rotation rate is 10^3 rpm, what is the peak output voltage?

10.14. Consider a betatron with these parameters: the radius of the electron orbit is 0.5 m; the kinetic energy of an electron injected into the accelerator is 2.0 MeV; and the rate of increase of the magnetic flux through the area of the toroidal vacuum chamber is 25 Wb/s. The electrons are ejected after 4 ms of acceleration.

(a) Compute the magnitude of the induced electric field.

(b) Compute the work done on an electron per revolution around the orbit.

(c) Compute the number of revolutions completed by an electron before ejection. (Approximate the electron speed by c.)

(d) Compute the final kinetic energy of the electron.

(e) In order to keep the radius r of the electron path constant, B at r must be one half of the average of B over the area enclosed by the circular orbit. What is dB/dt at r during the acceleration?

10.15. Show that a series LC circuit without resistance is a harmonic oscillator. Calculate the frequency if $L = 300\,\mathrm{mH}$ and $C = 1\,\mu\mathrm{F}$.

10.16. Consider a *coaxial cable* consisting of two long concentric hollow conducting cylinders with radii a and b. A current I travels up the inner cylinder, and down the outer cylinder. Determine the self-inductance per unit length, both from the definition $L = \Phi/I$, and from the magnetic energy $\frac{1}{2}LI^2$.

10.17. The circuit shown in Fig. 10.20 consists of a switch, inductor, light bulb, and battery. Assume $R = 10\,\Omega$ and $L = 10\,\mathrm{mH}$. The switch closes at time $t = 0$. At what time does the light reach 90% of its final brightness? (To answer this question one must make some assumption about the thermal conductivity of the bulb filament. Assume it is infinite, i.e., no time delay between Joule power and light intensity.)

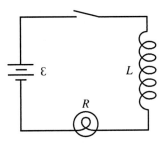

FIGURE 10.20   Exercise 17.

10.18. Determine the self-inductance per unit length of circumference, $L' = L/2\pi R$, for a *toroidal solenoid*—a wire wound around a torus. Show that in the limit $R \to \infty$ with A and n fixed, where R is the toroid radius, A the cross-sectional area, and n the number of turns per unit length, the result approaches the self-inductance per unit length of a cylindrical solenoid. (The self-inductance of the latter case is (10.25).)

Sec. 10.4. Mutual Inductance

10.19. Find the mutual inductance M between a long cylindrical solenoid, tightly wound with n_l turns per unit length, and a short solenoid inside it and far from the ends. The axis of the short solenoid is parallel to the axis of the long solenoid, and the short solenoid has cross-sectional area A and N_s turns. Notice that the mutual inductance is the same for any uniform cross-sectional shapes of the solenoids. Note also that M is the same whether the N_s turns of the inner solenoid are wound tightly or loosely.

10.20. Figure 10.21 shows an equilateral triangular loop whose altitude is a, and a long wire in the same plane. The wire is parallel to and at distance b from the base of the triangle. What is the mutual inductance? Assume the return path of the long wire is far away. [Answer:

$$M = \frac{\mu_0}{\pi\sqrt{3}}\left[(a+b)\ln\left(1 + \frac{a}{b}\right) - a\right]]$$

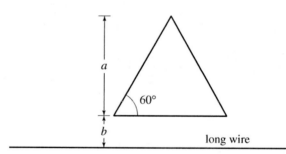

FIGURE 10.21   Exercise 20.

10.21. The ignition coil in a car supplies $20\,\text{kV}$ to the spark plugs. Suppose the maximum current in the primary coil is $4.0\,\text{A}$, and the current is interrupted 100 times per second (by the distributor points). Estimate the mutual inductance of the primary and secondary coils. (Hint: You will have to make a reasonable model of the function $I(t)$.)

Sec. 10.5. Magnetic Field Energy

10.22. Equal but opposite currents $+I$ and $-I$ flow in two long parallel plates, as shown in Fig. 10.22. The plates have width w and separation d, where d is small.

(a) Neglecting edge effects, find the magnetic field between the plates, from Ampère's Law.

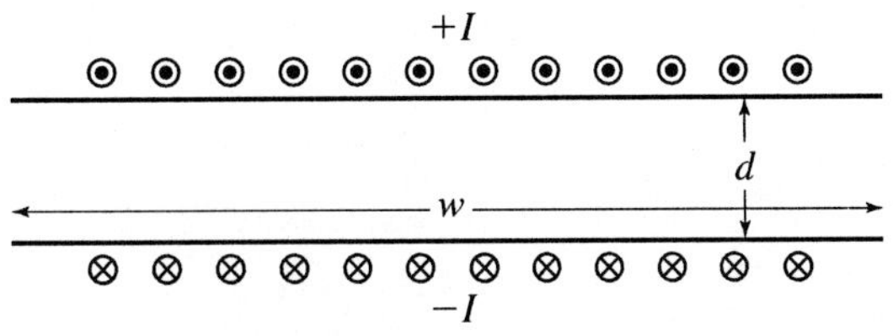

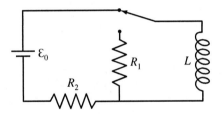

FIGURE 10.22 Exercise 22. End view of long parallel plates of width w and separation d, carrying total currents $+I$ and $-I$.

(b) Calculate the magnetic field energy per unit length.

(c) Use the result of (b) to show that the self-inductance per unit length is $\mu_0 d/w$.

10.23. Determine E/B if the electric and magnetic field energy densities are equal. Evaluate the ratio in m/s. What is this constant?

10.24. Consider the circuit shown in Fig. 10.23. Assume the switch has been in the position shown long enough that the steady state has been reached. Now at $t = 0$, flip the switch.

FIGURE 10.23 Exercise 24.

(a) Determine the current $I(t)$ for $t > 0$.

(b) Determine the total energy dissipated in the resistor after $t = 0$.

(c) Show that the result of (b) is equal to the energy stored in the inductor at $t = 0$.

10.25. Show that the magnetic field energy of *two current loops* is

$$\tfrac{1}{2}L_1 I_1^2 + \tfrac{1}{2}L_2 I_2^2 + M I_1 I_2$$

where L_1, L_2, and M are the self and mutual inductances. By convention we take M to be positive, and the signs of I_1 and I_2 determine the directions of the currents. (Reference: *The Feynman Lectures on Physics*, Vol. II, Sec. 17-8.)

10.26. Consider a collection of current loops $C_1, C_2, C_3, \dots$ fixed in space.

(a) Show that the total energy that must be supplied to establish currents I_1, I_2, $I_3, \dots$ is

$$U = \sum_i \frac{1}{2} L_i I_i^2 + \sum_{i<j} M_{ij} I_i I_j.$$

(Hint: The power is $-\sum I_i \mathcal{E}_i$, where $\mathcal{E}_i$ is the EMF induced by the changing flux through C_i.)

(b) Show that $U = \frac{1}{2} \sum_i I_i \Phi_i$.

10.27. Following are two exercises on astrophysical magnetic field energies.

(a) The average magnetic field in the interstellar space of our galaxy is about 3×10^{-10} T. If the galaxy is taken to be a disk-shaped volume, of diameter 10^{21} m and thickness 10^{19} m, what is the total magnetic energy of the galaxy? The total power radiated from all stars in the galaxy is about 10^{37} W. How many years of starlight are equivalent to the stored magnetic energy?

(b) The magnetic field at the surface of a neutron star, or pulsar, is about 10^8 T. What is the magnetic energy density for this field? Using the mass-energy relation $E = mc^2$, find the mass density that corresponds to the magnetic energy density. Compare this to the mass density of the neutron star, assuming it has the mass of the sun in a sphere of radius 10 km.

10.28. Assume the magnetic field of the Earth, in the region outside the Earth, is a pure dipole field corresponding to dipole moment $m = 8 \times 10^{22}$ J/T. (This is not completely accurate, because the solar wind cuts off the field at a distance of order 10 Earth radii, but it's a good enough approximation for this exercise.) Calculate the total energy of the magnetic field outside the Earth. Compare the result to the rotational kinetic energy of the Earth.

General Exercises

10.29. The solenoid in Fig. 10.24 is long, and has n turns per unit length and cross section A.

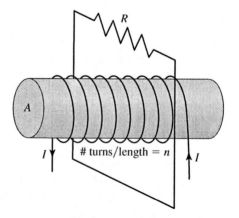

FIGURE 10.24   Exercise 29.

(a) If the current in the solenoid changes from I_1 to I_2, how much charge passes through the resistor?

(b) If the current in the solenoid, as a function of time t, is

$$I(t) = I_1 e^{-t/\tau} + I_2 \left(1 - e^{-t/\tau}\right),$$

what is the current $I_R(t)$ in the resistor?

(c) Use a computer to plot $I(t)$ and $I_R(t)$, as functions of t/τ, for $I_1 = 1$ A and $I_2 = 2$ A. Describe in words what is happening.

10.30. In Fig. 10.18 the conducting rod m is in contact with metal rails and the whole system is immersed in a uniform magnetic field of strength $B = 0.5$ T in the direction perpendicular to circuit. (The bar, rails and cross piece at the end of the rails form a continuous conducting path.)

(a) If the rod moves toward the right with a speed of 4.0 m/s, and the separation of the rails is $\ell = 0.10$ m, find the magnitude and direction of the induced emf.

(b) If the circuit has a resistance of 0.20 Ω when the rod is at a certain position, find the force exerted on the rod by the magnetic field.

(c) Compare the power of the external force and the power lost in Joule heating. Ignore friction.

10.31. Calculate the terminal velocity of a rectangular loop falling out of a region of **B**, as shown in Fig. 10.7. Take $B = 1$ T, $R = 0.1$ Ω, $M = 0.1$ kg, and $w = 0.1$ m.

10.32. For the Rowland Ring of Fig. 9.18 (Chapter 9):

(a) Show that if the primary current is changed by ΔI_p then the incremental change in H in the magnetic material is

$$\Delta H = N_p \, \Delta I_p / (2\pi R_0).$$

(b) Show that the resultant change in B in the magnetic material satisfies

$$\frac{\Delta B}{\Delta H} = -\frac{2\pi R_0 R_g \Delta Q_g}{A N_s N_p \Delta I_p},$$

where R_g is the resistance of the secondary circuit including the galvanometer and ΔQ_g is the amount of charge that passes through the galvanometer. (Assume that the cross-sectional dimension of the magnetic material is small compared to R_0.) This relation is used to construct magnetization curves of B versus H such as those in Fig. 9.19 of Sec. 9.6.

Computer Exercises

10.33. Consider damped oscillations in a series LRC circuit. The charge on the capacitor $Q(t)$, and the current in the circuit $I(t) = dQ/dt$, are functions of time t.

(a) Show that the differential equation satisfied by $Q(t)$ is

$$L\frac{d^2 Q}{dt^2} + R\frac{dQ}{dt} + \frac{Q}{C} = 0.$$

(b) Use a computer program to solve the equation numerically. As a template, here are Mathematica commands to solve the equation for parameter values $L = 1$,

$R = 1$, and $C = 1$ (in some system of units) and with initial values $Q(0) = 4$ and $I(0) = 0$:

```
{L,R,C}={1,1,1}
eqs={Q'[t]==I[t],
    I'[t]==-(R/L)I[t]-Q[t]/(L*C),
    Q[0]==4,I[0]==0}
soln=NDSolve[eqs,{Q[t],I[t]},{t,0,20}]
Qsol[t_]:=Evaluate[Q[t] /. soln]
Isol[t_]:=Evaluate[I[t] /. soln]
```

After these commands have been executed, `Qsol[t]` and `Isol[t]` are the functions $Q(t)$ and $I(t)$.

(c) Make plots of $Q(t)$ and $I(t)$ for overdamped oscillations ($R^2 > 4L/C$), underdamped oscillations ($R^2 < 4L/C$), and weakly damped oscillations ($R^2 \ll 4L/C$).

10.34. The differential equation for a harmonically driven series LRC circuit is

$$L\frac{d^2Q}{dt^2} + R\frac{dQ}{dt} + \frac{Q}{C} = \mathcal{E}_0 \sin \omega t.$$

(a) Let $L = 1$, $C = 1$, and $R = 0.1$, in some system of units. Solve the equation numerically, for initial values $Q(0) = 0$ and $I(0) = 0$, and driving frequency $\omega = 0.5$. After a transient period, the solution settles down to a steady state oscillation at the driving frequency ω.

(b) Now vary ω, in the range from 0.5 to 1.5. Show, by plots of $Q(t)$ and $I(t)$, that the steady state oscillations have maximum amplitude if $\omega \approx 1/\sqrt{LC}$. Explain this result in terms of *resonance*.

(c) Show graphically that there is a phase shift between the applied EMF and the charge on C.

CHAPTER
11

The Maxwell Equations

Before the development of field theory by Faraday and Maxwell, electric and magnetic phenomena were ascribed to "action at a distance" between charges and between currents. Action at a distance refers to the experimental observation that charges or currents exert forces on each other even though they are not in contact. The Coulomb force between static charges, and the force between currents explored by Ampère, both appear to be examples of action at a distance.

Faraday, from his extensive experience with experiments on electricity and magnetism, developed the idea that there must be something *physical* in the space around charges and currents, which he called *lines of force*. This idea was the kernel of field theory.

Maxwell expressed Faraday's idea in four partial differential equations—local relations—of fields and sources. Maxwell coined the phrase *electromagnetic field* as the name for his theory, because the action takes place not at a distance, but locally in the space around the sources. Maxwell's equations were published in 1864, and we still use the same equations today. This theory has really stood the test of time!

The interaction between charges in field theory is quite different from action at a distance. A charge creates an associated electromagnetic field throughout the surrounding space; another charge experiences a force exerted by the local field at its position. The fields have energy and momentum. They are not mere mathematical constructions, but have just as much physical reality as the charges. From this viewpoint, the answer to the question "What is the Universe made of?" is "particles and fields."

Maxwell's mathematical theory revealed that light is an electromagnetic phenomenon. Also, it predicted the existence of other forms of electromagnetic radiation, unknown at the time, such as radio waves and X rays. The theory led to great advances in technology. Recent applications of the theory are cellular telephones, positional devices for global navigation, and communication with spacecraft.

The purpose of this chapter is to introduce the Maxwell equations, both in vacuum and in the presence of matter, and to examine some of their consequences. In particular, we will begin the study of electromagnetic waves in Sec. 11.5.

11.1 ■ THE MAXWELL EQUATIONS IN VACUUM AND THE DISPLACEMENT CURRENT

The Maxwell equations express relations between the electromagnetic fields $\mathbf{E}(\mathbf{x}, t)$ and $\mathbf{B}(\mathbf{x}, t)$, and between the fields and their sources. We start our study by presenting the Maxwell equations *in vacuum*. Then in Sec. 11.3 we will consider how the equations should be rewritten in the presence of dielectric and magnetic materials.

In considering Maxwell's equations *in vacuum*, we don't necessarily mean in empty space. We have in mind a space in which there are sources, generally varying in time, of charge $\rho(\mathbf{x}, t)$ and of current $\mathbf{J}(\mathbf{x}, t)$. We may think of these as *ideal sources*, i.e., not associated with matter. In reality currents flow in wires, and charge resides in electrons or ions; real physical sources are associated with matter. But for many practical problems—and the Maxwell equations in vacuum have wide-ranging practical applications—the vacuum equations, with idealized sources, are an accurate description of electromagnetic phenomena. In such cases the material properties of the physical sources can be neglected. On the other hand, if materials are present with significant magnetic or electric susceptibilities, then bound charge and current must be taken into account, as explained in Sec. 11.3.

We have already studied three of the Maxwell equations,

$$\nabla \cdot \mathbf{E} = \rho/\epsilon_0 \tag{11.1}$$

$$\nabla \cdot \mathbf{B} = 0 \tag{11.2}$$

$$\nabla \times \mathbf{E} = -\partial \mathbf{B}/\partial t. \tag{11.3}$$

Equations (11.1) to (11.3) are correct and complete as they stand. They hold for time-dependent systems as well as for static systems. The first (11.1) is Gauss's Law for the electric field $\mathbf{E}$ and charge density ρ, which states that the electric field diverges from charges. The second (11.2) is Gauss's Law for the magnetic field $\mathbf{B}$, which states that there are no magnetic monopoles, i.e., no isolated pointlike magnetic sources. Equivalently, lines of $\mathbf{B}$ are closed loops. The third equation (11.3) is Faraday's Law of electromagnetic induction, which states that $\mathbf{E}$ curls around the rate of change of $\mathbf{B}$ if $\mathbf{B}$ varies in time.

Another necessary equation in electrodynamics is the continuity equation, discussed in Sec. 7.2, which expresses the local conservation of charge:

$$\nabla \cdot \mathbf{J} = -\frac{\partial \rho}{\partial t}. \tag{11.4}$$

In words, *electric charge is conserved throughout space*: the current flux away from a point is equal to the rate of decrease of the charge at that point.

In addition to the above equations, we have previously studied Ampère's Law for magnetostatics:

$$\nabla \times \mathbf{B} = \mu_0 \mathbf{J} \quad \text{(magnetostatics)}. \tag{11.5}$$

Equation (11.5) states that electric current is a source of magnetic field, and **B** curls around a constant current. Although (11.5) is true for time-independent currents, it is not complete and it must be modified for time-dependent problems. The modification consists of an additional term, called the *displacement current*, added to the right-hand side, which we will derive presently. Remarkably, it is possible to construct the complete equation by a purely theoretical analysis, which is how it was found originally by Maxwell.

Maxwell was the first theorist to construct differential equations such as (11.1) to (11.5) for electromagnetism. But he found that (11.1) to (11.5) are not mathematically self-consistent. Specifically, there is an inconsistency between (11.5) and (11.4) if the charge system is not static. To understand this point, let's first examine the divergence of (11.3), which will prove to be consistent, and then the divergence of (11.5), which will prove to be inconsistent for nonstatic electric fields. In the case of Faraday's Law, the divergence of the left-hand side of (11.3) is 0, because the divergence of the curl of any vector is 0; and the divergence of the right-hand side is also 0, by (11.2)

$$\mathbf{\nabla} \cdot \left(\frac{\partial \mathbf{B}}{\partial t} \right) = \frac{\partial}{\partial t} \left(\mathbf{\nabla} \cdot \mathbf{B} \right) = 0. \tag{11.6}$$

(It is important to understand that $\mathbf{\nabla}$ and $\partial/\partial t$ commute.) Thus equations (11.3) and (11.2) are consistent. But now consider the case of Ampère's Law from magnetostatics (11.5): Again the divergence of the left-hand side is 0. But the divergence of the right-hand side $\mu_0 \mathbf{\nabla} \cdot \mathbf{J}$ is in general *not* 0, by (11.4). $\mathbf{\nabla} \cdot \mathbf{J}$ is 0 if the charge density is static, but in general $\mathbf{\nabla} \cdot \mathbf{J}$ is $-\partial\rho/\partial t$. Equations (11.5) and (11.4) are thus inconsistent for time-dependent systems.

11.1.1 ■ The Displacement Current

Maxwell made the equations of electromagnetic fields consistent by modifying (11.5), for the general time-dependent case, adding another source term that he called the *displacement current*. The complete equation, in vacuum, is the fourth Maxwell equation

$$\mathbf{\nabla} \times \mathbf{B} = \mu_0 \mathbf{J} + \mu_0 \epsilon_0 \frac{\partial \mathbf{E}}{\partial t}. \tag{11.7}$$

We may call this equation the Ampère-Maxwell Law. It is consistent with the others. The divergence of the left-hand side of (11.7) is 0 as before. The divergence of the right-hand side is also 0: The divergence of **J** is $-\partial\rho/\partial t$ by conservation of charge, and the divergence of $\epsilon_0 \partial \mathbf{E}/\partial t$ is $+\partial\rho/\partial t$ by Gauss's Law. The four Maxwell equations, (11.1) to (11.3) and (11.7), along with the continuity equation (11.4), are a self-consistent theory of the electromagnetic field. These are the field equations in vacuum. They are tabulated in the second column of Table 11.1.

These equations speak to us if we understand the language of vector calculus. Faraday's Law tells us that a changing magnetic field acts as a source of electric field, as **E** curls around the change of **B**. The Ampère-Maxwell Law says that

TABLE 11.1   The Maxwell equations

Equation	In vacuum	In matter
Gauss	$\nabla \cdot \mathbf{E} = \rho/\epsilon_0$	$\nabla \cdot \mathbf{D} = \rho_f$
Gauss	$\nabla \cdot \mathbf{B} = 0$	$\nabla \cdot \mathbf{B} = 0$
Faraday	$\nabla \times \mathbf{E} = -\partial \mathbf{B}/\partial t$	$\nabla \times \mathbf{E} = -\partial \mathbf{B}/\partial t$
Ampère-Maxwell	$\nabla \times \mathbf{B} = \mu_0 \mathbf{J} + \mu_0\epsilon_0 \partial \mathbf{E}/\partial t$	$\nabla \times \mathbf{H} = \mathbf{J}_f + \partial \mathbf{D}/\partial t$

a changing electric field acts as a source of magnetic field, similarly. Maxwell named this source of $\mathbf{B}$ the displacement current. The *displacement current density in vacuum* is defined as

$$\mathbf{J}_D = \epsilon_0 \frac{\partial \mathbf{E}}{\partial t} = \frac{\partial \mathbf{D}}{\partial t}, \tag{11.8}$$

where $\mathbf{D} = \epsilon_0 \mathbf{E}$ is the *displacement field* in vacuum. That is, the Ampère-Maxwell Law is

$$\nabla \times \mathbf{B} = \mu_0 \left(\mathbf{J} + \mathbf{J}_D \right). \tag{11.9}$$

Both charge current $\mathbf{J}(\mathbf{x}, t)$ and displacement current $\mathbf{J}_D(\mathbf{x}, t)$ create the magnetic field. In fact, it is possible to have a magnetic field without any charge current at all, created only by an electric field that varies in time. We'll see later that this phenomenon occurs in electromagnetic waves. By integrating (11.9) over an open surface S with boundary curve C, and applying Stokes's theorem, we obtain the Ampère-Maxwell Law in *integral form*

$$\oint_C \mathbf{B} \cdot d\boldsymbol{\ell} = \mu_0 \int_S \left(\mathbf{J} + \mathbf{J}_D \right) \cdot d\mathbf{A}. \tag{11.10}$$

Written more simply,

$$\oint_C \mathbf{B} \cdot d\boldsymbol{\ell} = \mu_0 \left(I + I_D \right) \tag{11.11}$$

where I is the charge current and I_D the displacement current through S.

In laboratory systems of charges and currents, the displacement current is normally very small compared to typical charge currents, which explains why the displacement current was not discovered experimentally before Maxwell's theory. To observe $\mathbf{J}_D$ experimentally requires high frequencies, large $|\mathbf{E}|$, or both. Nevertheless, the displacement current is an essential part of electromagnetism because it is necessary to the propagation of electromagnetic waves, as we shall see. Indeed, the first experimental confirmation of the displacement current came with Hertz's laboratory demonstration of electromagnetic waves in 1887, more than 20 years after the publication of Maxwell's theory. A later *direct* laboratory observation of the displacement current, i.e., not in electromagnetic waves but in a system of charges and currents, was an ingenious experiment by M. R. Van Cauwenberghe in 1929.[1]

The Maxwell equations have wave solutions (Sec. 11.5) that describe light and other forms of electromagnetic radiation. The first production of electromagnetic waves in a laboratory experiment, which verified the Maxwell theory, was by Heinrich R. Hertz (1857–1894) in 1887. He discharged an induction coil through a spark gap, and observed small sparks in the spark gap of a receiver coil across the room. After studying this small effect in detail, he explained it as a consequence of electromagnetic waves traveling from the induction coil to the receiver. Hertz made this discovery as a result of *basic scientific research*, whose purpose was to clarify the role of electrical oscillations in Maxwell's theory. But Hertz's research had an immensely important *practical application* because it led to the development of radio communication. The waves he produced and whose properties he studied had wavelengths around 10 m; i.e., frequency 30 MHz, which falls between the AM and FM radio frequency bands and is now used for amateur hf-radio.[2]

The Maxwell equations are a unified field theory of electricity and magnetism. Furthermore, optics is a part of the theory, because light is an electromagnetic phenomenon described by Maxwell's equations.

EXAMPLE 1   Charging a capacitor, Part 1.   What are the fields $\mathbf{E}(\mathbf{x}, t)$ and $\mathbf{B}(\mathbf{x}, t)$ in and around a capacitor as it is being charged?

Figure 11.1 shows a parallel plate capacitor being charged by a current I that flows in wires along the z axis. The capacitor plates are disks with radius a. We will use the Ampère-Maxwell Law (11.7) to find $\mathbf{B}$ on the three circular paths C_1, C_2, and C_3 shown in the figure.

The charges on the left and right plates are, respectively, $Q(t)$ and $-Q(t)$; so the electric field between the plates, i.e., for $r < a$, is

$$\mathbf{E}(t) = \frac{\sigma(t)}{\epsilon_0}\hat{\mathbf{k}} = \frac{Q(t)}{\epsilon_0 \pi a^2}\hat{\mathbf{k}} \qquad (11.12)$$

by Gauss's Law, which holds for charge distributions that depend on time as well as for static charge distributions. Because both current densities $\mathbf{J}$ and $\mathbf{J}_D$ in (11.9) are in the z direction, so that $\nabla \times \mathbf{B}$ is in the z direction, the magnetic field is azimuthal, of the general form $\mathbf{B} = B_\phi(r, z)\hat{\phi}$ in each region. There are minor

---

[1] See Exercise 1.

[2] Richard P. Feynman (1918–1990) liked to use this history to emphasize how basic research can lead to the most important of applications, even more effectively than applied research. To paraphrase how Feynman put it: If Hertz had been hired by a company to find a way to communicate by voice over long distances, say between New York and California, then he would have experimented with "long tubes" instead of with inductors, oscillators, and sparks. Feynman in his lecture would demonstrate how Hertz might shout into the tube: "Hello! It's me, Hertz, in New York! Can you hear me yet in Pittsburgh?"

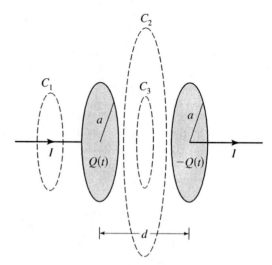

FIGURE 11.1   Charging a capacitor. The parallel plate capacitor is being charged by the current $I = dQ/dt$ along the z axis. The plates are disks of radius a separated by distance d with $d \ll a$. (The separation is exaggerated in the figure.) C_1, C_2, and C_3 are circular Amperian loops, on which $\mathbf{B}$ can be determined by (11.10).

corrections to these ideal fields, called *fringing fields*. The fringing fields are negligible between the plates if the plates are close together. They are only important near the edges.

Region 1:   What is $\mathbf{B}$ in the regions to the left and right of the capacitor?

We integrate (11.7) over the disk S_1 bounded by C_1, with radius r. By Stokes's theorem the flux of $\nabla \times \mathbf{B}$ through S_1 is equal to the circulation of $\mathbf{B}$ around C_1, so the field equation implies

$$\oint_{C_1} \mathbf{B} \cdot d\ell = \mu_0 \int_{S_1} \mathbf{J} \cdot \hat{\mathbf{n}} dA + \mu_0 \epsilon_0 \frac{\partial}{\partial t} \int_{S_1} \mathbf{E} \cdot \hat{\mathbf{n}} dA. \qquad (11.13)$$

Here the normal vector $\hat{\mathbf{n}}$ is $\hat{\mathbf{k}}$, and the direction of $d\ell$ along C_1 is defined by the right-hand rule. The charge current through S_1 is I so that the first term on the right side of (11.13) is $\mu_0 I$. The displacement current through S_1, expressed in the second term on the right, is zero in the quasistatic approximation.[3] The integral of E_n is not strictly zero because there is an $\mathbf{E}$ in the wire and an $E_\parallel$ near it, but unless the frequency is very high the displacement current is negligible compared

---

[3] The quasistatic approximation is the somewhat subtle issue that we have mentioned in previous chapters. In this approximation we use formulas derived from electrostatics or magnetostatics, to describe fields in systems that are changing in time. If the rate of change is slow, the approximation is accurate. In fact, the quasistatic approximation is quite accurate except in extreme cases. We will discuss the quasistatic approximation in more detail in Example 3.

to I. Therefore (11.13) becomes

$$B_\phi 2\pi r = \mu_0 I, \tag{11.14}$$

where the azimuthal symmetry has been used to evaluate the line integral; or,

$$B_\phi = \frac{\mu_0 I}{2\pi r}. \tag{11.15}$$

This calculation and its result are the same as for an infinite straight wire with current I. Here I may vary in time. Had we not made the quasistatic approximation there would be a small additional contribution to B_ϕ from displacement current.

Region 2:   What is **B** along C_2, i.e., between the plates in z, but outside the cylinder $(r > a)$?
   In this case we integrate (11.7) over the disk S_2 bounded by C_2, and obtain again an equation like (11.13), but now with integrals over C_2 and S_2. The charge current through S_2 is 0. The displacement current, using the electric field in (11.12), is

$$I_D(S_2) = \epsilon_0 \frac{\partial}{\partial t} \int_{S_2} E_z dA = \epsilon_0 \frac{dQ/dt}{\epsilon_0 \pi a^2} \pi a^2 = I \tag{11.16}$$

because $dQ/dt = I$ by conservation of charge. Therefore the magnetic field in Region 2 is again given by (11.15), the same as that of a straight wire. In calculating the magnetic field in this region we have, as a good approximation, ignored the fringing field of **E**.

Region 3:   What is **B** along C_3, i.e., between the plates in z, and inside the cylinder $(r < a)$?
   Again we integrate (11.7), this time over the disk S_3 bounded by C_3, and obtain an equation like (11.13), but now with integrals over C_3 and S_3. The charge current through S_3 is 0. The displacement current is

$$I_D(S_3) = \epsilon_0 \frac{\partial}{\partial t} \int_{S_3} E_z dA = \epsilon_0 \frac{dQ/dt}{\epsilon_0 \pi a^2} \pi r^2 = \frac{I r^2}{a^2}. \tag{11.17}$$

Thus, by Stokes's theorem,

$$B_\phi(r) = \frac{\mu_0 I r}{2\pi a^2}. \tag{11.18}$$

The magnetic field in Region 3 is the same as if the capacitor were replaced by a conducting cylinder, of radius a, carrying total current I uniformly distributed across its cross section. As a source of **B** the displacement current is equivalent to a charge current.

B(**x**, t) is continuous at the cylindrical surface $r = a$ because there is no surface current. However, **B** is discontinuous across either plate. The discontinuity of **B** is related to the *radial surface current* that flows in the plate as the capacitor is being charged, in accord with the general boundary condition at a current sheet.[4]

Example 1 can be used to understand in another way why the displacement current is logically necessary for consistency of the field equations. Figure 11.2 shows an Amperian loop C, in the form of a circle around the wire to the left of both capacitor plates. According to Stokes's theorem, the circulation of **B** around C is equal to the flux of $\mu_0(\mathbf{J} + \mathbf{J}_D)$ through *any surface S bounded by C*; this is (11.10) expressed in words. Therefore this flux must be the same for all such surfaces. First consider for S the disk S_a bounded by C. The flux through S_a is $\mu_0 I$, where I is the current in the wire, because **J** passes through S_a but $\mathbf{J}_D$ is 0 on S_a. Next consider instead the bag-shaped surface S_b, *whose boundary is also C*. The flux through S_b must also be $\mu_0 I$ for mathematical consistency. But there is no charge current through S_b, so without the displacement current there would be an inconsistency. There is displacement current through S_b given by $I_D = \partial(DA)/\partial t$ where $A = \pi a^2$ is the area of the plates. Now, by Gauss's law DA is equal to Q, the charge on the positive plate, so $I_D = dQ/dt = I$. In other words, the displacement current through S_b equals the charge current through S_a, both being I. The theory is self-consistent.

The displacement current is not a flow of charge, but a variation of **D**. The above discussion is based on the idea that both charge current and displacement

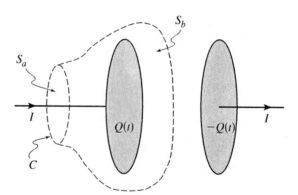

FIGURE 11.2   Why the displacement current is necessary. The capacitor is being charged by a current I. The circle C is the boundary curve of both the disk S_a and the bag-shaped surface S_b. By Stokes's theorem, the displacement current through S_b must be equal to the charge current through S_a.

[4] See Exercise 3.

current produce **B**. While this view is self-consistent, there are some subtle questions about "cause" and "effect." In a region where $\mathbf{J} = 0$, (11.7) describes a correlation between the spatial variation of **B** and the temporal variation of **E** (or **D**). But who is to say which is the cause and which the effect? For physical intuition it is helpful to think of $\partial \mathbf{D}/\partial t$ as a kind of current density, but a more precise statement is that $\partial \mathbf{D}/\partial t$ contributes to the curl of **B** in the same way as $\mathbf{J}$.[5]

11.2 ■ SCALAR AND VECTOR POTENTIALS

We now turn to an interesting and important general problem. How can we find the *time-dependent* fields $\mathbf{E}(\mathbf{x}, t)$ and $\mathbf{B}(\mathbf{x}, t)$ produced by *time-dependent* charge and current sources in vacuum? Much of modern communication technology, e.g., radio, television, and satellite communication, are applications of this problem.

The Maxwell equations contain all that is needed to solve the problem. However, calculating the fields directly from the sources is often difficult, and rarely the best approach. Rather, it is better to introduce *potential functions*—time-dependent scalar and vector potentials—whose mathematical connection to the sources is easier to handle. Once these potentials are known, the fields can be calculated by differentiation.

This situation is similar to what we found in electrostatics and magnetostatics. Recall the advantages of finding the potentials, namely $V(\mathbf{x})$ for electrostatics and $\mathbf{A}(\mathbf{x})$ for magnetostatics, and then calculating $\mathbf{E} = -\nabla V$ and $\mathbf{B} = \nabla \times \mathbf{A}$. First, the equations connecting the potentials and sources are simpler than the equations relating fields and sources. Also, the fields **E** and **B** have altogether six components, but the potentials V and **A** have only four. The same advantages exist for time-dependent problems.

It is straightforward to generalize the potentials to the case of time-dependent fields. Any vector function whose divergence is 0 can be written as the curl of another vector function. Therefore (11.2) implies that we can always find a vector potential $\mathbf{A}(\mathbf{x}, t)$ such that

$$\mathbf{B} = \nabla \times \mathbf{A}. \qquad (11.19)$$

Substituting this form of **B** into (11.3) gives

$$\nabla \times \left(\mathbf{E} + \frac{\partial \mathbf{A}}{\partial t} \right) = 0, \qquad (11.20)$$

where we have used the fact that ∇ and $\partial/\partial t$ commute. Now, any vector function whose curl is 0 can be written as the gradient of a scalar function. That is, (11.20) implies that we can always find a scalar potential $V(\mathbf{x}, t)$ such that

$$\mathbf{E} + \frac{\partial \mathbf{A}}{\partial t} = -\nabla V, \qquad (11.21)$$

[5] A. P. French, *The Physics Teacher*, **38**, 274 (2000).

or,

$$\mathbf{E} = -\nabla V - \frac{\partial \mathbf{A}}{\partial t}. \qquad (11.22)$$

(Note that the familiar electrostatics equation $\mathbf{E} = -\nabla V$ is *not true* for time-dependent fields.)

Equations (11.19) and (11.22) are a reflection of the Helmholtz theorem.[6] The theorem states that any vector field can be separated into a solenoidal part, which can be written as a curl, and an irrotational part, which can be written as a gradient. In (11.19) the purely solenoidal field $\mathbf{B}$ is written as the curl of $\mathbf{A}$. In (11.22) the purely irrotational function $\mathbf{E} + \partial \mathbf{A}/\partial t$ is written as $-\nabla V$.

With the fields $\mathbf{B}(\mathbf{x}, t)$ and $\mathbf{E}(\mathbf{x}, t)$ expressed in terms of the potentials $\mathbf{A}(\mathbf{x}, t)$ and $V(\mathbf{x}, t)$ by (11.19) and (11.22), two of the Maxwell equations are automatically satisfied, namely (11.2) and (11.3). To proceed we must rewrite the other two Maxwell equations in terms of $\mathbf{A}$ and V, obtaining the equations that will determine the potentials for whatever sources are present. But before deriving equations for $\mathbf{A}$ and V, we address the question of *uniqueness*.

11.2.1 ■ Gauge Transformations and Gauge Invariance

Are the potentials $\mathbf{A}(\mathbf{x}, t)$ and $V(\mathbf{x}, t)$ uniquely determined by the charge and current sources in a system? The answer is that these functions are *not* uniquely determined, a fact that creates some richness in the theory.

Of course the electric and magnetic *fields* are uniquely determined if the sources $\rho(\mathbf{x}, t)$ and $\mathbf{J}(\mathbf{x}, t)$ and boundary conditions of the system are established. After all, the fields are physical quantities, that could be measured by suitable experimental apparatus under the given conditions. In contrast, the potentials are introduced as a mathematical convenience; at least that is true in classical electromagnetism.[7] There are no classical instruments for measuring $V(\mathbf{x}, t)$ or $\mathbf{A}(\mathbf{x}, t)$, and the potentials do not have direct physical significance. They are not even unique for a given electromagnetic system.

If $\mathbf{A}(\mathbf{x}, t)$ and $V(\mathbf{x}, t)$ describe the fields of a certain system, then other potential functions, which also describe the same fields, can be constructed by transformations called *gauge transformations*. Let $f(\mathbf{x}, t)$ be an arbitrary scalar function. The gauge transformation associated with $f(\mathbf{x}, t)$ is

$$\mathbf{A} \longrightarrow \mathbf{A}' = \mathbf{A} + \nabla f \qquad (11.23)$$

$$V \longrightarrow V' = V - \frac{\partial f}{\partial t}. \qquad (11.24)$$

In words, the arrows ($\longrightarrow$) indicate that we replace the potentials $\mathbf{A}$ and V by new potentials $\mathbf{A}'$ and V' (the transformed potentials) that differ from $\mathbf{A}$ and V

---

[6]The Helmholtz theorem is presented in Appendix B.

[7]The Bohm-Aharanov effect in quantum electrodynamics is a physical effect that is directly sensitive to the vector potential $\mathbf{A}$.

by ∇f and $-\partial f/\partial t$, respectively. The *fields* $\mathbf{E}$ and $\mathbf{B}$ are the same for the primed potentials as for the unprimed potentials. The curl of $\mathbf{A}'$ is equal to the curl of $\mathbf{A}$ because $\nabla \times \nabla f = 0$, so $\mathbf{B}$ is the same for the two potentials. Similarly, $-\nabla V' - \partial \mathbf{A}'/\partial t$ is equal to $-\nabla V - \partial \mathbf{A}/\partial t$ because

$$\nabla \left(\frac{\partial f}{\partial t} \right) - \frac{\partial}{\partial t} (\nabla f) = 0,$$

so $\mathbf{E}$ is the same for the transformed and original potentials. Thus the transformed potentials, for any $f(\mathbf{x}, t)$, can be used as potentials for the same fields $\mathbf{B}$ and $\mathbf{E}$. In other words, $\mathbf{B}$ and $\mathbf{E}$ are invariant with respect to the gauge transformation (11.23) and (11.24). The theory is said to be *gauge invariant.*

We have previously encountered similar arbitrariness in defining potentials for static problems. In electrostatics, $V(\mathbf{x}) + C$ for any constant C gives the same electric field as $V(\mathbf{x})$. In magnetostatics, $\mathbf{A}(\mathbf{x}) + \nabla f(\mathbf{x})$ for any function $f(\mathbf{x})$ gives the same magnetic field as $\mathbf{A}(\mathbf{x})$. But the full gauge invariance with respect to the transformations (11.23) and (11.24) is more significant because it mixes $\mathbf{A}(\mathbf{x}, t)$ and $V(\mathbf{x}, t)$.

EXAMPLE 2 Scalar and vector potentials for a point charge at rest at the origin.   A natural choice of potentials for this static problem is

$$V(\mathbf{x}, t) = \frac{q}{4\pi \epsilon_0 r} \quad \text{and} \quad \mathbf{A}(\mathbf{x}, t) = 0. \tag{11.25}$$

The fields are $\mathbf{E} = q\hat{\mathbf{r}}/(4\pi \epsilon_0 r^2)$ and $\mathbf{B} = 0$.

But now make the gauge transformation with $f(\mathbf{x}, t) = qt/(4\pi \epsilon_0 r)$. The transformed potentials are

$$V'(\mathbf{x}, t) = 0 \quad \text{and} \quad \mathbf{A}'(\mathbf{x}, t) = \frac{-qt\hat{\mathbf{r}}}{4\pi \epsilon_0 r^2}. \tag{11.26}$$

The primed potentials give exactly the same fields as the unprimed potentials. However, the primed potentials look very surprising, because $\mathbf{A}'$ depends explicitly on t even though the sources and fields are all static! This example is additional evidence that the potentials do not have direct physical significance. The freedom to construct gauge transformations even permits time-dependent potentials for time-independent fields.

11.2.2 ▨ Gauge Choices and Equations for $\mathbf{A}(\mathbf{x}, t)$ and $V(\mathbf{x}, t)$

We are almost ready to substitute the fields, written in terms of $\mathbf{A}$ and V, into the Maxwell equations (11.1) and (11.7). However, there is one more preliminary step. It is generally useful to eliminate the gauge ambiguity by imposing some additional condition on the potentials $\mathbf{A}(\mathbf{x}, t)$ and $V(\mathbf{x}, t)$, rather than leaving their form completely arbitrary. The additional condition is called a *gauge choice*. Because of gauge invariance we have great freedom in the form of the potentials. We make use of this freedom to choose a convenient, or simple form.

Coulomb Gauge.   One example of a gauge choice is called the *Coulomb gauge*. The condition imposed in this case is

$$\nabla \cdot \mathbf{A} = 0. \qquad (11.27)$$

This condition makes $\mathbf{A}$ unique, because the gauge transformation (11.23) would in general produce a transformed potential that does not satisfy the condition (11.27). In other words, if $\mathbf{A}$ satisfies the condition (11.27) then $\mathbf{A}'$ of (11.23) would not satisfy the condition unless $\nabla^2 f = 0$; imposing a boundary condition on the vector potential, e.g., at infinity, would then require that f is constant, so that $\mathbf{A}'$ is just the same as $\mathbf{A}$. This means therefore that for given sources and boundary conditions there is *only one* $\mathbf{A}(\mathbf{x}, t)$ that gives the correct fields and also satisfies (11.27). The scalar potential for the problem is also uniquely determined, except for an arbitrary additive constant.

The scalar potential is particularly simple in the Coulomb gauge, because then Gauss's law (11.1) reduces to $-\nabla^2 V = \rho/\epsilon_0$. This is the same equation as in electrostatics—Poisson's equation—and we can solve it formally with the Green's function of $-\nabla^2$,

$$V(\mathbf{x}, t) = \frac{1}{4\pi\epsilon_0} \int \frac{\rho(\mathbf{x}', t)d^3x'}{|\mathbf{x} - \mathbf{x}'|}. \qquad (11.28)$$

But note that this function has an unphysical aspect: The potential V at an arbitrary field point $\mathbf{x}$ and time t depends on the charge density at all source points *at the very same time t*. Equation (11.28) implies that no matter how far the field point is from the source, there is an *instantaneous response* in $V(\mathbf{x}, t)$ to a change in $\rho(\mathbf{x}', t)$. This instantaneous response at arbitrary distance would be physically unacceptable if $V(\mathbf{x}, t)$ were measurable. In contrast the fields, which are measurable, do not respond instantaneously to a change in $\rho(\mathbf{x}', t)$; rather, there is a time delay—the time difference being the time for light to travel from $\mathbf{x}'$ to $\mathbf{x}$.[8] There is no inconsistency in the theory because $\mathbf{E}(\mathbf{x}, t)$ is determined by a combination of both $\mathbf{A}(\mathbf{x}, t)$ and $V(\mathbf{x}, t)$, as we see in (11.22). The contribution to $\mathbf{E}$ from $-\partial \mathbf{A}/\partial t$ cancels the instantaneous changes from V. The instantaneous response of $V(\mathbf{x}, t)$ in the Coulomb gauge is just an artifact of the gauge choice.

A disadvantage of the Coulomb gauge is that the vector potential $\mathbf{A}(\mathbf{x}, t)$ will in general be complicated if the sources vary in time. For this reason the Coulomb gauge is only useful for static systems, or to describe free electromagnetic waves. It is not a convenient gauge choice for problems involving the generation of radiation.

---

[8] We shall prove in Chapter 15 that the response of the fields to a change in the sources occurs at the retarded time.

Lorentz Gauge.   A different gauge choice, very useful for time-dependent problems like radiation, is called the *Lorentz gauge*[9] for which the gauge condition is

$$\mathbf{\nabla} \cdot \mathbf{A} = -\mu_0 \epsilon_0 \frac{\partial V}{\partial t}. \tag{11.29}$$

For this gauge choice the source equations (11.1) and (11.7) take a particularly attractive form, because V and $\mathbf{A}$ separate into independent equations. First, consider Gauss's law (11.1). From (11.22) $\mathbf{\nabla} \cdot \mathbf{E}$ is $-\nabla^2 V - \partial(\mathbf{\nabla} \cdot \mathbf{A})/\partial t$. Use the gauge condition (11.29) to eliminate $\mathbf{\nabla} \cdot \mathbf{A}$ for V. The result is a source equation involving V alone,

$$-\nabla^2 V + \mu_0 \epsilon_0 \frac{\partial^2 V}{\partial t^2} = \frac{\rho}{\epsilon_0}. \tag{11.30}$$

Similarly, the Ampère-Maxwell equation reduces to an equation for $\mathbf{A}$ alone,[10]

$$-\nabla^2 \mathbf{A} + \mu_0 \epsilon_0 \frac{\partial^2 \mathbf{A}}{\partial t^2} = \mu_0 \mathbf{J}. \tag{11.31}$$

Note that the equations for V and $\mathbf{A}$ are similar. The ith Cartesian component A_i with source $\mu_0 J_i$ satisfies an equation of the same form as V with source ρ/ϵ_0. We shall use the Lorentz gauge in Chapter 15 to analyze the radiation of electromagnetic waves by charges or currents that vary in time. There (11.30) and (11.31) will be solved. The results will be solidly intuitive.

It is also possible to write the *general* equations for $V(\mathbf{x}, t)$ and $\mathbf{A}(\mathbf{x}, t)$, i.e., without specifying any gauge condition.[11] However, the resulting equations are not very edifying—coupled scalar and vector partial differential equations—and not often used in practical calculations. The usual approach to finding V and $\mathbf{A}$ is to make a gauge choice as a first step. It should be verified at the end of the calculation that the solution does indeed satisfy the gauge condition.

EXAMPLE 2 Example 2 revisited.   Are the original potentials (11.25), for a point charge at rest at the origin, in the Coulomb or Lorentz gauge? Because $\mathbf{\nabla} \cdot \mathbf{A} = 0$ and $\partial V/\partial t = 0$ these potentials satisfy both (11.27) and (11.29). They are in *both* the Coulomb gauge and the Lorentz gauge.

Are the transformed potentials (11.26) in the Coulomb or Lorentz gauge? These potentials are in *neither* the Coulomb gauge *nor* the Lorentz gauge, because $\mathbf{\nabla} \cdot \mathbf{A}' = -(qt/\epsilon_0)\delta^3(\mathbf{x})$ and $\partial V'/\partial t = 0$. Recall that $q\delta^3(\mathbf{x})$ is the charge density of a point charge q at the origin.

---

[9] The first use of the Lorentz gauge may have been by Ludwig Lorenz of Austria, rather than Hendrik Lorentz of the Netherlands. We call (11.29) the Lorentz gauge because the condition is invariant under the Lorentz transformations of special relativity.

[10] See Exercise 4.

[11] See Exercise 5.

11.3 ■ THE MAXWELL EQUATIONS IN MATTER

When we wrote the Maxwell equations in vacuum (see Table 11.1) we allowed only isolated sources ρ and $\mathbf{J}$, but not dielectric or magnetic materials. We now consider the modification of the equations necessary to describe systems with such materials.

Equations (11.1) to (11.3) and (11.7) are still true for systems with matter, provided that we interpret ρ and $\mathbf{J}$ to include the bound charge and current—bound within atoms of the material. The underlying idea is that dielectric and magnetic materials, as indeed all matter in the Universe, consist of particles that are sources of, and interact with, the fields. The difficulty in solving Maxwell's equations from that starting point is that in order to find $\rho(\mathbf{x}, t)$ and $\mathbf{J}(\mathbf{x}, t)$ we'd need detailed knowledge about the microscopic nature of the particles in atoms and molecules, which would be very complicated. Therefore the vacuum equations are not a practical formulation of the theory in the presence of matter. Instead, we rewrite the equations by separating the free charge from the bound charge. We are not introducing new fundamental principles, but writing the equations in a form that is more convenient to account for the effects of the matter.

The way that matter interacts with the electromagnetic field is determined by the fact that *matter is atomic*. Atoms and molecules respond to electromagnetic fields mainly as electric and magnetic dipoles. As we learned in earlier chapters, the origin of the electric dipoles is the separation of positive and negative charges within atoms, and the origin of the magnetic dipoles is the current associated with electron orbital motion and spin. We have learned how to take these effects into account for static fields. In the case of a static electric field we introduced in Chapter 6 the *polarization field* $\mathbf{P}(\mathbf{x})$, the electric dipole moment density in matter as a function of position $\mathbf{x}$. In the case of a static magnetic field we introduced in Chapter 9 the *magnetization field* $\mathbf{M}(\mathbf{x})$, the magnetic dipole moment density of the matter.

The same matter fields $\mathbf{P}(\mathbf{x}, t)$ and $\mathbf{M}(\mathbf{x}, t)$, i.e., the dipole moment densities, account for the response of matter to time-dependent fields, but now the polarization and magnetization may depend on time. Recall from the static cases that $\mathbf{P}$ or $\mathbf{M}$ describes the *combined effect of many atoms* in a neighborhood of $\mathbf{x}$. The response of any single atom to a time-dependent field is complicated and quantum mechanical, but summing over the huge number of atoms in even a small volume yields the simple classical polarization fields.

11.3.1 ■ Free and Bound Charge and Current

Turning first to Gauss's Law (11.1) we know that when dielectrics are present the source of $\mathbf{E}$ will include bound charge ρ_b as well as free charge ρ_f. It is convenient to treat these sources separately, so we write

$$\nabla \cdot \mathbf{E} = \frac{\rho}{\epsilon_0} = \frac{1}{\epsilon_0}(\rho_f + \rho_b). \tag{11.32}$$

The field $\mathbf{E}(\mathbf{x}, t)$ and the densities $\rho_f(\mathbf{x}, t)$ and $\rho_b(\mathbf{x}, t)$ in (11.32) may depend on time. The bound charge density ρ_b describes the effect of atomic dipoles as sources of electric field. It is given by Eq. (6.12)

$$\rho_b = -\nabla \cdot \mathbf{P}, \tag{11.33}$$

which is valid for fields with arbitrary time dependence because Gauss's Law does not involve the time derivative. The displacement field is by definition (6.28)

$$\mathbf{D} = \epsilon_0 \mathbf{E} + \mathbf{P}, \tag{11.34}$$

so Gauss's Law may be written as

$$\nabla \cdot \mathbf{D} = \rho_f. \tag{11.35}$$

Equation (11.35) is Gauss's Law in the presence of dielectrics. This treatment of electric polarization for time-dependent fields has the same form as for static fields in Chapter 6.

The second and third Maxwell equations, (11.2) and (11.3), are universal—unchanged in the presence of dielectric or magnetic materials.

Finally we turn to the Ampère-Maxwell Law (11.7). Fundamentally, there are two source terms for $\mathbf{B}$, proportional to $\mathbf{J}$ and $\partial \mathbf{E}/\partial t$. To describe the interaction between matter and a magnetic field, we separate the current density $\mathbf{J}$ into free current, bound current, and polarization current

$$\mathbf{J} = \mathbf{J}_f + \mathbf{J}_b + \mathbf{J}_P. \tag{11.36}$$

The bound current, an effective current density equivalent to the atomic magnetic dipoles from the magnetization, is defined by Eq. (9.12)

$$\mathbf{J}_b = \nabla \times \mathbf{M}, \tag{11.37}$$

which we encountered when studying magnetic materials in Chapter 9. For time-dependent fields we must also include the *polarization current*, which does not appear in magnetostatics, defined by

$$\mathbf{J}_P = \frac{\partial \mathbf{P}}{\partial t}. \tag{11.38}$$

The polarization current is the current associated with change of electric polarization. Polarization is the separation of positive and negative charge, bound together in the atom. A *change* of polarization occurs when positive charge shifts one way and negative charge another. That motion of charge is a current—the polarization current.

Figure 11.3 explains the connection between $\mathbf{P}$, ρ_b, and $\mathbf{J}_P$. Because matter is electrically neutral, bound charge and polarization current satisfy a charge conservation equation. By their definitions (11.38) and (11.33) we have

$$\nabla \cdot \mathbf{J}_P = -\frac{\partial \rho_b}{\partial t}. \tag{11.39}$$

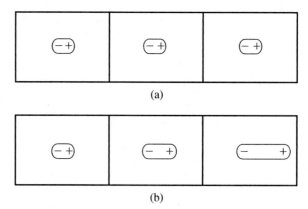

(a)

(b)

FIGURE 11.3   The polarization current. Three cells inside a dielectric are shown, and in each cell a typical atom polarized by the presence of an electric field. (There are *many* atoms in a cell.) In the state (a), **P(x)** is constant and ρ_b is 0. In the state (b), **P(x)** has nonzero divergence (**P** increases to the right so $\mathbf{\nabla} \cdot \mathbf{P} > 0$) and $\rho_b = -\mathbf{\nabla} \cdot \mathbf{P}$ is nonzero. The net charge in the center cell is negative because to go from state (a) to state (b) more positive charge passes out of cell through the right boundary than passes into the cell through the left boundary. If the material evolves in time from state (a) to state (b), then *during the change* there is a current $\mathbf{J}_P$ directed toward the right as positive charge moves to the right; that is the physical basis for (11.38). However, the individual charges remain bound in their atoms. The divergence of $\mathbf{J}_P$ is $-\partial \rho_b / \partial t$ by conservation of charge.

This result justifies the definition (11.38) for $\mathbf{J}_P$: A variation of the bound charge is a current, and the current out through a closed surface must equal the rate of decrease of charge enclosed, because charge is conserved. Furthermore, it is straightforward to show from (11.39) that the total charge $\rho = \rho_f + \rho_b$ and current (11.36) satisfy the full continuity equation (11.4), which is the fundamental equation.

Now, when (11.36) is substituted into the Ampère-Maxwell Law, the field equation becomes

$$\mathbf{\nabla} \times \mathbf{B} = \mu_0 \left(\mathbf{J}_f + \mathbf{\nabla} \times \mathbf{M} + \frac{\partial \mathbf{P}}{\partial t} \right) + \mu_0 \epsilon_0 \frac{\partial \mathbf{E}}{\partial t}, \tag{11.40}$$

from which, by rearranging terms,

$$\mathbf{\nabla} \times \left(\frac{\mathbf{B}}{\mu_0} - \mathbf{M} \right) = \mathbf{J}_f + \frac{\partial}{\partial t} (\epsilon_0 \mathbf{E} + \mathbf{P}). \tag{11.41}$$

Or, finally we have

$$\mathbf{\nabla} \times \mathbf{H} = \mathbf{J}_f + \frac{\partial \mathbf{D}}{\partial t}, \tag{11.42}$$

where **H** is the magnetic vector defined in (9.23)

$$\mathbf{H} = \frac{\mathbf{B}}{\mu_0} - \mathbf{M}. \tag{11.43}$$

Equation (11.42) is the Ampère-Maxwell Law in materials. $\mathbf{J}_f$ is the free current density and $\partial \mathbf{D}/\partial t$ is the displacement current density in matter.

The Maxwell equations in matter are (11.35) and (11.42), together with the universal equations (11.2) and (11.3), which are the same in matter or in vacuum. These are tabulated in the third column of Table 11.1. The sources ρ_f and $\mathbf{J}_f$ are just the *free* charge and current.

To use equations (11.35) and (11.42) we must know *from some independent information* how polarization and magnetization depend on **E** and **B**. In many applications of electromagnetism in matter, the polarization is proportional to **E** and the magnetization is proportional to **H**, for time-dependent fields if the frequency isn't too high, as well as for static fields. Materials with this property are called *linear materials*. The constants of proportionality, called the electric and magnetic susceptibilities χ_e and χ_m, are conventionally defined by

$$\mathbf{P} = \epsilon_0 \chi_e \mathbf{E}, \tag{11.44}$$

$$\mathbf{M} = \chi_m \mathbf{H}. \tag{11.45}$$

For linear materials **D** is proportional to **E**, and **B** is proportional to **H**; that is

$$\mathbf{D} = \epsilon \mathbf{E}, \tag{11.46}$$

$$\mathbf{B} = \mu \mathbf{H}, \tag{11.47}$$

where the constant ϵ is called the *permittivity* of the material, and μ the *permeability*. These relations specifying material properties are called *constitutive equations*. In terms of susceptibilites, $\epsilon = \epsilon_0(1 + \chi_e)$ and $\mu = \mu_0(1 + \chi_m)$. The constitutive equations are summarized in Table 11.2.

The most important application of Maxwell's equations in matter is to the theory of optics—the interaction between electromagnetic waves and matter—a subject we shall study in detail in Chapter 13.

11.3.2 ■ Boundary Conditions of Fields

The Maxwell equations are partial differential equations, and whenever partial differential equations occur in physics, the question of boundary conditions arises.

TABLE 11.2   Definitions and constitutive equations for linear materials

Material	Field definitions	Susceptibilities
Dielectrics	$\mathbf{D} = \epsilon_0 \mathbf{E} + \mathbf{P} = \epsilon \mathbf{E}$	$\mathbf{P} = \epsilon_0 \chi_e \mathbf{E}$
Magnetic materials	$\mathbf{B} = \mu_0(\mathbf{H} + \mathbf{M}) = \mu \mathbf{H}$	$\mathbf{M} = \chi_m \mathbf{H}$

What are the conditions that must hold at a boundary between different materials? Which fields are continuous? For field components that are discontinuous, what are the discontinuities? We can answer these questions by analyzing Maxwell's equations at the interface between the materials.

Let S be the surface separating two linear media, with material constants ϵ_1 and μ_1 on one side of S, and ϵ_2 and μ_2 on the other side. We have previously determined the following boundary conditions for static fields:

$$D_{2\perp} - D_{1\perp} = \sigma_f \tag{11.48}$$

$$B_{2\perp} - B_{1\perp} = 0 \tag{11.49}$$

$$\mathbf{E}_{2\|} - \mathbf{E}_{1\|} = 0 \tag{11.50}$$

$$\mathbf{H}_{2\|} - \mathbf{H}_{1\|} = \mathbf{K}_f \times \hat{\mathbf{n}} \tag{11.51}$$

where $\perp$ and $\|$ indicate vector components perpendicular and parallel to S, respectively, and $\hat{\mathbf{n}}$ is normal in the direction $1 \rightarrow 2$. *The very same boundary conditions are true for time-dependent fields.*

Recall how these boundary conditions were derived. Equations (11.48) and (11.49) were derived by integrating the flux of $\mathbf{D}$ or $\mathbf{B}$ over a closed Gaussian surface that cuts through the interface—a pill box with height δ that tends to 0 (see Fig. 11.4(a))—and applying Gauss's theorem. By (11.35) the flux of $\mathbf{D}$ is $\sigma_f dA$; hence (11.48). By (11.2) the flux of $\mathbf{B}$ is 0; hence (11.49). The same analysis applies to the time-dependent Maxwell equations, and leads to the same boundary conditions, because the field equations (11.35) and (11.2) do not involve the time derivative.

Equations (11.50) and (11.51) were derived for static fields by considering the circulation of $\mathbf{E}$ or $\mathbf{H}$ around a rectangular Amperian loop C cutting across S, and applying Stokes's theorem. Figure 11.4(b) shows such a loop, with tangential and normal dimensions ℓ and δ. In the time-dependent field equations there are additional terms involving the time derivative, but they do not contribute to the

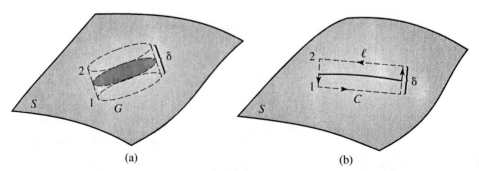

(a)   (b)

FIGURE 11.4   Boundary conditions of E and B. S is an interface between two regions. (a) A Gaussian pill box G of height δ cutting through S. (b) An Amperian loop C of height δ cutting through S.

discontinuity across S because the fields are nonsingular on S. For example, the circulation of **E** is $-d\Phi_B/dt$, where Φ_B is the magnetic flux through a surface bounded by C. As the height δ of C approaches 0, the flux $\Phi_B \to 0$ because the area $\ell\delta \to 0$ while **B** remains finite; hence (11.50). Similarly the circulation of **H** is $\mu_0(I_f + I_D)$. As $\delta \to 0$, the displacement current $I_D \to 0$ because again the area $\ell\delta \to 0$ while **D** remains finite; the free charge current approaches the surface current integrated across the length ℓ of the loop C; hence (11.51).

Where We Now Stand and How We Got Here

We've come quite far in our study of electromagnetism, so it's interesting and useful to look back on our path.

We now have Maxwell's equations in matter, written in the right column of Table 11.1. Taken together with the linear constitutive equations in Table 11.2, Ohm's law in the form $\mathbf{J} = \sigma\mathbf{E}$, and the boundary conditions (11.48) to (11.51), we therefore have everything necessary to solve, in principle, the following very general problem in electrodynamics: *What are the electric and magnetic fields for arbitrary time-dependent charge and current sources in the presence of linear dielectric, magnetic, and conducting materials?*

Recall that we started our work by first studying the equations $\nabla \cdot \mathbf{E} = \rho/\epsilon_0$ and $\nabla \times \mathbf{E} = 0$. With those PDEs, and boundary conditions, we could find **E** in vacuum for static charge sources. Second, we studied the equations $\nabla \cdot \mathbf{D} = \rho_f$ and $\nabla \times \mathbf{E} = 0$. With those, and a constitutive equation and boundary conditions, we could find **E** and **D** in the presence of dielectrics and static sources. Third, we turned to magnetostatics, and with $\nabla \cdot \mathbf{B} = 0$ and $\nabla \times \mathbf{B} = \mu_0\mathbf{J}$ we could find **B** in vacuum for static current sources. Fourth, we included magnetic materials, and found that with the equations $\nabla \cdot \mathbf{B} = 0$ and $\nabla \times \mathbf{H} = \mathbf{J}_f$, together with a constitutive equation and boundary conditions, we could find **B** and **H** in the presence of magnetic materials and static currents. With our work in this chapter, we are finally able to include arbitrary time dependence in our problems. Maxwell's equations in vacuum, as written in the middle column in Table 11.1, together with boundary conditions, enable us to find $\mathbf{E}(\mathbf{x}, t)$ and $\mathbf{B}(\mathbf{x}, t)$ for time-dependent sources $\rho(\mathbf{x}, t)$ and $\mathbf{J}(\mathbf{x}, t)$.

In the interest of completeness, we must remark that there are more general problems than the one in italics above. These would include materials with nonlinear electric and magnetic properties. In order to solve such problems we'd need more general constitutive equations, say of the form $\mathbf{D} = \mathbf{D}(\mathbf{E})$, $\mathbf{H} = \mathbf{H}(\mathbf{B})$, and $\mathbf{J} = \mathbf{J}(\mathbf{E}, \mathbf{B})$, or even more general relations. However, we will not consider such difficult problems.

11.4 ■ ENERGY AND MOMENTUM OF ELECTROMAGNETIC FIELDS

In general, electromagnetic fields possess both energy and momentum. The energy and momentum flow through space as the fields change. In previous chapters we deduced the equations for energy density in static fields, namely $u_E = \epsilon_0 E^2/2$

and $u_B = B^2/(2\mu_0)$ for fields in vacuum. The purpose of this section is to derive the equations for *energy flow* and *energy density* of arbitrary time-dependent fields. Field momentum is also described, although its derivation is postponed until Chapter 12. As we shall see in Chapter 12, it is natural to study energy and momentum together because they are linked by relativity, as are time and space.

11.4.1 ■ Poynting's Theorem

What is *energy*? Energy is a conserved quantity, defined by equations, associated with the ability to do work. A defining property of energy is that the total energy in an isolated system is constant. More precisely, if work is done by one part of the system on another part, then the first part loses energy equal to the work done and the second part gains an equal amount of energy.

In the case of field energy, an even stronger statement is true: Energy is not just conserved overall, but it is *locally conserved*, i.e., point-by-point in space. To derive the correct equations for the energy density and energy flux, we require that energy be locally conserved.

We studied in Chapter 7 another example of a local conservation law—conservation of electric charge. The equation for charge conservation is the continuity equation (11.4) where **J** is the flux and ρ is the density. Recall from Sec. 7.2 that local conservation of charge implies that if the net charge within *any* volume changes, then there must be a corresponding current flow, i.e., a net flux of charge, across the boundary surface. Equation (11.4) expresses this conservation. We will derive the formulas for energy flux and energy density by demanding that they satisfy a conservation equation analogous to (11.4).

For the case of energy, picture an arbitrary volume, either finite or infinitesimal, containing charges and fields as in Fig. 11.5. If the total energy inside this volume is to change then there must be a net energy flux across the boundary surface. It is necessary to account for both the field energy and the kinetic energy of the charged particles, because in general the fields and particles will exchange energy as they interact. That is, we seek an equation of the form

$$\mathbf{\nabla} \cdot \mathbf{S} = -\frac{\partial u}{\partial t} - \frac{\partial u_K}{\partial t}, \tag{11.52}$$

where **S** is the field energy flux (energy flow per unit time per unit area), u is the field energy density (energy per unit volume), and u_K is the *particle kinetic energy density*. (For simplicity we assume that the only forces on the particles are electromagnetic, so we do not need to account for other potential energies of the particles.) By Gauss's theorem, equation (11.52) says that the rate that energy flows out through the surface bounding a small volume at **x** equals the rate that energy decreases inside the volume.

If a particle with charge q moves, then the change of its kinetic energy K is equal to the work done on q; that is, $dK = \mathbf{F} \cdot \mathbf{v}dt$. This relation is called the work-kinetic energy theorem. The force is $q(\mathbf{E} + \mathbf{v} \times \mathbf{B})$ but the magnetic force

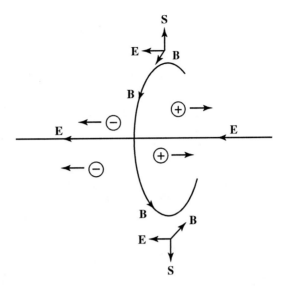

FIGURE 11.5   Poynting's theorem. Many kinds of energy transfer occur in a system of fields and charged particles, e.g., an antenna or a plasma. As charges are accelerated by $\mathbf{E}$ the kinetic energy density u_K changes. Also, the $\mathbf{E}$ and $\mathbf{B}$ fields change with time as the charges move, so the field energy density u changes. Waves may propagate away from the system with energy flux $\mathbf{S} = \mathbf{E} \times \mathbf{B}/\mu_0$. Poynting's theorem states that energy is conserved locally throughout the system.

does no work, so

$$\frac{dK}{dt} = q\mathbf{E} \cdot \mathbf{v}. \tag{11.53}$$

This result is for a single charged particle. To analyze a general continuous distribution of charge, with charge density $\rho(\mathbf{x}, t)$ and kinetic energy density $u_K(\mathbf{x}, t)$, consider the total kinetic energy in an infinitesimal volume d^3x at $\mathbf{x}$. In this case the work-kinetic energy theorem is

$$\frac{\partial u_K}{\partial t}d^3x = (\rho d^3x)\,\mathbf{E} \cdot \mathbf{v}, \tag{11.54}$$

completely analogous to (11.53), where $\mathbf{v}$ is the mean velocity of charges at $\mathbf{x}$. Now, $\rho\mathbf{v}$ is the current density $\mathbf{J}(\mathbf{x}, t)$. Thus the rate of change of particle kinetic energy *per unit volume* is

$$\frac{\partial u_K}{\partial t} = \mathbf{E} \cdot \mathbf{J}. \tag{11.55}$$

Using this formula for the kinetic energy term in (11.52) the continuity equation expressing conservation of energy is

$$\nabla \cdot \mathbf{S} = -\frac{\partial u}{\partial t} - \mathbf{E} \cdot \mathbf{J}. \tag{11.56}$$

This equation, together with the formulas for $\mathbf{S}(\mathbf{x}, t)$ and $u(\mathbf{x}, t)$ derived below, is called *Poynting's theorem*.[12]

Poynting determined the formulas for the energy flux $\mathbf{S}(\mathbf{x}, t)$ and energy density $u(\mathbf{x}, t)$ of electromagnetic fields; his results, which we shall verify, are

$$\mathbf{S} = \frac{1}{\mu_0}\mathbf{E} \times \mathbf{B}, \tag{11.57}$$

$$u = \frac{\epsilon_0}{2}E^2 + \frac{1}{2\mu_0}B^2. \tag{11.58}$$

The fields and $\mathbf{S}$ and u are all functions of position $\mathbf{x}$ and time t. In the case of the energy density u, (11.58) is the same as the equations we deduced previously for static fields; but here the derivation is more rigorous because it is based on local conservation of energy. The energy flux $\mathbf{S}(\mathbf{x}, t)$ is called the *Poynting vector*. The units of $\mathbf{S}$ are W/m^2. The meaning of $\mathbf{S}$ is that if $d\mathbf{A}$ is a small area in the field, then the energy passing through $d\mathbf{A}$ in the time dt is $(\mathbf{S} \cdot d\mathbf{A})dt$.

To verify (11.56) from Maxwell's equations is an exercise in vector calculus. Start with $\mathbf{E} \cdot \mathbf{J}$, and substitute for $\mathbf{J}$, from the Ampère-Maxwell law,

$$\mathbf{J} = \frac{1}{\mu_0}\nabla \times \mathbf{B} - \epsilon_0\frac{\partial \mathbf{E}}{\partial t}. \tag{11.59}$$

Now simplify $\mathbf{E} \cdot \mathbf{J}$ using various identities of vector calculus. For example, note that

$$\mathbf{E} \cdot (\nabla \times \mathbf{B}) = -\nabla \cdot (\mathbf{E} \times \mathbf{B}) + \mathbf{B} \cdot (\nabla \times \mathbf{E})$$

$$= -\nabla \cdot (\mathbf{E} \times \mathbf{B}) - \mathbf{B} \cdot (\partial \mathbf{B}/\partial t).$$

The first line is a vector identity, and the second follows from Faraday's law. Also,

$$\mathbf{E} \cdot \frac{\partial \mathbf{E}}{\partial t} = \frac{1}{2}\frac{\partial E^2}{\partial t} \quad \text{and} \quad \mathbf{B} \cdot \frac{\partial \mathbf{B}}{\partial t} = \frac{1}{2}\frac{\partial B^2}{\partial t}.$$

Therefore, putting it all together,

$$\mathbf{E} \cdot \mathbf{J} = -\nabla \cdot \left(\frac{1}{\mu_0}\mathbf{E} \times \mathbf{B}\right) - \frac{\partial}{\partial t}\left(\frac{1}{2}\epsilon_0 E^2 + \frac{1}{2\mu_0}B^2\right). \tag{11.60}$$

Poynting's theorem (11.56) follows immediately, with $\mathbf{S}$ and u in (11.57) and (11.58). Poynting used the same argument to obtain these results in 1884.

Equations (11.57) and (11.58) describe the field energy in vacuum. We used the vacuum form of Maxwell's equations to derive (11.56). If the calculation is redone for fields in matter the appropriate formulas are[13]

---

[12]J. H. Poynting was a professor of physics from 1880 until his death, at a school that became the University of Birmingham. Born 1852 in Monton, Lancashire. Died 1914 in Birmingham.
[13]See Exercise 7.

$$\mathbf{S} = \mathbf{E} \times \mathbf{H}, \tag{11.61}$$

$$u = \tfrac{1}{2}(\mathbf{E} \cdot \mathbf{D} + \mathbf{H} \cdot \mathbf{B}). \tag{11.62}$$

In this case, the energy transfer term $\partial u_K / \partial t$ is $\mathbf{E} \cdot \mathbf{J_f}$, the work on the free charge only, because $\mathbf{S}$ and u include the internal energy of the bound charge and current as well as field energy. Strictly speaking, (11.61) and (11.62) are only correct for linear materials.

EXAMPLE 3   Charging a capacitor, Part 2.   In Chapter 3 we argued that the energy of a capacitor belongs to the electric field. We can use Poynting's theorem to see how the electric field energy builds up in the capacitor as it becomes charged.

Recall the results of Example 1, and refer to Fig. 11.1. Again we use the quasistatic approximation, and neglect edge effects. On the cylindrical surface of radius a that encloses the volume between the plates, the electric field is $(Q/\pi a^2 \epsilon_0)\hat{\mathbf{k}}$ and the magnetic field is $(\mu_0 I/2\pi a)\hat{\boldsymbol{\phi}}$. The energy flux density on the surface is

$$\mathbf{S} = \frac{1}{\mu_0}\mathbf{E} \times \mathbf{B} = \frac{-QI}{2\pi^2 a^3 \epsilon_0}\hat{\mathbf{r}}. \tag{11.63}$$

Thus field energy flows radially *into* the cylinder as Q increases, i.e., when $I > 0$. The surface area is $2\pi a d$, where d is the distance between the plates, so the total rate at which field energy flows into the cylinder is

$$P = \frac{QId}{\pi a^2 \epsilon_0}. \tag{11.64}$$

The total electric field energy U_E between the plates is the density $\epsilon_0 E^2/2$ times the volume $\pi a^2 d$;

$$U_E = \frac{Q^2 d}{2\pi a^2 \epsilon_0}. \tag{11.65}$$

Therefore $dU_E/dt = P$, because $dQ/dt = I$ by conservation of charge. Field energy flowing in through the surface builds up in the field between the plates. This justifies the earlier statement that the energy stored in a capacitor resides in the electric field.

But what about magnetic field energy? While the capacitor is being charged, i.e., for $I \neq 0$, there is a magnetic field. The magnetic field energy between the plates is

$$U_M = \int_0^a \frac{B_\phi^2}{2\mu_0} 2\pi r \, dr \, d = \frac{\mu_0 I^2 d}{16\pi}, \tag{11.66}$$

where we have used the interior field $B_\phi(r)$ derived in Example 1. However, this magnetic field energy is negligible compared to the electric field energy. The ratio is

$$\frac{U_M}{U_E} = \mu_0\epsilon_0 \frac{a^2 I^2}{8Q^2}. \tag{11.67}$$

Because $\mu_0\epsilon_0$ is $1/c^2$, where c is the speed of light, the ratio is of order $a^2/(c\tau)^2$, where $\tau = Q/I$ is a characteristic time of the charging process. A realistic situation would have $\tau \gg a/c$, i.e., τ much larger than the time for light to cross the radius of the capacitor, so U_M is indeed negligible. This example is a good illustration of the quasistatic approximation. Because we used the quasistatic approximation for the fields when we calculated P, we must neglect U_M to be consistent, and this would be a good approximation in most real cases. On the other hand, if the time constant τ is not large compared to a/c, e.g., for a very high frequency alternating current, then a significant amount of electromagnetic radiation occurs, energy propagates away from the system in electromagnetic waves, and the quasistatic approximation breaks down.

■

EXAMPLE 4 Poynting vector for a current-carrying wire.   Consider a segment of conducting wire that extends from $z = 0$ to $z = \ell$, as shown in Fig. 11.6. The power supplied to the moving charges in the dc current I, by the potential difference V_0 from one end of the segment to the other, is $P = IV_0$. Although this power is ultimately supplied by a battery, or generator, which may be remote, it may be considered as transmitted by Poynting's vector across the surface of the wire at $r = a$, in the following way: At the surface the magnetic field is $\mathbf{B} = (\mu_0 I/2\pi a)\hat{\phi}$. The electric field inside the wire and on the surface is

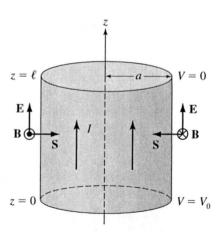

FIGURE 11.6   Example 4. There is a potential difference V_0 and resulting current I across the wire segment of radius a and length ℓ. The power flowing into the wire through the cylindrical surface, calculated from the Poynting vector, is equal to the power dissipated in resistance $P = IV_0$.

$\mathbf{E} = \mathbf{J}/\sigma = (V_0/\ell)\hat{\mathbf{k}}$. Thus at $r = a$ the Poynting vector $\mathbf{S}$ is $-(V_0 I/2\pi a\ell)\hat{\mathbf{r}}$. Because the surface area across which $\mathbf{S}$ flows is $2\pi a\ell$, the power inward through the surface is $V_0 I$.

In Examples 3 and 4 the Poynting vector is considered formally as transmitting power into a system. However, this viewpoint seems nonintuitive and abstract, especially in Example 4. Another, even more extreme example, is illustrated in Fig. 11.7, which shows a bar magnet and a point charge near the N pole. It is not difficult to see that the Poynting vector for this *static system* carries energy in closed circles in the $\hat{\phi}$ direction. However, this energy cannot be turned into heat, nor do work, nor be detected by any means!

When we use the Poynting vector to analyze energy flow in electromagnetic waves, in Sec. 11.5, the interpretation will be clear and natural.

11.4.2 ▨ Field Momentum

The electromagnetic field carries momentum as well as energy, and the momentum is also locally conserved. There is a continuity equation for momentum. But momentum is more complicated than energy. Energy is a scalar, so its flux $\mathbf{S}$ is a vector. But momentum is a *vector*, so its flux is a *tensor* of second rank. Here we are not going to derive the continuity equation for momentum conservation, which is a *vector* equation involving the divergence of the flux *tensor*, but we will examine the interesting result. The momentum density (momentum per unit volume, $d\mathbf{P}_{\text{em}}/dV$, a vector) of the electromagnetic field is[14]

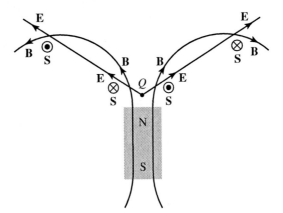

FIGURE 11.7   Magnet and charge. A bar magnet and a point charge Q have static fields **B** and **E**, respectively. With these fields there is a time-independent Poynting vector **S** in the directions shown, i.e., tangent to circles around the symmetry axis. The energy flow associated with **S** in this case is merely formal; it has no physical significance because it cannot be detected.

[14]For a derivation see Griffiths, Chapter 8, or Jackson, Chapter 6. We'll derive this result using relativity in Chapter 12.

$$\frac{d\mathbf{P}_{em}}{dV} = \mu_0 \epsilon_0 \mathbf{S}. \tag{11.68}$$

This result is consistent with the photon theory of electromagnetic waves. If the density of photons in a beam of light is n_γ, and each photon has energy E_γ and momentum $\mathbf{p}_\gamma$, then the energy density is $u = E_\gamma n_\gamma$, the energy flux is $\mathbf{S} = E_\gamma n_\gamma \mathbf{c}$, and the momentum density is $\mathbf{p}_\gamma n_\gamma$. Here c is the speed of light, or photons. The energy of a photon is $E_\gamma = p_\gamma c$, as for any massless particle in special relativity.[15] Thus the momentum density is $\mathbf{S}/c^2$, which agrees with (11.68) because $\mu_0 \epsilon_0 = 1/c^2$.

We will derive (11.68) in Chapter 12 using the relativistically covariant form of the theory. In relativity, energy and momentum are linked. The relativistic equation combines conservation laws of both energy and momentum.

There is also angular momentum in the electromagnetic field. The angular momentum density (angular momentum per unit volume) is

$$\frac{d\mathbf{L}_{em}}{dV} = \mathbf{r} \times \frac{d\mathbf{P}_{em}}{dV} = \mu_0 \epsilon_0 \mathbf{r} \times \mathbf{S}. \tag{11.69}$$

EXAMPLE 5   The Feynman disk paradox.   In Fig. 11.8 an insulator disk is free to rotate on its axis. Attached to the disk coaxially there are: (1) a solenoidal coil that can be energized by a battery on the disk, and (2) a ring of positive charge fixed on the disk. Initially the battery is not connected to the coil, no current flows, and the system is at rest.

Suppose the battery is now connected to the solenoid, by closing a switch, so that current flows in the solenoid in the direction shown. This causes a magnetic

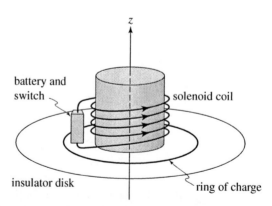

FIGURE 11.8   The Feynman disk paradox. When the switch is closed there is an impulsive torque on the disk. Is angular momentum conserved?

---

[15]Recall the energy-momentum relation in special relativity, $E = \sqrt{p^2 c^2 + m^2 c^4}$; if the particle mass m is 0, then $E = pc$.

field in the $+z$ direction through the coil. The changing flux that occurs while the current builds up to its final value produces an emf in the $-\hat{\phi}$ direction, by Lenz's law, which produces an impulsive torque on the ring of charge in the $-z$ direction. The disk begins to rotate. But the initial angular momentum was 0, and no external torques have been applied, so the final angular momentum should still be 0. How can this be?

The resolution to this apparent paradox is that when current is flowing in the solenoid there is angular momentum in the electromagnetic field; its density is $\mathbf{r} \times (\mathbf{E} \times \mathbf{B}) \, \epsilon_0$ by (11.69). This *field angular momentum* is equal but opposite to the mechanical angular momentum gained by the disk. Therefore the total angular momentum does remain 0. If the battery were now disconnected, by opening the switch, the magnetic field would decrease to 0 and there would be an emf in the $+\hat{\phi}$ direction. The resulting impulsive torque on the ring of charge, which would be in the $+z$ direction, would bring the system back to rest.

The fact that an electromagnetic field—something without mass—can have momentum and even angular momentum seems surprising. But from the viewpoint of *quantum electrodynamics* it is quite natural. A photon carries energy and momentum like a massless particle.

11.5 ■ ELECTROMAGNETIC WAVES IN VACUUM

In this section we consider electric and magnetic fields without charges or currents.

The Maxwell equations in vacuum, *without any matter at all*, i.e., with $\rho = 0$ and $\mathbf{J} = 0$, are

$$\nabla \cdot \mathbf{E} = 0 \tag{11.70}$$

$$\nabla \times \mathbf{E} = -\partial \mathbf{B}/\partial t \tag{11.71}$$

$$\nabla \cdot \mathbf{B} = 0 \tag{11.72}$$

$$\nabla \times \mathbf{B} = \mu_0 \epsilon_0 \partial \mathbf{E}/\partial t. \tag{11.73}$$

Although no charge sources are present in the space being considered, fields may exist there in the form of electromagnetic waves. Charge somewhere else in the universe may have created the waves in the first place, but the waves continue to exist, and propagate through the vacuum with constant energy into places where no charge is present.

There may also be static fields in a region where $\rho = 0$ and $\mathbf{J} = 0$, produced by static charges and currents outside the region. But we are now interested in time-dependent fields, propagating through a vacuum, independent of any charge, so we ignore possible static field components.

The purpose of this section is to find the mathematical properties of these wave solutions, and relate them to the physical properties of light and other forms of electromagnetic radiation. In Chapters 13–15 we will consider the inter-action of electromagnetic waves with matter.

11.5.1 ■ Derivation of the Wave Equation

The wave equation for a quantity $\varphi(\mathbf{x}, t)$ that has linear wave motion is

$$\nabla^2 \varphi - \frac{1}{c^2} \frac{\partial^2 \varphi}{\partial t^2} = 0 \tag{11.74}$$

where c is the wave speed. To see why c is the wave speed, consider propagation of plane waves in one dimension x, with φ independent of y and z. Then the equation is satisfied by any function of the form $\varphi(x, t) = F(x - ct)$. If the function $F(\xi)$ has a feature, e.g., an extremum or a node, at a certain value ξ_0 of its argument, then that feature in the solution $F(x - ct)$ is located at $x = \xi_0 + ct$; the feature moves in the x direction with speed c.

Equation (11.74) arises in many places in theoretical physics; for instance, φ could be the pressure or density in sound, the displacement from equilibrium of a point in an elastic medium, or the electric or magnetic field in light. All of these are quantities that may be described by the wave equation.[16]

To derive (11.74) from Maxwell's equations is rather straightforward. Use the "curl curl trick." Consider $\nabla \times (\nabla \times \mathbf{E})$. By the double cross product identity, it is

$$\nabla \times (\nabla \times \mathbf{E}) = \nabla(\nabla \cdot \mathbf{E}) - \nabla^2 \mathbf{E} = -\nabla^2 \mathbf{E}, \tag{11.75}$$

where the second equality follows from (11.70). But $\nabla \times \mathbf{E}$ is $-\partial \mathbf{B}/\partial t$ by Faraday's law, and making this replacement on the left-hand side of (11.75) we can also write

$$\nabla \times (\nabla \times \mathbf{E}) = \nabla \times (-\partial \mathbf{B}/\partial t) = -\frac{\partial}{\partial t}(\nabla \times \mathbf{B}) = -\mu_0 \epsilon_0 \frac{\partial^2 \mathbf{E}}{\partial t^2} \tag{11.76}$$

where the final equality follows from (11.73), the displacement current. Comparing (11.75) and (11.76) we see that $\mathbf{E}(\mathbf{x}, t)$ satisfies the wave equation

$$\nabla^2 \mathbf{E} - \mu_0 \epsilon_0 \frac{\partial^2 \mathbf{E}}{\partial t^2} = 0. \tag{11.77}$$

The result is actually a *vector equation*. Each Cartesian component of $\mathbf{E}$ satisfies the wave equation of the form (11.74). By a similar calculation[17] $\mathbf{B}(\mathbf{x}, t)$ satisfies the same equation

$$\nabla^2 \mathbf{B} - \mu_0 \epsilon_0 \frac{\partial^2 \mathbf{B}}{\partial t^2} = 0. \tag{11.78}$$

---

[16]We assume that the reader has encountered the wave equation and its solutions, at least in one dimension, in an introductory physics course. It might be advantageous to review the wave equation in an elementary textbook, such as those listed at the end of Chapter 1, or in one of the specialized books listed at the end of this chapter.

[17]See Exercise 8.

Equations (11.77) and (11.78) show that the fields obey the wave equation, but there is more to the story than that. What these equations do not show is the *relationship* between **E** and **B** in the wave. The waves in **E** and **B** are not independent, but must satisfy very strict conditions, which we will explore. For this reason, the waves are properly called *electromagnetic*.

The speed of electromagnetic waves is, according to (11.77) and (11.78),

$$c = \frac{1}{\sqrt{\mu_0 \epsilon_0}}. \tag{11.79}$$

When Maxwell derived this result he knew the values of the electric and magnetic parameters ϵ_0 and μ_0 from experimental measurements of electric and magnetic forces. He found that the speed c calculated from (11.79), is, within the experimental errors, the same as the measured speed of light. By this mathematical theory Maxwell discovered that light is an electromagnetic wave phenomenon.

Today we use (11.79) to determine the value of ϵ_0. In SI units μ_0 is assigned the value $4\pi \times 10^{-7}$ N/A^2, which defines the ampere (A) as we learned in Chapter 8. Then ϵ_0 is

$$\epsilon_0 = \frac{1}{\mu_0 c^2} = 8.85 \times 10^{-12} \frac{C^2}{Nm^2}.$$

The experimental verification of the full theory comes from a comparison of this ϵ_0 calculated from the wave speed, and the value of ϵ_0 measured from the Coulomb force between static charges. The two methods of evaluating ϵ_0 give the same result.

Maxwell's theory of electromagnetic waves—solutions of the field equations—predicted the existence of forms of radiation that were unknown in his time, having wavelengths outside the range of visible light. These predictions were verified years later. For example, in 1887 Hertz discovered what we now call radio waves, which are long-wavelength electromagnetic waves. Also, in 1895 Roentgen discovered X rays, which we now know are short-wavelength electromagnetic waves. These, and all forms of electromagnetic radiation, travel with the same speed c in vacuum.

The universality of the speed of light led to Einstein's theory of relativity in 1905.

There exist electromagnetic waves with any wavelength, or frequency. The wavelength λ and frequency ν are related by $\lambda\nu = c$, a relation that holds for any kind of harmonic waves. Table 11.3 lists the different parts of the *electromagnetic spectrum*, and how each part of the spectrum is used in practical applications of electromagnetic waves.

11.5.2 ■ An Example of a Plane Wave Solution

Because the wave equation is linear, electromagnetic waves satisfy the *Superposition Principle: Any linear combination of solutions is also a solution. A complete*

TABLE 11.3   The electromagnetic spectrum

Frequency (Hz)	Description	Wavelength
10^2	super low frequency (SLF) radio waves submarine communication	3000 km
10^3	ultra low frequency (ULF) radio waves	300 km
10^4	very low frequency (VLF) radio waves	30 km
10^5	low frequency (LF) radio waves marine radio	3 km
10^6	medium frequency (MF) radio waves AM radio is 0.53×10^6 to 1.60×10^6 Hz.	300 m
10^7	high frequency (HF) short-wave radio	30 m

10^8	(VHF) aircraft radio and navigation
	    FM radio is 0.87×10^8 to 1.08×10^8 Hz.
	    TV channels 2–13
		3 m

10^9	(UHF) cellular telephones, radar, microwave ovens, TV channels 14–83	30 cm
10^{10}	(SHF) microwaves, radar, mobile radio	3 cm

10^{11}	extremely high frequency (EHF)
	    Cosmic microwave background
	    maximum is at 3×10^{11} Hz.
		3 mm

10^{12}	far infrared	0.3 mm
10^{13}	far infrared	30 μm
10^{14}	near infrared Visible light is 3.9×10^{14} to 7.6×10^{14} Hz.	3 μm
10^{15}	near ultraviolet	0.3 μm
10^{16}	vacuum ultraviolet	30 nm
10^{17}	soft X rays	3 nm
10^{18}	soft X rays	0.3 nm
10^{19}	hard X rays	30 pm
10^{20}	gamma rays	3 pm
10^{21}	gamma rays	0.3 pm
10^{22}	cosmic gamma rays	30 fm

set of solutions—complete in the sense that any solution can be written as a super-position of these solutions—is the set of plane waves. A plane wave is a solution of the wave equation with definite values of wavelength λ and frequency ν, related by $c = \lambda \nu$. The *wave fronts* of a plane wave are infinite planes, perpendicular to the direction of propagation.

In Sec. 11.5.3 following we'll derive the general plane wave solution from the field equations. But before we go through the derivation it will be useful to examine a specific example of a plane wave, to understand its basic form. The reader will have to be patient to see where this form comes from; that will be derived in the next section. For now let's focus on understanding the geometrical characteristics of one specific example of an electromagnetic plane wave.

The electric field for a plane wave traveling in the z direction, polarized in the x direction, is

$$\mathbf{E}(\mathbf{x}, t) = E_0 e^{i(kz - \omega t)} \, \hat{\mathbf{i}}. \tag{11.80}$$

E_0 is the amplitude, $\hat{\mathbf{i}}$ is the polarization direction, the wavelength is $\lambda = 2\pi/k$, and the frequency is $\nu = \omega/2\pi$. The parameter k is called the wave number. The relation between ω and k, called the *dispersion relation*, is

$$\omega = ck, \tag{11.81}$$

which is equivalent to $\lambda \nu = c$. Equation (11.80) shows that the wave propagates with velocity ω/k, which is called the *phase velocity*. Then by (11.81) the phase velocity of the plane wave is c.

We've written the field at $\mathbf{x}$ and t as a complex number, but of course physically $\mathbf{E}$ must be real. When we write (11.80), using the complex exponential function, we mean that *the physical field is the real part of the right-hand side of the equation.* The use of complex numbers, though it will take some getting used to, is tremendously convenient.[18] This will be explained more in the subsection following.

There are several ways to see that (11.80) satisfies the wave equation. The simplest is just by direct substitution. ∇^2 acting on the exponential e^{ikz} becomes $-k^2$, and $\partial^2/\partial t^2$ acting on $e^{-i\omega t}$ becomes $-\omega^2$; so the wave equation is satisfied if $k = \omega/c$. Or, by a more roundabout argument, note that $\mathbf{E}(\mathbf{x}, t)$ can be written as

$$\mathbf{E}(\mathbf{x}, t) = F(z - ct) \hat{\mathbf{i}},$$

where $F(\zeta) = \exp(ik\zeta)$; the right-hand side has the familiar form of a one-dimensional wave, propagating in the z direction with speed c. The physical field $\mathbf{E}$ is $E_0 \cos(kz - \omega t) \hat{\mathbf{i}}$, the *real part* of (11.80). It oscillates in the $\pm \hat{\mathbf{i}}$ direction (the polarization direction) as the wave propagates in the $\hat{\mathbf{k}}$ direction.

Now, it is not enough that $\mathbf{E}(\mathbf{x}, t)$ satisfies the wave equation. The wave equation is necessary but not sufficient. $\mathbf{E}(\mathbf{x}, t)$ and $\mathbf{B}(\mathbf{x}, t)$ must together satisfy the four Maxwell equations. Therefore there must be a magnetic field as well, coupled with the electric field in ways specified by Maxwell's equations. The magnetic field, which *must be present* with the electric field (11.80), is

$$\mathbf{B}(\mathbf{x}, t) = B_0 e^{i(kz - \omega t)} \hat{\mathbf{j}}, \tag{11.82}$$

where

$$B_0 = E_0/c. \tag{11.83}$$

---

[18]"Something of the unreal is necessary to fecundate the real ...", Wallace Stevens in his preface to William Carlos Williams, *Collected Poems* (Objectivist Press, New York, 1934).

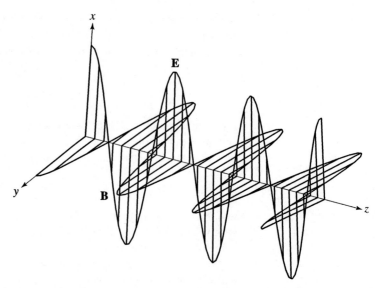

FIGURE 11.9   Electric and magnetic fields in an electromagnetic wave. The direction of propagation is $\hat{\mathbf{k}}$. The wave fronts are planes parallel to the xy plane. The figure is a snapshot of the fields along one line perpendicular to the xy plane. **E** oscillates in the $\pm\hat{\mathbf{i}}$ direction, and **B** in the $\pm\hat{\mathbf{j}}$ direction. Nodes of **E** and **B** coincide, as do the field maxima. $\mathbf{E} \times \mathbf{B}$ points in the direction of propagation.

Again, we will derive this result in complete generality in the next section, but first let's see what it means. Note that **E** and **B** oscillate in orthogonal directions ($\hat{\mathbf{i}}$ and $\hat{\mathbf{j}}$, respectively) and the wave in both cases travels in the third orthogonal direction ($\hat{\mathbf{k}}$). That is, **E**, **B**, and $\hat{\mathbf{k}}$ form an orthogonal triad of vectors.

Figure 11.9 illustrates the geometrical form of the plane-wave solution (11.80) and (11.82).

How to Treat Complex Waves

The solution (11.80) and (11.82) is a special case of the general solution to be derived in the next section, but as an exercise the reader should verify specifically that these functions satisfy all four of the Maxwell equations.[19]

The solution (11.80) and (11.82) is called a *complex wave*, because E_x and B_y are complex numbers at any point of space and time. Recalling Euler's identity, the complex exponential is

$$e^{i\theta} = \cos\theta + i\sin\theta, \tag{11.84}$$

where of course $i = \sqrt{-1}$. The complex number $e^{i\theta}$ has magnitude 1. Because $e^{i\theta}$ is periodic in θ with period 2π, the wave form $e^{i(kz-\omega t)}$ is periodic in z with

[19]See Exercise 9.

period $\lambda = 2\pi/k$ (the wavelength), and periodic in t with period $\tau = 2\pi/\omega$ (the reciprocal of the frequency).

But what is meant by a complex field, physically? Physical fields are defined by *forces* and surely these must be measured in *real numbers* ! The use of the complex wave is just a mathematical convenience—a kind of trick that we employ because calculations with complex exponentials are easier than with sinusoidal functions. When we write equations (11.80) and (11.82) *we mean that the real part of the complex wave is the physical field*. (The real part of $e^{i\theta}$ is $\cos\theta$.) The real part is just *understood*. Because Maxwell's equations are linear in the fields, both the real and imaginary parts of the complex wave satisfy the equations separately.

The constant coefficients E_0 and B_0 may also be complex. If $E_0 = |E_0|e^{i\alpha}$ then the physical field is

$$\mathrm{Re}\left\{|E_0|e^{i\alpha}e^{i(kz-\omega t)}\,\hat{\mathbf{i}}\right\} = |E_0|\cos(kz - \omega t + \alpha)\,\hat{\mathbf{i}}.$$

This shows that the phase angle α of the complex amplitude E_0 is the phase shift in the physical wave. That is, α determines the location of crests, troughs, and nodes for given t.

As long as we consider only expressions linear in the fields we may delay taking the real part. But before doing any calculation that is nonlinear in **E** and **B**, we must first take the real part, reducing the fields to real numbers. For example, the energy flux is $\mathbf{E} \times \mathbf{B}/\mu_0$. To calculate it we must first take the real parts of **E** and **B**. It is *understood* in the formula $\mathbf{S} = \mathbf{E} \times \mathbf{B}/\mu_0$ that **E** and **B** are the physical fields.

EXAMPLE 6 What are the energy flux, energy density, and momentum density in the polarized plane wave (11.80) and (11.82)?

The energy flux density $\mathbf{S}(\mathbf{x}, t)$, i.e., the Poynting vector (11.57), is quadratic in the fields, so to calculate **S** for the plane wave we must first take the real part in (11.80) and (11.82). Then the result is

$$\mathbf{S} = \frac{E_0^2}{\mu_0 c} \cos^2(kz - \omega t)\hat{\mathbf{k}}. \tag{11.85}$$

The direction of **S** is $\hat{\mathbf{k}}$, the direction of wave propagation. The *wave intensity* $\mathcal{I}$ is defined as the energy flux, in the direction of propagation, averaged in time over one period of oscillation. The units of $\mathcal{I}$, like **S**, are W/m^2. The average of $\cos^2\theta$ over one period is $\frac{1}{2}$, because $\cos^2\theta$ oscillates between 0 and 1 symmetrically about $\frac{1}{2}$. Therefore the wave intensity is

$$\mathcal{I} = \hat{\mathbf{k}} \cdot \mathbf{S}_{\mathrm{avg}} = \frac{E_0^2}{2\mu_0 c} = \frac{\epsilon_0 E_0^2 c}{2}. \tag{11.86}$$

The energy density u in the wave can be calculated from (11.58), which gives

$$u(\mathbf{x}, t) = \frac{\epsilon_0 E_0^2}{2} \cos^2(kz - \omega t) + \frac{B_0^2}{2\mu_0} \cos^2(kz - \omega t). \qquad (11.87)$$

The electric and magnetic contributions to u are equal in an electromagnetic wave, because $B_0 = E_0/c$. The equality of u_E and u_M means $\mathbf{E}$ and $\mathbf{B}$ have equal importance in a free wave; the wave is truly "electromagnetic"! We may also write

$$u(\mathbf{x}, t) = \epsilon_0 E_0^2 \cos^2(kz - \omega t). \qquad (11.88)$$

Note by comparing (11.85) and (11.88) that $S = uc$. This little equation has a big meaning—*energy conservation*. Picture a cylinder with cross section A and length $c\, dt$. The wave energy in the volume $A(c\, dt)$ will pass through the area A in time dt. That energy may be expressed either as $u A c\, dt$ or as $S A\, dt$; hence $S = uc$. It is also easy to verify Poynting's theorem for this free electromagnetic wave; i.e., that $\nabla \cdot \mathbf{S} = -\partial u/\partial t$.

The momentum density in the wave, calculated from (11.68), is

$$\frac{d\mathbf{P}_{em}}{dV} = \frac{\mathbf{S}}{c^2} = \frac{u}{c}\hat{\mathbf{k}}, \qquad (11.89)$$

where again u is the energy density. Equation (11.89) agrees with the photon theory of light: The energy-momentum relation for a photon is $E = pc$, as for any massless particle in special relativity. Therefore the momentum density in a beam of photons is equal to the energy density divided by c, in agreement with (11.89).

■————————————————

Radiation Pressure

Because an electromagnetic wave carries momentum, it exerts a pressure, called *radiation pressure*, when it hits a material surface. Suppose the material absorbs all the light incident on it. Then because momentum is conserved the material gains the momentum of the light. Recall from mechanics that $d\mathbf{p} = \mathbf{F}dt$ is the change of momentum during dt due to a force $\mathbf{F}$. Thus the pressure on an absorbing area A for normal incidence is $\mathcal{P}_{rad} = F/A = (dP)/(Adt)$, where dP is the wave momentum contained in the volume $A(cdt)$.[20] By (11.89) dP is $uAdt$, so the radiation pressure is $\mathcal{P}_{rad} = u$. In terms of intensity $\mathcal{I}$, the average pressure on an absorbing surface is $\langle \mathcal{P}_{rad} \rangle = \langle u \rangle = \mathcal{I}/c$.[21]

Radiation pressure is usually small (cf. the exercises) but in extreme cases it can be very large indeed. A dramatic example occurs in inertial confinement

[20] Please do not confuse $\mathcal{P}$ (pressure) with P or $\mathbf{P}$ (momentum).
[21] The brackets $\langle \ldots \rangle$ indicate the average of the enclosed quantity over a period of oscillation. For example, the intensity is $\mathcal{I} = \langle S \rangle$.

fusion experiments, such as are carried out at Lawrence Livermore National Laboratory, in which high-power lasers compress and heat small particles of matter.

EXAMPLE 7   Consider a high-power laser, operating at a wavelength of $1.6\,\mu$m, which puts out 10 kJ in 0.2 ns. The beam is focused on an area of diameter 0.5 mm.

First, what is the RMS electric field strength? The intensity (mean power per unit area) is $\mathcal{I} = 2.5 \times 10^{20}$ W/m^2, which by (11.86) implies $E_{rms} = E_0/\sqrt{2} = 3 \times 10^{11}$ V/m. Note that E_{rms} is far above the breakdown field strength of air (3 MV/m at room temperature and 1 atm) and even above that of fused silica (10^3 MV/m).

Second, what is the radiation pressure of the laser on an absorbing surface? We found above that $\langle \mathcal{P}_{rad} \rangle = \mathcal{I}/c$, so $\langle \mathcal{P}_{rad} \rangle = 8.5 \times 10^{11}$ Pa, which is about 8.5×10^6 atmospheres.

11.5.3 ■ Derivation of the General Plane Wave Solution

The particular solution (11.80) and (11.82) examined in the previous section was just pulled out of a hat. Let us now *derive mathematically* the general harmonic plane wave solution. By a harmonic wave we mean a solution with a definite frequency $\nu = \omega/2\pi$. By a plane wave we mean that the wave fronts are planes. Therefore the fields, as complex waves, have the form

$$\mathbf{E}(\mathbf{x}, t) = \mathbf{E}_0 e^{i(\mathbf{k}\cdot\mathbf{x} - \omega t)} \tag{11.90}$$

$$\mathbf{B}(\mathbf{x}, t) = \mathbf{B}_0 e^{i(\mathbf{k}\cdot\mathbf{x} - \omega t)}, \tag{11.91}$$

where $\mathbf{E}_0$ and $\mathbf{B}_0$ are constant vectors. The real parts are understood to be the physical fields. These functional forms describe an electromagnetic wave propagating in the direction of $\mathbf{k}$ with phase velocity ω/k. The vector $\mathbf{k}$, which determines the wavelength ($\lambda = 2\pi/k$) and the direction of propagation, is called the *wave vector*. In general the wave vector may be in any direction, of the form $k_x \hat{\mathbf{i}} + k_y \hat{\mathbf{j}} + k_z \hat{\mathbf{k}}$.

In a vacuum, the Maxwell equations (11.70) and (11.72) require

$$\mathbf{k} \cdot \mathbf{E}_0 = 0, \tag{11.92}$$

$$\mathbf{k} \cdot \mathbf{B}_0 = 0; \tag{11.93}$$

(The reason is that the $\mathbf{x}$ dependence of either $\mathbf{E}$ or $\mathbf{B}$ is in $e^{i\mathbf{k}\cdot\mathbf{x}}$, and when ∇ acts on the exponential it gives the exponential back again, times $i\mathbf{k}$ by the chain rule.) Thus an electromagnetic wave is a *transverse wave*: The fields oscillate in directions perpendicular to the direction of propagation. The Maxwell equations (11.71) and (11.73) require

$$\mathbf{k} \times \mathbf{E}_0 = \omega \mathbf{B}_0, \tag{11.94}$$

$$\mathbf{k} \times \mathbf{B}_0 = -\mu_0 \epsilon_0 \omega \mathbf{E}_0; \tag{11.95}$$

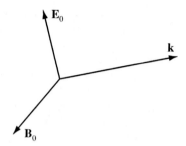

FIGURE 11.10   The triad $\mathbf{k}$, $\mathbf{E}_0$ and $\mathbf{B}_0$. Note that $\mathbf{k} \times \mathbf{E}_0$ is in the direction of $\mathbf{B}_0$, and $\mathbf{k} \times \mathbf{B}_0$ is in the direction of $-\mathbf{E}_0$.

because $\partial/\partial t$ acting on $e^{-i\omega t}$ gives the exponential times $-i\omega$. Therefore $\mathbf{E}_0$ and $\mathbf{B}_0$ are perpendicular to each other, as well as perpendicular to $\mathbf{k}$. Figure 11.10 shows the directions of $\mathbf{k}$, $\mathbf{E}_0$ and $\mathbf{B}_0$ implied by (11.94) and (11.95). The three vectors $\mathbf{E}_0$, $\mathbf{B}_0$, and $\mathbf{k}$ form an orthogonal triad of vectors, with $\mathbf{E}_0 \times \mathbf{B}_0$ parallel to $\mathbf{k}$. The Poynting vector is

$$\mathbf{S} = \frac{1}{\mu_0} \mathbf{E}_0 \times \mathbf{B}_0 \cos^2(\mathbf{k} \cdot \mathbf{x} - \omega t) \tag{11.96}$$

so the energy flux is in the direction of the wave vector $\mathbf{k}$.

The wave is transverse. Each field oscillates, in either space or time, in a direction perpendicular to the direction of propagation. An infinite number of transverse directions exist, but any transverse vector can be written as a linear combination of just two basis vectors. We say that plane waves with wave vector $\mathbf{k}$ have *two linearly independent polarizations*.[22] For example, if $\mathbf{k}$ is in the z direction, then $\hat{\mathbf{i}}$ and $\hat{\mathbf{j}}$ can be taken as the basis vectors for the transverse oscillations. Any plane wave is a superposition of the two basis polarizations. We will study polarization effects in optics in Chapter 13.

So far we have identified the directions of the vector fields. Now we turn to their magnitudes, and the wave speed. Since $\mathbf{E}_0$, $\mathbf{B}_0$, and $\mathbf{k}$ form an orthogonal triad, the cross products in (11.94) and (11.95) are simple. In (11.94) the magnitude of $\mathbf{k} \times \mathbf{E}_0$ is just kE_0, and the direction is parallel to $\mathbf{B}_0$, so (11.94) implies

$$\frac{E_0}{B_0} = \frac{\omega}{k}. \tag{11.97}$$

In (11.95) similarly $|\mathbf{k} \times \mathbf{B}_0| = kB_0$ and $\mathbf{k} \times \mathbf{B}_0$ is parallel to $-\mathbf{E}_0$, so (11.95) implies

$$\frac{E_0}{B_0} = \frac{k}{\mu_0 \epsilon_0 \omega}. \tag{11.98}$$

[22] See Exercise 19.

Equating the two ratios shows that

$$\frac{\omega^2}{k^2} = \frac{1}{\epsilon_0 \mu_0}. \tag{11.99}$$

This relation is required by Maxwell's equations. Now, ω/k is c by the dispersion relation $\omega = ck$. So the wave speed is $1/\sqrt{\mu_0\epsilon_0}$, as found before in (11.79). Then by (11.97) the ratio of amplitudes must be $E_0/B_0 = c$, as stated earlier in (11.83). Summarizing, for any harmonic plane wave,

$$\frac{\omega}{k} = c = \frac{1}{\sqrt{\mu_0\epsilon_0}} \quad \text{and} \quad B_0 = \frac{E_0}{c}. \tag{11.100}$$

Equations (11.90) and (11.91) are the most general linearly polarized harmonic plane wave. Any linear combination of plane waves is also a solution of Maxwell's equations, by the superposition principle. The converse is also true: A remarkable theorem of Fourier analysis states that any solution of the Maxwell equations in vacuum can be written as a superposition of plane waves. Common electromagnetic waves, such as light from the sun or from an electric bulb, are unpolarized and not monochromatic. Such waves are complicated superpositions of plane waves. For example, sunlight is white light—a superposition of all visible wavelengths (and also infrared and ultraviolet wavelengths) with no preferred direction of polarization. The electric field direction at any particular point in sunlight is constantly fluctuating this way and that, on a very short time scale, because there is no preferred polarization direction. An approximation of a linearly polarized harmonic wave could be produced by passing sunlight through a polarizer and a color filter, to "pick out" the component with the specified polarization and frequency.

Electromagnetic waves propagate through empty space. Neither matter nor charge is needed to maintain an electromagnetic wave. The wave carries its own energy and momentum, which are, of course, conserved. Electromagnetic waves may travel immense distances, e.g., the light arriving at Earth from distant galaxies. These waves of electromagnetic field are a part of the physical universe.

But field theory is a *local theory*, specified by differential equations. How are the electromagnetic fields at a particular point produced? What is the *source* of the fields at that point, if no charge is present?

How do electromagnetic waves work? Each field is produced by the variation in time of the other field, in a self-generating process: **E** is produced by the variation of **B**, by Faraday's law; **B** is produced by the variation of **E**, by the displacement current. By propagating through space as a transverse wave, the fields satisfy all four Maxwell equations at every point of space and time, self-consistently. The electromagnetic wave is a beautiful example of an extended geometric structure created by local mathematical rules.

The applications of electromagnetic waves are many, including all of optics. Applications usually involve interactions between waves and matter, or the generation of waves. These topics are covered in Chapters 13 to 15.

11.5.4 ■ A Spherical Harmonic Wave

A plane wave is very much a mathematical idealization. The wave fronts of the wave (11.90) and (11.91) are *infinite planes* perpendicular to $\mathbf{k}$. That is, the fields $\mathbf{E}(\mathbf{x}, t)$ and $\mathbf{B}(\mathbf{x}, t)$ are uniform over any infinite plane perpendicular to $\mathbf{k}$. The source of an ideal plane wave would have to be coherent over an infinite area. In contrast, a real physical wave must have finite extent.

In this section we'll construct another solution of the Maxwell equations in vacuum, that has the form of a *spherical wave*—a harmonic wave with spherical wave fronts propagating outward from the origin. Here we study only the fields in vacuum, so we are not yet in a position to characterize the source. The connection between waves and their sources is the subject of Chapter 15. For now it's enough to picture the spherical wave as emerging from a point source and propagating with equal speed in all directions.

An analogous example in two dimensions, familiar to everyone, is the circular wave on a water surface made by dropping a pebble into a pond. The circular wave fronts propagate isotropically away from the source point. However, in our example the electromagnetic wave is produced continuously, whereas the pebble makes a limited wave train. A better analogy would be the circular waves created in a ripple tank by a pointlike object oscillating up and down on the water surface.

To guarantee that all four Maxwell equations are satisfied, we shall construct the solution from a vector potential. The Cartesian components of $\mathbf{A}(\mathbf{x}, t)$ satisfy the wave equation (11.74), and it is sufficient to let $\mathbf{A}$ be everywhere in the z direction, $\mathbf{A} = A_z \hat{\mathbf{k}}$. Then a solution in spherical coordinates is[23]

$$\mathbf{A}(\mathbf{x}, t) = \Psi(r, t)\hat{\mathbf{k}} \tag{11.101}$$

where $\hat{\mathbf{k}}$ is the unit vector in the z direction, and

$$\Psi(r, t) = \frac{C}{r} e^{i(kr - \omega t)}. \tag{11.102}$$

The function $\Psi(r, t)$ is a spherically symmetric scalar solution of the wave equation (11.74). Using spherical coordinates to calculate the Laplacian ∇^2, it is straightforward to show that $\nabla^2 \Psi = -k^2 \Psi$; also, $\partial^2 \Psi / \partial t^2 = -\omega^2 \Psi$. The wave equation is satisfied for $\omega = ck$. The potential $\mathbf{A}(\mathbf{x}, t)$ is singular at the origin, i.e., at the source of the wave, because $\Psi \to \infty$ as $r \to 0$. We will not be able to determine the nature of the source until Chapter 15, but we can solve for the electromagnetic wave propagating away from the origin. It turns out that the source is a point-like electric dipole, oscillating in the z direction. For example, it could be a small charge undergoing simple harmonic motion on the z axis.

Now, the magnetic field is $\mathbf{B} = \nabla \times \mathbf{A}$. It is most convenient to use spherical coordinates to calculate the curl, and to express the fields in spherical components. Writing (11.101) in spherical coordinates, the vector potential is

---

[23] See Exercise 26.

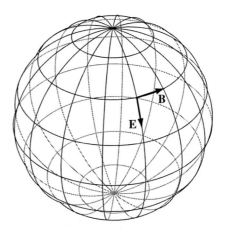

FIGURE 11.11   The asymptotic spherical wave. At any time and position the magnetic field is azimuthal—parallel to a line of latitude; and the electric field is orthogonal to **B**—parallel to a line of longitude. At the time and radius shown, **B** and **E** are in the $+\hat{\phi}$ and $+\hat{\theta}$ directions, respectively. One half period later at this radius **B** and **E** will be in the $-\hat{\phi}$ and $-\hat{\theta}$ directions. The Poynting vector points radially outward.

$$\mathbf{A}(\mathbf{x}, t) = \Psi(r, t)(\hat{\mathbf{r}} \cos\theta - \hat{\theta} \sin\theta) = \hat{\mathbf{r}} A_r + \hat{\theta} A_\theta. \tag{11.103}$$

Its curl is[24]

$$\mathbf{B}(\mathbf{x}, t) = \frac{\hat{\phi}}{r} \left[\frac{\partial}{\partial r} (r A_\theta) - \frac{\partial A_r}{\partial \theta} \right] = \hat{\phi} \left(-ik + \frac{1}{r} \right) \sin\theta \, \Psi(r, t). \tag{11.104}$$

The wave propagates in the $\hat{\mathbf{r}}$ direction, because of the factor $e^{i(kr - \omega t)}$ in $\Psi(r, t)$. The magnetic field at any point is in the azimuthal direction $\hat{\phi}$ as illustrated in Fig. 11.11. This direction is consistent with the fact that the wave comes from an electric dipole, oscillating along the z axis, because **B** curls around the current.

One approach to determining the electric field might be to construct the scalar potential $V(r, t)$, consistent with the Lorentz gauge choice. However, that is not necessary. It is easier to find $\mathbf{E}(\mathbf{x}, t)$ directly from $\mathbf{B}(\mathbf{x}, t)$, using the Ampère-Maxwell equation (11.73). $\mathbf{E}(\mathbf{x}, t)$ must have the same frequency as $\mathbf{B}(\mathbf{x}, t)$, i.e., $\mathbf{E}(\mathbf{x}, t)$ has the form $\mathbf{E}(\mathbf{x})e^{-i\omega t}$. Then the time derivative of **E** is equal to $-i\omega \mathbf{E}$, and (11.73) implies

$$\mathbf{E}(\mathbf{x}, t) = \frac{c^2 \nabla \times \mathbf{B}}{-i\omega}. \tag{11.105}$$

To calculate the curl of **B** is somewhat tedious, but the final result is not too complicated:

[24]See Exercise 27.

$$\mathbf{E}(\mathbf{x}, t) = \hat{\mathbf{r}} E_r + \hat{\boldsymbol{\theta}} E_\theta, \tag{11.106}$$

where

$$E_r = \frac{2c^2}{i\omega} \left(\frac{ik}{r} - \frac{1}{r^2} \right) \cos\theta \ \Psi(r, t) \tag{11.107}$$

$$E_\theta = \frac{c^2}{i\omega} \left(k^2 + \frac{ik}{r} - \frac{1}{r^2} \right) \sin\theta \ \Psi(r, t). \tag{11.108}$$

The electric field is everywhere orthogonal to the magnetic field.

It can be shown by direct substitution that $\mathbf{B}(\mathbf{x}, t)$ and $\mathbf{E}(\mathbf{x}, t)$ satisfy all four Maxwell equations in free space (for $r > 0$). But that rather long calculation is not really necessary because we can verify the equations by general considerations. $\mathbf{B}$ is a curl, so $\nabla \cdot \mathbf{B} = 0$. $\mathbf{E}$ is also a curl by (11.105) so $\nabla \cdot \mathbf{E} = 0$. Equation (11.105) was based on (11.73) so the latter equation is satisfied. All that remains is to check (11.71), and that is guaranteed by the fact that $\mathbf{B}$ satisfies the wave equation, because $\Psi(r, t)$ does.

This spherical wave has a complicated dependence on r. For small r, i.e., $r \ll \lambda$, the dominant field is the electric field, which is of order r^{-3}; that region is called the *near zone*, and the form of $\mathbf{E}$ is determined by proximity to the point source. For intermediate r, i.e., $r \simeq \lambda$, the fields do not have a simple power law dependence on r. For large r, i.e., $r \gg \lambda$, both fields are of order r^{-1}; that region is called the *radiation zone*. The asymptotic fields for large r are

$$\mathbf{B}_{\text{asy}}(\mathbf{x}, t) = -ik \sin\theta \ \Psi(r, t) \hat{\boldsymbol{\phi}}, \tag{11.109}$$

$$\mathbf{E}_{\text{asy}}(\mathbf{x}, t) = -i\omega \sin\theta \ \Psi(r, t) \hat{\boldsymbol{\theta}}. \tag{11.110}$$

(The real part of the right-hand side is understood.) These radiation fields, together with the direction of propagation $\hat{\mathbf{r}}$, form an orthogonal triad at every point in the radiation zone, with $\mathbf{E}_{\text{asy}} \times \mathbf{B}_{\text{asy}}$ in the direction of $\hat{\mathbf{r}}$. The magnetic field is azimuthal, i.e., parallel to lines of latitude, as shown in Fig. 11.11. The electric field is parallel to the lines of longitude, also shown in the figure. The Poynting vector is radially outward, so the wave propagating radially is carrying energy outward.

Note that the asymptotic fields are proportional to r^{-1}, and so the energy flux per unit area is proportional to r^{-2}. Therefore energy is conserved in the wave: The integrated power through a sphere of radius r is independent of r because the area grows as $4\pi r^2$. The average power flowing out through a small spherical surface enclosing the source is the same as the average power through any larger sphere. Of course energy is being fed into the wave by the source at $r = 0$, which is hidden from us because we have only used the field equations in the vacuum, outside the source. In Chapter 15 we will see that (11.104) and (11.106) are the fields of an oscillating electric dipole at the origin. Some external agent must be supplying energy to make the dipole oscillate.

The asymptotic fields (11.109) and (11.110) resemble a plane wave in these respects: (i) The spherical wave fronts are approximately planar. (ii) The fields and propagation direction form an orthogonal triad of vectors. (iii) The ratio $|\mathbf{B}_{asy}|/|\mathbf{E}_{asy}|$ is $1/c$. A plane wave is a mathematical idealization, which approximates a physical wave far from its source.

The reason plane wave solutions are so important is because they are mathematically *complete*; any solution of the wave equation can be written as a superposition of plane waves. Even the spherical wave is equal to a certain superposition of plane waves in all directions.

11.5.5 ■ The Theory of Light

In the late 17th century a dispute arose over the nature of light, between Newton who argued for a particle theory of light, and Huygens who favored a wave theory of light.

At the beginning of the the 19th century, Young observed interference of light, and so showed that light behaves as a wave. Fresnel explained that the optical phenomena of *light polarization* imply that light is a transverse wave. Later Maxwell developed the mathematical theory of the electromagnetic field, and found that solutions of the field equations describe the properties of light as an electromagnetic wave. These concepts were further developed by a group of British and Irish physicists, including Heaviside, FitzGerald, and Lodge, who called themselves *the Maxwellians*.

However, Maxwell's theory is not the end of the history of the science of light. Early in the 20th century, Planck and Einstein showed that light energy is quantized. Therefore light also behaves as if composed of particles, which we call photons. Compton did definitive experiments on X-ray scattering that verified the predictions of the photon theory. Dirac formulated a quantum field theory of electromagnetism. This *quantum theory of light* does not negate Maxwell's theory of electromagnetism. Indeed, Maxwell's equations are also true in quantum electrodynamics (QED), although their interpretation is different from the classical theory because in QED the fields have quantum uncertainty.

Classical electrodynamics—Maxwell's theory—is a valid description of electromagnetic phenomena, including light, in the limit of large field intensities. A large field intensity corresponds to large numbers of photons, so that the quantization unit (one photon) is small on the scale of the field. The classical theory is the limit of QED in the same sense that classical mechanics is the limit of quantum mechanics for large mechanical systems. The mean value of the quantum field obeys the classical field equations, and if the field intensity is large then the effects of uncertainty are negligible. The classical theory does break down for very small systems, such as a single photon or atom. But the classical theory remains important despite this limitation, for the many phenomena that occur on scales much larger than a single atom.

FURTHER READING

The following books discuss general properties of the wave equation, in a variety of applications.

F. S. Crawford, Jr., *Waves* (McGraw-Hill, New York, 1968).

A. P. French, *Vibrations and Waves* (Norton, New York, 1971).

W. C. Elmore and M. A. Heald, *Physics of Waves* (Dover, New York, 1969).

H. Georgi, *The Physics of Waves* (Prentice-Hall, Englewood Cliffs, NJ, 1993).

The interesting history of the electromagnetic theory of light is recounted in J. G. O'Hara and W. Pricha, *Hertz and the Maxwellians* (Peter Peregrinus, London, 1987).
The *Feynman Lectures on Physics*, Vol II, contain interesting insights on electromagnetic waves, especially in Chapters 18 and 21.

EXERCISES

Sec. 11.1. The Maxwell Equations in Vacuum and Displacement Current

11.1. In 1929 at the Free University of Brussels, M. R. Van Cauwenberghe measured, as directly as was then possible, the magnetic field caused by the displacement current between the plates of a parallel-plate capacitor. In his system the plates were specially shaped, but they were approximately parallel disks of diameter 1.5 m separated by 0.4 m. He applied across the plates an oscillating voltage, with amplitude 174 kV and frequency 50 Hz. For this system, with air between the plates, calculate the value of **B** at radial distance 0.4 m from the center.

To measure the small field, Van Cauwenberghe devised a magnetometer that consisted of a toroidal solenoid with an iron core, suspended parallel to the plates in the midplane of the capacitor. By using an iron core he increased the sensitivity by a large factor, probably more than 10^3. Also, the solenoid had 813 turns, which increased the sensitivity by another factor of almost 10^3. Ultimately he measured induced emfs in the solenoid of order 0.1 mV. His measurement agreed with Maxwell's theory of the magnetic field induced by the displacement current in the capacitor. (Reference: Van Cauwenberghe, *Journal de Physique et le Radium*, **10**, 303 (1929).)

11.2. A capacitor with circular parallel plates, with radius a and separation d, has potential difference $V(t)$.

  (a) Determine the magnetic field on the midplane of the capacitor, at radius r from the symmetry axis, for $r > a$. [Answer: $B(r) = (\mu_0 \epsilon_0 a^2 / 2rd)\, dV/dt$]

  (b) Show that B is the same as the field of a straight wire carrying current $I = dQ/dt$, where Q is the charge on the capacitor.

11.3. Show that the discontinuity of **B** across a capacitor plate as the capacitor is being charged with current I is equal to $\mu_0 \mathbf{K} \times \hat{\mathbf{n}}$, where **K** is

$$\mathbf{K} = \frac{\hat{\mathbf{r}} I}{2\pi} \left(\frac{1}{r} - \frac{r}{a^2} \right).$$

By the general boundary condition on the tangential field, $\mathbf{K}$ is the surface current density that flows radially as the plate becomes charged. Show that this functional form of the surface current implies that the charge remains uniformly distributed over the area of the plate. (Hint: Show that the rate of change of charge in an arbitrary annulus is proportional to the area of the annulus. Therefore the charge divided by the area is constant.)

Sec. 11.2. Scalar and Vector Potentials

11.4. Derive the equation (11.31) for $\mathbf{A}(\mathbf{x}, t)$ in the Lorentz gauge.

11.5. (a) Show that the *general* equations for the potentials $V(\mathbf{x}, t)$ and $\mathbf{A}(\mathbf{x}, t)$, i.e., without any gauge condition, are

$$\nabla^2 V + \frac{\partial}{\partial t} (\nabla \cdot \mathbf{A}) = -\frac{\rho}{\epsilon_0}$$

$$\nabla^2 \mathbf{A} - \mu_0 \epsilon_0 \frac{\partial^2 \mathbf{A}}{\partial t^2} - \nabla \left(\nabla \cdot \mathbf{A} + \mu_0 \epsilon_0 \frac{\partial V}{\partial t} \right) = -\mu_0 \mathbf{J}.$$

(b) The equations in a particular gauge can be obtained from the general ones by substituting the gauge condition. Derive, by this method, the field equations in the Coulomb gauge and the Lorentz gauge.

Sec. 11.3. Maxwell's Equations in Matter

11.6. (a) Show that the Maxwell equations in matter reduce to the Maxwell equations in vacuum if the polarization and magnetization are zero.

(b) Show that the Maxwell equations in vacuum are the same as the Maxwell equations in a uniform linear material with $\epsilon = \epsilon_0$ and $\mu = \mu_0$.

Sec. 11.4. Energy and Momentum of the Electromagnetic Field

11.7. Consider the electromagnetic field in a linear medium with material properties ϵ and μ. Calculate $\nabla \cdot \mathbf{S}$ and $\partial u / \partial t$ for the energy flux $\mathbf{S}$ and density u given in (11.61) and (11.62). Identify the energy transfer rate $\partial u_K / \partial t$ in (11.52) such that the energy continuity equation is satisfied, and interpret the result.

Sec. 11.5. Electromagnetic Waves in Vacuum

11.8. Use Maxwell's equations (11.70) through (11.73) to show that the magnetic field $\mathbf{B}(\mathbf{x}, t)$ satisfies the vector wave equation.

11.9. Verify explicitly that the specific plane wave solution (11.80) and (11.82) satisfies the four Maxwell equations.

11.10. (a) Determine the vector potential $\mathbf{A}(\mathbf{x}, t)$ and scalar potential $V(\mathbf{x}, t)$ in the Lorentz gauge, for the linearly polarized plane wave described by (11.80) and (11.82), with the boundary condition that the potentials must be finite at infinity. (Hint: Let $V = 0$.)

(b) Consider now the general linearly polarized plane wave, with wave vector **k**, polarization direction $\hat{\mathbf{n}}$, and frequency $\omega = c|\mathbf{k}|$. Determine the vector and scalar potentials for this case.

(c) Suppose the boundary condition in (a) is removed. Show that $\mathbf{A}(\mathbf{x}, t) = -B_0 x e^{i(kz - \omega t)}\hat{\mathbf{k}}$ and $V(\mathbf{x}, t) = -c B_0 x e^{i(kz - \omega t)}$ satisfy the Lorentz gauge condition and give the fields in (11.80) and (11.82).

11.11. Calculate $\nabla \cdot \mathbf{S}$ and $\partial u/\partial t$ for a linearly polarized plane wave propagating in the z direction and polarized in the x direction. Explain the meaning of the results. Show that the electric field and magnetic field have equal energy densities in the plane wave.

11.12. For a wave that depends only on z and t, the wave equation (11.74) becomes

$$\frac{\partial^2 \varphi}{\partial z^2} - \frac{1}{c^2}\frac{\partial^2 \varphi}{\partial t^2} = 0.$$

Show that a general solution of this equation may be written as $\varphi(z, t) = f(z - ct) + g(z + ct)$, where f and g are arbitrary functions of their arguments. Show by substitution in the equation that $\varphi(z, t) = C \cos kz \cos kct$ is a solution of the wave equation. Then determine f and g for this solution. Explain what is meant by the statement that a standing wave is the superposition of traveling waves in opposite directions.

11.13. Consider the electromagnetic field

$$\mathbf{E}(x, y, t) = E_0 \cos(\pi x/L)\cos(\pi y/L)\sin \omega t\ \hat{\mathbf{k}}$$

$$\mathbf{B}(x, y, t) = B_0\left[-\cos(\pi x/L)\sin(\pi y/L)\hat{\mathbf{i}} + \sin(\pi x/L)\cos(\pi y/L)\hat{\mathbf{j}}\right]\cos \omega t.$$

(a) Show that this field satisfies the Maxwell equations in vacuum if $\omega = \sqrt{2}\pi c/L$ and $B_0 = E_0/(\sqrt{2}c)$.

(b) This field represents a standing electromagnetic wave inside a box with metal walls and a square cross section of size $L \times L$ parallel to the xy plane, and very long in the z direction. This is an example of a *cavity oscillator*. Sketch the **E** and **B** fields. Note that the wave number k is π/L, so the wavelength is $2L$.

11.14. A field meter shows that the amplitude of the electric field oscillation in a certain radio wave is 5 millivolts per meter.

(a) What is the amplitude of the magnetic field oscillation, in T?

(b) What is the intensity in W/m^2?

11.15. Polarized light is incident on a perfect polarizer, and it is observed that 20% of the light intensity gets through. What is the angle between the polarizer axis and the polarization direction of the light? [Answer: 63 degrees]

11.16. A polarized plane electromagnetic wave moves in the y direction, with the electric field in the $\pm x$ direction. What is the direction of the magnetic field at a point where the electric field is in the $-x$ direction?

11.17. Suppose the electric field in an electromagnetic wave is

$$\mathbf{E}(\mathbf{x}, t) = \frac{E_0}{\sqrt{2}} \left(\hat{\mathbf{k}} - \hat{\mathbf{i}} \right) \sin(ky - \omega t).$$

(a) Determine the magnetic field. (Hint: $\nabla \times \mathbf{E} = -\partial \mathbf{B}/\partial t$.)

(b) Determine the Poynting vector.

11.18. *Circularly polarized waves.* Consider a superposition of waves traveling in the z direction, with fields

$$\mathbf{E}(\mathbf{x}, t) = \mathrm{Re} \left\{ \hat{\mathbf{i}} E_1 e^{i(kz - \omega t)} + \hat{\mathbf{j}} E_2 e^{i(kz - \omega t)} \right\}$$

$$\mathbf{B}(\mathbf{x}, t) = \mathrm{Re} \left\{ \hat{\mathbf{j}} \frac{E_1}{c} e^{i(kz - \omega t)} - \hat{\mathbf{i}} \frac{E_2}{c} e^{i(kz - \omega t)} \right\}$$

where E_1 and E_2 may be complex

$$E_1 = C_1 e^{i\phi_1}, \quad E_2 = C_2 e^{i\phi_2}.$$

(C_1 and C_2 are real.)

(a) Calculate the average energy flux $\mathbf{S}_{\mathrm{avg}}$.

(b) Suppose $E_1 = C$ and $E_2 = iC$, i.e., $C_1 = C_2 = C$ and $\phi_1 = 0$, $\phi_2 = \pi/2$. Determine the direction of $\mathbf{E}$ as a function of t, at a point on the xy plane. Describe the result in words and pictures.

(c) For the same field as (b), determine the direction of $\mathbf{E}$ as a function of z, for a snapshot of the field at $t = 0$. Describe the result in words and pictures.

11.19. Consider the fields

$$\mathbf{E}(\mathbf{x}, t) = \hat{\mathbf{i}} F_1(x - ct) + \hat{\mathbf{j}} F_2(x - ct) + \hat{\mathbf{k}} F_3(x - ct)$$

$$\mathbf{B}(\mathbf{x}, t) = \frac{1}{c} \left[\hat{\mathbf{i}} G_1(x - ct) + \hat{\mathbf{j}} G_2(x - ct) + \hat{\mathbf{k}} G_3(x - ct) \right]$$

where the functions $F_1, F_2, \ldots, G_3$ approach 0 in the limits $x \to \pm\infty$. These fields obviously satisfy the vector wave equation. They correspond to a pulse of radiation moving in the $+x$ direction. But the Maxwell equations place severe restrictions on the components $F_1, F_2, \ldots, G_3$.

(a) Show that the Maxwell equations require $F_1 = G_1 = 0$, $G_3 = F_2$, and $G_2 = -F_3$. (Thus there are only two independent polarizations.)

(b) Suppose $F_2(\xi) = G_3(\xi) = E_0 \exp(-\xi^2/a^2)$ and the other components are 0. (ξ stands for $x - ct$.) Make a sketch that shows a snapshot of the fields in space at time t.

11.20. (a) For an ideal plane electromagnetic wave in vacuum, show that the energy crossing an area A in 1 second, with A normal to the direction of propagation, is equal to the total field energy within a cylinder with base A and height 3×10^8 m.

(b) Show that $\mathcal{I} = cu_{\mathrm{av}}$ and explain this result in terms of conservation of energy.

11.21. (a) The intensity of sunlight at the Earth is $\mathcal{I}_{\text{solar}} = 1300 \, \text{W/m}^2$. Compute the RMS electric and magnetic field strengths (in V/m and T, respectively) for a plane wave with intensity $\mathcal{I}_{\text{solar}}$. Does the result depend on the wavelength of the plane wave?

(b) Estimate the RMS electric field strength in the light from a 100 W light bulb, at 1 m distance from the bulb.

(c) The total power of light from a small He-Ne laser pointer is 0.1 mW, in a beam of diameter 4 mm. Calculate the RMS electric field strength.

(d) The wavelength of light from a He-Ne laser is 633 nm. How many photons does the laser pointer emit in one second?

11.22. Consider a radio antenna radiating 20 kW of power in radio waves with frequency 100 MHz. Estimate the order of magnitude of the RMS electric field strength at a distance of 10 km, by assuming the radiation is uniformly distributed in a hemisphere around the transmitter. (More precisely, the intensity would depend somewhat on direction.)

11.23. Calculate the force due to radiation pressure by sunlight falling in the normal direction on a 1 cm^2 light-absorbing surface. The intensity of sunlight is 1.3 kW/m^2. What is the force if the surface reflects light? (Hint: $\mathbf{F}dt = d\mathbf{p}$.)

11.24. The dust tail of a comet points away from the sun because of radiation pressure by sunlight. Estimate the order of magnitude of the force on a dust grain with linear dimension 1 μm at the radius of the Earth's orbit, where the intensity of sunlight is 1300 W/m^2. Compare the radiation force to the force of solar gravity on the grain, assuming the grain has density 5 g/cm^3. Can the radiation affect significantly the orbit of the dust grain?

11.25. Why does light exert a pressure on a metal surface? By considering the directions of $\mathbf{E}$, $\mathbf{B}$, $\mathbf{J}$, and $\mathbf{F}$, show that there is a force on the metal, due to the *magnetic force* on the electrons, from light incident normal to the surface. The result is actually somewhat tricky. $\mathbf{J}$ is parallel to $\mathbf{E}$, and in phase with the oscillation of $\mathbf{E}$ (that's important!) because of *resistance*, which makes $\mathbf{J} = \sigma \mathbf{E}$.

The Spherical Wave

11.26. Show that $\mathbf{A}(\mathbf{x}, t)$ of (11.101) and (11.102) is a solution of the wave equation.

11.27. Carry out the calculations of $\mathbf{B}$ and $\mathbf{E}$ for the spherical wave in Section 11.5.4. First calculate $\mathbf{B} = \nabla \times \mathbf{A}$, where $\mathbf{A}$ is given by (11.101), using spherical polar coordinates. Then calculate $\mathbf{E}$ from (11.105).

The asymptotic vector potential at large r is proportional to r^{-1}, so one might guess that the asymptotic fields, involving derivatives of $\mathbf{A}$, would be of order r^{-2}. But the asymptotic fields are of order r^{-1}, a very important result because it implies there is an outgoing flux of energy. Where do the terms of order r^{-1} come from?

11.28. The wave fronts of the spherical wave in Sec. 11.5.4 are spheres. However, the energy flux is not isotropic.

(a) In the radiation zone show that the differential power $dP/d\Omega$, i.e., the average power per unit solid angle, is proportional to $\sin^2 \theta$, where θ is the polar angle.

(Hint: $dP/d\Omega$ is $r^2 \mathbf{S}_{av} \cdot \hat{\mathbf{r}}$; calculate that differential power in the radiation zone from the asymptotic fields.)

(b) Calculate the fraction of the total power within ± 10 degrees of the equatorial plane. Compare your result to the fractional solid angle of that range of directions (i.e., its fraction of the total solid angle 4π).

11.29. Show that for the spherical wave in Sec. 11.5.4, the asymptotic electric field as $r \to 0$, i.e., the *near field*, is an electric dipole field. (The dipole field is given in (3.101).)

General Exercises

11.30. Most of the electromagnetic energy in the Universe is in the cosmic microwave background radiation, a remnant of the Big Bang. This radiation was discovered by A. Penzias and R. Wilson in 1965, by observations with a radio telescope. The radiation is electromagnetic waves with wavelengths around 1.1 mm. The energy density is 4.0×10^{-14} J/m^3. (This is 2.5×10^5 eV/m^3, half the rest energy of an electron in each cubic meter of the Universe.)

(a) What is the RMS electric field strength of the cosmic microwave background radiation? [Answer: 0.067 V/m]

(b) How far from a 1000 W transmitter would you have to go to have the same field strength? Assume the power from the transmitter is isotropic. [Answer: 2.6 km]

11.31. What are the (real) electric and magnetic fields for a monochromatic plane wave whose polarization is parallel to the xy plane, and which is traveling in vacuum in the direction from the origin to the point $(0, 1, 1)$? Express your answers for both fields in terms of the amplitude E_0 of the electric field oscillations, the frequency ω, and the speed of light c. Use Cartesian coordinates and basis vectors.

11.32. Table 11.3 shows a wide range of frequencies and wavelengths for electromagnetic radiation. The relation $\lambda\nu = c$ holds over all time and length scales.

(a) What is ν if λ is the Earth-moon distance?

(b) What is ν if $\lambda = 2$ fm, the diameter of a He nucleus? What is the energy in eV for a photon of that frequency?

11.33. In 1991 the Fly's Eye detector in Utah observed a cosmic ray with energy 320 EeV (1 EeV=10^{18} eV). The identity of the primary particle is unknown, but it might have been a gamma ray. What is the photon wavelength for this energy? How does it compare to the size of a proton?

Computer Exercise

11.34. Consider an electromagnetic wave with vector potential $\mathbf{A}(\mathbf{x}, t) = \hat{\mathbf{j}}\, f(x - ct)$. (The scalar potential is 0.) The function $f(x - ct)$ approaches 0 as $x \to \pm\infty$, so the electromagnetic field is a *wave packet*. Suppose the wave hits an electron (charge $-e$, mass m) initially at rest at the origin.

(a) Derive the equations of motion for the electron velocity components v_x, v_y, v_z.

(b) Show that $v_z(t) = 0$.

(c) Show that $v_y(t) = (e/m)f(x - ct)$, where x is the x position of the electron at time t.

(d) Show that

$$v_x \left(1 - \frac{v_x}{2c}\right) = \frac{e^2}{2m^2c} [f(x - ct)]^2 .$$

(e) Describe in words and pictures the trajectory of the electron, assuming the wave packet has a short length. In particular, show that the electron will have a positive displacement in the x direction.

(f) Assume $v_x \ll c$ and $f(\xi) = K \exp(-\xi^2/d^2)$. Use a computer to solve the differential equation for $x(t)$ numerically and plot the result. (Let d be the unit of length, d/c the unit of time, and choose a relatively small value of the dimensionless constant $eK/(mc)$.)

(g) Solve for $y(t)$ and plot the trajectory in space. (For example, in *Mathematica* solve simultaneously the coupled equations for $x(t)$ and $y(t)$ using NDSolve and make the plot with ParametricPlot.)

CHAPTER

12

Electromagnetism and Relativity

"Everything should be made as simple as possible, but not simpler."

Albert Einstein

Einstein's theory of special relativity, published in 1905,[1] is the theory of coordinate transformations between inertial frames. It is based on two postulates:

1. **Postulate of relativity.** The laws of physics are the same in all inertial frames.
2. **Postulate of the absolute speed of light.** The speed of light in vacuum is the same in all inertial frames.

An *inertial frame* means a coordinate system with respect to which Newton's first law of motion—the law of inertia—is true. An inertial frame moves with constant velocity with respect to any other inertial frame. A noninertial frame is one that accelerates with respect to inertial frames. A reference frame has both a spatial coordinate system that locates positions by the distances x, y, z along Cartesian axes from an origin, and a time coordinate t that locates events in time.

Einstein was led to these postulates, especially the second, by the study of Maxwell's theory of the electromagnetic field. Maxwell's equations imply that light is an electromagnetic wave, with wave speed $c = 1/\sqrt{\mu_0 \epsilon_0}$ in vacuum. But this result raises an obvious question: *With respect to what frame of reference is the speed equal to* $1/\sqrt{\mu_0 \epsilon_0}$? In classical physics, if an object P moves with speed u in one inertial frame, then its speed in another frame is different. For example, if the second frame moves with speed v relative to the first frame, in the same direction as the velocity of P, then the speed of P in the second frame is $u - v$. However, there is no possibility within Maxwell's equations for the speed of an electromagnetic wave in vacuum to be anything but $1/\sqrt{\mu_0 \epsilon_0}$. To what frame of reference does that speed correspond? The answer is that the speed of light is $1/\sqrt{\mu_0 \epsilon_0}$ in *every* inertial frame, and the classical formula for combining velocities is wrong.

Before Einstein, theorists assumed that light propagates in a physical medium called the *aether*. Maxwell, and the field theorists of the next generation, thought

[1] A. Einstein, "On the Electrodynamics of Moving Bodies", *Ann. Phys.*, **17**, 891 (1905).

that the field equations would describe the electromagnetic field in the rest frame of the aether. Then the calculated wave speed would be the speed relative to the fixed aether. In a reference frame moving with respect to the aether, the speed of electromagnetic waves would be different than the speed relative to the aether.

In a famous series of experiments, culminating in a very precise experiment in 1887, Michelson and Morley attempted to observe a variation in the speed of light due to motion of the Earth through the aether. Using a very accurate interferometer, they measured the interference between light waves that had traveled in orthogonal directions. Varying the orientation of the apparatus would change the interference if the aether theory were true, but there was no observed change in the interference. The experiment had a *null result*. This was surely the most important null result in the history of physics! There is no detectable variation in the speed of light traveling in different directions with respect to the moving Earth. This striking result is explained in special relativity by overthrowing a basic assumption of classical physics—that lengths and time intervals are absolute quantities, i.e., the same for all observers.

The *Lorentz transformation* is the coordinate transformation between inertial frames constructed such that Postulate 2 is obeyed. We assume that the reader has some familiarity with this transformation,[2] although we will review it briefly. Our main interest in this chapter is to understand how Postulate 1 is satisfied in the theory of the electromagnetic field.

12.1 ■ COORDINATE TRANSFORMATIONS

12.1.1 ■ The Galilean Transformation

The first statement of the principle of relativity was by Galileo. He claimed that an observer, enclosed below decks in the hold of a large ship, would not be able to determine by any experiment whether the ship is at rest or moving with constant velocity. This passage from his writings (Ref. 1) is a statement clearly equivalent to Postulate 1:

> "Shut yourself up with some friend in the largest room below decks of some large ship and there procure gnats, flies, and such other small winged creatures. Also get a great tub full of water and put within it certain fishes; let also a certain bottle be hung up, which drop by drop lets forth its water into another narrow-necked bottle placed underneath. Then, the ship lying still, observe how those small winged animals fly with like velocity towards all parts of the room; how the fishes swim indifferently towards all sides; and how the distilling drops all fall into the bottle placed underneath. . . . Having observed all these particulars, though no man doubts that, so long as the vessel stands still, they ought to take place in this manner, make the

[2] See, e.g., Refs. 3–6.

ship move with what velocity you please, so long as the velocity is uniform and not fluctuating this way and that. You shall not be able to discern the least alteration in all the forenamed effects, nor can you gather by any of them whether the ship moves or stands still."

In modern terms we say that the laws of motion are the same in all inertial frames.

Consider two inertial frames, $\mathcal{F}$ and $\mathcal{F}'$, in relative motion. The two frames of reference are illustrated in Fig. 12.1. An event that occurs at coordinates x, y, z and time t of $\mathcal{F}$, occurs at coordinates x', y', z' and time t' of $\mathcal{F}'$. To be specific, suppose frame $\mathcal{F}'$ moves with velocity $v\,\hat{\mathbf{i}}$, in the x direction, relative to $\mathcal{F}$. For example, the origin O' of $\mathcal{F}'$ is at $(x', y', z') = (0, 0, 0)$. The coordinates of O' in $\mathcal{F}$ are $(x, y, z) = (vt, 0, 0)$ so that O' has velocity $v\,\hat{\mathbf{i}}$ in $\mathcal{F}$. Conversely, the origin O of $\mathcal{F}$ moves with velocity $-v\,\hat{\mathbf{i}}$ with respect to $\mathcal{F}'$. We assume that the origins of $\mathcal{F}$ and $\mathcal{F}'$ coincide at $t = 0$, which defines what we mean by "time zero."

The transformation from $\mathcal{F}$ to $\mathcal{F}'$ is a set of equations that relate the coordinates and time of an event observed in the two frames of reference. The *classical* transformation from $\mathcal{F}$ to $\mathcal{F}'$ is called the *Galilean transformation*. If $\mathcal{F}'$ moves with velocity $v\,\hat{\mathbf{i}}$ with respect to $\mathcal{F}$, then the Galilean transformation is

$$x' = x - vt, \quad y' = y, \quad z' = z, \quad t' = t. \qquad (12.1)$$

Note that O' has $\mathcal{F}'$ coordinate $x' = 0$, so by (12.1) it has $\mathcal{F}$ coordinate $x = vt$, as specified.

For an arbitrary velocity vector $\mathbf{v}$ of $\mathcal{F}'$ relative to $\mathcal{F}$, the Galilean transformation is

$$x'_\parallel = x_\parallel - vt, \quad \mathbf{x}'_\perp = \mathbf{x}_\perp, \quad t' = t, \qquad (12.2)$$

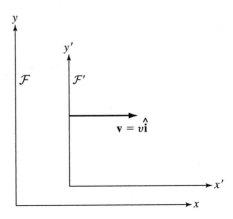

FIGURE 12.1   Two inertial frames of reference. $\mathcal{F}'$ moves with velocity $v\,\hat{\mathbf{i}}$ relative to $\mathcal{F}$.

where $\parallel$ and $\perp$ denote the vector components for directions parallel and perpendicular to $\mathbf{v}$. We emphasize that the relative velocity $\mathbf{v}$ must be constant.

Newton's equation of motion for a mass m is the same in $\mathcal{F}$ or $\mathcal{F}'$, provided the force is invariant under the transformation. The equation of motion in frame $\mathcal{F}$ is $m d^2\mathbf{x}/dt^2 = \mathbf{F}(\mathbf{x})$. Assume $\mathbf{F}'(\mathbf{x}') = \mathbf{F}(\mathbf{x})$ is the force in the coordinates of $\mathcal{F}'$. What is the equation of motion with respect to $\mathcal{F}'$? We need to write the acceleration vector in $\mathcal{F}'$ coordinates. First consider the velocity. The particle velocity in the frame $\mathcal{F}$ is $\mathbf{u} = d\mathbf{x}/dt$. The particle velocity in $\mathcal{F}'$ is

$$u'_\parallel = \frac{dx'_\parallel}{dt'} = \frac{d(x_\parallel - vt)}{dt} = u_\parallel - v, \tag{12.3}$$

$$\mathbf{u}'_\perp = \frac{d\mathbf{x}'_\perp}{dt'} = \frac{d\mathbf{x}_\perp}{dt} = \mathbf{u}_\perp; \tag{12.4}$$

this result is just the classical rule for combining velocities: $\mathbf{u}' = \mathbf{u} - \mathbf{v}$.[3] The Galilean transformation for the acceleration is

$$\mathbf{a}' = \frac{d\mathbf{u}'}{dt'} = \frac{d(\mathbf{u} - \mathbf{v})}{dt} = \mathbf{a}$$

because $\mathbf{v}$ is constant. Acceleration is a Galilean invariant. Therefore the equation of motion in the frame $\mathcal{F}'$ is $m d^2\mathbf{x}'/dt'^2 = \mathbf{F}'(\mathbf{x}')$, the same as in $\mathcal{F}$ as claimed.

We have just seen that Newtonian mechanics satisfies Postulate 1 if the transformation between inertial frames is Galilean and the force is invariant. However, Postulate 2 is not satisfied by the Galilean transformation. The coordinate transformation (12.2) implies the classical velocity addition formula, $\mathbf{u}' = \mathbf{u} - \mathbf{v}$. But if $\mathbf{u}$ is the velocity of a pulse of light in $\mathcal{F}$, then the speed $|\mathbf{u}'|$ in $\mathcal{F}'$ would not be equal to the speed $|\mathbf{u}|$ in $\mathcal{F}$. Einstein's Postulate 2 implies that (12.2) is *not* the correct transformation between inertial coordinates.

Equation (12.2) seems so obvious! How can it be wrong? There is an implicit assumption hiding in (12.2). These equations assume that lengths and time intervals are absolute quantities—the same for all observers.[4] However, on the contrary, length and time are *relative*. What is absolute is the speed of light.

12.1.2 ■ The Lorentz Transformation

The *relativistic* transformation between inertial frames is called the *Lorentz transformation*. Again, suppose that an inertial frame $\mathcal{F}'$ is moving with constant velocity $v\,\hat{\mathbf{i}}$ relative to another frame of reference $\mathcal{F}$, as illustrated in Fig. 12.1. The Lorentz transformation from coordinates x, y, z and time t of $\mathcal{F}$, to coordinates x', y', z' and t' of $\mathcal{F}'$, is

$$x' = \gamma(x - vt), \quad y' = y, \quad z' = z, \quad t' = \gamma(t - vx/c^2) \tag{12.5}$$

---

[3]We will usually use $\mathbf{u}$ to denote particle velocity, and $\mathbf{v}$ the relative velocity of different inertial frames. If $\mathcal{F}'$ is the rest frame of the particle, then $\mathbf{u}$ and $\mathbf{v}$ are the same.

[4]An "observer" in relativity theory means someone making measurements with respect to his laboratory, i.e., his rest frame.

where

$$\gamma = \frac{1}{\sqrt{1 - v^2/c^2}}. \qquad (12.6)$$

An event that occurs at x, y, z, t in frame $\mathcal{F}$ occurs at x', y', z', t' given by (12.5) in $\mathcal{F}'$.

It is a simple calculation to show that (12.5) implies Einstein's Postulate 2. Suppose that a flash of light is emitted from $\mathbf{x} = 0$ at time $t = 0$. As observed in the frame $\mathcal{F}$, there will be a spherical pulse of light traveling away from $\mathbf{x} = 0$ with speed c. The radius of the expanding light pulse is ct at time t, so the coordinates of any point on the pulse satisfy

$$x^2 + y^2 + z^2 = c^2 t^2. \qquad (12.7)$$

Now consider the same point on the pulse in $\mathcal{F}'$ coordinates, given by (12.5). The squared distance from the origin O' is

$$
\begin{aligned}
x'^2 + y'^2 + z'^2 &= \gamma^2 (x - vt)^2 + y^2 + z^2 \\
&= \gamma^2 (x - vt)^2 + c^2 t^2 - x^2 \\
&= (\gamma^2 - 1)x^2 - 2\gamma^2 xvt + (\gamma^2 v^2 + c^2)t^2 \\
&= c^2 \gamma^2 (t - vx/c^2)^2.
\end{aligned} \qquad (12.8)
$$

In the last step we used (12.6) for γ. The final result, using (12.5) for t', is that

$$x'^2 + y'^2 + z'^2 = c^2 t'^2. \qquad (12.9)$$

That is, the light pulse observed in the inertial frame $\mathcal{F}'$ is also a sphere traveling outward from $\mathbf{x}' = 0$ with speed c. Thus Postulate 2 is satisfied if the space and time coordinates of different inertial frames are related by (12.5). We could reverse this calculation to *derive* (12.5). If we demand (12.9), and require that the transformation be linear and not change the perpendicular coordinates (y and z), then the transformation must be (12.5).[5]

For an arbitrary relative velocity $\mathbf{v}$ the Lorentz transformation is

$$x'_\parallel = \gamma \left(x_\parallel - vt \right), \quad \mathbf{x}'_\perp = \mathbf{x}_\perp, \quad t' = \gamma \left(t - vx_\parallel/c^2 \right) \qquad (12.10)$$

where again $\parallel$ and $\perp$ denote vector components parallel and perpendicular to $\mathbf{v}$. This form of the transformation gives $\mathbf{x}'$ and t' in terms of $\mathbf{x}$ and t. The inverse transformation, which gives $\mathbf{x}$ and t in terms of $\mathbf{x}'$ and t', is

$$x_\parallel = \gamma \left(x'_\parallel + vt' \right), \quad \mathbf{x}_\perp = \mathbf{x}'_\perp, \quad t = \gamma \left(t' + vx'_\parallel/c^2 \right). \qquad (12.11)$$

---

[5] See Exercise 1.

The inverse is equivalent to exchanging $(\mathbf{x}, t)$ and $(\mathbf{x}', t')$ and replacing $\mathbf{v}$ by $-\mathbf{v}$, because $\mathcal{F}$ moves with velocity $-\mathbf{v}$ with respect to $\mathcal{F}'$.[6]

Now that we know how to satisfy Postulate 2, by the Lorentz transformation (12.10), what remains is to find theories that obey Postulate 1 for this transformation. The equations of physics must be the same whether written in coordinates of $\mathcal{F}$ or $\mathcal{F}'$. This property of the equations is called *covariance*. We shall analyze the covariance of electromagnetism in Sec. 12.3.

12.1.3 ■ Examples Involving the Lorentz Transformation

Before turning to electromagnetism, it is interesting to review some general consequences of the Lorentz transformation. In the examples below, $\mathcal{F}$ and $\mathcal{F}'$ denote inertial frames, and $\mathcal{F}'$ moves with constant velocity $\mathbf{v} = v\hat{\mathbf{i}}$ relative to $\mathcal{F}$, as in Fig. 12.1. In each example we analyze the spatial and temporal coordinates of *events* observed in the two frames.

EXAMPLE 1 Time dilation.   Consider two events that occur at the same spatial point in $\mathcal{F}'$, at times t_1' and t_2'. For example, these events could be two ticks of a clock at rest in $\mathcal{F}'$, i.e., moving with velocity $\mathbf{v}$ in $\mathcal{F}$. Or, the two events could be the creation and decay of an unstable elementary particle at rest in $\mathcal{F}'$. The time interval between the events, with respect to $\mathcal{F}'$, is $\Delta t' = t_2' - t_1'$. What is the time interval between the events as measured by an observer at rest in $\mathcal{F}$?

The time interval Δt in $\mathcal{F}$ is $t_2 - t_1$. To calculate times for $\mathcal{F}$ we need to use the inverse transformation (12.11), which gives $t = \gamma(t' + vx'/c^2)$. The time interval between the two events is $\Delta t = \gamma(\Delta t' + v\Delta x'/c^2)$ but $\Delta x' = 0$ because the events occur at the same location in $\mathcal{F}'$. Therefore,

$$\Delta t = \gamma \Delta t' = \frac{\Delta t'}{\sqrt{1 - v^2/c^2}}. \tag{12.12}$$

Note that $\Delta t > \Delta t'$; the time interval observed in $\mathcal{F}$ is longer than $\Delta t'$.

According to the observer at rest in $\mathcal{F}$, the time between the events is *dilated* compared to what it would be if the events were to occur at the same location in $\mathcal{F}$. The observer in $\mathcal{F}$ says "the moving clock runs slow." If the system is a clock that ticks every second (in its rest frame $\mathcal{F}'$), then the ticks are γ seconds apart in the time coordinate of $\mathcal{F}$. For example, if $v/c = 0.99$ then $\gamma \approx 7$. The observer in $\mathcal{F}$, using a clock of identical construction, will observe 7 ticks of his clock between two ticks of the moving clock. As v approaches c, the dilation factor γ tends to ∞.

Time dilation is an experimental fact. It is observed in the decay times of unstable elementary particles. Unstable particles, created in high-energy collisions,

[6] See Exercise 2.

travel mean distance $v\gamma\tau$ before decaying; here v is the speed of the particle, τ the mean lifetime *in its rest frame*, and $\gamma = 1/\sqrt{1 - v^2/c^2}$ the dilation factor. If v approaches the speed of light, then $\gamma \gg 1$ and the mean distance is much larger than $c\tau$.

EXAMPLE 2 Lorentz–FitzGerald contraction. Consider two points fixed on the x' axis. For example, these could be the endpoints of a ruler at rest in $\mathcal{F}'$, i.e., moving with velocity $v\hat{\mathbf{i}}$ in $\mathcal{F}$. Let $\Delta x'$ be the separation of the points in $\mathcal{F}'$. What is the separation between the *simultaneous positions of the two points* in the frame $\mathcal{F}$?

To calculate x values we again need to use the transformation (12.11), specifically $x = \gamma(x' + vt')$. Thus the distance between the two points is

$$\Delta x = \gamma\Delta x' + \gamma v\Delta t'. \tag{12.13}$$

We are interested in the simultaneous positions of the two points in $\mathcal{F}$, so $\Delta t = 0$. But that means $\Delta t'$ is *not* 0. Simultaneity is relative. By the Lorentz transformation (12.11), $\Delta t = 0$ implies $\Delta t' = -v\Delta x'/c^2$. Thus the separation of the points in $\mathcal{F}$ is

$$\Delta x = \gamma\left(1 - v^2/c^2\right)\Delta x' = \sqrt{1 - v^2/c^2}\,\Delta x'. \tag{12.14}$$

Note that $\Delta x < \Delta x'$; the length observed in $\mathcal{F}$ is shorter than $\Delta x'$.

For example, suppose a meter stick is laid along the x' axis and is at rest in $\mathcal{F}'$. If $v/c = 0.99$ then the instantaneous length of the meter stick in $\mathcal{F}$ is 0.14 m.

According to the observer at rest in $\mathcal{F}$, the length is *contracted* compared to what it would be if $\mathcal{F}'$ were at rest. The observer in $\mathcal{F}$ says "the moving ruler is short." This effect was first proposed by FitzGerald, and later independently by Lorentz, to explain the null result of the Michelson-Morley experiment. However, their idea was that motion through the aether caused a physical contraction of the apparatus. Einstein's theory is different. It was Einstein who first understood that the distance between points in space does not have an absolute measure; lengths are different for observers in relative motion.

EXAMPLE 3 Addition of velocities. Suppose a particle P has velocity $u'\hat{\mathbf{i}}$ with respect to the frame $\mathcal{F}'$. What is its velocity with respect to the frame $\mathcal{F}$? Recall that $\mathcal{F}'$ moves with velocity $v\hat{\mathbf{i}}$ with respect to $\mathcal{F}$.

In Newtonian physics the answer would be that the velocity of P in $\mathcal{F}$ is $(u' + v)\hat{\mathbf{i}}$. We know that is wrong. To determine the velocity of P in $\mathcal{F}$, we must apply the Lorentz transformation (12.11). Obviously the velocity is in the x direction, of the form $u\hat{\mathbf{i}}$. Let x_p and x'_p denote the particle's x coordinate in $\mathcal{F}$ and $\mathcal{F}'$, respectively. Then the x component of the velocity of P with respect to $\mathcal{F}$ is

$$u = \frac{dx_p}{dt} = \frac{\gamma \left(dx'_p + vdt' \right)}{\gamma \left(dt' + vdx'_p/c^2 \right)}; \qquad (12.15)$$

or, using that dx'_p/dt' is u',

$$u = \frac{u' + v}{1 + vu'/c^2}. \qquad (12.16)$$

Einstein called (12.16) the *theorem for addition of velocity*. If u' and v are small compared to c, then u is approximately the classical result $u' + v$. For example, if $u' = 0.1c$ and $v = 0.1c$ then $u = 0.198c$. But if u' or v approaches the speed of light then u also approaches the speed of light. For example, if $u' = 0.9c$ and $v = 0.9c$ then $u = 0.994c$. A massive particle does not move faster than c in any inertial frame. If the particle is massless, so that $u' = c$, then it moves at the speed of light in all inertial frames; indeed for $u' = c$ the equation gives $u = c$ for any v.

More generally, velocity is a vector $\mathbf{u}$. The transformations of velocity components can easily be derived by the same method as (12.16), with the results[7]

$$u_\parallel = \frac{u'_\parallel + v}{1 + vu'_\parallel/c^2} \qquad (12.17)$$

$$\mathbf{u}_\perp = \frac{\mathbf{u}'_\perp}{\gamma(1 + vu'_\parallel/c^2)}. \qquad (12.18)$$

12.2 ■ MINKOWSKI SPACE

12.2.1 ■ 4-vectors, Scalars, and Tensors

The Lorentz transformation is a linear transformation involving four coordinates. It is natural to combine the four coordinates in a 4-vector, and regard spacetime as a 4-dimensional space, called *Minkowski space*. We define the 4-vector x^μ by

$$x^\mu = \begin{pmatrix} x^0 \\ x^1 \\ x^2 \\ x^3 \end{pmatrix} = \begin{pmatrix} ct \\ x \\ y \\ z \end{pmatrix}. \qquad (12.19)$$

The notation here is important. The index μ, and all Greek indices ($\nu, \rho, \alpha, \ldots$), take the values 0, 1, 2, 3. We use "suffix notation" for 4-vectors, so x^μ stands for the vector itself in some contexts, and for the μ component of the vector in other contexts. It may be a little confusing at first, but we will regard x^μ as a symbol representing a 4-component vector.

------

[7] See Exercise 9.

The Lorentz transformation $x^\mu \rightarrow x'^\mu$ is linear, i.e., of the form

$$x'^\mu = \Lambda^\mu_\nu x^\nu, \tag{12.20}$$

where Λ^μ_ν is a matrix of constants. We use the Einstein summation convention throughout this chapter. Any repeated Minkowski index, such as ν in (12.20), is understood to be summed from 0 to 3. So, for example,

$$
\begin{aligned}
ct' = x'^0 &= \Lambda^0_\nu x^\nu \\
&= \Lambda^0_0 x^0 + \Lambda^0_1 x^1 + \Lambda^0_2 x^2 + \Lambda^0_3 x^3 \\
&= \Lambda^0_0 ct + \Lambda^0_1 x + \Lambda^0_2 y + \Lambda^0_3 z.
\end{aligned}
$$

The coefficients Λ^μ_ν of (12.20) may be written as a 4×4 matrix, since both μ and ν run from 0 to 3. If $\mathcal{F}'$ moves with velocity $v\,\hat{\mathbf{i}}$ relative to $\mathcal{F}$ then by (12.5) the transformation matrix is

$$\Lambda^\mu_\nu = \begin{pmatrix} \gamma & -\beta\gamma & 0 & 0 \\ -\beta\gamma & \gamma & 0 & 0 \\ 0 & 0 & 1 & 0 \\ 0 & 0 & 0 & 1 \end{pmatrix}, \tag{12.21}$$

where $\beta = v/c$ and $\gamma = 1/\sqrt{1-\beta^2}$. If (12.20) is written as a matrix equation, and then expanded in its four components, the result is the Lorentz transformation equations (12.5).[8]

Vectors and Tensors

Lorentz vectors and tensors are defined with respect to Lorentz transformations, in the same way that 3-dimensional vectors and tensors are defined with respect to rotations. A *4-vector* a^μ is a set of four components that transform in the same way as x^μ under Lorentz transformations,

$$a'^\mu = \Lambda^\mu_\nu a^\nu. \tag{12.22}$$

A tensor $T^{\mu\nu}$ is a set of 16 components that transform in the same way as $x^\mu x^\nu$,

$$T'^{\mu\nu} = \Lambda^\mu_\rho \Lambda^\nu_\sigma T^{\rho\sigma}. \tag{12.23}$$

(Note that ρ and σ are summed!)[9] It is also useful to define a quantity a_μ with *lower* index, associated with the arbitrary 4-vector a^μ, by

---

[8] See Exercise 5.

[9] Since μ and ν can each take four different values, (12.23) is a set of 16 equations; and the right-hand side of each equation is a sum of 16 terms.

$$a_\mu = \begin{pmatrix} a_0 \\ a_1 \\ a_2 \\ a_3 \end{pmatrix} = \begin{pmatrix} -a^0 \\ a^1 \\ a^2 \\ a^3 \end{pmatrix}. \qquad (12.24)$$

In this formalism there are two types of 4-vectors: A 4-vector with an upper index a^μ is called a contravariant vector, and one with a lower index a_μ is called a covariant vector.[10] The two types are different and must be distinguished by the upper or lower placement of the index. Tensors with lower indices are defined similarly. For example, $T_{\mu\nu}$ is defined from $T^{\mu\nu}$ by

$$T_{00} = T^{00}, \quad T_{ij} = T^{ij}, \quad T_{0i} = -T^{0i}, \quad T_{i0} = -T^{i0}. \qquad (12.25)$$

Raising or lowering a temporal index (0) changes the sign, but raising or lowering a spatial index makes no change. We use roman indices $(i, j, k, \ldots)$ to denote spatial indices. Any roman index takes the values 1, 2, 3, which correspond to the x, y, z Cartesian components, respectively. The Lorentz transformation rule for a covariant vector is $a'_\mu = a_\nu (\Lambda^{-1})^\nu{}_\mu$, where $(\Lambda^{-1})^\nu{}_\mu$ is the inverse of the matrix $\Lambda^\mu{}_\nu$.[11]

The Lorentz Product

The Lorentz product of two vectors, which plays the role of the scalar product in the Minkowski vector space, is defined by

$$a \cdot b = a_\mu b^\mu = -a^0 b^0 + \mathbf{a} \cdot \mathbf{b}. \qquad (12.26)$$

We shall prove the following crucial theorem: $a \cdot b$ is a Lorentz scalar, i.e., invariant under the Lorentz transformation.[12] The proof is just an algebraic exercise. It is sufficient to consider the transformation in (12.21):

$$
\begin{aligned}
a' \cdot b' &= -a'^0 b'^0 + a'^1 b'^1 + a'^2 b'^2 + a'^3 b'^3 \\
&= -\gamma \left(a^0 - \beta a^1\right) \gamma \left(b^0 - \beta b^1\right) + \gamma \left(a^1 - \beta a^0\right) \gamma \left(b^1 - \beta b^0\right) \\
&\quad + a^2 b^2 + a^3 b^3 \\
&= -a^0 b^0 + a^1 b^1 + a^2 b^2 + a^3 b^3 = a \cdot b.
\end{aligned}
$$

---

[10] The formalism of contravariant and covariant vectors has largely replaced the older formalism in which the fourth coordinate is an imaginary number ict.
[11] See Exercise 6.
[12] The analogue for 3-dimensional vectors is that the dot product $\mathbf{a} \cdot \mathbf{b}$ is invariant under coordinate rotations.

12.2.2 ■ Kinematics of a Point Particle

With the mathematical construction of Minkowski space in hand, we are now prepared to study the consequences of Postulate 1 for particle motion. First we need to introduce some more jargon.[13] The *worldline* means the trajectory of the particle (we'll call the particle P) in Minkowski space. That is, the worldline is a curve in four dimensions—the locus of points in spacetime touched by the particle. Also, the *proper time* of P, which we denote by τ, is defined by the differential relation

$$c d\tau = \sqrt{-dx_\mu dx^\mu} = \sqrt{c^2 (dt)^2 - (d\mathbf{x})^2}, \qquad (12.27)$$

where dx^μ is an infinitesimal displacement along the worldline. Note that $d\tau$, and therefore also τ, is a Lorentz scalar. We may also express $d\tau$ in terms of the particle velocity. Let $\mathbf{u}$ denote the velocity of P in the inertial frame $\mathcal{F}$, i.e., $\mathbf{u} = d\mathbf{x}/dt$. Then in terms of the time coordinate of frame $\mathcal{F}$, the proper time interval is

$$d\tau = dt \sqrt{1 - u^2/c^2}. \qquad (12.28)$$

Note that $d\tau$ is the time interval experienced by P *in its rest frame* $\mathcal{F}_p$, as P moves by dx^μ in $\mathcal{F}$. In the rest frame $\mathcal{F}_p$ the velocity $\mathbf{u}_p$ is 0, so by (12.28) $d\tau = dt_p$. In words, the proper time is the particle's own time.

4-velocity and 4-momentum
The *4-velocity* of P is defined by

$$\eta^\mu = \frac{dx^\mu}{d\tau}, \qquad (12.29)$$

where again dx^μ is along the worldline. The temporal and spatial components of 4-velocity are

$$\eta^0 = \frac{c}{\sqrt{1 - u^2/c^2}}, \qquad (12.30)$$

$$\boldsymbol{\eta} = \frac{\mathbf{u}}{\sqrt{1 - u^2/c^2}}, \qquad (12.31)$$

using the definition $\mathbf{u} = d\mathbf{x}/dt$. An important point is that η^μ is a 4-vector, because dx^μ is a vector and $d\tau$ a scalar. That is, the transformation of 4-velocity is simple: $\eta'^\mu = \Lambda^\mu_{\ \nu} \eta^\nu$. In other words, η^0 and $\boldsymbol{\eta}$ transform in the same way as ct and $\mathbf{x}$ under Lorentz transformations. The 3-dimensional vector $\mathbf{u}$ has a complicated transformation rule,[14] but the 4-vector η^μ transforms simply.

---

[13]Jargon is important in science because the words are defined precisely.
[14]See Exercise 9.

The *4-momentum* of P is defined by

$$p^\mu = m\eta^\mu \tag{12.32}$$

where m is the *rest mass*. The rest mass is a Lorentz scalar.[15] The 4-momentum is a 4-vector because η^μ is a 4-vector and m a scalar.

Energy and Momentum

The temporal component of p^μ is identified as E/c, where E is the total particle energy. The spatial components make up the spatial momentum $\mathbf{p}$. That is,

$$p^\mu = \begin{pmatrix} p^0 \\ p^1 \\ p^2 \\ p^3 \end{pmatrix} = \begin{pmatrix} E/c \\ p_x \\ p_y \\ p_z \end{pmatrix}. \tag{12.33}$$

Writing these components in terms of the particle velocity $\mathbf{u}$ in an arbitrary inertial frame $\mathcal{F}$, using (12.30) and (12.31), the relativistic formulas for particle energy and momentum are

$$E = m\eta^0 c = \frac{mc^2}{\sqrt{1 - u^2/c^2}}, \tag{12.34}$$

$$\mathbf{p} = m\eta = \frac{m\mathbf{u}}{\sqrt{1 - u^2/c^2}}. \tag{12.35}$$

By eliminating $\mathbf{u}$ we may relate energy and momentum,[16]

$$E = \sqrt{m^2c^4 + p^2c^2}. \tag{12.36}$$

If $\mathbf{p}$ is 0 then the energy is mc^2, the rest energy. If $|\mathbf{p}| \ll mc$ then E is approximately $mc^2 + p^2/2m$, the rest energy plus classical kinetic energy; this approximation is a Taylor expansion of (12.36) for small $|\mathbf{p}|$. If $p \gg mc$ then E is approximately pc, which is the energy of a massless particle.

Why are (12.34) and (12.35) the correct definitions of energy and momentum? Of course the reason is that they agree with experiment, e.g., collisions of high-energy particles. But why, *theoretically,* are they the correct equations? How did Einstein deduce them before there were any high-energy experiments? The answer is that the equations must satisfy Postulate 1 for the laws of conservation of energy and momentum. It is a basic law of physics that the total energy and momentum are conserved in a collision of particles; indeed, that is the defining

---

[15] In some discussions of special relativity, a velocity-dependent mass is introduced. In our treatment of the theory, m always denotes the *rest mass*, which is an invariant property of the particle, i.e., a scalar.

[16] Note that $E^2 - p^2c^2 = m^2c^4$.

property of energy and momentum. According to Postulate 1 this law must hold *in every inertial frame*. If E and $\mathbf{p}$ make up a 4-vector, as in (12.33), then the linearity of the Lorentz transformation guarantees that if energy and momentum are conserved in one inertial frame then they are conserved in every inertial frame. For example, consider a 2-body elastic collision. Conservation of energy and momentum in $\mathcal{F}$ is that $P^{\mu}_{\text{final}} = P^{\mu}_{\text{initial}}$, where P^{μ} denotes the *total* 4-momentum. Because the Lorentz transformation is linear, in any other frame $\mathcal{F}'$ an observer would find

$$P'^{\mu}_{\text{final}} = \Lambda^{\mu}_{\ \nu} P^{\nu}_{\text{final}} = \Lambda^{\mu}_{\ \nu} P^{\nu}_{\text{initial}} = P'^{\mu}_{\text{initial}}. \tag{12.37}$$

That is, energy and momentum are conserved in $\mathcal{F}'$. Equations (12.34) and (12.35) are correct because they make $p^{\mu} = (E/c, \mathbf{p})$ a 4-vector, so that P^{μ} is conserved in all inertial frames.

   Covariance of the conservation laws requires (12.34) and (12.35) for energy and momentum.

12.2.3 ■ Relativistic Dynamics

Postulate 1 states that the equations of physics must have the same form in all inertial frames. This property of the equations is called *covariance*. There is a method for writing equations that guarantees covariance, which is to write the equations as relations involving only 4-vectors, scalars, and tensors. That is, covariance is automatically true if the equations of a theory are written in Minkowski tensor form. The generic equation then has the form[17]

$$T^{\mu\nu}_1 = T^{\mu\nu}_2 \tag{12.38}$$

in some reference frame $\mathcal{F}$. In any other frame $\mathcal{F}'$ the equation has the same form, because by the definition of a tensor,

$$T'^{\mu\nu}_1 = \Lambda^{\mu}_{\ \alpha} \Lambda^{\nu}_{\ \beta} T^{\alpha\beta}_1 = \Lambda^{\mu}_{\ \alpha} \Lambda^{\nu}_{\ \beta} T^{\alpha\beta}_2 = T'^{\mu\nu}_2. \tag{12.39}$$

   Postulate 1 is equivalent to the statement that all the laws of physics can be written in Lorentz tensors. We have already seen an example of this fact, in the equation for energy and momentum conservation: $P^{\mu}_{\text{final}} = P^{\mu}_{\text{initial}}$. An equation written in tensors is said to be *manifestly covariant*.

The Equation of Motion of a Particle in Covariant Form

Newton's second law is written in the spatial coordinates and time of some inertial frame: $d\mathbf{p}/dt = \mathbf{F}(\mathbf{x}, t)$, where $\mathbf{p}$ is the spatial momentum and $\mathbf{F}$ the Newtonian force. But this equation is not manifestly covariant. To construct a manifestly covariant equation we must write the equation in 4-vectors and tensors. In analogy

---

[17] Tensors are defined with any number of indices. A rank-two tensor has two indices, as in (12.23), or as in the hypothetical generic equation (12.38). A rank-one tensor is a vector.

with the Newtonian equation, we write

$$\frac{dp^\mu}{d\tau} = K^\mu,$$

(12.40)

where K^μ is called the *Minkowski force*. In (12.40) the proper time is used as the independent variable to parametrize the position along the worldline in spacetime. The left-hand side is a 4-vector because τ is a scalar. Einstein's Postulate 1—covariance—requires that K^μ must also be a 4-vector, i.e., transform according to (12.22). Then (12.40) is the manifestly covariant equation of motion.

To appreciate the meaning of the Minkowski force, it is useful to relate its components **K** and K^0 to the Newtonian force **F**, which is *defined* as $d\mathbf{p}/dt$. The expressions are

$$\mathbf{K} \equiv \frac{d\mathbf{p}}{d\tau} = \frac{d\mathbf{p}/dt}{d\tau/dt} = \frac{\mathbf{F}}{\sqrt{1 - u^2/c^2}},$$

(12.41)

$$K^0 \equiv \frac{1}{c}\frac{dE}{d\tau} = \frac{\mathbf{F}\cdot\mathbf{u}/c}{\sqrt{1 - u^2/c^2}}.$$

(12.42)

In the second expression we have used the work-energy theorem, $dE = \mathbf{F}\cdot\mathbf{u}\,dt$. As in the case of 3-velocity, the Newtonian force **F** has a complicated transformation rule. But the Minkowski force transforms in a simple way—as a 4-vector.

Our goal in the next section is to put the equations of electromagnetism into manifestly covariant form.

12.3 ■ ELECTROMAGNETISM IN COVARIANT FORM

12.3.1 ■ The Lorentz Force and the Field Tensor

We know that in any particular frame $\mathcal{F}$, the equation of motion of a charge q moving under the influence of fields **E** and **B** is

$$\frac{d\mathbf{p}}{dt} = \mathbf{F} = q\mathbf{E} + q\mathbf{u} \times \mathbf{B}.$$

(12.43)

This **F** is called the *Lorentz force*. Indeed we used (12.43) to *define* the fields **E** and **B** in earlier chapters. But (12.43) is not written in Lorentz vectors and tensors. Our goal is to write the equation in tensors. In other words, we shall determine the Minkowski force K^μ that corresponds to (12.43), such that $dp^\mu/d\tau = K^\mu$. As a by-product, we shall learn how the fields **E** and **B** transform under a change of inertial coordinates.

The components of K^μ may be written concisely in terms of the 4-velocity η^μ, defined in (12.30) and (12.31). Using (12.41) the spatial components of K^μ are

$$\mathbf{K} = \frac{q\mathbf{E} + q\mathbf{u} \times \mathbf{B}}{\sqrt{1 - u^2/c^2}} = q\eta^0\frac{\mathbf{E}}{c} + q\eta \times \mathbf{B};$$

(12.44)

or, in suffix notation,

$$K^i = q\eta^0 \frac{E^i}{c} + q\epsilon_{ijk}\eta^j B^k, \tag{12.45}$$

where repeated roman indices (here j and k) are always summed from 1 to 3. (ϵ_{ijk} is the Levi-Civita tensor.) Using (12.42) the temporal component of K^μ is

$$K^0 = \frac{q\mathbf{E} \cdot \mathbf{u}/c}{\sqrt{1 - u^2/c^2}} = q\eta \cdot \frac{\mathbf{E}}{c}. \tag{12.46}$$

The brilliant insight of Minkowski was to combine $\mathbf{E}$ and $\mathbf{B}$ in an antisymmetric Lorentz tensor, which we call the *electromagnetic field tensor*, denoted by $F^{\mu\nu}(x)$. The spatial and temporal components of the tensor are defined by

$$F^{ij} = \epsilon_{ijk} B^k, \tag{12.47}$$

$$F^{0i} = -F^{i0} = \frac{E^i}{c}, \tag{12.48}$$

$$F^{00} = 0. \tag{12.49}$$

As usual, roman indices $(i, j, k, \ldots)$ are spatial, taking values $1, 2, 3$ that correspond to Cartesian components x, y, z; for example, $E^1 = E_x$ or $B^3 = B_z$. The field tensor is antisymmetric: $F^{\nu\mu} = -F^{\mu\nu}$. Written as a 4×4 matrix the field tensor is

$$F^{\mu\nu} = \begin{pmatrix} 0 & E_x/c & E_y/c & E_z/c \\ -E_x/c & 0 & B_z & -B_y \\ -E_y/c & -B_z & 0 & B_x \\ -E_z/c & B_y & -B_x & 0 \end{pmatrix}. \tag{12.50}$$

The Minkowski force takes very simple form in terms of $F^{\mu\nu}$. We shall verify that the Minkowski force on a point charge q is

$$K^\mu = q\eta_\nu F^{\mu\nu}. \tag{12.51}$$

We must show that (12.45) and (12.46) are correctly given by (12.51). First consider the ith spatial component of (12.51)

$$K^i = q\left(\eta_0 F^{i0} + \eta_j F^{ij}\right) = q\eta^0 \frac{E^i}{c} + q\epsilon_{ijk}\eta^j B^k;$$

to justify the second equality note that $\eta_0 = -\eta^0$ and $\eta_j = \eta^j$. The result is (12.45). Next consider the temporal component of (12.51)

$$K^0 = q\left(\eta_0 F^{00} + \eta_j F^{0j}\right) = q\eta^j \frac{E^j}{c};$$

this result is (12.46). Hence (12.51) is proven.

TABLE 12.1   Equations of electromagnetism in covariant and component form

Covariant form	Component form
$\dfrac{dp^{\mu}}{d\tau} = q\eta_{\nu}F^{\mu\nu}$	$\begin{cases} d\mathbf{p}/dt = q(\mathbf{E} + \mathbf{u} \times \mathbf{B}) \\ dE/dt = q\mathbf{u} \cdot \mathbf{E} \end{cases}$
$\dfrac{\partial J^{\mu}}{\partial x^{\mu}} = 0$	$\mathbf{\nabla} \cdot \mathbf{J} = -\dfrac{\partial \rho}{\partial t}$
$\dfrac{\partial F^{\mu\nu}}{\partial x^{\nu}} = \mu_0 J^{\mu}$	$\begin{cases} \mathbf{\nabla} \cdot \mathbf{E} = \rho/\epsilon_0 \\ \mathbf{\nabla} \times \mathbf{B} - (1/c^2)\partial \mathbf{E}/\partial t = \mu_0 \mathbf{J} \end{cases}$
$\dfrac{\partial G^{\mu\nu}}{\partial x^{\nu}} = 0$	$\begin{cases} \mathbf{\nabla} \cdot \mathbf{B} = 0 \\ \mathbf{\nabla} \times \mathbf{E} + \partial \mathbf{B}/\partial t = 0 \end{cases}$
$F_{\mu\nu} = \dfrac{\partial A_{\nu}}{\partial x^{\mu}} - \dfrac{\partial A_{\mu}}{\partial x^{\nu}}$	$\begin{cases} \mathbf{E} = -\mathbf{\nabla}V - \partial \mathbf{A}/\partial t \\ \mathbf{B} = \mathbf{\nabla} \times \mathbf{A} \end{cases}$

The 4-velocity η^{μ} and Minkowski force K^{μ} are Lorentz vectors. Therefore (12.51) implies that $F^{\mu\nu}$ must be a Lorentz tensor. That is, it must transform under a change of coordinates in the manner of a tensor (12.23). Just as the Lorentz product of two vectors $a_{\nu}b^{\nu}$ is a scalar, the similar product of a vector and a tensor $a_{\nu}T^{\mu\nu}$ is a 4-vector.[18] The right-hand side of (12.51) is a 4-vector if and only if $F^{\mu\nu}$ is a tensor. From this requirement we will deduce the transformations of $\mathbf{E}$ and $\mathbf{B}$, in Sec. 12.4.

In conclusion, the equation of motion of a particle of charge q is, in covariant form,

$$\frac{dp^{\mu}}{d\tau} = q\eta_{\nu}F^{\mu\nu}. \tag{12.52}$$

Table 12.1 lists the equations of electromagnetism in both covariant and component form. The first entry in the table is the equation of motion of a charged particle.

12.3.2 ■ Maxwell's Equations in Covariant Form

By writing the equation of motion of a point charge in covariant form, we have identified the electromagnetic field as an antisymmetric tensor $F^{\mu\nu}$. In accord with Einstein's Postulate 1, it must also be possible to write the Maxwell field equations in tensor form.

As a first step, we consider the charge and current. The densities $\rho(\mathbf{x}, t)$ and $\mathbf{J}(\mathbf{x}, t)$ are not covariant, but they combine to make a 4-vector current $J^{\mu}(x)$,

---

[18]In general, the product of a vector and a tensor of rank n is a tensor of rank $n - 1$.

defined by

$$J^\mu = \begin{pmatrix} c\rho \\ J_x \\ J_y \\ J_z \end{pmatrix}. \tag{12.53}$$

The continuity equation, which is an essential part of electrodynamics, is, in covariant form,

$$\frac{\partial J^\mu}{\partial x^\mu} = 0. \tag{12.54}$$

A subtle point here is that $\partial/\partial x^\mu$ transforms as a *covariant vector*, not a contravariant vector.[19] Therefore $\partial J^\mu/\partial x^\mu$ is, like the Lorentz product of vectors, a scalar. Local conservation of charge requires that this scalar is 0. It is a simple exercise to verify that in component form (12.54) is the familiar continuity equation, $\nabla \cdot \mathbf{J} = -\partial\rho/\partial t$.

Two of the Maxwell equations relate fields and sources—Gauss's law and the Ampère-Maxwell law. Gauss's law is a scalar equation, i.e., a scalar with respect to spatial rotations. The Ampère-Maxwell law is a vector equation with respect to rotations. The two equations together make one covariant 4-vector equation, i.e., covariant with respect to Lorentz transformations, which is

$$\frac{\partial F^{\mu\nu}}{\partial x^\nu} = \mu_0 J^\mu. \tag{12.55}$$

The temporal component ($\mu = 0$) of (12.55) is Gauss's law. Note that $\partial F^{0\nu}/\partial x^\nu$ is $\nabla \cdot \mathbf{E}/c$; the term in the sum over ν with $\nu = 0$ is 0 because $F^{00} = 0$. For $\mu = 0$ the right-hand side of (12.55) is $\rho/(\epsilon_0 c)$, because $\mu_0 c$ is equal to $1/(\epsilon_0 c)$. Hence the temporal component of (12.55) is equivalent to $\nabla \cdot \mathbf{E} = \rho/\epsilon_0$.

The spatial components of (12.55) are the Ampère-Maxwell law. Note that $\partial F^{i\nu}/\partial x^\nu$ is a sum of four terms. The term with $\nu = 0$ is $-(1/c^2)\partial E^i/\partial t$, and the sum over $\nu = 1, 2, 3$ is the i component of $\nabla \times \mathbf{B}$. Hence the spatial components of (12.55) are equivalent to $\nabla \times \mathbf{B} - (1/c^2)\partial\mathbf{E}/\partial t = \mu_0\mathbf{J}$. Table 12.1 lists side-by-side the field equations in covariant form and component form. The first two Maxwell equations appear as the third row in the table.

The other two Maxwell equations involve only the fields, not charge. With respect to spatial rotations, these equations are a scalar equation (Gauss's law for **B**) and a vector equation (Faraday's law). With respect to Lorentz transformations, the two equations together make one covariant vector equation, which may be written as

$$\frac{\partial G^{\mu\nu}}{\partial x^\nu} = 0. \tag{12.56}$$

[19]See Exercise 14.

$G^{\mu\nu}$ is another antisymmetric tensor, called *the dual field tensor*. It is defined in covariant form by the relation

$$G^{\mu\nu} = \tfrac{1}{2}\epsilon^{\mu\nu\alpha\beta} F_{\alpha\beta}, \tag{12.57}$$

where $\epsilon^{\mu\nu\alpha\beta}$ is the completely antisymmetric tensor in four dimensions. This tensor is the generalization to four dimensions of the Levi-Civita tensor in three dimensions: The component $\epsilon^{\mu\nu\alpha\beta}$ is 0 unless all four indices are different; and $\epsilon^{\mu\nu\alpha\beta}$ is $+1$ or -1, respectively, if $\mu\nu\alpha\beta$ is an even or odd permutation of 0123. So, for example, $\epsilon^{0ijk} = \epsilon_{ijk}$. Equation (12.57) is equivalent to these spacetime components of $G^{\mu\nu}$:

$$G^{ij} = \frac{1}{2}\epsilon^{ij\alpha\beta} F_{\alpha\beta} = -\epsilon_{ijk}\frac{E^k}{c}, \tag{12.58}$$

$$G^{0i} = -G^{i0} = \frac{1}{2}\epsilon^{0ijk} F_{jk} = B^i, \tag{12.59}$$

$$G^{00} = 0. \tag{12.60}$$

It is an interesting exercise to verify (12.58) and (12.59). Note that in (12.57) the nonzero contributions for $\mu\nu = ij$ have $\alpha\beta$ either $0k$ or $k0$, because $\epsilon^{\mu\nu\alpha\beta}$ is 0 if any two indices are equal; hence (12.58). Also, the nonzero contributions for $\mu = 0$ in (12.57) have $\nu\alpha\beta$ all spatial ijk; hence (12.59). In matrix form

$$G^{\mu\nu} = \begin{pmatrix} 0 & B_x & B_y & B_z \\ -B_x & 0 & -E_z/c & E_y/c \\ -B_y & E_z/c & 0 & -E_x/c \\ -B_z & -E_y/c & E_x/c & 0 \end{pmatrix}. \tag{12.61}$$

The dual tensor has the same form as $F^{\mu\nu}$, but with the replacements $\mathbf{E}/c \to \mathbf{B}$ and $\mathbf{B} \to -\mathbf{E}/c$.

As an exercise, it it straightforward to verify that the temporal component ($\mu = 0$) of (12.56) is Gauss's law ($\nabla \cdot \mathbf{B} = 0$); and the spatial components ($\mu = 1, 2, 3$) make up Faraday's law ($\nabla \times \mathbf{E} + \partial\mathbf{B}/\partial t = 0$).[20] These results are the fourth row in Table 12.1.

Maxwell's equations describe many physical phenomena. The fact that these equations reduce to the very compact form

$$\partial F^{\mu\nu}/\partial x^\nu = \mu_0 J^\mu \quad \text{and} \quad \partial G^{\mu\nu}/\partial x^\nu = 0,$$

indicates there is something really deep here!

12.3.3 ■ The 4-vector Potential

In earlier chapters we have often noted the importance of the potentials $V(\mathbf{x}, t)$ and $\mathbf{A}(\mathbf{x}, t)$. In the covariant form of electromagnetism, these potentials combine

---

[20]See Exercise 15.

to make a 4-vector potential $A^\mu(x)$, defined by

$$A^\mu = \begin{pmatrix} V/c \\ A_x \\ A_y \\ A_z \end{pmatrix}. \tag{12.62}$$

The field equation $\partial G^{\mu\nu}/\partial x^\nu = 0$ is automatically satisfied if $F_{\mu\nu}$ is written as

$$F^{\mu\nu} = \frac{\partial A^\nu}{\partial x_\mu} - \frac{\partial A^\mu}{\partial x_\nu}. \tag{12.63}$$

Note also that this formula automatically makes $F^{\mu\nu}$ antisymmetric. The components of the tensor equation (12.63) are given as the final row of Table 12.1. They are the familiar relations between the potentials V and $\mathbf{A}$ and the fields $\mathbf{E}$ and $\mathbf{B}$.

12.4 ■ FIELD TRANSFORMATIONS

The electromagnetic field is a Lorentz tensor $F^{\mu\nu}$, so the transformation of the field from one inertial frame $\mathcal{F}$ to another $\mathcal{F}'$ is prescribed by the general definition (12.23) of tensor transformations,

$$F'^{\mu\nu}(x') = \Lambda^\mu{}_\rho \Lambda^\nu{}_\sigma F^{\rho\sigma}(x) \tag{12.64}$$

where $x'^\mu = \Lambda^\mu{}_\lambda x^\lambda$. In this section we will split this field transformation into the separate parts for $\mathbf{E}'$ and $\mathbf{B}'$.

First, to be definite, assume that the velocity of $\mathcal{F}'$ relative to $\mathcal{F}$ is $v\,\hat{\mathbf{i}}$, in the x direction. Then the transformation matrix is (12.21). The electric field that would be measured by an observer at rest in $\mathcal{F}'$ is

$$E'^i = cF'^{0i} = c\Lambda^0{}_\rho \Lambda^i{}_\sigma F^{\rho\sigma}. \tag{12.65}$$

The right-hand side is a sum of 16 terms, because ρ and σ run from 0 to 3. However, most of the terms are 0 because of the form (12.21) of $\Lambda^\mu{}_\nu$. For $i = 1$ the only nonzero terms have $\rho\sigma = 01$ and 10; thus, since $E^1 = E_x$,

$$E'_x = c\gamma^2 F^{01} + c(-\beta\gamma)^2 F^{10} = cF^{01} = E_x. \tag{12.66}$$

For $i = 2$ the nonzero terms have $\rho\sigma = 02$ and 12, so

$$E'_y = c\gamma F^{02} - c\beta\gamma F^{12} = \gamma\left(E_y - vB_z\right); \tag{12.67}$$

and similarly, for $i = 3$,

$$E'_z = c\gamma F^{03} - c\beta\gamma F^{13} = \gamma\left(E_z + vB_y\right). \tag{12.68}$$

TABLE 12.2   Field transformations between frames $\mathcal{F}$ and $\mathcal{F}'$, where $\mathcal{F}'$ moves with velocity $v\,\hat{\mathbf{i}}$ with respect to $\mathcal{F}$

$E'_x = E_x$	$B'_x = B_x$
$E'_y = \gamma\left(E_y - vB_z\right)$	$B'_y = \gamma\left(B_y + vE_z/c^2\right)$
$E'_z = \gamma\left(E_z + vB_y\right)$	$B'_z = \gamma\left(B_z - vE_y/c^2\right)$
$E_x = E'_x$	$B_x = B'_x$
$E_y = \gamma\left(E'_y + vB'_z\right)$	$B_y = \gamma\left(B'_y - vE'_z/c^2\right)$
$E_z = \gamma\left(E'_z - vB'_y\right)$	$B_z = \gamma\left(B'_z + vE'_y/c^2\right)$

Table 12.2 lists the field transformation equations for this case in which the velocity of $\mathcal{F}'$ with respect to $\mathcal{F}$ is $v\,\hat{\mathbf{i}}$.

The magnetic field in $\mathcal{F}'$ is

$$B'^i = F'^{jk} = \Lambda^j{}_\rho \Lambda^k{}_\sigma F^{\rho\sigma}, \tag{12.69}$$

where ijk is a cyclic permutation of 123. For $i = 1$, jk is 23; then the only nonzero term in the sum over ρ and σ has $\rho\sigma = 23$; that is,

$$B'_x = F^{23} = B_x. \tag{12.70}$$

For $i = 2$, jk is 31; then the nonzero terms have $\rho\sigma = 30$ and 31, so

$$B'_y = -\beta\gamma F^{30} + \gamma F^{31} = \gamma\left(B_y + vE_z/c^2\right); \tag{12.71}$$

and similarly, for $i = 3$, $jk = 12$, and so

$$B'_z = -\beta\gamma F^{02} + \gamma F^{12} = \gamma\left(B_z - vE_y/c^2\right). \tag{12.72}$$

These magnetic field transformations are also listed in Table 12.2.

For an arbitrary relative velocity $\mathbf{v}$ the field transformations are, by generalizing the previous special case,

$$E'_\parallel = E_\parallel \quad \text{and} \quad E'_\perp = \gamma\left(\mathbf{E}_\perp + \mathbf{v} \times \mathbf{B}_\perp\right), \tag{12.73}$$

$$B'_\parallel = B_\parallel \quad \text{and} \quad B'_\perp = \gamma\left(\mathbf{B}_\perp - \mathbf{v} \times \mathbf{E}_\perp/c^2\right). \tag{12.74}$$

As before $\parallel$ and $\perp$ denote vector components parallel and perpendicular to $\mathbf{v}$. The inverse transformation equations are obtained from (12.73) and (12.74) by making the replacements $\mathbf{E} \leftrightarrow \mathbf{E}'$, $\mathbf{B} \leftrightarrow \mathbf{B}'$, and $\mathbf{v} \leftrightarrow -\mathbf{v}$. Given fields $\mathbf{E}$ and $\mathbf{B}$ in the inertial frame $\mathcal{F}$, however created, the fields that would be measured by

TABLE 12.3 Lorentz transformations of various quantities. The inertial frame $\mathcal{F}'$ moves with velocity $\mathbf{v}$ with respect to frame $\mathcal{F}$. The components denoted $\parallel$ and $\perp$ are parallel and perpendicular to $\mathbf{v}$.

Coordinates	
$t' = \gamma(t - vx_{\parallel}/c^2)$	$t = \gamma(t' + vx'_{\parallel}/c^2)$
$x'_{\parallel} = \gamma(x_{\parallel} - vt)$	$x_{\parallel} = \gamma(x'_{\parallel} + vt')$
$\mathbf{x}'_{\perp} = \mathbf{x}_{\perp}$	$\mathbf{x}_{\perp} = \mathbf{x}'_{\perp}$

Energy and momentum	
$E' = \gamma(E - vp_{\parallel})$	$E = \gamma(E' + vp'_{\parallel})$
$p'_{\parallel} = \gamma(p_{\parallel} - vE/c^2)$	$p_{\parallel} = \gamma(p'_{\parallel} + vE'/c^2)$
$\mathbf{p}'_{\perp} = \mathbf{p}_{\perp}$	$\mathbf{p}_{\perp} = \mathbf{p}'_{\perp}$

Velocity	
$u'_{\parallel} = (u_{\parallel} - v)/(1 - vu_{\parallel}/c^2)$	$u_{\parallel} = (u'_{\parallel} + v)/(1 + vu'_{\parallel}/c^2)$
$\mathbf{u}'_{\perp} = (1/\gamma)\mathbf{u}_{\perp}/(1 - vu_{\parallel}/c^2)$	$\mathbf{u}_{\perp} = (1/\gamma)\mathbf{u}'_{\perp}/(1 + vu'_{\parallel}/c^2)$

Electric and magnetic fields	
$E'_{\parallel} = E_{\parallel}$	$E_{\parallel} = E'_{\parallel}$
$\mathbf{E}'_{\perp} = \gamma(\mathbf{E}_{\perp} + \mathbf{v} \times \mathbf{B}_{\perp})$	$\mathbf{E}_{\perp} = \gamma(\mathbf{E}'_{\perp} - \mathbf{v} \times \mathbf{B}'_{\perp})$
$B'_{\parallel} = B_{\parallel}$	$B_{\parallel} = B'_{\parallel}$
$\mathbf{B}'_{\perp} = \gamma(\mathbf{B}_{\perp} - \mathbf{v} \times \mathbf{E}_{\perp}/c^2)$	$\mathbf{B}_{\perp} = \gamma(\mathbf{B}'_{\perp} + \mathbf{v} \times \mathbf{E}'_{\perp}/c^2)$

an observer at rest in $\mathcal{F}'$ are (12.73) and (12.74), where $\mathbf{v}$ is the relative velocity of $\mathcal{F}'$. Or, given $\mathbf{E}'$ and $\mathbf{B}'$, the inverse transformation determines $\mathbf{E}$ and $\mathbf{B}$. For archive purposes we record the general transformation for the fields in Table 12.3, along with the transformation rules for some other quantities.

The field transformations mix electric and magnetic fields. For example, suppose there is a set of charges that are at rest in $\mathcal{F}'$. Then the magnetic field $\mathbf{B}'$ is 0. An observer in $\mathcal{F}'$ experiences only an electric field. A bar magnet at rest in $\mathcal{F}'$ experiences no torque. However, an observer in $\mathcal{F}$ experiences both electric and magnetic fields due to the same set of charges. A bar magnet at rest in $\mathcal{F}$ experiences a torque. This is not surprising: The charges move with velocity $\mathbf{v}$ in $\mathcal{F}$, and their current creates a magnetic field. One important point is that if we want to calculate $\mathbf{B}(\mathbf{x}, t)$ and $\mathbf{E}(\mathbf{x}, t)$ for the inertial frame $\mathcal{F}$, it is not necessary to solve Maxwell's equations in the coordinates of $\mathcal{F}$. Instead, we only need to calculate the Lorentz transformation of $\mathbf{E}'(x')$, listed in Table 12.3. We shall use this method to determine the fields of a moving charge in the next section.

In special relativity, t and $\mathbf{x}$ combine into a 4-vector x^{μ}. Also, E and $\mathbf{p}$ combine into the 4-momentum p^{μ}. Quantities that were considered to be separate in classical physics, like time and space, or energy and momentum, are unified by the theory of special relativity. A change of inertial frame produces mixing of these quantities. Similarly, the fields **E** and **B** combine into the field tensor $F^{\mu\nu}$. The tensor structure specifies how the fields in one frame of reference are related to those in another frame.

EXAMPLE 4   Suppose in an inertial frame $\mathcal{F}$ there is a uniform electric field $\mathbf{E} = E_0\hat{\mathbf{k}}$, but no magnetic field, $\mathbf{B} = 0$. What are the fields in a frame $\mathcal{F}'$ that moves with velocity $\mathbf{v} = v\hat{\mathbf{i}}$ with respect to $\mathcal{F}$?

A physical realization of this example would be the space between parallel charged plates with surface charge densities $\pm\sigma_0$. See Figure 12.2; the lower plate is positive and the upper plate negative. In the rest frame of the plates ($\mathcal{F}$) the electric field is $(\sigma_0/\epsilon_0)\hat{\mathbf{k}}$ by Gauss's law. Now imagine a tiny observer moving between the plates with constant velocity $v\hat{\mathbf{i}}$ and making field measurements in her laboratory ($\mathcal{F}'$).

We'll use Table 12.2 to determine $\mathbf{E}'$ and $\mathbf{B}'$. The only nonzero field component in frame $\mathcal{F}$ is $E_z = E_0$. By Table 12.2

$$E_x' = 0, \ E_y' = 0, \quad \text{and} \quad E_z' = \gamma E_0; \tag{12.75}$$

and

$$B_x' = 0, \ B_y' = \frac{\gamma v}{c^2} E_0, \quad \text{and} \quad B_z' = 0. \tag{12.76}$$

In $\mathcal{F}'$ there is a uniform electric field $\gamma E_0\hat{\mathbf{k}}$ and a uniform magnetic field $\gamma v E_0/c^2\,\hat{\mathbf{j}}$.

The same results can also be derived by identifying the sources in $\mathcal{F}'$. In the primed coordinate system there are two charged plates with surface densities $\pm\sigma' = \pm\gamma\sigma_0$. The factor γ comes from the Lorentz–FitzGerald contraction of the plate in the x direction, which increases the density compared to $\mathcal{F}$. The electric field $\mathbf{E}'$ due to the surface charge is $(\sigma'/\epsilon_0)\hat{\mathbf{k}}$, which is $\gamma E_0\hat{\mathbf{k}}$. Also, in $\mathcal{F}'$ there are two current sheets because the charged plates move (in $\mathcal{F}'$) with ve-

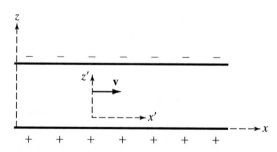

FIGURE 12.2   Example 4. $\mathcal{F}'$ moves with velocity $v\hat{\mathbf{i}}$ relative to $\mathcal{F}$, which is the rest frame of a pair of charged parallel plates.

locity $-v\hat{\mathbf{i}}$; the surface current densities in $\mathcal{F}'$ are $\mathbf{K}'_l = -\sigma' v\hat{\mathbf{i}}$ for the lower plate and $\mathbf{K}'_u = +\sigma' v\hat{\mathbf{i}}$ for the upper plate. The magnetic field between the plates is then $\mathbf{B}' = \mu_0\sigma' v\hat{\mathbf{j}}$ by Ampère's law, which is $(\gamma v E_0/c^2)\hat{\mathbf{j}}$.

Note that the fields observed in $\mathcal{F}'$ satisfy the relation

$$\mathbf{B}' = -\mathbf{v} \times \mathbf{E}'/c^2 \qquad \text{(for } \mathbf{B} = 0\text{)}. \tag{12.77}$$

This is a special case of a general theorem: If $\mathbf{B} = 0$ in some inertial frame $\mathcal{F}$, then (12.77) holds in any other inertial frame $\mathcal{F}'$; $\mathbf{v}$ is the constant velocity of $\mathcal{F}'$ with respect to $\mathcal{F}$.[21]

EXAMPLE 5   What does an electromagnetic wave look like in a moving frame of reference? More precisely, consider a plane wave traveling in the x direction and polarized in the z direction[22]

$$\mathbf{E}(\mathbf{x}, t) = E_0 \cos(kx - \omega t)\hat{\mathbf{k}},$$

$$\mathbf{B}(\mathbf{x}, t) = -\frac{E_0}{c}\cos(kx - \omega t)\hat{\mathbf{j}}. \tag{12.78}$$

What are the fields observed in a frame $\mathcal{F}'$ that moves with velocity $v\hat{\mathbf{i}}$ with respect to $\mathcal{F}$?

We can derive $\mathbf{E}'(\mathbf{x}', t')$ and $\mathbf{B}'(\mathbf{x}', t')$ using Table 12.2. Let's start by writing $kx - \omega t$ in primed coordinates, since that variable contains the spacetime dependence. By the Lorentz transformation,

$$kx - \omega t = k\gamma(x' + vt') - \omega\gamma(t' + vx'/c^2)$$
$$= k'x' - \omega't'$$

where we introduce new wave parameters k' and ω' for the primed frame,

$$k' = \gamma\left(k - v\omega/c^2\right) \tag{12.79}$$

$$\omega' = \gamma\left(\omega - vk\right). \tag{12.80}$$

(This transformation is equivalent to the statement that $(\omega/c, \mathbf{k})$ forms a 4-vector.) For an electromagnetic wave in vacuum, $\omega = ck$. Therefore the frequency observed by an observer at rest in $\mathcal{F}'$ is

$$\omega' = \gamma\left(1 - v/c\right)\omega = \sqrt{\frac{1 - v/c}{1 + v/c}}\,\omega. \tag{12.81}$$

---

[21] See Exercise 17.
[22] In (12.78) we have used the property of a plane wave that $B = E/c$.

This result is the Doppler shift for light. If $v > 0$ then the wave observed in $\mathcal{F}'$ has a lower frequency, i.e., a red shift, compared to $\mathcal{F}$; if $v < 0$ there is a blue shift. Note that the wave parameters for $\mathcal{F}'$ also obey $\omega' = ck'$, which implies that the wave speed is the same in $\mathcal{F}'$ as in $\mathcal{F}$.

Now we refer to Table 12.2 to find the field components observed in $\mathcal{F}'$. The nonzero components in $\mathcal{F}$ are E_z and B_y. According to the transformations in Table 12.2, the nonzero components in $\mathcal{F}'$ are E'_z and B'_y. For the electric field we find

$$E'_z = \gamma\,(E_0 - vE_0/c)\cos(kx - \omega t)$$
$$= E'_0 \cos(k'x' - \omega't') \tag{12.82}$$

where the amplitude in $\mathcal{F}'$ is

$$E'_0 = \gamma(1 - v/c)E_0 = \sqrt{\frac{1 - v/c}{1 + v/c}}\,E_0. \tag{12.83}$$

For the magnetic field we find

$$B'_y = \gamma\left(-E_0/c + vE_0/c^2\right)\cos(kx - \omega t)$$
$$= -\frac{E'_0}{c}\cos(k'x' - \omega't'); \tag{12.84}$$

again we have the relation $B' = E'/c$. The fields $\mathbf{E}'(x', t')$ and $\mathbf{B}'(x', t')$ form a transverse electromagnetic wave with the usual properties required by the Maxwell equations. The field transformations of an electromagnetic wave yield an electromagnetic wave. This is consistent with—indeed demanded by—Einstein's Postulate 1, covariance.

So, to answer the original question, the fields observed in $\mathcal{F}'$ form an electromagnetic wave. However, there are some differences compared to the wave in $\mathcal{F}$. In addition to the Doppler shift of frequency (or wavelength) there is a change in the amplitude of the field oscillations, given by (12.83). If $v > 0$, i.e., the observer moves in the direction of the wave, there is a red shift and a decrease in amplitude. In the limit $v \to c$ the wavelength tends to ∞ and the amplitude tends to 0. If $v < 0$, i.e., the observer moves in the opposite direction, there is a blue shift and an increase in amplitude.

12.5 ■ FIELDS DUE TO A POINT CHARGE IN UNIFORM MOTION

This section describes an example of the use of special relativity—the Lorentz transformation of coordinates and fields—to calculate fields. The example is very simple and basic: Consider a charge q moving with constant velocity with respect to an inertial frame $\mathcal{F}$. What are the electric and magnetic fields? The reader should appreciate that this very basic question has not been addressed in any previous chapter, except for the special case of a charge at rest.

To be specific, let the velocity of q be $v\,\hat{\mathbf{i}}$ in $\mathcal{F}$, along the x axis. Then the position of q as a function of time is $\mathbf{x}_q(t) = vt\,\hat{\mathbf{i}}$. We will derive $\mathbf{E}(\mathbf{x}, t)$ and $\mathbf{B}(\mathbf{x}, t)$ by Lorentz transformation from the rest frame of q, which we call $\mathcal{F}'$. The charge q is fixed at the origin of $\mathcal{F}'$. In $\mathcal{F}'$, the magnetic field $\mathbf{B}'$ is 0, and the electric field $\mathbf{E}'$ is simply the static Coulomb field

$$\mathbf{E}'(\mathbf{x}') = \frac{q}{4\pi\epsilon_0} \frac{\mathbf{x}'}{|\mathbf{x}'|^3}, \tag{12.85}$$

where $\mathbf{x}' = x'\,\hat{\mathbf{i}} + y'\,\hat{\mathbf{j}} + z'\,\hat{\mathbf{k}}$ is the field point in $\mathcal{F}'$ coordinates. The fields $\mathbf{E}(\mathbf{x}, t)$ and $\mathbf{B}(\mathbf{x}, t)$ observed in $\mathcal{F}$ are to be calculated from the inverse transformation (the lower rows) in Table 12.2. Note that $\mathcal{F}'$ moves with respect to $\mathcal{F}$ with the same velocity $\mathbf{v} = v\,\hat{\mathbf{i}}$ as q.

The field point has coordinates x, y, z and time t in frame $\mathcal{F}$, and coordinates x', y', z' and time t' in $\mathcal{F}'$. These coordinates and times are related by the Lorentz transformation (12.5). Therefore the vector $\mathbf{x}'$ in $\mathcal{F}'$, from q to the field point, but reexpressed in the coordinates of $\mathcal{F}$, is

$$\mathbf{x}' = \gamma(x - vt)\,\hat{\mathbf{i}} + y\,\hat{\mathbf{j}} + z\,\hat{\mathbf{k}}, \tag{12.86}$$

where $\gamma = 1/\sqrt{1 - v^2/c^2}$ as usual. The distance $|\mathbf{x}'|$ observed in $\mathcal{F}'$ is

$$|\mathbf{x}'| = \sqrt{\gamma^2(x - vt)^2 + y^2 + z^2}. \tag{12.87}$$

The electric field components in $\mathcal{F}$ are $E_x = E'_x$, $E_y = \gamma E'_y$, and $E_z = \gamma E'_z$, according to the field transformations in Table 12.2. Applying these transformations to the Coulomb field (12.85), but reexpressed in coordinates of $\mathcal{F}$, we obtain the electric field observed in $\mathcal{F}$,

$$\mathbf{E}(\mathbf{x}, t) = \frac{q\gamma}{4\pi\epsilon_0} \frac{(x - vt)\,\hat{\mathbf{i}} + y\,\hat{\mathbf{j}} + z\,\hat{\mathbf{k}}}{\left[\gamma^2(x - vt)^2 + y^2 + z^2\right]^{3/2}}. \tag{12.88}$$

Recall that in $\mathcal{F}$ the position of q at time t is $vt\,\hat{\mathbf{i}}$. Therefore $(x - vt)\,\hat{\mathbf{i}} + y\,\hat{\mathbf{j}} + z\,\hat{\mathbf{k}}$ is the vector in $\mathcal{F}$ from q to the field point $\mathbf{x}$. Equation (12.88) says that the direction of $\mathbf{E}(\mathbf{x}, t)$ points radially away from the *simultaneous* position of q.

If $v = 0$ then (12.88) is just the static Coulomb field, which is spherically symmetric, proportional to $\hat{\mathbf{r}}/r^2$, and centered at the origin. If v is small, i.e., $v \ll c$, then to first order in v/c we may approximate γ by 1. In this approximation $\mathbf{E}(\mathbf{x}, t)$ is the Coulomb field centered at the moving position $vt\,\hat{\mathbf{i}}$ of q. This limit is an example of the *quasistatic approximation*: The field has the same form as in electrostatics, but is centered at the slowly moving position of q.

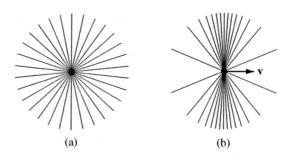

FIGURE 12.3   Electric field lines of a moving charge q. The speed of q in the two cases is (a) 0 and (b) $0.9\,c$. The density of lines is proportional to the field strength.

If v is large, i.e., approaching c, then the electric field is quite different from the Coulomb field. First, because of Lorentz–FitzGerald contraction in the x direction, the field lines are compressed into a disk normal to the direction of motion. An observer at rest at (x_0, y_0, z_0) in $\mathcal{F}$ would experience a sharp pulse of electric field as q moves past, with maximum strength when $t = x_0/v$. Second, the magnitude of the field is increased by the factor γ. The electric field lines for charged particles with speeds $v = 0$ and $v = 0.9\,c$ are compared in Fig. 12.3. Another way to visualize the electric field of a moving charge is shown in Fig. 12.4, which shows the electric field vectors of q for $v = 0$ and $v = 0.9$, at points equidistant from the charge.

The magnetic field components in $\mathcal{F}$ are $B_x = 0$, $B_y = -\gamma v E'_z/c^2$, and $B_z = \gamma v E'_y/c^2$, according to the field transformations in Table 12.2. Therefore we find

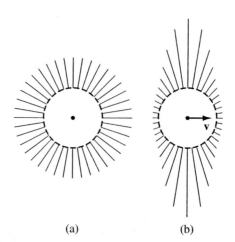

FIGURE 12.4   Electric field vectors of a moving charge q. The speed of q in the two cases is (a) 0 and (b) $0.9\,c$. The field vectors are plotted, for the same scale, at points equidistant from the moving charge.

that the magnetic field observed in $\mathcal{F}$ is

$$\mathbf{B}(\mathbf{x}, t) = \frac{q\gamma v}{4\pi\epsilon_0 c^2} \frac{\left(-\hat{\mathbf{j}}z + \hat{\mathbf{k}}y\right)}{\left[\gamma^2(x - vt)^2 + y^2 + z^2\right]^{3/2}}. \qquad (12.89)$$

The direction of $\mathbf{B}(\mathbf{x}, t)$ is tangent to a circle around the x axis, i.e., curling around the path of q in the sense given by the right-hand rule.[23] The ratio of magnetic and electric field strengths is

$$\frac{|\mathbf{B}|}{|\mathbf{E}|} = \frac{v}{c^2}\left(\frac{y^2 + z^2}{(x - vt)^2 + y^2 + z^2}\right)^{1/2}. \qquad (12.90)$$

If v is small compared to c, then $|\mathbf{B}| \ll |\mathbf{E}|/c$.

If v is large, approaching c, then the magnetic field is compressed into a sharp pulse coinciding with the electric field pulse. At the center of the pulse ($x = vt$) the ratio B/E is v/c^2, which is approximately $1/c$ if v approaches c. Thus the electric and magnetic pulse due to a relativistic charged particle resembles a pulse of light: $\mathbf{E}$ and $\mathbf{B}$ are perpendicular, with $B \approx E/c$.

To summarize the results we have obtained for $\mathbf{E}(\mathbf{x}, t)$ and $\mathbf{B}(\mathbf{x}, t)$ due to a charge moving with constant velocity, Fig. 12.5 represents a snapshot of the fields at an instant. In the space around a positive charge, $\mathbf{E}$ points away from the current position and $\mathbf{B}$ curls around the path of the particle.

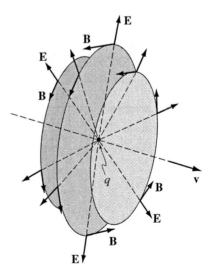

FIGURE 12.5   Fields of a moving charge. These are the fields at an instant. $\mathbf{E}$ points away from the position of q, and $\mathbf{B}$ curls around the trajectory.

[23]Note, too, that $\mathbf{B} = \mathbf{v} \times \mathbf{E}/c^2$, consistent with the general theorem in Exercise 17.

EXAMPLE 6   Suppose a point charge moves with velocity $\mathbf{v} = 0.9c\,\hat{\mathbf{i}}$ along the x axis. For a point P on the z axis, at distance d from the origin, we'll determine and plot the nonzero components of $\mathbf{E}$ and $\mathbf{B}$ as functions of time. The source point is at $(vt, 0, 0)$ and the field point P is at $(0, 0, d)$ as shown in Fig. 12.6(a).

Using (12.88) the nonzero components of the electric field at P are

$$E_x = \frac{q}{4\pi\epsilon_0 d^2}\,\frac{-\gamma\beta s}{\left[\gamma^2\beta^2 s^2 + 1\right]^{3/2}} \tag{12.91}$$

$$E_z = \frac{q}{4\pi\epsilon_0 d^2}\,\frac{\gamma}{\left[\gamma^2\beta^2 s^2 + 1\right]^{3/2}} \tag{12.92}$$

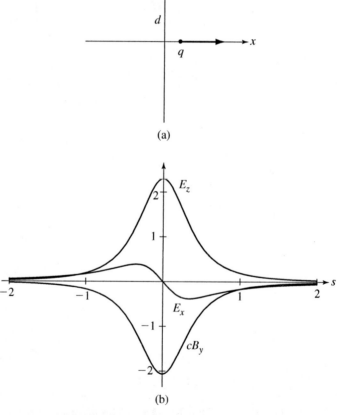

FIGURE 12.6   A moving charge. (a) The charge q moves along the x axis, and the field point P is on the z axis at distance d from the origin. (b) The nonzero field components, plotted as functions of the dimensionless variable $s \equiv ct/d$. (s is the time in units of d/c.)

where we have introduced a dimensionless variable $s \equiv ct/d$ proportional to the time. Using (12.89) the nonzero component of the magnetic field at P is

$$B_y = \frac{q}{4\pi\epsilon_0 cd^2} \frac{-\gamma\beta}{\left[\gamma^2\beta^2 s^2 + 1\right]^{3/2}}. \qquad (12.93)$$

These three functions are plotted in Fig. 12.6(b). In the graph the unit of electric field is $q/(4\pi\epsilon_0 d^2)$.

EXAMPLE 7 Fields of a moving line charge.   As another example of the use of field transformations, consider a long line of charge at rest with respect to the reference frame $\mathcal{F}'$, and lying along the x' axis. Let λ' be the linear charge density in $\mathcal{F}'$. To determine the field in $\mathcal{F}'$ is a problem in electrostatics. By Gauss's law,

$$\mathbf{E}'(\mathbf{x}') = \frac{\lambda'\hat{\mathbf{R}}'}{2\pi\epsilon_0 R'}, \qquad (12.94)$$

where $\mathbf{R}'$ is the perpendicular vector from the line to $\mathbf{x}'$. (We treat the charged line as infinite.) The magnetic field $\mathbf{B}'$ is 0.

Now let $\mathcal{F}$ be another inertial frame, relative to which the charged line (and $\mathcal{F}'$) moves with velocity $v\hat{\mathbf{i}}$. What are the electric and magnetic fields that would be observed in $\mathcal{F}$? According to the field transformation (12.73), or rather its inverse, the electric field observed in $\mathcal{F}$ has components $E_\| = E_\|' = 0$ and $\mathbf{E}_\perp = \gamma\mathbf{E}_\perp'$. This field $\mathbf{E}$ should be expressed in terms of coordinates of $\mathcal{F}$. Because $\mathbf{R}'$ is perpendicular to $\mathbf{v}$, and the perpendicular coordinates do not change under a Lorentz transformation, so the perpendicular vector from the line to $\mathbf{x}$ is $\mathbf{R} = \mathbf{R}'$. Thus the electric field is

$$\mathbf{E}(\mathbf{x}) = \frac{\lambda\hat{\mathbf{R}}}{2\pi\epsilon_0 R}, \qquad (12.95)$$

where $\lambda = \gamma\lambda'$. The direction of $\mathbf{E}$ is perpendicular to the moving line charge. The charge density in $\mathcal{F}$ is $\lambda = \gamma\lambda'$, increased by the factor γ compared to λ' because of Lorentz–FitzGerald contraction of the line. The electric field $\mathbf{E}$ is just the electrostatic field of a charged line with density λ.

But there is also a *magnetic field* in $\mathcal{F}$. By the field transformation (12.74) it is

$$\mathbf{B} = \gamma\mathbf{v}\times\mathbf{E}_\perp'/c^2 = \frac{\mu_0\lambda v\left(\hat{\mathbf{i}}\times\hat{\mathbf{R}}\right)}{2\pi R}. \qquad (12.96)$$

This field is just the magnetostatic field due to a steady current $\mathbf{I} = \lambda v\hat{\mathbf{i}}$. (Note that $\hat{\mathbf{i}}\times\hat{\mathbf{R}}$ is a unit vector $\hat{\psi}$ curling around the x axis.) Indeed, a charged line moving with constant velocity is a steady current $\mathbf{I}$. By the field transformations, we have merely reproduced the familiar static fields for a line charge and a line current.

As an interesting extension of Example 7, consider *two* charged lines, that have equal but opposite densities $\pm\lambda_0$ *in their rest frames.* Let the lines be parallel to the x axis, but slightly separated so that they can move independently. As a first instance, suppose the lines move in the lab frame with equal but opposite velocities $\pm\hat{\mathbf{i}}v/2$, respectively. Because they have opposite charges, the oppositely moving lines are both equivalent to currents in the $+\hat{\mathbf{i}}$ direction. Then the magnetic field is $\mu_0 I\hat{\psi}/(2\pi R)$, where $\hat{\psi}$ is azimuthal around the x axis, and $I = \lambda_0 v/\sqrt{1 - v^2/4c^2}$. The electric field is 0 because the positive and negative charges cancel. But as a second instance, suppose the positive line is at rest in the lab and the negative line moves with velocity $-\hat{\mathbf{i}}v$. In this case the magnetic field is $\mu_0 I\hat{\psi}/(2\pi R)$, where $I = \lambda_0 v/\sqrt{1 - v^2/c^2}$. The electric field is $(\lambda_+ - \lambda_-)\hat{\mathbf{R}}/(2\pi\epsilon_0 R)$, where $\lambda_+ - \lambda_-$ is the net charge density, which is $\lambda_0(1-\gamma)$. The negative line is Lorentz contracted, compared to the positive line, so there is a net negative charge and its corresponding electrostatic field.

A current in a metal wire is analogous to the second case in the previous paragraph, because the electrons move opposite to the current while the positive ions are at rest. Therefore a current carrying wire in its rest frame has a tiny electric field. The field is immeasurably small because the relevant electron velocity is the *drift velocity,* which is typically smaller than c by 12 orders of magnitude.

12.6 ■ MAGNETISM FROM RELATIVITY

In Chapters 3–11 we studied the field theory of electromagnetism one piece at a time, by considering particular electric and magnetic phenomena. We started with electrostatics, went on to steady currents and magnetostatics, then time-dependent fields, and finally electromagnetic waves. The theory that emerges from describing these phenomena mathematically—Maxwell's field theory—is consistent with the postulates of special relativity. Indeed it was the study of Maxwell's theory that led Einstein to special relativity. The presentation of electromagnetism in Chapters 3–11 resembles the historical development of the subject. We might call this presentation the *phenomenological approach,* because it emphasizes the physical phenomena. This approach is *inductive.*

We could, however, adopt a different presentation, which might be called the *axiomatic approach.* This approach is *deductive,* and more theoretical. The starting point would be to write down the minimal set of axioms from which the full structure of electromagnetic field theory can be deduced. The sufficient axioms are Coulomb's law, i.e., that the electric field of a point charge at rest is $\mathbf{E} = q\hat{\mathbf{r}}/(4\pi\epsilon_0 r^2)$; and the principles of special relativity. These lead inexorably to Maxwell's equations. In particular, magnetism can be derived as a necessary consequence of special relativity. The purpose of this section is to see how magnetism follows from relativity.

We'll start with an electric field. Consider a frame of reference $\mathcal{F}'$ in which there is a static electric field $\mathbf{E}'$, created by charges at rest. To an observer in $\mathcal{F}'$, a test charge q moves under the influence of the electric force $\mathbf{F}'$, according to the equation of motion

$$\frac{d\mathbf{p}'}{dt'} = \mathbf{F}' = q\mathbf{E}'. \tag{12.97}$$

To an observer in another frame $\mathcal{F}$, relative to which $\mathcal{F}'$ moves with velocity $\mathbf{v}$, the equation of motion is $d\mathbf{p}/dt = \mathbf{F}$. What is this force $\mathbf{F}$? We may determine $\mathbf{F}$ from the Lorentz transformation. Refer to Table 12.3. The component of $\mathbf{F}$ parallel to $\mathbf{v}$ is

$$F_{\parallel} = \frac{dp_{\parallel}}{dt} = \frac{\gamma\left(dp_{\parallel}' + vdE'/c^2\right)}{\gamma\left(dt' + vdx_{\parallel}'/c^2\right)}, \tag{12.98}$$

the first equality being the definition of $F_{\parallel}$, and the second following from the Lorentz transformation of 4-momentum and 4-position. According to the work-energy relation, $dE' = \mathbf{F}' \cdot \mathbf{u}'dt'$, where $\mathbf{u}' = d\mathbf{x}'/dt'$ is the velocity of q relative to frame $\mathcal{F}'$; thus,

$$F_{\parallel} = \frac{F_{\parallel}' + v\left(u_{\parallel}'F_{\parallel}' + \mathbf{u}_{\perp}' \cdot \mathbf{F}_{\perp}'\right)/c^2}{1 + vu_{\parallel}'/c^2} = F_{\parallel}' + \gamma v\mathbf{u}_{\perp} \cdot \mathbf{F}_{\perp}'/c^2. \tag{12.99}$$

In the second equality we have used the velocity transformation (see Table 12.3) to rewrite the expression in terms of the velocity $\mathbf{u}$ of q in $\mathcal{F}$. The final result is

$$F_{\parallel} = qE_{\parallel}' + \frac{q\gamma v}{c^2}\mathbf{u}_{\perp} \cdot \mathbf{E}_{\perp}'. \tag{12.100}$$

By a similar calculation, the components of $\mathbf{F}$ perpendicular to $\mathbf{v}$ are

$$\mathbf{F}_{\perp} = \frac{d\mathbf{p}_{\perp}}{dt} = \frac{q\mathbf{E}_{\perp}'}{\gamma\left(1 + vu_{\parallel}'/c^2\right)}. \tag{12.101}$$

Special relativity requires that the force on q, observed in frame $\mathcal{F}$, is given by (12.100) and (12.101). It is not obvious yet, but we shall show that this force is in fact the Lorentz force

$$\mathbf{F} = q\mathbf{E} + q\mathbf{u} \times \mathbf{B}, \tag{12.102}$$

where $\mathbf{E}(\mathbf{x}, t)$ and $\mathbf{B}(\mathbf{x}, t)$ are the fields in the frame $\mathcal{F}$. Our task is to determine the fields such that (12.102) is the same as (12.100) and (12.101). If we succeed then we have shown that magnetism—the force $q\mathbf{u} \times \mathbf{B}$—is a necessary consequence of relativity.

To proceed deductively is quite tedious. Instead, we shall state the result and prove that it is correct. The fields **E** and **B** are given by

$$E_\| = E_\|' \quad \text{and} \quad \mathbf{E}_\perp = \gamma \mathbf{E}_\perp' \tag{12.103}$$

$$B_\| = 0 \quad \text{and} \quad \mathbf{B}_\perp = \frac{\gamma}{c^2} \mathbf{v} \times \mathbf{E}_\perp'. \tag{12.104}$$

We need to prove that the Lorentz force with these fields is (12.100) and (12.101).

The parallel component.   Consider $\mathbf{v} \cdot \mathbf{F} = v F_\|$. For the Lorentz force,

$$v F_\| = q v E_\| + q \mathbf{v} \cdot (\mathbf{u} \times \mathbf{B}). \tag{12.105}$$

Now, $\mathbf{v} \cdot (\mathbf{u} \times \mathbf{B})$ is equal to $\mathbf{u} \cdot (\mathbf{B} \times \mathbf{v})$ by a vector identity. In the expression we may replace **B** by $\mathbf{B}_\perp$ because the cross product of the parallel component of **B** with **v** is 0. Similarly we may replace **u** by $\mathbf{u}_\perp$ because the parallel component would contribute nothing. Then inserting (12.104) for $\mathbf{B}_\perp$

$$v F_\| = q v E_\| + q \mathbf{u}_\perp \cdot \frac{\gamma}{c^2} \left[(\mathbf{v} \times \mathbf{E}_\perp') \times \mathbf{v} \right]. \tag{12.106}$$

The quantity in square brackets is $\mathbf{E}_\perp' v^2$ since by definition $\mathbf{E}_\perp'$ is perpendicular to **v**. Thus

$$v F_\| = v \left[q E_\|' + q \frac{v}{c^2} \gamma \mathbf{u}_\perp \cdot \mathbf{E}_\perp' \right] \tag{12.107}$$

and this is the same as (12.100).

The perpendicular components.   This time consider $\mathbf{v} \times \mathbf{F}$, which is the same as $\mathbf{v} \times \mathbf{F}_\perp$. For the Lorentz force,

$$\mathbf{v} \times \mathbf{F}_\perp = q \mathbf{v} \times \mathbf{E}_\perp + q \mathbf{v} \times (\mathbf{u} \times \mathbf{B}). \tag{12.108}$$

Now, the double cross product is $\mathbf{u} v B_\| - \mathbf{B} v u_\|$; or, $-\mathbf{B} v u_\|$ because $B_\|$ is 0. Inserting (12.103) and (12.104) we have

$$\mathbf{v} \times \mathbf{F}_\perp = q \mathbf{v} \times \left(\gamma \mathbf{E}_\perp' \right) - q v u_\| \frac{\gamma}{c^2} \mathbf{v} \times \mathbf{E}_\perp' = q \mathbf{v} \times \mathbf{E}_\perp' \gamma \left[1 - \frac{v u_\|}{c^2} \right]. \tag{12.109}$$

But by the addition of velocities the term in square brackets is

$$\gamma^{-2} \left(1 + v u_\|'/c^2 \right)^{-1}.$$

The final result is

$$\mathbf{v} \times \mathbf{F}_\perp = \mathbf{v} \times \frac{q \mathbf{E}_\perp'}{\gamma \left(1 + v u_\|'/c^2 \right)} \tag{12.110}$$

which is the same as (12.101). We have proven that the Lorentz force (12.102) is just what is required by relativity.

So, there must be a magnetic force in $\mathcal{F}$ if there is an electric force in $\mathcal{F}'$. Of course the fields (12.103) and (12.104) are precisely what we would obtain by applying the Lorentz transformation of fields, listed in Table 12.3, to compute **E** and **B** from **E**$'$. It follows that **E** and **B** form an antisymmetric tensor $F^{\mu\nu}$. Then for covariance the field equations must be Maxwell's equations.

In this sense magnetism is a relativistic effect: The Lorentz transformation implies the existence of a velocity-dependent force on q, perpendicular to **u**, in any reference frame where the charge sources are in motion. Furthermore, the structure of the field equations, including those of magnetism, is completely determined by the requirement of relativistic covariance. Although axiomitization is merely formalism, it is interesting that relativity *demands* certain interactions of particles and fields.

12.7 ■ THE ENERGY–MOMENTUM FLUX TENSOR

We will study one final theoretical development, relating to energy and momentum of the electromagnetic field. The covariant form of the electromagnetic theory is ideal for analyzing energy and momentum, because these quantities are combined by special relativity. The derivations in this section are difficult, but the results are rewarding.[24] They reveal the equations for *field momentum*.

The equation of motion for a single point charge is (12.52). We may *generalize* this equation, to describe any charge distribution in a volume. Let $J^{\mu}(x)$ be the 4-current $(c\rho, \mathbf{J})$. (x denotes the 4 spacetime coordinates.) The Minkowski force on the charge in a volume d^3x is obtained from (12.52) by replacing $q\eta_\nu$ by $J_\nu d^3x$. Therefore, if $\mathcal{P}^{\mu}(x)$ denotes the *momentum density* (momentum per unit volume) of the charged matter, then the equation of motion is

$$\frac{\partial \mathcal{P}^{\mu}}{\partial \tau} = J_\nu F^{\mu\nu}. \qquad (12.111)$$

In words, $J_\nu F^{\mu\nu}$ is the rate of change of 4-momentum, per unit volume, of the charged matter, with respect to proper time.

Now, local conservation of energy and momentum implies that $\partial \mathcal{P}^{\mu}/\partial t$ can be written as a divergence in Minkowski spacetime, i.e.,

$$J_\nu F^{\mu\nu} = -\frac{\partial T^{\mu\nu}}{\partial x^\nu}. \qquad (12.112)$$

Here $T^{\mu\nu}$ is a tensor, called the *energy-momentum flux tensor of the electromagnetic field*. We'll analyze this equation in detail, to understand the interpretation

---

[24]It is important to remember that the Einstein summation convention is used throughout this section. Greek indices run from 0 to 3, and Roman indices run from 1 to 3.

of $T^{\mu\nu}$. Later we'll write an equation for $T^{\mu\nu}$ in terms of the electromagnetic field $F^{\mu\nu}$.

Equation (12.112) is a kind of continuity equation. Recall that the continuity equation $\partial J^\nu/\partial x^\nu = 0$ expresses conservation of charge, a scalar quantity. In the absence of charge, i.e., for $J^\mu = 0$, the equation $\partial T^{\mu\nu}/\partial x^\nu = 0$ expresses conservation of the energy and momentum—a 4-vector quantity—of the electromagnetic field. In the *presence* of charged matter, there may be transfer of energy and momentum between the field and the matter. Then the local conservation of energy and momentum is described by (12.112).

To comprehend this equation, and learn the interpretation of $T^{\mu\nu}$, we inspect the temporal and spatial components of (12.112). The $\mu = 0$ component is

$$\frac{\mathbf{J} \cdot \mathbf{E}}{c} = -\frac{1}{c}\frac{\partial T^{00}}{\partial t} - \frac{\partial T^{0i}}{\partial x^i}. \tag{12.113}$$

$\mathbf{J} \cdot \mathbf{E}$ is the rate at which work is done, per unit volume, on the charged matter at x. In order for (12.113) to be a statement of conservation of energy, $\partial T^{00}/\partial t$ must be the rate of change of field energy, per unit volume, i.e., T^{00} is the field energy density. Also, $\partial(cT^{0i})/\partial x^i$ must be the *divergence* of the energy flux, i.e., cT^{0i} is the Poynting vector. In other words, (12.113) is just *Poynting's theorem*, which we encountered in Chapter 11, expressing the *local* conservation of energy.

The $\mu = i$ (spatial) component of (12.112) is

$$\rho E^i + \epsilon_{ijk} J^j B^k = -\frac{1}{c}\frac{\partial T^{i0}}{\partial t} - \frac{\partial T^{ij}}{\partial x^j}. \tag{12.114}$$

Equation (12.114) must express local conservation of 3-momentum. The left-hand side is the force per unit volume acting on the charged matter, i.e., the rate of change of its 3-momentum density. In order for (12.114) to be a statement of conservation of momentum, $\partial(T^{i0}/c)/\partial t$ must be the rate of change of field momentum density; that is, T^{i0}/c is the field momentum density. Also, $\partial T^{ij}/\partial x^j$ must be the divergence of the momentum flux. With these interpretations for the three parts of (12.114), the rate of change of momentum in a small volume at x equals the total flux of momentum into the volume. $T^{ij}(x)$ is the flux (momentum per unit area per unit time) in the jth direction, of the ith component of momentum.

Now that we have identified the *meaning* of $T^{\mu\nu}$, we must write $T^{\mu\nu}$ in terms of the electromagnetic field $F^{\mu\nu}$. The defining equation is (12.112). As we shall verify, the formula for $T^{\mu\nu}$ is

$$T^{\mu\nu} = \frac{1}{\mu_0}\left\{ F^{\mu\rho}F^\nu{}_\rho - \frac{1}{4}g^{\mu\nu}F^{\rho\sigma}F_{\rho\sigma} \right\}. \tag{12.115}$$

Here $g^{\mu\nu}$ is the *metric tensor*:

$$g^{00} = -1, \quad g^{11} = g^{22} = g^{33} = +1, \quad \text{and} \quad g^{\mu\nu} = 0 \quad \text{for} \quad \mu \neq \nu. \tag{12.116}$$

It can be shown that $g^{\mu\nu}$ transforms as a tensor, i.e., $\Lambda^{\mu}_{\rho}\Lambda^{\nu}_{\sigma}g^{\rho\sigma} = g^{\mu\nu}$. Therefore (12.115) is a covariant equation. All of our considerations of field energy and momentum in this section start from tensor equations, so the conclusions are consistent with Einstein's postulate of relativity.

To verify that (12.115) does satisfy the continuity equation (12.112), we note that the 4-space divergence of (12.115) can be written in the form

$$\frac{\partial T^{\mu\nu}}{\partial x^{\nu}} = \frac{1}{\mu_0}\left\{F^{\mu\rho}\left(\frac{\partial F^{\nu}_{\rho}}{\partial x^{\nu}}\right) - \frac{1}{2}F_{\rho\sigma}\left[\frac{\partial F^{\rho\sigma}}{\partial x_{\mu}} + \frac{\partial F^{\mu\rho}}{\partial x_{\sigma}} + \frac{\partial F^{\sigma\mu}}{\partial x_{\rho}}\right]\right\}. \quad (12.117)$$

(We have used the fact that $g^{\mu\nu}\partial/\partial x^{\nu} = \partial/\partial x_{\mu}$.) By Maxwell's equation (12.55) the quantity $\partial F^{\nu}_{\rho}/\partial x^{\nu}$, which appears in the first term on the right, is equal to $-\mu_0 J_{\rho}$. By Maxwell's equation (12.56) the quantity in square brackets is 0. Hence (12.112) is satisfied.

Now we may go back to the interpretation of $T^{\mu\nu}$ and write the energy and momentum density and flux in terms of **E** and **B**. The field energy density, denoted by $u(\mathbf{x}, t)$ in previous chapters, is

$$u = T^{00} = \frac{1}{\mu_0}\left[\frac{E^2}{2c^2} + \frac{B^2}{2}\right], \quad (12.118)$$

where T^{00} has been evaluated in terms of **E** and **B** from (12.115). The result agrees with our earlier analyses of field energy. The energy flux is

$$\mathbf{S} = \hat{\mathbf{e}}_i c T^{0i} = \frac{1}{\mu_0}\mathbf{E} \times \mathbf{B} \quad (12.119)$$

where again T^{0i} has been evaluated from (12.115). The result agrees with the equation for the Poynting vector derived in Chapter 11.

The electromagnetic field carries momentum as well as energy. The field momentum density must be T^{i0}/c, as we argued above. By (12.115) the energy-momentum flux tensor is *symmetric*: $T^{\nu\mu} = T^{\mu\nu}$. Therefore $T^{i0} = T^{0i} = S^i/c$. *The momentum density* $\mathbf{\Pi}(\mathbf{x}, t)$ *in the electromagnetic field is* $\mathbf{S}(\mathbf{x}, t)/c^2$. We have used this result previously (in Chapter 11) but without deriving it.

The relation $\mathbf{\Pi} = \mathbf{S}/c^2$ is consistent with the photon theory of light. In the photon theory, the energy flux is $\mathbf{S} = E_{\gamma}nc\hat{\mathbf{k}}$, where n is the photon density, E_{γ} the energy of a photon, and $\hat{\mathbf{k}}$ the direction of photon velocity. The momentum density is $\mathbf{p}_{\gamma}n$, where the single photon momentum is $\mathbf{p}_{\gamma} = (E_{\gamma}/c)\hat{\mathbf{k}}$ because a photon is a massless particle. Thus the momentum density in the photon theory is $\mathbf{S}/c^2$, the same as in the classical theory.

FURTHER READING

1. Galileo, *Dialogue on the Two World Systems*.

2. A. Einstein, *The Meaning of Relativity*, 5th ed. (Princeton Univ. Press, Princeton, 1955). [This little book is quite readable. The first half is on special relativity.]

3. K. Krane, *Modern Physics*, 2nd ed. (Wiley, New York, 1996). [Chapter 2 is an introduction to special relativity, including a discussion of the experimental verification of the theory.]

4. S. T. Thornton and A. Rex, *Modern Physics*, 2nd ed. (Saunders, Philadelphia, 2000). [Chapter 2 is on special relativity.]

5. R. Resnick, *Introduction to Special Relativity* (Wiley, New York, 1968).

6. E. F. Taylor and J. A. Wheeler, *Spacetime Physics* (W. H. Freeman, San Francisco, 1966).

7. E. M. Purcell, *Electricity and Magnetism*, 2nd ed. (McGraw-Hill, New York, 1985). [This famous introductory book uses the axiomatic presentation of electromagnetism, deriving magnetism from the Lorentz transformation.]

8. M. Schwartz, *Principles of Electrodynamics* (Dover, New York, 1985). [This book, too, uses the axiomatic presentation. It is more advanced than Ref. [7].]

EXERCISES

Sec. 12.1. Coordinate Transformations

12.1. Derive the Lorentz transformation by assuming that the transformation is linear, and does not change the perpendicular coordinates. Write the transformation as

$$x' = A_1(x - vt), \quad y' = y, \quad z' = z, \quad t' = A_2 t + A_3 x.$$

Determine A_1, A_2, A_3 by requiring that a flash of light produces an outgoing spherical wave, with velocity c, in either frame $\mathcal{F}$ or $\mathcal{F}'$.

12.2. If $\mathcal{F}'$ moves with velocity $v\,\hat{\mathbf{i}}$ in $\mathcal{F}$, then the Lorentz transformation relating coordinates is

$$x' = \gamma \, (x - vt), \quad y' = y, \quad z' = z, \quad t' = \gamma \left(t - vx/c^2 \right),$$

where $\gamma = 1/\sqrt{1 - v^2/c^2}$. Find the inverse of this transformation, i.e., solve for x, y, z, t in terms of x', y', z', t'. Verify that the inverse transformation has the same form as the Lorentz transformation except with the replacement $v \to -v$. This form is necessary for self-consistency since $\mathcal{F}$ moves with velocity $-v\,\hat{\mathbf{i}}$ in the frame $\mathcal{F}'$.

12.3. The mean lifetime of muons is 2×10^{-6} s in their rest frame. Muons are produced in the upper atmosphere, as cosmic-ray secondaries.

   (a) Calculate the mean distance traveled by muons with speed $0.99\,c$, assuming classical physics, i.e., without special relativity.

   (b) Calculate the mean distance with special relativity.

(c) What percentage of muons produced at an altitude of 10 km reach the ground, assuming they travel downward with speed $0.99 c$? [Answer: 9%]

12.4. Suppose an inertial frame $\mathcal{F}'$ moves with velocity $v\,\hat{\mathbf{i}}$ with respect to another frame $\mathcal{F}$. The *rapidity* of $\mathcal{F}'$ is the variable θ defined by

$$\tanh \theta = v/c.$$

(a) Sketch a plot of θ versus v, for $-c \leq v \leq c$. Use computer graphics if you are unfamiliar with the hyperbolic tangent, but then explain analytically the form of the graph.

(b) Write the Lorentz transformation matrix Λ^{μ}_{ν} in terms of $\sinh \theta$ and $\cosh \theta$. Compare the result to the transformation matrix for a coordinate rotation about the z axis.

(c) Define similarly the rapidity θ_p of a particle, moving with velocity $u\,\hat{\mathbf{i}}$ in $\mathcal{F}$, by $\tanh \theta_p = u/c$. From Einstein's velocity addition theorem, derive the transformation of particle rapidity from $\mathcal{F}'$ to $\mathcal{F}$. [Answer: $\theta_p = \theta'_p + \theta$]

Sec. 12.2. Minkowski Space

12.5. Write (12.20) as a matrix equation, for the transformation matrix in (12.21), and verify that the result is equivalent to the Lorentz transformation (12.5).

12.6. (a) Show that $a^{\mu} b_{\mu} \equiv -a^0 b^0 + \mathbf{a} \cdot \mathbf{b}$ is invariant under Lorentz transformations.

(b) Show that the invariance of $a^{\mu} b_{\mu}$ implies that b_{μ} transforms according to $b'_{\mu} = b_{\sigma} (\Lambda^{-1})^{\sigma}_{\mu}$, where Λ^{-1} in the inverse of the matrix in (12.20).

12.7. The set of Lorentz transformations in a given direction ("boosts") forms a group. Consider boosts in the x direction. Let $\Lambda^{\mu}_{\nu}(v)$ denote the Lorentz transformation matrix for a boost of velocity $v\,\hat{\mathbf{i}}$.

(a) Show that the product of two boosts, of velocities $v_1\,\hat{\mathbf{i}}$ and $v_2\,\hat{\mathbf{i}}$, is also a boost, corresponding to velocity $(v_1 + v_2)/(1 + v_1 v_2/c^2)$.

(b) What is the inverse of the boost of velocity $v\,\hat{\mathbf{i}}$?

(The group of Lorentz transformations in all directions, and rotations, is called the Lorentz group.)

12.8. Sketch a spacetime diagram, i.e., the (x, ct) plane, showing: (i) the *light cone* (locus of points touched by light emitted at $x = 0$ at time $t = 0$, (ii) the trajectory of an observer at rest at the origin, and (iii) the trajectory of an observer who travels from $x = 0$ at $t = 0$ to $x = L$, and then back to $x = 0$.

What parts of the spacetime diagram are impossible for a traveler to reach, leaving $x = 0$ at $t = 0$?

12.9. Derive the transformation equations for particle velocity. Let $\mathbf{u} = d\mathbf{x}/dt$ be the particle velocity in $\mathcal{F}$, and $\mathbf{u}' = d\mathbf{x}'/dt'$ that in $\mathcal{F}'$. Determine $u'_{\parallel}$ and $u'_{\perp}$ in terms of $u_{\parallel}$ and $\mathbf{u}_{\perp}$.

12.10. An inertial frame $\mathcal{F}'$ moves in the x direction with speed $v = 0.9c$ relative to another frame $\mathcal{F}$. A traveler is moving with velocity $\mathbf{u}' = 0.5c\,\hat{\mathbf{i}} + 0.5c\,\hat{\mathbf{j}}$ in the frame $\mathcal{F}'$. Determine the traveler's velocity in the frame $\mathcal{F}$.

12.11. **(a)** Let η^μ be the 4-velocity of a particle. Evaluate $\eta^\mu \eta_\mu$. [Answer: $-c^2$]

   (b) Let p^μ be the 4-momentum of a particle. Evaluate $p^\mu p_\mu$. [Answer: $-m^2 c^2$]

12.12. Sketch a graph of $E - mc^2$ versus particle speed u, where E is the particle energy. On the same graph sketch the classical kinetic energy. Explain why the relativity curve implies that a particle cannot accelerate to a speed greater than c.

12.13. **(a)** Sketch a graph of particle energy E versus particle momentum p. What is the *name* of this mathematical curve?

   (b) How does E depend on p if $p \ll mc$? What is the curve in this limit?

   (c) How does E depend on p if $p \gg mc$? What is the curve in this limit?

Sec. 12.3. Electromagnetism in Covariant Form

12.14. **(a)** Prove that $\partial/\partial x^\mu$ transforms as a covariant vector. For example, if $\varphi(x)$ is a scalar function (x means the 4-vector x^μ) then $\partial\varphi/\partial x^\mu$ is a covariant vector. Before you try to construct the proof, check that the result makes sense, by considering $\partial\varphi/\partial x^\mu$ for $\varphi = C \cdot x$, where C^μ is a constant 4-vector, and for $\varphi = (C \cdot x)^2$, and for $\varphi = x^2$. (Hint: Use the chain rule of differentiation: the derivative with respect to x' is the derivative with respect to x times the derivative of x with respect to x'.)

   (b) Prove that the covariant equation $\partial J^\mu/\partial x^\mu$ is the same as the continuity equation $\nabla \cdot \mathbf{J} = -\partial\rho/\partial t$.

12.15. Verify that the temporal component ($\mu = 0$) of (12.56) is the same as $\nabla \cdot \mathbf{B} = 0$, and the spatial components ($\mu = i = 1, 2, 3$) are the same as $\nabla \times \mathbf{E} = -\partial\mathbf{B}/\partial t$.

12.16. **(a)** From the requirement that $(c\rho, \mathbf{J})$ is a Lorentz vector J^μ, determine the Lorentz transformations of ρ and $\mathbf{J}$.

   (b) Suppose in a frame $\mathcal{F}'$, ρ' is nonzero, but $\mathbf{J}'$ is 0. What are ρ and $\mathbf{J}$ in another frame $\mathcal{F}$? Show that $\mathbf{J} = \rho\mathbf{v}$. Explain why ρ is greater than ρ'.

Sec. 12.4. Field Transformations

12.17. Prove the following theorem: If $\mathbf{B} = 0$ in some inertial frame $\mathcal{F}$, then $\mathbf{B}' = -\mathbf{v} \times \mathbf{E}'/c^2$ in any other inertial frame $\mathcal{F}'$; $\mathbf{v}$ is the constant velocity of $\mathcal{F}'$ with respect to $\mathcal{F}$.

12.18. In 1952 Einstein wrote: "What led me more or less directly to the special theory of relativity was the conviction that the electromotive force acting on a body in motion in a magnetic field was nothing else but an electric field."

   Suppose in an inertial frame $\mathcal{F}$ there is a magnetic field $\mathbf{B} = b\hat{\mathbf{j}}$, but the electric field is 0. If a charge q moves with velocity $\mathbf{u} = u\hat{\mathbf{i}}$ in $\mathcal{F}$, then it experiences a force $qub\hat{\mathbf{k}}$ (the magnetic force). Now let $\mathcal{F}'$ be the rest frame of q.

   (a) Find $\mathbf{E}'$ and $\mathbf{B}'$, the fields in $\mathcal{F}'$.

   (b) Find $\mathbf{F}'$, the force on q in $\mathcal{F}'$.

   (c) How does the result relate to what Einstein wrote?

12.19. An observer, Stephanie, measures the electric and magnetic fields of a large charged plate, which is at rest in the xy plane with uniform charge density σ_S, in a region far from the edges of the plate. What are her measured fields $\mathbf{E}_S$ and $\mathbf{B}_S$?

Another observer, Loretta, is moving with velocity $v\,\hat{\mathbf{i}}$ with respect to Stephanie, and Loretta also measures the electric and magnetic fields. What are her measured fields, $\mathbf{E_L}$ and $\mathbf{B_L}$?

Loretta attributes the fields to a surface charge density σ_L and current density $\mathbf{K_L}$. Relate these quantities to σ_S.

12.20. Michele, an astronomy student, observes that the Lyman α spectral line emitted from hydrogen in a distant quasar has a wavelength of 790 nm, in the infrared. The wavelength of the Lyman α line from terrestrial hydrogen is 122 nm, in the ultraviolet. How fast is the quasar moving away from Earth? How did Michele recognize this line considering how far it has been shifted?

Sec. 12.5. Fields of a Moving Charge

12.21. The fields $\mathbf{E}(\mathbf{x}, t)$ and $\mathbf{B}(\mathbf{x}, t)$ due to a charge q moving with constant velocity $\mathbf{v} = v\,\hat{\mathbf{i}}$ are given in (12.88) and (12.89). Show that these functions satisfy the vacuum form of Maxwell's equations, apart from the singularity at $\mathbf{x} = vt\,\hat{\mathbf{i}}$.

12.22. A proton at the origin exerts a force on the nucleus of a gold atom at rest on the z axis at $z = 100\,\text{fm}$. (The atomic electrons are irrelevant.) Compute the force on the Au nucleus if the proton is (a) at rest, and (b) moving along the x axis with speed $0.99\,c$. Also, in the latter case, plot the magnitude of the force on the Au nucleus as a function of time. In making the plot it is convenient to measure time in zeptoseconds (zs) where $1\,\text{zs} = 10^{-21}\,\text{s}$. Express the forces in MeV/fm. (Neglect the motion of the heavy Au nucleus.)

General Exercises

12.23. A parallel plate capacitor is at rest in an inertial frame $\mathcal{F}'$. The plates have area A', are located at $z' = d/2$ and $z' = -d/2$, parallel to the xy plane, and have charge $-Q$ and $+Q$ respectively. The capacitor moves in the x direction with speed v in another frame $\mathcal{F}$.

Neglecting edge effects, find the electric and magnetic fields in the frame $\mathcal{F}$.

12.24. Consider a particle (call it P) with rest mass m moving under the influence of a constant force $\mathbf{F} = F\,\hat{\mathbf{i}}$. For example, P could be a proton in a constant electric field. The equation of motion is $d\mathbf{p}/dt = \mathbf{F}$ where $\mathbf{p} = m\mathbf{u}/\sqrt{1 - u^2/c^2}$. Here $\mathbf{u}$ is the velocity of P. Assume P is at rest at $t = 0$. Then $\mathbf{u} = u\,\hat{\mathbf{i}}$ at later times.

   (a) The solution of the equation of motion is $\mathbf{p} = \mathbf{F}t$. Determine u as a function of t. Sketch a graph of u versus t.

   (b) Let τ be the proper time of P. From the definition $(d\tau = dt\sqrt{1 - u^2/c^2})$ show that $Ft/mc = \sinh(F\tau/mc)$. Sketch a plot of τ versus t.

   (c) Show that $x(\tau) = (mc^2/F)\,[\cosh(F\tau/mc) - 1]$.

   (d) Use computer graphics to make an accurate plot of the worldline of P in the plane of points (x, x^0). Use mc^2/F as the unit of length. (Hint: Make a parametric plot.)

12.25. We showed that the Newtonian force $\mathbf{F}$ must be invariant under a change of inertial frame if the transformation between inertial frames is Galilean. But the transformation between inertial frames is *not* Galilean, but Lorentzian. So how does the Newtonian force $\mathbf{F}$, defined to be $d\mathbf{p}/dt$, transform under a change of inertial frame?

(a) By transforming the equation of motion $\mathbf{F} = d\mathbf{p}/dt$ to $\mathcal{F}'$ coordinates, i.e., writing $\mathbf{F}' = d\mathbf{p}'/dt'$, prove that

$$F'_\| = F_\| - \frac{v\mathbf{u}_\perp \cdot \mathbf{F}_\perp}{c^2 \left(1 - \mathbf{v} \cdot \mathbf{u}/c^2\right)}$$

$$\mathbf{F}'_\perp = \frac{\mathbf{F}_\perp}{\gamma \left(1 - \mathbf{v} \cdot \mathbf{u}/c^2\right)}$$

where $\mathbf{u}$ is the velocity of the particle. ($\|$ and $\perp$ refer to vector components parallel and perpendicular to $\mathbf{v}$.)

(b) From the Lorentz transformations of $\mathbf{u}$, $\mathbf{E}$ and $\mathbf{B}$, prove that the Lorentz force $q(\mathbf{E} + \mathbf{u} \times \mathbf{B})$ transforms in this manner.

12.26. A charge q moves with constant velocity $\mathbf{u} = u\,\hat{\mathbf{i}}$. Define the antisymmetric tensor $r^{\mu\nu}$ by

$$r^{\mu\nu} = \frac{1}{c}\left(\eta^\mu x^\nu - \eta^\nu x^\mu\right)$$

where η^μ is the 4-velocity of q. Also, define $r^2 = -r^{\mu\nu}r_{\mu\nu}/2$.

(a) Prove that

$$r^2 = \frac{(x - ut)^2}{1 - u^2/c^2} + y^2 + z^2.$$

(b) From equations (12.88) and (12.89) for $\mathbf{E}$ and $\mathbf{B}$ of the moving charge, show that the electromagnetic field tensor is

$$F^{\mu\nu} = \frac{q}{4\pi\epsilon_0 c}\frac{r^{\mu\nu}}{(r^2)^{3/2}}.$$

This is the electromagnetic field due to a moving charge, in covariant form. Such a fundamental quantity has a beautiful form when written covariantly.

CHAPTER
13

Electromagnetism and Optics

We have learned that the field theory of electromagnetism is a unified theory including both time-independent and time-dependent phenomena of electricity and magnetism. As part of this study we saw in Chapter 11 that the Maxwell equations for $\mathbf{E}(\mathbf{x}, t)$ and $\mathbf{B}(\mathbf{x}, t)$ in vacuum have wave solutions. We explored the properties of electromagnetic waves, valid over a wide range of frequencies. In vacuum, all electromagnetic waves travel at speed $c = 3.00 \times 10^8$ m/s.

In this chapter we will apply and extend the same principles to consider electromagnetic waves interacting with matter. An important part of this subject is the branch of physics called Optics, the study of light. Among their many interesting properties, light waves from the Sun bathe the Earth and support life. They make vision possible—the ability to see our surroundings (including this textbook). From the electromagnetic theory, we'll come to understand how light propagates in matter, how it reflects from materials (dielectrics and conductors), the role of polarization, and the basis for many familiar optical phenomena.

It's useful to bear in mind that a lot of what we'll study in this chapter holds not only for light (visible, infrared, ultraviolet) but also for other kinds of electromagnetic waves (radio, microwaves, X rays). The behavior of the fields depends on the material with which the wave is interacting. The idea is that wave solutions *in vacuum* are similar for all frequencies and wavelengths, because in vacuum $\mathbf{E}(\mathbf{x}, t)$ and $\mathbf{B}(\mathbf{x}, t)$ interact only with each other. But when electromagnetic waves interact with *matter*—it's usually the interaction between the electric field and the electrons that determines what happens—then atomic and molecular properties become important.

13.1 ■ ELECTROMAGNETIC WAVES IN A DIELECTRIC

Recall that by a *dielectric* we mean a *linear material* with electric permittivity ϵ and magnetic permeability μ but which has negligible conductivity, $\sigma = 0$. Examples of dielectrics are glass, water, plastics, and other insulators. In this section we'll assume, for simplicity, that ϵ and μ are real constants, independent of the wave frequency. The constitutive equations are $\mathbf{D}(\mathbf{x}, t) = \epsilon\mathbf{E}(\mathbf{x}, t)$ and $\mathbf{H}(\mathbf{x}, t) = \mathbf{B}(\mathbf{x}, t)/\mu$. Electromagnetic waves in dielectrics are similar in many ways to those in free space, except that they travel more slowly.

The Maxwell equations in a uniform dielectric with neither free charge ($\rho_{\text{free}} = 0$) nor free current ($\mathbf{J}_{\text{free}} = 0$) are

$$\nabla \cdot \mathbf{D} = 0 \quad \text{and} \quad \nabla \cdot \mathbf{B} = 0, \tag{13.1}$$

$$\nabla \times \mathbf{E} = -\frac{\partial \mathbf{B}}{\partial t} \quad \text{and} \quad \nabla \times \mathbf{H} = \frac{\partial \mathbf{D}}{\partial t}. \tag{13.2}$$

These have the same form as Maxwell's equations in vacuum, with the replacements $\epsilon_0 \rightarrow \epsilon$ and $\mu_0 \rightarrow \mu$. Therefore (13.1) and (13.2) have transverse wave solutions with the same form as electromagnetic waves in vacuum. However, the wave speed in the dielectric is

$$v = 1/\sqrt{\epsilon\mu}, \tag{13.3}$$

smaller than the wave speed in vacuum $c = 1/\sqrt{\epsilon_0\mu_0}$. The *index of refraction* of the dielectric, denoted by n, is defined as the ratio of wave speeds

$$n = \frac{c}{v}, \tag{13.4}$$

so that

$$n = \sqrt{\frac{\epsilon\mu}{\epsilon_0\mu_0}}. \tag{13.5}$$

For common dielectrics the magnetization is small, so $\mu \approx \mu_0$; and $\epsilon > \epsilon_0$. Then v is less than c, and $n > 1$. Table 13.1 lists values of the index of refraction for some common dielectrics. Much of classical optics is concerned with the effects of the index of refraction. The fact that n is directly related to the electric and magnetic parameters ϵ and μ relates optics to electricity and magnetism.

In applying (13.5) it is important to note that ϵ for real dielectrics may depend on frequency.[1] The ratio ϵ/ϵ_0 must be measured at the frequency of the light whose index of refraction is being calculated. It is too naive to substitute into (13.5) the static dielectric constant $\kappa = \epsilon/\epsilon_0$, which we studied in Chapter 6.

TABLE 13.1   Index of refraction of various materials for light of wavelength 589 nm in vacuum

Air	1.0003
Water	1.33
Crown glass	1.52
Diamond	2.42

[1] A theoretical model for the frequency dependence of ϵ is described in Sec. 13.4.

(However, most dielectrics are approximately nonmagnetic at all frequencies, so in (13.5) we can set $\mu/\mu_0 = 1$.) The frequency dependence $\epsilon = \epsilon(\omega)$ is usually negligible for nonpolar gases (e.g., air or inert gases), nonpolar liquids (e.g., benzene), and nonpolar solids (e.g., diamond). For those substances (13.5) gives a good approximation to the index of refraction of light even if the static value for κ is used. But clearly this formula does not give the correct index of refraction for water, if one compares the value $n = 1.33$ in Table 13.1 with $\sqrt{\kappa} = \sqrt{80} \approx 9$ from Table 6.3. For water, and other polar materials, the ratio $\epsilon(\omega)/\epsilon_0$ is smaller at optical frequencies than for the static case, because the molecular dipoles in the material cannot keep up with the rapidly changing $\mathbf{E}$ field in light, so the net polarization is smaller. In contrast, in a microwave oven, where the frequency (typically 3×10^9 Hz) is much lower than optical frequencies (typically 6×10^{14} Hz) the dipoles do oscillate in phase with the field. It is the thermalization of this motion that cooks the food.

The fields of a polarized plane wave in the dielectric are

$$\mathbf{E}(\mathbf{x}, t) = \mathbf{E}_0 e^{i(\mathbf{k}\cdot\mathbf{x}-\omega t)}$$

$$\mathbf{B}(\mathbf{x}, t) = \mathbf{B}_0 e^{i(\mathbf{k}\cdot\mathbf{x}-\omega t)} \tag{13.6}$$

where the *real part* is understood to be the physical field on the right-hand side of these equations. The fields in (13.6) must satisfy Maxwell's equations in the form (13.1) and (13.2). Applying the divergence equations (13.1) to these fields gives

$$\epsilon \nabla \cdot \mathbf{E} = \epsilon i \mathbf{k} \cdot \mathbf{E}_0 e^{i(\mathbf{k}\cdot\mathbf{x}-\omega t)} = 0$$

$$\nabla \cdot \mathbf{B} = i \mathbf{k} \cdot \mathbf{B}_0 e^{i(\mathbf{k}\cdot\mathbf{x}-\omega t)} = 0 \tag{13.7}$$

which imply that the amplitudes $\mathbf{E}_0$ and $\mathbf{B}_0$ are orthogonal to the wave vector $\mathbf{k}$. Hence the electromagnetic wave is transverse. Applying Faraday's Law (the first equation of (13.2)) to the fields gives

$$i\mathbf{k} \times \mathbf{E}_0 e^{i(\mathbf{k}\cdot\mathbf{x}-\omega t)} = i\omega \mathbf{B}_0 e^{i(\mathbf{k}\cdot\mathbf{x}-\omega t)}, \tag{13.8}$$

which relates the two field amplitudes,

$$\mathbf{B}_0 = \frac{\mathbf{k} \times \mathbf{E}_0}{\omega}. \tag{13.9}$$

Similarly, from Ampère's Law (the second equation of (13.2)),

$$\mathbf{E}_0 = \frac{-\mathbf{k} \times \mathbf{B}_0}{\mu\epsilon\omega}. \tag{13.10}$$

Equations (13.6) to (13.10) show that electromagnetic waves in a dielectric have the same form as in vacuum. In both cases the vectors $\mathbf{E}_0$, $\mathbf{B}_0$, and $\mathbf{k}$ are an orthogonal triad, and the wave propagates in the direction of the wave vector $\mathbf{k}$ with velocity ω/k.

One difference between electromagnetic waves in vacuum and those in a dielectric is the wave speed. In vacuum there is a unique propagation velocity c, and the dispersion relation is $\omega = ck$ for all frequencies. The vacuum phase velocity $v_{\mathrm{phase}} = \omega/k$ and group velocity $v_{\mathrm{group}} = d\omega/dk$ are both equal to c. So in vacuum all electromagnetic waves, whether a monochromatic laser beam or a polychromatic sun beam, travel at speed c. In a real dielectric, however, the propagation velocity v given by (13.3) depends on the frequency because ϵ varies with ω. The dispersion relation has the form $\omega = v(\omega)k$; it follows that v_{phase} and v_{group} are in general different and not equal to c. So, for example, the wave speed of a laser beam in a dielectric depends on the laser frequency; also, a pulse will spread in time because $v_{\mathrm{group}} \neq v_{\mathrm{phase}}$. Or, a sun beam will be *dispersed* as it passes through the material because its constituent frequencies propagate with different velocities. We will discuss phase and group velocities more later.

The *intensity* of a plane wave is defined as the mean energy flux. The Poynting vector (energy flux vector) in a linear material is $\mathbf{S} = \mathbf{E} \times \mathbf{H} = \mathbf{E} \times \mathbf{B}/\mu$. In applying this equation we must use the real parts of $\mathbf{E}$ and $\mathbf{B}$. To calculate $\mathbf{S}$ for the plane wave (13.6), assume for now that $\mathbf{E}_0$, $\mathbf{B}_0$, and $\mathbf{k}$ are all real. In that case we have

$$\mathbf{S} = \frac{\mathbf{E}_0 \times \mathbf{B}_0}{\mu} \cos^2(\mathbf{k} \cdot \mathbf{x} - \omega t) = \mathbf{k}\frac{E_0^2}{\mu\omega}\cos^2(\mathbf{k} \cdot \mathbf{x} - \omega t). \tag{13.11}$$

The second equality follows from the fact that $\mathbf{E}_0$, $\mathbf{B}_0$ and $\mathbf{k}$ form an orthogonal triad, and $B_0 = kE_0/\omega$. The intensity of the plane wave is therefore

$$I = \left\langle \frac{\mathbf{k}}{k} \cdot \mathbf{S} \right\rangle = \frac{1}{2}\epsilon v E_0^2, \tag{13.12}$$

where the factor $\frac{1}{2}$ is the time average of $\cos^2(\mathbf{k} \cdot \mathbf{x} - \omega t)$. We have derived (13.12) for the case of real $\mathbf{E}_0$ and $\mathbf{B}_0$ but the result is more general.[2]

13.2 ■ REFLECTION AND REFRACTION AT A DIELECTRIC INTERFACE

The fact that the speed of light is different in different dielectrics leads to the important optical phenomena of reflection and refraction at a dielectric interface, as, for example, the reflection of the sky in the surface of a lake. The geometrical laws that govern reflection and refraction are a direct consequence of the physical boundary conditions that $\mathbf{E}$ and $\mathbf{B}$ must satisfy at the interface. First we will derive these laws, and later we will determine the reflectivity itself.

Consider the wave solution illustrated in Fig. 13.1. The yz plane is the interface between two dielectrics, with parameters ϵ_1, μ_1 for $x \leq 0$ and ϵ_2, μ_2 for $x > 0$. The x axis is normal to the interface. In the region $x \leq 0$ the wave is the superposition of an incident plane wave with wave vector $\mathbf{k}$ and a reflected plane wave with wave vector $\mathbf{k}''$; in the region $x > 0$ is a transmitted wave with wave

---

[2] See Exercise 3.

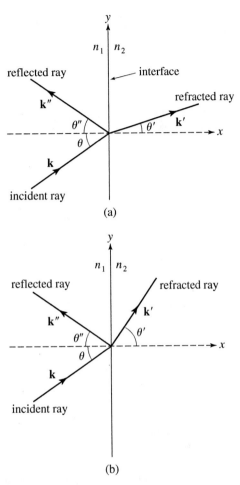

FIGURE 13.1   Wave vectors for reflection and refraction of light at the interface $x = 0$ between two dielectrics. Dielectric 1 is to the left of the interface, where $x < 0$. Dielectric 2 is to the right of the interface, where $x > 0$. The arrows point in the directions of the wave vectors $\mathbf{k}$, $\mathbf{k}'$, and $\mathbf{k}''$. (a) Light incident from an optically less dense dielectric onto an optically denser dielectric, $n_1 < n_2$. (b) Light incident from an optically more dense dielectric onto a less dense dielectric, $n_1 > n_2$.

vector $\mathbf{k}'$. The electric field is

$$\mathbf{E}(\mathbf{x}, t) = \begin{cases} \mathbf{E}_0 e^{i(\mathbf{k}\cdot\mathbf{x} - \omega t)} + \mathbf{E}_0'' e^{i(\mathbf{k}''\cdot\mathbf{x} - \omega'' t)} & \text{for} \quad x \le 0 \\ \mathbf{E}_0' e^{i(\mathbf{k}'\cdot\mathbf{x} - \omega' t)} & \text{for} \quad x > 0. \end{cases} \tag{13.13}$$

The magnetic field can be determined by Faraday's Law, as in (13.9), for each term. Figure 13.1 shows the *rays* of the three waves, which are lines in the directions of the wave vectors $\mathbf{k}$, $\mathbf{k}'$, and $\mathbf{k}''$. In each wave the electric and magnetic fields are everywhere orthogonal to the light ray.

Normally there is neither free surface charge nor free surface current on the interface. In that case the four boundary conditions that must be satisfied are

$$\epsilon_1 E_{1\perp} = \epsilon_2 E_{2\perp} \quad \text{and} \quad B_{1\perp} = B_{2\perp}, \tag{13.14}$$

$$\mathbf{E}_{1\parallel} = \mathbf{E}_{2\parallel} \quad \text{and} \quad \mathbf{B}_{1\parallel}/\mu_1 = \mathbf{B}_{2\parallel}/\mu_2, \tag{13.15}$$

where 1 and 2 indicate the two dielectric regions. The subscripts $\perp$ and $\parallel$ denote the directions perpendicular and parallel to the interface, also called the *normal* and *tangential* directions, respectively.

13.2.1 ■ Wave Vectors

The boundary conditions (13.14) and (13.15) must hold for all t. Therefore the incident, reflected, and transmitted waves must have the same frequency

$$\omega = \omega' = \omega''. \tag{13.16}$$

For a plane wave, ω/k is equal to the wave speed c/n. The index of refraction n is the same for the incident and reflected waves because they are both in dielectric 1, so the equality $\omega'' = \omega$ implies $k'' = k$. But the refracted wave is in dielectric 2, so that $\omega' = \omega$ implies

$$k' = \frac{n_2}{n_1}k. \tag{13.17}$$

If $n_2 > n_1$ (the case illustrated in Fig. 13.1(a)) as for example the case of a wave in air incident on water or glass, the wavelength of the refracted wave $(2\pi/k')$ is less than that of the incident wave $(2\pi/k)$. Although the frequency remains unchanged when a light wave enters an optically denser medium, the wavelength and speed both decrease.

The incident wave in Fig. 13.1 is a plane wave and therefore illuminates the entire interface, i.e., the yz plane ($x = 0$). Similarly the reflected and refracted waves come from the entire yz plane. The boundary conditions (13.14) and (13.15) must hold for all y and z at $x = 0$. Therefore the variation with y or z must be the same for all three waves. In other words, the y and z components of the three wave vectors (incident, reflected, and refracted) must be equal:

$$k_y = k_y' = k_y'', \tag{13.18}$$

$$k_z = k_z' = k_z''. \tag{13.19}$$

Suppose the wave vector of the incident wave has no z component, $k_z = 0$. In that case k_z' and k_z'' are also both zero, according to (13.19). This means that if the incident ray lies in the xy plane then the reflected and refracted rays will also lie in the xy plane. In our idealized solution (13.6) the wave fronts are infinite. In practice, a beam from a laser might be used for demonstrations or laboratory experiments on reflection and refraction. The laser beam is not infinite, but its diameter is *many* wavelengths, so the plane wave approximation (13.6) is valid.

If the incident beam direction is in the xy plane then the reflected and refracted beams are in that same plane, called *the plane of incidence*.

Equation (13.18) will be used to connect the angles of incidence θ, reflection θ', and refraction θ''. The relations are familiar but remarkable. The three angles, shown in Fig. 13.1, are defined as the angles between the respective wave vectors and the *normal* to the interface. If the xy plane is the plane of incidence then the wave vectors are

$$\mathbf{k} = n_1 \frac{\omega}{c} \left(\hat{\mathbf{i}} \cos\theta + \hat{\mathbf{j}} \sin\theta \right), \tag{13.20}$$

$$\mathbf{k}' = n_2 \frac{\omega}{c} \left(\hat{\mathbf{i}} \cos\theta' + \hat{\mathbf{j}} \sin\theta' \right), \tag{13.21}$$

$$\mathbf{k}'' = n_1 \frac{\omega}{c} \left(-\hat{\mathbf{i}} \cos\theta'' + \hat{\mathbf{j}} \sin\theta'' \right). \tag{13.22}$$

The conditions (13.18) then imply

$$\sin\theta = \sin\theta'', \tag{13.23}$$

$$n_1 \sin\theta = n_2 \sin\theta'. \tag{13.24}$$

Equation (13.23) is the *law of reflection* for specular reflection; in words, the angle of incidence equals the angle of reflection. This behavior of light was known to the ancient Greeks. Equation (13.24) is the *law of refraction*, or Snell's law.[3]

Total Internal Reflection

Figure 13.1(a) illustrates reflection and refraction for the case $n_2 > n_1$. In that case, (13.24) implies $\theta' < \theta$, which means that the incident beam is refracted toward the normal as shown in the figure. Figure 13.1(b) illustrates the case for $n_1 > n_2$, so that $\theta' > \theta$, which means that the beam is refracted away from the normal. A critical case occurs in Fig. 13.1(b) when $\theta' = \pi/2$. For that special case the angle of incidence is called the *critical angle* θ_c, and (13.24) becomes $n_1 \sin\theta_c = n_2$, so that

$$\theta_c = \arcsin\left(\frac{n_2}{n_1}\right). \tag{13.25}$$

If light is incident at an angle equal to or greater than the critical angle, $\theta \geq \theta_c$, then the incident beam is totally reflected in the sense that all the energy in the incident beam goes into the reflected beam. This phenomenon is called *total internal reflection*. For reflection from a glass-air interface the critical angle is $\theta_c = \arcsin(1.0003/1.52) = 41°$. Total internal reflection can only occur for light incident from an optically dense material to a less dense material, i.e., from a negative change of n.

---

[3] In France, Snell's law has been called the Snell–Descartes law. Snell discovered the relation of sines in 1621 but did not publish it. Descartes rediscovered the relation and published it in a book on optics in 1637.

It is interesting to ask, for the case of total internal reflection, what happens to the transmitted wave in (13.13). For any angle of incidence, the transmitted wave can be written, by inserting (13.21) into (13.13), as

$$\mathbf{E}'(\mathbf{x}, t) = \mathbf{E}'_0 \exp \left\{ i \left(n_2 \frac{\omega}{c} \cos \theta' x + n_2 \frac{\omega}{c} \sin \theta' y - \omega t \right) \right\}. \qquad (13.26)$$

In writing (13.26) we've taken the plane of incidence to be the xy plane as before, and set $\omega' = \omega$ by (13.16). Equation (13.26) is *formally* correct for any angle of incidence. By Snell's law, the factor $n_2 \sin \theta'$ in the middle term in the parentheses can be replaced by $n_1 \sin \theta$. But if the angle of incidence is greater than θ_c, then $\cos \theta'$ is purely imaginary,

$$\cos \theta' = \sqrt{1 - \sin^2 \theta'} = i\xi, \qquad (13.27)$$

where

$$\xi = \sqrt{\frac{n_1^2}{n_2^2} \sin^2 \theta - 1} = \sqrt{\frac{\sin^2 \theta}{\sin^2 \theta_c} - 1}. \qquad (13.28)$$

Thus the electric vector of the transmitted wave for the supercritical case $\theta > \theta_c$ is

$$\mathbf{E}'(\mathbf{x}, t) = \mathbf{E}'_0 \exp \left(-n_2 \frac{\omega}{c} \xi x \right) \exp \left\{ i \left(n_1 \frac{\omega}{c} \sin \theta y - \omega t \right) \right\}. \qquad (13.29)$$

The exponential in y and t tells us that the "transmitted" wave travels parallel to the interface, i.e., in the y direction, with phase velocity $v_p = c/(n_1 \sin \theta)$. But the exponential in x shows that as this wave penetrates into dielectric 2 (the less dense medium) its amplitude falls off exponentially with distance x. The attenuation length δ, given by

$$\delta = \frac{c}{n_2 \omega \xi}, \qquad (13.30)$$

is the distance in x over which the amplitude of $\mathbf{E}'$ decreases by the factor $1/e$.

For $\theta < \theta_c$ there is nothing unusual. In that case we would observe, in addition to the reflected wave in medium 1, a transmitted wave in medium 2 bent away from the normal, i.e., toward the surface. But now let θ increase, tending to θ_c. As θ approaches θ_c, the direction of the transmitted wave approaches the surface, and when $\theta = \theta_c$ it is parallel to the surface. For $\theta > \theta_c$ the wave in medium 2 propagates parallel to the surface but it does not extend far into medium 2 because its amplitude is attenuated exponentially, as we see from (13.29). Because the fields decrease with distance x into dielectric 2 (the less dense dielectric), the electric and magnetic fields in dielectric 2 are called an *evanescent wave*.

Associated with the electric field (13.29) is a magnetic field $\mathbf{B}(\mathbf{x}, t)$, given by (13.9). We then ask the natural question: What energy is associated with the evanescent wave? The answer is surprising. There is no net energy flow *across the boundary* because the time averaged Poynting vector in the x direction is zero. But there is a finite flow in the y direction. This flow of energy parallel to the boundary is consistent with total internal reflection because in this ideal case the incident and reflected waves are plane waves with infinite wave fronts.[4]

If another refractive medium with index n_1 is brought very close to the interface, say to within $x \le \delta$, then a transmitted wave will appear in it. This effect is usually offered as support for the existence of the evanescent wave. But the argument is somewhat indirect because the presence of the new medium changes the boundary conditions by introducing a reflected wave at the new interface.

Near-field Scanning Optical Microscopy (NSOM) is a technique that uses evanescent waves to obtain optical images of surface details much smaller than the wavelength of light. Ordinary optical microscopy using lenses is limited, because of diffraction, to resolving distances $\ge \lambda/2$, where λ is the wavelength of the incident light. In NSOM the evanescent waves are produced by light emitted from a subwavelength aperture, of diameter, say, $\lambda/5 \approx 100\,\mathrm{nm}$ or smaller, at the tip of a tapered optical fiber. There are evanescent waves in the neighborhood outside the tip. These waves are exponentially attenuated with distance, but if the probe is scanned over a surface very close to the tip, say within $\lambda/20 \approx 25\,\mathrm{nm}$, then light is reflected from the scanned surface. Analysis of the reflected light reveals surface details at resolutions less than $100\,\mathrm{nm}$. Among the advantages of NSOM is that it can be carried out on samples in air or water, which makes it useful for studying biological materials.

Total internal reflection has many applications. At the glass-air interfaces of the prisms in binoculars, total internal reflection makes erect the inverted image formed by the objective lenses. Total internal reflection is also the basis for efficiently guiding light along optical fibers. Figure 13.2 is an illustration of an optical fiber; the light internally reflects at points along the glass-air interface of the fiber

FIGURE 13.2   Illustration of an optical fiber. Light undergoes total internal reflection at the glass-air interfaces. There is a small field for a short distance outside the fiber—the evanescent wave.

[4]See Ref. [1].

surface. Optical fibers of this kind are widely used in communication, where they make possible transmission of information at optical frequencies over long distances. In medicine they are used to view inaccessible tissues and, with lasers, to deliver heat energy there.

13.2.2 ■ Reflectivity for Normal Incidence

A basic problem in the electromagnetic theory of optics is to calculate the reflectivity, i.e., reflected intensity, at the interface between two dielectrics, in terms of the electric and magnetic parameters ϵ_1, μ_1 and ϵ_2, μ_2. We ask now: If light is incident normally on the interface, how much of the energy is reflected and how much is transmitted? This is the special case with $\theta = 0$, and therefore $\theta' = \theta'' = 0$ in Figs. 13.1(a) or (b). In this case then, $\mathbf{k}$ is directed normally toward the interface, and both $\mathbf{k}'$ and $\mathbf{k}''$ are directed normally away from the interface.

As we know, in an electromagnetic wave $\mathbf{E}$ and $\mathbf{B}$ are perpendicular to the propagation direction. For normal incidence, therefore, $\mathbf{E}$ and $\mathbf{B}$ for the incident, reflected, and transmitted waves, are all parallel to the dielectric interface. That will make it simple to solve the boundary conditions (13.14) and (13.15). First, the conditions (13.14), on the perpendicular components, are satisfied trivially because all the perpendicular field components are zero. The tangential conditions (13.15) would be the same for any polarization, but to be definite let the polarization direction be $\hat{\mathbf{j}}$. Then $\mathbf{E}_0 = E_0 \hat{\mathbf{j}}$, $\mathbf{E}_0' = E_0' \hat{\mathbf{j}}$, and $\mathbf{E}_0'' = E_0'' \hat{\mathbf{j}}$. Also, by Faraday's Law (13.9)[5]

$$\mathbf{H}_0 = \frac{E_0}{\mu_1 v_1}\hat{\mathbf{z}}, \quad \mathbf{H}_0' = \frac{E_0'}{\mu_2 v_2}\hat{\mathbf{z}} \quad \text{and} \quad \mathbf{H}_0'' = \frac{-E_0''}{\mu_1 v_1}\hat{\mathbf{z}}. \tag{13.31}$$

Thus the tangential continuity conditions (13.15) are

$$E_0 + E_0'' = E_0', \tag{13.32}$$

$$\frac{E_0 - E_0''}{\mu_1 v_1} = \frac{E_0'}{\mu_2 v_2}. \tag{13.33}$$

Here $v_1 = c/n_1$ and $v_2 = c/n_2$ are the wave speeds in the two dielectrics. Also, $\mathbf{k}'' = -\mathbf{k}$ for normal incidence, leading to the minus signs in (13.31) and (13.33). From these boundary conditions we solve for the electric field amplitudes of the transmitted and reflected waves. The results are

$$E_0' = \frac{2\mu_2 n_1}{\mu_2 n_1 + \mu_1 n_2} E_0, \tag{13.34}$$

$$E_0'' = \frac{\mu_2 n_1 - \mu_1 n_2}{\mu_2 n_1 + \mu_1 n_2} E_0. \tag{13.35}$$

---

[5]In this chapter we will use the notation $\hat{\mathbf{z}}$ for the unit vector in the z direction, rather than $\hat{\mathbf{k}}$ which might be confused with the wave vectors.

Phase Change on Reflection

In most dielectrics we may approximate $\mu = \mu_0$. Then if $n_1 < n_2$, as in the case of light incident from air into water or glass, E_0'' and E_0 have opposite signs because the fraction in (13.35) is negative. That is, the electric field in the reflected wave is 180 degrees out of phase with that in the incident wave. (But the magnetic fields of the reflected and incident waves are in phase.) On the other hand, if $n_1 > n_2$ there is no phase change on reflection. In either case the transmitted wave is in phase with the incident wave.

Reflectivity

We tend to think of glass or water as transparent to light, but in fact these dielectrics do reflect some of the incident light. We define the *reflectivity*, or reflection coefficient, by the ratio of reflected and incident intensities,

$$R = \frac{I''}{I}. \tag{13.36}$$

The wave intensity I is given by (13.12). By (13.35) the reflectivity for normal incidence is

$$R = \frac{E_0''^2}{E_0^2} = \left(\frac{\mu_2 n_1 - \mu_1 n_2}{\mu_2 n_1 + \mu_1 n_2}\right)^2. \tag{13.37}$$

Similarly, the transmissivity, or transmission coefficient, T is defined as I'/I; for normal incidence, by (13.34),

$$T = \frac{\epsilon_2 v_2 E_0'^2}{\epsilon_1 v_1 E_0^2} = \frac{4\mu_1 \mu_2 n_1 n_2}{(\mu_2 n_1 + \mu_1 n_2)^2}. \tag{13.38}$$

Note that $R + T = 1$, which is a consequence of energy conservation: The sum of the energy fluxes in the reflected and transmitted waves is equal to the incident energy flux.

For light incident normally from air ($n_1 = 1$) into glass ($n_2 = 1.5$) the reflectivity is $R = 0.04$; that is, 4% of the incident light intensity is reflected. For light incident from air into water ($n_2 = 1.33$), $R = 0.02$; from air into diamond ($n_2 = 2.42$), $R = 0.17$. We approximate μ by μ_0 for these materials.

The expressions (13.37) and (13.38) are unchanged if the subscripts 1 and 2 are interchanged. This means that the reflectivity for light incident from dielectric 2 into dielectric 1 is just the same as the reflectivity for light incident from dielectric 1 into dielectric 2. The same is true for the transmissivity.

Nonreflecting Lens Coatings

An important application of wave optics is to design a lens-air interface that transmits 100% of the light incident on it. As we will now see, this can be achieved by

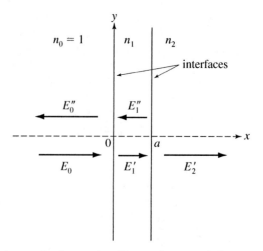

FIGURE 13.3   A nonreflective coating. The coating extends from $x = 0$ to $x = a$ and has index of refraction n_1. The directions of propagation of the plane waves are indicated by arrows. The electric vectors are all taken to point in the $\hat{\mathbf{j}}$ direction.

coating the lens with a layer of dielectric having just the right thickness and index of refraction.

Figure 13.3 shows the geometry. Monochromatic light is incident normally from air ($n_0 = 1$) onto a dielectric layer of thickness a (index of refraction n_1) that is on the surface of a glass lens (n_2).

In the air there is an incident wave propagating in the $+\hat{\mathbf{i}}$ direction and, in general, a reflected wave in the $-\hat{\mathbf{i}}$ direction. In the layer there are waves in both the $+\hat{\mathbf{i}}$ and $-\hat{\mathbf{i}}$ directions. In the glass there is only a transmitted wave in the $+\hat{\mathbf{i}}$ direction. If we assume $\mu = \mu_0$ and take the direction of polarization of the waves to be $\hat{\mathbf{j}}$, then the fields in the three regions are

$$
\mathbf{E}(\mathbf{x}, t) = \begin{cases}
E_0 \hat{\mathbf{j}} e^{i(kx - \omega t)} + E_0'' \hat{\mathbf{j}} e^{-i(kx + \omega t)} & \text{for} \quad x \leq 0 \\
E_1' \hat{\mathbf{j}} e^{i(k_1 x - \omega t)} + E_1'' \hat{\mathbf{j}} e^{-i(k_1 x + \omega t)} & \text{for} \quad 0 \leq x \leq a \\
E_2' \hat{\mathbf{j}} e^{i(k_2 x - \omega t)} & \text{for} \quad x \geq a
\end{cases}
\qquad (13.39)
$$

and

$$
c\mathbf{B}(\mathbf{x}, t) = \begin{cases}
E_0 \hat{\mathbf{z}} e^{i(kx - \omega t)} - E_0'' \hat{\mathbf{z}} e^{-i(kx + \omega t)} & \text{for} \quad x \leq 0 \\
n_1 E_1' \hat{\mathbf{z}} e^{i(k_1 x - \omega t)} - n_1 E_1'' \hat{\mathbf{z}} e^{-i(k_1 x + \omega t)} & \text{for} \quad 0 \leq x \leq a \\
n_2 E_2' \hat{\mathbf{z}} e^{i(k_2 x - \omega t)} & \text{for} \quad x \geq a.
\end{cases}
\qquad (13.40)
$$

The wave numbers are $k = \omega/c$, $k_1 = n_1 k$, and $k_2 = n_2 k$. The waves that propagate in the $-x$ direction are written as $e^{-i(kx + \omega t)}$ and $e^{-i(k_1 x + \omega t)}$; then all terms have the same time dependence $e^{-i\omega t}$. Also, we have used the transversality of $\mathbf{E}$ and $\mathbf{B}$ in a plane wave.

The boundary conditions (13.15) at the interface $x = 0$ between air and dielectric are

$$E_0 + E_0'' = E_1' + E_1'' \tag{13.41}$$

$$E_0 - E_0'' = n_1 \left(E_1' - E_1'' \right), \tag{13.42}$$

and those at $x = a$ between dielectric and glass are

$$E_1' e^{ik_1 a} + E_1'' e^{-ik_1 a} = E_2' e^{ik_2 a} \tag{13.43}$$

$$n_1 \left(E_1' e^{ik_1 a} - E_1'' e^{-ik_1 a} \right) = n_2 E_2' e^{ik_2 a}. \tag{13.44}$$

From these equations the four amplitudes E_0'', E_1', E_1'' and E_2' can be found in terms of the incident wave amplitude E_0 for any n_1, n_2 and a.

Zero reflectivity.   We seek now a solution to the boundary conditions for which there is no reflected wave, i.e., with $E_0'' = 0$. To be more precise, we ask: What are the requirements on n_1 and a such that $E_0'' = 0$? The algebra is straightforward. Set $E_0'' = 0$ in (13.41) and (13.42), and solve the resulting equations for E_1' and E_1''. Substitute these into (13.43) and (13.44), and form the ratio of the resulting equations. After some simplification the result is

$$\frac{n_2}{n_1} = \frac{(n_1 + 1)e^{ik_1 a} - (n_1 - 1)e^{-ik_1 a}}{(n_1 + 1)e^{ik_1 a} + (n_1 - 1)e^{-ik_1 a}} = \frac{n_1 i \sin k_1 a + \cos k_1 a}{n_1 \cos k_1 a + i \sin k_1 a}. \tag{13.45}$$

Now, the right-hand side of (13.45) must be real (because n_1 and n_2 are real) and there are only two possibilities; either $\cos k_1 a = 0$ or $\sin k_1 a = 0$. It is the first case that interests us now. For $\cos k_1 a = 0$ the possible values of a are $a = \lambda_1/4, 3\lambda_1/4, \ldots, (2j + 1)\lambda_1/4, \ldots$, where λ_1 is the wavelength in the dielectric coating. The thinnest coating has $a = \lambda_1/4$. Then by (13.45) the ratio n_2/n_1 must be n_1, so the index of refraction of the dielectric coating must be $n_1 = \sqrt{n_2}$. We have done the calculation for $n_0 = 1$, but it is easy to generalize to $n_0 \neq 1$. In that case the conditions for zero reflectivity are again that the layer thickness must be $a = (2j + 1)\lambda_1/4$, and that the index of refraction of the layer must be $n_1 = \sqrt{n_0 n_2}$. The ideal index of refraction is the geometric mean of the indices on either side of the coating.

Camera lenses are often coated with a layer of magnesium fluoride MgF_2, for which $n_1 = 1.38$, therefore approximately satisfying the condition $n_1 = \sqrt{n_2} = \sqrt{1.5}$. The thickness of the layer is designed to minimize reflection in the middle of the range of optical wavelengths. In cameras the purpose of nonreflective coatings is to maximize the transmitted light.

Stealth aircraft technology also uses the technology of nonreflective coatings. In this case the purpose is to minimize the reflection of radar waves in order to render the aircraft invisible to radar. Because the incident radar may have a variety of frequencies, multiple coatings are used.

Another way for the right-hand side of (13.45) to be real is if $\sin k_1 a = 0$. This implies fulfillment of the two conditions $a = j\lambda_1/2$ and $n_2 = 1$ (or, more generally, $n_2 = n_0$). We picture, for example, two layers of air separated by a small thickness a of glass, or two layers of glass separated by a small thickness a of air. The dark bands observed for monochromatic light in Newton's rings, or from light scattered by a thin wedge of air between two optically flat pieces of glass, are examples of zero reflectivity at locations where the air gap has thickness $j\lambda/2$. In such cases the zero reflectivity is properly understood as the solution of the *complete* boundary value problem. It is not simply the result of interference between waves reflected from different surfaces, because while the reflected waves have phase difference of π they have different amplitudes so that the destructive interference between them is only partial. To understand the complete cancellation of the waves requires the full solution of the boundary value problem.

So far we have limited the discussion to light that is normally incident on the surface, i.e., $\theta = 0$. We next consider light incident at an arbitrary angle.

13.2.3 ■ Reflection for Incidence at Arbitrary Angles: Fresnel's Equations

For light incident at an arbitrary angle θ, the intensities of the transmitted and reflected waves depend on the direction of polarization of the wave. Therefore we will consider separately the reflectivity for TE (transverse electric) and TM (transverse magnetic) polarization.

TE Polarization
The definition of TE polarization is that the electric field is *perpendicular to the plane of incidence*. The plane of incidence is the plane spanned by the normal to the surface and the incident wave vector **k**. In Fig. 13.4 the plane of incidence is the xy plane. The TE electric fields of the three waves—incident, transmitted, and reflected—are

$$\mathbf{E} = \hat{\mathbf{z}} E_0 e^{i(\mathbf{k} \cdot \mathbf{x} - \omega t)}, \tag{13.46}$$

$$\mathbf{E}' = \hat{\mathbf{z}} E_0' e^{i(\mathbf{k}' \cdot \mathbf{x} - \omega t)}, \tag{13.47}$$

$$\mathbf{E}'' = \hat{\mathbf{z}} E_0'' e^{i(\mathbf{k}'' \cdot \mathbf{x} - \omega t)}. \tag{13.48}$$

The magnitudes of the wave vectors are $k = k'' = \omega/v_1$ and $k' = \omega/v_2$. In the TE case the electric fields are all parallel to the boundary, so the boundary condition $D_{1\perp} = D_{2\perp}$ is trivially satisfied. The boundary condition $\mathbf{E}_{1\parallel} = \mathbf{E}_{2\parallel}$ implies

$$E_0 + E_0'' = E_0'. \tag{13.49}$$

In Fig. 13.4 the three electric fields are drawn pointing in the $+z$ direction. The choice of direction $+z$ or $-z$ is arbitrary, because the electric vectors oscillate in both space and time. Also, there may be a phase change on reflection, so that E_0'' has the opposite sign as E_0. In any case, it is necessary to use (13.9), i.e., $\mathbf{B} = \mu\mathbf{H} = \mathbf{k} \times \mathbf{E}/\omega$, to find the magnetic field directions. From Fig. 13.4,

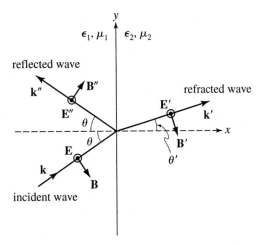

FIGURE 13.4   Electric fields, magnetic fields, and wave vectors for light incident on the interface between two dielectrics, and polarized perpendicular to the plane of incidence (TE polarization). The xy plane is the plane of incidence and the plane $x = 0$ is the interface, so that the **E** vectors are parallel to the interface. For each wave the vectors **E**, **B**, and **k** form an orthogonal triad that satisfies $\mathbf{B} = \mathbf{k} \times \mathbf{E}/\omega$.

which shows the directions of **E** and **H** for each of the three waves, we find the amplitudes of the magnetic fields to be

$$\mathbf{H}_0 = \frac{E_0}{\mu_1 v_1} \left[\hat{\mathbf{i}} \sin\theta - \hat{\mathbf{j}} \cos\theta \right] \tag{13.50}$$

$$\mathbf{H}_0' = \frac{E_0'}{\mu_2 v_2} \left[\hat{\mathbf{i}} \sin\theta' - \hat{\mathbf{j}} \cos\theta' \right] \tag{13.51}$$

$$\mathbf{H}_0'' = \frac{E_0''}{\mu_1 v_1} \left[\hat{\mathbf{i}} \sin\theta + \hat{\mathbf{j}} \cos\theta \right], \tag{13.52}$$

where $v_1 = c/n_1$ and $v_2 = c/n_2$. From (13.46) to (13.48) and (13.50) to (13.52) it is straightforward to verify that $\mathbf{E} \times \mathbf{H}$ is in the direction of the corresponding wave vector for each wave.

The $\hat{\mathbf{i}}$ component of **B** is normal to the interface; the boundary condition $B_{1x} = B_{2x}$ is the same as (13.49) by Snell's law. The $\hat{\mathbf{j}}$ component of **H** is parallel to the interface, so the boundary condition $\mathbf{H}_{1\parallel} = \mathbf{H}_{2\parallel}$ implies $H_{0y} + H_{0y}'' = H_{0y}'$, i.e.,

$$\frac{n_1 \left(E_0 - E_0'' \right)}{\mu_1} \cos\theta = \frac{n_2 E_0'}{\mu_2} \cos\theta'. \tag{13.53}$$

Equations (13.49) and (13.53) make two equations for the two unknowns E_0' and E_0''. The solutions for these amplitudes are

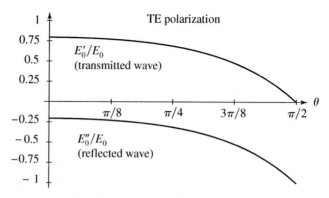

FIGURE 13.5   Graphs of E_0''/E_0, the ratio of amplitudes of the reflected to the incident wave, and E_0'/E_0, the ratio of amplitudes of the transmitted to the incident wave, for the case of TE polarization. The abscissa is the angle of incidence θ. The light is taken to be incident on the interface from air ($n_1 = 1$) to glass ($n_2 = 1.5$).

$$E_0' = \frac{2\mu_2 n_1 \cos\theta}{\mu_2 n_1 \cos\theta + \mu_1 n_2 \cos\theta'} E_0, \tag{13.54}$$

$$E_0'' = \frac{\mu_2 n_1 \cos\theta - \mu_1 n_2 \cos\theta'}{\mu_2 n_1 \cos\theta + \mu_1 n_2 \cos\theta'} E_0. \tag{13.55}$$

For most dielectrics we may approximate $\mu_1 = \mu_2 = \mu_0$; then the second result may be simpified, using Snell's law, to

$$E_0'' = \frac{\sin(\theta' - \theta)}{\sin(\theta' + \theta)} E_0, \tag{13.56}$$

which is *Fresnel's equation* for the amplitude of the reflected wave with TE polarization.[6]

Figure 13.5 shows plots of E_0''/E_0 and E_0'/E_0 as functions of the angle of incidence θ for TE waves incident from air into glass. The sign of E_0''/E_0 is always negative, so the reflected wave is 180° out of phase with the incident wave.

TM Polarization

The definition of TM polarization is that the magnetic field is perpendicular to the plane of incidence. The magnetic fields in terms of $\mathbf{H} = \mathbf{B}/\mu$ are

$$\mathbf{H} = \hat{\mathbf{z}} \frac{E_0}{\mu_1 v_1} e^{i(\mathbf{k}\cdot\mathbf{x} - \omega t)}, \tag{13.57}$$

$$\mathbf{H}' = \hat{\mathbf{z}} \frac{E_0'}{\mu_2 v_2} e^{i(\mathbf{k}'\cdot\mathbf{x} - \omega t)}, \tag{13.58}$$

[6] See Exercise 6.

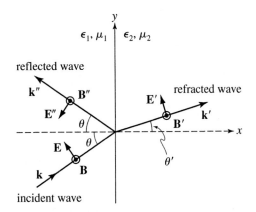

FIGURE 13.6   Electric fields, magnetic fields, and wave vectors for light incident on the interface between two dielectrics and polarized parallel to the plane of incidence (TM polarization). The xy plane is the plane of incidence and the plane $x = 0$ is the interface, so that the **B** vectors are parallel to the interface.

$$\mathbf{H}'' = \hat{\mathbf{z}}\frac{E_0''}{\mu_1 v_1}e^{i(\mathbf{k}''\cdot\mathbf{x}-\omega t)}. \tag{13.59}$$

Figure 13.6 shows the relations among **E**, **H**, and **k**, where the **H** fields have been shown pointing in the $+\hat{\mathbf{z}}$ direction. The boundary condition $\mathbf{H}_{1\parallel} = \mathbf{H}_{2\parallel}$ implies

$$\frac{n_1}{\mu_1}\left(E_0 + E_0''\right) = \frac{n_2}{\mu_2}E_0'. \tag{13.60}$$

The electric field amplitudes for TM polarization are

$$\mathbf{E}_0 = E_0\left[-\hat{\mathbf{i}}\,\sin\theta + \hat{\mathbf{j}}\,\cos\theta\right] \tag{13.61}$$

$$\mathbf{E}_0' = E_0'\left[-\hat{\mathbf{i}}\,\sin\theta' + \hat{\mathbf{j}}\,\cos\theta'\right] \tag{13.62}$$

$$\mathbf{E}_0'' = E_0''\left[-\hat{\mathbf{i}}\,\sin\theta - \hat{\mathbf{j}}\,\cos\theta\right]. \tag{13.63}$$

Referring to Fig.13.6 the $\hat{\mathbf{j}}$ component of **E** is parallel to the surface, so the condition $\mathbf{E}_{1\parallel} = \mathbf{E}_{2\parallel}$ implies

$$\left(E_0 - E_0''\right)\cos\theta = E_0'\cos\theta'. \tag{13.64}$$

Equations (13.60) and (13.64) can be solved for the reflected and refracted amplitudes. The results are

$$E_0' = \frac{2\mu_2 n_1\cos\theta}{\mu_2 n_1\cos\theta' + \mu_1 n_2\cos\theta}E_0 \tag{13.65}$$

$$E_0'' = \frac{\mu_1 n_2 \cos\theta - \mu_2 n_1 \cos\theta'}{\mu_2 n_1 \cos\theta' + \mu_1 n_2 \cos\theta} E_0. \tag{13.66}$$

Again, if $\mu_1 = \mu_2 = \mu_0$ then the second result may be simplified by the use of Snell's law; after some algebra (see Exercise 6)

$$E_0'' = \frac{\tan(\theta - \theta')}{\tan(\theta + \theta')} E_0, \tag{13.67}$$

which is Fresnel's equation for TM polarization.

Figure 13.7 shows plots of E_0''/E_0 and E_0'/E_0 as functions of the angle of incidence for TM waves incident from air ($n_1 = 1$) into glass ($n_2 = 1.5$). At grazing incidence ($\theta = 90°$) all the light is reflected. For those angles for which $E_0''/E_0 > 0$, the reflected wave is $180°$ out of phase with the incident wave, but when $E_0''/E_0 < 0$ the reflected and incident waves are in phase.[7]

In solving this problem we used the two boundary conditions expressed in (13.15), namely, the continuity of $\mathbf{H}_\parallel$ and of $\mathbf{E}_\parallel$. However, four boundary conditions, including those of (13.14), must be satisfied at the interfacial plane $x = 0$. What became of the conditions (13.14) on the normal components? The answer is that the continuity of $B_\perp$ is automatically satisfied for TM polarization because $B_\perp$ is zero for all three waves. The continuity of $\epsilon E_\perp$ is equivalent to (13.60) by Snell's law.

It is interesting to carry the analysis of TM polarization further by calculating the reflectivity R and transmissivity T. The reflectivity is again defined as the

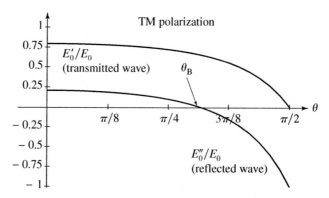

FIGURE 13.7 Graphs of E_0''/E_0, the ratio of amplitudes of the reflected to the incident wave, and E_0'/E_0, the ratio of amplitudes of the transmitted to the incident wave, for the case of TM polarization. The abscissa is the angle of incidence θ. The light is taken to be incident on the interface from air ($n_1 = 1$) to glass ($n_2 = 1.5$).

[7] The sign conventions for E_0'' of TE and TM waves, defined respectively in Figs. 13.4 and 13.6, are different.

ratio of reflected intensity to incident intensity,[8]

$$R = \frac{\epsilon_1 v_1 E_0''^2/2}{\epsilon_1 v_1 E_0^2/2} = \frac{(n_2 \cos\theta - n_1 \cos\theta')^2}{(n_1 \cos\theta' + n_2 \cos\theta)^2}. \tag{13.68}$$

Similarly, the transmissivity is the ratio of transmitted and incident intensities,

$$T = \frac{\epsilon_2 v_2 E_0'^2/2}{\epsilon_1 v_1 E_0^2/2} = \frac{4 n_1 n_2 \cos^2\theta}{(n_1 \cos\theta' + n_2 \cos\theta)^2}. \tag{13.69}$$

The conservation of energy for TM polarization (and also for TE polarization) is expressed by the equation

$$\frac{\cos\theta'}{\cos\theta} T + R = 1, \tag{13.70}$$

and it is straightforward to verify that this equation is satisfied.[9]

In writing (13.70) it is necessary to multiply T by the factor $(\cos\theta'/\cos\theta)$, which is the ratio of the cross-sectional area of the transmitted beam to that of the incident beam. The reason for this factor is that T is the ratio of the transmitted *intensity* to the incident *intensity*. Intensity is power per unit area, with units W/m². It is not intensity, but integrated power, that is conserved. The transmitted beam has a different cross-sectional area from the incident beam. Figure 13.8 explains

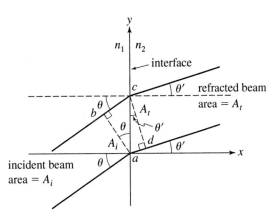

FIGURE 13.8   Areas of the incident and transmitted beams. If the incident beam has cross-sectional area A_i, shown at ab, then the area illuminated on the $x = 0$ dielectric interface is $A_d = A_i / \cos\theta$, shown at ac. The cross-sectional area of the transmitted beam is $A_t = A_i (\cos\theta'/\cos\theta)$, shown at dc; note that the area of the transmitted beam is greater than the area of the incident beam. The angle of incidence is θ, the angle of refraction is θ', and $n_1 \sin\theta = n_2 \sin\theta'$.

[8] We have set $\mu_1 = \mu_2 = \mu_0$.
[9] See Exercise 7.

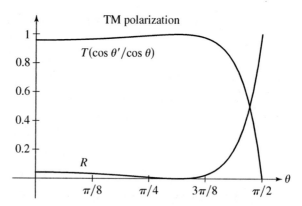

FIGURE 13.9   The reflectivity R and transmissivity T plotted as functions of incident angle θ for TM polarization. The transmissivity has been scaled by the factor $(\cos\theta'/\cos\theta)$ as explained in the text. The light is taken to be incident on the interface from air ($n_1 = 1$) to glass ($n_2 = 1.5$).

the geometry. If the cross-sectional area of the incident beam is A_i then the area that it illuminates on the dielectric interface is $A_d = A_i/\cos\theta$. The cross-sectional area of the transmitted beam is $A_t = A_d \cos\theta' = A_i(\cos\theta'/\cos\theta)$. If the transmitted beam is refracted toward the normal, as in Fig. 13.8, then $A_t > A_i$; or, if the transmitted beam is refracted away from the normal, $A_t < A_i$. In any case the fact that the areas are different must be factored into the comparison of the integrated power. It is not necessary to include a similar factor for the reflected wave, because the areas of the reflected and incident beams are equal. In the case of normal incidence, $\theta = \theta' = 0$, the areas of all three beams are the same.

The transmissivity and reflectivity for TM polarization are plotted in Fig. 13.9.

Brewster's Angle

For TM polarization, the intensity of the reflected wave is 0 if $\theta + \theta' = \pi/2$, because then the denominator in (13.67) is infinite. (We assume $\mu_1 = \mu_2$.) The angle of incidence at which this condition occurs is called Brewster's angle θ_B. By Snell's law, with $\theta = \theta_B$ and $\theta' = \pi/2 - \theta_B$, $n_1 \sin\theta_B$ is equal to $n_2 \cos\theta_B$; therefore,

$$\tan\theta_B = \frac{n_2}{n_1}. \tag{13.71}$$

If the angle of incidence of a TM wave is θ_B then there is no reflected wave. The reflectivity tends to 0 as θ approaches θ_B.

The condition (13.71) for Brewster's angle, together with Snell's Law, has a simple geometrical interpretation: For a TM wave with $\theta = \theta_B$, the vectors $\mathbf{k}''$ and $\mathbf{k}'$ are orthogonal, because $\theta_B + \theta' = \pi/2$. (Note from Fig. 13.6 that the angle between $\mathbf{k}'$ and $\mathbf{k}''$ is $\pi - \theta - \theta'$, which is $\pi/2$ if $\theta = \theta_B$.) Or, because $\mathbf{E}_0'$ is also perpendicular to $\mathbf{k}'$, at Brewster's angle the polarization of the transmitted wave

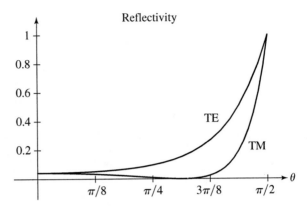

FIGURE 13.10   Reflectivities versus angle of incidence for TE and TM polarization for light incident on the interface from air ($n_1 = 1$) to glass ($n_2 = 1.5$).

($\mathbf{E}_0'$) is parallel to the direction of specular reflection ($\mathbf{k}''$). When this occurs the intensity of the reflected wave is 0.

If *unpolarized* light is incident on a dielectric surface at angle θ_B then the reflected light is TE polarized because the amplitude of the TM reflected component is 0, as we have seen. The reflected light is 100% polarized at $\theta = \theta_B$, and it is partially polarized at other angles. Figure 13.10 shows the reflectivities for the two polarizations, as functions of the angle of incidence, for light incident from air ($n_1 = 1$) into glass ($n_2 = 1.5$). In this case Brewster's angle is 56°. The difference between the two curves is a measure of the degree of polarization of reflected unpolarized light. This graph shows why polarized sunglasses reduce glare: When unpolarized light reflects from a dielectric surface, the reflected light is highly polarized. If the sunglasses' polarization axis is orthogonal to the polarization of the reflected light, then the intensity of reflected light that will pass through the sunglasses is low. For instance, sunglasses for people who fish are designed to absorb horizontally polarized light, so as to reduce glare from sunlight reflected from the water surface.

Grazing Incidence
It is interesting to note from Fig. 13.10 that $R \rightarrow 1$ as $\theta \rightarrow \pi/2$, for either polarization. That is, at grazing incidence the light is 100% reflected. This agrees with our everyday experience of light hitting a glass surface at grazing incidence.

13.3 ■ ELECTROMAGNETIC WAVES IN A CONDUCTOR

In this chapter until now we've considered how electromagnetic waves interact with dielectrics. The dominant underlying physical process is the interaction between $\mathbf{E}$ of the electromagnetic wave and electrons bound in atoms. Next we consider how electromagnetic waves interact with conductors, materials with free electrons. The dominant interaction for this case is between $\mathbf{E}$ and the free electrons. This interaction has many interesting consequences. For example, metals are shiny—that's why a silvered mirror reflects light so well. Also, light propagating in a metal is rapidly attenuated—that's why metals are opaque. The speed of propagation of electromagnetic waves in metals is much smaller than in vacuum or dielectrics. The magnetic field component lags behind the electric field for an electromagnetic wave in a conductor, and much more of the energy resides in the magnetic field than in the electric field. We shall explore how these phenomena follow from the Maxwell equations.

In applying Maxwell's equations to metals, we will describe the metal as a linear material with ohmic conductivity. The free current in the material is

$$\mathbf{J}_{\text{free}}(\mathbf{x}, t) = \sigma \mathbf{E}(\mathbf{x}, t). \tag{13.72}$$

Also, we assume the metal has linear dielectric and magnetic properties, so $\mathbf{D} = \epsilon \mathbf{E}$ and $\mathbf{H} = \mathbf{B}/\mu$. These depend on the bound electrons of the metal atoms.

Inside a metal we may set $\rho_{\text{free}} = 0$, because any excess charges repel each other and move rapidly to the surface. We showed in Chapter 7 that the characteristic time for decay of excess charge is ϵ_0/σ. Strictly speaking we should restrict the discussion to frequencies less than σ/ϵ_0, for which it is a justifiable approximation to set $\rho_{\text{free}} = 0$. Naively, the frequency at which this criterion breaks down is, for Cu, about 10^{18} Hz. However, a detailed analysis yields the more stringent limit of 10^{14} Hz, which is the frequency of electron collisions in a metal. Still, we get a qualitatively correct picture even for optical frequencies.

With these model simplifications, the Maxwell equations in a metal become

$$\nabla \cdot \mathbf{E} = 0 \quad \text{and} \quad \nabla \cdot \mathbf{B} = 0,$$

$$\nabla \times \mathbf{E} = -\partial \mathbf{B}/\partial t \quad \text{and} \quad \nabla \times \mathbf{B} = \mu \sigma \mathbf{E} + \mu \epsilon \partial \mathbf{E}/\partial t. \tag{13.73}$$

The new term, $\mu \sigma \mathbf{E}$, is the conduction current, and will have interesting and important consequences.

Before we construct the wave solutions of (13.73), let's think about this problem physically, and anticipate the behavior of the solution. With $\mathbf{J}_{\text{free}} = \sigma \mathbf{E}$ there is ohmic resistance, and therefore dissipation of energy. The energy of the currents and fields is transferred to random molecular energy by heat, through the interactions responsible for the resistance. Consequently an electromagnetic wave will lose energy as it propagates through the conductor. This situation is unlike a wave in vacuum, or in an insulator, which propagates without loss of intensity.

To analyze (13.73) we eliminate $\mathbf{B}$ by the *curl curl trick*: Take the curl of both sides of Faraday's Law. On one side of the equation we then have $\nabla \times (\nabla \times \mathbf{E})$. By the double curl identity from Chapter 2 this quantity is $-\nabla^2 \mathbf{E}$, because $\nabla \cdot \mathbf{E} = 0$. On the other side of the equation we have $-\partial \nabla \times \mathbf{B}/\partial t$, and we may replace $\nabla \times \mathbf{B}$ using the Ampère–Maxwell Law. The result is a wave equation for $\mathbf{E}(\mathbf{x}, t)$,

$$\nabla^2 \mathbf{E} = \mu \sigma \frac{\partial \mathbf{E}}{\partial t} + \mu \epsilon \frac{\partial^2 \mathbf{E}}{\partial t^2}. \tag{13.74}$$

It can be shown that $\mathbf{B}$ satisfies the same equation, by taking the curl of both sides of the Ampère–Maxwell equation. Now, construct a solution of (13.74) in the form of a plane wave propagating in the x direction

$$\mathbf{E}(\mathbf{x}, t) = \mathbf{E}_0 e^{i(\kappa x - \omega t)}. \tag{13.75}$$

As usual when using complex functions, the real part of the right-hand side is understood to be the physical field. Because $\nabla \cdot \mathbf{E} = 0$, the constant amplitude vector $\mathbf{E}_0$ must again be transverse; that is, $\hat{\mathbf{i}} \cdot \mathbf{E}_0 = 0$. The wave equation (13.74) requires that κ and ω must be related according to the dispersion relation

$$\kappa^2 = +i\mu\sigma\omega + \mu\epsilon\omega^2. \tag{13.76}$$

If the conductivity σ were 0, as it is in dielectrics, then the dispersion relation would imply that κ is real, and equal to the wave number $k \equiv 2\pi/\lambda$. In metals, however, $\sigma \neq 0$ so that κ^2 and κ are complex. In this case it is necessary to write κ in terms of its real and imaginary parts, κ_1 and κ_2 respectively, as

$$\kappa = \kappa_1 + i\kappa_2. \tag{13.77}$$

Substituting into (13.76) and equating the real and imaginary terms, we obtain two real equations,

$$\kappa_1^2 - \kappa_2^2 = \mu\epsilon\omega^2 \quad \text{and} \quad 2\kappa_1\kappa_2 = \mu\sigma\omega. \tag{13.78}$$

These may be solved for the real and imaginary parts of κ, and the result is

$$\left.\begin{array}{c} \kappa_1 \\ \kappa_2 \end{array}\right\} = \omega\sqrt{\frac{\epsilon\mu}{2}}\left[\sqrt{1 + \left(\frac{\sigma}{\epsilon\omega}\right)^2} \pm 1\right]^{1/2}, \tag{13.79}$$

where the $+$ sign is for κ_1 and the $-$ sign is for κ_2. The general expression (13.79) is rather complicated, but we shall use it only in limiting cases where it is simpler. The quantities κ_1 and κ_2 are both important physically. The imaginary part κ_2 of κ leads to exponential *decrease* of the wave amplitude with distance x in the propagation direction, because if we substitute (13.77) into the original wave expression (13.75) it becomes

$$\mathbf{E}(\mathbf{x}, t) = \mathbf{E}_0 e^{-\kappa_2 x} e^{i(\kappa_1 x - \omega t)}. \tag{13.80}$$

This exponential *attenuation* of the wave with x is what we anticipated by physical arguments: The intensity decreases because electric resistance robs the wave of energy.

Equation (13.80) shows that κ_1, the real part of κ, determines the wavelength and phase velocity in the conductor. As usual, these are $\lambda = 2\pi/\kappa_1(\omega)$ and $v_{\text{phase}} = \omega/\kappa_1(\omega)$, where we have emphasized the dependence of these quantities on ω. A medium in which v_{phase} depends on frequency is called *dispersive* because an electromagnetic signal with several component frequencies will undergo *dispersion* as it traverses the medium; that is to say it will broaden and in general change shape. The motion of waves in a dispersive medium is character-ized by the *group velocity*, defined by $v_{group} = d\omega/d\kappa_1$. (In terms of (13.79) this derivative is $v_{group} = (d\kappa_1/d\omega)^{-1}$.) v_{phase} is the velocity of a point of constant phase in the wave. v_{group} is the velocity of the envelope of a wave pulse.[10] The group velocity is important because it is the velocity of information carried in a wave signal.

A material is called a *good conductor* in the context of its interaction with an electromagnetic wave if the charge current (σE) is large compared to the displace-ment current ($\epsilon\partial E/\partial t$). The criterion for a good conductor is therefore

$$\sigma \gg \epsilon\omega. \quad \text{(good conductor)} \tag{13.81}$$

In the case of a good conductor we may neglect the term $\mu\epsilon\omega^2$ in (13.76) so that κ^2 is purely imaginary. Then, from the identity $\sqrt{i} = (1 + i)/\sqrt{2}$, the real and imaginary parts of κ are equal, and given by

$$\kappa_1 = \kappa_2 = \sqrt{\frac{\mu\sigma\omega}{2}}. \tag{13.82}$$

Note that for a good conductor

$$\frac{c\kappa_1}{\omega} = \sqrt{\frac{\mu_0 c^2 \sigma}{2\omega}} = \sqrt{\frac{\sigma}{2\epsilon_0\omega}} \tag{13.83}$$

so $c\kappa_1 \gg \omega$. Therefore the phase velocity is small compared to c. In (13.83) we have assumed $\mu = \mu_0$.

Equation (13.81) is a quantitative criterion for conduction in a material to be important. For example, a typical metal has σ of order $10^8 \ \Omega^{-1} \ m^{-1}$. For infrared light the frequency ω is of order $10^{14} \ s^{-1}$ and ϵ is of order $\epsilon_0 = 8.85 \times 10^{-12}$ F/m. Thus (13.81) suggests that a metal is a good conductor for interactions with in-frared light.

According to (13.80), a wave penetrates into a good conductor by only a very small distance, of order $\delta \equiv 1/\kappa_2 = \sqrt{2/(\mu\sigma\omega)}$. This characteristic length—the attenuation length of E in the conductor—is called the *skin depth*. To estimate δ for infrared light in a metal, let $\mu = \mu_0$, $\sigma = 10^8 \ \Omega^{-1} \ m^{-1}$, and $\omega = 10^{14} \ s^{-1}$; these imply $\delta = 10^{-8}$ m. Although this is a very small distance, corresponding to a thickness of, say, 20 atomic layers, the continuum approximation for the metal is still adequate. The result is consistent with our experience that light does not penetrate into a metal. Even very thin metal foils, like tinfoil or Al foil, are opaque to light.

In making the estimates of the preceding paragraph we have casually approx-imated σ at high frequencies by its dc or low-frequency value. In fact σ is a function of ω, so it is important to ask: Up to how high in frequency is it rea-sonable to approximate σ by its dc value? The answer to this question involves,

[10]See Exercise 12.

again, the electron collision time which, for Cu, is about $\tau = 2.4 \times 10^{-14}$ s. The approximation we made is a good one up to frequencies for which $\omega\tau = 1$. For frequencies less than a few times 10^{12} Hz, we can confidently replace $\sigma(\omega)$ with the dc value. For higher frequencies we must ultimately rely on experimental data for $\sigma(\omega)$.

The magnetic field associated with the electric field in (13.75) can be deduced from Faraday's Law. The frequency of oscillation of **B** and **E** must be equal, so $\partial\mathbf{B}/\partial t$ is $-i\omega\mathbf{B}$, and Faraday's Law is

$$i\omega\mathbf{B} = \nabla \times \mathbf{E} = i\kappa\,\hat{\mathbf{i}} \times \mathbf{E}; \tag{13.84}$$

that is,

$$\mathbf{B} = \hat{\mathbf{i}} \times \mathbf{E}_0\frac{\kappa}{\omega}e^{i(\kappa x - \omega t)}. \tag{13.85}$$

We must remember that the real part of the right-hand side is understood to be the physical field. But κ has an imaginary part, so there is a *phase difference* between the magnetic and electric field oscillations. For a good conductor, $\kappa = \sqrt{\mu\sigma\omega/2}(1 + i) = \sqrt{2}\kappa_1\,e^{i\pi/4}$. In that case the magnetic field lags the electric field by $\pi/4$ radians, i.e., $\frac{1}{8}$ of a cycle.

Taking the magnitude of both sides of (13.85), the ratio of the amplitude of the magnetic field to that of the electric field is

$$\frac{|B_0|}{|E_0|} = \frac{|\kappa|}{\omega} = \sqrt{\frac{\mu\sigma}{\omega}}. \tag{13.86}$$

For Cu and a frequency of 1 MHz, a typical AM radio frequency, this is $|B_0|/|E_0| = 10^6/c$. Comparing this to the ratio of the amplitudes in vacuum, which is $1/c$, we see that in a good conductor there is an enhancement by a factor of 10^6 in the magnitude of the magnetic field relative to the electric field. Also, the energy density of the magnetic field is much larger than that of the electric field; the ratio of the energy densities is

$$\frac{B_0^2/2\mu}{\epsilon E_0^2/2} = \frac{\sigma}{\epsilon\omega} \gg 1. \tag{13.87}$$

Evaluating this ratio for Cu at 1 MHz gives 10^{12}. For an electromagnetic wave in vacuum, by comparison, the magnetic and electric energy densities are equal. The physical interpretation of these results is easy to understand. For a good conductor a small electric field drives a large electric current density, which produces a large magnetic field. The magnetic field of an electromagnetic wave in a metal comes predominantly from the charge current rather than the displacement current.

13.3.1 ■ Reflectivity of a Good Conductor

From daily experience we know that metal surfaces are shiny. The basis of this effect in classical electrodynamics, as we have already discussed, is that when light shines on a metal the electric field of the electromagnetic wave interacts with the free electrons. The field exerts a force, of frequency ω, on the electrons and causes them to accelerate. What we see as the reflected light is the combined radiation emitted by all the accelerating electrons. The statement that metals are shiny, means that the reflectivity R is close to unity for metals.

In this section we calculate the reflectivity for normal incidence at a conducting surface. Consider a wave traveling in the x direction, incident from vacuum or air for $x \leq 0$, into a conductor for $x > 0$. We wish to calculate E_0''/E_0, the ratio of the amplitudes of the reflected (E_0'') and incident (E_0) waves. The method of calculation is the same that we used earlier to calculate the reflectivity for normal incidence at a dielectric surface, by solving the boundary conditions of the fields. In fact, the algebra for the conductor problem is exactly the same as for the earlier dielectric problem, so we do not need to set up the problem all over again. We can just use the final result (13.35) with appropriate reinterpretation of the parameters.[11]

Region 1, the region of the incident and reflected waves, will be taken to be vacuum, so $\mu_1 = \mu_0$ and $n_1 = 1$. Region 2, the region of the transmitted wave, is the conductor, which we assume to have negligible magnetization, so $\mu_2 = \mu_0$. When (13.35) is applied to a metal it is necessary to think carefully what to substitute for n_2 in the equation. For the dielectric problem, n_2 can be written in terms of k' by (13.17) which, if medium 1 is vacuum, gives $n_2 = ck'/\omega$. This makes sense: It is just c/v_{phase}. In a metal it is the complex quantity κ of (13.75) that corresponds to k', so to apply (13.35) to this case we must make the substitution $n_2 \to c\kappa/\omega$. Thus the ratio of the amplitudes is

$$\frac{E_0''}{E_0} = \frac{1 - n_2}{1 + n_2} = \frac{\omega - c\kappa}{\omega + c\kappa}. \tag{13.88}$$

Now, for the conductor κ is a complex number $\kappa_1 + i\kappa_2$. The reflectivity, i.e., the ratio of intensities, is $|E_0''/E_0|^2$, the *squared magnitude* of the complex number in (13.88); that is,

$$R = \frac{I''}{I} = \frac{(\omega - c\kappa_1)^2 + c^2\kappa_2^2}{(\omega + c\kappa_1)^2 + c^2\kappa_2^2}. \tag{13.89}$$

But for a good conductor we saw in (13.82) that $\kappa_1 = \kappa_2$; and furthermore, $c\kappa_1 \gg \omega$, so we may treat ω as a small parameter in (13.89). To first order in ω,

$$R \approx 1 - \frac{4\omega c\kappa_1}{c^2(\kappa_1^2 + \kappa_2^2)} = 1 - \sqrt{\frac{8\omega\epsilon_0}{\sigma}}. \tag{13.90}$$

where the second equality follows from (13.82). Thus the reflectivity of a good conductor is only slightly less than 1.

Very little power flows into the conductor, to be dissipated in resistance, because the Poynting vector $\mathbf{E} \times \mathbf{H}$ is very small at the surface. The electric field in

[11]See Exercise 9.

the conductor is $\mathbf{J}/\sigma$, which is small because σ is large; so $\mathbf{E} \times \mathbf{H}$ is small inside the conductor. The electric field just outside the conductor is small because the $\mathbf{E}$ fields of the incident and reflected waves nearly cancel. $\mathbf{E}$, being tangential, must be continuous at the surface; so if $\mathbf{E}$ is small on one side of the surface it must be small on the other side. Although it is an oversimplification to assume that the conductivity of a metal is a real constant in the frequency range of visible light, the calculation does explain qualitatively the basic reason why metals are shiny.

13.4 ■ A CLASSICAL MODEL OF DISPERSION: THE FREQUENCY DEPENDENCE OF MATERIAL PROPERTIES

In our previous calculations we assumed that ϵ, μ, and σ are simply real constants. For high-frequency oscillations, however, such as encountered in visible or ultraviolet light, this assumption is not correct physically. *Dispersion*, i.e., the frequency dependence of material properties, is an important aspect of optics. The fact that the index of refraction of water or glass varies with frequency explains rainbows, why prisms spread light into its spectral components, and chromatic aberration in lenses.

Dispersion—frequency dependence—occurs because the response to change, by atoms or electrons in a material, is not instantaneous. For example, if the permittivity is a function of frequency ω then the constitutive equation is not simply $\mathbf{D}(t) = \epsilon \mathbf{E}(t)$; rather, there are *time delays* between a change of $\mathbf{E}$ and the responsive change of polarization, or $\mathbf{D}$.

A complete physical theory describing the response of an atom to a varying field requires quantum mechanics. However, we can get some insight by studying a classical model. Consider an electron moving under the influence of atomic forces and an electric field $\mathbf{E}(t) = \mathbf{E}_0 e^{-i\omega t}$. A simple model equation for the motion of the electron is

$$m\frac{d^2\mathbf{x}}{dt^2} = -K\mathbf{x} - \gamma\frac{d\mathbf{x}}{dt} - e\mathbf{E}_0 e^{-i\omega t}. \qquad (13.91)$$

Here $\mathbf{x}$ is the electron position, with the nucleus at the origin. The right-hand side has various forces on the electron: (i) $-K\mathbf{x}$ represents the restoring force that keeps the electron in the atom. (ii) $-\gamma d\mathbf{x}/dt$ is a dissipative force, which in this model we take to be proportional, but opposite in direction, to the electron velocity; the model parameter γ expresses the strength of the dissipation. (iii) $-e\mathbf{E}$ is the electric force due to the electromagnetic wave. It is understood that the *real part* of (13.91) describes the electron motion, but we use the complex equation during intermediate stages of the calculation for convenience. At the end of the calculation, Re $\mathbf{x}(t)$ is the position of the electron.

The *steady-state solution* of (13.91) has the form $\mathbf{x}(t) = \mathbf{x}_0 e^{-i\omega t}$. The electron undergoes harmonic motion with the same frequency as the driving force $-e\mathbf{E}$, although in general there is a phase shift from the phase of the complex amplitude $\mathbf{x}_0$. It is straightforward to show that the steady-state solution is

$$\mathbf{x}(t) = \frac{-e\mathbf{E}_0}{K - m\omega^2 - i\omega\gamma} e^{-i\omega t}. \tag{13.92}$$

To obtain (13.92) we just substitute $\mathbf{x} = \mathbf{x}_0 e^{-i\omega t}$ into (13.91) and solve for $\mathbf{x}_0$. For any specified initial condition the solution to (13.91) also includes a transient term, but that decays in time and (13.92) remains.

13.4.1 ■ Dispersion in a Dielectric

The electric dipole moment of an atom, due to the displacement from equilibrium of one electron, is

$$\mathbf{p}(t) = -e\mathbf{x}(t) = \alpha\mathbf{E}(t), \tag{13.93}$$

assuming the dipole moment is 0 at the position of static equilibrium ($\mathbf{x} = 0$). Using (13.92) shows that the parameter α, called the *polarizability* of the atom, is

$$\alpha = \frac{e^2}{K - m\omega^2 - i\omega\gamma}. \tag{13.94}$$

Note that α depends on ω and is complex. In this microscopic calculation α is the polarizability of a single atom. The polarization $\mathbf{P}(\mathbf{x}, t)$ is the dipole moment per unit volume of bulk material induced by a macroscopically averaged electric field. If the atomic density ν is small, so that the polarization of one atom doesn't affect the others, then the local field experienced by an atom is equal to the macroscopic field, and the polarization $\mathbf{P}$ due to the atomic dipole moments is $\mathbf{P} = \nu\mathbf{p} = \nu\alpha\mathbf{E}$. The permittivity ϵ of the material is defined by $\epsilon\mathbf{E} = \epsilon_0\mathbf{E} + \mathbf{P}$, so

$$\epsilon = \epsilon_0 + \nu\alpha. \tag{13.95}$$

Note that this permittivity is complex.

There is an interesting correction that must be applied to (13.95) if the material is dense, because then the force in (13.91) must also include a contribution due to the other polarized atoms in the neighborhood of the electron. Under those conditions the local field at the atom is larger than the macroscopic average field, i.e., $\mathbf{E}_{\text{local}} = \mathbf{E} + \mathbf{P}/3\epsilon_0$, a result from Section 6.3.2. Then the dielectric constant ϵ/ϵ_0 is given by

$$\frac{\epsilon}{\epsilon_0} = \frac{3\epsilon_0 + 2\nu\alpha}{3\epsilon_0 - \nu\alpha}, \tag{13.96}$$

which is another way to write the Clausius-Mossotti equation (6-39).

Now, how does an electromagnetic wave propagate in this model dielectric? The electric field of a plane wave, propagating in the x direction, polarized in the y direction, and with frequency ω is

$$\mathbf{E}(\mathbf{x}, t) = E_0 e^{i(\kappa x - \omega t)} \hat{\mathbf{j}}. \tag{13.97}$$

The dispersion relation is the same as for a nonmagnetic dielectric,

$$\kappa = \omega\sqrt{\epsilon\mu_0}. \tag{13.98}$$

But here $\epsilon(\omega)$ is complex, so $\kappa(\omega) = \kappa_1(\omega) + i\kappa_2(\omega)$. Therefore the electric field is

$$\mathbf{E}(\mathbf{x}, t) = E_0 e^{-\kappa_2 x} e^{i(\kappa_1 x - \omega t)}\,\hat{\mathbf{j}}. \tag{13.99}$$

This result displays both absorption and dispersion. The energy absorption length d is $1/2\kappa_2$; i.e., the energy density decreases exponentially with distance, by the factor $1/e$ for each length d. The index of refraction is $n = c/v = c\kappa_1/\omega$, where v is the phase velocity ω/κ_1. Both d and n depend on frequency through the ω dependence of α in (13.94).

Figure 13.11 illustrates the phenomena of absorption and dispersion associated with the electron dynamics within this simple model. Assuming $v\alpha/\epsilon_0$ is small, the index of refraction is

$$n = \frac{c\kappa_1}{\omega} = 1 + \frac{ve^2}{2\epsilon_0 m\omega_0^2}\frac{\omega_0^2(\omega_0^2 - \omega^2)}{(\omega_0^2 - \omega^2)^2 + (\omega\gamma/m)^2} \tag{13.100}$$

and the inverse absorption length is

$$d^{-1} = 2\kappa_2 = \frac{ve^2}{\epsilon_0 mc}\frac{\omega^2\gamma/m}{(\omega_0^2 - \omega^2)^2 + (\omega\gamma/m)^2}. \tag{13.101}$$

The quantity $\omega_0 \equiv \sqrt{K/m}$ is the resonant frequency for zero damping. The graph in Fig. 13.11 shows $n - 1$ in units of $ve^2/(2\epsilon_0 m\omega_0^2)$ and d^{-1} in arbitrary units. The dissipation parameter γ has been set equal to $0.1 m\omega_0$ for illustration purposes only. The behavior is an example of *resonance*, because the absorption becomes strong when the driving frequency ω is near the natural frequency ω_0.

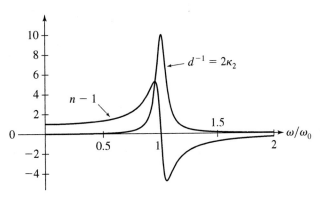

FIGURE 13.11   The inverse absorption length $d^{-1} = 2\kappa_2(\omega)$ and the index of refraction $n = c\kappa_1(\omega)/\omega$ versus ω/ω_0, where $\omega_0 = \sqrt{K/m}$.

Below the resonant frequency ω_0, the index of refraction $n(\omega)$ increases with ω. For many transparent liquids and solids the characteristic atomic frequencies are in the ultraviolet, so in the visible range $n(\omega)$ increases with ω. Consequently violet light refracts more than red light in glass or water, which is certainly consistent with our knowledge of prisms or rainbows. At higher frequencies, in the region of strong absorption, there occurs *anomalous dispersion*, in which the index of refraction decreases as a function of frequency.

Our classical model has only a single atomic frequency, but a real atom has many resonant frequencies, corresponding to quantum transitions between electron states with frequencies $\Delta E/\hbar$. If there are no transitions in the visible part of the spectrum, then the material is colorless and transparent. If there are transitions in the visible part of the spectrum, then the absorption determines the color of the material. In real materials $n(\omega)$ is a complicated function.

13.4.2 ■ Dispersion in a Plasma

A plasma is an ionized gas consisting at least partly of free electrons and positive ions; it is therefore a conducting material. We have not discussed plasma properties previously, although it is a fascinating subject. Plasmas are interesting for themselves, important for applications, and ubiquitous—sparks are plasmas, illuminated gas tubes contain plasmas, and plasmas play a key role in fusion power research. Plasmas are also an important part of the Universe—the Sun and stars are largely plasma. We will explore the dispersion of conductivity in a plasma using the classical electron model.

Because the electrons in a plasma are not bound in atoms, we set $K = 0$ in the equation of motion (13.91). The electron velocity $\mathbf{v} = d\mathbf{x}/dt$ may then be calculated for the steady-state solution,

$$\mathbf{v}(t) = \frac{-i\omega e}{m\omega^2 + i\omega\gamma}\mathbf{E}(t). \tag{13.102}$$

Letting v_e be the electron density in the plasma, the current density is $\mathbf{J} = -ev_e\mathbf{v} = \sigma\mathbf{E}$, and so the conductivity σ is

$$\sigma(\omega) = \frac{i\omega e^2 v_e}{m\omega^2 + i\omega\gamma}. \tag{13.103}$$

Again the material parameter is a complex number. We have previously obtained the wave solution of Maxwell's equations in a conducting medium. The fields $\mathbf{E}(\mathbf{x}, t)$ and $\mathbf{B}(\mathbf{x}, t)$ are given by (13.75) and (13.85), with wave dependence $e^{i(\kappa x - \omega t)}$. The dispersion relation for $\kappa(\omega)$ is (13.76).

For a *dilute plasma*, the electron collisions that produce dissipation of electron energy are rare, so the dissipation is small. In this case we may set $\gamma = 0$. (Also, the polarization and magnetization are negligible for a dilute plasma, so $\mu = \mu_0$ and $\epsilon = \epsilon_0$.) Then the conductivity σ is purely imaginary

$$\sigma(\omega) = \frac{ie^2 v_e}{m\omega}. \tag{13.104}$$

Equation (13.104) is a remarkable relation because σ in other contexts is so easily understood and intuitive. What does it mean that σ is *purely imaginary*? Recall that the power per unit volume lost to ohmic heating is

$$\frac{dP}{dV} = \mathbf{J} \cdot \mathbf{E}. \tag{13.105}$$

The real parts of $\mathbf{J}$ and $\mathbf{E}$, i.e., the physical fields, must be used in this relation. But since σ is imaginary, $\mathbf{J}$ and $\mathbf{E}$ oscillate 90° out of phase and the time average of $\mathbf{J} \cdot \mathbf{E}$ is 0. That is, there is no energy loss in the dilute plasma. This is not really surprising, because we set the only dissipation in the model—the damping parameter γ—equal to 0.

The dispersion relation (13.76) becomes

$$\kappa^2 = \mu_0 \epsilon_0 \omega^2 - \frac{\mu_0 e^2 v_e}{m} = \frac{\omega^2 - \omega_p^2}{c^2}. \tag{13.106}$$

In the second equality we have defined ω_p, which is called the *plasma frequency*, by

$$\omega_p = \sqrt{\frac{e^2 v_e}{m\epsilon_0}}. \tag{13.107}$$

If ω is greater than ω_p, then the wave vector κ is real, so the wave propagates without attenuation through the plasma. There is dispersion, because the phase velocity ω/κ depends on frequency, $v_{\text{phase}} = c/\sqrt{1 - (\omega_p/\omega)^2}$; but there is no attenuation. An interesting feature for $\omega > \omega_p$ is that the phase velocity is greater than c.[12] The group velocity is less than c.[13]

On the other hand, if ω is less than ω_p, then κ is purely imaginary, so the wave does not propagate at all; its amplitude just dies out exponentially with decay length $d = c/\sqrt{\omega_p^2 - \omega^2}$. But if waves with $\omega < \omega_p$ cannot propagate in the plasma, what happens if a low-frequency wave from vacuum impinges upon a plasma boundary? Energy is conserved, but there is no dissipation in the plasma, and no propagation into the plasma; so the wave can only be reflected back into the vacuum. When the plasma is disturbed by a low-frequency field, charges move to screen out the field. If $\omega < \omega_p$ then the electrons can move fast enough to prevent the field from penetrating far into the plasma, and so the electromagnetic wave reflects from the boundary.

---

[12] In Chapter 14 we'll see another example where the phase velocity is greater than c, for an electromagnetic wave in a wave guide, and discuss the implications.
[13] See Exercise 18.

Radio and the Ionosphere

An important application of the interaction between plasmas and electromagnetic waves is to radio propagation. The upper atmosphere is a plasma, called the ionosphere, with a typical number density of electrons $v_e = 10^{11}$ free electrons/m³. These electrons come from atoms that have been ionized by ultraviolet radiation from the sun. There is an equal charge density of positive ions but their motion can be neglected because their mass is much larger than the electron mass. According to (13.107), the plasma frequency for the ionosphere is $\omega_p \approx 2 \times 10^7 \, s^{-1}$, which corresponds to $f_p \approx 3 \times 10^6$ Hz. The decay length in the ionosphere for low frequency waves, those with $\omega \ll \omega_p$, is $c/\omega_p \approx 15$ m.

AM radio waves have frequencies from 0.55 to 1.60 MHz, below the plasma frequency f_p, and they therefore reflect from the ionosphere. The reason an AM signal may be received at distances of hundreds of miles from the transmitter is that the radio waves can bounce back and forth multiple times from the ionosphere and the Earth's surface. For AM and shortwave radio propagation the ionosphere acts as a global reflecting layer.

The existence of the ionosphere was predicted independently by Heaviside and Kennelly, in 1901, from the observation that Marconi's radio signals reached Nova Scotia from London with far greater intensity than expected. Indeed some theorists had claimed that radio signals could not be received at such large distances because of the curvature of the Earth. For many years people referred to the ionosphere as the Heaviside layer.[14] Today it is known that the ionosphere has a complex and dynamic structure with several layers. The reason why the range of an AM radio station becomes extended at night, as is commonly observed, is that the outer atmosphere cools after sunset and the warmer ionosphere rises. Therefore radio waves that reflect from the bottom of the ionosphere come back to Earth farther away than during the day.

FM radio ($\approx 10^8$ Hz) and TV (10^8–10^9 Hz) have $\omega > \omega_p$, so these signals pass through the ionosphere without being reflected back. FM and TV signals cannot be received from a transmitter far beyond the horizon, because without ionospheric reflections, the receiving antenna must be within the line of sight of the transmitting antenna. Also, signals at high frequencies ($f > 10^8$ Hz) must be used for communication between Earth and satellites, because these signals can pass through the ionosphere.

FURTHER READING

1. M. Born and E. Wolf, *Principles of Optics*, 7th expanded edition (Cambridge University Press, New York, 1999). This is a classic book on the electromagnetic theory of propagation, interference, and diffraction of light.

[14]Oliver Heaviside (1850–1925) was a brilliant mathematical physicist, though eccentric. He had only an elementary formal education and beyond that was largely self-taught. He did extensive research on field theory, and published a three-volume work entitled "Electromagnetic Theory," not unlike this book but longer and more original. Another scientist complained to Heaviside that his papers were very difficult to read. His retort: "That may well be but they were much more difficult to write."

EXERCISES

Sec. 13.1. Electromagnetic Waves in a Dielectric

13.1. Starting with Maxwell's equations in a linear dielectric, and equation (13.6) for the fields of an electromagnetic wave:

(a) Prove that $\mathbf{E}_0$ and $\mathbf{B}_0$ are perpendicular to $\mathbf{k}$.

(b) Prove that $\mathbf{B}_0$ is perpendicular to $\mathbf{E}_0$.

(c) Prove that $B_0/E_0 = k/\omega$ and $B_0/E_0 = \mu\epsilon\omega/k$.

(d) Prove the dispersion relation $\omega = vk$ and determine v.

13.2. (a) Show that the mean energy density of the plane wave (13.6), averaged over a period of oscillation, is $\langle u \rangle = \epsilon E_0^2/2$.

(b) Show that the intensity (energy flux) is $I = \langle u \rangle v$, and explain why this means that the energy flows with velocity v.

13.3. Suppose the complex wave amplitude in (13.6) is $\mathbf{E}_0 = e^{i\Phi}\mathbf{E}_r$ where Φ and $\mathbf{E}_r$ are real. What is the magnetic wave amplitude $\mathbf{B}_0$? Determine the intensity for this electromagnetic wave, and show that it is the same as (13.12). Explain the physical meaning of complex $\mathbf{E}_0$ and $\mathbf{B}_0$.

Sec. 13.2. Reflection and Refraction at a Dielectric Surface

13.4. Imagine a lucite prism ($n = 1.5$) whose cross section is a quarter circle of radius a. As shown in Fig. 13.12, one flat side rests on a table, and light is incident normal to the other flat side. The region from P to Q on the table is not illuminated by light from the prism. Determine the position of Q. [Answer: $\overline{OQ} = 3a/\sqrt{5}$]

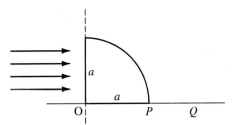

FIGURE 13.12   Exercise 4. Light incident on a lucite prism whose cross section is a quarter circle.

13.5. Consider light traveling from $x = -\infty$, incident normally on a plate of glass with thickness a. The plate is parallel to the yz plane, with one face at $x = 0$ and the other at $x = a$. The index of refraction is $n_0 = 1$ for $x < 0$ and $x > a$, and $n = 1.5$ for $0 \le x \le a$. The electromagnetic field in the region $x < 0$ is a superposition of right and left traveling waves (where right means $\hat{\mathbf{i}}$ and left means $-\hat{\mathbf{i}}$), which are the incident and reflected waves. In the region $0 \le x \le a$ there are both right and left traveling waves, and in the region $x > a$ there is only the transmitted right traveling wave.

(a) Write $\mathbf{E}(x, t)$ and $\mathbf{B}(x, t)$ in the three regions, letting $\hat{\mathbf{j}}$ be the polarization direction. Write the four boundary conditions on the wave amplitudes.

(b) Solve for the transmission coefficient T, i.e., the ratio of transmitted intensity to incident intensity.

(c) Plot T as a function of ka, where k is the incident wave vector.

13.6. (a) Derive Fresnel's equation (13.56) for the reflected wave amplitude of TE polarized light, from (13.55) and Snell's law.

(b) Do the same for (13.67), TM polarization, by deriving it from (13.66) and Snell's law.

13.7. From the amplitudes E_0, E_0', and E_0'' for light scattering from a dielectric surface, show that the conservation of energy equation (13.70) holds at arbitrary angles for (a) TE polarization and (b) TM polarization.

13.8. *Brewster's angle.* Consider light incident on the plane interface between two dielectrics, from index of refraction n_1 to n_2, at angle of incidence θ. For incident TM polarized light the reflection is 0 at $\theta = \theta_B$, where $\tan \theta_B = n_2/n_1$. (The dielectrics are assumed to have $\mu_1 = \mu_2 = \mu_0$.) Therefore, for *unpolarized light* incident at $\theta = \theta_B$ the reflected light is TE polarized.

(a) Prove that for a TM polarized incident wave, the electric field of the transmitted wave is parallel to the direction of the reflected ray if $\theta = \theta_B$. (The direction of the reflected ray is well-defined, although the intensity of the reflected wave is 0.)

The Rainbow Caustic

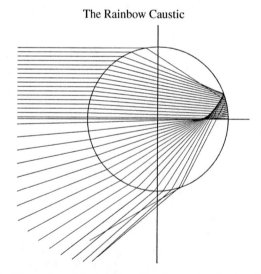

FIGURE 13.13 Illustration of how light refraction and reflection in a water droplet causes a rainbow. Light incident from the left is refracted on entering the spherical drop, reflects partially from the back surface, and is refracted on leaving the front surface. (Only rays incident above the midplane are shown.) The caustic—the region where the exiting rays are most concentrated—is the rainbow. The colors of the rainbow are the result of dispersion; the angle of the caustic varies with wavelength.

(b) Calculate Brewster's angle for light incident from air into water.

(c) Calculate Brewster's angle for light incident from water into air.

An example of polarization by reflection is the light in a rainbow. Rainbow light is produced by a three-step process of light scattering from a spherical drop of water: refraction into the drop, internal reflection at the back surface, and refraction back out of the drop. Figure 13.13 illustrates the paths of the rays. At the internal reflection of the caustic, the angle of incidence is approximately 39°. Because this angle is close to Brewster's angle for water, rainbow light is highly polarized—about 96% polarized.

Sec. 13.3. Electromagnetic Waves in Conductors

13.9. (a) From first principles, set up the boundary conditions for a plane wave incident normally on the surface of a conductor. Let the incident wave travel in the x direction, and be polarized in the y direction. The reflected and transmitted waves travel in the $-x$ and $+x$ directions, respectively. The wave vector κ in the conductor is complex, and the dispersion relation is (13.76). Solve the boundary conditions, and show that the amplitude of the reflected wave is correctly given by (13.88).

(b) For a good conductor, i.e., $\sigma \gg \epsilon\omega$, derive (13.90).

(c) Show that for a good conductor, at the surface the field of the reflected wave is approximately equal but opposite to the field of the incident wave.

13.10. In this chapter we have discussed reflection and transmission for dielectrics, in which $\sigma/\epsilon\omega \approx 0$, and for conductors, in which $\sigma/\epsilon\omega \gg 1$. But many interesting systems are intermediate between these extremes. In the current exercise, it is a fairly good approximation to use the dielectric expressions for R and T.[15]

For biological tissues with high water content, including skin, ϵ and σ depend on frequency. At 2450 MHz their values are $\epsilon = 47\epsilon_0$ and $\sigma = 2.21\,\Omega^{-1}\text{m}^{-1}$. What is the value of $\sigma/\epsilon\omega$ for this system? Evaluate κ_2 from (13.79) to estimate the attenuation length.

If a radar signal at 2450 MHz impinges on skin and other similar tissues, from air, how much of the incident power is absorbed? Assume the tissue thickness is 15 cm. (The precise answer is $T = 0.43$.)

13.11. Calculate the radiation pressure exerted by light incident normally on the surface of a good conductor. (Hint: The force is equal to the rate of change of momentum of the light.) Express the result in terms of the intensity of the incident light. Explain qualitatively how the pressure results from the Lorentz force on the electrons in the conductor.

13.12. As an example of *group velocity*, consider a Gaussian wave pulse of some quantity $\phi(x, t)$ that undergoes wave motion, given by

$$\phi(x, t) = \int_{-\infty}^{\infty} e^{i(kx - \omega t)} f(k)\, \frac{dk}{2\pi}$$

[15]The general case is treated in detail in U.S. Inan and A.S. Inan, *Engineering Electromagnetism* (Addison-Wesley, Menlo Park, California, 1998).

where $f(k) = f_0 e^{-(k-k_0)^2 a^2}$. The k range is $(-\infty, \infty)$, but the integrand is peaked at $k = k_0$, and the width of the peak is of order $1/a$. Assume that within the peak $\omega(k)$ may be approximated by

$$\omega(k) = \omega(k_0) + (k - k_0)\omega'(k_0).$$

(a) Evaluate the integral, and obtain explicitly the function $\phi(x, t)$. (Hint: Let $k = k_0 + q$, change the variable of integration to q, and use a table of integrals or an analytic computer program (e.g., Mathematica or Maple) to evaluate the integral.)

(b) Show explicitly that the phase velocity is $\omega(k_0)/k_0$ and the group velocity is $\omega'(k_0)$.

13.13. The dispersion relation for deep water gravity waves is $\omega = \sqrt{gk}$. Show that the group velocity is one-half the phase velocity. (This leads to an interesting effect that may be observed in water waves produced by dropping a pebble into a pond. The outgoing wave group travels slower than the individual ripples. Wave crests in the group are born at the back, travel through the group at twice the group velocity, reach maximum amplitude near the middle, and die out at the front.)

13.14. Prove and explain why $\sqrt{i} = (1+i)/\sqrt{2}$. Also, why is the other root $(-(1+i)/\sqrt{2})$ not used in calculating κ_1 and κ_2 in (13.82)?

Sec. 13.4. Dispersion

13.15. Typical laboratory plasmas have number densities of $10^{18} - 10^{22}$ electrons/m^3. What is the corresponding range of plasma frequencies? What is the corresponding range of attenuation lengths for low frequency electromagnetic waves? (Low frequency means much less than the plasma frequency.)

13.16. As a simple model, the conduction electrons in a metal may be considered to be a plasma (together with the ion cores that make the plasma neutral).

(a) Calculate the plasma frequency in copper, assuming one conduction electron per atom. The density of copper is 8.93×10^3 kg/m^3.

(b) According to this model, for what wavelengths is copper transparent to electromagnetic waves? To what part of the electromagnetic spectrum do these wavelengths belong?

13.17. In their interaction with electromagnetic radiation, the alkali metals behave approximately like plasmas.

(a) What is the plasma frequency for Na?

(b) Sodium is transparent for $\lambda \leq 210$ nm. Show that this agrees with the answer to (a).

13.18. For an electromagnetic wave in a dilute plasma, sketch a graph showing both the phase velocity and the group velocity as functions of frequency ω.

General Exercises

13.19. Consider a model of the conductivity of a plasma in which the conductivity is purely imaginary $\sigma = i\sigma_2$; that is, $\mathbf{J} = i\sigma_2 \mathbf{E}$ in the complex phase representation of the wave.

(a) From Maxwell's equations, derive the relation between κ and ω for an electromagnetic wave propagating in the plasma, with fields given by

$$\mathbf{E} = E_0 \hat{\mathbf{j}} e^{i(\kappa x - \omega t)},$$
$$\mathbf{B} = B_0 \hat{\mathbf{k}} e^{i(\kappa x - \omega t)}.$$

(b) Show that $\mathbf{J}$ lags behind $\mathbf{E}$ by 90 degrees.

13.20. What are the reflectivity and transmissivity for AM radio waves, frequency 1 MHz, incident normally on the surface of a lake? (Assume water is a dielectric with $\epsilon/\epsilon_0 = 81$ at this frequency.)

Theoretical physics often uses simple models to explain physical phenomena, at least qualitatively. The next two problems concern the classical electron theory that we used to estimate the frequency dependence of electromagnetic parameters.

13.21. Using the classical electron theory we derived a formula for the polarizability of a molecule

$$\alpha(\omega) = \frac{n_e e^2/m}{\omega_0^2 - \omega^2 - i(\gamma/m)\omega} \approx \frac{n_e e^2}{m\omega_0^2}.$$

This is essentially (13.94) but we have extended it by calling n_e the number of electrons that contribute to the dipole moment of the molecule, and in the approximate form we have assumed $\omega \ll \omega_0$. Here $\omega_0 = \sqrt{K/m}$ is the natural frequency, assumed to be the same for all electrons. This model should give a reasonable estimate of the dielectric constant κ ($\equiv \epsilon/\epsilon_0$) according to the Clausius-Mossotti relation

$$\frac{\kappa - 1}{\kappa + 2} = \frac{N\alpha}{3\epsilon_0},$$

where N is the molecular density.

(a) Estimate $\hbar\omega_0$ using this model for water ($\kappa = 1.8$ for visible light).
(b) From what you know about atomic and molecular physics, does your result make sense?

13.22. Using the classical electron theory we derived a formula for the (complex) conductivity of a plasma

$$\sigma(\omega) = \frac{\nu_e e^2/m}{(\gamma/m) - i\omega} = \frac{\epsilon_0 \omega_p^2}{(\gamma/m) - i\omega},$$

where ω_p is the plasma frequency. For a dilute plasma the damping factor γ is small, so for the purposes of illustration assume $\gamma = 0.01 m\omega_p$. Consider a typical laboratory plasma, with $\nu_e = 10^{15}$ cm^{-3}.

(a) For what frequencies is $\mathrm{Re}\,\sigma \gg \mathrm{Im}\,\sigma$? Describe in words the behavior of an electromagnetic wave in the plasma if the frequency is in this range.
(b) For what frequencies is $\mathrm{Re}\,\sigma \ll \mathrm{Im}\,\sigma$? Describe in words the behavior of an electromagnetic wave in the plasma if the frequency is in this range.

13.23. Figure 13.13 shows the light scattering process that creates the primary rainbow. Light rays at varying impact parameter refract into a spherical water drop, reflect from the back surface, and refract out of the drop. (In the figure, only the rays entering the upper half of the drop are shown. The rays shown are the rays that would reach the ground.) At a scattering angle of about 42° there is a concentration of scattered rays, called the *caustic*; and that somewhat more intense scattered light is the rainbow.

(a) Explain why the ordering of colors (ROYGBIV) is red at the outer edge of the arc, and violet at the inner edge.

(b) A secondary rainbow, in which the order of colors is reversed, is sometimes visible at a higher angle than the primary. Explain this second arc.

(c) Explain why the area inside the primary rainbow is brighter that the area outside.

13.24. The criterion (13.81) for a good conductor depends on frequency as well as conductivity. If the frequency is low enough, matter that we intuitively consider insulating can behave as conducting. This exercise uses that idea.

(a) What are the skin depths of 10 kHz electromagnetic waves for: dry earth ($\epsilon = 3\epsilon_0, \sigma = 10^{-4}$ S/m), wet earth ($\epsilon = 10\epsilon_0, \sigma = 10^{-2}$ S/m), and sea water ($\epsilon = 81\epsilon_0, \sigma = 4$ S/m)? Frequencies in this range are used for naval communication.

(b) What are the reflectivities of these materials for 10 kHz waves?

Computer Exercises

13.25. Use computer graphics to plot the reflection and transmission amplitudes, E_0''/E_0 and E_0'/E_0 respectively, as functions of the angle of incidence θ, for TM polarized light incident from air into diamond. (Transverse magnetic (TM) polarization has **B** perpendicular, and therefore **E** parallel, to the plane of incidence.) The index of refraction of diamond is 2.42.

13.26. Calculate the reflectivity R and the transmissivity T for TE waves obliquely incident on the interface between two dielectrics.

Use computer graphics to plot R and T as functions of the angle of incidence θ for the case $n_1 = 1$ and $n_2 = 1.5$.

13.27. Consider the boundary conditions (13.41) and (13.44) for light incident normally on a dielectric coating on glass. The reflectivity is $R \equiv |E_0''/E_0|^2$.

Plot R as a function of $k_1 a$ in the range from 0 to 2π, for various values of n_1. (Let $n_2 = 1.5$ for glass.) Note that $n_1 = 1$ and $n_1 = n_2$ are both equivalent to no coating. For what value of n_1 is it possible to have $R = 0$? What happens if $n_1 > n_2$?

CHAPTER

14

Wave Guides and Transmission Lines

Microwaves are electromagnetic waves with frequencies from 300 MHz to 300 GHz. The range of wavelengths is 1 mm to 1 m. This part of the electromagnetic spectrum is very important in modern electrical engineering. The ability to generate microwaves, and to control their direction and intensity, has created new technologies that we use every day, in some cases without even being aware of it.

The first use of microwaves was for radar. Efficient microwave generators—the magnetron and the klystron—were developed during World War II for military radar.[1] Today radar has many commercial applications, including air traffic control, weather observation, and enforcement of speed limits. Detecting an object by scattering of waves—the basic idea of radar—requires that the wavelength be smaller, or at most comparable, to the size of the object. Microwaves have the wavelengths appropriate for radar.

Another significant application of microwaves is in communications. Sending large amounts of information rapidly and economically over large distances requires many separate channels. So, for example, in 1981 the cellular phone system in North America was established by the Federal Communications Commission to operate in the microwave frequency ranges 824–849 MHz and 869–894 MHz. The cellular system has grown so that now there are more than 30×10^6 portable telephones, the latest operating in the GHZ range, over 20×10^3 cell sites in the United States, and many more worldwide. As another example, microwave links between antennas on high towers carry television or telephone signals over thousands of miles. (The UHF channels of commercial television are in the microwave range 470–890 MHz.) Microwaves are not reflected by the ionosphere, so each tower in the microwave network must be in the line-of-sight of its neighbors. The concept of microwave links has been extended by using satellites orbiting the Earth as the repeater stations.

A variety of other applications of microwaves include research in atomic and molecular physics (electron spin resonance), the use of microwave cavities in high-energy charged-particle accelerators, and microwave ovens (2.45 GHz). On a cosmic scale, the radiation remnant of the big bang is a blackbody spectrum at temperature 3 K, which is maximum in the microwave range.

---

[1] Much of the early development of radar was done at the M. I. T. Radiation Laboratory. "It's simple," [I. I.] Rabi told the theorists who were staring at the disassembled parts of the [magnetron] tube. "It's just a kind of whistle." "Okay, Rabi," Edward Condon responded, "how does a whistle work?" (History of Physics Collection, Niels Bohr Library, American Institute of Physics)

523

An essential component of microwave circuits is the *wave guide*—a hollow metal tube with, most simply, a rectangular cross section—in which the electromagnetic waves propagate with little attenuation from one part of the device to another. For example, in a radar station microwaves generated in a resonant cavity travel through a wave guide to a horn antenna from which they emerge into free space to be scattered by the detected objects.

The propagation of electromagnetic waves in a finite volume containing either vacuum or loss-free dielectric, and bounded by conductors, is thus an important problem in electromagnetic field theory. The propagation is rather different from propagation in free space, because the fields are affected by charge and current distributions induced in the conducting walls. Heuristically we can picture it this way: A beam of radiation sent into the end of a wave guide will reflect from any conducting wall that it hits, and so it will travel down the guide bouncing back and forth between opposite walls. Physically, the fields in the guide and the charges in the walls influence each other and so affect the propagation. Mathematically, the propagation characteristics are determined by partial differential equations and boundary conditions, the whole set of equations being satisfied in a self-consistent way.

We'll analyze several basic examples involving wave motion in bounded volumes. We'll find that for a given frequency there are a finite number of discrete *modes of propagation*, the number depending on the frequency; and there is *dispersion*, i.e., the group velocity is different from the phase velocity.

14.1 ■ ELECTROMAGNETIC WAVES BETWEEN PARALLEL CONDUCTING PLANES

Our first example is a very idealized case—propagation of waves in the space between infinite parallel planes. Although not realistic,[2] this simple model will provide some useful insights, which will later help us understand the realistic problem of a rectangular wave guide.

Figure 14.1 shows the geometry of this example. The planes $y = 0$ and $y = b$ are boundaries of perfect conductors in the regions $y < 0$ and $y > b$. The region between the planes $0 \leq y \leq b$ is vacuum, or some weak dielectric like air for which the polarization is negligible. The question is, how do electromagnetic waves propagate in this bounded space? We shall solve the field equations and boundary conditions for harmonic waves propagating in the $+z$ direction. (Because the planes are infinite the propagation is the same in any direction orthogonal to the y axis.) For simplicity we shall only consider waves that are translation invariant in the x direction, a further idealization.

By a perfect conductor we mean that the conductivity of the material is infinite. Then the skin depth is 0, and the electric and magnetic fields do not penetrate into the material. Another way to understand why there are no fields in the material

---

[2]This "parallel plane transmission line" does resemble in some respects a device in microwave engineering called a microstrip transmission line.

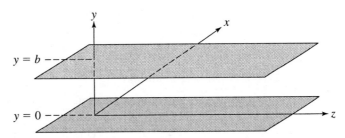

FIGURE 14.1 Parallel planar boundaries of a conducting material. For $y < 0$ and $y > b$ the conductivity is infinite, and for $0 \leq y \leq b$ the conductivity is 0.

is to recall that $\mathbf{E} = 0$ inside a perfect conductor. It then follows from Faraday's law that $\partial \mathbf{B}/\partial t = 0$, so any magnetic field must be constant in time. We set $\mathbf{B} = 0$ initially, so it remains 0. We will use this simple fact several times in the discussions in this chapter.

The normal component of $\mathbf{B}$ and the tangential components of $\mathbf{E}$, being continuous at any surface, must be 0 at the planes $y = 0$ and $y = b$. Also, there will be surface charge σ and surface current $\mathbf{K}$ on the boundary planes, induced by the fields. Approximating a metal surface by a perfect conductor gives a reasonable description of the fields and surface charge distributions, but it cannot be used to determine the ohmic loss in the walls.

A harmonic wave propagating in the z direction and uniform in the x direction has fields of the form[3]

$$\mathbf{E}(\mathbf{x}, t) = \left[\hat{\mathbf{i}} \, E_x(y) + \hat{\mathbf{j}} \, E_y(y) + \hat{\mathbf{z}} E_z(y) \right] e^{i(kz - \omega t)} \tag{14.1}$$

$$\mathbf{B}(\mathbf{x}, t) = \left[\hat{\mathbf{i}} \, B_x(y) + \hat{\mathbf{j}} \, B_y(y) + \hat{\mathbf{z}} B_z(y) \right] e^{i(kz - \omega t)}. \tag{14.2}$$

As usual when using complex functions, the real part of (14.1) or (14.2) is understood to be the physical field. We could insert these forms into the Maxwell equations and deduce the general solutions, but it will be simpler to anticipate some properties of the basic solutions, and pick specific cases of (14.1) and (14.2). The boundary conditions of the fields will place additional constraints on the solutions. Indeed we will find that the boundaries have significant and interesting effects on wave propagation.

In an infinite vacuum the plane electromagnetic wave is transverse in *both* $\mathbf{E}$ *and* $\mathbf{B}$. Such a wave, for which $\mathbf{E}$ and $\mathbf{B}$ are both orthogonal to the direction of propagation, is called a TEM wave ("transverse electric and magnetic"). A TEM wave propagating in the z direction has $E_z = 0$ and $B_z = 0$. In a bounded volume there are also solutions for which only one of E_z and B_z is 0. Such solutions with $E_z = 0$ are called TE waves ("transverse electric"), and those with $B_z = 0$ are called TM waves ("transverse magnetic"). Superpositions of any of these waves are also solutions. Furthermore, the harmonic waves are *complete*: any solution

---

[3] In this chapter we use $\hat{\mathbf{z}}$ for the unit vector in the z direction.

can be written as a superposition of harmonic waves. For Fig. 14.1 we shall begin by deriving the TEM solution, which is rather simple, and then analyze the more complicated TE and TM waves.

14.1.1 ■ The TEM Solution

There exists a solution of the field equations for which $E_x = E_z = 0$ in (14.1). As we shall see, this solution is a TEM wave. Gauss's Law for this case is $\nabla \cdot \mathbf{E} = \partial E_y / \partial y = 0$, so E_y is a constant E_0; the electric field between the planes is

$$\mathbf{E}(\mathbf{x}, t) = \hat{\mathbf{j}} E_0 e^{i(kz - \omega t)}. \tag{14.3}$$

We can determine the magnetic field from Faraday's Law, $\nabla \times \mathbf{E} = -\partial \mathbf{B} / \partial t$. The frequency of oscillation of $\mathbf{B}$ must be the same as $\mathbf{E}$, so $-\partial \mathbf{B} / \partial t = i\omega \mathbf{B}$; thus

$$\mathbf{B}(\mathbf{x}, t) = \frac{-i}{\omega} \nabla \times \mathbf{E} = -\hat{\mathbf{i}} \frac{k E_0}{\omega} e^{i(kz - \omega t)}. \tag{14.4}$$

The Ampère–Maxwell Law $c^2 \nabla \times \mathbf{B} = \partial \mathbf{E} / \partial t$ must also be satisfied; a short calculation shows that this field equation requires

$$\omega = ck. \tag{14.5}$$

(Alternatively we could arrive at (14.5) from the wave equation.) Thus the wave propagates in the z direction with speed c, for any frequency. The real parts of the complex fields in (14.3) and (14.4) are the physical fields. As the electric and magnetic fields at a fixed point oscillate in time, their oscillations remain perfectly *in phase*.

The TEM wave resembles a plane wave in free space, but chopped off below $y = 0$ and above $y = b$. The fields are illustrated in Fig. 14.2. The wave fronts are planar strips orthogonal to the direction of propagation ($\hat{\mathbf{z}}$) with $0 \le y \le b$. At each point, $\mathbf{E}$, $\mathbf{B}$, and $\hat{\mathbf{z}}$ form an orthogonal triad, and $|\mathbf{B}| = |\mathbf{E}|/c$. The Poynting vector is

$$\mathbf{S}(\mathbf{x}, t) = \frac{1}{\mu_0} \mathbf{E} \times \mathbf{B} = \hat{\mathbf{z}} \frac{E_0^2}{\mu_0 c} \cos^2(kz - \omega t), \tag{14.6}$$

so energy flows only in the direction of propagation, with intensity $E_0^2 / (2\mu_0 c)$.

However, there is another aspect of this problem—the charge and current on the boundary surfaces. Their densities, σ and $\mathbf{K}$ respectively, can be calculated from general boundary conditions. The tangential components of $\mathbf{E}$ are 0 at the surfaces $y = 0$ and $y = b$, in accord with the continuity of E_t because $\mathbf{E} = 0$ in the perfect conductor. However, the normal component E_n is discontinuous, and its discontinuity is σ / ϵ_0; thus the surface charge densities are

$$\sigma = \epsilon_0 E_y = \epsilon_0 E_0 \cos(kz - \omega t) \quad \text{at} \quad y = 0;$$

$$\sigma = -\epsilon_0 E_y = -\epsilon_0 E_0 \cos(kz - \omega t) \quad \text{at} \quad y = b.$$

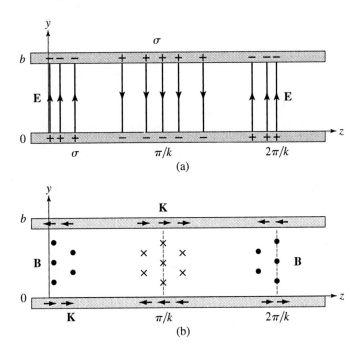

FIGURE 14.2 Snapshot of the TEM mode, for waves between parallel conducting planes. The diagram shows schematically (a) the electric field and charge density, and (b) the magnetic field and current density, at an instant of time. A dot indicates a vector pointing out of the page, and a cross, one into the page. The charges and currents reside on the boundary surfaces at $y = 0$ and $y = b$. The fields and densities move in the z direction as time passes.

The normal component of **B** is 0 at $y = 0$ and b, in accord with the continuity of B_n. However, the tangential component B_t is discontinuous, corresponding to a surface current density $\mathbf{K} = \hat{\mathbf{n}} \times \mathbf{B}/\mu_0 = \hat{\mathbf{z}} K_z$ with

$$K_z = -\frac{B_x}{\mu_0} = +\frac{E_0}{\mu_0 c} \cos(kz - \omega t) \quad \text{at} \quad y = 0;$$

$$K_z = +\frac{B_x}{\mu_0} = -\frac{E_0}{\mu_0 c} \cos(kz - \omega t) \quad \text{at} \quad y = b.$$

Figure 14.2 is a schematic diagram of the fields, charge, and current distributions, in the form of a *snapshot* at $t = 0$ of one wavelength of the TEM mode. (The wavelength λ is $2\pi/k$.) Figure 14.2(a) shows the electric field and the surface charge, and Fig. 14.2(b) shows the magnetic field and the surface current. Unlike TEM modes in an infinite vacuum, for which **E** can point in any direction orthogonal to the direction of propagation, the TEM mode in the bounded volume must have **E** pointing across the gap.

The TEM mode is rather featureless. It propagates with velocity c for any frequency. The wave fronts are planar cross sections of the gap, orthogonal to the direction of propagation, and the fields are independent of x and y. The other modes of propagation (TE and TM) are more interesting.

14.1.2 ■ TE Waves

What we mean by a TE wave is that $\mathbf{E}$ is transverse to the direction of propagation. But $\mathbf{B}$ will have components in both transverse and longitudinal directions.

For a TE wave, $E_z = 0$ in (14.1). Then Gauss's law in the bounded volume is

$$\mathbf{V} \cdot \mathbf{E} = \partial E_y / \partial y = 0,$$

so E_y is constant. In the previous section we considered $E_y = E_0$, a nonzero constant, and $E_x = 0$, which led to the TEM mode. Now we take $E_0 = 0$ and consider the orthogonal polarization with E_x nonzero. That is, we seek solutions for which the electric field is polarized in the x direction,

$$\mathbf{E}(\mathbf{x}, t) = \hat{\mathbf{i}}\, E_x(y) e^{i(kz - \omega t)}. \tag{14.7}$$

Our first task is to determine $E_x(y)$ and the relation between the frequency ω and wave vector k.

We can immediately determine the form of the magnetic field from Faraday's Law, $\partial \mathbf{B}/\partial t = -\mathbf{V} \times \mathbf{E}$. The frequency of oscillation of $\mathbf{B}$ must be the same as $\mathbf{E}$, so $\partial/\partial t$ may be replaced by $-i\omega$. Therefore, evaluating $\mathbf{V} \times \mathbf{E}$,

$$\mathbf{B}(\mathbf{x}, t) = \frac{-i}{\omega} \mathbf{V} \times \mathbf{E} = \left[\hat{\mathbf{j}} \frac{k E_x(y)}{\omega} + \hat{\mathbf{z}} \frac{i E_x'(y)}{\omega} \right] e^{i(kz - \omega t)}. \tag{14.8}$$

(The prime denotes the derivative with respect to y.) Note that Gauss's law $\mathbf{V} \cdot \mathbf{B} = 0$ is already satisfied by the field in (14.8)—a nontrivial result.

One more Maxwell equation must be satisfied—the Ampère–Maxwell Law, $\partial \mathbf{E}/\partial t = c^2 \mathbf{V} \times \mathbf{B}$. Substituting the functions (14.7) and (14.8) this field equation implies

$$-i\omega E_x(y) = \frac{c^2}{\omega} \left[i E_x''(y) - i k^2 E_x(y) \right];$$

or,

$$E_x''(y) = -\left(\frac{\omega^2}{c^2} - k^2 \right) E_x(y). \tag{14.9}$$

This differential equation will be easy to solve, but the solution depends on the boundary conditions. The most general solution of (14.9) is

$$E_x(y) = c_1 \sin \nu y + c_2 \cos \nu y, \tag{14.10}$$

where c_1 and c_2 are constants, and $\omega^2/c^2 = k^2 + \nu^2$. The boundary conditions place constraints on c_1, c_2, and ν.

Boundary Conditions

The transverse components of **E** are continuous at any surface, so in particular at the boundary planes $y = 0$ and $y = b$. Because we are assuming perfectly conducting walls, the electric field is 0 in the material. Thus the boundary conditions for the function $E_x(y)$ in (14.7) are

$$E_x(0) = E_x(b) = 0. \tag{14.11}$$

The first condition implies that $c_2 = 0$ in (14.10); the second condition implies that $\sin \nu b = 0$, so νb is an integer multiple of π. Therefore the solution to the differential equation (14.9) with boundary values (14.11) is

$$E_x(y) = E_0 \sin\left(\frac{n\pi y}{b}\right), \tag{14.12}$$

where n is a positive integer, and

$$\frac{\omega^2}{c^2} = k^2 + \left(\frac{n\pi}{b}\right)^2. \tag{14.13}$$

Equation (14.13), or more specifically the function $\omega = \omega(k)$, is called the *dispersion relation* for these waves. We conclude that the electric and magnetic fields of a TE solution are

$$\mathbf{E}(\mathbf{x}, t) = \hat{\mathbf{i}} \, E_0 \sin\left(\frac{n\pi y}{b}\right) e^{i(kz - \omega t)}, \tag{14.14}$$

$$\mathbf{B}(\mathbf{x}, t) = \left[\hat{\mathbf{j}} \frac{k E_0}{\omega} \sin\left(\frac{n\pi y}{b}\right) + \hat{\mathbf{z}} \frac{i n\pi E_0}{b\omega} \cos\left(\frac{n\pi y}{b}\right)\right] e^{i(kz - \omega t)}. \tag{14.15}$$

We derived these solutions requiring that $\mathbf{E}(\mathbf{x}, t)$ is translation invariant in x. In fact it can be shown[4] that all solutions that are harmonic in time, traveling in the z direction and polarized in the x direction, satisfy this requirement. Also, the *real parts* of the complex functions in (14.14) and (14.15) are the physical fields.

Another boundary condition is that the normal component of **B** must be continuous at $y = 0$ and $y = b$. That is, since $\mathbf{B} = 0$ in the conducting walls,

$$B_y(0) = B_y(b) = 0 \tag{14.16}$$

using the notation of (14.2). These conditions are indeed satisfied by (14.15).

Propagation Characteristics

The solution (14.14) and (14.15) is called the TE(n) mode. Each field is a traveling wave in the z direction and a standing wave in the y direction. The electric field is

[4] See Exercise 3.

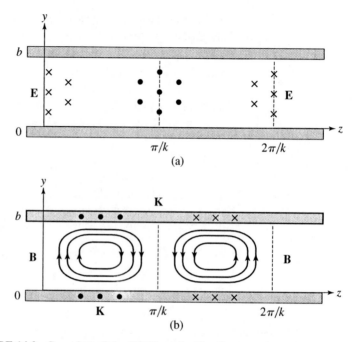

FIGURE 14.3   Snapshot of the TE(1) mode. The diagrams show schematically (a) the electric field, and (b) the magnetic field and current density, for a TE(1) wave between planar boundaries of perfect conductors, at an instant of time. The wave travels in the z direction, and the wavelength is $\lambda = 2\pi/k$. A dot indicates a vector pointing out of the page, and a cross, one into the page. The curves in (b) are tangent curves of the magnetic field.

polarized in the x direction. The magnetic field has two components: a transverse component in the y direction, which oscillates in phase with the electric field, and a longitudinal component in the z direction, which oscillates with a 90 degree phase shift relative to the electric field.

Figure 14.3 is a schematic diagram of a snapshot at $t = 0$ of the electric and magnetic fields for one wavelength of the TE(1) mode. (A quantitatively accurate plot is provided in Fig. 14.5.) In Fig. 14.3 the wavelength in the z direction is $\lambda = 2\pi/k$. The electric field alternates in the $\pm x$ direction, and the magnetic field curls around the changing electric field. The wave propagates in the z direction, so at a later time t the fields in Fig. 14.3 will be translated in the z direction by $\omega t/k$. What this diagram cannot illustrate is the *field magnitudes* as functions of y and z. E_x is maximum at the center and 0 at the boundaries; B_y has the same pattern. B_z is maximum at the boundaries and 0 at the center.

The dispersion relation (14.13) determines the wave velocities. For electromagnetic waves in an infinite vacuum the phase velocity and group velocity are both equal to $c = 3 \times 10^8$ m/s. The TEM mode for Fig. 14.1 is the same. But the TE modes are much different, because the frequency ω is not a linear function of the wave vector k; the phase and group velocities differ, and depend on the frequency.

The *phase velocity* is ω/k, because a constant phase point in the factor $e^{i(kz-\omega t)}$ moves in the z direction with velocity $\delta z/\delta t = \omega/k$; thus

$$v_{\text{ph}} = \frac{\omega}{k} = \frac{c\omega}{\sqrt{\omega^2 - (n\pi c/b)^2}}. \tag{14.17}$$

The phase velocity is always greater than c, the speed of light in free space.[5] The phase velocity approaches c in the limit of high frequencies, but it approaches ∞ as ω decreases to the value $n\pi c/b$. Also, there is no propagation in the TE(n) mode for $\omega < n\pi c/b$, because for $\omega < n\pi c/b$ the wave vector k determined from (14.13) would be imaginary; in this case the fields would decay exponentially as a function of z rather than propagating as a wave. The parameter $\omega_n \equiv n\pi c/b$ is called the *cutoff frequency* of the mode TE(n). The absolute cutoff frequency of TE waves, below which no TE wave can propagate between the planes, is the cutoff frequency ω_1 of the TE(1) mode,

$$\omega_{\text{cutoff}} = \frac{\pi c}{b}. \tag{14.18}$$

(The TEM mode propagates for any ω, all the way down to 0.)

The *group velocity* of TE waves in the space between the conducting planes is

$$v_{\text{gr}} = \frac{d\omega}{dk} = \frac{c^2 k}{\omega} = c\sqrt{1 - (n\pi c/\omega b)^2}. \tag{14.19}$$

(Note from (14.13) that $\omega d\omega = c^2 k dk$.) The group velocity is the signal velocity, and also the velocity of energy transport. It is less than c for any frequency, and it approaches 0 as ω approaches the cutoff frequency $n\pi c/b$. Figure 14.4 shows plots of v_{ph} and v_{gr} versus ω.

The Surface Current

The longitudinal ($\hat{\mathbf{z}}$) component of $\mathbf{B}(\mathbf{x}, t)$ is discontinuous at the conducting surfaces. The discontinuity of this tangential component implies the existence of surface currents in the conducting planes. Recall the boundary condition on $\mathbf{B}_t$,

$$\mathbf{B}_{t2} - \mathbf{B}_{t1} = \mu_0 \hat{\mathbf{n}} \times \mathbf{K}, \tag{14.20}$$

where $\mathbf{K}(\mathbf{x}, t)$ is the surface current density and $\hat{\mathbf{n}}$ is the unit normal vector pointing from $2 \rightarrow 1$, i.e., from the conductor into the space between. For the case of a perfect conductor, we set $\mathbf{B}_{t2} = 0$ in (14.20) and operate with $\hat{\mathbf{n}}\times$ on both sides. The result is $\mathbf{K} = \hat{\mathbf{n}} \times \mathbf{B}_{t1}/\mu_0$, a convenient form for calculating the surface currents in this chapter. The normal vector $\hat{\mathbf{n}}$ is $+\hat{\mathbf{j}}$ at $y = 0$ and $-\hat{\mathbf{j}}$ at $y = b$; the

---

[5]At this point it is natural to ask whether the result $v_{\text{ph}} > c$ conflicts with the theory of relativity. It does not. We shall explore this question later.

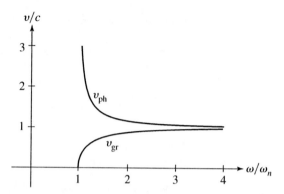

FIGURE 14.4   Phase and group velocities, for TE(n) or TM(n) modes of electromagnetic waves propagating between parallel planes, as functions of frequency ω. The cutoff frequency is $\omega_n = n\pi c/b$. For $\omega < \omega_n$ the modes do not propagate.

direction of **B** is $\pm\hat{\mathbf{z}}$ at the inner surfaces, the sign depending on z and t. Therefore surface currents flow in the $\pm\hat{\mathbf{i}}$ direction. Because **B** is 0 in the perfect conductor, the discontinuity of **B** is just the value of **B** at the surface in the interior. Thus the surface current densities are given by

$$\mathbf{K} = +\frac{B_z(0)}{\mu_0}\,\hat{\mathbf{i}} = -\frac{n\pi E_0}{\mu_0 b\omega}\sin(kz - \omega t)\hat{\mathbf{i}} \quad \text{at} \quad y = 0,$$

$$\mathbf{K} = -\frac{B_z(b)}{\mu_0}\,\hat{\mathbf{i}} = (-1)^n\frac{n\pi E_0}{\mu_0 b\omega}\sin(kz - \omega t)\hat{\mathbf{i}} \quad \text{at} \quad y = b. \quad (14.21)$$

(The imaginary parts have been dropped.) In other words, current waves propagate on the boundary surfaces along with the field wave in the space between the planes. The existence of a longitudinal component of **B** is not surprising once we see that there are surface currents, because **B** must curl around **K**. Figure 14.3(b) shows the surface currents for a snapshot of one wavelength of the mode TE(1).

Energy Transport
The final part of the analysis is to find the energy transport of the wave. The energy flux is the Poynting vector $\mathbf{S} = \mathbf{E} \times \mathbf{B}/\mu_0$, which has both z and y components. To calculate **S** we must use the physical fields—the real part of the complex wave. It is straightforward to show from the fields (14.14) and (14.15) that the Poynting vector for the mode TE(n) is

$$\mathbf{S}(\mathbf{x}, t) = \frac{kE_0^2}{\mu_0\omega}\left[\hat{\mathbf{z}}\sin^2\left(\frac{n\pi y}{b}\right)\cos^2(kz - \omega t)\right. \quad (14.22)$$

$$\left. + \hat{\mathbf{j}}\frac{n\pi}{kb}\sin\left(\frac{n\pi y}{b}\right)\cos\left(\frac{n\pi y}{b}\right)\sin(kz - \omega t)\cos(kz - \omega t)\right].$$

The y component of $\mathbf{S}$, which describes energy transport in the direction from one plane to the other, is proportional to $\sin(kz - \omega t)\cos(kz - \omega t)$. This function oscillates about 0, and averages to 0 over one period of oscillation, so it corresponds to energy transfer back and forth across the gap. In contrast, there is a net transport of energy in the z direction, i.e., the direction of wave propagation. The average power per unit area is

$$\mathbf{S}_{\text{avg}} = \frac{\omega}{2\pi} \int_0^{2\pi/\omega} \mathbf{S}(\mathbf{x}, t)\, dt = \frac{kE_0^2}{2\mu_0\omega} \sin^2\left(\frac{n\pi y}{b}\right) \hat{\mathbf{z}}. \tag{14.23}$$

Integrating $\mathbf{S}_{\text{avg}}$ over y yields the power per unit of length in the x direction carried by the electromagnetic wave

$$\frac{dP}{dx} = \int_0^b \mathbf{S}_{\text{avg}} \cdot \hat{\mathbf{z}}\, dy = \frac{bE_0^2}{4\mu_0 c} \sqrt{1 - (n\pi c/\omega b)^2}. \tag{14.24}$$

The power carried by the wave is proportional to the group velocity,[6] so it tends to 0 as ω decreases to the cutoff frequency $n\pi c/b$. This result shows that the group velocity is the velocity of energy transport.

Figure 14.5 is a quantitatively accurate diagram of the energy flow for the TE(1) mode, in the form of a snapshot at $t = 0$ for one wavelength of the

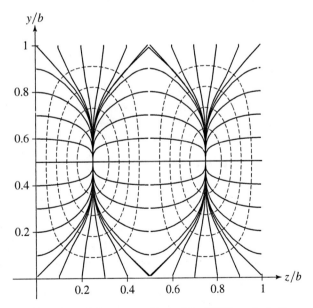

FIGURE 14.5 Snapshot of the energy flux of the TE(1) mode in the space between parallel planar boundaries of perfect conductors. The wavelength λ is equal to the separation between the planes b in this figure. The solid curves are everywhere tangent to the Poynting vectors, and the dashed curves are tangent to the magnetic field vectors.

[6]See Exercise 1.

mode. The solid curves are the tangent curves of the field of Poynting vectors. The magnetic field is also illustrated; the dashed curves are the tangent curves of the magnetic field vectors. (Note that **S** is everywhere orthogonal to **B**.) This plot is for the special case $\lambda = b$.

Phase Velocity Greater Than c

It is sometimes stated, incorrectly, that "nothing can travel faster than c, the speed of light in vacuum." We have just seen an example of something that does travel faster than c: the constant phase point of a guided wave. It is true that no massive particle can travel faster than c, but that's different. Also, no signal carrying information can travel faster than c, but that's also different. The constant phase point is not a carrier of information.

To pursue this idea a little further, consider a pulse of electromagnetic radiation. The pulse carries information, because it might be, for example, part of a message in Morse code; or it might be a signal sent at some specific time. By Fourier analysis, the pulse may be considered to be composed of infinite trains of waves with a broad spectrum of frequencies. If the pulse is emitted into free space then all the component waves travel at speed c (the phase velocity) so that the pulse also travels at speed c (the group velocity) and arrives at its destination with the same shape with which it was emitted. In this case the information associated with the pulse has traveled with speed c. But suppose instead that the pulse is emitted into a parallel-plane transmission line such as we have been considering. Then the pulse will change as it travels, because of *dispersion*. Low frequencies cannot propagate in the bounded region. Modes with frequency above the cut-off propagate in the waveguide, but each mode travels with a different phase velocity that depends on frequency. All the phase velocities are greater than c, as we have seen. However, no *information* is carried by the individual infinite wave trains. Information can only be transmitted by some identifiable feature of the pulse, such as the point of maximum amplitude. Such features travel with the group velocity, which is less than c.[7]

There is no conflict between the theory of relativity and the fact that the constant phase point travels faster than c. Figure 14.6 shows an analogous situation—ocean waves breaking on a beach. The waves travel at speed v and approach the shore at angle α. The point on the wave where the wave encounters the shore travels down the beach with speed $v/\cos\alpha$, which is greater than the speed of the wave in open water. This example is analogous to the motion of the constant phase point of a TE wave between parallel planes; in both cases a certain phase point travels faster than the free wave speed.

The TE Wave as a Sequence of Reflections

We derived the TE wave fields (14.14) and (14.15) by a straightforward mathematical approach—solving the boundary value problem for Maxwell's equations in the region between the planes. It is also interesting to understand the solution in another way. The electric field (14.14) is a traveling wave in z and a standing

---

[7]Good discussions of group and phase velocities can be found in Refs. [3] and [4].

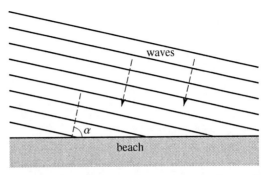

FIGURE 14.6   Waves on a beach. The wave speed is v, the distance between crests is λ, and the waves approach the beach at an angle α. In the time it takes for the wave to travel a distance λ, the point where the wave breaks travels distance $\lambda/\cos\alpha$ to the left along the beach. Therefore the point where a wave encounters the shore travels down the beach at speed $v/\cos\alpha$.

wave in y. But a standing wave is a superposition of traveling waves in opposite directions. Therefore (14.14) is a superposition of two waves with wave vectors $\mathbf{k}_+$ and $\mathbf{k}_-$ given by

$$\mathbf{k}_\pm = k\hat{\mathbf{z}} \pm \frac{n\pi}{b}\hat{\mathbf{j}}; \tag{14.25}$$

specifically,

$$\mathbf{E}(\mathbf{x}, t) = \mathbf{E}_+ e^{i(\mathbf{k}_+ \cdot \mathbf{x} - \omega t)} + \mathbf{E}_- e^{i(\mathbf{k}_- \cdot \mathbf{x} - \omega t)}, \tag{14.26}$$

where $\mathbf{E}_\pm = \pm \hat{\mathbf{i}}\, E_0/2i$. In any small section of the wave guide the two terms in (14.26) may be interpreted as an incident wave plus a reflected wave. The wave vectors $\mathbf{k}_+$ and $\mathbf{k}_-$ have equal components along $\hat{\mathbf{z}}$ and equal but opposite components along $\hat{\mathbf{j}}$. This relationship between $\mathbf{k}_+$ and $\mathbf{k}_-$ describes *specular reflection* from the planes $y = 0$ or $y = b$.

The individual terms in (14.26) have the same properties as waves in free space. The dispersion relation (14.13) implies that

$$\omega = c|\mathbf{k}_+| \quad \text{and} \quad \omega = c|\mathbf{k}_-|. \tag{14.27}$$

The vector amplitudes $\mathbf{E}_+$ and $\mathbf{E}_-$ are orthogonal to the respective wave vectors $\mathbf{k}_+$ and $\mathbf{k}_-$. We may also write the magnetic field as a superposition of waves with the wave vectors $\mathbf{k}_\pm$, as

$$\mathbf{B}(\mathbf{x}, t) = \mathbf{B}_+ e^{i(\mathbf{k}_+ \cdot \mathbf{x} - \omega t)} + \mathbf{B}_- e^{i(\mathbf{k}_- \cdot \mathbf{x} - \omega t)}. \tag{14.28}$$

By then applying $\nabla \times \mathbf{E} = -\partial\mathbf{B}/\partial t$ we obtain for the vector amplitudes

$$\mathbf{B}_\pm = \left[\hat{\mathbf{z}}\frac{n\pi}{b} \mp \hat{\mathbf{j}}k\right]\frac{iE_0}{2\omega}. \tag{14.29}$$

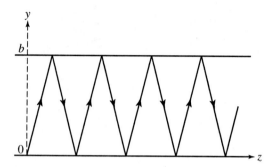

FIGURE 14.7   Rays for the superposed fields in (14.26) and (14.28).

The amplitudes $\mathbf{B}_+$ and $\mathbf{B}_-$ are orthogonal to $\mathbf{E}_+$ and $\mathbf{E}_-$, and also orthogonal to the respective wave vectors $\mathbf{k}_+$ and $\mathbf{k}_-$. Furthermore, the ratios $|\mathbf{B}_+|/|\mathbf{E}_+|$ and $|\mathbf{B}_-|/|\mathbf{E}_-|$ are both $1/c$. So in all respects the separate electromagnetic waves with wave vectors $\mathbf{k}_+$ and $\mathbf{k}_-$ are like waves in free space.

By writing the guided wave fields as the superposition in (14.26) and (14.28) we justify the heuristic picture of an electromagnetic wave reflecting back and forth between the planes, as illustrated in Fig. 14.7. However, the "ray" picture in Fig. 14.7 does not do complete justice to the phenomenon of wave motion in a wave guide, because it does not illustrate the *interference* between the superposed waves. Nevertheless, a nice point of this development is that it explains why the phase velocity of the guided wave is greater than c. Either of the superposed waves in (14.26) has phase speed c, but the phase velocities are in the directions of $\mathbf{k}_+$ or $\mathbf{k}_-$ rather than the direction $\hat{\mathbf{z}}$. The superposition is a traveling wave in the z direction with phase speed $c/\cos\alpha$, where α is the angle between $\mathbf{k}_+$ or $\mathbf{k}_-$ and $\hat{\mathbf{z}}$.[8]

Figure 14.5 shows a snapshot of the energy flow for the pure TE(1) mode, for the case $\lambda = b$. The energy flows back and forth between the planes, with the average flux in the z direction. This plot is consistent with the heuristic picture of the wave motion as a sequence of reflections.

Equation (14.22) shows that the ratio of the component of energy flux across the gap S_y, to the component in the direction of propagation S_z, is proportional to λ/b, where $\lambda = 2\pi/k$ is the wavelength parallel to the planes. If the wavelength is much shorter than b then the energy flow is strongly directed parallel to the planes; this is the case of high frequencies. In contrast, for long wavelengths there is strong energy flow back and forth across the gap. The cutoff frequency has $\lambda = \infty$, and in this limit all the energy flow is perpendicular to the bounding planes.

---

[8]See Exercise 2.

14.1.3 ■ TM Waves

For the TEM mode, **E** is normal to the conducting planes and **B** is tangential; see Fig. 14.2. For the TE modes **E** is tangential and **B** has both normal and tangential components; see Fig. 14.3. For the TM modes **B** is again purely tangential, as well as transverse to the direction of propagation. So, in this case the solutions take the form

$$\mathbf{B}(\mathbf{x}, t) = \hat{\mathbf{i}}\, B_x(y) e^{i(kz - \omega t)}. \tag{14.30}$$

However, unlike the TEM mode, **B** varies with y.

Gauss's Law ($\nabla \cdot \mathbf{B} = 0$) is satisfied by (14.30) for any $B_x(y)$. In a region of vacuum **B** must also obey the wave equation $\nabla^2 \mathbf{B} - c^{-2} \partial^2 \mathbf{B}/\partial t^2 = 0$. Substituting (14.30) into the wave equation we find that $B_x(y)$ must satisfy

$$B_x''(y) = -\gamma^2 B_x(y) \tag{14.31}$$

where $\gamma^2 = \omega^2/c^2 - k^2$. (Again, the prime indicates differentiation with respect to y.) Therefore $B_x(y)$ is a linear combination of $\cos \gamma y$ and $\sin \gamma y$. Another restriction on **B** is the boundary condition that the normal component must be 0 at the planes $y = 0$ and $y = b$; but this is always true for (14.30) because **B** is parallel to the planes.

The further constraints on $B_x(y)$ come from the boundary condition on the tangential component of the *electric field*. The form of **E** is determined by the Ampère–Maxwell Law

$$\nabla \times \mathbf{B} = \frac{1}{c^2} \frac{\partial \mathbf{E}}{\partial t} = \frac{-i\omega}{c^2} \mathbf{E},$$

which implies

$$\mathbf{E}(\mathbf{x}, t) = \left[-\hat{\mathbf{j}} k B_x(y) - \hat{\mathbf{z}} i B_x'(y) \right] \frac{c^2}{\omega} e^{i(kz - \omega t)}. \tag{14.32}$$

The tangential component E_z must be zero at the boundary surfaces, so

$$B_x'(0) = B_x'(b) = 0. \tag{14.33}$$

The solution of the differential equation (14.31) and boundary conditions (14.33) is

$$B_x(y) = B_0 \cos\left(\frac{n\pi y}{b}\right), \tag{14.34}$$

where n is an integer and $\gamma = n\pi/b$. The solution with $n = 0$ is precisely the TEM mode derived earlier. The TM modes have $n = 1, 2, 3, \ldots$.

Summarizing, the electric and magnetic fields of the TM(n) mode are

$$\mathbf{B}(\mathbf{x}, t) = \hat{\mathbf{i}}\, B_0 \cos\left(\frac{n\pi y}{b}\right) e^{i(kz - \omega t)} \tag{14.35}$$

$$\mathbf{E}(\mathbf{x}, t) = \left[-\hat{\mathbf{j}} \cos\left(\frac{n\pi y}{b}\right) + \hat{\mathbf{z}} \frac{in\lambda}{2b} \sin\left(\frac{n\pi y}{b}\right) \right] \frac{c^2 k B_0}{\omega} e^{i(kz - \omega t)}. \tag{14.36}$$

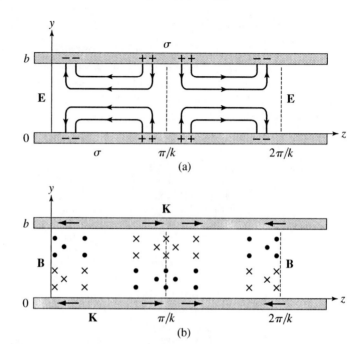

FIGURE 14.8   Snapshot of the TM(1) mode, for propagation between parallel conducting planes. The diagram shows schematically (a) the electric field and charge density, and (b) the magnetic field and current density. The charges and currents reside on the bounding surfaces at $y = 0$ and $y = b$.

The fields of the TM(1) mode are shown schematically in Fig. 14.8. This figure is a snapshot at $t = 0$ of one wavelength ($\lambda = 2\pi/k$) of the wave. As for TE waves, the TM waves may be regarded as a sequence of reflections.

The dispersion relation for the TM(n) mode is $\omega^2 = c^2 k^2 + (n\pi c/b)^2$, which is just the same as for the TE(n) mode. Therefore the phase and group velocities are also the same, as shown in Fig. 14.4.

From Fig. 14.8(b) we see that $\mathbf{B}_t$ is discontinuous at the boundary planes. Therefore there are surface currents with density $\mathbf{K}$ given, for the TM(1) mode, by

$$\mu_0 \mathbf{K} = +\hat{\mathbf{j}} \times \mathbf{B} = -\hat{\mathbf{z}} B_0 \cos(kz - \omega t) \quad \text{at} \quad y = 0,$$

$$\mu_0 \mathbf{K} = -\hat{\mathbf{j}} \times \mathbf{B} = -\hat{\mathbf{z}} B_0 \cos(kz - \omega t) \quad \text{at} \quad y = b;$$

these surface currents are shown in Fig. 14.8(b). From Fig. 14.8(a) we see that E_n is discontinuous at the boundary planes. Therefore there are surface charges with density σ given, for the TM(1) mode, by

$$\sigma/\epsilon_0 = +\hat{\mathbf{j}} \cdot \mathbf{E} = -\frac{c^2 k B_0}{\omega} \cos(kz - \omega t) \quad \text{at} \quad y = 0,$$

$$\sigma/\epsilon_0 = -\hat{\mathbf{j}} \cdot \mathbf{E} = -\frac{c^2 k B_0}{\omega} \cos(kz - \omega t) \quad \text{at} \quad y = b;$$

these surface charges are shown at $t = 0$ in Fig. 14.8(a). A wave of charge and current propagates on the surfaces along with the electromagnetic fields in the interior.

The Poynting vector for the TM(n) mode is

$$\mathbf{S}(\mathbf{x}, t) = \frac{c^2 k B_0^2}{\mu_0 \omega} \left[\hat{\mathbf{z}} \cos^2 \left(\frac{n\pi y}{b} \right) \cos^2(kz - \omega t) \right. \tag{14.37}$$

$$\left. - \hat{\mathbf{j}} \frac{n\lambda}{2b} \cos \left(\frac{n\pi y}{b} \right) \sin \left(\frac{n\pi y}{b} \right) \cos(kz - \omega t) \sin(kz - \omega t) \right].$$

Figure 14.9 shows a snapshot at $t = 0$ of the energy flux for one wavelength of the mode TM(1) for parameters with $\lambda = b$. The average energy flux is

$$\mathbf{S}_{\text{avg}} = \frac{c^2 k B_0^2}{2\mu_0 \omega} \cos^2 \left(\frac{n\pi y}{b} \right) \hat{\mathbf{z}}. \tag{14.38}$$

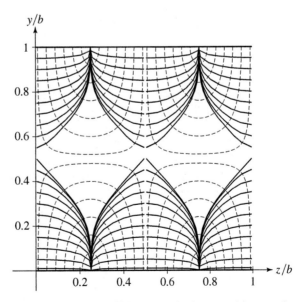

FIGURE 14.9 Energy flux of a TM(1) wave in the space between planar boundaries of perfect conductors. The wavelength λ is equal to the separation between the planes b in this figure. The solid curves are everywhere tangent to the Poynting vectors, and the dashed curves are tangent to the electric field vectors.

This power density is similar to that of the TE(n) mode (14.23), but there is this difference: In a TE mode the power density is 0 at the boundaries whereas in a TM mode it is maximum there.

14.1.4 ▓ Summary

The example in this section is very idealized; the boundary planes are infinite and we have only considered waves that are coherently uniform in the x direction. Though unrealistic, this example has the value that it is simple enough to analyze in detail, and yet it demonstrates some important ways that propagation of electromagnetic waves in a space bounded by conductors is different from propagation in free space. The different properties of the TE and TM waves in the bounded space are these:

- There exist discrete modes of propagation, and a given mode is only possible if the frequency is higher than a certain minimum value—the cutoff frequency—associated with that mode. For a given frequency the number of modes is finite.
- Either **B** or **E** may have a nonzero longitudinal component.
- The dispersion relation is nonlinear, so the phase and group velocities differ and depend on the frequency.
- The Poynting vector is not always parallel to the direction of propagation, although $\mathbf{S}_{avg}$ is.
- There are nonzero charge and current densities on the boundary surfaces.

The same properties also hold for electromagnetic waves in a rectangular wave guide, a device with practical applications in microwave technology, to which we turn now.

14.2 ■ THE RECTANGULAR WAVE GUIDE

We saw in the previous section that electromagnetic waves in a space bounded by conductors have particular *modes of propagation*, each mode having a characteristic vector functional form and specific phase and group velocities. In the present section we shall determine the complete set of harmonic modes in a wave guide with a rectangular cross section. There are two classes of guided waves—TE and TM. There is no TEM mode in a hollow rectangular wave guide.

We assume the wave guide is much longer in one dimension, which we take to be the z direction, than in the other two; then it is a good approximation to imagine the guide infinitely long in the z direction. Figure 14.10 shows the bounded space in which the electromagnetic waves travel. The cross section has dimensions $a \times b$, where a is the length in the x direction and b that in the y direction. In the regions $x < 0$ or $x > a$, and $y < 0$ or $y > b$, there is a perfect conductor—a material with electric conductivity $\sigma = \infty$. This means that the electric and magnetic fields are 0 in these exterior regions. We will assume that the interior of the wave

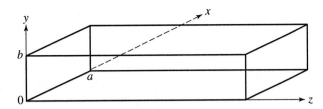

FIGURE 14.10   Geometry of a rectangular wave guide. The exterior is a conductor, and the interior is a dielectric. Waves propagate in the z direction. The cross section is a rectangle of dimensions $a \times b$.

guide is vacuum, but air or some other weak dielectric would be essentially the same.[9] The boundaries of the wave guide are four planar strips:

$$x = 0 \quad \text{or} \quad x = a \quad \text{and} \quad 0 \le y \le b,$$

$$y = 0 \quad \text{or} \quad y = b \quad \text{and} \quad 0 \le x \le a. \tag{14.39}$$

On these surfaces B_n and $\mathbf{E}_t$ must be 0, by the continuity of these field components. The boundary conditions play a major role in determining the modes of propagation.

14.2.1 ▥ Transverse Electric Modes TE(m, n)

First we consider electromagnetic waves propagating down the guide in the $+z$ direction with $\mathbf{E}(\mathbf{x}, t)$ orthogonal to the direction of propagation—the *transverse electric modes*. The electric field has the form

$$\mathbf{E}(\mathbf{x}, t) = \left[\hat{\mathbf{i}} E_x(x, y) + \hat{\mathbf{j}} E_y(x, y) \right] e^{i(kz - \omega t)}. \tag{14.40}$$

The various field equations and boundary conditions will determine the functions $E_x(x, y)$ and $E_y(x, y)$. First, Gauss's Law $\nabla \cdot \mathbf{E} = 0$ implies

$$\frac{\partial E_x}{\partial x} + \frac{\partial E_y}{\partial y} = 0. \tag{14.41}$$

It is convenient to satisfy this equation by introducing a scalar function $\psi(x, y)$ such that

$$E_x = -\frac{\partial \psi}{\partial y} \quad \text{and} \quad E_y = +\frac{\partial \psi}{\partial x}; \tag{14.42}$$

obviously then (14.41) is satisfied. In other words, we write

$$\mathbf{E} = \nabla \times \left(-\hat{\mathbf{z}} \psi e^{i(kz - \omega t)} \right); \tag{14.43}$$

that $\nabla \cdot \mathbf{E}$ is 0 follows from the general identity that the divergence of a curl is 0.

[9] If the interior is a dielectric with permittivity ϵ then replace ϵ_0 in the equations by ϵ, and c by $c\sqrt{\epsilon_0/\epsilon}$.

The electric field in a region of vacuum also satisfies the wave equation,

$$\nabla^2 \mathbf{E} - \frac{1}{c^2} \frac{\partial^2 \mathbf{E}}{\partial t^2} = 0,$$

where $c = 1/\sqrt{\epsilon_0 \mu_0}$. Substituting (14.40) and (14.42) into the wave equation we find that the equation is satisfied if

$$\nabla^2 \psi - k^2 \psi + \frac{\omega^2}{c^2} \psi = 0; \tag{14.44}$$

that is,

$$\nabla^2 \psi = -\gamma^2 \psi \quad \text{where} \quad \gamma^2 = \frac{\omega^2}{c^2} - k^2. \tag{14.45}$$

Equation (14.45) is the *Helmholtz equation* in two dimensions (2D). We can solve this partial differential equation (p.d.e.) by the familiar technique of separation of variables. Write $\psi(x, y) = f(x)g(y)$. Then the p.d.e. in (14.45) becomes, after dividing both sides of the equation by ψ,

$$\frac{1}{f} \frac{d^2 f}{dx^2} + \frac{1}{g} \frac{d^2 g}{dy^2} = -\gamma^2. \tag{14.46}$$

The two terms on the left-hand side depend independently on x and y, whereas the right-hand side is a constant. Therefore each term on the left-hand side will be a constant (call them $-\mu^2$ and $-\nu^2$ respectively) so

$$\frac{d^2 f}{dx^2} = -\mu^2 f \quad \text{and} \quad \frac{d^2 g}{dy^2} = -\nu^2 g, \tag{14.47}$$

and $\gamma^2 = \mu^2 + \nu^2$. The general solutions of (14.47) are

$$f(x) = c_1 \cos \mu x + c_2 \sin \mu x \tag{14.48}$$

$$g(y) = c_3 \cos \nu y + c_4 \sin \nu y; \tag{14.49}$$

the constants $c_1 \ldots c_4$ and parameters μ and ν must be determined from the boundary conditions.

Boundary Conditions

The tangential components of $\mathbf{E}$ must be 0 at the four boundaries listed in (14.39). First, E_y must be 0 at $x = 0$ and $x = a$, for all y between 0 and b. Here $E_y \propto \partial \psi / \partial x = f'(x)g(y)$ (where the prime means the derivative) so the boundary conditions on the walls at $x = 0$ and $x = a$ are

$$f'(0) = 0 \quad \text{which implies} \quad c_2 = 0,$$

$$f'(a) = 0 \quad \text{which implies} \quad \sin \mu a = 0.$$

Similarly E_x must be 0 at $y = 0$ and $y = b$. In this case $E_x \propto -\partial \psi / \partial y = -f(x)g'(y)$ so the boundary conditions on the walls at $y = 0$ and $y = b$ are

$$g'(0) = 0 \quad \text{which implies} \quad c_4 = 0,$$

$$g'(b) = 0 \quad \text{which implies} \quad \sin \nu b = 0.$$

The boundary conditions restrict the possible values of μ and ν; both μa and νb must be an integer multiple of π,

$$\mu = \frac{m\pi}{a} \quad \text{and} \quad \nu = \frac{n\pi}{b}, \tag{14.50}$$

where m and n are non-negative integers. The parameter γ in (14.45) is therefore

$$\gamma = \sqrt{\left(\frac{m\pi}{a}\right)^2 + \left(\frac{n\pi}{b}\right)^2}. \tag{14.51}$$

Finally then the function $\psi(x, y)$ is

$$\psi(x, y) = \Psi_0 \cos\left(\frac{m\pi x}{a}\right) \cos\left(\frac{n\pi y}{b}\right). \tag{14.52}$$

The constant Ψ_0 determines the wave intensity. The corresponding electric field in the guide is, according to (14.40) and (14.42),

$$\mathbf{E}(\mathbf{x}, t) =$$
$$\left[\hat{\mathbf{i}} \frac{n\pi}{b} \cos\left(\frac{m\pi x}{a}\right) \sin\left(\frac{n\pi y}{b}\right) - \hat{\mathbf{j}} \frac{m\pi}{a} \sin\left(\frac{m\pi x}{a}\right) \cos\left(\frac{n\pi y}{b}\right)\right] \Psi_0 e^{i(kz - \omega t)}. \tag{14.53}$$

The electric field is the product of a traveling wave in z and standing waves in x and y. In other words, the boundary condition $\mathbf{E}_t = 0$ requires that an integer number (m) of half waves must fit between the walls at $x = 0$ and a, and similarly (n) between the walls at $y = 0$ and b. In general $\mathbf{E}(\mathbf{x}, t)$ has both components E_x and E_y transverse to the direction of wave motion; but for the special cases with m or n equal to 0 the electric field has just one of these components. The mode of propagation with electric field (14.53) is called the TE(m, n) mode.

The Magnetic Field of the TE(m, n) Mode
Given the electric field it is now straightforward to derive the magnetic field from Faraday's Law, $\nabla \times \mathbf{E} = -\partial \mathbf{B} / \partial t = i\omega \mathbf{B}$, i.e.,

$$\mathbf{B} = -\frac{i}{\omega} \begin{vmatrix} \hat{\mathbf{i}} & \hat{\mathbf{j}} & \hat{\mathbf{z}} \\ \partial/\partial x & \partial/\partial y & \partial/\partial z \\ E_x & E_y & 0 \end{vmatrix}. \tag{14.54}$$

After a short calculation using (14.40) and (14.42) we find

$$\mathbf{B}(\mathbf{x}, t) = \left[-\frac{k}{\omega} \nabla \psi + \frac{i \gamma^2}{\omega} \hat{\mathbf{z}} \psi \right] e^{i(kz - \omega t)}. \tag{14.55}$$

Or, substituting the solution (14.52) for $\psi(x, y)$

$$\begin{aligned}
\mathbf{B}(\mathbf{x}, t) = \Bigg[& \hat{\mathbf{i}} \frac{m\pi}{a} \sin\left(\frac{m\pi x}{a}\right) \cos\left(\frac{n\pi y}{b}\right) \\
& + \hat{\mathbf{j}} \frac{n\pi}{b} \cos\left(\frac{m\pi x}{a}\right) \sin\left(\frac{n\pi y}{b}\right) \\
& + \hat{\mathbf{z}} \frac{i\gamma^2}{k} \cos\left(\frac{m\pi x}{a}\right) \cos\left(\frac{n\pi y}{b}\right) \Bigg] \frac{k \Psi_0}{\omega} e^{i(kz - \omega t)}. \tag{14.56}
\end{aligned}$$

The magnetic field of a TE mode is not transverse. The ratio of the longitudinal component B_z to a transverse component B_x or B_y is proportional to $1/k = \lambda/2\pi$, where λ is the wavelength of the propagating wave. So, the longitudinal component is small for short wavelengths but large for long wavelengths.

Propagation Characteristics

The dispersion relation for the TE(m, n) wave is found from (14.45) and (14.51),

$$\omega^2 = c^2 k^2 + \left(\frac{m\pi c}{a}\right)^2 + \left(\frac{n\pi c}{b}\right)^2. \tag{14.57}$$

Because the dispersion relation is nonlinear, the phase and group velocities are different. The phase velocity is

$$v_{\text{ph}} = \frac{\omega}{k} = \frac{c\omega}{\sqrt{\omega^2 - \left(\frac{m\pi c}{a}\right)^2 - \left(\frac{n\pi c}{b}\right)^2}}. \tag{14.58}$$

Note that the phase velocity is always greater than c and that it diverges as ω decreases to the *cutoff frequency* ω_{mn} defined by

$$\omega_{mn} = \sqrt{\left(\frac{m\pi c}{a}\right)^2 + \left(\frac{n\pi c}{b}\right)^2}. \tag{14.59}$$

For $\omega < \omega_{mn}$ there is no propagation in the TE(m, n) mode, because then the dispersion relation (14.57) would require k to be imaginary. The group velocity of the TE(m, n) wave is

$$v_{\text{gr}} = \frac{d\omega}{dk} = \frac{c^2 k}{\omega} = c\sqrt{1 - \omega_{mn}^2/\omega^2}. \tag{14.60}$$

The group velocity is always less than c, and it approaches 0 as ω decreases to the cutoff ω_{mn}. For high frequencies $\omega \gg \omega_{mn}$ both the phase and group velocity approach c. An interesting relation is that

$$v_{\mathrm{ph}} v_{\mathrm{gr}} = c^2. \tag{14.61}$$

A plot of the phase and group velocities as functions of frequency ω would look exactly like the earlier case of TE or TM waves between parallel conducting planes, shown in Fig. 14.4.

For any given value of the angular frequency ω there are only a finite number of propagating TE modes. The mode numbers m and n of a propagating wave are limited by the inequality

$$\omega_{mn} < \omega. \tag{14.62}$$

Discreteness of wave states in a bounded volume is a common feature of waves of all kinds. The mode with lowest cutoff frequency is $(m, n) = (1, 0)$ if $a > b$, or $(m, n) = (0, 1)$ if $b > a$. So, for example, if $a > b$ there can be no wave propagation through the wave guide with frequency $\omega < \omega_{\mathrm{cutoff}}$, where

$$\omega_{\mathrm{cutoff}} = \omega_{10} = \frac{\pi c}{a}. \tag{14.63}$$

If $a > b$ then the next lowest TE mode is either $(m, n) = (2, 0)$ or $(1, 1)$, depending on the relative size of a and b. If the frequency of oscillation ω is between ω_{10} and the next cutoff frequency then only the mode TE(1,0) can propagate down the wave guide. It is common in microwave applications to choose dimensions of the wave guide for a given frequency of excitation such that only the mode with lowest cutoff frequency can propagate. Any higher modes that might be excited by some irregularity will decay exponentially to negligible levels before reaching the end of the guide.

Why do electromagnetic waves with $\omega < \omega_{\mathrm{cutoff}}$ decay when they enter the wave guide? (By the dispersion relation (14.57) k is imaginary if $\omega < \omega_{mn}$, so then e^{ikz} decreases exponentially with z.) There is no dissipation in this system because the conductivity of the walls is ∞, so the attenuation is not due to dissipation. Rather, the charge and current densities induced in the conducting walls, which we will calculate presently, shield the interior of the wave guide from fields outside the guide if the frequency is low. For the special case $\omega = 0$, i.e., electrostatics, this result is the familiar Faraday cage phenomenon. But we have found more generally that for $\omega < \omega_{mn}$ there is exponential attenuation with distance in the wave guide.

If a radiating antenna is placed in a wave guide, but the frequency is below the cutoff, then the energy supplied to the antenna is not converted to propagation of electromagnetic waves. Oscillating fields occur in the neighborhood of the antenna, and these create oscillating charge and current densities on the conducting walls. But the field and source oscillations are not in phase, and each opposes wave propagation (in z) of the other. If ω is large then the displacement current dominates the charge current and the wave propagates. If $\omega < \omega_{\mathrm{cutoff}}$ then there is destructive interference, increasing with z, between the fields of the antenna and those of the wall currents, and the wave decays.

The fact that low frequency electromagnetic waves do not penetrate into a region bounded by conductors explains why an AM radio signal cannot be picked up by a car radio inside a metal bridge or tunnel.[10]

EXAMPLE 1   Consider a wave guide with rectangular cross section of dimensions 10 cm × 5 cm. Which TE modes can propagate for angular frequency 4π GHz? What are the 10 lowest cutoff frequencies for TE modes, and their corresponding mode numbers?

In this example, $a = 0.10$ m and $b = 0.05$ m. The cutoff frequency of the mode TE(m, n) is

$$\omega_{mn} = (3\pi \text{ GHz}) \sqrt{m^2 + 4n^2}. \tag{14.64}$$

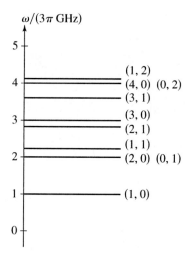

FIGURE 14.11   Cutoff frequencies of the TE modes for the rectangular wave guide in Example 1.

The mode TE(m, n) propagates in the guide only if $\omega_{mn} < \omega$. The only TE mode that can propagate with angular frequency 4π GHz is the TE(1,0) mode. Figure 14.11 shows the cutoff frequencies of the 10 lowest TE modes, which are, in order: $(mn) = $ (10), (20) and (01) (degenerate), (11), (21), (30), (31), (40) and (02) (degenerate), (12).

---

[10]See Exercise 6.

14.2.2 ■ Transverse Magnetic Modes TM(m, n)

In the previous section we found a set of modes in a rectangular wave guide for which the electric field is transverse, i.e., orthogonal to the axis of the guide. These modes are complete for transverse electric waves: Any TE wave can be expressed as a superposition of the TE modes. But there is another class of waves, for which the magnetic field is transverse. In this section we describe the TM modes.

Following the same line of analysis that we used to construct the TE modes, now we write the magnetic field as

$$\mathbf{B}(\mathbf{x}, t) = \left(-\hat{\mathbf{i}} \frac{\partial \psi}{\partial y} + \hat{\mathbf{j}} \frac{\partial \psi}{\partial x} \right) e^{i(kz - \omega t)}, \qquad (14.65)$$

where again $\psi(x, y)$ satisfies the 2D Helmholtz equation (14.45). This form guarantees that $\nabla \cdot \mathbf{B} = 0$, and that $\mathbf{B}$ satisfies the wave equation. The associated electric field is deduced from the Ampère–Maxwell Law,

$$\mathbf{E}(\mathbf{x}, t) = \frac{ic^2}{\omega} \nabla \times \mathbf{B} = \frac{-c^2 k}{\omega} \left[-\nabla \psi + \frac{i\gamma^2}{k} \hat{\mathbf{z}} \psi \right] e^{i(kz - \omega t)}. \qquad (14.66)$$

All that remains is to impose the boundary conditions.

It is sufficient to consider the boundary conditions on the normal component of $\mathbf{B}$. At the boundaries listed in (14.39) B_n must be continuous, which is to say 0 because $\mathbf{B}$ is zero in the perfectly conducting walls. In terms of the function $\psi(x, y)$ the boundary conditions are

$$\frac{\partial \psi}{\partial y} = 0 \quad \text{at} \quad x = 0 \quad \text{and} \quad x = a,$$

$$\frac{\partial \psi}{\partial x} = 0 \quad \text{at} \quad y = 0 \quad \text{and} \quad y = b.$$

A complete set of solutions to (14.45) subject to these boundary conditions consists of

$$\psi(x, y) = \Psi_0 \sin\left(\frac{m\pi x}{a} \right) \sin\left(\frac{n\pi y}{b} \right), \qquad (14.67)$$

where m and n are both positive integers. This mode of propagation is called the TM(m, n) mode, and its fields are

$$
\begin{aligned}
\mathbf{B}(\mathbf{x}, t) = \Big[&-\hat{\mathbf{i}} \frac{n\pi}{b} \sin\left(\frac{m\pi x}{a} \right) \cos\left(\frac{n\pi y}{b} \right) \\
&+ \hat{\mathbf{j}} \frac{m\pi}{a} \cos\left(\frac{m\pi x}{a} \right) \sin\left(\frac{n\pi y}{b} \right) \Big] \Psi_0 e^{i(kz - \omega t)},
\end{aligned}
\qquad (14.68)
$$

and

$$E(\mathbf{x}, t) = \left[\hat{\mathbf{i}}\,\frac{m\pi}{a}\cos\left(\frac{m\pi x}{a}\right)\sin\left(\frac{n\pi y}{b}\right)\right.$$

$$+ \hat{\mathbf{j}}\,\frac{n\pi}{b}\sin\left(\frac{m\pi x}{a}\right)\cos\left(\frac{n\pi y}{b}\right)$$

$$\left. - \hat{\mathbf{z}}\,\frac{i\gamma^2}{k}\sin\left(\frac{m\pi x}{a}\right)\sin\left(\frac{n\pi y}{b}\right)\right]\frac{c^2 k\Psi_0}{\omega}e^{i(kz-\omega t)}. \quad (14.69)$$

The dispersion relation for the TM(m, n) mode is the same as for the TE(m, n) mode, given in (14.57). However, there is a major difference between the TM and TE modes: The TM mode with lowest cutoff frequency has $(mn) = (11)$ because there is no TM mode with either m or n equal to 0, while the lowest TE mode is either $(m, n) = (1, 0)$ or $(0, 1)$.

EXAMPLE 2   What are the surface charge densities for the mode TM(1, 1)?

The normal component of $\mathbf{E}$ on a conducting boundary surface is σ/ϵ_0, where σ is the surface charge density. On the horizontal surfaces, i.e., those parallel to the xz plane, the normal unit vector is $\hat{\mathbf{n}} = +\hat{\mathbf{j}}$ for $y = 0$ and $\hat{\mathbf{n}} = -\hat{\mathbf{j}}$ for $y = b$; at those boundaries

$$\sigma = \frac{\epsilon_0\pi c^2 k\Psi_0}{b\omega}\sin\left(\frac{\pi x}{a}\right)\cos(kz - \omega t) \quad \text{(horizontal surfaces).} \quad (14.70)$$

On the vertical surfaces, i.e., those parallel to the yz plane, $\hat{\mathbf{n}} = +\hat{\mathbf{i}}$ for $x = 0$ and $\hat{\mathbf{n}} = -\hat{\mathbf{i}}$ for $x = a$; at those boundaries

$$\sigma = \frac{\epsilon_0\pi c^2 k\Psi_0}{a\omega}\sin\left(\frac{\pi y}{b}\right)\cos(kz - \omega t) \quad \text{(vertical surfaces).} \quad (14.71)$$

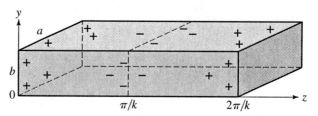

FIGURE 14.12   Surface charge distributions for the transverse magnetic mode TM(1, 1) of a rectangular wave guide.

A snapshot of the surface charge distributions for one wavelength of the TM(1, 1) mode is shown in Fig. 14.12. This charge distribution propagates as a wave along the surfaces in the $+z$ direction. One can easily figure out the surface current densities from the relation $\mathbf{K} = \sigma v_{\text{ph}}\hat{\mathbf{z}}$.

14.3 ■ WAVE GUIDE OF ARBITRARY SHAPE

The rectangular wave guide is most common, but other shapes are possible. In this section we analyze harmonic electromagnetic waves in a wave guide with an arbitrary cross section. The wave guide is infinitely long in the z direction, and the shape of the cross section is independent of z. The boundary surface is a cylinder parallel to the z axis. A cross section of the interior region is bounded by a closed curve C parallel to the xy plane. For simplicity we assume ideal conditions: The exterior of the guide is a perfect conductor and the interior is vacuum. The basic equations for the rectangular wave guide can be generalized to an arbitrary cross section.

TE modes.   The fields for a TE wave propagating in the $+z$ direction may be written in the form

$$\mathbf{E}(\mathbf{x}, t) = -\nabla \times \left(\hat{\mathbf{z}} \psi e^{i(kz - \omega t)} \right) = - \left[\nabla \times \left(\hat{\mathbf{z}} \psi \right) \right] e^{i(kz - \omega t)} \qquad (14.72)$$

$$\mathbf{B}(\mathbf{x}, t) = \frac{k}{\omega} \left[-\nabla \psi + \frac{i \gamma^2}{k} \hat{\mathbf{z}} \psi \right] e^{i(kz - \omega t)}, \qquad (14.73)$$

where $\psi(x, y)$ is a scalar function independent of z. These forms are the same as (14.43) and (14.55) used in the analysis of the rectangular wave guide. All four Maxwell equations are satisfied if $\psi(x, y)$ is a solution of the 2D Helmholtz equation

$$\nabla^2 \psi = -\gamma^2 \psi \qquad (14.74)$$

with $\omega^2 / c^2 = k^2 + \gamma^2$.[11] All that remains is to impose the boundary conditions.

The normal component of $\mathbf{B}$ must be 0 on C, the boundary curve of a cross section of the guide. What is this condition in terms of $\psi(x, y)$? By (14.73) the normal component of $\mathbf{B}$ at a point on the surface is proportional to $\hat{\mathbf{n}} \cdot \nabla \psi$, where $\hat{\mathbf{n}}$ is the unit normal vector at the point in the plane of C; thus

$$\hat{\mathbf{n}} \cdot \nabla \psi = 0 \quad \text{on} \quad C. \qquad (14.75)$$

The tangential components of $\mathbf{E}$ must also be 0 on C, but that requirement leads to the same condition (14.75).

For the TE modes then, $\psi(x, y)$ obeys the 2D Helmholtz equation in the region enclosed by C, with normal derivative 0 on the boundary. Equation (14.74) is an eigenvalue problem, with operator ∇^2 and eigenvalue $-\gamma^2$. That is, $\psi(x, y)$ is the eigenfunction of the 2D Laplacian with the Neumann boundary condition (14.75). There are an infinite number of discrete eigenstates. If the angular frequency ω of the field oscillations is greater than $c\gamma$ then the solution describes a propagating wave, because k is real in that case. But if ω is less than $c\gamma$ then the solution

[11] See Exercise 10.

decays exponentially with z, and the fields in that case are called an *evanescent wave*.

TM modes.   The fields of the TM modes are

$$\mathbf{B}(\mathbf{x}, t) = -\nabla \times \left(\hat{\mathbf{z}} \psi e^{i(kz-\omega t)} \right) \tag{14.76}$$

$$\mathbf{E}(\mathbf{x}, t) = \frac{c^2 k}{\omega} \left[\nabla \psi - \frac{i\gamma^2}{k} \hat{\mathbf{z}} \psi \right] e^{i(kz-\omega t)} \tag{14.77}$$

where again $\psi(x, y)$ is a solution of (14.74). The tangential component of $\mathbf{E}$ must be 0 on the boundary curve C. In particular, the $\hat{\mathbf{z}}$ component in (14.77) must be 0, so for TM modes

$$\psi = 0 \quad \text{on} \quad C. \tag{14.78}$$

The same condition implies that the other tangential component of $\mathbf{E}$, and the normal component of $\mathbf{B}$, are also 0 on C; these components are both proportional to $(\hat{\mathbf{z}} \times \hat{\mathbf{n}}) \cdot \nabla \psi$, i.e., the tangential derivative of ψ, which is 0 because ψ is constant along C.

The function $\psi(x, y)$ for a TM mode is an eigenfunction of the Laplacian for the Dirichlet boundary condition (14.78) with eigenvalue $-\gamma^2$. This eigenvalue problem is mathematically equivalent to the transverse vibrations of a membrane. For the analogue membrane, $\psi(x, y)$ is the displacement from equilibrium of the point on the membrane at (x, y); the boundary condition (14.78) is that the membrane is fixed along the curve C like a drumhead. As in the TE case there are an infinite number of discrete eigenstates.

TEM modes.   Under what circumstances does a TEM mode exist? Looking at $\mathbf{B}$ for the TE modes (14.73) or at $\mathbf{E}$ for the TM modes (14.77) we see that these fields are transverse if and only if $\gamma = 0$. So the partial differential equation for a TEM mode is

$$\nabla^2 \psi = 0, \tag{14.79}$$

i.e., Laplace's equation.

We can now prove the following theorem: There are no TEM modes in a hollow conducting wave guide of the kind we have been considering—one that has a single closed curve C as the boundary of the cross section. This theorem is a corollary of the uniqueness theorem for Laplace's equation. The *unique* solution to (14.79) for either boundary condition (14.75) or (14.78) is $\psi(x, y) = 0$; that is, the fields are 0.

On the other hand there can be TEM modes if there is more than one conductor in the system, because then Laplace's equation and the boundary conditions can be satisfied with $\psi \neq 0$. We saw an example of this case in Sec. 14.1, TEM waves between parallel conducting planes. There the boundary of a cross section of the propagation space consists of two infinite lines, $y = 0$ and $y = b$. The TEM solution (14.3) and (14.4) corresponds to the TE fields (14.72) and (14.73) with $\psi(x, y) = E_0 x$ and $\gamma = 0$.[12] This $\psi(x, y)$ obeys Laplace's equation and the Neumann boundary condition (14.75) at $y = 0$ and b. The uniqueness theorem is circumvented by the fact that the region is unbounded in x. Another example with a TEM mode, in which the propagation space is finite, is the coaxial cable transmission line, to be discussed in the next section.

Finally, recall that the dispersion relation for waves in the cylinder is $\omega^2 = c^2(k^2 + \gamma^2)$. A TEM mode ($\gamma = 0$) has $\omega = ck$, the same dispersion relation as electromagnetic waves in free space. Therefore any TEM mode propagates with wave speed c and without dispersion.

14.4 ■ THE TEM MODE OF A COAXIAL CABLE

A transmission line is used to carry a signal in the radio or microwave frequency range from a transmitter to a receiver, or to an antenna, or to some other device. In analyzing a transmission line one can either focus on the current and voltage signal, or the wave signal, but the two are intricately coupled. One of the most common transmission lines is the coaxial cable, familiar in computer networks and cable television. The cable consists of two concentric conducting cylinders separated by an isotropic dielectric insulator. We can either analyze the electromagnetic wave in the dielectric, or the current and voltage on the conductors.

The geometry is defined in Fig. 14.13. The axis of the cable is defined to be the z axis. So that end effects are negligible, the cable is taken to be infinitely long. The radius of the inner cylinder is a, and the inner radius of the outer cylinder is b. The regions $r < a$ and $r > b$ are conductors, which we'll take to have infinite conductivity. The waves propagate in the region $a \leq r \leq b$, where we'll say the dielectric permittivity is ϵ.

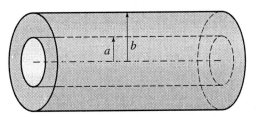

FIGURE 14.13   Coaxial cable. Electromagnetic waves propagate in the space between the cylinders, which is filled with a uniform dielectric. Charge and current waves move on the cylinder surfaces.

[12] Alternatively the TEM mode corresponds to the TM fields (14.77) and (14.76) with $\psi(x, y) = B_0 y$ and $\gamma = 0$.

The coaxial cable has a TEM mode, and this mode of propagation is the most important for signal propagation in practical applications. The purpose of this section is to derive this solution of the field equations. It resembles closely the TEM mode between parallel conducting planes described earlier, but with the planes rolled up into concentric cylinders. We shall use this similarity to help construct the solution. To solve the field equations it is most natural to use cylindrical coordinates.

By the cylindrical symmetry, the electric or magnetic field of a wave propagating in the z direction is a vector function of r and ϕ times $e^{i(kz-\omega t)}$. The field directions of the TEM mode are orthogonal to $\hat{z}$, so $\mathbf{E}$ and $\mathbf{B}$ may have radial and azimuthal components. In analogy with the TEM mode of parallel conducting planes, the electric field points from one conductor to the other, i.e., in this case radially. Surface charges on the cylinders will satisfy the boundary condition of the normal component of $\mathbf{E}$. Also, $\mathbf{B}$ must be purely azimuthal, i.e., circling around the z axis, because from the general analysis in Sec. 14.3 $\mathbf{B}$ and $\mathbf{E}$ are orthogonal. Surface currents will satisfy the boundary condition of the tangential components of $\mathbf{B}$. By the cylindrical symmetry, the field magnitudes must be independent of the azimuthal angle ϕ. Putting together all these symmetry considerations we are led to a solution of the form

$$\mathbf{E}(\mathbf{x}, t) = \hat{\mathbf{r}} E_r(r) e^{i(kz-\omega t)}, \qquad (14.80)$$

$$\mathbf{B}(\mathbf{x}, t) = \hat{\boldsymbol{\phi}} B_\phi(r) e^{i(kz-\omega t)}. \qquad (14.81)$$

We may proceed in various ways, but perhaps the most straightforward is to return to first principles by substituting (14.80) and (14.81) directly into Maxwell's equations.[13] The divergence of $\mathbf{E}$ is

$$\nabla \cdot \mathbf{E} = \frac{1}{r} \frac{d}{dr} (r E_r) = 0. \qquad (14.82)$$

The divergence of $\mathbf{B}$ is always 0 for any field of the form (14.81). From the divergence equations we learn that $r E_r$ is constant,

$$E_r(r) = \frac{C}{r}. \qquad (14.83)$$

The curl of $\mathbf{E}$ is

$$\nabla \times \mathbf{E} = \hat{\boldsymbol{\phi}} i k E_r(r) e^{i(kz-\omega t)} \qquad (14.84)$$

which, by Faraday's Law, must equal

$$-\partial \mathbf{B}/\partial t = \hat{\boldsymbol{\phi}} i \omega B_\phi(r) e^{i(kz-\omega t)}; \qquad (14.85)$$

---

[13] Alternatively we could use the general results of Sec. 14.3; see Exercise 12.

that is,

$$B_\phi(r) = \frac{kC}{\omega r}. \tag{14.86}$$

The curl of **B** is

$$\nabla \times \mathbf{B} = -\hat{\mathbf{r}} i k B_\phi(r) e^{i(kz-\omega t)} \tag{14.87}$$

which, by the Ampère–Maxwell Law, must equal

$$\mu_0\epsilon \frac{\partial \mathbf{E}}{\partial t} = -\hat{\mathbf{r}} i \mu_0\epsilon\omega E_r(r) e^{i(kz-\omega t)}; \tag{14.88}$$

that is,

$$B_\phi(r) = \frac{\mu_0\epsilon\omega C}{kr}. \tag{14.89}$$

Together the curl equations show that the dispersion relation for the TEM mode is the same as for electromagnetic waves in an unbounded dielectric,

$$\omega = vk \quad \text{where} \quad v = 1/\sqrt{\mu_0\epsilon}. \tag{14.90}$$

Also, the ratio $|\mathbf{B}|/|\mathbf{E}|$ is $1/v$. Of course these results are in accord with the discussion in Sec. 14.3 of the TEM solution for a wave guide with arbitrary shape.

The potential difference $V(z, t)$ between the inner and outer cylinders at position z along the cable is, by definition,

$$V(z, t) = \int_a^b \mathbf{E} \cdot \hat{\mathbf{r}} dr. \tag{14.91}$$

Substituting the TEM electric field, we find

$$V(z, t) = V_0 e^{i(kz-\omega t)} \quad \text{where} \quad V_0 = C \ln(b/a). \tag{14.92}$$

In terms of the amplitude V_0 of oscillation of the potential, the fields of the TEM mode are

$$\mathbf{E}(\mathbf{x}, t) = \frac{\hat{\mathbf{r}} V_0}{r \ln(b/a)} e^{i(kz-\omega t)}, \tag{14.93}$$

$$\mathbf{B}(\mathbf{x}, t) = \frac{\hat{\boldsymbol{\phi}} V_0}{vr \ln(b/a)} e^{i(kz-\omega t)}. \tag{14.94}$$

So far we have ignored the boundary conditions. Taking the cylinders to be perfect conductors, **E** and **B** are 0 in the conducting regions. Therefore B_n and E_t must vanish at the boundary surfaces. But these components of **B** and **E** are 0 everywhere for (14.80) and (14.81), so these boundary conditions are automatically satisfied. On the other hand, E_n and $\mathbf{B}_t$ are discontinuous at the boundaries. Their discontinuities determine the surface charge and current densities:

$$\sigma = \epsilon E_n = \begin{cases} \dfrac{+\epsilon V_0}{a \ln(b/a)} e^{i(kz-\omega t)} & \text{for} \quad r = a \\[3mm] \dfrac{-\epsilon V_0}{b \ln(b/a)} e^{i(kz-\omega t)} & \text{for} \quad r = b, \end{cases} \tag{14.95}$$

and

$$\mathbf{K} = \frac{1}{\mu_0} \hat{\mathbf{n}} \times \mathbf{B}_t = \begin{cases} \dfrac{+\hat{\mathbf{z}} v \epsilon V_0}{a \ln(b/a)} e^{i(kz-\omega t)} & \text{for} \quad r = a \\[3mm] \dfrac{-\hat{\mathbf{z}} v \epsilon V_0}{b \ln(b/a)} e^{i(kz-\omega t)} & \text{for} \quad r = b. \end{cases} \tag{14.96}$$

Thus charge/current waves propagate on the surfaces of the coaxial cylinders. Figure 14.14(a) shows the electric field and charge distribution, and Fig. 14.14(b) shows the magnetic field and current distribution for the TEM mode of a coaxial cable.

Current and Impedance

The total currents in the inner and outer cylinders, calculated by integrating the surface current density $\mathbf{K}$ around the circumference, are equal but opposite. Letting $I(z, t)$ be the total current at z on the inner cylinder,

$$I(z, t) = \int_0^{2\pi} \mathbf{K} \cdot \hat{\mathbf{z}} a d\phi = \frac{2\pi v \epsilon V_0}{\ln(b/a)} e^{i(kz-\omega t)}. \tag{14.97}$$

At a fixed position z on the cable there is an alternating current in the $\pm z$ direction with angular frequency ω. At the same position the potential difference between the cylinders is given by (14.92). The *impedance* Z of the coaxial cable is defined as the ratio of $V(z, t)$ over $I(z, t)$, so

$$Z = \frac{V(z, t)}{I(z, t)} = \sqrt{\frac{\mu_0}{\epsilon}} \frac{\ln(b/a)}{2\pi}. \tag{14.98}$$

For example, if the polarizability of the dielectric material between the cylinders is negligible, so that $\epsilon = \epsilon_0$, then the impedance is

$$Z = (60\,\Omega) \ln(b/a). \tag{14.99}$$

The coaxial cable also has TE and TM modes, which could be determined from the general considerations of Sec. 14.3. However, in common applications the cable is designed to carry only the TEM mode, so its dimensions are small enough that only the TEM mode is a propagating mode.

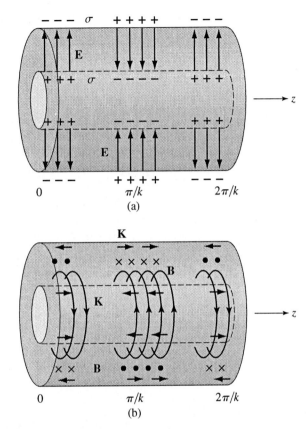

FIGURE 14.14   The TEM mode of a coaxial cable. (a) Electric field and charge densities on the conducting surfaces. (b) Magnetic field and current densities.

14.5 ■ CAVITY RESONANCE

If a finite length L of wave guide is closed at both ends by reflecting walls, the result is an example of a resonant cavity. Suppose the cross section of the wave guide is bounded by a closed curve C. We already know solutions for an unbounded wave guide—TE or TM waves traveling in the $+z$ or $-z$ direction. At an end wall the fields can be resolved into incident waves and reflected waves, which interfere with each other. For simplicity assume the ends are perfect reflectors. Then if the length L is an integer number of half wavelengths, standing waves will be set up along the z direction with nodes of some field at the walls.

For example, consider the TE modes. The fields for traveling waves are (14.72) and (14.73)

$$\mathbf{E}(\mathbf{x}, t) = -\left[\nabla \times (\hat{\mathbf{z}}\psi)\right] e^{\pm ikz} e^{-i\omega t} \qquad (14.100)$$

$$\mathbf{B}(\mathbf{x}, t) = \frac{k}{\omega}\left[\mp\nabla\psi + \frac{i\gamma^2}{k}\hat{\mathbf{z}}\psi\right] e^{\pm ikz} e^{-i\omega t} \qquad (14.101)$$

where $\psi(x, y)$ is the eigenfunction of ∇^2 with eigenvalue $-\gamma^2$. Here $\mathbf{E}$ is orthogonal to $\hat{\mathbf{z}}$ so it is *tangential* at the end walls. To satisfy the boundary condition $\mathbf{E}_t = 0$ for perfectly conducting walls at $z = 0$ and $z = L$ we need the standing wave that results from the superposition

$$e^{+ikz} - e^{-ikz} = 2i\sin kz.$$

(The factor $2i$ can be ignored, because it can be absorbed into the amplitude.) This function vanishes at $z = 0$ for any k, and it vanishes at $z = L$ provided kL is an integer multiple of π. The cavity fields are

$$\mathbf{E}(\mathbf{x}, t) = \left[\hat{\mathbf{z}} \times \nabla\psi\right]\sin\left(\frac{N\pi z}{L}\right) e^{-i\omega t} \qquad (14.102)$$

$$\mathbf{B}(\mathbf{x}, t) = \frac{ik}{\omega}\left[\nabla\psi\cos kz + \frac{\gamma^2}{k}\hat{\mathbf{z}}\psi\sin kz\right] e^{-i\omega t} \qquad (14.103)$$

where N is a positive integer and $k = N\pi/L$. The relation between ω and k must be $\omega^2 = c^2(\gamma^2 + k^2)$, the same as for traveling waves, because that relation came from Maxwell's equations independent of the boundary conditions. So the frequency of this cavity mode is

$$\omega_N = \sqrt{c^2\gamma^2 + (N\pi c/L)^2}. \qquad (14.104)$$

There is a set of *resonant frequencies* $\omega_1, \omega_2, \omega_3, \ldots$ for each eigenstate $\psi(x, y)$ of the 2D Laplacian. The resonant frequencies are all greater than the cutoff frequency $c\gamma$ for the corresponding wave guide mode. Associated with the fields (14.102) and (14.103) are charge and current distributions on the sides and ends of the cavity. In a manner consistent with all the field equations, charge oscillates back and forth over the surfaces, and $\mathbf{E}$ and $\mathbf{B}$ oscillate inside the cavity.

A section of wave guide closed at both ends is just one form of resonant cavity, and other forms are possible. Resonant modes of oscillation exist for a cavity of any shape. Analytic methods can be used to calculate the modes for simple symmetric shapes, like a spherical cavity, but for a complex shape the equations must be solved by numerical methods.

The importance of resonant cavities is that they are used in microwave generators. Various devices have been developed to produce microwaves. Historically the first were vacuum tube resonators—the magnetron and klystron—containing electron beams. In a magnetron a bunched electron beam circling in a toroidal cavity with conducting walls excites a resonance with the electromagnetic field. The resonant frequency depends on the geometry of the cavity and the orbital period of the electron bunches. A magnetron is the source of microwaves in a microwave oven.

Today there are solid-state microwave generators, which convert dc power to microwave power. Think of a hand-held cellular phone: The power is provided by a battery, but the phone transmits information by microwaves. The common design of solid-state microwave generators consists of a diode or transistor oscillator circuit that produces the oscillation, combined with a cavity resonator, which locks the frequency to the desired value. For radio wave frequencies the resonator could be a "lumped circuit," i.e., made of separate capacitive and inductive components (the "lumps"). But for higher frequencies, in the microwave range, the resonator must be an integrated device like a cavity resonator, e.g., a shorted wave guide.

FURTHER READING

1. J. C. Slater and N. H. Frank, *Electromagnetism* (Dover, New York, 1969). Chapter XI is a brief but clear exposition of wave guides.

2. M. A. Heald and J. B. Marion, *Classical Electromagnetic Radiation*, 3rd ed. (Saunders, Philadelphia, 1995).

3. W. C. Elmore and M. A. Heald, *Physics of Waves* (Dover, New York, 1969).

4. H. Georgi, *The Physics of Waves* (Prentice-Hall, Englewood Cliffs, NJ, 1993).

5. Microwave electronics is a major industry, and some engineering textbooks are entirely devoted to this subject: R. S. Elliott, *Guided Waves and Microwave Circuits* (Prentice-Hall, Englewood Cliffs, NJ, 1993).

EXERCISES

Sec. 14.1. Waves Between Conducting Planes

14.1. For the TE(n) mode of propagation in the space between parallel conducting planes, show that $\mathbf{S}_{avg} = u_{avg} v_{gr} \hat{\mathbf{z}}$, where u is the field energy density. Here the subscript avg implies an average over both t and y. Explain why this result means that the group velocity is the velocity of energy transport.

14.2. Show that the superposition of waves in (14.26) has phase speed $v_{ph} = c/\cos\alpha$ with phase velocity in the direction of $\hat{\mathbf{z}}$. (α is the angle between $\mathbf{k}_+$ or $\mathbf{k}_-$ and $\hat{\mathbf{z}}$.)

14.3. Consider waves between parallel conducting planes, traveling in the z direction and polarized in the x direction. Show that $\mathbf{E}(\mathbf{x}, t)$ must be independent of x.

14.4. For electromagnetic waves between parallel planar boundaries of perfect conductors, show that as ω approaches the cutoff frequency of a TE or TM mode the wavelength for that mode approaches ∞.

14.5. Verify that the field (14.26) is the same as (14.14).

Sec. 14.2. Rectangular Wave Guide

14.6. Consider a car entering a tunnel of dimensions 15 m wide and 4 m high. Assuming the walls are good conductors, can AM radio waves propagate in the tunnel?

14.7. Suppose the dielectric in a rectangular wave guide undergoes breakdown (sparking) at 10 kV/m. Calculate the maximum power that can be carried by the TE(1,0) mode without breakdown if the dimensions of the waveguide are 5 cm × 5 cm. Assume the frequency is 2× the cutoff frequency.

14.8. Consider a wave guide with a square cross section of dimensions $a \times a$. Let the z axis be the axis of the wave guide. Suppose the region $z < 0$ is vacuum, and the region $z > 0$ is a dielectric with permittivity ϵ. Write a solution of the wave equations and boundary conditions such that there is an incident and reflected wave for $z < 0$ and a transmitted wave for $z > 0$. All three waves are TE(10) waves. Determine the transmitted power as a fraction of the incident power. [Answer: $S_{\text{trans}}/S_{\text{inc}} = 4kk'/(k + k')^2$, where k is the wave vector in the region $z < 0$ and k' is the wave vector in the region $z > 0$.]

14.9. Determine the 10 TE modes of a wave guide with a square cross section with lowest cutoff frequencies. Also, what are the lowest 10 TM modes?

Sec. 14.3. Wave Guide of Arbitrary Shape

14.10. Prove that all four Maxwell equations are satisfied for the fields in (14.72) and (14.73) if $\psi(x, y)$ is a solution of the 2D Helmholtz equation. (Hint: In calculating $\nabla \cdot \mathbf{B}$, $\nabla \times \mathbf{B}$, and $\nabla \times \mathbf{E}$, be careful to take account of the factor $e^{i(kz-\omega t)}$.)

14.11. Determine the TM and TE modes of propagation in a cylindrical wave guide with a circular cross section. (Hint: Solve (14.74) by separation of variables in plane polar coordinates. The radial functions are Bessel functions regular at $r = 0$.) Determine the first few cutoff frequencies of each type of mode, and note any degeneracies.

Sec. 14.4. Coaxial Cable

14.12. Use the general results of Sec. 14.3 to derive the TEM mode of a coaxial cable. Start with (14.72) and (14.73), with $\gamma = 0$. $\psi(x, y)$ is a solution of Laplace's equation with appropriate boundary conditions.

General Exercises

14.13. Consider electromagnetic waves inside a perfectly conducting metal wave guide. Assume the interior cross section of the wave guide (parallel to the xy plane) is a square of size $a \times a$, and the length of the wave guide along the z axis is infinite. For a particular set of TE modes, the electric field oscillates in the x direction

$$\mathbf{E}(\mathbf{x}, t) = \hat{\mathbf{i}} \, E(y) e^{i(kz-\omega t)}$$

where $E(y)$ depends only on y.

(a) For a given value of k determine the possible values of ω.

(b) Sketch a graph of phase velocity as a function of k for the first few modes. Mark the speed of light in vacuum c on the velocity axis.

(c) Sketch a graph of group velocity as a function of k for the first few modes. Mark c on the velocity axis.

CHAPTER
15

Radiation of
Electromagnetic Waves

In this final chapter we study the process of *radiation*. Radiation means the *generation* of electromagnetic waves, which then propagate away from the source to infinity. Classically, electromagnetic waves are created by charged particles accelerating, or, equivalently, currents varying in time. In Chapter 12 we determined the electric and magnetic fields of a point charge moving with constant velocity. There is no radiation in that case; the energy in the electromagnetic field is transported along with the particle, but does not propagate away toward infinity. But if the charge velocity is not constant there must be radiation.

As we start the study of radiation, it is interesting to remember that the classical field theory was developed from experiments and observations on macroscopic objects such as charged bodies, insulators, conductors, current-carrying wires, and magnets. The classical theory is therefore at its best when applied to macroscopic electromagnetic phenomena. So, for example, the theory applies to radiation of radio waves by an antenna, or microwaves by a cavity resonator. In this book we concentrate mainly on systems of that kind.

However, much of the radiation around us—in fact much of the radiation in the Universe—comes from *microscopic* systems and must be described by quantum electrodynamics. For example, sunlight, light from the filament of an incandescent bulb, fluorescent light, or laser light can only be explained by quantum considerations of individual atoms, molecules, or systems of atoms. For these systems the classical theory does not really apply, except perhaps in a qualitative, heuristic way. Only quantum electrodynamics can properly describe radiation by an atom. Despite its limited applicability, the classical theory of radiation is an important part of electromagnetism. Quantum electrodynamics, which has limitations of its own, relies on a foundation of classical electrodynamics.

A significant application of radiation theory is the analysis and design of antennas. A common radio antenna carries an alternating current. This time-dependent current, composed of charges with nonzero acceleration, creates electromagnetic waves. Some type of *modulation* of the alternating current encodes the information carried by the wave. In AM radio the amplitude is modulated, while in FM radio and in many cell phones the frequency is modulated. The receiver circuit decodes the modulation when the radio wave is picked up by the receiving antenna.

In the first section of this chapter we construct the potential functions for arbitrary time-dependent charge and current densities. In the later sections we use the general results to analyze some specific radiating systems.

15.1 ■ THE RETARDED POTENTIALS

The behavior of electromagnetic fields interacting with charged particles is governed by the Maxwell equations. It is sufficient for our purposes to consider fields in vacuum, i.e., systems without polarizable materials. Then the field equations are

$$\nabla \cdot \mathbf{E} = \rho/\epsilon_0 \quad \text{and} \quad \nabla \times \mathbf{E} = -\partial \mathbf{B}/\partial t \tag{15.1}$$

$$\nabla \cdot \mathbf{B} = 0 \quad \text{and} \quad \nabla \times \mathbf{B} = \mu_0 \left(\mathbf{J} + \epsilon_0 \partial \mathbf{E}/\partial t \right). \tag{15.2}$$

We simplify the mathematics by introducing vector and scalar potentials, $\mathbf{A}(\mathbf{x}, t)$ and $V(\mathbf{x}, t)$. Recall from Sec. 11.2 that the fields are written as

$$\mathbf{B} = \nabla \times \mathbf{A} \quad \text{and} \quad \mathbf{E} = -\partial \mathbf{A}/\partial t - \nabla V. \tag{15.3}$$

Also, it is most convenient in radiation theory to impose upon $\mathbf{A}(\mathbf{x}, t)$ and $V(\mathbf{x}, t)$ the *Lorentz gauge condition*,

$$\nabla \cdot \mathbf{A} = -\frac{1}{c^2} \frac{\partial V}{\partial t}; \tag{15.4}$$

there is no loss of generality in making this requirement, because it can always be satisfied by a suitable gauge transformation if necessary. Following the usual notation c is the speed of electromagnetic waves in vacuum, which is equal to $1/\sqrt{\mu_0 \epsilon_0}$. With the Lorentz gauge choice, Maxwell's equations imply that $\mathbf{A}(\mathbf{x}, t)$ and $V(\mathbf{x}, t)$ satisfy the differential equations

$$-\nabla^2 V + \frac{1}{c^2} \frac{\partial^2 V}{\partial t^2} = \frac{\rho}{\epsilon_0}, \tag{15.5}$$

$$-\nabla^2 \mathbf{A} + \frac{1}{c^2} \frac{\partial^2 \mathbf{A}}{\partial t^2} = \mu_0 \mathbf{J}. \tag{15.6}$$

We met these equations earlier as (11-30) and (11-31). The scalar equation (15.5) is called *d'Alembert's equation*; the differential operator $-\nabla^2 + (1/c^2)\partial^2/\partial t^2$ is called the d'Alembertian. Each Cartesian component of $\mathbf{A}(\mathbf{x}, t)$ also satisfies d'Alembert's equation. We shall solve (15.5) for $V(\mathbf{x}, t)$ and (15.6) for $\mathbf{A}(\mathbf{x}, t)$ by constructing the *Green's function* of the d'Alembertian.

15.1.1 ■ Green's Functions

Our goal is to develop a formula that solves (15.5) for $V(\mathbf{x}, t)$ for an arbitrary source function $\rho(\mathbf{x}, t)$. The same formula will apply to (15.6). To construct the solution we shall analyze (15.5) in several steps. To begin, suppose the charge density is constant in time, $\rho = \rho(\mathbf{x})$. Then $V(\mathbf{x})$ is independent of time, assuming $\rho(\mathbf{x})$ is the only source of the potential, and so (15.5) reduces to Poisson's equation

$$-\nabla^2 V = \rho/\epsilon_0. \tag{15.7}$$

In Chapter 3 we studied Green's solution of Poisson's equation,

$$V(\mathbf{x}) = \frac{1}{\epsilon_0} \int \frac{\rho(\mathbf{x}')d^3 x'}{4\pi |\mathbf{x} - \mathbf{x}'|} \quad \text{(static case)}. \tag{15.8}$$

This formula solves Poisson's equation because $1/4\pi r$ is the Green's function of $-\nabla^2$; that is, in terms of the Dirac delta function

$$-\nabla^2 \frac{1}{4\pi |\mathbf{x} - \mathbf{x}'|} = \delta^3(\mathbf{x} - \mathbf{x}'). \tag{15.9}$$

When $-\nabla^2$ acts on the integral in (15.8) it gives $\int \delta^3(\mathbf{x} - \mathbf{x}')\rho(\mathbf{x}')d^3 x'$, which is $\rho(\mathbf{x})$ because the delta function is an infinitely sharp spike at $\mathbf{x}' = \mathbf{x}$. Hence (15.7) is satisfied. In other words, $1/(4\pi r)$ is the potential (aside from the constant factor $1/\epsilon_0$) at distance r from a unit point charge. The integral in (15.8) is the superposition of the potentials due to elements of the charge distribution, integrated over the charge distribution.

As the next step toward solving (15.5), suppose the charge density is *harmonic in time* with constant angular frequency ω. That is, let $\rho(\mathbf{x}, t)$ have the form

$$\rho(\mathbf{x}, t) = \tilde{\rho}(\mathbf{x})e^{-i\omega t}. \tag{15.10}$$

As usual we use complex fields and densities for mathematical convenience, but it is understood that the real part of a complex function is the physical quantity. If $\rho(\mathbf{x}, t)$ is harmonic for all times then so is $V(\mathbf{x}, t)$, and we may write

$$V(\mathbf{x}, t) = \tilde{V}(\mathbf{x})e^{-i\omega t}. \tag{15.11}$$

Now we substitute these forms for V and ρ into d'Alembert's equation. Note that the operator $\partial/\partial t$ acting on $V(\mathbf{x}, t)$ is equivalent to multiplying by $-i\omega$. Thus we find that $\tilde{V}(\mathbf{x})$ must obey

$$-\nabla^2 \tilde{V} - k^2 \tilde{V} = \tilde{\rho}/\epsilon_0, \tag{15.12}$$

where the scalar parameter k is ω/c. Equation (15.12) is the *Helmholtz equation*, and it, too, can be solved by the Green's function method. We just need the Green's function of the operator $-(\nabla^2 + k^2)$.

We shall show that the Green's function of $-(\nabla^2 + k^2)$ is $e^{ikr}/(4\pi r)$. That is,

$$-\left(\nabla^2 + k^2\right)\frac{e^{ikr}}{r} = 4\pi\delta^3(\mathbf{x} - \mathbf{x}'), \tag{15.13}$$

where $r = |\mathbf{x} - \mathbf{x}'|$. Note that if $k = 0$ this reduces to the static case. To prove (15.13) we must verify that the left-hand side has the properties of the delta function. Without loss of generality let $\mathbf{x}'$ be at the origin. We must show that the left-hand side of (15.13) is 0 for $r \neq 0$ but has integral 4π over a volume that includes the origin. It is straightforward to show, in spherical coordinates, $\nabla^2(e^{ikr}/r) = -k^2(e^{ikr}/r)$ for $r \neq 0$; so the left-hand side of (15.13) is 0 away from the singular point. But by Gauss's theorem the integral of the left-hand side over a spherical volume of radius R around $r = 0$ is

$$-\frac{d}{dR}\left[\frac{e^{ikR}}{R}\right]4\pi R^2 - k^2 4\pi \left[\frac{-1}{(ik)^2} + \frac{r}{ik}\right]e^{ikr}\Bigg|_0^R = 4\pi.$$

The integral is 4π for any R. Thus the left-hand side of (15.13) does have exactly the properties of a delta function, and the equation is proven. Using the Green's function, $\widetilde{V}(\mathbf{x})$ is

$$\widetilde{V}(\mathbf{x}) = \frac{1}{\epsilon_0}\int \frac{e^{ik|\mathbf{x}-\mathbf{x}'|}}{4\pi |\mathbf{x} - \mathbf{x}'|}\widetilde{\rho}(\mathbf{x}')d^3x'. \tag{15.14}$$

The operator $-(\nabla^2 + k^2)$ acting on the integral gives $\widetilde{\rho}(\mathbf{x})$ because the delta function is an infinitely sharp spike at $\mathbf{x}' = \mathbf{x}$.

The formula (15.14) solves the potential problem for a source harmonic in time. There are many interesting problems with harmonic fields. Equation (15.14) and its partner for $\widetilde{\mathbf{A}}(\mathbf{x})$ may be used to solve such problems. But now we are interested in general time-dependence.

As the final step in the analysis of d'Alembert's equation, let $\rho(\mathbf{x}, t)$ have *arbitrary* time dependence. For any time dependence we may express $\rho(\mathbf{x}, t)$ and $V(\mathbf{x}, t)$ as Fourier integrals, i.e., as superpositions of harmonic functions:

$$\rho(\mathbf{x}, t) = \int_{-\infty}^{\infty} \widetilde{\rho}(\mathbf{x}, \omega)e^{-i\omega t}d\omega, \tag{15.15}$$

$$V(\mathbf{x}, t) = \int_{-\infty}^{\infty} \widetilde{V}(\mathbf{x}, \omega)e^{-i\omega t}d\omega. \tag{15.16}$$

The fact that any time-dependent function can be written as a Fourier integral (with reasonable assumptions about smoothness and finiteness that would be true for any physical field) is a powerful theorem of mathematics. Now, by the superposition principle—applicable here because (15.5) is a linear relation between V and ρ—the component of $V(\mathbf{x}, t)$ with frequency ω is determined by the component of $\rho(\mathbf{x}, t)$ with the same frequency. That is, $\widetilde{V}(\mathbf{x}, \omega)$ is given in terms of $\widetilde{\rho}(\mathbf{x}, \omega)$ by (15.14) with $k = \omega/c$. Substituting this solution for $\widetilde{V}(\mathbf{x}, \omega)$ into (15.16) we have

$$V(\mathbf{x}, t) = \frac{1}{\epsilon_0}\int_{-\infty}^{\infty}\int \frac{e^{i\omega|\mathbf{x}-\mathbf{x}'|/c}}{4\pi |\mathbf{x} - \mathbf{x}'|}\widetilde{\rho}(\mathbf{x}', \omega)d^3x'e^{-i\omega t}d\omega. \tag{15.17}$$

The ω integration in (15.17) can be evaluated. Combine the exponentials $e^{-i\omega t}$ and $e^{i\omega r/c}$ as $e^{-i\omega t'}$ where

$$t' = t - |\mathbf{x} - \mathbf{x}'|/c. \tag{15.18}$$

Now the ω integral can be done by (15.15), giving $\rho(\mathbf{x}', t')$. Therefore the scalar potential is

$$V(\mathbf{x}, t) = \frac{1}{\epsilon_0} \int \frac{\rho(\mathbf{x}', t - |\mathbf{x} - \mathbf{x}'|/c)}{4\pi |\mathbf{x} - \mathbf{x}'|} d^3 x'. \tag{15.19}$$

This is our general formula for the scalar potential due to a time-dependent charge density in the Lorentz gauge. d'Alembert's equation for the vector potential is solved in the same way, by

$$\mathbf{A}(\mathbf{x}, t) = \mu_0 \int \frac{\mathbf{J}(\mathbf{x}', t - |\mathbf{x} - \mathbf{x}'|/c)}{4\pi |\mathbf{x} - \mathbf{x}'|} d^3 x'. \tag{15.20}$$

The potentials (15.19) and (15.20) are called the *retarded potentials*.

The meaning of (15.19) or (15.20) is that a potential at point $\mathbf{x}$ and time t is a superposition of terms, integrated over all source points $\mathbf{x}'$. However, the contribution to the potential from the element of the source at position $\mathbf{x}'$, is determined by the density $\rho(\mathbf{x}', t')$ or $\mathbf{J}(\mathbf{x}', t')$ not at time t but at the *earlier time* t' given by $t - r/c$. This time t' is called the *retarded time*. The time difference $t - t'$ is equal to the time it would take light to travel from $\mathbf{x}'$ to $\mathbf{x}$. In other words, there is a time delay, equal to the light travel time, for any action at the source point $\mathbf{x}'$ to affect the field at $\mathbf{x}$. Field theory is a local theory—the antithesis of action at a distance—and it takes time for a change at one point in the system to influence the system at another point. The effect of a change in the charge or current density travels at the speed of light.

The idea of the retarded potential was first stated by Riemann for scalar waves. It was first applied to electromagnetism by Ludwig Lorenz. By the way, (15.19) and (15.20) imply that the Green's function $G(\mathbf{r}, \tau)$ of the d'Alembertian operator, defined by

$$-\nabla^2 G + \frac{1}{c^2} \frac{\partial^2 G}{\partial \tau^2} = \delta^3(\mathbf{r})\delta(\tau), \tag{15.21}$$

is

$$G(\mathbf{r}, \tau) = \frac{\delta(\tau - r/c)}{4\pi r}. \tag{15.22}$$

In terms of this time-dependent Green's function, the potential is

$$V(\mathbf{x}, t) = \frac{1}{\epsilon_0} \int G(\mathbf{x} - \mathbf{x}', t - t')\rho(\mathbf{x}', t')d^3 x' dt'. \tag{15.23}$$

As an illustration of the use of retarded potentials, we consider the following interesting, although unrealistic, example.

EXAMPLE 1 Suppose at $t = 0$ a uniform surface current $\mathbf{K}(t) = K_0(t)\hat{\mathbf{i}}$ is suddenly established on the entire xy plane. Figure 15.1 illustrates the source and waves. What are the radiated wave fields $\mathbf{E}(\mathbf{x}, t)$ and $\mathbf{B}(\mathbf{x}, t)$?

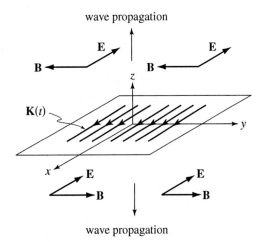

FIGURE 15.1   Example 1. The xy plane carries a time-dependent surface current density $\mathbf{K}(t)$ in the x direction. Waves are radiated in the $\pm z$ directions.

First, note that $V(\mathbf{x}, t) = 0$ because there is no net charge. Second, calculate the vector potential by replacing $\mathbf{J} d^3 x'$ in (15.20) by $\mathbf{K} dS'$

$$\mathbf{A}(\mathbf{x}, t) = \frac{\mu_0}{4\pi} \int \frac{1}{r} \mathbf{K}(\mathbf{x}', t - r/c) dS'. \qquad (15.24)$$

Here r denotes the distance $|\mathbf{x} - \mathbf{x}'|$ from source point to field point. The integral may be evaluated by integrating over infinitesimal annuli of current, as shown in Fig. 15.2. For a point P at distance z from the plane of current

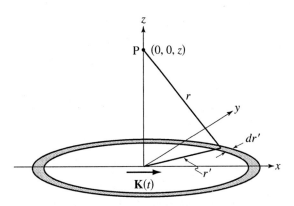

FIGURE 15.2   An integration element on a current sheet. The xy plane has surface current density $\mathbf{K}$ in the x direction. The shaded region is an elemental ring.

$$\mathbf{A}(\mathbf{x}, t) = \frac{\mu_0}{4\pi} \hat{\mathbf{i}} \int \frac{1}{r} K_0(t - r/c) 2\pi r' dr', \tag{15.25}$$

where $r^2 = r'^2 + z^2$.

Now change the variable of integration from r' to r, using $r'dr' = rdr$. The lower limit of r is z for $z > 0$, or $-z$ for $z < 0$; therefore we write $|z|$. The upper limit of r is ct because $K_0(t - r/c)$ is 0 for $t - r/c < 0$. Thus

$$\mathbf{A}(\mathbf{x}, t) = \frac{\mu_0}{2} \hat{\mathbf{i}} \int_{|z|}^{ct} K_0(t - r/c) dr. \tag{15.26}$$

The effect of a change of the current propagates at the speed of light. For $t < |z|/c$ there is no field at P. At $t = |z|/c$ the field due to the nearest source, at the origin, arrives at P. For $t > |z|/c$ all sources with $r \leq ct$ contribute to the field at P.

To facilitate the calculation of $\mathbf{E}$ and $\mathbf{B}$ we make another change of variables, to $\xi \equiv t - r/c$,

$$\mathbf{A}(\mathbf{x}, t) = \frac{\mu_0 c}{2} \hat{\mathbf{i}} \int_0^{t - |z|/c} K_0(\xi) d\xi. \tag{15.27}$$

The fields at P are

$$\mathbf{E}(\mathbf{x}, t) = -\frac{\partial \mathbf{A}}{\partial t} = -\frac{\mu_0 c}{2} K_0(t \mp z/c) \hat{\mathbf{i}} \tag{15.28}$$

$$\mathbf{B}(\mathbf{x}, t) = \nabla \times \mathbf{A} = \hat{\mathbf{j}} \frac{\partial A_x}{\partial z} = \mp \frac{\mu_0}{2} K_0(t \mp z/c) \hat{\mathbf{j}} \tag{15.29}$$

where the upper sign is for $z > 0$ and the lower sign is for $z < 0$. These fields are illustrated in Fig. 15.1.

As an exercise it can be shown that (15.28) and (15.29) represent outgoing plane waves, and that the proper boundary conditions are satisfied.[1]

![black rectangle marker]

One final comment should be made about the retarded potentials. There are also *advanced potentials*, for which the potential at the spacetime point $(\mathbf{x}, t)$ depends on the source at $(\mathbf{x}', t'')$, where $t'' = t + |\mathbf{x} - \mathbf{x}'|/c$. Note that t'' is a *later* time than t; the time t'' is called the *advanced time*. Going back through our derivation of (15.19) and (15.20) we would generate the advanced potentials by using $e^{-ikr}/(4\pi r)$ as the Green's function of $-(\nabla^2 + k^2)$. In (15.14) we used $e^{+ikr}/(4\pi r)$, an *outgoing* spherical wave, as the effect of the harmonic source; whereas $e^{-ikr}/(4\pi r)$ describes an *incoming* spherical wave. Either retarded or advanced potentials satisfy the differential equations (15.5) and (15.6). But the retarded potentials satisfy the proper *boundary condition in time* for radiation problems: Causes must precede their effects, so the response to a change in the

[1] See Exercise 1.

source occurs later, at the retarded time. The advanced potentials are unphysical solutions, the use of which would imply a violation of causality. Nevertheless there are theoretical speculations about electromagnetism in which advanced potentials play a role.[2]

15.2 ■ RADIATION FROM AN ELECTRIC DIPOLE

Equations (15.19) and (15.20) determine the potentials, and hence the fields, for any densities $\rho(\mathbf{x}, t)$ and $\mathbf{J}(\mathbf{x}, t)$. At this point we may specify a system of current and charge and calculate the fields. At large distances from the source the asymptotic fields will be electromagnetic waves, propagating outward with some angular distribution of intensity. Such a calculation is an example in *antenna theory*. Whole books are devoted to this important topic in electrical engineering.

Antenna calculations are somewhat technical and specialized. In this section we study the simplest example—radiation from a pointlike electric dipole. Although this case has limited applicability, it is a good pedagogical example because it is simple enough that the field calculation can be done analytically, but complete enough to illustrate the general features of radiation problems.

In general a source of electromagnetic radiation has both charge density $\rho(\mathbf{x}, t)$ and current density $\mathbf{J}(\mathbf{x}, t)$. It turns out to be simplest to analyze the current density. $\mathbf{J}(\mathbf{x}', t')$ denotes the volume current density as a function of position $\mathbf{x}'$ in the source and time t'. We shall assume that the source occupies a finite volume around the origin, as shown in Fig. 15.3. Then we shall calculate the *asymptotic* potentials and fields, i.e., at points far from the source. As we shall see, the asymptotic fields are of order $1/r$ where $r = |\mathbf{x}|$ is the distance from the origin. Therefore *we systematically neglect terms of order $1/r^2$ or smaller* in calculating

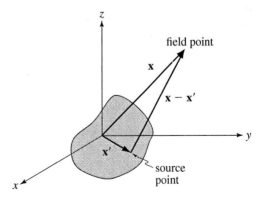

FIGURE 15.3   Asymptotics. Far from the source $|\mathbf{x} - \mathbf{x}'|$ approaches r. A better approximation is $|\mathbf{x} - \mathbf{x}'| \approx r - \hat{\mathbf{r}} \cdot \mathbf{x}'$.

[2]J. A. Wheeler and R. P. Feynman, *Rev. Mod. Phys.* **17**, 157 (1945), and **21**, 425 (1949); J. L. Anderson, *Am. J. Phys.* **60**, 465 (1992).

the fields. It is characteristic of radiation fields that they fall off as $1/r$ at large distances. The asymptotic region is called the *radiation zone*.

The exact vector potential for any field point $\mathbf{x}$ is (15.20). Asymptotically, $|\mathbf{x}|$ is much larger than $|\mathbf{x}'|$ for any source point $\mathbf{x}'$. Therefore $|\mathbf{x}-\mathbf{x}'|$ approaches $r \equiv |\mathbf{x}|$, the distance to the origin, as demonstrated in Fig. 15.3. This approximation is accurate enough to determine the asymptotic fields of a pointlike electric dipole.[3] Thus we start with the relation

$$\mathbf{A}(\mathbf{x}, t) \sim \frac{\mu_0}{4\pi r} \int \mathbf{J}(\mathbf{x}', t - r/c)d^3 x', \qquad (15.30)$$

where the symbol "$\sim$" is to be read as "approaches asymptotically".[4] Again, $r = |\mathbf{x}|$ is the distance from the origin to the field point $\mathbf{x}$, which is to say the radial coordinate of $\mathbf{x}$. To simplify the notation, let t_r denote the retarded time for the source point at the origin

$$t_r = t - r/c. \qquad (15.31)$$

In the approximation (15.30) all source points have the same retarded time t_r, which is what we mean by a pointlike source. For this approximation to be valid the source size a must be small compared to any relevant length in the system. For one thing, a must be small compared to r. But more importantly, a must also be small compared to $c\tau$, where τ is the characteristic time constant of the variation of $\mathbf{J}(\mathbf{x}', t')$. For example, if $\mathbf{J}(\mathbf{x}', t')$ varies harmonically in time, with angular frequency ω, then the characteristic time constant τ would be the period $2\pi/\omega$. In that case, because $c\tau$ would be the wavelength λ of the radiation, we require $a \ll \lambda$. This statement makes precise what we mean by a pointlike source. Or, looking at it another way, the condition for validity of (15.30) is that the changes occur slowly, with a long characteristic time $\tau \gg a/c$.

Having made the approximation (15.30) the rest of the calculation is just a matter of simplifying the expression. First we relate $\mathbf{A}(\mathbf{x}, t)$ to the electric dipole moment of the source. Recall from Chapter 3 the definition of the electric dipole moment of a charge distribution,

$$\mathbf{p}(t) = \int \mathbf{x}\rho(\mathbf{x}, t)d^3 x, \qquad (15.32)$$

which we assume is nonzero. (We temporarily drop the prime on the source points to simplify the notation; for now $\mathbf{x}$ denotes a point in the source.) We shall prove that $\int \mathbf{J}d^3 x$ is $d\mathbf{p}/dt$.

Consider the integral, over all space, of $\nabla \cdot (x_i \mathbf{J})$ where i is a Cartesian index.[5] By Gauss's theorem the integral is 0, because $\mathbf{J}(\mathbf{x}, t)$ is 0 on the sphere at ∞. The divergence is $J_i + x_i \nabla \cdot \mathbf{J}$ by a vector identity; and $\nabla \cdot \mathbf{J}$ is $-\partial\rho/\partial t$ by charge

---

[3] We will use a better approximation to calculate magnetic dipole radiation in Sec. 15.2.3.

[4] The mathematics of asymptotic expansion is used often in radiation theory. We approximate functions of $\mathbf{x}$ in the limit $r \to \infty$, keeping terms of order r^{-1} but neglecting terms of order r^{-2}.

[5] The index values $i = 1, 2, 3$ correspond to the x, y, z components, as usual.

conservation. Thus the all-space integral of the current density is

$$\int J_i d^3 x = -\int x_i \nabla \cdot \mathbf{J} d^3 x = \frac{d}{dt} \int x_i \rho d^3 x = \frac{dp_i}{dt}, \tag{15.33}$$

as claimed. Using this result in (15.30) the asymptotic potential is

$$\mathbf{A}(\mathbf{x}, t) \sim \frac{\mu_0}{4\pi r} \frac{d\mathbf{p}}{dt}, \tag{15.34}$$

where the derivative of $\mathbf{p}$ is evaluated at the retarded time t_r.

The magnetic field is $\nabla \times \mathbf{A}$. Computing the curl of (15.34) gives two terms, but one is negligible asymptotically. If ∇ acts on the factor $1/r$ the result is of order $1/r^2$, which we must neglect as before. But there is also an r dependence in $d\mathbf{p}/dt$ because this function is evaluated at $t_r = t - r/c$. The curl of $d\mathbf{p}/dt$ is $\nabla t_r \times d^2\mathbf{p}/dt^2$ by the chain rule; and ∇t_r is $-\hat{\mathbf{r}}/c$. Thus the asymptotic magnetic field is

$$\mathbf{B}(\mathbf{x}, t) \sim \frac{-\mu_0}{4\pi r c} \hat{\mathbf{r}} \times \frac{d^2\mathbf{p}}{dt^2} \equiv \mathbf{B}_{\text{rad}}. \tag{15.35}$$

The notation $\mathbf{B}_{\text{rad}}$ emphasizes that the asymptotic field is the electromagnetic radiation.

The asymptotic scalar potential may be determined from the Lorentz gauge condition (15.4),

$$\frac{\partial V}{\partial t} = -c^2 \nabla \cdot \mathbf{A} \sim -\frac{c^2 \mu_0}{4\pi r} \frac{d^2\mathbf{p}}{dt^2} \cdot \left(\frac{-\hat{\mathbf{r}}}{c} \right). \tag{15.36}$$

The *asymptotic part* of $\nabla \cdot \mathbf{A}$ comes from ∇ acting on the r dependence of $\mathbf{p}(t_r)$, i.e.,

$$\nabla \cdot \frac{d\mathbf{p}(t_r)}{dt} = \frac{d^2\mathbf{p}}{dt^2} \cdot \nabla t_r = \frac{d^2\mathbf{p}}{dt^2} \cdot \left(\frac{-\hat{\mathbf{r}}}{c} \right). \tag{15.37}$$

By (15.36) the asymptotic potential is

$$V \sim \frac{\mu_0 c}{4\pi r} \hat{\mathbf{r}} \cdot \frac{d\mathbf{p}}{dt}. \tag{15.38}$$

The electric field is $-\partial \mathbf{A}/\partial t - \nabla V$. We could calculate the asymptotic form directly, which we leave as an exercise.[6] However, the result may be derived more easily by using properties of electromagnetic waves. Far from the source, $\mathbf{E}$ and $\mathbf{B}$ propagate as an outgoing spherical wave. In any direction the wave approaches asymptotically a plane wave. Therefore $\mathbf{E}_{\text{rad}}$, $\mathbf{B}_{\text{rad}}$, and $\hat{\mathbf{r}}$ form an orthogonal triad of vectors; and the magnitudes of the electric and magnetic fields are related by $B_{\text{rad}} = E_{\text{rad}}/c$, as for plane waves. These properties of the outgoing waves follow

[6] See Exercise 4.

directly from Maxwell's equations in the vacuum outside the source. Therefore the asymptotic electric field $\mathbf{E}_{rad}$ is $c\mathbf{B}_{rad} \times \hat{\mathbf{r}}$. Substituting (15.35), and simplifying the double cross product, we find the electric field of the outgoing radiation

$$\mathbf{E}_{rad} = \frac{\mu_0}{4\pi r} \left[\hat{\mathbf{r}} \left(\hat{\mathbf{r}} \cdot \frac{d^2\mathbf{p}}{dt^2} \right) - \frac{d^2\mathbf{p}}{dt^2} \right]. \tag{15.39}$$

Again, $d^2\mathbf{p}/dt^2$ is evaluated at the retarded time $t_r = t - r/c$.

It is important to note that $\mathbf{E}_{rad}$ and $\mathbf{B}_{rad}$ are of order $1/r$. Indeed we have neglected any other r dependence. *Static sources* always produce fields that decrease with a higher power of $1/r$. For example, the electric field of a static dipole is of order $p/(\epsilon_0 r^3)$. But the electric field of the *radiation* produced by a time-varying dipole is of order $\ddot{p}/(\epsilon_0 c^2 r)$. It is also important to understand the implications of this asymptotic behavior. The fields decrease as $1/r$, so the energy flux decreases as $1/r^2$, as required by conservation of energy.

The energy flux $\mathbf{S}$ (power per unit area) in the radiation zone is $\mathbf{E}_{rad} \times \mathbf{B}_{rad}/\mu_0$. The calculation of $\mathbf{E}_{rad} \times \mathbf{B}_{rad}$ requires several steps. $\mathbf{B}_{rad}$ is proportional to $\hat{\mathbf{r}} \times \alpha$, where $\alpha \equiv d^2\mathbf{p}/dt^2$, and $\mathbf{E}_{rad}$ is proportional to $\hat{\mathbf{r}}\alpha_r - \alpha$. The cross-product reduction proceeds via vector product identities as follows:

$$\left(\hat{\mathbf{r}}\alpha_r - \alpha \right) \times \left(\hat{\mathbf{r}} \times \alpha \right) = \alpha_r \hat{\mathbf{r}} \times \left(\hat{\mathbf{r}} \times \alpha \right) - \alpha \times \left(\hat{\mathbf{r}} \times \alpha \right)$$

$$= \alpha_r \left[\hat{\mathbf{r}}\alpha_r - \alpha \right] - \hat{\mathbf{r}}\alpha^2 + \alpha_r \alpha$$

$$= \hat{\mathbf{r}} \left(\alpha_r^2 - \alpha^2 \right).$$

Therefore

$$\mathbf{S} = \frac{\mu_0 \hat{\mathbf{r}}}{16\pi^2 r^2 c} \left[\left(\frac{d^2\mathbf{p}}{dt^2} \right)^2 - \left(\hat{\mathbf{r}} \cdot \frac{d^2\mathbf{p}}{dt^2} \right)^2 \right]. \tag{15.40}$$

The energy flux is in the direction of $\hat{\mathbf{r}}$, i.e., radially outward. The magnitude includes the factor $1/r^2$ as it must by energy conservation: The power through a sphere cannot depend on the radius if the radiation rate is constant. Or, whatever power passes a radius r_1 at time t_1 passes any larger radius r_2 at $t_2 = t_1 + (r_2 - r_1)/c$, carried outward by the fields of the propagating wave. The surface area is $4\pi r^2$ so the power *per unit area* must decrease as $1/r^2$.

The total power of radiation passing through a sphere of radius R is

$$P(R, t) = \oint \mathbf{S} \cdot \hat{\mathbf{r}} R^2 d\Omega.$$

To evaluate the integral over solid angles, set up a coordinate system with the z direction parallel to $d^2\mathbf{p}/dt^2$ and use polar coordinates; note that $\oint d\Omega = 4\pi$ and

$\oint \cos^2 \theta \, d\Omega = 4\pi/3$. Thus the total power through the sphere is

$$P(R, t) = \frac{1}{6\pi \epsilon_0 c^3} \left| \frac{d^2 \mathbf{p}}{dt^2} \right|^2. \tag{15.41}$$

It is worth stating again that $d^2\mathbf{p}/dt^2$ in (15.40) and (15.41) is evaluated at $t - R/c$, the time when the radiation was emitted from the origin. So $P(R, t)$ is the power at radius R of the radiation that was generated at the retarded time.

The results (15.40) and (15.41) are general for any pointlike dipole moment $\mathbf{p}(t)$. For example, if the dipole changes only during a short time interval, then the radiation is a pulse propagating outward. Another example is continuous harmonic radiation, which we consider next.

15.2.1 ▪ The Hertzian Dipole

As a simple example of a radiation source, consider the harmonically oscillating dipole shown in Fig. 15.4, called the *Hertzian dipole*. An alternating current $I(t)$ carries charge between two small spheres on the z axis at $z = \pm d/2$. The charges on the two spheres are $\pm Q(t)$, where $Q(t) = Q_0 \cos \omega t$. The current in the wire connecting the spheres is $I = dQ/dt = -\omega Q_0 \sin \omega t$. Thus the electric dipole moment, as a function of time t, is

$$\mathbf{p}(t) = \hat{\mathbf{k}} Q(t) d = \hat{\mathbf{k}} p_0 \cos \omega t \tag{15.42}$$

where $p_0 = Q_0 d$. The derivatives of $\mathbf{p}$, which we need to calculate the radiation fields, are

$$d\mathbf{p}/dt = -\hat{\mathbf{k}} p_0 \omega \sin \omega t$$
$$d^2\mathbf{p}/dt^2 = -\hat{\mathbf{k}} p_0 \omega^2 \cos \omega t.$$

A physical realization of the Hertzian dipole is a small parallel plate capacitor in a circuit with an alternating emf.

The electromagnetic waves from the Hertzian dipole have frequency ω and wavelength $\lambda = 2\pi c/\omega$. We assume $\lambda \gg d$ so that the dipole is pointlike.

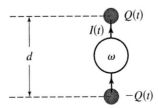

FIGURE 15.4 The Hertzian dipole. The size of the dipole is assumed to be much smaller than one wavelength of the radiation.

Then the radiated power is (15.41) and its angular distribution can be calculated from (15.40).

Let r, θ, ϕ be spherical polar coordinates, with the origin at the dipole. The radiation zone for this small dipole is the region $r \gg \lambda$. The radiation fields at (r, θ, ϕ) are, by (15.35) and (15.39),

$$\mathbf{B}_{\text{rad}} = B_{\text{rad}}\hat{\boldsymbol{\phi}} \quad \text{and} \quad \mathbf{E}_{\text{rad}} = cB_{\text{rad}}\hat{\boldsymbol{\theta}} \qquad (15.43)$$

where

$$B_{\text{rad}} = \frac{-\mu_0 p_0 \omega^2}{4\pi c} \frac{\sin\theta}{r} \cos\left[\omega(t - r/c)\right]. \qquad (15.44)$$

The magnetic field oscillates in the azimuthal direction, and the electric field oscillates in the meridional direction, as illustrated in Fig. 15.5. The Poynting vector is

$$\mathbf{S} = \frac{cB_{\text{rad}}^2}{\mu_0}\,\hat{\mathbf{r}}, \qquad (15.45)$$

which is in the radial direction. The energy flux at a point in the radiation zone oscillates between 0 and a maximum value S_{max}, with frequency 2ω. The *radiation intensity* $\mathcal{I}$ is defined as the average energy flux—averaged over a period of oscillation—so the intensity is $\mathcal{I} = S_{\text{max}}/2$. The differential power, i.e., the average power per unit solid angle, is

$$\frac{dP_{\text{avg}}}{d\Omega} = r^2\hat{\mathbf{r}} \cdot \mathbf{S}_{\text{max}}/2 = \frac{p_0^2 \omega^4}{32\pi^2 \epsilon_0 c^3} \sin^2\theta. \qquad (15.46)$$

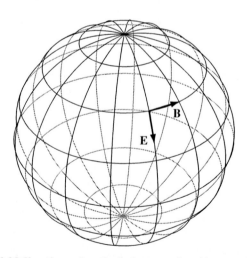

FIGURE 15.5   Field directions of a spherical wave radiated by a small electric dipole at the origin. The Poynting vector is radially outward. One half cycle later the directions of **E** and **B** at the point shown are opposite to the directions shown.

(We have substituted $\mu_0 = 1/(\epsilon_0 c^2)$.) An important result is that the radiation intensity is proportional to ω^4. If the frequency is doubled, the intensity increases by a factor of 16.

The total power passing through a sphere in the radiation zone is the integral of $\mathbf{S} \cdot \hat{\mathbf{r}}$ over the surface, or equivalently the integral of $dP/d\Omega$ over all directions. The relevant solid-angle integral is $\oint \sin^2 \theta \, d\Omega = 8\pi/3$. Thus the integrated power, averaged over a period of oscillation, is

$$P_{\text{avg}} = \frac{p_0^2 \omega^4}{12\pi \epsilon_0 c^3}. \tag{15.47}$$

The result is independent of the radius of the sphere, as expected.

The angular distribution of the radiation is proportional to $\sin^2 \theta$, i.e.,

$$\frac{dP_{\text{avg}}}{d\Omega} = \frac{3 P_{\text{avg}}}{8\pi} \sin^2 \theta. \tag{15.48}$$

The intensity is independent of the azimuthal angle, as it must be by symmetry. Figure 15.6 shows the $\sin^2 \theta$ distribution in the form of a polar plot. On this plot the distance from the origin to a point on the curve is proportional to the intensity of radiation in that direction. (θ is the angle with the z axis.) The $\sin^2 \theta$ distribution is characteristic of electric dipole radiation. The intensity is greatest in the plane perpendicular to the oscillating moment ($\theta = \pi/2$); and the intensity is 0 in the direction of motion of the oscillating charge ($\theta = 0$ or π). Electromagnetic waves are *transverse*; that is, the fields oscillate in directions orthogonal to the direction of propagation. Current, alternating up and down on the z axis as shown in Fig. 15.4, and the varying charge on the spheres, create radiation fields with greatest magnitude in the xy plane, and with zero magnitude in the longitudinal direction z.

Electrical engineers conventionally express the power characteristics of an antenna in terms of a quantity called the *radiation resistance*, or *antenna resistance*. The radiation fields are linear functions of the current I, so the average power is proportional to I_{rms}^2, the mean squared current. For a current $I_0 \sin \omega t$ the rms

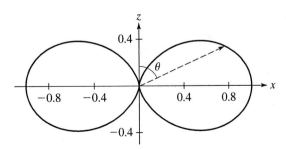

FIGURE 15.6   Angular distribution of the intensity of electric dipole radiation, as a polar plot. The distance on the graph from the origin to a point on the curve is proportional to the intensity in that direction.

current is $I_0/\sqrt{2}$. The antenna resistance is defined by

$$R_{ant} = P_{avg}/I_{rms}^2. \tag{15.49}$$

That is, $I_{rms}^2 R_{ant}$ is the average power radiated by the antenna. (Recall that the average power dissipated in a resistor R is $I_{rms}^2 R$.) R_{ant} has units of ohms. The radiation resistance of an antenna depends on its geometry, and is a measure of its efficiency as a radiator.

For the Hertzian dipole the power is (15.47) and the rms current is $I_{rms} = I_0/\sqrt{2} = \omega Q_0/\sqrt{2}$, where $Q_0 = p_0/d$. Therefore the antenna resistance is

$$R_{ant} = \frac{\omega^2 d^2}{6\pi \epsilon_0 c^3} = \frac{1}{6\pi} \sqrt{\frac{\mu_0}{\epsilon_0}} (kd)^2, \tag{15.50}$$

where k in the second equality is ω/c. The quantity $\sqrt{\mu_0/\epsilon_0} = 377\,\Omega$ is called the impedance of free space. Numerically, we calculate

$$R_{ant} = (20\,\Omega)(kd)^2. \tag{15.51}$$

The Hertzian dipole is small; we required $d \ll \lambda$ to justify the pointlike source approximation leading to (15.47). Therefore $kd \ll 1$ because $k = 2\pi/\lambda$. The radiation resistance is small, so a small dipole is not an efficient radiator. For example, if $d = 0.1/k$ then R_{ant} is only $0.2\,\Omega$. By comparison, the radiation resistance of a half-wave antenna is $73\,\Omega$. The half-wave antenna has $d = \lambda/2$, and is not accurately described by the pointlike source approximation. For any antenna the radiation resistance can be written as $\sqrt{\mu_0/\epsilon_0}$ times a dimensionless factor that is determined by the geometry of the source.

15.2.2 ■ Atomic Transitions

Radiation by an atom is quite different from radiation by a macroscopic source. An atom in an excited state emits a *single photon* and jumps to a state of lower energy. The quantum process is discrete and probabilistic. We cannot say exactly when the transition will occur, but we can calculate the *transition rate* (= transition probability per unit time) in terms of a transition matrix element.

For a hydrogen atom, if the transition satisfies the electric-dipole selection rules then the relevant transition element is the dipole moment operator $e\mathbf{x}$ between the inital and final atomic states $|i\rangle$ and $|f\rangle$. ($\mathbf{x}$ is the electron position with respect to the nucleus.) The matrix element may be written as

$$\langle f|e\mathbf{x}|i\rangle = ed_{fi}\hat{\mathbf{k}}$$

letting the z axis be in the direction of $\langle f|\mathbf{x}|i\rangle$. Then the transition rate per unit solid angle of the emitted photon direction, calculated in quantum theory, is

$$\text{transition rate} = \left(\frac{dR}{d\Omega}\right)_{qu} = \frac{e^2 d_{fi}^2 \omega^3}{(4\pi\epsilon_0)2\pi\hbar c^3}\sin^2\theta,$$

where θ is the angle between the z axis and the photon direction. The photon has energy $\hbar\omega$ so the mean power radiated per unit solid angle is

$$\text{power radiated} = \hbar\omega \left(\frac{dR}{d\Omega}\right)_{\text{qu}} = \frac{e^2 d_{fi}^2 \omega^4}{8\pi^2 \epsilon_0 c^3} \sin^2\theta.$$

This quantum result has the same form as the classical power (15.46) for a dipole moment $p_0 = 2ed_{fi}$.

The mean energy flux, as a function of direction, is the same in both theories, with a suitable matching of the dipole moments. But the nature of the radiation is very different in the two theories. It is discrete in quantum theory but continuous in classical theory. The radiation from an atomic transition, if observed in a detector, is a single packet of energy $\hbar\omega$ with a definite direction. In contrast, the radiation from a macroscopic antenna arrives at a detector continuously and spreads out as a wave moving in all directions from the source. A single photon also has an associated wave—a *probability wave*—but it cannot be detected simultaneously at different points like a classical wave. The radiation from a macroscopic antenna is a state with a large and indeterminate number of photons streaming in all directions, and this state behaves just like the classical Maxwellian theory.

The quantum theory of atomic transitions is an important part of electromagnetism but outside the scope of this book. We return now to the classical theory of radiation.

15.2.3 ■ Magnetic Dipole Radiation

The radiation fields in (15.35) and (15.39) were derived in a general way, and are the dominant fields for a small source ($a \ll \lambda$) unless the electric dipole moment happens to be 0 because of some symmetry of the source. If $\mathbf{p}(t) = 0$ (for all t) then the dominant radiation may be magnetic dipole, electric quadrupole, or some higher multipole, depending on the geometry of the source. For example, consider a current loop in the xy plane, with alternating current $I(t) = I_0 e^{-i\omega t}$, as shown in Fig. 15.7. This source has zero electric dipole moment, and indeed zero charge density. But it radiates because of the varying current. The dominant radiation, if the current loop is small compared to $\lambda = 2\pi c/\omega$, is magnetic dipole radiation. The magnetic moment is $\mathbf{m}(t) = \hat{\mathbf{k}} I_0 A e^{-i\omega t}$, where A is the area of the loop.

The vector potential can be calculated from (15.20). For simplicity we will only calculate the fields in the radiation zone. Asymptotically we may approximate $1/|\mathbf{x} - \mathbf{x}'|$ by $1/r$, where $r = |\mathbf{x}|$ is the radial coordinate of the field point. However, we need to be more accurate in evaluating the retarded time. A better approximation of $|\mathbf{x} - \mathbf{x}'|$ is $r - \mathbf{x}' \cdot \hat{\mathbf{r}}$; then by Taylor series expansion

$$\mathbf{J}(\mathbf{x}', t - |\mathbf{x} - \mathbf{x}'|/c) \approx \mathbf{J}(\mathbf{x}', t_r) + \frac{\mathbf{x}' \cdot \hat{\mathbf{r}}}{c} \frac{\partial}{\partial t_r} \mathbf{J}(\mathbf{x}', t_r). \tag{15.52}$$

Again t_r is $t - r/c$, the retarded time of a source at the origin. In the second term of (15.52) $\partial/\partial t_r$ may be replaced by $-i\omega$. To use (15.20) we need the integral of

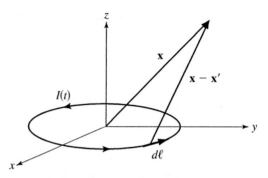

FIGURE 15.7   Magnetic dipole radiator. The magnetic moment is $I\,A\hat{\mathbf{k}}$, where A is the area of the planar loop. If $I(t)$ oscillates harmonically then electromagnetic waves with the same frequency are generated.

$\mathbf{J}(\mathbf{x}', t')$ over all $\mathbf{x}'$. The integral of the first term is 0 by (15.33), because $\rho(\mathbf{x}, t) = 0$. (This is why we needed a more precise retarded time; the lowest approximation is 0.) The integral of the second term is $(-i\omega/c)\mathbf{m} \times \hat{\mathbf{r}}$, by (8.72), which was derived from the definition (8.73) of the magnetic dipole moment of a current distribution. Thus the asymptotic vector potential is

$$\mathbf{A}(\mathbf{x}, t) \sim \frac{-\mu_0 i \omega}{4\pi rc}\mathbf{m}(t_r) \times \hat{\mathbf{r}}. \qquad (15.53)$$

It is convenient to define the xy plane as the plane of the current loop; then using polar coordinates

$$\mathbf{A}(\mathbf{x}, t) \sim \frac{-\mu_0 i \omega I_0 A}{4\pi c}\frac{\sin\theta}{r}e^{-i\omega(t-r/c)}\hat{\boldsymbol{\phi}}. \qquad (15.54)$$

The asymptotic vector potential for magnetic dipole radiation is $\propto m_0/c$, while for electric dipole radiation it is $\propto p_0$. In radiation from atoms, magnetic dipole radiation is weaker than electric dipole radiation because typically m_0/c is small compared to p_0.

The electric field is $\mathbf{E} = -\partial \mathbf{A}/\partial t = i\omega\mathbf{A}$, because $V = 0$. Thus $\mathbf{E}_{\text{rad}}$ is azimuthal. In the radiation zone the magnetic field is orthogonal to $\mathbf{E}$ and $\hat{\mathbf{r}}$, with magnitude E/c, so $\mathbf{B} = \hat{\mathbf{r}} \times \mathbf{E}/c$. Thus $\mathbf{B}_{\text{rad}}$ is meridional. The directions of $\mathbf{E}_{\text{rad}}$ and $\mathbf{B}_{\text{rad}}$ are interchanged relative to radiation from an electric dipole. We leave it as an exercise[7] to show that the magnetic dipole results are equivalent to the following *dual transformation* of the electric dipole results:

$$p_0 \rightarrow m_0/c, \qquad \mathbf{E}_{\text{rad}} \rightarrow -c\mathbf{B}_{\text{rad}}, \qquad \mathbf{B}_{\text{rad}} \rightarrow \mathbf{E}_{\text{rad}}/c, \qquad (15.55)$$

where $m_0 = I_0 A$ is the amplitude of oscillation of the magnetic moment. That is, starting from the equations for the electric dipole fields, and making the replacements (15.55), yields the magnetic dipole fields.

[7] See Exercise 7.

The mean power radiated by the magnetic dipole is

$$P_{avg} = \frac{m_0^2 \omega^4}{12\pi \epsilon_0 c^5}, \qquad (15.56)$$

which is just the dual transformation of the power (15.47) from an oscillating electric dipole. The dipole moment for a current loop is $m_0 = I_0 A$, so the power may be written as

$$P_{avg} = \frac{I_0^2 A^2 k^4}{12\pi \epsilon_0 c}, \qquad (15.57)$$

where $k = \omega/c$. The antenna resistance for an oscillating pointlike magnetic dipole is

$$R_{ant} = \frac{1}{6\pi} \sqrt{\frac{\mu_0}{\epsilon_0}} \left(k^2 A\right)^2 = (20\,\Omega)\left(k^2 A\right)^2. \qquad (15.58)$$

Let a be the linear dimension of the loop, so that $A \propto a^2$. Then R_{ant} is proportional to $(ka)^4$, whereas for electric dipole radiation R_{ant} is proportional to $(ka)^2$. Thus, since $ka \ll 1$ for small sources, radiation from a magnetic dipole is much less intense than radiation from an electric dipole, assuming the dimensions and currents are comparable. A related result in atomic physics is that magnetic dipole transition rates are much smaller than electric dipole.[8]

15.2.4 ■ Complete Fields of a Hertzian Dipole

In Sec. 15.2.1 we calculated the *radiation fields* of a Hertzian dipole, i.e., a small electric dipole whose direction oscillates on the z axis. These are the asymptotic fields propagating away from the dipole. Now we will calculate the complete fields, for all positions in space. By small dipole we mean that the size d is much less than the wavelength of the radiation, which is the only other scale in the problem.

Our starting point is the exact retarded potential (15.20). Because the source is pointlike this reduces to

$$\mathbf{A}(\mathbf{x}, t) = \frac{\mu_0}{4\pi r}\frac{d\mathbf{p}}{dt}$$

as in (15.34), where $\mathbf{p}$ is the electric dipole moment at the retarded time $t - r/c$. This is the vector potential that we used earlier to calculate the asymptotic fields, i.e., for $r \gg \lambda$. But here we will not assume that r is large compared to λ. We just assume that r is much larger than the dimensions of the source.

Let the dipole moment point in the z direction, so that $\mathbf{p}(t) = p(t)\hat{\mathbf{k}}$. Then

$$\mathbf{A}(\mathbf{x}, t) = \frac{\mu_0}{4\pi r} p'(t - r/c)\hat{\mathbf{k}},$$

[8] See Exercise 9.

where p' denotes the derivative of p with respect to its argument. Note that $\mathbf{A}$ is in the z direction, $A_z\hat{\mathbf{k}}$. The magnetic field is

$$\mathbf{B} = \nabla \times \mathbf{A} = \nabla A_z \times \hat{\mathbf{k}} = \frac{\mu_0}{4\pi}\left[\frac{p'}{r^2} + \frac{p''}{cr}\right]\sin\theta\,\hat{\boldsymbol{\phi}}. \tag{15.59}$$

In evaluating (15.59) use spherical coordinates to calculate ∇A_z, and note that $\hat{\mathbf{r}} \times \hat{\mathbf{k}} = -\sin\theta\,\hat{\boldsymbol{\phi}}$. The magnetic field is *everywhere* azimuthal, curling around the dipole axis.

The electric field can be calculated most easily from the field equation

$$\frac{\partial \mathbf{E}}{\partial t} = c^2 \nabla \times \mathbf{B},$$

which holds in the vacuum outside the dipole source. Because $\mathbf{B} = B_\phi\hat{\boldsymbol{\phi}}$ is azimuthal the curl in polar coordinates gives

$$\frac{\partial \mathbf{E}}{\partial t} = c^2\left\{\frac{\hat{\mathbf{r}}}{r\sin\theta}\frac{\partial}{\partial\theta}\left(B_\phi\sin\theta\right) - \frac{\hat{\boldsymbol{\theta}}}{r}\frac{\partial}{\partial r}\left(rB_\phi\right)\right\}$$

$$= \frac{\mu_0 c^2}{4\pi}\left[\hat{\mathbf{r}}2\cos\theta\left(\frac{p'}{r^3} + \frac{p''}{cr^2}\right) + \hat{\boldsymbol{\theta}}\sin\theta\left(\frac{p'}{r^3} + \frac{p''}{cr^2} + \frac{p'''}{c^2r}\right)\right].$$

This can immediately be integrated in t, and we find the electric field everywhere,

$$\mathbf{E} = \frac{\mu_0 c^2}{4\pi}\left[\hat{\boldsymbol{\theta}}\sin\theta\left(\frac{p}{r^3} + \frac{p'}{cr^2} + \frac{p''}{c^2r}\right) + \hat{\mathbf{r}}\cos\theta\left(\frac{2p}{r^3} + \frac{2p'}{cr^2}\right)\right]. \tag{15.60}$$

Near, intermediate, and far fields. If we examine (15.59) and (15.60) we see three regions. For large r the dominant terms are of order p''/r. These are the asymptotic radiation fields propagating away from the origin, that we calculated before, also called the *far fields*. They have a relatively simple form: Both fields are transverse to the direction of propagation, $\mathbf{B}$ in the $\hat{\boldsymbol{\phi}}$ direction and $\mathbf{E}$ in the $\hat{\boldsymbol{\theta}}$ direction, with $B = E/c$. The other limit is small r, where the dominant field is $\mathbf{E}$, which is of order p/r^3. This is called the *near field*, and it, too, has a simple form. This field is precisely the electrostatic field of a point dipole (see (3-101)) with instantaneous dipole moment $p(t)\hat{\mathbf{k}}$. The near field is also a good approximation for all r if the dipole moment changes very slowly, so that p' and p'' are small, which is an example of the quasistatic approximation. Both near and far fields have simple forms, but the *intermediate field*, which interpolates from one to the other, is more complicated.

EXAMPLE 2 A harmonically oscillating dipole moment. Suppose now that $p(t) = p_0 e^{-i\omega t}$. What are the fields?

By our previous results the magnetic field is $B_\phi \hat{\phi}$, where

$$B_\phi = \frac{\mu_0 p_0}{4\pi} \left[\frac{-i\omega}{r^2} - \frac{\omega^2}{cr} \right] \sin\theta e^{-i\omega(t-r/c)}; \qquad (15.61)$$

and the electric field is $E_\theta \hat{\theta} + E_r \hat{r}$, where

$$E_\theta = \frac{\mu_0 p_0 c^2}{4\pi} \left[\frac{1}{r^3} - \frac{i\omega}{cr^2} - \frac{\omega^2}{c^2 r} \right] \sin\theta e^{-i\omega(t-r/c)} \qquad (15.62)$$

$$E_r = \frac{\mu_0 p_0 c^2}{4\pi} \left[\frac{2}{r^3} - \frac{2i\omega}{cr^2} \right] \cos\theta e^{-i\omega(t-r/c)}. \qquad (15.63)$$

The phase of the oscillation in time is a complicated function of r. It depends on r because of the retardation, and because of the factors of $i = e^{i\pi/2}$ in the terms of order $1/r^2$. Also, we can be more precise about the validity of the near and far field approximations in this example. The near zone is the region with $r \ll c/\omega$; or, since the wavelength is $\lambda = 2\pi c/\omega$, the near zone is $r \ll \lambda$. Similarly the far zone is the region $r \gg \lambda$.

In the far zone **E** and **B** are orthogonal, as for a plane wave. There the Poynting vector is always radially outward. In the intermediate zone **E** and **B** are not orthogonal and the Poynting vector has an oscillating component in the θ direction, corresponding to field energy moving back and forth along meridians.

Although the complete fields are complicated they satisfy the *free field equations*. The dipole is pointlike, so there is no charge or current for $r > 0$. In fact, we have previously derived the solution (15.61) to (15.63) from the vacuum field equations.[9]

15.3 ■ THE HALF-WAVE LINEAR ANTENNA

Many radio transmitters are linear antennas with length ℓ equal to a multiple of $\lambda/2$, where λ is the wavelength of the radiation. In this section we study the radiation fields and power from a *half-wave antenna*, $\ell = \lambda/2$. Figure 15.8(a) shows the antenna schematically. Alternating current $I_0 \sin\omega t$ feeds into the center of the antenna from a transmission line. (There is no radiation from the transmission line because its two currents cancel.) Let the z axis be along the antenna, with the center of the antenna at $z = 0$. Let $I(z, t)$ be the current at distance z from the center and time t. At $z = 0$ the antenna current is $I_0 \sin\omega t$, oscillating harmonically between I_0 and $-I_0$. At the ends $z = \pm\ell/2$ the current must be 0, because there is nowhere for the charges to go to or come from.

[9] See Exercise 8.

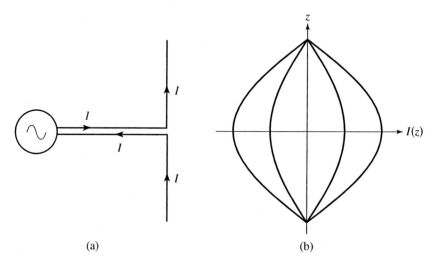

(a)   (b)

FIGURE 15.8   The half-wave linear antenna. (a) Alternating current is fed into the center of the antenna by a transmission line. (b) The electric current $I(z, t)$ of (15.64) as a function of z for several values of t. The length of the antenna is $\lambda/2$.

A reasonable approximation of the antenna current as a function of z and t is

$$I(z, t) = I_0 \cos kz \sin \omega t, \tag{15.64}$$

where $k = \pi/\ell$. This function has the correct behavior at $z = 0$ and at $z = \pm\ell/2$. Also it is harmonic in time, and all points in the antenna oscillate in phase. The function $I(z, t)$ is shown in Fig. 15.8(b). The parameter k is π/ℓ, so that $I = 0$ at $z = \pm\ell/2$. For a half-wave antenna the frequency must match the length of the antenna. The length ℓ is $\lambda/2$, so $\omega = 2\pi c/\lambda = \pi c/\ell$. (It also follows that $k = \omega/c$.) For example, if the antenna length is $10\,\mathrm{m}$, then the frequency must be $f = c/(2\ell) = 15\,\mathrm{MHz}$. The condition $\ell = \lambda/2$ is a *resonance* condition. If the antenna length and the wave length satisfy this relation, the antenna is an efficient radiator.

We shall calculate only the asymptotic fields, i.e., terms of order $1/r$ for $r \gg \ell$. The exact formula for $\mathbf{A}(\mathbf{x}, t)$ is (15.20). For the one-dimensional current in the antenna the current element $\mathbf{J}(\mathbf{x}', t')d^3x'$ must be replaced by $\hat{\mathbf{z}}I(z', t')dz'$, and the integral is over the length of the antenna as illustrated in Fig. 15.9, so

$$\mathbf{A}(\mathbf{x}, t) = \mu_0 \int_{-\ell/2}^{\ell/2} \frac{I(z', t - R/c)}{4\pi R} dz'\, \hat{\mathbf{z}}. \tag{15.65}$$

The distance R from the antenna element dz' to the field point $\mathbf{x}$ approaches r in the limit $r \gg \ell$. The approximation $R \approx r$ is good enough for the factor $1/R$ in the integrand. But for the phase of the current we need a better approximation because the antenna length is comparable to the wavelength of the radiation; as

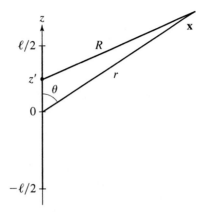

FIGURE 15.9   Geometry of the half-wave antenna. **x** is the field observation point, and $r = |\mathbf{x}|$. Asymptotically $R \approx r - z' \cos\theta$, where z' is a point on the antenna.

z' varies along the length of the antenna the phase of the current varies over a significant part of π. By Taylor series expansion

$$R = \sqrt{r^2 - 2rz'\cos\theta + z'^2} \approx r - z'\cos\theta.$$

With these approximations the asymptotic potential is

$$\mathbf{A}(\mathbf{x}, t) \sim \frac{\mu_0}{4\pi r} \int_{-\ell/2}^{\ell/2} I\left(z', t - (r - z'\cos\theta)/c\right) dz'\, \hat{\mathbf{z}}$$

$$= \frac{\mu_0 I_0}{4\pi r} \int_{-\ell/2}^{\ell/2} \cos kz' \sin\left(\omega t - kr + kz'\cos\theta\right) dz'\, \hat{\mathbf{z}}. \quad (15.66)$$

The calculation of the z' integral is an interesting exercise in calculus. First, change the variable of integration from z' to $\psi = kz'$. The endpoints of ψ are $\pm\pi/2$. Then

$$\mathbf{A}_{\mathrm{rad}}(\mathbf{x}, t) = \frac{\mu_0 I_0}{4\pi kr} \int_{-\pi/2}^{\pi/2} \cos\psi \sin(\psi\gamma - q)d\psi\, \hat{\mathbf{z}},$$

where we have introduced the symbols $\gamma \equiv \cos\theta$ and $q \equiv kr - \omega t$. The integration range can be separated into two parts, $(-\pi/2, 0)$ and $(0, \pi/2)$; then, with a change of variable, the ψ integral becomes

$$\int_0^{\pi/2} \cos\psi\, [\sin(\psi\gamma - q) - \sin(\psi\gamma + q)] d\psi = -2\sin q \int_0^{\pi/2} \cos\psi \cos(\psi\gamma)d\psi$$

$$= -2\sin q \frac{\cos(\pi\gamma/2)}{1 - \gamma^2}.$$

Substituting back $\gamma = \cos\theta$ we have

$$\mathbf{A}_{\text{rad}}(\mathbf{x}, t) = \frac{-\mu_0 I_0}{2\pi k r} \frac{\cos\left[\frac{\pi}{2}\cos\theta\right]}{\sin^2\theta} \sin(kr - \omega t)\hat{\mathbf{z}}. \tag{15.67}$$

Next we determine the fields. The magnetic field is obtained from the curl,

$$\mathbf{B}_{\text{rad}} = \nabla \times \mathbf{A}_{\text{rad}} = \nabla A_z \times \hat{\mathbf{z}}.$$

However, because we are only calculating the asymptotic field, i.e., the radiation, we must neglect terms of order $1/r^2$. So, for example, ∇ acting on the factor $1/r$ gives terms that are negligible in the radiation zone. Also, ∇ acting on the θ dependence is down by a power of r, and so is negligible. The only contribution to the field of order $1/r$ comes from ∇ acting on $\sin(kr - \omega t)$. Thus

$$\mathbf{B}_{\text{rad}} = \frac{-\mu_0 I_0}{2\pi k r} \frac{\cos\left[\frac{\pi}{2}\cos\theta\right]}{\sin^2\theta} k\cos(kr - \omega t)\hat{\mathbf{r}} \times \hat{\mathbf{z}}$$

$$= \frac{\mu_0 I_0}{2\pi r} \frac{\cos\left[\frac{\pi}{2}\cos\theta\right]}{\sin\theta} \cos(kr - \omega t)\hat{\boldsymbol{\phi}}. \tag{15.68}$$

The magnetic field is azimuthal. This direction makes sense because the current is along the z axis and $\mathbf{B}$ curls around the current. The electric field in the radiation zone is obtained in the manner of previous examples from the requirement that the electromagnetic wave is transverse, with $B = E/c$; therefore $\mathbf{E}_{\text{rad}} = c\mathbf{B}_{\text{rad}} \times \hat{\mathbf{r}} = c B_{\text{rad}}\hat{\boldsymbol{\theta}}$.

The energy flux of the electromagnetic waves radiated by the half-wave antenna is the Poynting vector

$$\mathbf{S} = \frac{\mu_0 I_0^2 c}{4\pi^2 r^2} \left[\frac{\cos\left(\frac{\pi}{2}\cos\theta\right)}{\sin\theta} \right]^2 \cos^2(kr - \omega t)\hat{\mathbf{r}}. \tag{15.69}$$

Figure 15.10 shows the angular distribution of the radiated power, in the form of a polar plot. The radiation pattern is qualitatively like a Hertzian dipole, but somewhat more focused in the xy plane.

The average power through a solid angle $d\Omega$ is $\mathbf{S}_{\text{avg}} \cdot \hat{\mathbf{r}} r^2 d\Omega$, and the average total power, i.e., integrated over all solid angles, is

$$P_{\text{avg}} = \frac{\mu_0 I_0^2 c}{4\pi} \int_0^\pi \frac{\cos^2\left(\frac{\pi}{2}\cos\theta\right)}{\sin\theta} d\theta. \tag{15.70}$$

The θ integral cannot be reduced to elementary functions but the definite integral can be evaluated numerically;[10] its value is approximately 1.22. The antenna

[10]See Exercise 12.

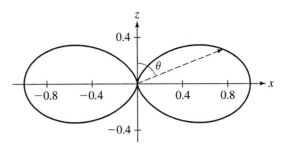

FIGURE 15.10   Angular distribution of the radiation intensity from a half-wave dipole antenna, expressed as a polar plot. The dipole moment oscillates in the $\pm z$ direction. On the graph, the distance from the origin to a point on the curve is proportional to the intensity in that direction.

resistance R_{ant} is defined as $P_{\text{avg}}/I_{\text{rms}}^2$, where I_{rms}^2 is $I_0^2/2$, so

$$R_{\text{ant}} = \frac{1.22}{2\pi} \sqrt{\frac{\mu_0}{\epsilon_0}} = 73 \, \Omega. \tag{15.71}$$

Exercise 12 explores the strength of radiation from a half-wave antenna.

Radio and Cellular Phones

The radiation and antenna requirements of cellular phones are relatively straight-forward, although the associated engineering problems, which must be solved to make the system operate properly, are very challenging. Cellular phones operate in reserved ultrahigh frequency bands near 800 MHz and near 1900 MHz. These bands are divided into channels. Each channel has a specific frequency and an associated frequency width within which the signal transfer between cellular phones actually takes place. Originally the signals were frequency modulated (FM), and a typical 25 MHz wide band accommodates about 400 channels; but digital modulation, which is a more complex but more efficient way to send information, is now often used.

A large geographical area is divided into cells, each with its own antenna and frequency channels. An important feature of cellular telephony is that the radiation intensity must be low enough so that signals from one cell do not interfere with those in other cells; this also implies that all receivers must be sensitive. To help avoid interference, adjacent cells use different channels. Each cell uses one-seventh of the available channels. Channels are reused by being assigned to many, but nonadjacent, cells; every seventh cell has the same channels. Figure 15.11 shows an idealized arrangement of hexagonal cells, in which cells with the same letter are assigned the same channels. Some important technical difficulties are assigning channels, maintaining contact between cellular telephones—sometimes a continent apart—as they move from cell to cell, and, of course, billing.

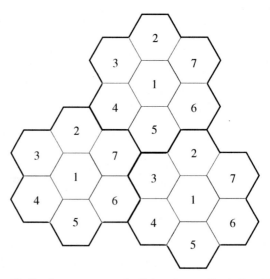

FIGURE 15.11   Idealized arrangement of cellular phone cells. Cells with the same number are assigned the same channels.

15.4 ■ THE LARMOR FORMULA: RADIATION FROM A POINT CHARGE

Electric charge is a property of electrons and protons. Any radiating current is a collection of moving charged particles. Therefore a basic problem in electrodynamics is to determine the radiation from a single charged particle.

The charge and current densities associated with a moving charge q are

$$\rho(\mathbf{x}, t) = q\delta^3(\mathbf{x} - \mathbf{x}_q(t)) \tag{15.72}$$

$$\mathbf{J}(\mathbf{x}, t) = q\mathbf{v}(t)\delta^3(\mathbf{x} - \mathbf{x}_q(t)) \tag{15.73}$$

where $\mathbf{x}_q(t)$ is the position of q as a function of time, and $\mathbf{v} = d\mathbf{x}_q/dt$ is its velocity. The delta function implies that $\rho(\mathbf{x}, t)$ and $\mathbf{J}(\mathbf{x}, t)$ vanish everywhere in space *except* at the position of q. Note that the integral of $\rho(\mathbf{x}, t)$ over a volume V is equal to q if the charge is inside V and 0 otherwise. Therefore (15.72) is the charge density function of a single point charge at $\mathbf{x}_q(t)$. Also, the charge and current densities must obey the continuity equation, $\nabla \cdot \mathbf{J} = -\partial\rho/\partial t$. The time derivative of $\rho(\mathbf{x}, t)$ is, by the chain rule of differentiation,

$$\frac{\partial\rho}{\partial t} = q\left(-\mathbf{v}(t)\right) \cdot \nabla\delta^3(\mathbf{x} - \mathbf{x}_q).$$

It might seem very daring to differentiate such a singular function as the delta function, but the derivative is well defined if the delta function is regarded as the limit of a sharply peaked distribution. The divergence of $\mathbf{J}(\mathbf{x}, t)$ is

$$\nabla \cdot \mathbf{J} = q\mathbf{v} \cdot \nabla \delta^3(\mathbf{x} - \mathbf{x}_q)$$

so indeed the continuity equation is satisfied.

Given the trajectory of q, the potentials associated with q may be calculated by evaluating the retarded potentials (15.19) and (15.20) for the densities in (15.72) and (15.73). In a later section (Sec. 15.6) we shall investigate the exact fields of q. But in this section our goal is simpler: To determine only the *radiation fields*, i.e., the asymptotic fields propagating away from q as a wave. Furthermore, we will assume that the motion of q is nonrelativistic, $|\mathbf{v}| \ll c$.

The radiation fields can be deduced from (15.35) and (15.39), which are the asymptotic fields for a small charge and current distribution. For the case of an isolated moving charge q, for which the current density is (15.73), the dipole moment is $\mathbf{p}(t) = q\mathbf{x}_q(t)$, and its derivative is $d\mathbf{p}/dt = q\mathbf{v}$. Therefore in (15.35) and (15.39), $d^2\mathbf{p}/dt^2$ is $q\mathbf{a}$, where $\mathbf{a} = d\mathbf{v}/dt$ is the acceleration of q. So, applying the general formulas to the special case of an accelerating particle, the radiation fields are

$$\mathbf{B}_{\text{rad}}(\mathbf{x}, t) = \frac{-\mu_0 q}{4\pi c} \frac{\hat{\mathbf{R}} \times \mathbf{a}}{R}, \tag{15.74}$$

$$\mathbf{E}_{\text{rad}}(\mathbf{x}, t) = c\mathbf{B}_{\text{rad}} \times \hat{\mathbf{R}}. \tag{15.75}$$

Here $\mathbf{R}$ is the vector from the position of q at the retarded time t_r to the field point $\mathbf{x}$; also, $R = |\mathbf{R}|$ and $\hat{\mathbf{R}} = \mathbf{R}/R$. As usual the radiation fields are inversely proportional to the distance from the source. The retarded time t_r is determined by an *implicit equation*,

$$t_r = t - |\mathbf{x} - \mathbf{x}_q(t_r)|/c. \tag{15.76}$$

Also, in (15.74) and (15.75) the acceleration $\mathbf{a}$ is evaluated at the retarded time, because in the general formulas all mention of the source is at the retarded time. The above equations for $\mathbf{B}_{\text{rad}}$ and $\mathbf{E}_{\text{rad}}$ are valid if $|\mathbf{v}| \ll c$, a point that will be discussed presently.

If the charge q is moving with constant velocity ($\mathbf{a} = 0$) then the radiation fields are 0. This is necessary from the theory of relativity: In the rest frame there are no propagating waves produced by the charge, so in any Lorentz frame there can be no propagating waves.

The Poynting vector in the radiation zone is $\mathbf{S} = \mathbf{E}_{\text{rad}} \times \mathbf{B}_{\text{rad}}/\mu_0$. Substituting the radiation fields, and simplifying the cross products we obtain

$$\mathbf{S} = \frac{q^2}{16\pi^2 \epsilon_0 c^3 R^2} \left[a^2 - \left(\mathbf{a} \cdot \hat{\mathbf{R}} \right)^2 \right] \hat{\mathbf{R}}. \tag{15.77}$$

(We have substituted $\mu_0 = 1/(\epsilon_0 c^2)$.) The energy flux direction is radially away from the position $\mathbf{x}_q(t_r)$ of q at the retarded time. The total power of electromagnetic radiation passing through a sphere of radius R at time t, the center of the sphere being at the position $\mathbf{x}_q(t_r)$, is

$$P = \oint \mathbf{S} \cdot \hat{\mathbf{R}} \, R^2 \, d\Omega. \tag{15.78}$$

This P is the instantaneous power that was radiated by q at the retarded time t_r. The integral over solid angles is computed most conveniently by setting up a co-ordinate system with the z axis in the direction of $\mathbf{a}$; then $a^2 - (\mathbf{a} \cdot \hat{\mathbf{R}})^2$ is $a^2 \sin^2 \theta$, and the relevant solid-angle integral is $\int \sin^2 \theta \, d\Omega = 8\pi/3$. Our final result, then, for the instantaneous power radiated by a nonrelativistic point charge is

$$P = \frac{1}{4\pi \epsilon_0} \frac{2q^2 a^2}{3c^3}. \tag{15.79}$$

This celebrated formula was derived by Larmor in 1897.

The Larmor formula is valid for *nonrelativistic motion* of q; that is, in the limit $v \ll c$. Where did this condition enter the derivation of (15.79)? The general formulas (15.35) and (15.39) for the asymptotic fields of a localized distribution of charge and current were originally derived for a pointlike source, so that the retarded time may be approximated as a constant throughout the source. More precisely, the condition for validity of those equations is that $\ell \ll c\tau$, where ℓ is the size of the source and τ is the time constant of the variation—a characteristic time over which the charge density changes significantly. For example, for an oscillating source τ is the period of oscillation and (15.35) is valid if $\ell \ll \lambda$. In the case of a radiating particle the time ℓ/v plays the role of the characteristic time τ, so the condition for (15.35) and (15.39) to be accurate is $v \ll c$.

The Larmor formula emphasizes that *acceleration* of a charged particle pro-duces radiation. The accelerating particle must either lose energy, or else an ex-ternal agent must do work on the particle to maintain the particle energy. The next example shows that a classical model of atomic structure, based on planetary or-bits of electrons, is not physically reasonable because of energy loss by radiation.

EXAMPLE 3   Instability of the classical atom.   A classical, "planetary" model of a hydrogen atom has an electron in a circular orbit of radius r around the nucleus. Neglecting radiation, the equation of motion is

$$\frac{mv^2}{r} = \frac{e^2}{4\pi \epsilon_0 r^2} \tag{15.80}$$

and the particle energy is

$$E = \frac{1}{2}mv^2 - \frac{e^2}{4\pi \epsilon_0 r} = \frac{-e^2}{8\pi \epsilon_0 r}. \tag{15.81}$$

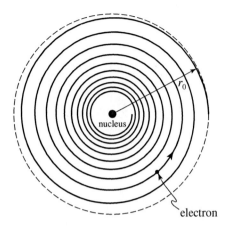

FIGURE 15.12   Example 3. Instability of the classical atom. The circular orbit is unstable because the electron radiates. As it loses energy, the electron spirals into the nucleus.

Now take into account the radiation. The electron loses energy, and spirals into the nucleus as illustrated in Fig. 15.12. How long does it take for the electron to hit the nucleus?

The electron loses energy slowly enough that the orbit remains almost circular, and we may continue to use (15.80) and (15.81) for the electron orbit as the electron spirals in.[11] By conservation of energy the electron must lose orbital energy at the same rate that it radiates electromagnetic waves; so, since the latter rate is the Larmor power,

$$\frac{dE}{dt} = -P = \frac{-1}{4\pi\epsilon_0}\frac{2e^2a^2}{3c^3} \tag{15.82}$$

where E is the particle energy (15.81). The acceleration in circular motion is $a = v^2/r$, and we can eliminate v^2 using (15.80). This leads to an equation for the decreasing radius $r(t)$,

$$\frac{dr}{dt} = -\frac{4cr_e^2}{3r^2}. \tag{15.83}$$

We have introduced the constant r_e, called the *classical radius of the electron*, defined by

$$r_e = \frac{e^2}{4\pi\epsilon_0 mc^2} = 2.8 \times 10^{-15} \text{ m}. \tag{15.84}$$

It is left as an exercise[12] to show that the solution of this differential equation is

[11] See Exercise 13.
[12] See Exercise 14.

$$r(t) = \left(r_0^3 - 4cr_e^2 t\right)^{1/3}, \tag{15.85}$$

where r_0 is the orbit radius at $t = 0$. The electron hits the nucleus at time $t = r_0^3/(4cr_e^2)$. This time is of order 10^{-11} s.

The instability of the classical atom is an example of the breakdown of classical physics at the atomic scale. What makes the atom stable in reality, i.e., in the quantum world, is that there is a lower bound on the electron orbital energy, of -13.6 eV in a hydrogen atom. The electron cannot radiate from this state because it has no lower energy state into which it can go.

EXAMPLE 4 An α particle with energy 1 Mev collides head-on with a uranium nucleus. How much of the initial kinetic energy is converted to energy of radiated electromagnetic waves?

The energy radiated is small, so a good approximation is to apply the Larmor formula to the trajectory that is calculated ignoring radiation. The total radiated energy is the integral of P over the entire (one-dimensional) trajectory. Let the U nucleus ($Q = Ze$ with $Z = 92$) be at rest at the origin and the α ($q = 2e$) travel along the x axis. The α starts at $x = \infty$ with $v = -v_\infty$, decelerates on its incoming path from ∞ to x_0, the distance of closest approach, then reverses direction and accelerates on its outgoing path to $x = \infty$, where $v = v_\infty$. For simplicity we'll neglect the recoil of the U nucleus, which is reasonable because the mass is much greater than the α mass m.

First we must determine v_∞ and x_0. It is specified that $mv_\infty^2/2 = 1$ Mev; the numerical value is $v_\infty = 6.9 \times 10^6$ m/s, or $v_\infty = 0.023c$. At closest approach the kinetic energy is 0 and the total energy is the Coulomb potential energy; thus $mv_\infty^2/2 = qQ/(4\pi\epsilon_0 x_0)$, which may be solved for x_0.

The energy radiated on the inward trajectory equals the energy radiated on the outward trajectory, by symmetry, so, using the Larmor formula, the total energy radiated is

$$U_{rad} = 2\int_{outward} P(t)dt = \frac{q^2}{3\pi\epsilon_0 c^3}\int_{outward} a^2(t)dt. \tag{15.86}$$

Or, changing the variable of integration from t to x, using $dt = dx/v(x)$, we have

$$U_{rad} = \frac{q^2}{3\pi\epsilon_0 c^3}\int_{x_0}^{\infty} \frac{a^2(x)}{v(x)}dx. \tag{15.87}$$

The acceleration is given by Newton's second law, $a = qQ/(4\pi\epsilon_0 mx^2)$. The velocity may be obtained from conservation of energy, $mv_\infty^2/2 = mv^2/2 + qQ/(4\pi\epsilon_0 x)$; after some algebra,

$$v(x) = v_\infty(1 - x_0/x)^{1/2}. \tag{15.88}$$

Substituting these expressions into the integral gives

$$U_{\text{rad}} = \frac{q^4 Q^2}{48\pi^3 \epsilon_0^3 c^3 m^2 v_\infty} \int_{x_0}^\infty \frac{dx}{x^4(1 - x_0/x)^{1/2}}. \tag{15.89}$$

For evaluation we reduce the integral to nondimensional form by changing the variable to $\xi \equiv x_0/x$. After some final simplifications the result is

$$U_{\text{rad}} = \frac{1}{12Z}\left(\frac{mv_\infty^2}{2}\right)\left(\frac{v_\infty}{c}\right)^3 \int_0^1 \frac{\xi^2 d\xi}{(1-\xi)^{1/2}}, \tag{15.90}$$

where $Z = 92$. The integral may be determined numerically as 1.07. Thus $U_{\text{rad}} = 1.2 \times 10^{-8}$ MeV. Only a very small fraction of the initial kinetic energy is radiated away.

The transition between classical theory and quantum electrodynamics is not clear cut, and there are essentially microscopic phenomena to which classical models apply surprisingly well. Example 4 is such a case, and others are treated by Jackson.[13, 14]

15.5 ■ CLASSICAL ELECTRON THEORY OF LIGHT SCATTERING

When an electromagnetic wave impinges upon a system of charges, they accelerate in response to the electric and magnetic forces, and radiate. Energy from the incident wave is transferred to an outgoing spherical wave centered at the radiating charges. This phenomenon is a kind of *scattering* because light intensity is taken from the incident direction and transferred to other directions. For example, if a pulse of light traveling in the z direction hits a hydrogen atom at the origin, there will be an outgoing spherical pulse of light—the scattered light—with some angular distribution of intensity. In the classical picture energy is transferred from the incident wave to the electron, and simultaneously from the electron to the scattered wave. (Radiation by the proton is negligible, as we'll see later in the discussion.) The goal of a theory of light scattering is to calculate the *scattering cross section*, defined below. Although quantum theory is necessary for a complete description of the interaction between photons and atoms, some interesting aspects of this physics can be explored using the classical theory. The classical theory is an important first step toward understanding the scattering of light by electrons, atoms, or molecules.

As a simple model, we shall describe an atomic electron as a point charge $-e$ bound to the origin by a Hooke's law force. We used this classical electron

---

[13] J. D. Jackson, *Classical Electrodynamics*, Chapter 15.

[14] Another example is found in D. R. Stump and G. L. Pollack, "Radiation from a neutron in a magnetic field," *Eur. J. Phys.* **19**, 59 (1998).

theory previously in Chapter 13 to understand the frequency dependence of the index of refraction of a dielectric. The equation of motion (13-91) for the electron interacting with an electromagnetic wave is

$$m\frac{d^2\mathbf{x}}{dt^2} = -K\mathbf{x} - \gamma\frac{d\mathbf{x}}{dt} - e\mathbf{E}_0 e^{-i\omega t}. \tag{15.91}$$

For simplicity we neglect the magnetic force on the electron, which would be small compared to the electric force. Equation (15.91) is an inhomogeneous linear equation, describing a driven damped harmonic oscillator.

For an electromagnetic wave traveling in the z direction and linearly polarized in the x direction, and with a wavelength much larger than the atom,[15] the electric field experienced by the electron is $E_0 e^{-i\omega t}\,\hat{\mathbf{i}}$, independent of the electron position $\mathbf{x}$. There is no driving force in the y and z directions so we may set the coordinates y and z equal to their equilibria $y(t) = z(t) = 0$. The steady-state solution to (15.91) for the x coordinate of the electron is

$$x(t) = \frac{-eE_0 e^{-i\omega t}}{m\left(\omega_0^2 - \omega^2\right) - i\gamma\omega}, \tag{15.92}$$

where $\omega_0 = \sqrt{K/m}$ is the natural frequency of the electron oscillation. Equation (15.92) is the electron motion—simple harmonic motion.

The electron has nonzero acceleration, so it radiates. This radiation is the scattered light in the classical picture, as illustrated by Figure 15.13. The Larmor formula (15.79) gives the power. The acceleration is $\mathbf{a} = d^2x/dt^2\,\hat{\mathbf{i}} = -\omega^2 x\,\hat{\mathbf{i}}$, where the *real part* is understood to be the physical acceleration, as usual; thus, discarding the imaginary part,

$$\mathbf{a} = [C_1\cos\omega t + C_2\sin\omega t]\,\hat{\mathbf{i}} \tag{15.93}$$

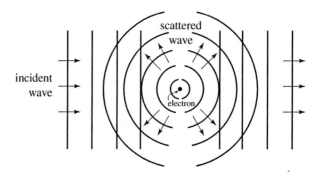

FIGURE 15.13   The classical picture of light scattering by an electron. Incident plane waves cause the electron to oscillate, and radiate outgoing waves. This radiation is the scattered light.

[15]Treating the electric field as constant over the volume of the atom is a good approximation. One wavelength of visible light is of order 10^3 atomic radii.

where

$$C_1 = \frac{e E_0 \omega^2 m \left(\omega_0^2 - \omega^2\right)}{m^2 \left(\omega_0^2 - \omega^2\right)^2 + \gamma^2 \omega^2} \tag{15.94}$$

$$C_2 = \frac{e E_0 \gamma \omega^3}{m^2 \left(\omega_0^2 - \omega^2\right)^2 + \gamma^2 \omega^2}. \tag{15.95}$$

The Larmor power is proportional to a^2. To calculate the *average power*, averaged over a period of oscillation, note that the average of $\cos^2 \omega t$ or $\sin^2 \omega t$ is $\frac{1}{2}$, while that of $\cos \omega t \sin \omega t$ is 0. The average power radiated by the electron is

$$P_{\text{avg}} = \frac{e^2 (C_1^2 + C_2^2)}{12 \pi \epsilon_0 c^3} = \frac{e^4 E_0^2 \omega^4}{12 \pi \epsilon_0 c^3 \left[m^2 (\omega_0^2 - \omega^2)^2 + \gamma^2 \omega^2 \right]}. \tag{15.96}$$

Note that the radiated power is proportional to E_0^2; the reason is that the electron is accelerated by the electric field of the incident light, so $a \propto E_0$. Therefore the power is proportional to the incident *intensity*. The scattering cross section σ is defined as the ratio of the radiated power to the incident power per unit area, i.e., to the incident intensity;

$$\sigma = \frac{P_{\text{avg}}}{S_{\text{inc}}} \tag{15.97}$$

where S_{inc} is the average of the Poynting vector of the incident wave. The cross section has units of area, and is a measure of the scattering efficiency of the atom. For the incident plane wave S_{inc} is $\epsilon_0 c E_0^2 / 2$. Therefore the cross section for light scattering by the bound electron in this classical theory is

$$\sigma = \frac{8 \pi r_e^2}{3} \frac{\omega^4}{\left(\omega_0^2 - \omega^2\right)^2 + (\gamma \omega / m)^2}. \tag{15.98}$$

Again r_e is the parameter $e^2 / (4 \pi \epsilon_0 m c^2)$, called the classical radius of the electron, defined earlier in (15.84).

Figure 15.14 is a graph of σ versus ω. In this graph γ has been set rather arbitrarily to $0.1\, m\omega_0$, for illustration purposes. There are three distinct regions: $\omega \ll \omega_0$, $\omega \approx \omega_0$, and $\omega \gg \omega_0$. At low frequencies, $\omega \ll \omega_0$, the cross section approaches the *Rayleigh cross section*

$$\sigma_{\text{Rayleigh}} = \frac{8 \pi r_e^2}{3} \left(\frac{\omega}{\omega_0} \right)^4. \tag{15.99}$$

In this limit the cross section varies as ω^4, or λ^{-4} where λ is the wavelength. Long waves are scattered less than short waves, and the dependence on λ is strong.

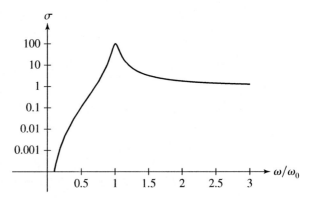

FIGURE 15.14   The cross section $\sigma(\omega)$ for scattering of light by a bound electron, calculated from the classical theory, as a function of the angular frequency ω of the light. ω_0 is the natural frequency $\sqrt{K/m}$.

Rayleigh scattering explains why the sky is blue. The sky is a big light, produced by scattering of sunlight from atmospheric molecules, predominantly N_2 and O_2. A characteristic natural frequency ω_0 for the electrons in an atom or molecule might be in the ultraviolet, so the Rayleigh cross section applies to scattering of visible light. The incident sunlight is white—a uniform superposition of visible wavelengths. But the sky is more intense at short wavelengths (violet and blue) than long (orange and red) because the cross section increases with frequency; the sky is blue. In contrast, clouds are white. A cloud consists of liquid droplets of water, which are much larger than a wavelength of light. A water droplet scatters light as a macroscopic dielectric, i.e., by reflection and refraction at the surface, rather than by radiation by a single electron. This scattering is approximately independent of frequency so the scattered light from a cloud is white, like the incident sunlight.

Why is $\sigma_{\text{Rayleigh}} \propto \omega^4$? It is because for a harmonic oscillation the acceleration goes as ω^2, and the Larmor power goes as the acceleration squared.

Scattering for ω near ω_0 is called *resonant scattering*. The cross section is sharply peaked near ω_0, because the damping is small, $\gamma/m \ll \omega_0$. When the electron is driven near its natural frequency, the motion and acceleration are large, leading to a large cross section. This case is analogous to resonance fluorescence in atoms, or to the absorption line spectrum: Incident light is absorbed if the frequency is near a spectral frequency.

For high frequencies, $\omega \gg \omega_0$, the electron behaves as if it were a *free particle*: ω_0 and γ/m can be neglected compared to ω, so the atomic forces in (15.91) are negligible. Then the cross section approaches the *Thomson cross section*

$$\sigma_{\text{Thomson}} = \frac{8\pi r_e^2}{3} = 6.6 \times 10^{-29}\,\text{m}^2, \tag{15.100}$$

which is the cross section for scattering of light, or photons, from a free electron. This cross section is independent of ω. Thomson scattering is observed from plasmas. For example, in experimental plasma physics Thomson scattering is used as a diagnostic tool for plasma production. Or, the corona of the sun, seen during a total eclipse as a large bright region around the sun, is sunlight scattered from the tenuous plasma around the sun, which extends to distances comparable to the solar radius.

For very high frequencies, $\hbar\omega \approx mc^2$ or larger, the classical theory breaks down. This range of hard X-ray scattering is called *Compton scattering*. For high frequency, i.e., high photon energy, the scattered light has a different frequency from the incident light. The frequency shift depends on the scattering angle in a way that can be calculated in the photon theory of light. The existence of a frequency shift shows that the classical theory cannot explain the phenomenon. In the classical theory the scattered light is light radiated by the accelerating electron, so it has the same frequency as the electron oscillation; and because the electron oscillation is driven by the incident light, that is the same frequency as the incident light. Compton's experiments verified the photon theory of light, because the frequency shift measured in the experiment agrees with that predicted by the photon theory.

In quantum theory, Compton scattering is the process $\gamma_1 + e_1 \rightarrow \gamma_2 + e_2$. In a frame where the initial electron e_1 is at rest, the outgoing photon γ_2 has less energy than the incoming photon γ_1, because of the recoil of the electron. Thus γ_2 has a lower frequency than γ_1. For $\hbar\omega \ll mc^2$ the frequency shift is small, and the classical theory gives the correct cross section. But for $\hbar\omega$ of order mc^2 or larger the scattering process can only be described by quantum electrodynamics.

15.6 ■ COMPLETE FIELDS OF A POINT CHARGE: THE LIÉNARD–WIECHERT POTENTIALS

We end this book with a very basic problem—to determine the complete electromagnetic field of a charged particle in arbitrary motion. If the particle is at rest the field is just the electrostatic Coulomb field. If the particle moves with constant velocity, the field can be calculated by the Lorentz transformation of the static field. But what if the particle moves on a trajectory $\mathbf{x}_q(t)$, with accelerations and decelerations, and changes of direction? What is the electromagnetic field then?

For any charge density, the scalar potential in the Lorentz gauge is

$$V(\mathbf{x}, t) = \frac{1}{\epsilon_0} \int \frac{\rho(\mathbf{x}', t - R/c)}{4\pi R} d^3x', \tag{15.101}$$

where $R = |\mathbf{x} - \mathbf{x}'|$; this is (15.19). For a point charge q moving on the trajectory $\mathbf{x}_q(t)$ the charge density is

$$\rho(\mathbf{x}', t') = q\delta^3\left(\mathbf{x}' - \mathbf{x}_q(t')\right) \tag{15.102}$$

as we saw before in (15.72). The δ-function requires in (15.101)

$$\mathbf{x}' = \mathbf{x}_q(t - |\mathbf{x} - \mathbf{x}'|/c). \tag{15.103}$$

That is, $\mathbf{x}'$ is a point on the trajectory at a certain time t_r such that

$$\mathbf{x}_q(t_r) = \mathbf{x}_q(t - |\mathbf{x} - \mathbf{x}_q(t_r)|/c). \tag{15.104}$$

The time t_r must be determined by the *implicit equation*

$$t_r = t - |\mathbf{x} - \mathbf{x}_q(t_r)|/c. \tag{15.105}$$

Or, in words, t_r is the retarded time for the observation point $(\mathbf{x}, t)$ if the particle moves on the trajectory $\mathbf{x}_q(t)$. Then the potential is

$$V(\mathbf{x}, t) = \frac{q}{4\pi \epsilon_0 R_q} \int \delta^3 \left[\mathbf{x}' - \mathbf{x}_q \left(t - |\mathbf{x} - \mathbf{x}'|/c\right)\right] d^3x', \tag{15.106}$$

where $R_q = |\mathbf{x} - \mathbf{x}_q(t_r)|$. What makes this problem difficult is that the retarded time t_r depends on the observation point $\mathbf{x}$.

The integral in (15.106) can be evaluated, because the δ-function picks out one point in space, namely $\mathbf{x}' = \mathbf{x}_q(t_r)$. But the evaluation is nontrivial because of the complicated way that the δ-function depends on $\mathbf{x}'$. It is tempting, but wrong, to say $\int \delta^3(\mathbf{x}' - \mathbf{x}_0)d^3x' = 1$; that evaluation is correct if $\mathbf{x}_0$ is a constant, but not so if $\mathbf{x}_0$ depends on $\mathbf{x}'$. Before figuring out the 3-dimensional integral, let's practice on the following 1-dimensional integral,

$$I_1 = \int_{-\infty}^{\infty} \delta \left[x' - f(x')\right] dx', \tag{15.107}$$

where $f(x')$ is some well-behaved function. The only way to evaluate this integral is to change the variable of integration from x' to $x'' \equiv x' - f(x')$. Note that

$$dx'' = \left[1 - f'(x')\right] dx' \tag{15.108}$$

the prime on f meaning the derivative with respect to its argument; thus

$$I_1 = \int_{-\infty}^{\infty} \frac{\delta(x'')dx''}{|1 - f'(x')|} = \frac{1}{|1 - f'(x_0)|}. \tag{15.109}$$

In the second equality we have used the defining property of the δ-function: $\delta(x'')$ is an infinitely sharp spike at $x'' = 0$, so $\int \delta(x'')g(x'')dx'' = g(0)$. In (15.109) x_0 is the value of x', where $x'' = 0$; that is, $x_0 = f(x_0)$. The key to the evaluation of I_1 is to change the variable of integration to the argument of the δ-function.

Now consider the 3-dimensional integral

$$I_3 = \int \delta^3 \left[\mathbf{x}' - \mathbf{x}_q \left(t - |\mathbf{x} - \mathbf{x}'|/c\right)\right] d^3x', \tag{15.110}$$

which appears in (15.106). The key to evaluating I_3 is to change the variable of integration to $\mathbf{x}''$, the argument of the δ-function,

$$\mathbf{x}'' = \mathbf{x}' - \mathbf{x}_q \left(t - |\mathbf{x} - \mathbf{x}'|/c \right). \tag{15.111}$$

Then

$$I_3 = \int \delta^3(\mathbf{x}'') \frac{d^3 x''}{\mathcal{J}(\mathbf{x}'')} = \frac{1}{\mathcal{J}(0)}; \tag{15.112}$$

here $\mathcal{J}$ is the *Jacobian* of the change of variables,

$$\mathcal{J}(\mathbf{x}'') = \mathrm{Det} \left(\frac{\partial x_i''}{\partial x_j'} \right). \tag{15.113}$$

By the chain rule of differentiation, the transformation matrix for the change of variables is

$$\frac{\partial x_i''}{\partial x_j'} = \delta_{ij} - \frac{\partial x_{qi}}{\partial t} \frac{\partial \left(t - |\mathbf{x} - \mathbf{x}'|/c \right)}{\partial x_j'}$$

$$= \delta_{ij} - \frac{v_i}{c} \frac{(x - x')_j}{|\mathbf{x} - \mathbf{x}'|}. \tag{15.114}$$

The Jacobian is to be evaluated at $\mathbf{x}'' = 0$, i.e., at $\mathbf{x}' = \mathbf{x}_q(t_r)$. Therefore in (15.114), $\mathbf{x} - \mathbf{x}'$ is $\mathbf{x} - \mathbf{x}_q(t_r) \equiv \mathbf{R}_q$. Combining these results we have

$$\mathcal{J}(0) = \mathrm{Det} \left[\delta_{ij} - \beta_i n_j \right] \tag{15.115}$$

where we define

$$\boldsymbol{\beta} = \mathbf{v}/c \quad \text{and} \quad \hat{\mathbf{n}} = \mathbf{R}_q / R_q. \tag{15.116}$$

The unit vector $\hat{\mathbf{n}}$ points in the direction from the position of q *at the retarded time* to the observation point $\mathbf{x}$. Also, $\boldsymbol{\beta}$ is the velocity in units of c. Finally, a short calculation[16] reveals the value of the Jacobian to be

$$\mathcal{J}(0) = 1 - \hat{\mathbf{n}} \cdot \boldsymbol{\beta}. \tag{15.117}$$

Thus the scalar potential of the point charge q is

$$V(\mathbf{x}, t) = \frac{q}{4\pi \epsilon_0 R_q} \frac{1}{1 - \hat{\mathbf{n}} \cdot \boldsymbol{\beta}}. \tag{15.118}$$

---

[16]See Exercise 20.

To find the vector potential, consider (15.20) with $\mathbf{J}(\mathbf{x}', t') = \rho(\mathbf{x}', t')\mathbf{v}(t')$, where t' is the retarded time (15.18); thus

$$\mathbf{A}(\mathbf{x}, t) = \frac{\mathbf{v}}{c^2}V(\mathbf{x}, t) = \frac{q\mathbf{v}}{4\pi\epsilon_0 c^2 R_q}\frac{1}{1 - \hat{\mathbf{n}}\cdot\boldsymbol{\beta}}. \tag{15.119}$$

Again, $\mathbf{R}_q = \mathbf{x} - \mathbf{x}_q = R_q\hat{\mathbf{n}}$ is the vector from q at the retarded time to the observation point; and $\boldsymbol{\beta} = \mathbf{v}/c$ is the velocity of q in units of c. In all expressions the source variables ($\mathbf{x}_q$ and $\mathbf{v}$) are evaluated at the retarded time t_r defined in (15.105). The potentials (15.118) and (15.119) are called the *Liénard–Wiechert potentials*.

15.6.1 ■ A Charge with Constant Velocity

An important special case is for constant velocity $\mathbf{v} = \mathbf{v}_0$. We can calculate the potential and fields for this case in another way, by making the Lorentz transformation from the rest frame of q (the frame $\mathcal{F}'$ in which the potentials are $V' = q/(4\pi\epsilon_0 r')$ and $\mathbf{A}' = 0$) to the observation frame in which q has velocity $\mathbf{v}_0$. This method only works for constant velocity. On the other hand, the Liénard–Wiechert potentials are general, so they must give the same answer for this special case, as we shall now verify.

For convenience set up a coordinate system with the z axis in the direction of motion, so $\mathbf{v}_0 = v_0\hat{\mathbf{k}}$; and let the origin be the position of q at $t = 0$. Then the trajectory is $\mathbf{x}_q(t) = v_0 t\hat{\mathbf{k}}$. The distance between q at the observation time t and the observation point $\mathbf{x}$ at time t is $r = \sqrt{x^2 + y^2 + (z - v_0 t)^2}$. The geometry is shown in Fig. 15.15.

The scalar potential $V(\mathbf{x}, t)$ in (15.118) is expressed in terms of the position of q at the *retarded time* t_r. To compare with the potential calculated by the Lorentz transformation, we must rewrite $V(\mathbf{x}, t)$ in terms of the position of q at the *observation time* t. Refer to Fig. 15.15. P is the observation point at (x, y, z). Q is the position of q at the observation time, $(0, 0, v_0 t)$; and $\mathrm{Q_r}$ is the position of q at the retarded time, $(0, 0, v_0 t_r)$. The distance $\mathrm{QQ_r}$ is $v_0(t - t_r) = \beta R_q$, where $\beta = v_0/c$. The distance QN is $z - v_0 t$, and the distance NP is $\sqrt{x^2 + y^2}$. Now here is the crucial step in the analysis: The distance $\mathrm{Q_r M}$ can be written as

$$\mathrm{Q_r M} = \beta R_q \cos\theta = R_q\hat{\mathbf{n}}\cdot\boldsymbol{\beta};$$

Therefore the quantity $R_q(1 - \hat{\mathbf{n}}\cdot\boldsymbol{\beta})$, which appears in the Liénard–Wiechert potentials, is

$$R_q(1 - \hat{\mathbf{n}}\cdot\boldsymbol{\beta}) = \text{the distance MP.} \tag{15.120}$$

But the distance MP can be determined using the right triangles QPM, QPN, and $\mathrm{Q_r PN}$:

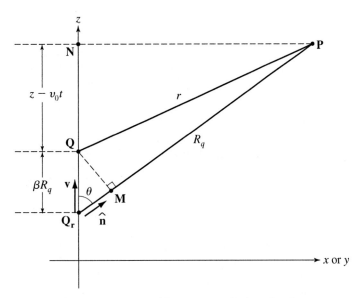

FIGURE 15.15 Charged particle with constant velocity. The charge moves on the z axis with velocity $v_0\hat{\mathbf{k}}$. P is the observation point at (x, y, z). Q is the position of q at the observation time, $(0, 0, v_0 t)$; and Q_r is the position of q at the retarded time, $(0, 0, v_0 t_r)$. R_q is the distance Q_rP.

$$\mathrm{MP} = \sqrt{r^2 - \left(\beta R_q \sin\theta\right)^2}$$

$$= \sqrt{x^2 + y^2 + (z - v_0 t)^2 - \beta^2(x^2 + y^2)}$$

$$= \frac{1}{\gamma}\sqrt{x^2 + y^2 + \gamma^2(z - v_0 t)^2} \qquad (15.121)$$

where $\gamma = 1/\sqrt{1 - \beta^2}$ is the familiar γ parameter in the Lorentz transformation. Substituting this result for MP into (15.118) we find

$$V(\mathbf{x}, t) = \frac{q}{4\pi\epsilon_0} \frac{\gamma}{\left[x^2 + y^2 + \gamma^2(z - v_0 t)^2\right]^{1/2}}. \qquad (15.122)$$

Similarly,

$$\mathbf{A}(\mathbf{x}, t) = \frac{q}{4\pi\epsilon_0 c^2} \frac{\gamma v_0\hat{\mathbf{k}}}{\left[x^2 + y^2 + \gamma^2(z - v_0 t)^2\right]^{1/2}}. \qquad (15.123)$$

The same potentials result from making the Lorentz transformation from the rest frame of q to the observation frame, because $A^\mu = (V/c, \mathbf{A})$ is a Lorentz 4-vector. The corresponding fields $\mathbf{E}(\mathbf{x}, t)$ and $\mathbf{B}(\mathbf{x}, t)$ are the same fields of a particle with constant velocity that we calculated by the Lorentz transformation in Sec. 12.5.

A charged particle moving with constant velocity does not radiate electromagnetic waves. The Poynting vector can be analyzed to show that the field energy is carried along with the particle.[17]

15.6.2 ■ The Complete Fields

Now we return to the general case. We have the potentials (15.118) and (15.119). To calculate the fields we only need to evaluate some derivatives

$$\mathbf{E} = -\nabla V - \partial \mathbf{A}/\partial t \quad \text{and} \quad \mathbf{B} = \nabla \times \mathbf{A}. \tag{15.124}$$

However, calculating these derivatives is fiendishly difficult because the retarded time t_r depends on $\mathbf{x}$ and t. In other words, ∇ may act on the *time dependence* of $\mathbf{R}_q = \mathbf{x} - \mathbf{x}_q(t_r)$ or $\boldsymbol{\beta}(t_r)$. For example, consider the quantity ∇t_r. From (15.105)

$$\nabla t_r = -\frac{1}{c} \nabla R_q$$

$$= -\frac{1}{c} \frac{(x_i - x_{qi})}{|\mathbf{x} - \mathbf{x}_q|} \nabla \left(x_i - x_{qi} \right)$$

$$= -\frac{1}{c} \left[\hat{\mathbf{n}} - \hat{\mathbf{n}} \cdot \mathbf{v} \, \nabla t_r \right] ;$$

thus

$$\nabla t_r = \frac{-\hat{\mathbf{n}}}{c \left(1 - \hat{\mathbf{n}} \cdot \boldsymbol{\beta} \right)} \quad \text{and} \quad \nabla R_q = \frac{\hat{\mathbf{n}}}{1 - \hat{\mathbf{n}} \cdot \boldsymbol{\beta}}. \tag{15.125}$$

Also, even more complicated, ∇ may act on $\hat{\mathbf{n}} = \mathbf{R}_q/R_q$.

The result of the very intricate calculations of the derivatives in (15.124) is, for the electric field,

$$\mathbf{E}(\mathbf{x}, t) = \frac{q}{4\pi \epsilon_0 R_q^2} \frac{(\hat{\mathbf{n}} - \boldsymbol{\beta})\left(1 - \beta^2\right)}{\left(1 - \hat{\mathbf{n}} \cdot \boldsymbol{\beta}\right)^3} + \frac{q}{4\pi \epsilon_0 c^2 R_q} \frac{\hat{\mathbf{n}} \times \left[(\hat{\mathbf{n}} - \boldsymbol{\beta}) \times \mathbf{a}\right]}{\left(1 - \hat{\mathbf{n}} \cdot \boldsymbol{\beta}\right)^3},$$
$$\tag{15.126}$$

where $\mathbf{a} = d\mathbf{v}/dt$ is the acceleration of the particle; and the result for the magnetic field is very simply related,

$$\mathbf{B}(\mathbf{x}, t) = \hat{\mathbf{n}} \times \mathbf{E}(\mathbf{x}, t)/c. \tag{15.127}$$

The first term in (15.126) is proportional to R_q^{-2}, like the field of a static charge. If q is static ($\beta = 0$) then this first term is just the Coulomb field. This term is negligible at large R_q, and is not part of a propagating wave. The second term in

---

[17] See Exercise 21.

(15.126) is proportional to R_q^{-1} and is nonzero if the charge accelerates. Asymptotically this term, together with the associated term in **B**, is the radiation field, a wave propagating away from the charge.

For nonrelativistic velocities, β is much less than 1 and many factors in (15.126) simplify. In particular, if $\beta \ll 1$ then the asymptotic fields at large distances ($r \equiv |\mathbf{x}| \to \infty$) are

$$\mathbf{E}_{\text{rad}} = \frac{q\hat{\mathbf{n}} \times (\hat{\mathbf{n}} \times \mathbf{a})}{4\pi\epsilon_0 c^2 r} \quad \text{for} \quad \beta \ll 1, \tag{15.128}$$

$$\mathbf{B}_{\text{rad}} = \frac{-\mu_0 q\hat{\mathbf{n}} \times \mathbf{a}}{4\pi cr} \quad \text{for} \quad \beta \ll 1. \tag{15.129}$$

(Asymptotically, $R_q \sim r$.) These fields agree with the radiation fields of a nonrelativistic charge that we derived in Sec. 15.4, which led to the Larmor formula for the radiated power.

15.6.3 ■ Generalization of the Larmor Formula

Knowing the exact fields (15.126) and (15.127) we can now calculate the power radiated by an accelerating charge moving with arbitrary velocity. For *nonrelativistic* motion the power is given by the Larmor formula (15.79). The exact formula is a generalization of that result.

The calculations that must be done are first to find the asymptotic fields for large $r \equiv |\mathbf{x}|$, and then to integrate the Poynting vector over a sphere in the radiation zone. Rather than go through the details of these long calculations, it will be more interesting to consider one special case, and then state and analyze the general result.

Acceleration Parallel to Velocity
As a special case, suppose **a** and **v** are in the same direction; this is the case if q moves along a straight line. Then $\beta \times \mathbf{a} = 0$ and the asymptotic electric field is

$$\mathbf{E}_{\text{rad}} = \frac{q}{4\pi\epsilon_0 c^2 r} \frac{\hat{\mathbf{n}} \times (\hat{\mathbf{n}} \times \mathbf{a})}{\kappa^3}, \tag{15.130}$$

where

$$\kappa = 1 - \hat{\mathbf{n}} \cdot \beta. \tag{15.131}$$

The asymptotic magnetic field $\mathbf{B}_{\text{rad}}$ is $\hat{\mathbf{n}} \times \mathbf{E}_{\text{rad}}/c$. The Poynting vector is

$$\mathbf{S} = \frac{1}{\mu_0} \mathbf{E}_{\text{rad}} \times \mathbf{B}_{\text{rad}} = \frac{\hat{\mathbf{n}}}{\mu_0 c} E_{\text{rad}}^2. \tag{15.132}$$

Note that the energy flux is directed away from the retarded position of the charge.

Now, from (15.130)

$$E_{rad}^2 = \left(\frac{q}{4\pi\epsilon_0 c^2 r}\right)^2 \frac{\left(a^2 - (\hat{\mathbf{n}}\cdot\mathbf{a})^2\right)}{\kappa^6}. \tag{15.133}$$

Thus the energy that passes through a sphere of radius r during a small time interval dt is

$$dU = \frac{q^2}{16\pi^2\epsilon_0 c^3}\int \frac{a^2 - (\hat{\mathbf{n}}\cdot\mathbf{a})^2}{\kappa^6}d\Omega\, dt. \tag{15.134}$$

The energy in (15.134) was emitted by the particle during the time interval dt_r, where t_r is the retarded time. Equation (15.105), which relates the times t and t_r, implies

$$dt_r = dt - \frac{1}{c}\frac{\mathbf{x}-\mathbf{x}_q}{|\mathbf{x}-\mathbf{x}_q|}\cdot d\left[\mathbf{x} - \mathbf{x}_q(t_r)\right] = dt + \hat{\mathbf{n}}\cdot\boldsymbol{\beta}dt_r; \tag{15.135}$$

or, rearranging,

$$dt = (1 - \hat{\mathbf{n}}\cdot\boldsymbol{\beta})dt_r = \kappa dt_r. \tag{15.136}$$

Therefore the *particle's* rate of energy emission, per steradian of the radiation direction, is

$$\frac{dP_{rad}}{d\Omega} = \frac{d^2U}{d\Omega dt_r} = \frac{q^2}{16\pi^2\epsilon_0 c^3}\frac{a^2 - (\hat{\mathbf{n}}\cdot\mathbf{a})^2}{\kappa^5}. \tag{15.137}$$

To analyze the angular distribution, let θ be the angle between $\hat{\mathbf{n}}$ and the velocity or acceleration. For example, set up a coordinate system with q at the origin and the velocity in the z direction; θ is the polar angle of the field point. Then the differential power as a function of θ is

$$\frac{dP_{rad}}{d\Omega} = \frac{1}{4\pi\epsilon_0}\frac{q^2 a^2}{4\pi c^3}\frac{\sin^2\theta}{(1 - \beta\cos\theta)^5}. \tag{15.138}$$

Figure 15.16 shows the angular distribution of the intensity in the form of a polar plot, for $\beta = 0.1$, $\beta = 0.3$, and $\beta = 0.5$. For small β the distribution is approximately $\sin^2\theta$, which is symmetric for $\theta < \pi/2$ and $\theta > \pi/2$. As β approaches 1 the distribution becomes peaked forward, with maximum intensity in a small angular ring around the path of the particle.

The total power of electromagnetic radiation from the particle is

$$P_{rad} = \int_0^\pi \frac{dP_{rad}}{d\Omega}2\pi\sin\theta d\theta = \frac{1}{4\pi\epsilon_0}\frac{2q^2 a^2}{3c^3}\frac{1}{\left(1-\beta^2\right)^3}. \tag{15.139}$$

For $\beta \ll 1$ the result approaches the Larmor formula. For β approaching 1, the power for a given acceleration is much larger than Larmor's formula.

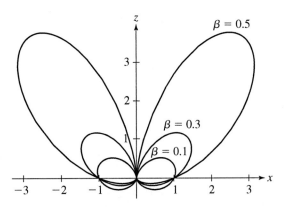

FIGURE 15.16   Polar plot of the intensity of radiation from a particle with acceleration and velocity in the same direction. The particle moves along the z axis. The power
distribution is shown for $\beta = 0.1, 0.3$ and 0.5.

Liénard's Formula

The results of the previous section were for a special case, in which the acceleration and velocity of the particle are in the same direction. The general result for
the total power radiated by a charged particle is[18]

$$P_{\text{rad}} = \frac{1}{4\pi\epsilon_0}\frac{2q^2a^2}{3c^3}\frac{(1 - \beta^2\sin^2\psi)}{(1 - \beta^2)^3}. \qquad (15.140)$$

Here ψ is the angle between the velocity and acceleration. This result was first
derived by Liénard in 1898.

EXAMPLE 5 Synchrotron radiation.   For circular motion the acceleration is
centripetal, orthogonal to the velocity, so $\psi = \pi/2$. The acceleration is $a = v^2/R$,
where R is the radius of the orbit, so by Liénard's formula the power is

$$P_{\text{rad}} = \frac{1}{4\pi\epsilon_0}\frac{2q^2}{3c^3R^2}\frac{v^4}{(1 - \beta^2)^2}. \qquad (15.141)$$

For example, consider an electron ($q = -e$) circulating in a synchrotron of
radius R and magnetic field B. The relativistic equation of motion is

$$\frac{mv}{\sqrt{1 - v^2/c^2}} = eBR. \qquad (15.142)$$

[18]To derive (15.140) by integrating the Poynting vector is quite a challenge. (Using an algebraic
manipulation program is a big help.) It can also be derived by Lorentz transformation of the Larmor
formula; see Heald and Marion, Sec. 14.9.

In the high-energy limit we have $v \approx c$ and

$$\frac{1}{1 - \beta^2} \approx \left(\frac{eBR}{mc}\right)^2 . \qquad (15.143)$$

Thus the radiated power approaches

$$P_{\text{rad}} = \frac{1}{4\pi\epsilon_0} \frac{2e^4 B^2 \gamma^2}{3m^2 c}, \qquad (15.144)$$

where $\gamma = 1/\sqrt{1 - \beta^2}$. Numerically,

$$P_{\text{rad}} = (9.9 \times 10^4 \,\text{eV/s}) \left(\frac{B}{1\,\text{T}}\right)^2 \gamma^2 \quad \text{(high energy limit).} \qquad (15.145)$$

An electron in a synchrotron loses a lot of energy by radiation.[19] Therefore the next high-energy electron accelerator, which will be built sometime in the future to study elementary particles and their interactions, will be a linear accelerator rather than a synchrotron. Proton accelerators have no problem with synchrotron radiation at current energies. The power of synchrotron radiation is proportional to γ^2/m^2, so the power radiated by a proton is less than 10^{-13} times that from an electron of the same energy.

Applications of synchrotron radiation.   Synchrotron radiation is not just an unwanted by-product of high-energy accelerators. Many electron synchrotrons—over 20 worldwide with more planned—are in use as high-intensity light sources for scientific and engineering studies. The electrons travel in large circular orbits at ultra-relativistic speeds, such as $1 - v/c \lesssim 5 \times 10^{-9}$. The radiation is strongly concentrated in the direction of motion, like a searchlight beam. The radiation extends over a broad range of frequencies, and is intense, collimated, and polarized. With monochromators, radiation studies can be carried out from 0.01 eV (far infrared) to above 1 GeV. Synchrotron radiation is used for research on atoms, molecules, solids, X-ray diffraction studies of protein structure, and X-ray lithography.

The path followed by a charged particle in a synchrotron is controlled by magnetic fields. In addition to the main magnets that determine the overall circular trajectory, other magnets, called wiggler and undulator magnets, cause short-range accelerations of the particles. The purpose of these magnets is to enrich the spectrum of emitted radiation so that it can be used to probe efficiently the materials or processes studied by scattering of the synchrotron light.

---

[19] See Exercise 23.

FURTHER READING

M. A. Heald and J. B. Marion, *Classical Electromagnetic Radiation*, 3rd ed. (Saunders, Ft. Worth, 1995). This book is a thorough treatment of waves and radiation in classical electromagnetism, at the intermediate level.

W. L. Stutzmann and G. A. Thiele, *Antenna Theory and Design* (Wiley, New York, 1981). This book is intended for students of electrical engineering, and it has many interesting examples of the physics of antennas.

The Feynman Lectures on Physics, Vol. II, Sec. 21. Feynman discusses interesting equations for the fields of a charged particle with arbitrary motion, called the Heaviside-Feynman equations. In the notation of Sec. 15.6.2, these equations are

$$\mathbf{E}(\mathbf{x}, t) = \frac{q}{4\pi\epsilon_0}\left[\frac{\hat{\mathbf{n}}}{R_q^2} + \frac{R_q}{c}\frac{d}{dt}\left(\frac{\hat{\mathbf{n}}}{R_q^2}\right) + \frac{1}{c^2}\frac{d^2\hat{\mathbf{n}}}{dt^2}\right]$$

and

$$\mathbf{B}(\mathbf{x}, t) = \hat{\mathbf{n}} \times \mathbf{E}(\mathbf{x}, t)/c.$$

A derivation and discussion are given in A. R. Janah, T. Padmanabhan, and T. P. Singh, *Am. J. Phys.* **56**, 1036 (1988).

EXERCISES

Sec. 15.1. The Retarded Potentials

15.1. Show that (15.28) and (15.29) represent plane waves propagating in the $+z$ and $-z$ directions at speed c. Show also that the discontinuity in $\mathbf{B}$ at the xy plane satisfies the appropriate boundary condition.

15.2. Suppose that at $t = 0$ a current I is suddenly established in an infinite wire that lies on the z axis. What are the resulting electric and magnetic fields? Show that after a long time, $t \gg r/c$, the magnetic field is the same as the static field of a long wire with constant current I. What is the electric field for $t \gg r/c$? [Answer:

$$E_z(r, t) = \frac{-\mu_0 I c}{2\pi\sqrt{c^2 t^2 - r^2}}$$

$$B_\phi(r, t) = \frac{\mu_0 I}{2\pi r}\frac{ct}{\sqrt{c^2 t^2 - r^2}}]$$

15.3. Starting from Maxwell's equations with sources $\rho(\mathbf{x}, t)$ and $\mathbf{J}(\mathbf{x}, t)$ in vacuum, show that

$$-\nabla^2\mathbf{E} + \frac{1}{c^2}\frac{\partial^2\mathbf{E}}{\partial t^2} = -\frac{1}{\epsilon_0}\nabla\rho - \mu_0\frac{\partial\mathbf{J}}{\partial t}$$

$$-\nabla^2\mathbf{B} + \frac{1}{c^2}\frac{\partial^2\mathbf{B}}{\partial t^2} = \mu_0\nabla \times \mathbf{J}.$$

In comparison with (15.5) and (15.6) the left-hand sides of these equations are the d'Alembertians of the fields themselves rather than of the potentials. The source terms, on the right-hand sides, are more complicated than ρ/ϵ_0 and $\mu_0 \mathbf{J}$. By using (15.22), the Green's function of the d'Alembertian, the equations can be solved for $\mathbf{E}$ and $\mathbf{B}$ as integrals involving the retarded source terms, analogous to (15.19) and (15.20). The integral expressions, which are known as Jefimenko's equations, can be used to discuss interesting problems in electrodynamics.[20]

Sec. 15.2. Dipole Radiation

15.4. Calculate the asymptotic electric field $\mathbf{E}_{rad}$ produced by a small time-dependent electric dipole, by differentiating the vector and scalar potentials in the Lorentz gauge.

15.5. For the Hertzian dipole, i.e., $\mathbf{p}(t') = p_0 \cos \omega t' \hat{\mathbf{k}}$, show that the asymptotic potentials are

$$\mathbf{A} \sim \frac{-\mu_0 p_0}{4\pi r} \omega \sin \omega (t - r/c) \hat{\mathbf{k}}$$

$$V \sim \frac{-\mu_0 c p_0}{4\pi} \left(\frac{\cos \theta}{r} \right) \omega \sin \omega (t - r/c).$$

Verify the Lorentz gauge condition asymptotically.

15.6. Figure 15.6 shows the angular distribution of electric dipole radiation as a polar plot. One-half of the intensity is emitted in the angular range

$$\frac{\pi}{2} - \frac{\alpha}{2} \leq \theta \leq \frac{\pi}{2} + \frac{\alpha}{2},$$

i.e., in a sector with angular measure α centered at the xy plane. Determine α, which is called the half-intensity angle.

15.7. (a) From the potentials show that the magnetic dipole radiation fields are

$$\mathbf{E}_{rad}(\text{mag.dipole}) = -\frac{\mu_0}{4\pi c r} \frac{d^2 \mathbf{m}}{dt^2} \times \hat{\mathbf{r}}, \tag{15.146}$$

$$\mathbf{B}_{rad}(\text{mag.dipole}) = \frac{\mu_0}{4\pi c^2 r} \left\{ \hat{\mathbf{r}} \left(\hat{\mathbf{r}} \cdot \frac{d^2 \mathbf{m}}{dt^2} \right) - \frac{d^2 \mathbf{m}}{dt^2} \right\}. \tag{15.147}$$

(b) Show that the dual transformation (15.55) converts the electric dipole radiation fields into the magnetic dipole radiation fields.

15.8. Show that the full fields of a harmonically oscillating pointlike electric dipole are identical to the spherical wave discussed in Section 11.5.4.

15.9. Show for typical atomic dimensions that $m_0/c \ll p_0$.

---

[20]O. D. Jefimenko, *Electromagnetic Retardation and Theory of Relativity* (Electret Scientific Company, Star City, W.Va., 1997).

Sec. 15.3. Half-Wave Antenna

15.10. The current $I(z, t)$ in the half-wave linear antenna is given by (15.64). Determine the *linear charge density*, i.e., the charge per unit length $\lambda(z, t)$, in the antenna. Show that the charge density is maximum at times when the current is 0. What is the maximum charge density at the end of the antenna if $I_0 = 10 \, A$ and $\omega = 2\pi$ MHz?

15.11. Calculate the half-intensity angle (see Exercise 6) for the half-wave dipole antenna.

15.12. **(a)** Use a computer program to evaluate the θ integral in (15.70).

   (b) An FM radio station has frequency 10 MHz and power 10 kW from a half-wave antenna. How long is the antenna? What is the rms current?

Sec. 15.4. Larmor Formula

15.13. Use the results of Example 3 to estimate the number of revolutions made by the radiating electron in the classical model of the atom as it spirals into the nucleus from initial radius $r_0 = 10^{-10}$ m.

15.14. Derive the solution (15.85) of the differential equation for the orbit radius as a function of time in the classical model of an atom.

15.15. The average power radiated by an oscillating dipole, with dipole moment $p(t) = p_0 \cos \omega t$, is

$$P_{\text{avg}} = \frac{p_0^2 \omega^4}{12\pi \epsilon_0 c^3}.$$

   Derive this result from the Larmor formula, treating the dipole as an oscillating pair of charges $\pm q_0$, which oscillate 180 degrees out of phase with amplitude of oscillation $d/2$. (Note that $p_0 = q_0 d$.) But be careful! The waves radiated by q_0 and $-q_0$ *interfere*, so in Larmor's formula you must add the qa's and then square, rather than adding the squares of the qa's!

15.16. **(a)** Energy loss by radiation is an insignificant contribution to the resistance of an ordinary wire. Consider a conduction electron with velocity v_0 (moving in one dimension) that is brought to rest, because of a collision, over a distance D. Derive the ratio f of the energy radiated to the initial energy, assuming constant deceleration.

   (b) Let v_0 be a typical Fermi velocity, say for Cu ($\frac{1}{2}mv_F^2 = kT_F$) and D a typical lattice dimension, say 4×10^{-10} m. Calculate f.

15.17. Suppose the velocity $\mathbf{v}$ and acceleration $\mathbf{a}$ of a charged particle are collinear. Let $dP/d\Omega$ be the differential power radiated, i.e., the power per *steradian*. Call the total power $P_T = \int (dP/d\Omega) d\Omega$.

   Use computer graphics to plot the angular distribution of the power

$$f(\theta) = \frac{1}{P_T} \frac{dP}{d\Omega},$$

   as a function of the angle θ with the direction of motion, for these speeds (superimposed): $0.01 \, c$, $0.5 \, c$, and $0.99 \, c$. Use either a polar plot or an ordinary function plot.

Sec. 15.5. Light Scattering

15.18. Compute the numerical value of the ratio $\sigma(\text{violet})/\sigma(\text{red})$, where $\sigma(\omega)$ is the Rayleigh cross section for scattering, by a bound electron, of light with frequency $\omega \ll \omega_0$.

15.19. The electron density n in the solar corona is roughly 10^9 cm^{-3}. Let $\mathcal{I}_{\text{direct}}$ be the intensity of direct sunlight (at the Earth) and $\mathcal{I}_{\text{corona}}$ the intensity of coronal light. Use dimensional analysis to estimate the ratio $\mathcal{I}_{\text{corona}}/\mathcal{I}_{\text{direct}}$. (Hint: The differential cross section for Thomson scattering is

$$\frac{d\sigma}{d\Omega} = \frac{r_e^2}{2}\left(1 + \cos^2\theta\right).)$$

Sec. 15.6. Liénard–Wiechert Potentials

15.20. Evaluate the Jacobian determinant in (15.115).

15.21. Calculate the Poynting vector and energy density of the electromagnetic field of a charged particle moving with constant velocity. Show that the field energy is carried along with the particle.

15.22. Calculate the integral of $d P_{\text{rad}}/d\Omega$ for radiation from a charged particle with acceleration and velocity in the same direction.

15.23. Calculate the energy radiated per revolution by an electron in a synchrotron accelerator with magnetic field 1 T if the energy of the electron is 50 GeV.

APPENDIX

Electric and Magnetic Units

In this book we use the MKSA (for meter, kilogram, second, and ampere) system of units, also called SI (after *Système Internationale*) units. MKSA units are now used in most books on electromagnetism. There are, however, other systems of units. Of these, the most common is the cgs (for centimeter, gram, second) system, also called Gaussian units. Gaussian units are commonly used in books on quantum mechanics. It is often desirable to convert equations and quantities between these systems. The two tables following are intended to help with making these conversions.

TABLE I   Equations in the MKSA and Gaussian systems

Quantity	MKSA	Gaussian
Fields		
$\mathbf{E}$	$\dfrac{1}{4\pi\epsilon_0}\dfrac{q\,\hat{\mathbf{r}}}{r^2}$	$\dfrac{q\,\hat{\mathbf{r}}}{r^2}$
$d\mathbf{B}$	$\dfrac{\mu_0}{4\pi}\dfrac{I d\boldsymbol{\ell}\times\hat{\mathbf{r}}}{r^2}$	$\dfrac{1}{c}\dfrac{I d\boldsymbol{\ell}\times\hat{\mathbf{r}}}{r^2}$
Maxwell's Equations (in vacuum)		
$\nabla\cdot\mathbf{E}$	$\dfrac{\rho}{\epsilon_0}$	$4\pi\rho$
$\nabla\cdot\mathbf{B}$	0	0
$\nabla\times\mathbf{E}$	$-\dfrac{\partial\mathbf{B}}{\partial t}$	$-\dfrac{1}{c}\dfrac{\partial\mathbf{B}}{\partial t}$
$\nabla\times\mathbf{B}$	$\mu_0\mathbf{J}+\epsilon_0\mu_0\dfrac{\partial\mathbf{E}}{\partial t}$	$\dfrac{4\pi}{c}\mathbf{J}+\dfrac{1}{c}\dfrac{\partial\mathbf{E}}{\partial t}$
Maxwell's Equations (in matter)		
$\nabla\cdot\mathbf{D}$	ρ_f	$4\pi\rho_f$
$\nabla\cdot\mathbf{B}$	0	0
$\nabla\times\mathbf{E}$	$-\dfrac{\partial\mathbf{B}}{\partial t}$	$-\dfrac{1}{c}\dfrac{\partial\mathbf{B}}{\partial t}$
$\nabla\times\mathbf{H}$	$\mathbf{J}_f+\dfrac{\partial\mathbf{D}}{\partial t}$	$\dfrac{4\pi}{c}\mathbf{J}_f+\dfrac{1}{c}\dfrac{\partial\mathbf{D}}{\partial t}$
Constitutive relations		
$\mathbf{D}$	$\epsilon_0\mathbf{E}+\mathbf{P}$	$\mathbf{E}+4\pi\mathbf{P}$
$\mathbf{H}$	$\dfrac{1}{\mu_0}\mathbf{B}-\mathbf{M}$	$\mathbf{B}-4\pi\mathbf{M}$
Linear media		
$\mathbf{D}$	$\epsilon\mathbf{E}$	$\epsilon\mathbf{E}$
$\mathbf{H}$	$\dfrac{1}{\mu}\mathbf{B}$	$\dfrac{1}{\mu}\mathbf{B}$
$\mathbf{P}$	$\epsilon_0\chi_e\mathbf{E}$	$\chi_e\mathbf{E}$
$\mathbf{M}$	$\chi_m\mathbf{H}$	$\chi_m\mathbf{H}$
Lorentz force $\mathbf{F}$	$q\,(\mathbf{E}+\mathbf{v}\times\mathbf{B})$	$q\left(\mathbf{E}+\dfrac{\mathbf{v}}{c}\times\mathbf{B}\right)$
Energy density (vacuum) u	$\dfrac{\epsilon_0}{2}E^2+\dfrac{1}{2\mu_0}B^2$	$\dfrac{1}{8\pi}\left(E^2+B^2\right)$
Poynting vector (vacuum) $\mathbf{S}$	$\dfrac{1}{\mu_0}\mathbf{E}\times\mathbf{B}$	$\dfrac{c}{4\pi}\mathbf{E}\times\mathbf{B}$
Larmor power P	$\dfrac{1}{4\pi\epsilon_0}\dfrac{2q^2a^2}{3c^3}$	$\dfrac{2q^2a^2}{3c^3}$

TABLE II   Conversion factors between MKSA and Gaussian units.

In the conversion table below, the speed of light c is an important factor. In the table it has been approximated as $c = 3 \times 10^8$ m/s $= 3 \times 10^{10}$ cm/s; the approximation is indicated by writing (3). For greater accuracy, replace (3) by 2.99792.

Quantity	MKSA	Gaussian
Length	1 meter (m)	10^2 centimeters (cm)
Mass	1 kilogram (kg)	10^3 grams (g)
Force	1 newton (N)	10^5 dynes
Energy, Work	1 joule (J)	10^7 ergs
Power	1 watt (W)	10^7 ergs/second
Charge q	1 coulomb (C)	$(3) \times 10^9$ statcoulombs (esu[†])
Charge density ρ	1 C/m^3	$(3) \times 10^3$ statcoulombs/cm^3
Current I	1 ampere (A)	$(3) \times 10^9$ statamperes
Current density $\mathbf{J}$	1 A/m^2	$(3) \times 10^5$ statamperes/cm^2
Potential V	1 volt (V)	$\dfrac{1}{(3)} \times 10^{-2}$ statvolts
Electric field $\mathbf{E}$	1 V/m	$\dfrac{1}{(3)} \times 10^{-4}$ statvolts/cm
Polarization $\mathbf{P}$	1 C/m^2	$(3) \times 10^5$ statcoulombs/cm^2
Displacement $\mathbf{D}$	1 C/m^2	$(3) \times 4\pi \times 10^5$ statcoulombs/cm^2
Conductivity σ	1 siemens/m (S/m)	$(3)^2 \times 10^9$ (seconds)$^{-1}$
Capacitance C	1 farad (F)	$(3)^2 \times 10^{11}$ cm
Magnetic field $\mathbf{B}$	1 tesla (T)	10^4 gauss
Magnetic flux Φ	1 weber (Wb*)	10^8 maxwells*
Magnetic field $\mathbf{H}$	1 A-turn/m	$4\pi \times 10^{-3}$ oersted (Oe)
Magnetization $\mathbf{M}$	1 A/m	10^{-3} Oe
Inductance L	1 henry (H)	$\dfrac{1}{(3)^2} \times 10^{-11}$ stathenrys (s^2/cm)

[†]esu is an abbreviation for electrostatic unit.

*1 Wb=1 T m^2; 1 maxwell=1 gauss cm^2

Further Reading
Thorough discussions of conversions among units are in

J. D. Jackson, *Classical Electrodynamics*, 3rd ed. (Wiley, New York, 1999);

R. K. Wangsness, *Electromagnetic Fields*, 2nd ed. (Wiley, New York, 1986).

E. M. Purcell, *Electricity and Magnetism*, 2nd ed. (McGraw-Hill, New York, 1985). This classic book uses Gaussian units.

For further discussion of units and dimensions see R. T. Birge, *Am. Phys. Teacher* (now *Am. J. Phys.*) **2**, 41 (1934); **3**, 102 and 171 (1935).

APPENDIX

B

The Helmholtz Theorem

The field equations of electromagnetism are partial differential equations involving the curl and divergence of $\mathbf{E}(\mathbf{x}, t)$ and $\mathbf{B}(\mathbf{x}, t)$. For example, the electric field produced by a static charge density $\rho(\mathbf{x})$ obeys

$$\nabla \times \mathbf{E} = 0 \quad \text{and} \quad \nabla \cdot \mathbf{E} = \rho/\epsilon_0. \tag{B.1}$$

Or, the magnetic field produced by a static current density $\mathbf{J}(\mathbf{x})$ obeys

$$\nabla \times \mathbf{B} = \mu_0 \mathbf{J} \quad \text{and} \quad \nabla \cdot \mathbf{B} = 0. \tag{B.2}$$

These relations hold at all $\mathbf{x}$ and each is a precise statement about the *variation* of the field at $\mathbf{x}$.

The idea of field theory is that *local equations* determine the fields *globally*. For example, given a charge at $\mathbf{x}_1$ and a grounded conducting sphere nearby, the differential equations (B.1) should determine the field at a point $\mathbf{x}_2$ on the other side of the sphere.

But is a vector function really determined by its curl and divergence? The answer is yes, provided the sources are finite in extent. Any real system must be finite so the foundation of field theory is solid.

The Helmholtz theorem states that any vector field can be decomposed into an irrotational part (i.e., with curl = 0) and a solenoidal part (i.e., with divergence = 0), and these components are determined by the divergence and curl of the field, respectively. If the field tends to 0 at infinity, faster than r^{-1} as $r \to \infty$, then the decomposition is unique. This abstract mathematical theorem is an important part of the foundation of field theory.

Statement of the Theorem

Let $\mathbf{F}(\mathbf{x})$ be a differentiable vector function, for which the divergence $\nabla \cdot \mathbf{F} = d(\mathbf{x})$ and curl $\nabla \times \mathbf{F} = \mathbf{c}(\mathbf{x})$ are specified at all points in space. To be consistent, $\nabla \cdot \mathbf{c} = 0$. Assume $d(\mathbf{x})$ and $\mathbf{c}(\mathbf{x})$ approach 0 faster than r^{-2} as $r \to \infty$, and $\mathbf{F}(\mathbf{x})$ approaches 0 faster than r^{-1}. Then $\mathbf{F}(\mathbf{x})$ may be written in the form

$$\mathbf{F}(\mathbf{x}) = -\nabla\psi + \nabla \times \mathbf{A}, \tag{B.3}$$

and the components $-\nabla\psi$ and $\nabla \times \mathbf{A}$ are unique. Explicit functions $\psi(\mathbf{x})$ and $\mathbf{A}(\mathbf{x})$ for the representation (B.3) are

$$\psi(\mathbf{x}) = \int \frac{d(\mathbf{x}')d^3x'}{4\pi |\mathbf{x} - \mathbf{x}'|}, \tag{B.4}$$

$$\mathbf{A}(\mathbf{x}) = \int \frac{\mathbf{c}(\mathbf{x}')d^3x'}{4\pi\,|\mathbf{x} - \mathbf{x}'|}, \tag{B.5}$$

in which the integrals extend over all space.

Note that $\mathbf{F}(\mathbf{x})$ is expressed in (B.3) as the sum of an irrotational function $(-\nabla\psi)$ and a solenoidal function $(\nabla \times \mathbf{A})$. The functions ψ and $\mathbf{A}$ are called *potentials*.

Proof

Because $d(\mathbf{x}')$ and $\mathbf{c}(\mathbf{x}')$ tend to 0 at ∞, the integrals in (B.4) and (B.5) are convergent and so the functions defined in those equations are well defined. To prove the theorem we must first verify that $\nabla \cdot \mathbf{F}$ is the specified divergence $d(\mathbf{x})$. By (B.3), $\nabla \cdot \mathbf{F} = -\nabla^2\psi$. But $1/\left(4\pi\,|\mathbf{x} - \mathbf{x}'|\right)$ is the Green's function of $-\nabla^2$ (cf. Section 3.5) so (B.4) is Green's solution to Poisson's equation, $-\nabla^2\psi = d(\mathbf{x})$. That means the divergence of $\mathbf{F}$ is $d(\mathbf{x})$, as required.

Also, we must verify that $\nabla \times \mathbf{F}$ is the specified curl $\mathbf{c}(\mathbf{x})$. By (B.3),

$$\nabla \times \mathbf{F} = \nabla \times (\nabla \times \mathbf{A}) = \nabla (\nabla \cdot \mathbf{A}) - \nabla^2\mathbf{A}. \tag{B.6}$$

Again, $\mathbf{A}$ is Green's solution to Poisson's equation, $-\nabla^2\mathbf{A} = \mathbf{c}$. Also, $\nabla \cdot \mathbf{A} = 0$ by the following calculation:

$$\nabla \cdot \mathbf{A} = \int \left(\nabla \frac{1}{4\pi\,|\mathbf{x} - \mathbf{x}'|}\right) \cdot \mathbf{c}(\mathbf{x}')d^3x'$$

$$= \int \left(-\nabla' \frac{1}{4\pi\,|\mathbf{x} - \mathbf{x}'|}\right) \cdot \mathbf{c}(\mathbf{x}')d^3x'$$

$$= \int \left\{-\nabla' \cdot \left[\frac{\mathbf{c}(\mathbf{x}')}{4\pi\,|\mathbf{x} - \mathbf{x}'|}\right] + \frac{\nabla' \cdot \mathbf{c}(\mathbf{x}')}{4\pi\,|\mathbf{x} - \mathbf{x}'|}\right\}d^3x' = 0.$$

In making the last step, the first term in the line above is 0 by Gauss's theorem because $\mathbf{c} \to 0$ on the boundary sphere at ∞; the second term is 0 because $\nabla \cdot \mathbf{c} = 0$. Thus (B.6) implies that the curl of $\mathbf{F}$ is $\mathbf{c}(\mathbf{x})$, as required. ∎

Is the decomposition (B.3) unique? This question is the same as asking whether we could add to $\mathbf{F}$ a function $\mathbf{F}'$ with zero curl and zero divergence and that also satisfies the boundary condition $\mathbf{F}' \to 0$ at ∞. There exist functions with $\nabla \times \mathbf{F}' = 0$ and $\nabla \cdot \mathbf{F}' = 0$, the simplest being $\mathbf{F}' = $ constant; but they do not approach 0 at ∞. So the boundary condition that $\mathbf{F} \to 0$ as $r \to \infty$ makes (B.3) unique.

Let's prove the statement that $\mathbf{F}'(\mathbf{x})$ must be 0. We have specified that $\nabla \times \mathbf{F}' = 0$, so $\mathbf{F}'$ can be written as $\mathbf{F}' = \nabla f$ for some scalar function $f(\mathbf{x})$. Without loss of generality we may demand that $f \to 0$ as $r \to \infty$. (If f approaches a constant C, then $f - C$ approaches 0 and has the same gradient as f.) The condition $\nabla \cdot \mathbf{F}' = 0$ requires $\nabla^2 f = 0$. Now consider the integral

$$\mathcal{J} \equiv \int f\nabla^2 f d^3x, \tag{B.7}$$

where the integration region is all of space. Of course $\mathcal{J} = 0$ because $\nabla^2 f = 0$. But now rewrite $\mathcal{J}$ as

$$\mathcal{J} = \int \left[\nabla \cdot (f \nabla f) - (\nabla f)^2 \right] d^3x = - \int (\nabla f)^2 \, d^3x; \qquad \text{(B.8)}$$

the first term in square brackets has integral 0 by Gauss's theorem because $f \nabla f \to 0$ on the boundary surface at ∞. Thus $\int (\nabla f)^2 \, d^3x = 0$. But this means that $\nabla f(\mathbf{x}) = 0$ for all $\mathbf{x}$, i.e., $\mathbf{F}'(\mathbf{x}) = 0$ as claimed.

In summary, $\mathbf{F}(\mathbf{x})$ is determined by its curl and divergence, assuming the functions tend to 0 at infinity. The irrotational component $(-\nabla \psi)$ depends on the divergence $d(\mathbf{x})$ and the solenoidal component $(\nabla \times \mathbf{A})$ depends on the curl $\mathbf{c}(\mathbf{x})$.

The simplest application of the Helmholtz theorem is to the field $\mathbf{E}(\mathbf{x})$ in electrostatics, for which $\mathbf{c} = 0$ and $d = \rho/\epsilon_0$. But the general theorem (B.3) has a myriad of other applications, including magnetostatics, hydrodynamics, aerodynamics, elasticity, acoustics and seismology. It can also be extended to waves, in which $-\nabla \psi$ describes a longitudinal wave and $\nabla \times \mathbf{A}$ a transverse wave.

Index

Physical Constants

$c = 2.998 \times 10^8$ m/s   Speed of light

$\mu_0 = 4\pi \times 10^{-7}$ N/A^2 (or H/m)   Permeability constant in vacuum

$\epsilon_0 = \dfrac{1}{\mu_0 c^2} = 8.854 \times 10^{-12}$ C^2/Nm^2 (or F/m)   Permittivity constant in vacuum

$\dfrac{1}{4\pi\epsilon_0} = 10^{-7} c^2 = 8.988 \times 10^9$ Nm^2/C^2

$e = 1.602 \times 10^{-19}$ C   Magnitude of electron charge

$m_e = 0.9109 \times 10^{-30}$ kg   Electron mass

Useful Integrals

$$\int \frac{dx}{\sqrt{a^2 + x^2}} = \ln\left(x + \sqrt{a^2 + x^2}\right)$$

$$\int \frac{dx}{a^2 + x^2} = \frac{1}{a}\arctan\frac{x}{a}$$

$$\int \frac{dx}{\left(a^2 + x^2\right)^{3/2}} = \frac{x}{a^2\sqrt{a^2 + x^2}}$$

Binomial Expansion

$$(1 + \epsilon)^p = 1 + p\epsilon + \frac{p(p-1)}{2!}\epsilon^2 + \frac{p(p-1)(p-2)}{3!}\epsilon^3 + \ldots$$

Notation for Position Vector

$$\mathbf{x} = \hat{\mathbf{i}}\,x + \hat{\mathbf{j}}\,y + \hat{\mathbf{k}}\,z$$

$$r = |\mathbf{x}| = \sqrt{x^2 + y^2 + z^2} \quad \text{and} \quad \hat{\mathbf{r}} = \frac{\mathbf{x}}{r}$$